Oxidative Stability and Shelf Life of Foods Containing Oils and Fats

Editors

Min Hu
DuPont Nutrition and Health
New Century
KS, USA

Charlotte Jacobsen
National Food Institute
Technical University of Denmark
Kongens Lyngby, Denmark

ELSEVIER

AMSTERDAM • BOSTON • HEIDELBERG • LONDON
NEW YORK • OXFORD • PARIS • SAN DIEGO
SAN FRANCISCO • SINGAPORE • SYDNEY • TOKYO

Academic Press is an imprint of Elsevier

AOCS PRESS

Academic Press and AOCS Press
Academic Press is an imprint of Elsevier
125 London Wall, London EC2Y 5AS, UK
525 B Street, Suite 1800, San Diego, CA 92101-4495, USA
50 Hampshire Street, Cambridge, MA 02139, USA
The Boulevard, Langford Lane, Kidlington, Oxford OX5 1GB, UK

Published in cooperation with American Oil Chemists' Society www.aocs.org
Director, Content Development: Janet Brown

ISBN: 978-1-63067-056-6

British Library Cataloguing-in-Publication Data
A catalogue record for this book is available from the British Library

Library of Congress Cataloging-in-Publication Data
A catalog record for this book is available from the Library of Congress

For information on all Academic Press publications
visit our website at www.elsevier.com

Publisher: Nikki Levy
Acquisition Editor: Nancy Maragioglio
Editorial Project Manager: Jaclyn Truesdell
Production Project Manager: Caroline Johnson
Designer: Ines Cruz

Typeset by TNQ Books and Journals
www.tnq.co.in

Oxidative Stability and Shelf Life of Foods Containing Oils and Fats

Contents

6. Oxidative Stability and Shelf Life of Bulk Animal Fats and Poultry Fats

Poulson Joseph

7. Oxidative Stability and Shelf Life of Frying Oils and Fried Foods

Hong-Sik Hwang and Jill K. Winkler-Moser

8. Oxidative Stability and Shelf Life of Food Emulsions

Charlotte Jacobsen

List of Contributors

Monica Anese Dipartimento di Scienze degli Alimenti, Università di Udine, Udine, Italy

Marit Aursand Department of Processing Technology, SINTEF Fisheries and Aquaculture, Trondheim, Norway

Robin A. Boyle Kalsec®, Inc., Kalamazoo, MI, USA

Sonia Calligaris Dipartimento di Scienze degli Alimenti, Università di Udine, Udine, Italy

Ana K. Carvajal Department of Processing Technology, SINTEF Fisheries and Aquaculture, Trondheim, Norway

Hall Clifford Department of Plant Sciences, North Dakota State University, Fargo, ND, USA

Etsehiwot Gebreselassie Department of Plant Sciences, North Dakota State University, Fargo, ND, USA

Min Hu DuPont Nutrition and Health, New Century, KS, USA

Hong-Sik Hwang United States Department of Agriculture, Agricultural Research Service, National Center for Agricultural Utilization Research, Functional Foods Research, Peoria, IL, USA

Charlotte Jacobsen National Food Institute, Technical University of Denmark, Kgs. Lyngby, Denmark

Poulson Joseph Kalsec®, Inc., Kalamazoo, MI, USA

John L. Koontz FDA–CFSAN, Division of Food Processing Science and Technology, Bedford Park, IL, USA

Vera Kristinova Department of Processing Technology, SINTEF Fisheries and Aquaculture, Trondheim, Norway

Lara Manzocco Dipartimento di Scienze degli Alimenti, Università di Udine, Udine, Italy

Revilija Mozuraityte Department of Processing Technology, SINTEF Fisheries and Aquaculture, Trondheim, Norway

Maria C. Nicoli Dipartimento di Scienze degli Alimenti, Università di Udine, Udine, Italy

K.M. Schaich Department of Food Science, Rutgers University, New Brunswick, NJ, USA

Fereidoon Shahidi Department of Biochemistry, Memorial University of Newfoundland, St. John's, NL, Canada

Inger B. Standal Department of Processing Technology, SINTEF Fisheries and Aquaculture, Trondheim, Norway

Asim Syed Dow AgroSciences LLC, Indianapolis, IN, USA

Ingrid Undeland Biology and Biological Engineering – Food and Nutrition Science, Chalmers University of Technology, Gothenburg, Sweden

Jill K. Winkler-Moser United States Department of Agriculture, Agricultural Research Service, National Center for Agricultural Utilization Research, Functional Foods Research, Peoria, IL, USA

Xiaoqing Yang Kalsec®, Inc., Kalamazoo, MI, USA

Preface

The oxidative stability and shelf-life of foods and pet foods containing oils and fats are critical for food and pet food companies to develop successful new food products and improve existing food products that deliver the quality expected by the end consumer in terms of nutrition, appearance, texture, taste, and aroma. The professionals and managers in the food and pet food industries have been expecting a book on evaluation of shelf-life of oils/fats and oil/fat-containing foods and pet foods to be published. The book *Oxidative Stability and Shelf-life of Foods Containing Oils and Fats* is now published to meet the need. The book is dedicated not only to the readers working in the food and pet food industries but also to students, researchers, and professors in universities and research institutes who may use it as a useful reference.

Although there are some published books that deal with the stability and shelf-life of foods, the microbial stability, and shelf-life of foods as the main focus, the oxidative stability of the foods is discussed in only few chapters of other books. The principal focus of this book, however, is the oxidative stability and shelf-life of a variety of foods containing oils and fats, the applications of various analytical and evaluating methods to studies of oxidative stability and shelf-life of bulk oils and fats, frying oils and fried foods, food emulsions, low-moisture (dry) foods including pet foods, meat and meat products, as well as seafood.

The authors of the book are experts who have been working in the food industry or research institutes and have rich experience and expertise in the field of oxidative stability and shelf-life of foods containing fats and oils or professors who have conducted in-depth research in the area of lipid and protein oxidation and antioxidants in foods. The book chapters are summaries of their research experience, expertise, and knowledge, as well as published literature. The newest advances in the studies of oxidative stability and shelf-life of fat/oil-containing foods are presented in the book.

The book is composed of 13 chapters plus an introduction. The introduction briefly summarizes oxidation (rancidity) processes in foods containing fats and oils, measurements of lipid oxidation, definitions of oxidative stability and shelf-life, methods to evaluate oxidative stability and shelf-life, challenges of assessing oxidative stability and shelf-life, particularly in heterogeneous and multiple-phase food systems, factors influencing and determining oxidative stability, and strategies of minimizing lipid oxidation and extending the shelf-life of foods containing oils and fats. Chapters 1 to 3 cover a variety of commonly used methods of analyzing lipid oxidation and protein oxidation, the methods of determining and predicting the shelf-life of food products, and the sensory evaluation of oils/fats and oil/fat-containing food products. Chapters 4 to 12 mainly deal with the oxidative stability and shelf-life of various food systems such as vegetable oils, fish oils, animal and poultry fats, frying oils and fried foods, food emulsions such as mayonnaise, salad dressing, margarine and spreads, low-moisture (dry) foods such as milk powders, egg powders, microencapsulated oil powders, breakfast cereals, snack foods and dry pet foods, meat and meat products, seafood, as well as crackers, cookies, and biscuits. The oxidative stability and active packaging are also highlighted in the last chapter.

Finally, we are indebted to all the authors contributing to the book. The book would not be published without their hard work and support. We are also indebted to Janet Brown, director of AOCS (American Oil Chemists' Society), for initiating a project at AOCS to publish the book and for organizing and executing the project, and to Lori Weidert, production editor at AOCS, for editing the book, as well as all the staff members who worked on the book at Elsevier. We hope that readers of the book will find it useful.

Min Hu and Charlotte Jacobsen

Introduction

Min Hu
DuPont Nutrition and Health, New Century, KS, USA

MAIN FOCUS OF THE BOOK

Food stability and shelf-life are critical in the development of successful new food products that deliver the quality expected by the end consumer in terms of nutrition, appearance, texture, taste, and aroma. In general, food stability may include chemical stability (e.g., oxidative stability), physical stability, and microbial stability. There are several published books (Eskin and Robinson, 2000; Kilcast and Subramaniam, 2000, 2011, Man and Jones, 2000) focusing primarily on microbial stability and the shelf-life of various foods or dealing with microbiological spoilage, chemical deterioration, and physical instability of a variety of foods, including fruits, vegetables, bread and bakery products, beverages, milk and milk products, seafood, meat, and poultry, as well as oils and fats. The central focus of this book, however, is the oxidative stability and shelf-Life of oils/fats and oils/fats-containing foods—in particular, the application of various evaluation methods to studies of oxidative stability and shelf-life in oils/fats and oils/fats-containing foods in the food and pet food industries. In addition, lipid co-oxidation with protein is also discussed, since a number of food products contain both lipids and proteins.

This book is directed mainly toward readers working in the food and pet food industries. It may also serve as a useful reference for students, researchers, and professors in universities and research institutes.

OXIDATION (RANCIDITY) PROCESSES IN OILS/FATS AND OILS/FATS-CONTAINING FOODS AND MEASUREMENTS OF LIPID OXIDATION

One of the most important tasks for a manufacturer of edible oils/fats and oils/fats-containing foods is to control the lipid oxidation process in order for those foods to have an acceptable oxidative stability and shelf-life over a certain period of storage time. A better understanding of lipid oxidation processes is particularly helpful in controlling these processes and developing new oils/fats-containing food products with acceptable oxidative stability and shelf-life.

Triacylglycerols (triacylglycerides), or glycerol esters of fatty acids, make up to 99% of lipids of plant and animal origin (Fennema, 1996). Thus, triacylglycerides are major components (around 99%) in edible oils and fats. Phospholipids (phosphoglycerides) are important structural lipids in foods and cell membranes, in addition to being a minor component in crude vegetable oils and crude animal fats. Because a large portion of phospholipids is removed by degumming in the refining of vegetable oils (Frankel, 2005), oxidation of refined edible vegetable oils primarily involves the oxidation of triacylglycerides. Oxidation of crude vegetable oils and crude animal fats—such as animal fats produced with animal byproducts or poultry byproduct meals—may involve the oxidation of both triacylglycerides and phospholipids. However, oxidation of whole egg and egg yolk powders, as well as meat and meat products, involves the oxidation of triacylglycerides, phospholipids, and cholesterol (Meynier et al., 2014; Mazalli and Bragagnalo, 2007).

Triacylglycerols are subject to hydrolysis by either enzymes (e.g., lipase) or alkalis/acids to produce a mixture of diacylglycerols, monoacylglycerols, free fatty acids, and glycerol. Thus, the substrate of lipid (oils and fats) oxidation could be free fatty acids and mono-, di-, or triacylglycerides. However, free fatty acids are more easily oxidized than others.

There are three major reaction pathways responsible for lipid oxidation in foods (Figure 1): auto-oxidation, photo-oxidation, and enzymatic oxidation (Carlson et al., 2005). Auto-oxidation is a free radical chain reaction, involving chain initiation, propagation, and termination, and is a primary pathway for the formation of lipid hydroperoxides. In the presence of initiators such as light, trace transition metals, or free radicals, unsaturated fatty acids (LH) lose a hydrogen atom to form an alkyl radical ($L^\bullet$) that reacts rapidly with molecular oxygen (O_2) to yield a peroxide radical ($LOO^\bullet$). The peroxide radical then reacts with another LH to form hydroperoxide (LOOH), a primary oxidation product. Hydroperoxide is tasteless and odorless and has no significant impact on the sensory quality of the oils/fats. It is, however, generally unstable and can react with ferrous iron (Fe^{2+}) to yield lipid alkoxyl radical ($LO^\bullet$), thus leading to the formation of secondary lipid oxidation products by β scission such as aldehydes (e.g., hexanal and propanal), ketones, alcohols, organic acids (e.g., hexanoic and propanoic acid), epoxides, and hydrocarbons. These secondary oxidation products would contribute flavor notes that significantly impact the sensory quality of oils/fats and oils/fats-containing foods. For instance, one can sense the rancid notes of a vegetable oil or animal fat after the oil/fat was stored for a long time, which means that the oil/fat is considered rancid.

Photo-oxidation occurs when an oil or a fat is exposed to light in the presence of sensitizers such as chlorophyll and pheophytine for Type II photo-oxidation or riboflavin for Type I photo-oxidation. When activated by light, chlorophyll reacts with triplet oxygen ($^3O_2$) to yield reactive singlet oxygen ($^1O_2$), which then reacts with LH to form lipid LOOH. The last reaction takes place when

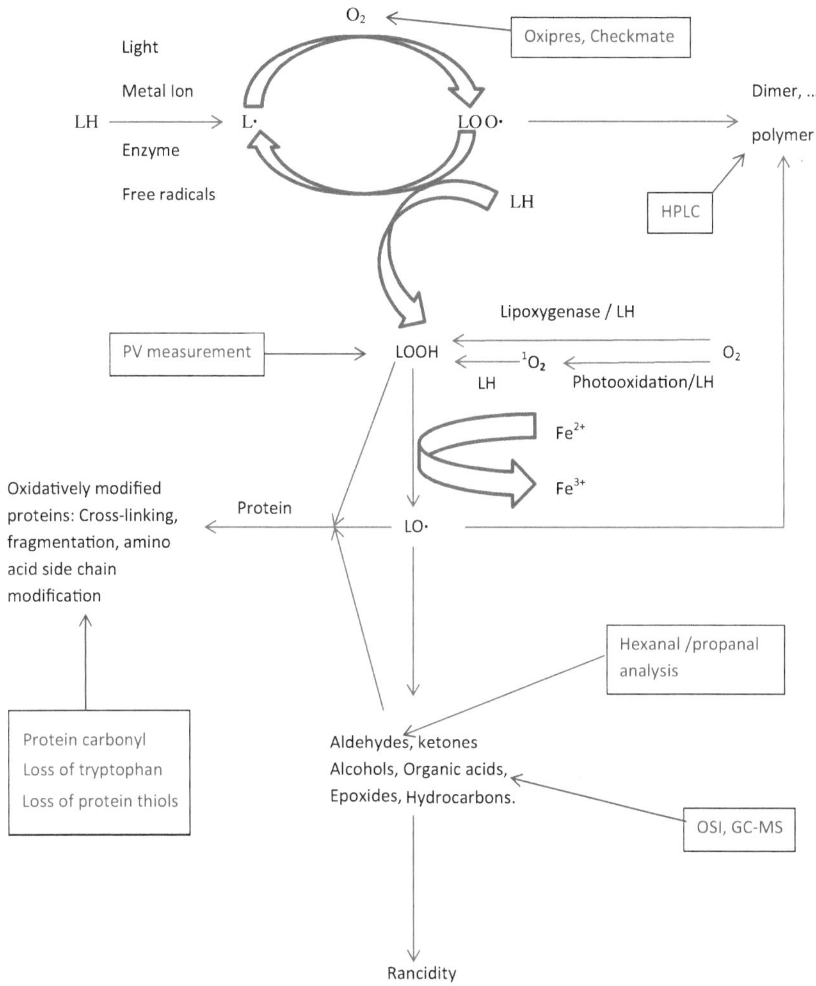

FIGURE 1 Three major reaction pathways responsible for lipid oxidation in foods.

electrophilic singlet oxygen directly reacts with a high–electron-density double bond without activation of LH (Choe and Min, 2006). Thus, oxidative rancidity of the oil or fat occurs very rapidly in the presence of light without an induction period (Frankel, 1998; Mattaus, 2010). In the presence of riboflavin, however, the light-activated riboflavin transfers energy to LH to form an alky radical ($L^•$), which reacts with triple oxygen to form a peroxide radical ($LOO^•$) via auto-oxidation as shown in Figure 1.

The third pathway is enzymatic oxidation. Lipo-oxygenases can incorporate oxygen into LH to yield lipid LOOH, which commonly occurs in chicken meat and soybeans (Skibsted, 2010).

The free radicals generated through auto-oxidation—e.g., peroxide radical (LOO•) and alkoxyl radical (LO•)—could polymerize to form dimers, trimers, and polymers after lipid oxidation occurs at a high temperature for a longer time, as with deep frying oil and the oil or fat at an advanced stage of accelerated storage.

A number of food products contain a variety of complex matrices with both oils/fats and protein as major components, for instance, milk powder, lipoproteins, tortilla chips, roasted peanuts, peanut butter, meat and meat products, oil-in-water emulsions, as well as dry pet foods and ingredients. More and more studies are focusing on lipid co-oxidation with protein. Protein oxidation can be triggered by reactive oxygen species such as alkoxyl radical (LO•), primary oxidation product (LOOH), or secondary oxidation products such as aldehydes (Genot et al., 2003). These can react with proteins and amino acids, either free or bound in protein or peptides, resulting in oxidatively modified proteins through protein cross-linking, amino acid side chain modification and protein fragmentation (Pokorny et al., 2005; Skibsted, 2010).

Lipid co-oxidation with protein in lipid-and-protein–containing foods moves system oxidation away from lipids to other molecules, since reactive lipid hydroperoxide, aldehydes, ketones, and free radicals (e.g., LO•) may react with protein, thereby resulting in an appreciable decrease of reactive substances such as lipid hydroperoxide and aldehydes. Therefore, measuring only lipid oxidation in the foods containing both lipids and proteins may underestimate lipid oxidation in complex food systems (Schaich, 2014; Skibsted, 2010). Measuring both lipid oxidation and protein oxidation in lipid-and-protein–containing foods might be critical for accurately assessing the extent of oxidative rancidity in the foods.

In an oxidative stability and shelf-life study, PV (hydroperoxide value) measures the concentration of primary oxidation product lipid LOOH formed during lipid oxidation (Figure 1). Oxipres and Checkmate measure the headspace level of molecular oxygen (O_2) consumed in the reaction to form peroxide radicals (LOO•) in a sealed vessel containing bulk oils/fats or a dry food (e.g., encapsulated fish oil or milk powder), dry pet food, or food emulsion. An oxygen electrode could be used to analyze the molecular oxygen level consumed in the reaction forming peroxide radicals (LOO•) in an oil-in-water emulsion (Hu and Skibsted, 2002). GC-MS (Gas Chromatography - Mass Spectrometry) or GC-FID (Gas Chromatography - Flame Ionization Detector) headspace analysis measures concentrations of secondary volatile oxidation products such as hexanal, 2,4-decadienal, propanal, 2,4-heptadienal, hexanoic, and propanoic acid, formed during the lipid oxidation. GC–MS or GC-FID can also be used to measure Strecker aldehydes and volatile nonenzymatic browning products formed from reactions between aldehydes and amino groups in krill oils (Lu et al., 2014). Hexanoic or propanoic acid was found to be suitable for assessing oxidative stability and shelf-life of dry foods such as breakfast cereals containing $n-6$ or $n-3$ fatty acids after long-term storage. The correlation between hexanoic acid and sensory data was established

(Paradiso et al., 2008). The oxidative stability index (OSI) measures levels of organic acids, secondary oxidation products. High performance size-exclusive chromatography analyzes levels and distribution of dimers, trimers, and polymers formed during the frying of oil. Measurements of protein carbonyl, loss of tryptophan, and protein thiols provide information about protein oxidation and lipid co-oxidation with protein (Figure 1).

WHAT ARE THE OXIDATIVE STABILITY (RANCIDITY) AND SHELF-LIFE OF OILS/FATS AND OILS/FATS-CONTAINING FOODS?

Lipid oxidation in foods can be described as oxidative stability, oxidative rancidity, or oxidative deterioration. Oxidative stability refers to the ability of oils/fats and oils/fats-containing foods to resist oxidative rancidity (or deterioration) over processing and storage periods. Oxidative rancidity is essentially the oxidative instability or oxidative deterioration of oils/fats and oils/fats-containing foods during processing and storage. Oxidative deterioration in foods may involve oxidation in both the lipid phase (e.g., polyunsaturated lipids) and the aqueous phase (e.g., proteins), depending on the matrices in the foods (Skibsted, 2010).

Published definitions of food and beverage shelf-life tend to vary somewhat. According to IFST (Institute of Food Science and Technology, UK.) Guidelines (1993), shelf-life is defined as the time during which the food product will (1) remain safe; (2) be certain to retain desired sensory, chemical, physical, and microbiological characteristics; and (3) comply with any label declaration of nutritional data, when stored under recommended conditions.

According to the New Zealand Food Safety Authority (2005), shelf-life is considered a guide for setting consumer expectations regarding the period of time a food product can be kept before it starts to deteriorate, assuming that any recommended storage conditions have been adhered to.

According to Fu and Labuza (1997), the shelf-life of a food is the time period within which the food is safe to consume and/or is of a quality acceptable to consumers.

For oils/fats and oils/fats-containing food products, shelf-life also could be defined as the time period before the oxidative rancidity and overall quality (safety, sensory, and physicochemical characteristics) of the product reaches an unacceptable (rancid) degree under specific storage conditions.

WHAT IS CRITICAL ABOUT EVALUATING OXIDATIVE STABILITY AND SHELF-LIFE?

First, food manufacturers have the responsibility to conduct shelf-life study and sensory evaluation for a new food product containing oils/fats in order to ensure that the product with an expected shelf-life and high quality is accepted by consumers during the shelf-life period. A "best before" date that is based on a shelf-life study is needed on the product label. Ideally, a food manufacturer will conduct a real-time (or ambient shelf-life) study of an oils/fats-containing food product if time permits. However, if new product development time is short or if the new product is oxidatively stable, the manufacturer will have to conduct an accelerated shelf-life study. In such cases, manufacturers must make a determination regarding the oxidative stability and shelf-life of the new food product in a short period of time. Assessment is usually based on new accelerated shelf-life data, previous shelf-life data, the experience of the evaluators, and sensory tests. Therefore, previously accumulated oxidative stability and shelf-life data and experience on evaluating oils/fats and oils/fats-containing foods and a well-trained sensory panel are essential for a food or pet food company to successfully develop various new products.

Second, lipid oxidation may decrease the quality of foods in terms of color, taste, odor, and palatability. It may also decrease the levels of nutrients such as essential amino acids, essential fatty acids, and vitamins and lead to the formation of toxic compounds. Therefore, it is critical to evaluate the oxidative stability, shelf-life, and quality of oils/fats and oils/fats-containing foods and to add appropriate antioxidants to prevent lipid oxidation. Third, analytical methods of evaluating oxidative stability can be used to assess the efficacy of both synthetic and natural antioxidants and antioxidant blends, as well as to study antioxidant mechanisms. The research and application of these analytical evaluation methods can help food companies develop oxidatively stable products and new and efficacious antioxidants.

WHY IS EVALUATING THE OXIDATIVE STABILITY AND SHELF-LIFE OF FOODS AND FOOD INGREDIENTS A BIG CHALLENGE FOR THE FOOD AND PET FOOD INDUSTRIES?

Professors may order pure ingredients from chemical companies to conduct various tests in university laboratories. However, food and pet food companies have a harder time obtaining pure ingredients for their production lines. Many ingredients contain a number of "impurities" or minor components that may negatively impact the finished products. It can be difficult to accurately assess the composition and history of oils/fats or oils/fats-containing ingredients. For instance, the level of mixed tocopherols may not be found on the label of a commercial vegetable oil, and those levels can vary. So it is important to understand when and how much additional antioxidants containing tocopherols are added to the vegetable oils. Oxidative stability and shelf-life may not be increased if a relatively high level of mixed tocopherols is added to a vegetable oil that already contains higher levels of naturally occurring tocopherols (Akoh, 1994). In general, animal fats contain a relatively low level of mixed tocopherols. Adding mixed tocopherols would increase the oxidative stability of animal fats provided that no antioxidant is added to those commercial animal fat products. Obviously, each antioxidant has an appropriate use range. The antioxidant may not work if the usage is over the range.

Further, if one uses different oils/fats or oils/fats-containing ingredients with varying oxidative stability and different degrees of freshness to produce a new product, it is difficult to evaluate the oxidative stability and shelf-life of the finished product and to explain the oxidative stability data, since the oxidation of the new product is not that of a pure bulk oil but is that of mixed oils/fats with different degree of freshness and varying oxidative stability.

In addition, although a number of analytical evaluation methods are available, only a few are used in the food industry, because each has its own limitations. Some may be good for theoretical research but may not be suitable for routine analysis in the food industry. For example, ESR (Electron Spin Resonance), NMR (Nuclear Magnetic Resonance), DSC (Differential Scanning Calorimetry), chemiluminescence, and fluorescence methods are widely used in academia, but are not widely used as routine analysis in the food and pet food industries. Further, while a given method may work well in one food system, it may not work well in another. For instance, PV, Schaal oven, and active oxygen method (AOM) tests could be working well in bulk oil during the early stage of lipid oxidation. But they may be less effective in complicated multiphase food systems, such as dry foods and certain food emulsions. PV and p-anisidine vale (*p*-AnV) work in evaluating the quality of newly produced bulk oils and fats, or the oxidative stability of neat oils and fats at early stage of oxidation, but may not work for evaluating the shelf-life of oils/fats-containing multiphase food products at later stage of oxidation. *p*-AnV often does not correlate well with sensory data despite the fact that it evaluates the level of volatile oxidation products. TBARS (Thiobarbituric Acid Reactive Substances) works well in meat and meat products but may not work well in some dry foods.

Finally, timing and complexity can be complicating factors. Accelerated development of new oils/fats-containing products can make it impractical to conduct real-time or ambient shelf-life studies. Food products that contain lipids, protein, and other components often involve a degree of complexity due to lipid co-oxidation with protein, oxidized aldehydes such as hexanal and pentanal further oxidizing to produce hexanoic acid and pentanoic acid, and interactions of the components, making it difficult to explain oxidative stability data, and even harder to draw final conclusions.

COMMON METHODS FOR EVALUATING OXIDATIVE STABILITY AND SHELF-LIFE IN THE FOOD AND PET FOOD INDUSTRIES: ADVANTAGES AND DISADVANTAGES

Usually, physical and chemical analytical methods and the sensory test are used to assess the oxidative stability and shelf-life of oils/fats and oils/fats-containing foods. As mentioned previously, although numerous analytical methods exist, a limited array of methods is used routinely due to the convenience and effectiveness they provide for the food and pet food industries. Table 1 summarizes the various analytical methods commonly used in the food and pet food industries,

TABLE 1 Analytical Methods Available: Applications and Limitation

Method	Applications	Limitation	Application in Food and Pet Food Industries
Oxipres (oxygen bomb) Oxidography (oils/fats)	Mainly for dry foods/ ingredients, but can be used for bulk oils/fats. Bath temp. 80–100 °C: 80 °C for products rich in polyunsaturated fatty acids, e.g., fish meal; 100 °C for extruded foods, baked foods, fried foods, dry pet foods. In some cases, bath temp. may be adjusted to 50–60 °C. Oxidography is mainly used for measuring bulk oils/fats.	• High measuring temperature • May not be suitable for some of powder samples	Widely used
Checkmate (headspace oxygen measurement)	For both liquid and solid samples. Analyze headspace oxygen level at ambient temperature. Used for shelf-life study of samples stored at different temperatures over time	Low validity (some ingredients such as protein) may absorb oxygen	Used
OSI	• Mainly for bulk oils/fats Measuring temperature: 100–120 °C for saturated, mono-, di-unsaturated oils/ fats, like sunflower oil, chicken fat; 60–80 °C for polyunsaturated oils, like fish oil • Modified method with AMVN at lower temperature, i.e., 50 °C	• High measuring temperature • Extraction is needed for solid samples • Low sensitivity	Widely used

TABLE 1 Analytical Methods Available: Applications and Limitation—cont'd

Method	Applications	Limitation	Application in Food and Pet Food Industries
PV (hydroperoxide value)	• Accurate, widely used, simple, economical, suitable for evaluation of the early stages of lipid oxidation • AOCS titration: Need extracted oils/fats for analysis and solvent removal • Modified FOXII (xylenol orange): Liquid/solid samples. High sensitivity. Do not need extracted oils/fats • IDF (using thiocyanate): Liquid/solid samples. High sensitivity. Do not need extracted oils/fats	• Good for "fresh" pure bulk oils/fats, emulsions, seafood • Many factors impacting measurement • Low sensitivity and validity. • Unit: From mEq/kg to μmol/kg	Widely used
Conjugated dienes	Oils/fats or extracted lipids	Not suitable for quantitative comparison	Used but widely used in academia
p-Anisidine value (*p*-AV)	• Simple and mainly used for bulk oils/fats • Routine analysis • TOTOX = 2 PV + p-AV	• Only good for pure bulk oils/fats • Low validity	Widely used
Hexanal or 2,4- decadienal, hexanoic acid	• Used for samples rich in ω-6 fatty acids. May be correlated to sensory evaluation • GC static headspace: Rapid, routine analysis • Dynamic headspace: Higher sensitivity, lower re-productivity. May not be good for routine analysis • SPME: Higher sensitivity, routine analysis	• Different pretreatments may have different impact on analytical result • May need a standard method • For SPME, problems with competition for space on the fiber	Widely used

Continued

TABLE 1 Analytical Methods Available: Applications and Limitation—cont'd

Method	Applications	Limitation	Application in Food and Pet Food Industries
Propanal or 1-penten-3-one or 2,4-heptadienal, propanoic acid	• Used for samples rich in $n-3$ fatty acids, such as fish oil, algal oil, and Krill oil	Same as above	Used
Schaal Oven Test	• Mainly for bulk oils/fats • Measure PV over time at 60 °C and sensory evaluation	May not be suitable for solid samples	Used
TBA	• Used in fat-containing foods, in particular, meat and meat products • Used in samples with polyunsaturated fatty acids with 3 and more double bonds	• Lacking of specificity and sensitivity • Different analytical procedures may result in different analytical results • Interfered when analyzing oxidized samples with color	Widely used
Polymer determination	Using high-performance, size-exclusive chromatography, mainly for frying oil, encapsulated oil	Can be used for routine analysis?	Used
Polar compounds	Using absorption chromatography, mainly for frying oil	Time consuming	Widely used
HPLC for measuring antioxidant level	Measuring antioxidant level change (loss, residue) in oils/fats and oils/fats-containing foods over processing or storage to evaluate lipid oxidation over storage period, e.g., level changes in tocopherols, carnosic acid, EGCG, BHA, BHT, and ethoxyquin	Challenge for analyzing very low level of residue antioxidants	Used

TABLE 1 Analytical Methods Available: Applications and Limitation—cont'd

Method	Applications	Limitation	Application in Food and Pet Food Industries
ESR (EPR), NMR, DSC, NIR, chemiluminescence fluorescence	Widely used in academia	Not widely used in food industry as routine analysis, except NIR	Not widely used but widely used in Academia
• Protein–carbonyl compounds • Loss of tryptophan fluorescence • Loss of protein thiols	• Simple • Used for protein oxidation • Indicating co-oxidation of lipid with protein	Only certain imine compounds showing fluorescence	Used and will be widely used

as well as their primary applications and associated advantages and disadvantages. The most widely used analytical methods in food and pet food industries could be: Oxipres (or oxygen bomb), OSI, PV, P-anisidine, hexanal, or 2.4-decadienal and hexanoic acids, propanal or 2,4-heptadienal and propanoic acids, thiobarbituric acid (TBA), polar compounds (for frying oil), and protein carbonyl measurement (for protein oxidation). In Chapter 1, analytical methods will be discussed in detail.

Sensory evaluation remains one of the primary methods for assessing oxidative stability and shelf-life in oils/fats and oils/fats-containing foods. An experienced panel assesses the aroma (smell), taste, appearance and mouthfeel of the samples and scores and describes the attributes of the tested samples. Sensory evaluation correlates closely to customer acceptance. It is highly sensitive and valid and widely used in the food and pet food industries. However, the usefulness of this method is limited because it tends to be very expensive, requires a well-trained panel, and delivers a low level of precision (Frankel, 1998). Optimally, both sensory and oxidative stability data would be used to evaluate product shelf-life, and it is critical to correlate sensory data with oxidative stability data. Both agreement and conflict between sensory data and analytical data of oxidative stability of oils/fats and oils/fats-containing foods have been reported (Maisuthisakul et al., 2007; Lloyd et al., 2009), but there is usually consistency between sensory data and levels of secondary oxidation products such as hexanal and TBARS. The food and pet food industries may need to establish a clear correlation between sensory and oxidative stability data for various food products. Some companies may have a trained panel for conducting sensory tests of their products, but most do not, instead sending samples out for sensory testing, or they may train their employees in sensory testing. Employees in most of companies could be asked to do the sensory test for their own products.

GC-olfactometry (GCO) uses a combination of GC with olfactometric detection (essentially a sniff test) to establish which volatile secondary oxidation compounds contribute to rancid and other off-odor notes and at what concentration. Electric nose and electronic tongue technology have been used to evaluate oxidative rancidity in food products (Tikk et al., 2008).

HOW TO EVALUATE THE OXIDATIVE STABILITY AND SHELF-LIFE OF OILS/FATS AND OILS/FATS-CONTAINING FOODS

Selecting appropriate analytical methods and oxidative indicators (markers) is of critical importance. It is recommended that at least two methods are employed—one to analyze primary oxidation product quantified by hydroperoxide value (PV) and the other to measure secondary oxidation products, such as hexanal or 2,4-decadienal and hexanoic acids for foods containing $n-6$ fatty acids, such as corn, soybean, rapeseed, sunflower, olive, peanut, and sesame oils, as well as lard (or pork fat), chicken fat and tallow, and propanal, 1-penten-3-one or 2,4-heptadienal and propanoic acids for foods containing $n-3$ fatty acids, such as fish, algal, flax seed, and krill oils. For krill oil, it is also recommended to measure nonenzymatic browning products such as pyrroles and/or Strecker aldehydes such as 3-methylbutanal. In the food and pet food industries, commonly used oxidative indicators (markers) may include PV, *p*-AnV, TOTOX (total oxidation), hexanal or 2,4-decadienal, hexanoic acid, propanal or 2,4-heptadienal, propanoic acid, TBARS, IP (induction period) of OSI or Oxipres, and polar compounds (for frying oil). In order to evaluate the oxidative stability and shelf-life of a food product, one should first establish a correlation between an oxidative indicator, such as hexanal or TOTOX (usually TOTOX is defined as $2\,PV + p\text{-}AnV$, which will be discussed in detail later) and product quality based on the sensory data during storage. Hexanal or TOTOX can be a rapid and useful tool for evaluating the oxidative stability and shelf-life of the products once the correlation is set up.

It also is essential to select different analytical methods based on the matrices of the foods tested. For instance, if a sample is bulk oil or fat, OSI can be used to roughly evaluate the oxidative stability and shelf-life of the bulk oil or fat at different temperatures, such as 110, 100, 90, and 80 °C. Then, the Arrhenius equation or Q_{10} is used to extrapolate the actual (or ambient) IP (induction period) value at lower temperature from the IP at high temperatures in order to predict shelf-life of the oil or fat at lower temperature. A modified Schaal oven test may be used to assess the shelf-life of the oil or fat at 50–60 °C using changes in both PV and hexanal levels during storage as well as sensory test data. If a sample is solid or powder, Oxipres can be used to roughly evaluate the shelf-life of the sample at different temperatures. Then, the Arrhenius equation or Q_{10} is used to extrapolate the actual IP value at lower temperature based on the IP at high temperatures. An ambient storage test at 22 °C, or an accelerated storage test at 37 °C, may be used to assess the oxidative stability and shelf-life of solid or powder and food emulsion samples using PV and hexanal data over the shelf-life period. The correlation between IP values and hexanal and PV values in the tested temperatures must then be determined. If the sample is an emulsion, Oxipres or oxidography may be used to roughly evaluate the shelf-life of the sample at relatively lower temperatures, such as 70 °C. PV and hexanal measurements at ambient or accelerated storage test (i.e., 37 °C) over storage may be used for more accurate assessment.

Third, in order to use primary and secondary oxidation products to comprehensively evaluate oxidative stability and shelf-life of lipids, TOTOX value was defined as follows (Allen and Hamilton, 1994):

$$\text{TOTOX value A} = 2\text{PV} + p\text{-AnV}$$

TOTOX value A can be used to evaluate bulk oils/fats quality, but may not be suitable for complex multiphase food systems. Thus a new TOTOX value has been defined (Shahidi and Wanasundara, 2008):

$$\text{Totox value B} = 2\text{PV} + \text{TBARS}$$

TOTOX value B may be appropriate for meat and meat products, but may not be suitable for other dry food matrices due to the limitation of the TBARS method. Therefore, yet another TOTOX is defined as:

$$\text{Totox value C} = 2\text{PV} + \text{hexanal}$$

TOTOX value C may be widely preferred over TOTOX value A and B, since hexanal is a very useful oxidative indicator or marker for assessing secondary oxidation product for bulk oils/fats, emulsions and dry foods. However, TOTOX Value C may not be well correlated to sensory data, because the ratio of PV to hexanal (the contribution of PV and hexanal values to lipid oxidation) is 2:1. For pure bulk oil, TOTOX value C may be suitable if the AOCS titration method is used to measure PV and a GC headspace method used to analyze hexanal. However, for complicated multiphase food systems such as dry foods, TOTOX value C may not be appropriate for evaluation. Thus, for different food systems, emulsions and dry foods, the ratio of PV to hexanal should be determined by correlation between PV and hexanal as well as sensory data, as varying food matrices and different measuring methods may have appreciable impact on PV and hexanal values. Accordingly, a widely used TOTOX value could be defined as:

$$\text{Totox value} = a\ \text{PV} + b\ \text{hexanal}$$

where a and b can be determined by experimental data to match different food matrices (systems).

It should be pointed out that there are at least three different methods for measuring PV and three for analyzing hexanal, as well as a number of pretreatment methods for hexanal measurement when using GC headspace method. Varying analytical methods and pretreatment methods would result in different PV and hexanal data. Therefore, the same PV and hexanal methods must be used when calculating and comparing TOTOX value. We have a standard PV titration method from AOCS, but no standard hexanal method. In order to compare TOTOX values determined in different laboratories, a standard hexanal analytical method is needed. In the TOTOX value equation, hexanal may be replaced by propanal, or 1-penten-3-one, 1-penten-3-ol, or 2,4-heptadienal if oils or fats contain a large amount of $n-3$ fatty acids, such as fish oil and algal oil. In the food and pet food industries, a cut-off (end point) value for oxidative indicator (marker) could be required for a specific product in order to quickly evaluate shelf-life and compare the oxidative stability of similar products by using TOTOX or hexanal. A cut-off (end point) value may be defined as a value over which the product will be considered rancid and not acceptable to consumers.

In addition, selecting appropriate actual and accelerated storage methods for shelf-life studies is essential for the food and pet food industries. The disadvantage of actual storage testing is that it is time-consuming and expensive; a normal ambient shelf-life test may need a year for an oxidatively stable food product. But data generated in actual (e.g., ambient) storage tests are much more reliable than those generated at high temperature. Food companies may need shelf-life data and sensory data in a short period of time, however—a month or even 10 days—in order to make a business decision. Because of this, accelerated storage tests to predict shelf-life of oils/fats-containing food products is often required for food and pet food industries.

The Arrhenius equation:

$$k = Ae^{-Ea/RT} \text{ or } \ln\ k = \ln A - Ea/RT$$

and temperature coefficient Q_{10} are the basis of the prediction. In order to predict the actual shelf-life (e.g., in ambient temperature) of a food product using rate constant obtained at high temperatures, such as 80, 90, 100 and 110 °C, the Arrhenius equation is used to plot (ln k versus $1/T$) and extrapolate to the rate constant at lower (ambient) temperature from the rate constants at high temperatures. A measurable quality factor, such as IP (induction period), is usually employed in a shelf-life study. For instance, OSI or Oxipres is selected to measure IP of a tested food product at several different temperatures for shelf-life prediction. An actual induction period (shelf-life in hours, days, weeks, and months) would be evaluated by an Arrhenius plot of log induction period versus accelerated temperature and extrapolation from IP at high temperatures to IP at low (ambient) temperature.

Labuza and Riboh (1982), Labuza and Schmidl (1985) and Labuza (1979) have conducted a lot of research on predicting shelf-life of food products using

Q_{10} and the Arrhenius equation. Q_{10} is the factor by which the reaction rate increases as a consequence of increasing the temperature by 10 °C. Any measurable quality factor of the progress of a process, i.e., the process of oxidative rancidity, may be used to represent the reaction rate. Q_{10} is calculated from the measurable quality factor (e.g., IP) obtained at two or more storage temperatures in accelerated storage tests and usually has a value around 2 to 3 for many chemical reactions. It can be also used to predict expected shelf-life of a food product, with shelf-life being defined as follows (Frankel, 1998):

$$Q_{10} = \text{shelf life at } T/\text{shelf life at } T + 10 = Q_t/Q_{t+10\,°\mathrm{C}} = 2 - 3$$

For instance, if a food product has a shelf-life of 3 months at 42 °C, then it has a shelf-life of 6 (3×2) and 12 (6×2) months at 32 °C and 22 °C, respectively, when Q_{10}=2. Different food products may have different Q_{10} values since they may have differing oxidation mechanisms, reaction order and activation energy. Accurately determining Q_{10} is critical since the effect of an increased Q_{10} is very large. In general, the reaction activation energy of oxidative rancidity is 10–25 kcal/mol.

Theoretically, one can use the Arrhenius equation and Q_{10} to predict the shelf-life of various food products. Actually, it is relatively easy to predict the shelf-life of bulk oils/fats, but it is a great challenge to predict the shelf-life of emulsions and dry food products using accelerated methods such as OSI and Oxipres. This is because high temperatures, such as 100 °C, may cause physical and chemical changes, including emulsion structural changes, protein denaturation, fats melting, moisture evaporation, heat-sensitive antioxidant evaporation, or degradation (Frankel, 1998; Frankel and Meyer, 2000), volatilization of secondary oxidation products, interactions among lipids, lipid oxidation products, active oxygen species and proteins, small peptides, amino acids, and other molecules. Thus, the lipid oxidation mechanism and reaction activation energy may change, and the linear correlation between the accelerated shelf-life test and actual shelf-life test may be lost. Accordingly, lower measuring temperatures are recommended for the prediction of shelf-life.

Generally, actual and accelerated shelf-life studies may be required for the food products to be tested. First, collect the samples that represent the whole food products. Then conduct actual (such as ambient) and accelerated shelf-life studies, as well as a sensory evaluation. Subsequently, the correlation between ambient testing data and accelerated testing data should be established using regression analysis. If the product launch timeline allows too little time for ambient testing, an accelerated shelf-life study must be conducted. Accelerated data (i.e., IP data from OSI or Oxipres measured at different temperatures), sensory test data and previous experience in evaluating the shelf-life of similar products should be employed to comprehensively evaluate the shelf-life of the tested product under actual (or ambient) storage conditions. When extrapolating ambient data from accelerated test data, prior experience can be invaluable if the Arrhenius plot is nonlinear.

Finally, we may establish an acceptable cut-off level for the shelf-life of a food product by determining the correlation between oxidative stability data and sensory evaluation data generated by a trained sensory panel.

CHALLENGES POSED BY HETEROGENEOUS, MULTIPHASE FOOD SYSTEMS

In the past few decades, the oxidative stability and shelf-life of bulk oils/fats, oil-in-water, and water-in-oil food emulsions have been widely investigated (Chaiyasit et al., 2007; Choe and Min, 2006. McClements and Decker, 2000). A great progress has recently been made (Laguerre et al., 2015). In bulk oil, the crucial site of oxidation is not the air-oil interface but the association colloids formed with trace of water and surface active molecules. In oil-in-water emulsion, the nonlinear influence (or cut-off effect) of the hydrophobicity on antioxidant capacity was found. Currently, the oxidative stability and shelf-life of neat oils/fats are relatively well understood (Skibsted, 2010); however, those of low moisture or dry food products have not been studied thoroughly. Methods for assessing lipid oxidation in complex food systems containing oils and fats are still in their infancy (Eldin, 2010), although emulsified and dry food products are a significant category in food and pet food markets. Food matrix has appreciable impact on oxidative stability of low moisture foods. An antioxidant has been shown to be efficacious in rapeseed oil, but not so in cauliflower soup powder containing the rapeseed oil (Ratio et al., 2011). Lipid oxidation in solid phases may not be exactly same as that in liquid phase and further study is needed (Hu, 2014). Further, some low-moisture food products contain both lipids and proteins. Thus, lipid co-oxidation with protein may need to be considered when evaluating such products (Schaich, 2014; Wanibadullah, 2013). Finally, the oxidized product aldehydes such as hexanal in the dry food may further oxidize to form hexanoic acid. Therefore, it may be difficult to assess the oxidative stability using the evolution of hexanal.

As previously mentioned, it is highly challenging to evaluate the shelf-life and oxidative stability of dry foods containing oils/fats or oils/fats containing ingredients with varying oxidative stability and different degrees of freshness. PV and hexanal level curves over storage time, for instance, may not be normal. In particular, the initiate PV and hexanal levels could be higher—even much higher—than those tested in a bulk oil (Figure 2(A) and (B)). This could be because one or two oils/fats-containing ingredient(s) may have begun oxidation while these ingredients were used to produce the dry food. Actually, the oxidation of the dry food involves mixed oils/fats with varying oxidative stability and different degree of freshness, which is more complicated than the oxidation of bulk oil and fat. In addition, a dry food may also contain both lipids and proteins. Lipid co-oxidation with protein may decrease or change PV and hexanal

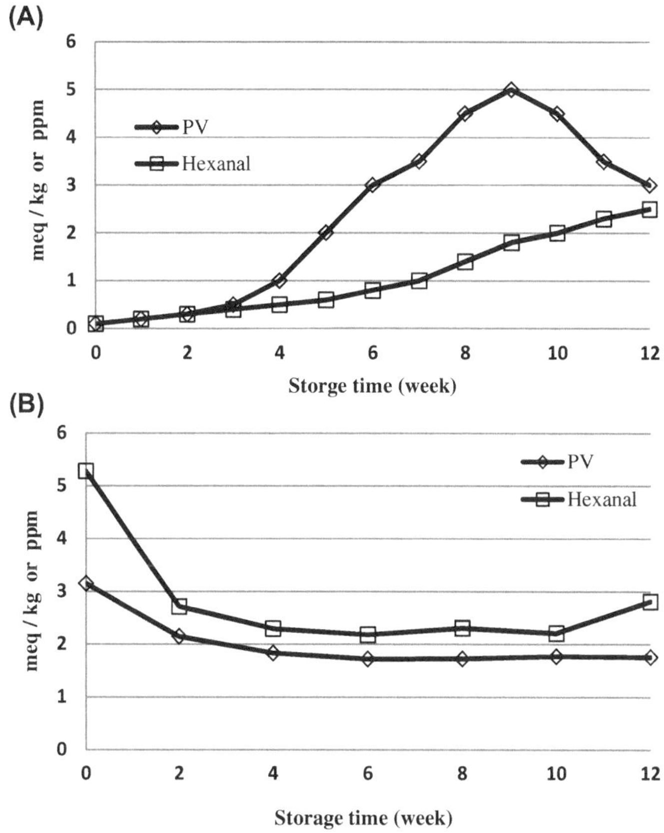

FIGURE 2 (A) The course of oxidation of bulk fat. (B) The course of oxidation of fat in a dry food containing both fat and protein.

levels during storage, while protein oxidation tends to increase the concentration of protein carbonyl compound (Dong, 2011; Ratio et al., 2011). The formation of hexanoic acid may also decrease the hexanal level over the storage.

Figure 2(A) shows the changes of PV and hexanal in bulk fat or oil during storage. Figure 2(B) exhibits the change of PV and hexanal in a dry food containing both fat and protein. Lipid co-oxidation with protein may explain unusual PV and hexanal changes in dry foods containing both lipids and proteins during storage, because the reactions between hydroperoxide or hexanal and protein may result in the decrease of PV and hexanal during storage compared with pure bulk fat. In addition, unusual PV and hexanal changes also may be attributed to oils/fats with varying oxidative stability and different degrees of freshness, as well as the further oxidation of secondary oxidation products such as hexanal in a low-moisture food.

FACTORS INFLUENCING AND DETERMINING OXIDATIVE STABILITY

There are internal (intrinsic) and external (extrinsic) factors that impact oxidative stability and shelf-life of oils/fats and oils/fats-containing foods. Internal factors for edible bulk oils/fats include oil/fat types, degree of unsaturation, fatty acid profiles, free fatty acid content, mono- and diacylglycerols, free radicals and minor components, such as chlorophylls, carotenoids, tocopherols, tocotrienols, phospholipids, sterols, and polyphenols (Choe and Min, 2006). For instance, bulk animal fats, such as poultry fat and lard, are generally more oxidatively stable than plant oils such as corn, sunflower, and canola oils, which are more oxidatively stable than fish, algal, and flax seed oils. This is primarily due to the different fatty acid profiles and varying degrees of unsaturation of fatty acids in those oils and fats. Minor components in oils/fats also impact oxidative stability and shelf-life (Abuzaytoun and Shahidi, 2006). For example, mixed tocopherols, tocotrienols, carotenoids, and polyphenols in vegetable oils may increase the oxidative stability of vegetable oils. However, moisture, transition metal iron (e.g., ferrous iron), free radicals, free fatty acids, and chlorophylls in plant oils may decrease the oxidative stability of the plant oils. As for oils/fats-containing foods, internal factors may also include the type of food matrices, water activity, moisture content and other components, such as protein, as well as the quality of ingredients containing oils/fats. In general, a bulk oil may be more oxidatively stable than a dry food containing the bulk oil, which, in turn, may be more oxidatively stable than an oil-in-water emulsion containing the bulk oil. The optimal water activity range for oxidative stability in some dry food products may be from 0.2 to 0.4 according to food stability map, but other dry foods do not always follow the food stability map (Ratio et al., 2011). Hence polymer glass transition theory was put forwarded to explain lipid oxidation and predict oxidative stability of lower and intermediate moisture foods (Nelson and Labuza, 1992). Interactions between lipids and other components, like protein and amino acids, have an appreciable impact on oxidative stability of the oils/fats-containing food products. Negative impacts can be attributed to lipid co-oxidation with protein, as in tortilla chips, peanut butter, dry pet foods, and protein meals, while positive impacts might be seen when protein and amino acids produce an antioxidant effect to increase the oxidative stability of food emulsions. These proteins may include whey protein isolate, soy protein isolate, and casein in oil-in-water emulsions (Hu et al., 2003a,b). The quality of ingredients (raw materials) containing oils/fats is also a critically important factor in the production of oxidatively stable oils/fats-containing food products.

External factors influencing lipid oxidation may include temperature, light, transition metals, enzymes, air (oxygen), processing and storage conditions, packaging, antioxidants, ratio of surface area–to–oils/fats volume, and environmental and systematic pHs. The presence of oxygen in air is indispensable to lipid oxidation since there is no auto-oxidation without oxygen. Thus food packaging is important in extending the shelf-life of oils/fats-containing food products. It is well understood that transition metal, such as ferrous iron, reacts with hydroperoxide to form lipid alkoxyl radical (LO·), leading to the formation of secondary lipid oxidation products. Iron and copper therefore significantly decrease oxidative stability of oils/fats-containing foods. In the food industry, antioxidants are required to prevent lipid oxidation in food products by free radical scavenging and transition metal ion chelation. Enzymes also have an impact on lipid oxidation. For instance, lipo-oxygenase in food products may react with LHs to form hydroperoxide. Lipase may react with triglycerides in the presence of water to yield free fatty acids. The ratio of surface area–to–oils/fats volume would appreciably influence the oxidative stability of bulk oils/fats and oils/fats-containing food products. In general, a small ratio of surface area–to–volume ratio of bulk oil may make it much more stable than oil-in-water emulsions with a larger such ratio, although the opposite relationship has been observed. Finally, environmental pH values impact the oxidative stability of food products and the efficacy of antioxidants. For example, baicalein and quercetin show a strong ability to bind transition metal ions at around neutral pH, but would not work well or efficiently at a lower, acidic pH (i.e., <4). This is because pH has a significant impact on protonation and deprotonation of the phenolic hydroxyls in flavonoids, which in turn influences chelating activity of those flavonoids.

MINIMIZING LIPID OXIDATION AND EXTENDING SHELF-LIFE OF OILS/FATS AND OILS/FATS-CONTAINING FOOD PRODUCTS

Controlling lipid oxidation and extending the shelf-life of food products are essential for most sectors of the food industry and pet food industry. Strategies for minimizing lipid oxidation should focus on factors that influence oxidative stability.

Decreasing processing and storage temperatures is critical for most food products since temperature is a key factor affecting the rate of lipid oxidation. Minimizing exposure to light and oxygen during processing and storage is also highly important. Modified atmosphere package and new active packaging offering the controlled release of antioxidants can help further extend the shelf-life of oils/fats-containing foods (Estaca et al., 2014; Silva et al., 2014). Oxygen absorbents, too, have been useful in increasing the shelf-life of packaged foods. It is also important to remove pro-oxidant transition metals, free radicals, and oxidized compounds by using chain-breaking antioxidants to scavenge free radicals as well as chelators to chelate transition metals (Decker et al., 2001). Commonly used natural and synthetic chain-breaking antioxidants in food industry include tocopherols, catechins in green tea extract, carnosic acid and carnosol in rosemary extract, proanthocyanidins in grape seed extract, and punicalagins in pomegranate extract, as well as BHA (butylated hydroxyanisole), BHT (butylated hydroxytoluene), PG (propyl gallate), TBHQ (tertiary butylhydroquinone), and ascorbic palmitate. Commonly used chelators are citric acid,

EDTA (ethylenediaminetetraacetic acid), phosphate, polyphosphates, hexanetaphosphate, pyrophosphate, quercetin, plant extracts and proteins, such as lactoferrin, casein, soy protein isolate, and whey protein isolate, as well as small peptides like carnosine. A big challenge for the food industry is that natural, effective chelators are limited, and the application range of natural chelators from plant extracts is similarly limited. Matrices may vary widely among different food product types including edible bulk oils, oil-in-water emulsions (such as mayonnaise), water-in-oil emulsions (such as margarine), dry powders (such as encapsulated fish oil powder or milk powder), or dry foods (such as extruded chips). Lipid oxidation can be minimized, however, if the appropriate antioxidant or antioxidant blend is selected for a food product with a specific matrix.

Moreover, food companies should try to obtain ingredients with low levels of hydroperoxides, transition metals, other pro-oxidants and impurities. Unfortunately, most food ingredients contain at least some impurities that can appreciably increase challenges in meeting oxidative stability and shelf-life goals. It is critical for food scientists and industry managers to gain practical experience and know-how in the evaluation of oxidative stability and shelf-life in commercial food products made with unpurified ingredients. Further, oxidatively unstable oils or fats may be blended with oxidatively stable oils or fats in food industry practices. Finally, designing a food product with a small ratio of lipid surface area-to-volume can in some cases substantially increase oxidative stability and shelf-life.

WHAT YOU WILL FIND IN THIS BOOK

The first three chapters of this book introduce basic analytical methods for measuring lipid oxidation in edible oils/fats and oils/fats-containing food products, methods of determining and predicting the shelf-life of food products, and the sensory evaluation of oils/fats and oils/fats-containing food products. Chapters 4 to 12 cover oxidative stability and shelf-life in different food systems using the various methods described in the first three chapters. The food systems include vegetable oils, fish and algal oils, and other oils with long-chain PUFAs, animal and poultry fats, frying oils and fried foods, food emulsions, complex dry foods, meat and meat products, sea foods, as well as crackers, cookies, and biscuits. The final chapter reviews oxidative stability and active packaging.

REFERENCES

Abuzaytoun, R., Shahidi, F., 2006. Oxidative stability of algal oils as affected by their minor components. J. Agric. Food Chem. 54, 8253–8260.

Akoh, C.C., 1994. Oxidative stability of fat substitutes and vegetable oils by the oxidative stability index method. J. Am. Oil Chem. Soc. 71, 211–216.

Allen, J.C., Hamilton (Eds.), 1994. Rancidity in Food, third ed. Springer, London, p. 31.

Carlson, C.U., Møller, J.K.S., Skibsted, L.H., 2005. Heme-iron in lipid oxidation. Coord. Chem. Rev. 249, 465–498.

Chaiyasit, W., Elias, R.J., McClements, D.J., Decker, E.A., 2007. Role of physical structure in bulk oils on lipid oxidation. Crit. Rev. Food Sci. Nutr. 47, 299–317.

Choe, E., Min, D.B., 2006. Mechanisms and factors for edible oil oxidation. Compr. Rev. Food Sci. Food Saf. 5, 169–186.

Decker, E.A., Elias, R.J., McClements, D.J. (Eds.), 2001. Woodhead Publishing Limited, Cambridge, Cambridgeshire, UK, pp. 3–35.

Dong, Y., October 2011. Protein Modifications in Baked versus Fried Tortilla Chips (M.S. thesis). The State University of New Jersey, Rutgers.

Eldin, A.K., 2010. Methods to determine the extent of lipid oxidation in foods. In: Decker, E.A., Elias, R.J., McClements, D.J. (Eds.), Understanding mechanisms of Oxidation and Antioxidant Activity, vol. 1. Woodhead Publishing Limited, Cambridge, Cambridgeshire, UK, p. 191.

Eskin, M., Robinson, D.S., 2000. Food Shelf-Life Stability: Chemical, Biochemical, and Microbiological Changes. CRC Press, Boca Raton, Florida.

Estaca, J.G., Dicastillo, C.L.O., Munoz, P.H., Catala, R., Gavara, R., 2014. Advances in antioxidant active food packaging. Trends Food Sci. Tech. 35, 42–51.

Fennema, O.R., 1996. Food Chemistry. Marcel Dekker, Inc., New York, p. 226.

Frankel, E.N., 1998. Lipid Oxidation, first ed. The Oily Press, Bridgwater, England, pp. 43–54, 79–81, 99–113.

Frankel, E.N., 2005. Lipid Oxidation, second ed. PJ Barnes and Associates, Bridgwater, England.

Frank, E.N., Meyer, A., 2000. The problem using one-dimensional methods to evaluate multifunctional food and biological antioxidants. J. Sci. Food Agric. 48, 1925–1941.

Fu, B., Labuza, T.P., 1997. Shelf life testing: procedures and prediction methods for frozen foods. In: Erickson, M.C., Hung, Y.C. (Eds.), Quality in Frozen Foods. Kluwer Academic/Plenum Publishers, London.

Genot, C., Meynier, A., Riaublanc, A., Chobert, J.M., 2003. Protein alterations due to lipid oxidation in multiphase systems. In: Kamal-Eldin, A. (Ed.), Lipid Oxidation Pathways. AOCS Press, Champaign, Illinois, pp. 265–292.

Hu, M., 2014. Lipid oxidation in dry pet foods. Inform 25 (1), 7–10.

Hu, M., Skibsted, L.H., 2002. Antioxidative capacity of Rhizome extract and Rhizome knot extract of edible Lotus (*Nelumbo nuficera*). Food Chem. 76, 327–333.

Hu, M., McClements, D.J., Decker, E.A., 2003a. Lipid oxidation in corn oil-in-water emulsions stabilized by casein, whey protein isolate and soy protein isolate. J. Agric. Food Chem. 51, 1696–1700.

Hu, M., McClements, D.J., Decker, E.A., 2003b. Impact of whey protein emulsifiers on the oxidative stability of Salmon oil-in-water emulsions. J. Agric. Food Chem. 51, 1435–1439.

Institute of Food Science and Technology, 1993. UK, Shelf Life of Foods: Guidelines for Its Determination and Prediction, London. .

Kilcast, D., Subramaniam, P., 2000. The Stability and Shelf Life of Food. Woodhead Publishing Limited, Cambridge, England.

Kilcast, D., Subramaniam, P., 2011. Food and Beverage Stability and Shelf Life. Woodhead Publishing Limited, Cambridge, England.

Labuza, T.P.A., 1979. Theoretical comparison of losses in foods under fluctuating telerature sequences. J. Food Sci. 44, 1162–1168.

Labuza, T.P., Riboh, D., 1982. Theory and application of Arrhenius kinetics to the prediction of nutrient losses in foods. Food Technol. 32, 66–74.

Labuza, T.P., Schmidl, M.K., 1985. Accelerated shelf life testing of foods. Food Technol. 39, 57–64, 134.

Laguerre, M., Bayrasy, C., Panya, A., Weiss, J., McClments, D.J., Lecomte, J., Decker, E.A., Villeneuve, P., 2015. What makes good antioxidants in lipid-based systems? The next theories beyond the polar paradox. Crit. Rev. Food Sci. Nutr. 55, 183–201.

Lloyd, M.A., Drake, M.A., Gerard, P.D., 2009. Flavor variability and flavor stability of US-produced whole milk powder. J. Food Sci. 74 (7), 334–343.

Lu, F.S.H., Bruheim, I., Haugsjerd, B.O., Jacobsen, C., 2014. Effect of temperature towards lipid oxidation and non-enzymatic browning reactions in krill oil upon storage. Food Chem. 157 (2014), 398–407.

Mazalli, M.R., Bragagnolo, N., 2007. Effect of storage on cholesterol oxide formation and fatty acid alterations in egg powder. J. Agric. Food Chem. 55, 2743–2748.

Maisuthisakul, P., Gordon, M.H., pongsawatmanit, R., Suttajit, M., 2007. Enhancing the oxidative stability of rice crackers by addition of the ethanolic extract of phytochemicals from Cratoxylum formosum Dyer. Asia Pacific J. Clin. Nutr. 16 (Suppl. 1), 37–42.

Man, D., Jones, A., 2000. Shelf-Life Evaluation of Foods. Aspen Publishers, Inc., Gaithersburg, Maryland.

Matthaus, B., 2010. Oxidation of edible oil. In: Decker, E.A., Elias, R.J., McClements, D.J. (Eds.), Oxidation in Foods and Beverage and Antioxidant Applications. Management in Different Industry Sectors, vol. 2. Woodhead Publishing Limited, Cambridge, Cambridgeshire, pp. 183–237.

McClments, D.J., Decker, E.A., 2000. Lipid oxidation in oil-in-water emulsions: impact of molecular environment on chemical reactions in heterogeneous food systems. J. Food Sci. 65, 1270–1282.

Meynier, A., Leborgne, C., Viau, M., Schuck, P., Guichardant, M., Rannou, C., Anton, M., 2014. *n*-3 fatty acid enriched eggs and production of egg yolk powders: an increased risk of lipid oxidation? Food Chem. 153, 94–100.

Nelson, K.A., Labuza, T.P., 1992. Relationship between water and lipid oxidation rates: water activity and glass transition theory. In: St. Angelo, A.J. (Ed.), Lipid Oxidation in Food. ACS, Washington, DC, pp. 93–103.

New Zealand Food Safety Authority, 2005. A Guide to Calculate the Shelf Life of Foods, Information Booklet for the Food Industry, Wellington, New Zealand. .

Paradiso, V.M., Summo, C., Antonio Trani, A., Caponio, F., 2008. An effort to improve the shelf life of breakfast cereals using natural mixed tcopherols. J. Cereal Sci. 47, 322–330.

Pokorny, J., Kolakowska, A., Bienkiewicz, G., 2005. Analysis of interaction products of oxidized lipids with amino acids, proteins and carbohydrates. In: Kamal-Eldin, A., Pokorny, J. (Eds.), Analysis of Lipid Oxidation. AOCS Press, Urbana, Illinois, pp. 263–280.

Raitio, R., Orlien, V., Skibsted, L.H., 2011. Storage stability of cauliflower soup powder: the effect of lipid oxidation and protein degradation reactions. Food Chem. 128, 371–379.

Sanches-Silva, A., Costa, D., Albuquerque, T.G., Buonocore, G.G., Ramose, F., Castilho, M.C., Machado, A.V., Costa, H.S., 2014. Trends in the use of natural antioxidants in active food packaging: a review. Food Addit. Contam. Part A 31 (3), 374–395.

Schaich, K.M., 2014. Lipid co-oxidation of proteins: one size does not fit all. Inform 25 (3), 134–139.

Shahidi, F., Wanasundara, U.N., 2008. Methods for measuring oxidative rancidity in fats and oils. In: Casimir, C.A., David, B.M. (Eds.), Food Lipids: Chemistry, Nutrition and Biotechnology. CRC Press, Taylor & Francis Group, Boca Raton, F.L.

Skibsted, L.H., 2010. Understanding oxidation processes in foods. In: Decker, E.A., Elias, R.J., McClements, D.J. (Eds.), Oxidation in Foods and Beverage and Antioxidant Applications. Understanding Mechanisms of Oxidation and Antioxidant Activity, vol. 1. Woodhead Publishing Limited, Cambridge, Cambridgeshire, UK, pp. 3–35.

Tikk, K., Haugen, J.E., Aaslyng, M.D., 2008. Monitoring of warmed-over flavour in pork using the electronic nose-correlation of sensory attributes and secondary lipid oxidation products. Meat Sci. 80, 1254–1263.

Wanibadullah, W., October 2013. Lipid Protein Interactions in Peanut Butter (Ph.D. dissertation). The State University of New Jersey, Rutgers.

Chapter 1

Analysis of Lipid and Protein Oxidation in Fats, Oils, and Foods

K.M. Schaich
Department of Food Science, Rutgers University, New Brunswick, NJ, USA

1.1 INTRODUCTION

During nearly 20 years of the low-/no-fat era, lipid oxidation was essentially forgotten as a stability problem. Now, the era of healthy fats focuses on reformulating foods with highly unsaturated (polyunsaturated) fatty acids (PUFAs). PUFAs are essential – they cannot be synthesized by humans and they are required for many physiological processes – but also are extremely sensitive to oxidation and can very rapidly degrade food quality, function, and nutrition by production of off-flavors and odors, loss of essential fatty acids, and co-oxidation of proteins and vitamins. Thus, the need for accurate and sensitive tracking of lipid oxidation has once again moved to the forefront in food analyses and quality control.

A book published by the American Oil Chemists' Society, *Lipid Oxidation: Challenges in Food Systems* (Logan et al., 2013), addressed issues that pose stumbling blocks to understanding, analyzing, and stabilizing lipid oxidation. The current book moves forward to practical problems in dealing with lipid oxidation in foods. Chapter 2 in the preceding book presented a number of conceptual and practical challenges to analyzing lipid oxidation (Schaich, 2013b). Readers are referred to that chapter because it provides a critical base for this chapter, which moves forward to consider options in analyzing lipid oxidation.

Deciding how to measure lipid oxidation and its associated co-oxidations is much more complicated than just selecting an assay from a list or following procedures of standardized methods. The assay selected must match the chemistry of lipid oxidation, the information needed, and the equipment and time available. Some key questions that must be considered include:

- Are you measuring continuously over time, infrequently (weekly or monthly), or only once on isolated samples?
- Are you most interested in early or secondary degradation products, or do you need both for a more accurate picture?

Oxidative Stability and Shelf Life of Foods Containing Oils and Fats
http://dx.doi.org/10.1016/B978-1-63067-056-6.00001-X

- Do you need detailed data for research or just markers that correlate with sensory responses or known rejection points?
- How fast do you need the information?
- Are co-oxidations likely to re-route lipid oxidation radicals and secondary products to other molecules, particularly proteins, which must then also be monitored to determine system oxidation?

Research on mechanisms requires more detailed product analysis over longer time periods from initial to extended oxidation, whereas strategies to assess effects of processing or antioxidants on stability must follow changes rapidly, and quality control analyses are more often conducted as isolated analyses on samples with unknown extent of oxidation.

Now add to these issues new information indicating that lipid oxidation pathways and formation of lipid oxidation products may not follow the sequence that has dictated most oxidation analyses in current use (Schaich, 2005) but have addition, internal rearrangement, scission, and other reactions that compete with the well-known hydrogen abstractions. Research and industrial food laboratories alike have encountered situations in which standard lipid oxidation products such as hydroperoxides or carbonyls are measured but the results do not correlate with fatty acid loss, sensory evaluations, or off-odors and yellowing in the product. Further, products expected to be secondary frequently begin forming at the same time as conjugated dienes (CDs) and hydroperoxides rather than after observable hydroperoxide decomposition. Thus, old "rules" of how to measure lipid oxidation do not always provide accurate pictures of lipid oxidation. Most critically, they often underestimate total oxidation levels and misrepresent active pathways.

To address these challenges, this chapter argues for a new paradigm in analysis of lipid oxidation. Rather than focusing on single products, analytical strategies must move toward analyzing multiple classes of products to detect all active pathways, to avoid missing some oxidation, and to more accurately assess the extent and pathways of lipid oxidation. The main focus is on analyses used or likely to be applied in the food industry. Highly specialized methods such as electron paramagnetic resonance, nuclear magnetic resonance, and mass spectrometry (MS) requiring expensive instrumentation are not included.

1.1.1 Fundamental Processes of Lipid Oxidation

For more than 70 years, lipid oxidation has been viewed as a radical chain reaction driven by hydrogen abstractions. Hydroperoxides are the first stable product and their decomposition radicals serve as precursors for all secondary products that form at the end of oxidation chains (Figure 1.1). Many books and reviews have presented details of initiation processes, kinetics, and reactions leading to secondary products (Lundberg, 1961; Scott, 1965; Chan and Coxon, 1987; Frankel, 2005b; Schaich, 2005). In the context of oxidation analysis, a key characteristic of the chain reaction is that there is a specific sequence in

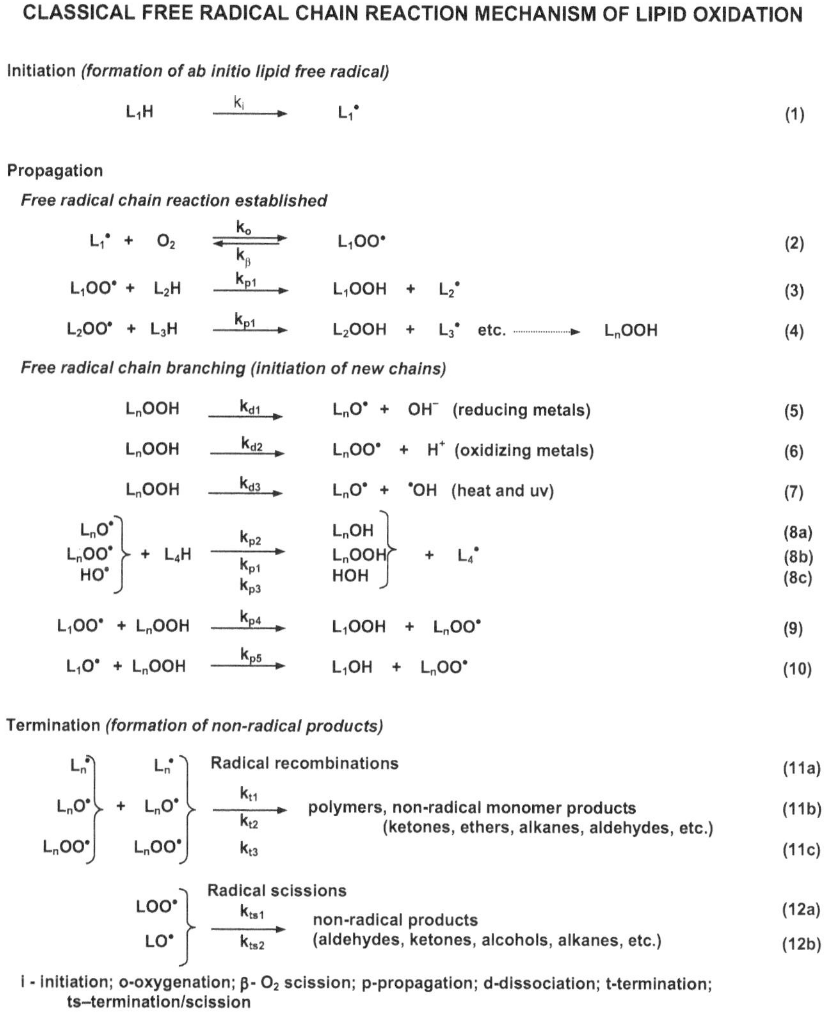

FIGURE 1.1 Traditional free radical chain reaction of lipid oxidation. *From Schaich (2005); used with permission.*

formation of products – first CDs, then hydroperoxides, and secondary products only evolve after hydroperoxide decomposition. Assumptions also have been that lipid oxidation is relatively slow, so products need only be monitored weekly or monthly in shelf-life studies. Indeed, most analytical strategies have been designed to fit with this scenario.

However, experimental results for both kinetics and products often have been inconsistent with the traditional reaction scheme. Attempts to find explanations identified a number of reactions of peroxyl and alkoxyl radicals that compete with hydrogen abstraction and have the potential for greatly complicating lipid oxidation. These alternate reactions were combined with the traditional sequence of hydrogen abstractions to develop an integrated reaction scheme that may more fully and more accurately describe the complex reactions of lipid oxidation, shown in Figure 1.2 (Schaich, 2005, 2012, 2013a).

The existence of simultaneous alternate pathways has several critical consequences that must be considered when analyzing lipid oxidation. Products do not wait until the end of oxidation to form. Addition, rearrangement, and dismutation reactions of lipid peroxyl radicals (LOO•) generate dimers, epoxides, and aldehydes from the beginning of oxidation in parallel with hydroperoxides, and they can divert radicals from the formation of hydroperoxides. Indeed, we have recently observed epoxides forming before and at higher levels than hydroperoxides in oxidizing methyl linoleate (Xie, 2015). Lipid alkoxyl radicals, LO•, also undergo alternate reactions of internal rearrangement, scission, and addition in competition with hydrogen abstraction, so expected aldehydes are not always present. As a consequence, measuring only the traditional scission products hexanal and decadienal may miss major proportions of products and significantly

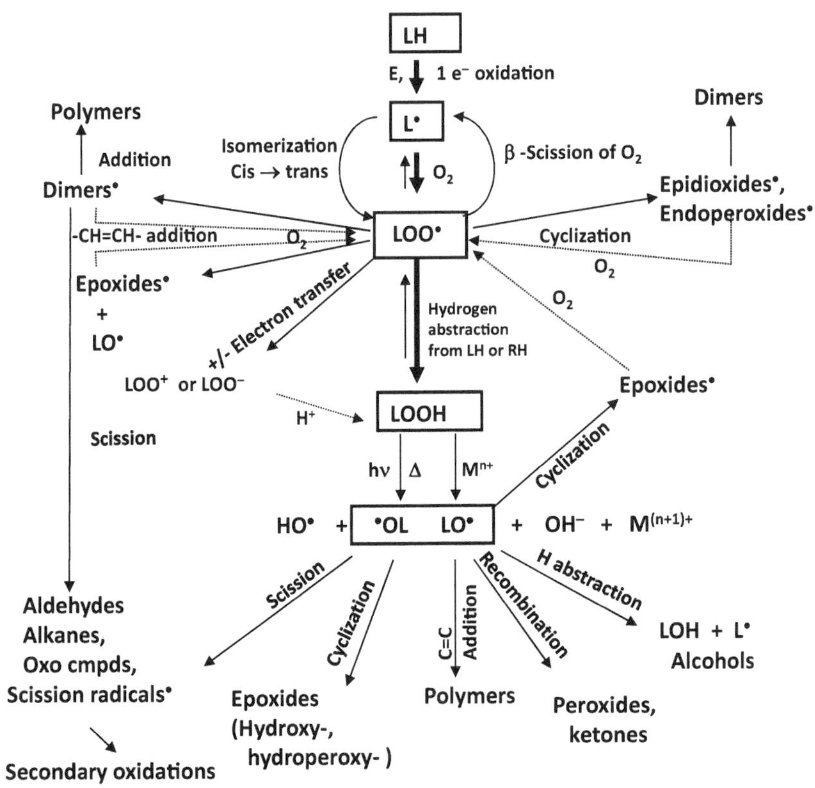

FIGURE 1.2 Reaction scheme integrating competing alternate reactions (side loops) with traditional hydrogen abstraction (boxes center). *From Schaich (2005); used with permission.*

underestimate the extent of lipid oxidation. Activation energies of the alternate pathways vary, so the balance between pathways shifts with conditions. This means the same products will not be appropriate markers for oxidation in every oil or food product during production, after normal versus accelerated storage, in early oxidation versus in late oxidation, and in wet foods versus dry foods. Altogether, these observations demonstrate that multiple pathways and products must be tracked to accurately assess the extent of lipid oxidation.

In complex foods, the situation is complicated even further by the presence of co-oxidations in which lipid oxidation radicals and secondary products react with other system components, particularly proteins. Co-oxidations have the effect of moving oxidation to other molecules in the food system while at the same time consuming lipid oxidation intermediates and products. Under these circumstances, lipid oxidation can appear to be low or inhibited when actually it is actively and rapidly broadcasting oxidation to other molecules. Hence, measures of co-oxidations appropriate for the food system must be included to assess the full extent of oxidative degradation and to avoid underestimating lipid oxidation.

Strategies for analyzing lipid oxidation in the past have usually been rather simplistic, monitoring only a few products (e.g., CDs, hydroperoxides, and perhaps aldehydes as hexanal or anisidine reactants) or using only a single assay (e.g., Oxidative Stability Index [OSI] or gas chromatography [GC] of volatiles). With the challenges of today's food formulations and complex matrices, it is time to move forward with more comprehensive and more accurate analyses of lipid oxidation in foods. Thus, in presenting an overview of analytical methods and critical considerations for measuring lipid oxidation in foods, this chapter seeks to provide a bridge between major methods traditionally employed in the food industry and methods that are perhaps not as well known but are becoming necessary to provide cleaner analyses and a broader, more accurate picture of full system oxidation in foods. Included in the bridging is an introduction to methods for analyzing protein co-oxidations to provide a guide to tracking the footprints of lipid oxidation in other molecules.

1.2 HANDLING CONSIDERATIONS CRITICAL FOR LIPID OXIDATION ANALYSES

The practices outlined in this section may seem obvious but their effects should never be underestimated. Most of the considerations raised apply across the board when handling lipids, but they are absolutely mandatory for preventing inaccuracies and artifacts when analyzing lipid oxidation. Indeed, it will be emphasized repeatedly throughout this section that extraction and handling procedures adequate for analysis of lipid composition or other properties cannot always be applied when oxidation analyses are the endpoint. Unless the handling considerations described here are the cornerstone of all lipid handling and lipid oxidation analyses, results are erratic and inconsistent and have high-oxidation backgrounds.

1.2.1 Glassware Washing

Washing of glassware may seem like a mundane, routine laboratory procedure but it is absolutely critical when handling lipids, especially for oxidation analyses. Borosilicate glass found in most beakers, flasks, and test tubes contains metals and also binds and traps metals in surface pores. The metals then provide very active sites to catalyze lipid oxidation and to transform products. Traces of lipids not fully removed during cleaning also provide radicals and secondary products to catalyze oxidation of subsequent samples. Thus, both prevention of oxidation during normal handling and accurate analyses of lipid oxidation require removal of metals, lipids, and other contaminants in all glassware being used.

A simple but effective protocol for routine glassware cleaning uses the following steps: (1) Wash glassware with an acidic, phosphate-free detergent to remove surface residues, rinse three times with tap water followed by three times with double-distilled, deionized water purified to 18-MΩ resistivity, such as that obtained with Milli-Q or Barnstead purification systems; (2) soak glassware overnight in denatured alcohol saturated with potassium hydroxide to saponify lipid traces and hydrolyze other contaminants, including proteins; (3) repeat step 1; (4) soak overnight in 1 N low-metal hydrochloric acid prepared with 18 MΩ water to dissolve metals on glass; and (5) rinse three times with 18-MΩ water and dry upside-down in hot oven to eliminate traces of bound water. Since background metal levels in laboratory reagent acids and alkalis can be very high, all reagents used in these steps should be the highest purity (especially lowest metal content) reasonably available to avoid adding further contamination. Ultrex™ level reagents are desirable but horrendously expensive for glass washing except for highly specialized chemistry.

Special applications and highly sensitive reactions or assays will require soaking glassware for short times in stronger acid solutions such as 6 mol/L hydrochloric acid or aqua regia (1:3 v/v concentrated nitric and hydrochloric acids) after the two cleaning steps described in the previous paragraph. For example, we use aqua regia cleaning before and after lipid reactions with metal catalysts. Other procedures are available for removing trace metals but either have the potential to leave oxidizing components on the glass (chromic, sulfuric, and nitric acids), are toxic (chromic acid) or are not useful for large scale cleaning (McCormick, 2006). Aqua regia solutions are extremely corrosive and must be handled in a hood with great caution to avoid explosions or skin burns. Guidelines for handling strong acids and aqua regia are available online (Princeton University, 2014; University of Illinois, 2015).

Issues with glassware washers. Questions have been raised about use of laboratory glassware washers for lipid oxidation applications. This author has never had the luxury of using a laboratory glassware washer so cannot speak from direct experience. Nevertheless, I will share my concerns about glassware washers for lipid oxidation analyses and individual laboratories can decide for themselves which method is acceptable. First, nearly all laboratory glassware washers are made of stainless steel, which sounds good, but all stainless steel, even the high-purity SS 316, leaches metals in acidic environments. The acidic detergents recommended for applications where proteins and metals must be dissolved increase potential leaching from the stainless steel, and metals dissolved in the wash water then become deposited on the glass. Each washer model must be carefully inspected to assure that water supplies or circulation never contact non–stainless steel metal parts. One copper or galvanized steel connector can be disastrous for lipid oxidation applications. Detergents must be free of phosphates, surfactants, and other components that may remain as residues on the glassware. Washer models claiming water volume reduction recycle small volumes of wash and rinse water. This may be acceptable for general applications, but not when sensitive chemistry is involved because it concentrates contaminants that may adhere to the glass. Rinsing with tap water is totally verboten due to high contamination with metals and other potential oxidation catalysts. Even rinsing with doubly distilled deionized water is questionable since there are many levels of remaining metals in deionized water, depending on generation method and handling. Water purified to 18-MΩ resistivity in systems such as Milli-Q™ or Barnstead™ should be used, but connecting the laboratory cartridge systems to the washer and providing large volumes of this water rapidly would require modifications. Some detergents claim ability to remove metals, proteins, and greases during short wash times that clean by high-pressure water circulation rather than filling the washer and allowing for soaking. This procedure is adequate for most laboratory applications that are not supersensitive to trace contaminants but is untested for lipid oxidation and other redox applications. Overall, this author is skeptical that automatic glassware washers can remove residues and contaminants adequately for lipid oxidation studies, but as noted above, this is without personal experience.

1.2.2 Solvents

Like glassware, solvents are mostly taken for granted in general laboratory analyses, but solvent purity, stability, and oxygen solubility are critical when dealing with lipids and lipid oxidation products.

1.2.2.1 General Solvent Properties and Selectivity

Which solvent to use is always a quandary, particularly with current pressures to replace the traditional chlorinated hydrocarbons due to their carcinogenicity and neurotoxicity. A number of comparisons of solvent extraction capabilities have shown that, in general, solvent composition has a relatively small effect on the extraction of predominant lipid classes (triacylglycerides, cholesterol esters, and phosphatidylcholines) but strongly influences solubilization of the more polar lipid classes (Hara and Radin, 1978; Reis et al., 2013).

This gives analysts considerable flexibility in solvent selection for tailoring extractions. Seldom considered during solvent selection, however, is that solvents have different effects on lipid oxidation than on lipid yields because they intervene in oxidation and analytical reactions, and solvents giving high lipid yields may either miss lipid oxidation products or enhance oxidation during extraction. Individual laboratories have probably observed solvent artifacts and influences, but organized information about specific effects of different solvents on solubility, reactions, and stability of lipid oxidation products is not available. Thus, the section that follows is intended not to rehash well-known solvent properties, but to emphasize their effects on lipid oxidation and lipid analyses.

Chloroform. Decades ago, Schmid demonstrated that chloroform was the most efficient solvent for extraction of all but the most polar lipid classes (Schmid and Hunter, 1971; Schmid, 1973) and indeed, chloroform now is considered the quintessential universal lipid solvent. Chloroform disrupts hydrogen bonding between lipid and protein thus promoting release of all lipids. It also extracts proteolipids, which are active oxidation targets, and it inactivates lipoxygenases in muscle foods and plants, thus protecting lipids from enzymatic oxidation during extraction. However, chloroform also has a donatable hydrogen that can be abstracted by lipids, and the $CCl_3OO^•$ radicals that result can intervene in both oxidation chains and analyses (cf., hydroperoxide analyses).

In biological tissues, methanol is usually added to chloroform in 2:1 or 1:1 proportions to increase solubilization of very polar lipid fractions such as sphingholipids and gangliosides in biological tissues and to improve solvent miscibility with the high water materials. Although adding methanol is general practice also in foods, is not always necessary since chloroform alone can extract the important lipid fractions in most foods. Methanol addition is beneficial in muscle tissues, dairy products, and high-moisture foods but can increase extraction of nonlipid materials, particularly starch, in dry foods.

As an important practical point, chloroform is easy to remove by vacuum evaporation.

Methylene chloride (dichloromethane). Methylene chloride is similar to chloroform in extraction capabilities yet has a number of properties that affect oxidation, particularly slightly higher polarity (Reichardt, 2003), lower hydrophobicity, higher water solubility, and mixed gas solubilities (see Sections 1.2.2.3 and 1.2.3). Thus, it does not behave the same as chloroform in extractions of all materials or in all oxidation analyses. That methylene chloride adsorbs moisture and does not separate from water phases as cleanly as chloroform causes particular problems for oxidation applications.

Although less toxic than chloroform, methylene chloride is still a potent neurotoxin. Like chloroform, it is on the Substance Priority List of Hazardous Substances of the Agency for Toxic Substances and Disease Registry (ASTDR, 2014), so should not be considered a "safe" alternative to chloroform.

Hexane. Among the most hydrophobic nonpolar solvents, hexane is good for removing lipid in layers or isolated bodies. However, it does not dissolve polar lipids, particularly phospholipids and many oxidation products. Such behavior is very helpful in oilseed and food oil extractions before refining, but in foods, hexane by itself misses components important for lipid oxidation analyses. For total lipid extractions, hexane must be mixed with a polar solvent such as an alcohol. Hexane is immiscible with methanol, but the two solvents can be used together in pressurized solvent extraction where they are added to the sample under pressure separately but simultaneously. Hexane:isopropanol 3:2 mixtures were originally demonstrated by Hara and Radin (1978) to give high yields and clean extracts; the isopropanol eliminates heme contamination in tissue extracts, including meat (Rose and Oklander, 1965). Although it took some time to attract attention, the mixture is now increasingly being used as one substitute for chloroform:methanol. Despite its solvent comparability, hexane is not the best substitute for lipid oxidation analyses due to its high oxygen solubility (see Section 1.2.3).

Iso-octane. With even greater hydrophobicity and lower volatility and water solubility than hexane, iso-octane is also being considered as replacement for chlorinated hydrocarbons in extractions and analyses. Like hexane, iso-octane must be mixed with an alcohol to extract polar lipids and products. Iso-octane:isopropanol mixtures have been suggested for extractions but may not be the best choice for lipid oxidation analyses due to high oxygen solubility in iso-octane (Table 1.2) and peroxide formation in isopropanol. The very high boiling point of iso-octane also makes it difficult to evaporate from lipids after extraction. At the same time, this solvent extracts very little, if any, nonlipid material such as starches and proteins, so iso-octane extracts may be used directly in analyses.

Petroleum ether. This classic lipid solvent has been used extensively to extract neutral lipids from many types of materials. Being very nonpolar, it does not dissolve phospholipids, polar nonlipids such as starch, or oxidation products. Its main application is extraction of free or surface lipids as well as fatty acids released after acid hydrolysis.

Diethyl ether. A classic fat solvent, diethyl ether dissolves all fats and oxidation products. However, it also dissolves nonlipid materials and picks up water during extractions, it promotes phospholipase D activity in tissues and forms peroxides readily, so it is not a solvent of choice for oxidation analyses. It is useful mainly for removing surface lipids or fatty acids after acid digestion. Diethyl ether should not be used when an alternative is available due to its volatility, anesthetic effects, high flammability, and explosion risks.

Methyl-tert-butyl ether (MTBE). This synthetic solvent, initially used as an oxygenate in unleaded gasoline to improve combustion efficiency, has been proposed as a chloroform substitute due to lack of evidence for human toxicity (Matyash et al., 2008). However, MTBE is carcinogenic in rats (Anonymous, 1998). MTBE may have limitations as a lipid solvent because it is flammable and forms azeotropes with water (52.6 °C; 96.5% MTBE) and methanol

(51.3 °C; 68.6% MTBE). There is as yet relatively little experience base with lipids extractions or oxidation using this solvent.

Methanol. Methanol has long been the preferred solvent for phospholipids, as well as the major polar modifier for very hydrophobic solvents. It dissolves fatty acids and oxidation products but not triacylglycerols unless enough phospholipid is present as a co-solvent. Because methanol co-extracts high concentrations of hemes, which catalyze rapid oxidation in extracts and also interfere with analyses, it should be replaced with isopropanol to eliminate hemes from extracts of muscle foods (Rose and Oklander, 1965).

Isopropanol. Isopropanol is an interesting solvent with mixed effects. It is a good solvent for complex plant and animal tissues because it solubilizes both polar and nonpolar lipids while limiting extraction of pro-oxidant pigments chlorophyll (Ramluckan et al., 2014) and heme compounds (Rose and Oklander, 1965). It also inhibits phospholipase D and lipases (Christie, 1976), thus minimizing degradation of lipids by tissue enzymes. However, oxygen solubility in isopropanol is double that in methanol (Section 1.2.3; Table 1.2) so extra care is needed to prevent oxidation during handling. Also, isopropanol forms peroxides (Redemann, 1942) and these must be removed by distillation or other treatment before use in extractions or analyses.

Water-saturated butanol. This is the traditional solvent for disrupting starch complexes and releasing bound lipids in cereals but the heat required for expanding starch helices also modifies oxidation products. Butanol is not a good solvent for studying lipid oxidation because it promotes phospholipase D activity in plant and animal tissues, extracts significant amounts of nonlipid materials and is difficult to evaporate at temperatures that do not promote lipid oxidation, even in vacuum.

1.2.2.2 Solvent Stability

Many solvents used for extracting or dissolving lipids form peroxides that interfere with assays and can catalyze lipid oxidation as samples stand. Of these, ethyl ether, tetrahydrofuran, isopropanol, and butanol form peroxides most readily and must be protected by keeping all solvent bottles sparged (liquid and headspace) with argon and tightly sealed after opening, blocking out light and maintaining minimum headspace (Clark, 2001), and adding peroxide decomposers such as XPell commercial pellets that are inert and act much like molecular sieves in solvents (XploSafe, 2012). In addition, ethyl ether must be stored frozen because of its volatility.

Peroxides in water-immiscible solvents may be detected by shaking 2 mL solvent with 1 mL freshly prepared 10% potassium iodide in 18-MΩ water and adding a drop of starch indicator solution. Peroxides are indicated by development of color -- light yellow-brown for low levels of peroxides or blue for active peroxides (Cyberlipid, 2014). Solvents testing strongly positive for peroxides should be discarded; solvents with low levels of peroxides should be distilled immediately before use in lipid oxidation assays or extractions.

Ethanol is usually considered to be a stable solvent. However, it oxidizes to aldehydes during storage. Thus, ethanol used in oxidation analyses should be stored under argon. Presence of aldehydes can be detected by reaction with 2,4 dinitrophenylhydrazine. Carbonyl-free ethanol for extractions and assays can be prepared by adding 1% dinitrophenylhydrazine and a few drops of concentrated hydrochloric acid, refluxing for 3 h, then distilling. Solvent blanks must be run in all oxidation analyses to eliminate contributions from contaminants.

Chloroform, the solvent used most often with lipids, oxidizes to form $CH_3^{\bullet}$ and $CH_3OO^{\bullet}$ radicals (Cunningham et al., 1985; Connor et al., 1994; Rosen et al., 1999), so should never be stored with oxidizing lipids. Chloroform also forms toxic phosghene ($COCl_2$) and hydrochloric acid over time (Sigma-Aldrich, 2012). This can be prevented by addition of stabilizers and by storing under argon after opening. The best stabilizer to use is dictated by the material and analytical application. Ethanol is the preferred chloroform stabilizer from a toxicity standpoint, but must be added at high concentrations (1–2%) (Sigma-Aldrich, 2012). This increases solvent polarity, which can alter extractions and reactions. Ethanol also often complicates incubations and analytical reactions because it is susceptible to radical attack by oxidizing lipids, forming reactive radicals that accelerate oxidation (Schaich et al., 1993), and it complexes with lipid and other radicals (Wu et al., 1977; Rockenbauer and Tudos, 1979; Gardner et al., 1985). Short-chain alkenes added at about 100 ppm similarly stabilize solvents by scavenging radicals, but in the process, form radicals that can interfere with analyses. Cyclohexene interferes with gas chromatographic analysis of products (Sigma-Aldrich, 2012), and amylene (1-pentene) generates high backgrounds in the ferric thiocyanate hydroperoxide assay (Richards and Feng, 2000). Alkenes also are less effective in preventing phosgene formation over time (Turk, 1998). Thus, we have found it necessary to screen solvents for each assay to maximize protection and minimize interferences.

1.2.2.3 Water Solubility in Solvents and Vice Versa

Although methylene chloride (dichloromethane) is increasingly substituted for chloroform in lipid extractions and solvation for lower toxicity, it does not match chloroform properties exactly. As already mentioned, one particularly irksome issue is that methylene chloride has higher solubility than chloroform in water, and water is more soluble in methylene chloride (Table 1.1). This makes extracts more difficult to dry, definitely complicates phase separations in procedures like the Folch wash, and also allows some diffusion of catalysts between phases. In addition to these physical effects, water in solvents can alter chemistry – it stabilizes some products and directs shifts in degradation pathways which change both levels of oxidation and distribution of products. Thus, chemically methylene chloride is not the solvent of first choice for lipid extractions and analyses when oxidation is the analytical end point. All solvents used for all lipid oxidation analyses should be dried for accurate detection of

TABLE 1.1 Water-in-Solvent and Solvent-in-Water Solubilities of Organic Solvents Used with Lipids

Solvent	Solvent Solubility in Water (%)	H_2O Solubility in Solvent (%)
n-Octane	6.6×10^{-7}	0.0095
n-Hexane	0.00123	0.0111
1-Octanol	0.0538	(—)
n-Hexanol	0.706	7.42
1,2-Dichloroethane	0.81	0.32
Chloroform	0.815	0.0135
Dichloromethane	1.3	0.093
Methyl isobutyl ketone	1.7	1.9
1,1-Dichloroethane	5.03	0.187
Diethyl ether	6.04	1.47
n-Butanol	7.45	16.9
Isobutanol	10	7.46
Methyl ethyl ketone	24	10
Methyl *t*-butyl ether	5.44	4.2

Extracted from Kislik (2012). MBTE data from UN Environment Program (1998).

chemistry and maximum reproducibility (Burfield et al., 1977; DeLloyd, 2006; Bradley et al., 2010).

1.2.3 Gas Solubility in Solvents

Both air and solvents contain oxygen that drives lipid oxidation, so the obvious first rule of thumb in handling lipids is to maintain test materials under inert atmosphere at all times. Obviously, vacuum affords most protection, but few laboratories actually handle samples on vacuum lines except for exacting basic research. As an alternative for protecting lipids during normal handling, headspaces and solvents must always be sparged with inert gas, and whenever possible the materials themselves must also be saturated to eliminate dissolved or entrapped oxygen.

Nitrogen sparging or blanketing, most commonly used in lipid research due to habit and lower cost, is widely assumed to prevent lipid oxidation. Not so! Oxygen solubility in most solvents is significantly higher than that of

TABLE 1.2 Solubility of Oxygen, Nitrogen, and Argon in Common Lipid Solvents at Room Temperature (293 K)

	Mol Fraction × 10^4		
Solvent	**Oxygen**[a]	**Nitrogen**	**Argon**
Octane	21.22		
Iso-octane	26.81	14.13[c]	29.2[d]
Heptane	19.98		25.0[d]
Hexane	19.60	14.0[d]	25.3[d]
Cyclohexane	12.31	7.55[d]	14.9[d]
Toluene	9.17		10.95[d]
Acetone	8.34	5.19[c]	
Tetrahydrofuran	8.03		
Butanol	8.03	4.52[c]	
Chloroform	7.23[b]	7.89[b]	3.82[b]
Dichloromethane	5.61[b]	3.27[b]	5.02[b]
Ethanol	5.87	3.55[c]	
Methanol	4.15	2.75[c]	
Isopropanol	7.88	4.66	
Olive oil	0.51		

[a]*Unless otherwise noted, data compiled from Battino et al. (1983).*
[b]*Shirono et al. (2008).*
[c]*Kretschmer et al. (1946).*
[d]*Jolley and Hildebrand (1958).*
From Schaich (2013b); used with permission.

nitrogen (Table 1.2), nitrogen is less dense than air so it escapes more rapidly, and the oxygen in even high-purity nitrogen is sufficient to drive lipid oxidation (Schaich, 2013b). Rather, all lipid research should use argon, preferably high purity.[1] Argon is a much more protective atmosphere because it is heavier than air and thus does not escape from solutions or vessels readily, it does not contain the traces of oxygen present in nitrogen, and argon solubility in both water and oil is double (or more) that of nitrogen (Gunstone and Padley, 1997).

1. Vendors vary, but for our source standard pre-purified nitrogen has <5 ppm O_2 while high purity argon has <1 ppm O_2. Multiple purities of each gas are available. Individual vendors should be consulted for O_2 contents of specific product grades.

Argon is more soluble than oxygen in water, and it is 2.5 times more soluble than nitrogen in water up to 50 °C (Battino and Clever, 1966). More importantly for lipids, argon is more soluble than both oxygen and nitrogen in most organic solvents except dichloromethane (Table 1.2). The disadvantageous balance between oxygen and inert gas solubility is one reason why dichloromethane is not the best choice of solvents for lipid oxidation analyses. Our laboratory uses high-purity argon exclusively in lipid handling. The extra protection it affords more than justifies its higher cost.

To be most effective, gas sparging (especially with nitrogen) must be alternated with vacuum and carried out for at least three 15- to 30-min cycles (Rollie et al., 1987).

Oxygen solubility in all the major solvents used in lipid research is substantial (Table 1.2), so solvents should not be overlooked as major sources of catalytic levels of oxygen. One of the reasons why chloroform, and now also dichloromethane, are preferred as lipid solvents is that they have nearly the lowest oxygen solubilities (Table 1.2). We became sensitized to this effect when hexane always caused more oxidation than chloroform in lipid extractions. In contrast, iso-octane, which is used in many oxidation assays, has triple the oxygen solubility of chloroform. Hence, it should be standard practice to significantly reduce or eliminate oxygen in all solvents used with lipids by distillation (and immediate use) or sparging thoroughly with inert gas. Most laboratories opt for the latter, but here solubilities of inert gases in the sample become the determining issue.

1.2.4 Light

Oils and food materials destined for oxidation analysis must be protected from light during storage, handling, and analyses. The main effect of UV light is to decompose hydroperoxides, while visible light catalyzes singlet oxygen–mediated oxidation when pigments or other photosensitizers are present. The effect of light is usually unpredictable since it simultaneously catalyzes formation and destruction of the same intermediates, but it can also be disastrous. For example, when purifying lipid hydroperoxides synthesized in high concentrations by lipoxygenase, we found that all hydroperoxides were lost going through a silica gel column in the laboratory. Yields improved when the column was wrapped with aluminum foil to block laboratory light, but expected yields were only achieved when the column was moved to a cold room under red light for full protection. We have attempted to track light effects during sample preparation in the laboratory and observed considerable variation depending on the food or oil form and composition.

Light also affects stability of nearly all reagents used in many oxidation analyses (cf., warnings in Material Safety Data Sheets). Thus, all protocols should include mandatory procedures to protect samples, extracts, and reactions from light, unless light is a required catalyst. This can be achieved by turning off laboratory lights when possible (assuming there is diffuse light from windows), wrapping test tubes or flasks in aluminum foil, incubating reactions in a closed drawer or dark box, or working under a red light. When in doubt, handle in the dark.

1.2.5 Handling and Storage

Temperature. It is a general rule of thumb when handling lipids, especially for oxidation analyses, to keep both handling and storage temperatures as low as possible. For storage, −80 °C is preferable to standard freezers (−10 to −20 °C), and with very sensitive materials we have sometimes found it necessary to conduct isolations and analyses in a cold room. Hydroperoxides are reasonably stable up to ~40 °C but decompose rapidly at higher temperatures. This means extractions and solvent evaporations should be limited to 40 °C or slightly lower. At 60 °C, which is frequently used in accelerated shelf-life studies, reactions change dramatically and do not follow Arrhenius kinetics, thus the conditions should not be construed as just rapid models of normal oxidations.

Time. A common practice is to extract samples during incubations and then store them frozen until analyses can be completed. However, extracts change rapidly even when they are stored at −80 °C under argon or vacuum, and analytical variability increases with time of storage. In kinetic studies, methyl linoleate most often showed a surge of oxidation during the first day or two followed by stabilization or decrease before resuming oxidation (Xie, 2015). In closed systems such as laminate packaging and sealed vials, we have observed iterative cycling of products (i.e., increase then decrease repeatedly over short periods), as well as shifting of individual products (e.g., aldehydes) even when total class analyses show relatively little change. The cycling largely disappeared when samples were analyzed with longer intervals between testing. Altogether, these observations indicate that new approaches to lipid oxidation analysis must focus on short time scales – analyze within hours of extractions, if not possible immediately, and with short intervals between sampling. Short, in this case, means no longer than daily for early oxidation and weekly for extended incubations in order to avoid missing key oxidation patterns.

Packaging. A critical line of defense against air is using glass containers that can be sealed rather than plastic materials that are oxygen-permeable. The common practice of handling and storing samples in plastic bags or containers before and during analyses allows at least slow diffusion of oxygen into test materials; with some plastics, samples might as well be sitting open on the laboratory bench. Polyethylene, polypropylene, polystyrene, and polycarbonate all have relatively high oxygen permeabilities, ranging from 1 to 4 mL/sq. inch (Massey, 2003).

Moisture permeability through these plastics is also high. Moisture permeability is lowest in polypropylene and high-density polypropylene (0.2–0.5 g/100 sq. inch), and it increases with polarity of the packaging polymer (e.g., polyethylene terephthalate, nylon, ethylene vinyl alcohol) (Massey, 2003). Thus, we often are faced with the conundrum of whether to block oxygen or moisture. Admission of either changes oxidation rates, pathways, and products (Schaich, 2013b).

Plastic packaging (bags and containers) also contains various plasticizers and additives, some of which are soluble in and reactive with lipids, and they can potentially interfere with product analyses. It is also almost impossible to eliminate all contaminants on the surfaces of plastics. Thus, for numerous reasons, glass containers are usually the best option for handling, storing, and analyzing materials and extracts destined for lipid (or protein) oxidation analyses. Teflon containers are acceptable alternatives but are expensive.

1.3 EXTRACTION OF LIPIDS FOR OXIDATION ANALYSES

Nearly all chemical analyses of lipid oxidation require extraction of lipids as a first step. Extractions, for this application, have three requirements—provide a representative lipid sample with maximum yield, avoid new lipid oxidation and degradation/transformation of existing products, and be able to handle large numbers of samples simultaneously. Unfortunately, these requirements impose stringent handling standards and limitations that make most standard extraction methods unacceptable when lipid oxidation is the analytical endpoint. The discussion that follows emphasizes extraction and handling considerations that are unique to lipid oxidation.

1.3.1 Pretreatments

Lipids adsorbed on the surface are relatively easy to remove, while lipids entrapped in food matrices or bound to other molecules require special treatments to increase their availability. Common pretreatments to disrupt physical matrices and facilitate lipid release include drying, homogenization or grinding to small particles, and digestion with enzymes such as proteases or amylases. Although necessary, each approach presents significant challenges for preventing adventitious oxidation and oxidation artifacts.

Drying. Water is often detrimental in extractions because it limits access of hydrophobic solvents to lipid regions in tissues and food matrices, it mobilizes catalysts and supports activity of degradative enzymes, and it mediates shifts in oxidation reactions, particularly alkoxyl radical scissions. Water plasticizes foods, making them more difficult to disperse during extraction, and it mixes with solvents to form azeotropes, which are difficult to remove by evaporation. To avoid these problems, samples may be dried prior to extraction but all precautions discussed above regarding handling definitely must be applied to avoid oxidation artifacts. In particular, no dehydration methods using heat can be used if oxidation is to be analyzed. Lyophilization is an acceptable alternative, but resulting matrices will be very porous and exposed, so samples should be analyzed immediately after drying.

Grinding. Grinding may seem to be a straightforward task, yet it presents significant challenges for lipid oxidation analysis because it must not degrade any food molecules and it must be performed under conditions that protect samples from air, heat, and light. First, the practice of grinding individual samples by any method and allowing them to sit on the counter under laboratory lights until all samples are prepared is unacceptable. Samples must proceed to extraction immediately or be frozen under vacuum until analyzed, preferably within a few hours at most. Second, nearly every method presents some limitations. One vexing problem is finding a method that does not result in the lipid plugging the grinder sieves, as occurs with Wiley mills. Mills with high shear, such as centrifugal mills, require cryogenic cooling (e.g., liquid nitrogen) to remove frictional heat as it is generated, and they should not be used at all when protein co-oxidation is to be measured because they break protein disulfide bonds and generate extensive sulfur radicals (Schaich, 2002). Hand grinding with mortar and pestle presents the least physical stress but requires long times that expose samples to greater adventitious oxidation, and it is often difficult to reduce particle sizes sufficiently. Food processors with argon blown in over the sample can be a useful alternative if grinding times are limited and samples are extracted immediately afterward. Early design cryomills that froze and pulverized products directly in liquid nitrogen often caused freezing damage to materials. However, new designs protect samples by freezing in containers cooled by liquid nitrogen or solid carbon dioxide (dry ice) and by pulverizing under inert gas. Although there is relatively little experience base with cryomilling at the present time, we recommend that increased attention be given to this method and feel that it could well become the grinding method of choice for lipid oxidation analyses.

Enzymatic digestion. Grinding does not always guarantee increased access to lipids entrapped in complex molecular matrices such as extruded products, baked goods, and other foods composed mainly of starches and proteins. In such cases, selective enzymatic digestion of the molecules that are blocking release of lipids can be beneficial. For example, incubation of 1 g extruded corn-soy product with 25 mg α-amylase overnight released starch-bound lipids quantitatively, and they could then be collected by extraction into 2:1 chloroform-methanol (Strange and Schaich, 2000). Careful argon saturation of the enzyme solution and sparging of the flask headspace as well as incubation in the dark were necessary to minimize lipid oxidation during the process. Predigestion with amylases or proteases can disrupt matrices of extruded products (Wang, 2000), thereby making lipids more accessible. Predigestion with trypsin or snailase (mixture of enzymes that digest most structural biomolecules) increased lipid extraction from microalgae from <10% to about 35% (Liang et al., 2012). Cellulases also increased lipid yields, but conditions required for enzyme activity (pH 4, 50 °C) would degrade oxidation products.

Acid hydrolysis. Acid hydrolysis is the method of choice for quantitative determination of total fatty acid content expressed as triacylglycerol equivalents. However, the lipids generated by this method cannot be analyzed for lipid oxidation because the required sample refluxing in 6N HCl or 5–10% HCl in methanol for 4–30h depending on food matrix and lipid target causes extensive degradation of lipid oxidation products. When measures of both lipid composition and lipid oxidation are required, separate but parallel extractions and sample preparation must be applied.

1.3.2 Extraction Methods

Many methods have been developed to enhance removal of lipids from biological tissues, both plant and animal, with the goal of maximizing yields for compositional analysis (Schaich, 2013b). Typically, the same methods have been adapted to extract lipids from foods, but the practice is not always valid, even for quantitation, due to differences in lipid levels and compartmentalization. Tissue extraction methods were designed for low lipid levels and become increasingly inaccurate as lipid levels increase above 2%, which is a very low lipid content for foods (APJCN-NHRI, 2015). Nearly all enhanced extraction methods have significant limitations when lipid oxidation is the issue because they cause some tissue damage to increase lipid yields (Schaich, 2013b). This both exaggerates levels and obscures mechanisms of native oxidation present. Loss of 1% or 2% of total lipids may not matter in quantitation, but extensively oxidized foods involve less than 1% of the total lipids. Unfortunately, this means that either separate extractions must be used for lipid quantitation and lipid oxidation analyses, or a single method that is efficient for both goals must be developed.

There is also the issue of extracting free versus bound lipids. Exhaustive extraction with acid hydrolysis to account for both free and bound lipids may be required for compositional analysis and nutrition labeling, but whether this is warranted for routine oxidation evaluation is questionable. Oxidation products in the free lipids – those readily extractable – are of greatest interest for industrial quality control analyses because they will be most strongly and immediately perceived by consumers. Bound lipids certainly oxidize and probably contribute significantly to co-oxidations in both foods and biological tissues. To be fair, the distinction between oxidation of free versus bound lipids has barely been studied, for good reason, so little definitive information is available. However, the chemical stresses required to disrupt the molecular associations should also degrade and modify oxidation products, which would invalidate subsequent analyses of oxidation. Thus, methods described below focus on extraction of lipids that are free (unbound) or associated weakly by forces that can be disrupted by solvent interactions.

Limitations of extraction methods for lipid oxidation were discussed previously (Schaich, 2013b). Only a general overview of the problems will be presented here. Needless to say, all of the solvent and inert gas issues discussed in Sections 1.2.2 and 1.2.3 apply.

Manual extraction as a baseline for comparison. The simplest extraction method is manual extraction, where food samples are allowed to sit in solvent for varying periods of time with or without shaking, then filtered, and the solvent is evaporated from the lipid residue. Manual extraction works best with foods in which the lipids are mostly coated or adsorbed on the surface or are distributed in the food matrix without complexation, entrapment, or binding; it is much less effective in removing lipids from complex matrices. Unfortunately, despite easy application, so much handling is required that preventing adventitious lipid oxidation is almost impossible. Manual extractions also require extensive amounts of glassware, and thus add significantly to cleaning requirements and cycling times for experiments. Thus, manual extraction is typically reserved for teaching students basic extraction concepts, for range-finding, and for situations where other methods are not available. Alternative methods are applied for increased lipid yields, for protection against oxidation, or both.

1.3.2.1 Methods that Improve Extraction Yields and Efficiency

Soxhlet, microwave, ultrasound, and supercritical CO_2 extractions have been developed to improve lipid yields over manual extractions by increasing solvent contact while breaking down molecular matrices that bind or entrap lipids. They are particularly helpful when lipid composition (identification or quantitation) is the analytical goal. Unfortunately, all of these methods apply stresses (heat, radiation, radicals) that degrade lipids and alter oxidation products, they require long times, and samples must be extracted singly or in limited numbers (Schaich, 2013b). Consequently, they are not suitable when analysis of lipid oxidation is the endpoint.

Acid hydrolysis as an extraction method must be distinguished from acid hydrolysis as an analytical method for determining fatty acid content. In extraction, the entire food sample is hydrolyzed to release lipids to determine fat content, while in analysis, lipid extracts are hydrolyzed to release fatty acids to determine fatty acid composition or to reconstruct total equivalent fat content (triacylglycerols plus all other esterified fatty acids). Acid hydrolysis for lipid extraction is used in two types of applications (Min and Ellefson, 2010):

1. Where lipid fractions are strongly bound to proteins and/or starches; in this case, samples are typically digested 1 h in 3 N HCl before extracting released lipids into solvents by any standard method;
2. To break emulsions, digest proteins, and release internal phase lipids which are then collected physically by centrifugation, such as in dairy products.

It is easy to see that both methods destroy all useful information about oxidation because heat and acid both induce some hydrolysis of triacylglycerols and also strongly modify lipid oxidation products.

Alkaline hydrolysis has been the preferred digestion method when determining lipid content for nutrition labels because it can harvest all metabolizable fatty acids wherever they are esterified with minimal degradation. These can then be extracted, analyzed by GC and reconstructed into fat (i.e., triacylglycerol) equivalents for nutritional accounting.

Acids and alkalies both contain high levels of metal contaminants, particularly iron and copper, which rapidly initiate lipid oxidation and also transform existing products. Thus, if hydrolysis cannot be avoided for particular food products, ultrapure reagents such as Baker Ultrex™ must be used.

The Folch method (Folch et al., 1951, 1957) and its Bligh–Dyer simplified version (Bligh and Dyer, 1959) have become universally accepted (almost required) methods for lipid extractions of tissues for biological and medical research. The Folch procedure was developed specifically to remove total lipids, including very polar but trace lipids such as gangliosides, from homogenized fresh tissues (animal or vegetable) and to eliminate nonlipid contaminants in the same procedure. Briefly, 1 g tissue is homogenized for 3 min with 19 mL chloroform:methanol 2:1 to extract lipids. The mixture is then filtered and the extract solvent phase is shaken with 0.2 volume of water or 0.05 mol/L potassium chloride to remove nonlipid contaminants such as proteins, pigments, antioxidants, and other alcohol-soluble compounds. Final solvent proportions (counting tissue water content) are chloroform:methanol:water 8:4:3. After standing or centrifugation, solvents separate into two layers, with nonlipid components in the top aqueous layer (C:M:H_2O = 3:48:47) and lipids in the bottom organic layer (C:M:H_2O = 86:14:1). Water washes loose some polar lipids and results in accumulation of a proteolipid foam at the phase interface (Folch et al., 1951). This foam layer is eliminated and lipid yields are increased when salt washes replace water.

Adding chloroform directly to tissues often leads to clumping and aggregation of proteins and occlusions of lipids. To avoid this, Bligh and Dyer (1959) homogenized tissue first with methanol:chloroform 2:1 to disrupt hydrogen bonding networks and electrostatic forces between lipids and proteins, then added more chloroform to dissolve lipids, and completed the procedure with water washes. Critical chloroform:methanol:water proportions here are 1:2:0.8 for the first stage and 2:2:1.8 for the second stage after addition of water.

Both methods have been applied extensively to plant as well as animal tissues with many modifications, and reportedly have also been applied almost indiscriminately to a wide variety of foods and food materials even though they were designed for high water materials (e.g., >80%) with very low lipid (2% lipid for Folch and 1% for Bligh–Dyer). "Reportedly" is used here because examination of actual conditions shows that many extractions listed as one of the two classics are actually modified chloroform–methanol extractions and do not follow exact solvent proportions or omit the washes. Authentic application of true Folch or Bligh–Dyer conditions to materials with compositions outside the targeted range requires adjustment of the water added, the solvent amounts and proportions, or sample size in order to maintain required sample–solvent proportions and solvent:water ratios. Even with adjustments, however, the solvent–sample proportions of Folch/Bligh–Dyer become increasingly inefficient (i.e., less than 100% lipid extracted) as lipid contents increase above 2%, with the degree of underestimation of lipid content increasing with lipid content of the material (Iverson et al., 2001). Perhaps yields are not considered when Folch/Bligh–Dyer extractions are applied to higher-lipid foods.

Folch and Bligh–Dyer procedures for extracting lipids are basically traditional chloroform–methanol extractions with water or salt washes added, mostly to remove nonlipid contaminants such as carbohydrates and proteins. One alternative for cleaning up food extracts while eliminating water and inducing fewer oxidation artifacts is filtration on Sephadex G-25 columns (Wells and Dittmer, 1963). Sephadex G-25 is first washed with 1N high-purity hydrochloric acid (e.g., Baker Ultrex™ or equivalent) to remove contaminant metals, then rinsed three times with 18 MΩ water, dried, and packed into a pipette or column (washed as described in Section 1.2.1). Lipids are loaded onto the support, then selectively removed with solvents in sequence, such as $CHCl_3$:MeOH (19:1) to elute most lipids, then $CHCl_3$:MeOH (19:1), water-saturated, with added acetic acid to elute gangliosides and acidic phosphatides. Polar nonlipid materials remain on the column.

As with the Folch/Bligh–Dyer extractions, the extra handling required for the Sephadex columns is a significant disadvantage for lipid oxidation analyses. It is difficult to avoid exposure to air (even with argon sparging), and light, solid-phase surfaces induce oxidation and alter oxidation pathways (e.g., both facilitate new epoxide formation and enhance decomposition of preformed epoxides); only a few samples can be handled simultaneously. Sephadex is useful for small extract volumes, but large extract volumes are not easily protected against oxidation and modification.

The clean-up procedure we prefer to minimize handling and artifacts in lipid oxidation analyses is removing initial solvent under vacuum followed by redissolving the extract in argon-saturated chloroform. This leaves behind residues of starch, protein, etc. although it occasionally retains some artificial pigments and additives along with the lipid fractions in the chloroform (these can be detected by thin layer or high-pressure chromatography). Most importantly, it minimizes handling and adventitious oxidation.

Restrictions on the use of chlorinated solvents have stimulated development of alternative solvent mixtures that can work with Bligh–Dyer procedures. A mixture of propan-2-ol–cyclohexane–water (8:10:11 v/v/v) gave lipid yields of seafoods (plaice, mussel, and herring) that were comparable to chloroform:methanol but less sensitive to sample-phase volume ratios (Smedes, 1999). Heptane/ethanol/water/sodium dodecyl sulfate (SDS) mixtures gave good yields if the SDS content remained low but tended to underextract phospholipids. Very low oxidation in the extracts suggested that these solvents did not extract lipid oxidation products (Undeland et al., 1998). Ethanol-water-hexane mixtures gave reasonable yields (80% of expected) with microalga (Fajardo et al., 2007). Hexane:isopropanol (3:2 v/v) mixtures were not very effective, probably due to the insolubility of hexane in water and loss of polar fractions (Khor and Chan, 1985).

This section cannot close without emphasizing issues of lipid oxidation in extractions. Despite their popularity, Folch and Bligh–Dyer extractions have critical disadvantages for oxidation endpoints. Water induces formation of lysolipids and high losses of polar lipids (including oxidation products) in the water phase, enhances transformation of some oxidation products, and facilitates scission of alkoxyl radicals. Metals in the salts and water catalyze lipid oxidation. Perhaps most importantly, it is exceedingly difficult to complete all the handling steps under inert atmosphere so adventitious lipid oxidation can be substantial. As noted previously, loss of a few percent of the lipids by oxidation is barely visible in analyses of lipid content and composition, but it is disastrous when extent of lipid oxidation must be accurately ascertained. Thus, Folch and Bligh–Dyer extractions may be appropriate for detailed studies of lipid content in meat tissues or fresh fruit and vegetable tissues but their strict application to processed foods or when lipid oxidation of the extracts will be measured is questionable.

1.3.2.2 Methods that Increase Extraction Efficiency While also Limiting Lipid Oxidation

So far, all the extraction methods described have increased lipid yields but posed serious problems for lipid oxidation analyses. Are there any extraction methods that *do not* compromise lipid stability yet efficiently remove lipids from complex matrices?

Pressurized solvent extraction (also known as accelerated solvent extraction [ASE]) is a relatively new method based on the principles of supercritical extraction, but it uses solvents instead of gases under pressure. Pressurized solvent extraction lacks the experience base and methodological development of the other techniques, but it offers protections and conditions that, in the experience and opinion of this author, make it the method of choice for lipid oxidation analyses. To increase permeation of solvent into the sample, degassed solvents are applied to the sample in sealed stainless steel cells under pressure (typically about 1500 psi) and elevated temperature. Elevated temperature decreases solvent viscosity, which makes it easier for pressure to force the solvent through even dense molecular matrices. Extraction efficiency is increased further by combining static and dynamic steps. In static extractions, solvent is held under pressure in contact with the sample for selected periods of time to force solvent through the molecular matrix to reach lipid and dissolve it. Dynamic extractions then flow fresh solvent continually through the sample to remove loosely bound lipids or lipids extracted in previous static steps. Static and dynamic extractions can be linked in many combinations and with repeated cycles to tailor the process to specific food compositions and structures (Singleton and Stikeleather, 1999; Mendiola et al., 2007; Herraro et al., 2013).

Pressurized solvent extraction offers a number of advantages for lipid oxidation analyses in foods because it limits handling while facilitating lipid removal from complex matrices without extreme treatments. Static extractions (periods where solvent is held in contact with foods) use pressure to drive solvent into the food matrix to dissolve lipids, then dynamic extraction steps wash the dissolved lipid out into collection flasks. Time, temperature, and pressure of static extractions, volume of dynamic extractions, and number of extraction cycles can be varied to optimize lipid removal from different types of materials. Indeed, these adjustments are mandatory for full lipid extraction. The samples are sealed in cells protected from light and pressurized under inert gas, the solvents used are degassed, and large numbers of samples (e.g., up to 24) can be extracted in a single programmed operation. Typically, extractions can be completed in 15–30 min per sample but can be extended when necessary.

In testing a broad range of food matrices, we have found that if extraction temperature is limited to 40 °C, lipids can be extracted with minimal new oxidation and no hydroperoxide decomposition (Yao and Schaich, 2014). Pressurized solvent extraction often eliminates the need for matrix digestions by amylases (e.g., extruded cereal products) and proteases (meat-based products), although grinding of samples to about 250-μm particle size is still necessary to allow efficient solvent penetration. Multiple extraction steps with different conditions (e.g., solvent) can be programmed and controlled automatically to extract specific fractions. Extracts can be collected separately with sequences of different solvents for individual class analyses or combined for assays of total lipids. High-purity diatomaceous earth in a special expanded form (commercially available as Hydromatrix™) disperses samples to allow more efficient solvent access and binds water and water-soluble contaminants, which reduces clean-up steps. Substantially lower solvent volumes are required than other extraction methods. Clearly capital equipment cost is an impediment to adapting this methodology, but the extracts are so much cleaner and so much easier to obtain that we recommend pressurized solvent extraction as the method of choice for lipid oxidation analysis of most food materials.

As a disadvantage, pressurized solvent extraction is relatively new so guidelines for extracting lipids from a wide range of materials are not yet available yet. Every matrix and lipid composition is different so optimum solvent and extraction conditions – solvent, static time, number of cycles, temperature – must be determined for every material. However, once procedures are optimized, efficient extractions can be automated.

Examples of successful lipid extractions by pressurized solvent extraction that showed lipid yields comparable to or better than other standard methods include baked and extruded pet foods (Yao and Schaich, 2014), ground corn kernels and freshly ground rolled oats (Moreau et al., 2003), fish tissue (Dodds et al., 2004); cereal, egg yolk, and chicken breast muscle (Schäfer, 1998); poultry meat (Toschi et al., 2003): and peanut meal (Singleton and Stikeleather, 1999). The method needs to be tested with emulsion products such as margarines and mayonnaise.

1.4 CHEMICAL ANALYSES OF OXIDATION PRODUCTS IN OILS AND EXTRACTS

The "meat and potatoes" of lipid oxidation analyses are, of course, the chemical assays for individual classes of products. While CDs and hydroperoxides are almost universally assayed, secondary products are followed variably or not at all, at least in part because assays that are sensitive, quantitative, and easy to handle are not available. Another quandary is which assay to use.

A consequence of the competing alternate pathways shown in Figure 1.2 is that one or two assays cannot be guaranteed to provide an accurate picture of oxidation. There is a very high risk of underestimating oxidation, or missing oxidation altogether, if an inactive or degraded pathway is analyzed. This is especially critical when discard decisions are made based on results tracking oxidation on production lines or during storage.

The time for defining oxidation by set levels of a single product is past. We have tested samples that were yellowed and smelled very bad yet showed low peroxide values (PVs) and little or no hexanal but had very high epoxide levels. In methyl linoleate, total carbonyls remain low but often shift in distribution of individual aldehydes with different oxidation conditions, altering odor impacts and potential for secondary reactions (Xie, 2015). Studies of oils oxidized at room temperature versus accelerated storage (60 °C) versus frying temperatures (150–180 °C) showed very different reaction mechanisms and product patterns (Qin, 2011; Repko Reader, 2013; Tian, 2013; Bogusz, 2015; Xie, 2015). Thus, no single assays would accurately reflect differences in the chemistry occurring under all three conditions.

Rather, we need to adopt a new analytical paradigm in which multiple products are assayed with as much detail as possible. At a minimum, CDs, hydroperoxides, epoxides, total carbonyls, and hopefully also alcohols should be monitored just to track activity in alternate pathways. Even better is determining individual products within classes, such as positions of hydroperoxides and specific epoxides and aldehydes. This information is of more than just academic interest. Consistent analyses with this level of detail will begin to develop a data base of more accurate oxidation patterns that can explain deterioration of product quality and suggest key points for intervention.

To provide some guidelines for understanding the various assays and for developing effective combinations of oxidation analyses, this section will cover principles, advantages, disadvantages and limitations, and applications of assays for individual classes of products. Detailed protocols are available in numerous resources elsewhere, such as Wrolstad et al. (2005), the Methods in Molecular Biology series, and standard methods of AOCS, AOAC, and IUPAC.

1.4.1 Conjugated Dienes

Principle of analysis. Lipids have no innate chromophores for optical detection. However, formation of radicals in polyunsaturated fatty acids (linoleic acid and above) causes a shift in double bond position, converting nonconjugated to conjugated fatty acids (Reaction (1)):

$$\text{–CH=CH–CH}_2\text{–CH=CH–} \longrightarrow \text{–CH=CH–CH=CH–C}^{\bullet}\text{H– and –C}^{\bullet}\text{H–CH=CH–CH=CH–} \quad (1)$$

Hence, CDs are the first relatively stable chemical marker of oxidation in polyunsaturated fatty acids (Parr and Swoboda, 1976). Because of the resonance, conjugated fatty acids absorb UV light in the range of 231–235 nm. In general, wavelength of maximum absorption for CDs increases with chain length and decreases with conversion of double bonds to *trans* configuration: ~235 nm for *cis-cis*, ~233 nm for *cis-trans*, *trans-cis*, and ~231 for *trans–trans* (Recknagel and Glende, 1984). There can be interference from nonlipid components absorbing near 233 nm, especially antioxidant phenols (Gruszka et al., 2008; Hornero-Méndez et al., 2001; Dobarganes and Velasco, 2002; Frankel, 2005b), so nonspecific absorbance (as shown by very high absorbance) is sometimes corrected for by measuring ratio of absorbance at 233 and 215 nm (oxidation index) (Klein, 1970).

Standardized procedures (AOCS method Ti 1a-64) report CDs as percent lipid weight to eliminate differences in lipid composition:

$$\%\,\text{CDs} = 0.84\left[(A_s/bc) - k_0\right]$$

where A_s is the sample absorbance at 234 nm, b is the cuvette length in cm, c is the lipid sample weight (g/L) in the final reaction mixture, and k_0 is the absorptivity by acids (0.03) and esters (0.07). The 0.84 factor is undefined. However, this method for reporting CDs is rather outdated.

For mass balance with other products, concentration of CDs can also be calculated on a molar basis from Beer's Law using known extinction coefficients (molar $\varepsilon = 29{,}500$ for iso-octane, 28,000 for ethanol, 25,200 for cyclohexane, and 30,000 for tetracyanoethylene) and normalizing to moles of lipid analyzed (Tian, 2013; Xie, 2015). When a single oil is being studied, direct comparisons of CD molarities can be made because the density of the oil samples is constant. Densities of most vegetable oils are sufficiently alike (0.91–0.94) that variation in CD concentrations occurs at the sub-micromolar level if variations in densities are ignored. Thus, general comparisons generally can be made without adjustments. However, when detailed differences between samples must be determined, the CD molarity determined from Beer's Law calculations must be corrected for density differences between oils (Reaction (2)):

$$\frac{(\text{Absorbance 234 nm}/\varepsilon) * \text{Rxn vol (ml)}}{\left[\dfrac{\text{Volume oil sample (ml)} * \text{oil density (g/ml)}}{\text{Oil or fatty acid mol wt}}\right]} = \frac{\text{mmol conjugated dienes}}{\text{mol lipid}} \qquad (2)$$

where ε is the extinction coefficient (molar absorptivity) in L/(mol cm) for the solvent being used.

Conjugated trienes form in higher polyunsaturated fatty acids during later stages of oxidation by simultaneous oxidation at both ends (C9 and C16) of linolenic acid and by reduction of hydroperoxides to alcohols followed by dehydration (Reaction (3)).

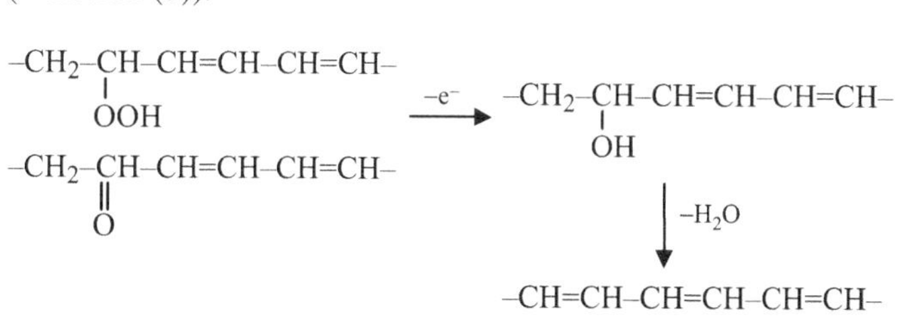

(3)

Conjugated trienes are detected at 268–275 nm (Fishwick and Swoboda, 1977). Note: carbonyls also absorb in this range, so absorbance should not be interpreted as conjugated trienes without secondary verification of structure.

Advantages. Being the first transformed product, CDs are good indicators of initiation and early oxidation in polyunsaturated lipids, and they are good markers for following initiation and progression of oxidation continuously, as in antioxidant testing. CDs are perhaps the simplest assay for lipid oxidation – lipids are dissolved in solvent and the absorbance is read at ~234 nm.

Disadvantages and precautions. The "peak" of CDs is often just a shoulder for which the actual absorbance maximum is difficult to determine. Absorbance at 234 nm is used as an average and recorded mostly by rote without appropriate blanks and corrections. Many solvents and contaminants absorb in the UV range, so high-purity solvents must be used to minimize background absorbance, and absorbance must be recorded over the range 210–250 nm for each set of samples to verify that an authentic peak exists between 230 and 235 nm. This is particularly important for extracts that may contain phenolic or carotenoid antioxidants, as noted earlier.

Corongiu and Milia (1983) have shown that converting the absorption curve into its second derivative gives more accurate determination of the actual peak. This, in turn, may improve differentiation of the diene species while also increasing accuracy of the analyses. Derivatizing spectra is simple with contemporary instrumentation and computer capabilities, so incorporating this modification into standard analyses deserves serious attention.

CD structures are retained in hydroperoxides but not when epoxides, addition, or scission products are formed. Thus, this assay becomes more inaccurate as a single marker as oxidation progresses, and it should be used with caution as a quantitative measure in spot analyses of products, such as checking oxidation of stored foods. Similarly, CDs are not an accurate quantitative measure of thermal degradation, where thermal scissions are a major driving force and hydroperoxides and other products are rapidly transformed. To avoid misinterpreting low CD levels, this assay should be combined with measures of other products in most applications.

Oxidation continues and CDs are transformed even during analysis so handling must be minimized, solvents must be sparged with argon before adding lipid, and samples must be mixed singly and analyzed immediately. Batch analyses are not feasible.

The lipid and solvent volumes specified in AOCS Ti 1a-64 (70–100 mg and 75 mL isooctane, respectively), designed primarily for industrial quality control laboratories, must be micronized for research and academic laboratories where only small samples are available and solvent use must be limited. AOCS Ti 1a-64 also specifies weighing, dissolving, warming, cooling, and diluting lipid samples. Progressive oxidation is very difficult to control under these conditions. Better protection is offered if oil samples (usually 10–30 μL depending on extent of oxidation) are pipetted directly into argon-saturated solvent, mixed by inversion, and absorbance is measured immediately. Results can easily be converted to lipid weight using oil density. Extracts that are dried and weighed after solvent removal can be analyzed directly by adding 10 mL argon-saturated iso-octane to the evaporator flask, swirling to dissolve lipid residue, and analyzing immediately; results are calculated on a weight basis. It can be argued that pipetting oil samples leads to inaccuracies and variability from inconsistent sample delivery. However, low retention pipette tips improve pipetting accuracy and more than counterbalance adventitious oxidation caused by excess handling.

Analyses can be slow because expensive quartz cuvettes are required. Most laboratories have only a few, and these must be thoroughly cleaned and dried between samples.

Applications. The assay can only be used with fatty acids containing two or more double bonds – it is not applicable to monounsaturated oils. Otherwise, CDs are routinely analyzed as early markers in most oils and lipid extracts. CDs can be analyzed readily in emulsions if water and oil phases are separated first, such as by solvent extraction, centrifugation, freezing (Alemán et al., 2015), or lyophilization (Belhaj et al., 2010). However, it has also been reported that dilute oil-in-water emulsions can be analyzed directly by recording spectra from 200 to 300 nm and analyzing second derivatives with multivariate partial least squares chemometrics (Baron et al., 1997). The main issue in all applications is limiting or removing background and interferences by computer subtraction of reagent and sample controls, sample clean-up, etc.

1.4.2 Hydroperoxides

Hydroperoxides are considered to be the first stable product in lipid oxidation so they are the most prevalent measure of lipid oxidation experimentally and are the best-understood product quantitatively. Indeed, PVs[2] have been correlated with sensory response, and a scale of "acceptability" has been developed where fresh oils have PVs of <1, cut-off values for industrial use is 5, and oils with 10 mEq hydroperoxide/kg lipid are considered rancid and rejected by taste panels. Even so, it must be stressed that hydroperoxides alone cannot accurately evaluate the actual level of lipid oxidation in an oil or extract. Low PVs can indicate little oxidation has occurred, but they also occur in active oxidation when hydroperoxides decompose to other products (Figure 1.1), such as at elevated temperatures or in UV light, and when peroxyl radicals generate other products in parallel or preference to hydroperoxides (Figure 1.2). On the other hand, disproportionately high PVs can result when hydroperoxides are stabilized, such as

2. Peroxide values have been expressed variously as milliequivalents of either oxygen or hydroperoxides per kg fat. Hydroperoxide concentrations determined from standard curves are almost always in molarity. Hence, the basis for calculations must be specified since oxygen equivalents are two times the hydroperoxide molarity.

by low temperatures or hydrogen bonding to phenolic antioxidants. Hence, at a minimum, hydroperoxides, CDs, epoxides, and carbonyls should be measured. Sensitive assays for all these products are now available.

Hydroperoxide assays were tested in detail for chemical accuracy, required handling, reproducibility, and sensitivity. Detailed evaluation of these assays may be found in Steltzer (2012). Only general information and guidelines for those most used are provided here.

1.4.2.1 Iodometric Titration (AOCS Cd 8–53)

Principle of analysis. Potassium iodide, KI, dissociates in saturated acidic solution to provide a source of electrons (Reaction (4)), which then reduce lipid hydroperoxides (LOOH) to hydroxyl derivatives (LOH) (Reaction (5)) (Pokorný, 2005). In the net reaction (Reaction (6)), reduction of one LOOH releases one I_2:

$$2KI \leftrightarrow 2K^+ + 2I^- \leftrightarrow I_2 + 2e^- \quad (4)$$

$$LOOH + 2H^+ + 2e^- \leftrightarrow LOH + H_2O \quad (5)$$

$$\underset{\text{(clear)}}{LOOH + 2H^+ + 2KI} \rightarrow \underset{\text{(yellow)}}{I_2} + LOH + H_2O + 2K^+ \xrightarrow{\text{starch}} \underset{\text{(purple)}}{\text{starch-}I_2} \quad (6)$$

The released I_2 is then quantitated by compleximetric titration with sodium thiosulfate (Reaction (7)).

$$\underset{\text{(yellow)}}{I_2} + 2Na_2S_2O_3 \rightarrow \underset{\text{(colorless)}}{Na_2S_4O_6} + 2NaI \quad (7)$$

In practice, lipid is dissolved in [chloroform or iso-octane]:acetic acid 3:2 v/v and KI is added for reaction. Water is added to stop the reaction and extract the molecular iodine into an aqueous phase for titration with sodium thiosulfate. Starch added as an indicator during titration forms a blue–purple complex with free iodine molecules. Transition from purple to clear marks the end point of the reaction when all the iodine has been reacted (Skoog et al., 1996).

Advantages. Although this assay has been much maligned, it is the only hydroperoxide assay we have tested that is chemically accurate and absolutely quantitative, dependent entirely on the peroxide functionality and independent of the R group of the hydroperoxides (Steltzer, 2012). Because of this accuracy, iodometric titration has been recommended for primary quantification of hydroperoxide standards used in other assays (Gay et al., 1999b). Iodometric titration is also the only assay that can handle highly oxidized lipids, and it can be performed in even minimal laboratories, requiring only an inert gas supply, a good magnetic stirrer, and a burette.

Limitations and precautions. Iodometric titration has two major limitations: low sensitivity (minimum detection about 0.5 mEq/kg lipid) and difficult handling. The assay is notoriously "empirical," requiring special care and exact conditions to yield reproducible results. Nevertheless, we have found that when attention is paid to the chemistry of the reaction, this titration has good reproducibility and achieves the most accurate and absolutely quantitative measures of hydroperoxides (Steltzer, 2012). Scrupulous oxygen removal is required to prevent oxidation of I^- and generation of erroneously high PVs (Mehlenbacher, 1960). This can be achieved by thoroughly sparging solvents with argon before reaction and maintaining active argon sparging throughout the titration. The reaction should take place in the dark or in diffused daylight to prevent decomposition of hydroperoxides (Pokorný, 2005). The reaction has two immiscible phases – lipids in chloroform (or iso-octane):methanol and potassium iodide in dilute acetic acid. For reaction with iodide ions to occur, the hydroperoxides must partition from the organic to the aqueous phase, and this requires active stirring or agitation (e.g., magnetic stirring). Gentle hand-swirling is inadequate (Steltzer, 2012). Adsorption or addition of iodine to unsaturated bonds of fatty acids, which decreases PVs, can be avoided by rapid titration.

The titration and special handling requirements make this assay somewhat tedious and very time consuming, so that large numbers of samples are difficult to handle. Autotitrators can facilitate high throughput but argon sparging, active stirring, and protection from light still must be maintained.

The factor most limiting sensitivity of this reaction is difficulty in visually distinguishing the color change marking the actual end point determination. The purple color fades gradually in many shades rather than colored to clear instantaneously, so reactions are often overtitrated. Visual end point detection may be obscured altogether with colored lipids or extracts. As an alternative, potentiometric titration can take the guesswork out of end point detection and reduce the detectable concentration of peroxides to nanoequivalents (Hara and Totani, 1988; Oishi et al., 1992), although deaeration requirements here are even more rigid than for the standard assay (Oette et al., 1963). Nevertheless, given the chemical accuracy of the KI reaction with hydroperoxides and the greatly increased sensitivity of electrochemical endpoints, this end point detection certainly merits more attention.

Finally, it has been claimed that large samples are required. If following the standard AOCS method, 5 g of sample are recommended for PVs <5, and about 1 g of sample for PVs >10 just to be able to see the end point clearly. However, smaller quantities of lipids are routinely analyzed with standard procedures, especially in research laboratories, and the method is still valid (Crowe and White, 2001). General guidelines for amounts of lipid required are given in Table 1.3 (Wrolstad et al., 2005). Less than 10 mg lipid is required when potentiometric end points are used (Hara and Totani, 1988).

Applications. This is the only assay that can detect high concentrations of hydroperoxides accurately without dilution. It is equally applicable to oils, extracts, reagents, solvents, and standard pure hydroperoxides as long as precautions outlined above are followed.

1.4.2.2 Xylenol Orange (XO)

Principle of analysis. The XO (or FOX) assay was developed to greatly increase hydroperoxide detection levels, reduce sample size requirements, and eliminate oxygen sensitivity in analyses of biological tissues (Jiang et al., 1991;

TABLE 1.3 Sample (Oil or Lipid Extract) Requirements for Iodometric Titration of Lipid Hydroperoxides

Expected Peroxide Value (mEq Active O_2/kg Lipid)	Weight Lipid Required (g)
0–2	5
2–10	2
10–25	1
25–50	0.5
50–100	0.3

From Wrolstad et al. (2005); used with permission.

Gay et al., 1999a). Ferrous ions in an acidic medium reduce hydroperoxides (Reaction (8)), and the resulting ferric ions are complexed by XO dye to produce a blue–purple complex that has an absorbance maximum at 550–600 nm (Reaction (9)) (Wolff, 1994; Gay et al., 1999b).

$$ROOH + Fe^{2+} \rightarrow Fe^{3+} + RO^{\bullet} + {}^{-}OH \quad (8)$$

$$Fe^{3+} + XO \rightarrow (Fe^{3+}...XO)\text{ complex} \quad (9)$$

Hydroperoxide concentrations are determined by comparison of absorbance values to a standard curve and normalization to lipid weight in the sample.

Two procedures for the Fe^{3+}-XO (or FOX) assay are in use. FOX1 is a water-soluble version containing sorbitol as a reaction enhancer (Wolff, 1994); it is used with protein and other nonlipid hydroperoxides. A second version (FOX2) omits sorbitol and adds butylated hydroxytoluene (BHT) and methanol (Wolff, 1994) for detection of lipid hydroperoxides. A number of variations using different solvents to solubilize lipids have evolved.

Advantages. The XO assay has high sensitivity, detecting nanomolar to micromolar levels of hydroperoxides in very small samples of oil or extract (typically 25 μL diluted) (Steltzer, 2012). It is a simple assay to set up, larger numbers of samples can be handled in batches, disposable test tubes can be used, and the assay is adaptable to plate readers for high throughput. The assay is available in commercial kits from several companies.

Limitations and precautions. Absolute quantitation of hydroperoxides such as that obtained with iodometric titrations is not possible because assays with optical detection depend on accurate extinction coefficients or comparison to standard curves for quantitation. Unfortunately, extinction coefficients vary not only with the reaction system (solvent and reagent proportions) but also with the hydroperoxide being analyzed, as shown in Table 1.4 (Gay et al., 1999b; Bou et al., 2008). It has now been shown in numerous studies that responses

TABLE 1.4 Variation in Extinction Coefficient of Xylenol Orange with Commercial Source, Hydroperoxide Structure, and Solvent; Unless Otherwise Noted, Measured at 560 nm

		ε (mol/(L cm)) Xylenol Orange Source[a]		
Hydroperoxide	**Solvent**	**Unspecified**	**Sigma**	**Aldrich**
Ferric sulfate[b]	25 mmol/L H_2SO_4	13,400		
Ferric nitrate[c]	50 mmol/L $HClO_4$	26,600		
Ferric chloride	25 mmol/L H_2SO_4	15,000		
H_2O_2	25 mmol/L H_2SO_4	43,000, 45,600, 45,000	44,000	34,360
H_2O_2[b]	25 mmol/L H_2SO_4	26,800		
H_2O_2	50% acetic acid		59,800	45,360
Cumene-OOH	25 mmol/L H_2SO_4	43,000	99,300	79,000
	50% acetic acid		116,200	88,750
t-Butyl-OOH	25 mmol/L H_2SO_4	43,000	98,000	78,900
	50% acetic acid		115,700	84,500
Linoleic acid-OOH	25 mmol/L H_2SO_4/90% MeOH	47,000	60,000	44,000
LDL-OOH	25 mmol/L H_2SO_4/90% MeOH	43,000		
Lipoproteins	25 mmol/L H_2SO_4	45,000		
Fats	25 mmol/L H_2SO_4	45,000		

[a]*The two chemical companies listed have merged since the initial publication of these data. It is not clear which product is currently marketed, but the differences in extinction coefficient listed illustrate the inaccuracies that can be introduced when published ε values are applied indiscriminantly to all systems. Standard curves must be run for each batch of xylenol orange and each analyte system.*
[b]*Measured at 540 nm.*
[c]*Measured at 550 nm.*
Excerpted from Gay et al. (1999b).

vary with hydroperoxide structure and with the batch and source of XO. Thus, separate standard curves must be generated for the specific conditions of each test system and each dye batch. Standard hydroperoxides should be quantitated independently by iodometric titration before plotting quantitation curves. As much as possible, use of lipid hydroperoxide standards generated from lipoxygenase and standardized by iodometric titration are preferable to cumene hydroperoxide. When standard hydroperoxides that match components of complex systems cannot be synthesized or are not available commercially, results should be expressed as equivalents of the hydroperoxide standard used (except H_2O_2 is not a good match for lipids) (Gay et al., 1999b).

A second major limitation is that stoichiometry of the reaction and accurate quantitation are both complicated by secondary reactions of the alkoxyl radicals with solvents. Normal stoichiometry of hydroperoxide reactions with XO should be 2 mol of iron oxidized per mole of hydroperoxide; variance from this value indicates involvement of additional reactions. H_2O_2, cumene, and *t*-butyl hydroperoxides each have interfering reactions that yield different stoichiometry and make them less than optimal standards for lipid hydroperoxides (Gay et al., 1999b). H_2O_2 gives a reaction stoichiometry of about 2.5 because the hydroxyl radicals generated from H_2O_2 reduction react with XO, yielding a second species that oxidizes Fe^{2+}.

$$Fe^{2+} + H_2O_2 \longrightarrow Fe^{3+} + HO^{\bullet} + HO^{-} \quad (10)$$

$$HO^{\bullet} + XO \longrightarrow HOXO^{\bullet} \quad (11)$$

$$HOXO^{\bullet} + Fe^{2+} \longrightarrow HOXO^{-} + Fe^{3+} \quad (12)$$

$$HO^{\bullet} + Fe^{2+} \longrightarrow Fe^{3+} + OH^{-} \quad (13)$$

Cumene and *t*-butyl hydroperoxides exhibit stoichiometries of about 5 in both aqueous (Gay et al., 1999b) and lipophilic (Steltzer, 2012) reactions because both hydroperoxides decompose to methyl radicals, which oxidize to additional radicals that also participate in the analysis.

Stoichiometry greater than 2 has also been reported for lipid hydroperoxides (Yildiz et al., 2003; Osawa et al., 2007; Steltzer, 2012) and most likely results from secondary contributions of solvent radicals generated by lipid alkoxyl radical reactions. Variation in measured PVs between studies arise in part from differences in solvents used to solubilize the lipids being tested. Differences in reactivity of solvent radicals and presence of preformed solvent hydroperoxides both contribute to PVs observed. Thus, it is critical to test solvent blanks for each run and subtract these from sample experimental values. In addition, since is it not possible to determine stoichiometry from analytes with unknown hydroperoxide structures and levels, Osawa et al. (2007) have recommended use of "adjustment factors" to eliminate stoichiometry variations by normalizing each system to iodometric titration, i.e., for each experiment, analyze some samples by both methods to determine the proportionality between the two and apply this factor to all analyses of that system.

The XO assay is sensitive to oxygen, which oxidizes ferrous ions and forms new peroxides from any secondary radicals generated during the reactions (Kolthoff and Medalia, 1949, 1951; Mihaljevic et al., 1996; Steltzer, 2012). Thus, the reaction should be run with all solutions carefully sparged with argon. Similarly, reagent ferrous iron oxidizes over time, so blanks of reagent iron need to be run with each analysis.

Light has no effect on XO or iron, but hydroperoxides decompose under UV light, giving two radicals ($LOOH \rightarrow LO^{\bullet} + {}^{\bullet}OH$) instead of the one $LO^{\bullet}$ generated by iron reduction. We have found more consistent results and more accurate stoichiometry when reactions were held in the dark until optical analyses.

Perhaps the most insidious problem with the assay when working with food analyses is that at high oxidation levels, excess hydroperoxides react with the XO–Fe complex and bleach the color (Steltzer, 2012). This bleaching explains observations that XO results are comparable to iodometric titrations at very low hydroperoxide concentrations but increasingly diverge as concentrations increase (Osawa et al., 2007). If analysts are not aware of this potential, low absorbances can be very much misinterpreted and oxidation underestimated. When samples have very low absorbance readings, bleaching can be distinguished from low peroxide levels by diluting samples sequentially and reanalyzing. Increase in absorbance with sample dilution indicates bleaching from high levels of hydroperoxides, in which case the assay must be conducted with smaller samples or higher dilutions of original oils or extracts.

Bou has reviewed a broad range of factors affecting performance of the XO reaction (Bou et al., 2008).

Applications. The XO assay has been used to determine hydroperoxides in a wide range of systems including oils (Nourooz-Zadeh et al., 1995; Yildiz et al., 2003; Tian, 2013); lipid extracts from plants, foods, and food materials (Hermes-Lima et al., 1995; Grau et al., 2000; DeLong et al., 2002; Navas et al., 2004); phospholipids in solution, liposomes, low-density lipoproteins and membranes (Jiang et al., 1991; Yin and Porter, 2003; Fukuzawa et al., 2006); and biological fluids (Södergren et al., 1998; Banerjee et al., 2004). It is applicable to any samples with low levels of oxidation as long as precautions noted above are followed. It is not useful for even moderately oxidized lipids without extensive dilution.

Comments on two kits available commercially. We have found it exceedingly difficult to run XO assays directly from independent reagents, due partly to differences in XO from different sources and partly to ubiquitous iron contamination in all reagents, particularly acids. Kits provide standardization of reagents which appears to be critical with this assay, but at a cost – exact reagents and reagent concentrations are not known so it is difficult to test accuracy and adjust reaction conditions when necessary. Some observations on two commercial kits are offered here.

PeroxySafe assay (MP Biomedical). Detection range: claimed by manufacturer – 0.01–50 mEq/kg; actual measured 0.1–0.5 mEq/kg direct linear detection range (Steltzer, 2012). The standard assay comes with reagents in bottles equipped with calibrated pump pipettors. We found these units to be inaccurate, irreproducible, and extremely wasteful of expensive reagents due to the priming required. Replacing pump pipettors with direct pipetting of reagents with micropipettors improved assay results tremendously. The assay is run nominally at room temperature; we found that use of a heating block set at 25 °C to maintain a constant "room temperature" was necessary for acceptable reproducibility. Contrary to claims of the supplier and reactions in aqueous systems, we found that no clear endpoint was reached in the reaction run in proprietary stabilized isopropanol solvent, so timing of mixing and absorbance reading is critical. Bleaching of the XO–iron complex at high peroxide concentrations is a significant problem, so food oils and lipid extracts of foods usually require dilution by several orders of magnitude to put the samples in linear response range. Absorbance of a reaction measured in the MicroChem optical detector designed for the assay generates a "number" supposedly in mEq/kg lipid, but the calculations are blind and algorithms are unavailable from the company so what the results actually represent is not clear. Converting the assay to a conventional UV/VIS spectrophotometer gives more comprehensible and controllable results. The assay uses cumene hydroperoxide to generate standard curves; as discussed above, these may not be appropriate standards for lipid hydroperoxides. When standards are fresh and undegraded, the slope of the regression equation for the standard curve is very close to 1. However, the slope decreases as the controls age, so control reaction curves should be monitored frequently and controls replaced when the slope drops below a limit point (e.g., 0.90–0.95) or PVs of samples are erroneously low (Steltzer, 2012). We agree with the recommendation of Osawa et al. (2007) to calculate adjustment factors for each reaction system by normalizing to iodometric titration results. The kits are available in standard and microplate versions.

Despite these limitations, we have found the PeroxySafe assay is useful and can be made sufficiently accurate for most analyses of lipid hydroperoxides in foods.

PeroxoQuant (Pierce, Thermo Scientific). Detection range: cumene hydroperoxides (0.01–0.30 mEq/kg), *t*-butyl hydroperoxide (0.07–0.50 mEq/kg) (Steltzer, 2012). Concentration response curves were nonlinear over the entire concentration range and quite different for all hydroperoxides tested. This overall response pattern raises serious questions about appropriate hydroperoxide standards to use with lipids. The kit does not supply standards but suggests use of hydrogen peroxide, which is very inaccurate for lipid hydroperoxides. Because factors controlling reactivity appeared to be more complex than just redox potentials of the peroxides, we substituted authentic Fe^{3+} for standard curves. Bleaching of the XO complex was a significant problem. Until reaction chemistry is better defined and issues of standards can be resolved, we consider this assay not acceptable for quantitative analyses of lipid oxidation.

1.4.2.3 Ferric Thiocyanate (FeSCN)

Principle of analysis. In acidic media, ferrous iron reduces lipid hydroperoxides (LOOH) and becomes oxidized to ferric iron (Reaction (14)). The lipid alkoxyl radical (LO•) generated from LOOH oxidizes a second ferrous ion (Reaction (15)). The ferric ions then react with thiocyanate to form a red FeSCN complex that absorbs at 500 nm (Kolthoff and Medalia, 1949, 1951).

$$LOOH + Fe^{2+} \longrightarrow LO^{\bullet} + OH^{-} + Fe^{3+} \quad (14)$$

$$LO^{\bullet} + Fe^{2+} + H^{+} \longrightarrow LOH + Fe^{3+} \quad (15)$$

Hydroperoxide concentrations are determined from standard curves prepared from Fe^{3+} or standard hydroperoxides (e.g., cumene hydroperoxide), adjusted for the assumed stoichiometry of 2 mol Fe^{3+} generated per mole LOOH. Fe^{3+} standard curves are more accurate because iron salts are available in higher purity and are more stable than hydroperoxides. If hydroperoxide standards are used, actual reaction stoichiometry can be determined by dividing iron concentration generated by hydroperoxide concentration reacted. Peroxide values for lipid samples are determined by dividing determined hydroperoxide molarity by lipid weight to normalize, then converting to (mEq LOOH/kg fat).
Advantages. High sensitivity. Working concentration range: 5–90 nmol organic hydroperoxide per reaction; linear response range 1–40 µmol/L in oil or lipid extract (Steltzer, 2012).

Limitations and precautions. The assay has many chemical limitations and interferences that can create havoc with its use (Mihaljevic et al., 1996). Starting with the most critical issues, quantitative assays require a constant stoichiometry for all analytes and conditions. Huge problems arise in the FeSCN assay because the stoichiometry of the reaction varies with the hydroperoxide structure, the hydroperoxide concentration, the solvent, and the presence of oxygen (Steltzer, 2012). Reaction curves for different hydroperoxides may superimpose at very low concentrations, but they increasingly diverge above hydroperoxide concentrations of about 10 µmol/L (in the total reaction mixture) due to differences in hydroperoxide reactivity with both iron and solvents. Since oils and lipid extracts contain hydroperoxides of different structures, and none of these are available in high purity commercially, selection of an appropriate hydroperoxide standard is a challenge and use of any standard leads to inaccuracies in absolute quantitation. Perhaps the best options under such circumstances are to either use an Fe^{3+} standard curve and report results as iron equivalents, or to report lipid hydroperoxides as equivalents of whatever hydroperoxide standard is used.

Solvent interferences arise because alkoxyl radicals formed in hydroperoxide decomposition abstract hydrogen atoms from solvents, forming solvent radicals in competition with oxidizing ferrous ions. The net effect on the assay varies with the solvent and the redox potential of its radicals and is not always predictable. For example, with chloroform and methanol, the solvents commonly used in this assay, chloroform produces trichloromethyl radicals ($Cl_3C^\bullet$) (Reaction (16)), which are strongly oxidizing and peroxidizable. $Cl_3C^\bullet$ oxidize ferrous ions directly (Reaction (17)) and after oxidation to $^\bullet OOCCl_3$ radicals (Reactions (18) and (19)):

$$LO^\bullet + CHCl_3 \rightarrow LOH + {}^\bullet CCl_3 \quad (16)$$

$$Fe^{2+} + {}^\bullet CCl_3 + H^+ \rightarrow CHCl_3 + Fe^{3+} \quad (17)$$

$$^\bullet CCl_3 + O_2 \rightarrow {}^\bullet OOCCl_3 \quad (18)$$

$$^\bullet OOCCl_3 + Fe^{2+} \rightarrow Fe^{3+} + {}^- OOCCl_3 \quad (19)$$

The $LO^\bullet$ attack on methanol, on the other hand, produces hydroxymethyl radicals (Reaction (20)), which are strong reducing agents and recycle Fe^{3+} to Fe^{2+} (Reaction (21)):

$$LO^\bullet + CH_3OH \rightarrow LOH + CH_2OH \quad (20)$$

$$^\bullet CH_2OH + Fe^{3+} \rightarrow HCHO + Fe^{2+} + H^+ \quad (21)$$

These are all competing reactions, so the resulting amount of Fe^{3+} generated in the reaction will be determined by nature of the solvent, levels and type of hydroperoxide, oxygen availability, and pH. In other words, the stoichiometry will depend on the analytical solution composition (Mihaljevic et al., 1996).

For most samples, accurate results require analyses of a range of concentrations; testing a single concentration of sample has a high likelihood of erroneous results, particularly significant underestimation of peroxides. The assay can only analyze low concentrations of hydroperoxides for two reasons. Due to the high extinction coefficients of FeSCN complexes, even very low concentrations will push optical absorbances to >1, the dependable detection limit on most spectrometers. Hence, except in early oxidation, oils and extracts usually must be diluted and/or very small aliquots analyzed to stay in range of the assay. In addition, even more than in the XO assay, excess hydroperoxides bleach the FeSCN complex, so highly oxidized lipids can appear to have little or no reaction. Since the hydroperoxide levels of test oils or food samples are unknown, the only way to tell whether low apparent reactivity is due to low lipid oxidation versus competitive bleaching is to dilute the sample sequentially, even over several orders of magnitude. If high hydroperoxide levels are present, reaction will be observed at sufficient dilution.

Oxygen interferences arise from slow oxidation of the ferrous iron (especially important at low hydroperoxide concentrations) and from promotion of secondary reactions in lipid and solvent after hydroperoxide reduction.

These interferences can be eliminated by sparging solutions with argon and running the reaction under a headspace of argon.

Finally, for this assay it is important to use chloroform stabilized with ethanol rather than amylene (Richards and Feng, 2000). Chloroform stabilized with amylene causes very high blank readings and nonlinear calibration curves with cumene hydroperoxide. Although amylene is added as a radical scavenger, it does not prevent formation of trichloromethyl radicals in chloroform, nor does it prevent trichloromethyl radicals from oxidizing ferrous ion. In contrast, ethanol as preservative eliminates these interferences.

Applications. In either the basic form developed to monitor oxidation in dairy lipids (International Dairy Federation, 1991) or the modified form (Shantha and Decker, 1994), this assay has been applied to determine hydroperoxides in a wide range of oil and food systems. However, the inconsistency of stoichiometry and problems with FeSCN bleaching raise serious questions about validity and accuracy of hydroperoxide quantitation with this assay, even for comparing similar samples.

1.4.3 Epoxide Assays

Epoxides have not been analyzed systematically in oxidized oils or lipid extracts, in part because they were not previously considered important, and in part because assays with sufficient sensitivity were not available. However, epoxides have been detected at higher levels than hydroperoxides in neat methyl linoleate (Haynes and Vonwiller, 1990; Xie, 2015), they are suspected of toxicity, and they are more reactive with proteins and DNA than are than aldehydes. This makes epoxides important potential mediators of co-oxidations. Thus, analysis of epoxides should be given prime attention in new protocols for lipid oxidation.

Four epoxide assays (hydrogen bromide AOCS Cd 9–57, 4-*p*-nitrobenzylpyridine, *N,N*-diethyldithiocarbamate [DETC], and nuclear magnetic resonance [NMR]) were recently evaluated for chemical accuracy, sensitivity, reproducibility, handling requirements, and applicability to analysis of oxidized lipids (Liao, 2013). Numerous limitations were found with the nitrobenzyl pyridine assay. NMR shows promise as a research tool but is not readily available to quality control laboratories in the food industry. The hydrobromic acid titration has problems but is an AOCS standardized assay that has been applied in the food industry. The DETC-high-pressure liquid chromatography (HPLC) assay showed the greatest sensitivity and flexibility, plus it is equally applicable to quality control and basic research. Thus, only the HBr and DETC assays will be discussed here.

1.4.3.1 Hydrobromic Acid Titration (AOCS Standard Method Cd 9–57, 1997)

Principle of the assay. Compounds with epoxides are dissolved in glacial acetic acid then titrated with HBr that has been standardized with potassium hydrogen phthalate. The acid H^+ catalyzes ring opening by an SN1 mechanism and the nucleophilic bromide ion (Br^-) adds to the epoxy ring at the less substituted end (Reaction (22)) (Clayden et al., 2006).

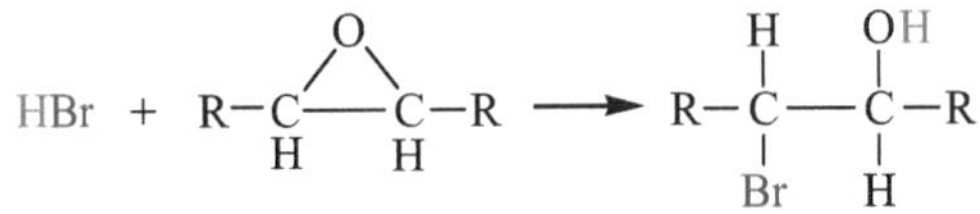

(22)

Epoxide concentration is calculated as % oxirane oxygen = [mL HBr * N * 1.6]/mass oil (g).where N is the normality of the HBr (Durbetaki, 1956).

Advantages. The assay is simple and requires no special equipment, and standardized procedures are available.

Disadvantages and precautions. This assay has a number of serious limitations. It detects only relatively high epoxide concentrations; the detection range is 0.0075–0.1 mol/L, but response is nonlinear below 10 mmol/L epoxide (Liao, 2013). The 0.1 mol/L HBr prescribed by the standard assay (designed for high epoxide concentrations in modified oils) considerably overtitrates low epoxide concentrations, but more dilute HBr is too volatile and autoxidizes readily, so gives erratic and inaccurate titrations. To limit HBr oxidation, the reaction mixture must be stirred well during titration to ensure maximum contact between reactants, and also must be continuously sparged with a flow of argon that is sufficient to prevent oxidation of the HBr but not enough to facilitate its volatilization.

Due to the instability and volatility of HBr, its solutions must be standardized frequently, at least daily and sometimes more often.

As with the thiosulfate titration in the hydroperoxide assay, the reaction endpoint is difficult to determine accurately and reproducibly by eye. The crystal violet indicator has a purple-violet color. As the hydrobromic acid is consumed by addition to the epoxide during titration, the color gradually fades and turns into blue-green. That the color does not change completely and instantly increases likelihood of overtitration; variation in color perception among individuals also contributes to high variability in results.

Perhaps most problematic for analyzing oxidized lipids is that the HBr reaction rate and stoichiometry vary with epoxide structure, increasing with epoxide chain length (Liao, 2013). This adds a large factor of uncertainty for absolute quantitation and creates particular difficulties for comparing results between oils and extracts of differing composition, where multiple epoxides may be present and their structures are not known. Also, the reaction is not totally specific for epoxides. β-Unsaturated ketones, conjugated dienols, and other oxygenated secondary products of lipid oxidation compete with epoxides for HBr and may react in addition to epoxides; whether the net effect is synergistic or antagonistic varies with the samples. At any rate, results in complex systems such as food oils or extracts (as opposed to pure materials) are questionable.

Applications. This assay, originally designed to analyze industrially epoxidized oils, is too insensitive to detect levels of epoxides expected in foods, especially during early stages of oxidation. However, if great care is maintained to limit oxidation and volatilization of the HBr, this assay is suitable for determining epoxidation levels in oils modified for lubricants and other nonfood uses.

1.4.3.2 DETC Assay with HPLC Separation and Quantification of Epoxy Adducts

Principle of assay: DETC reacts with epoxides in warm (60 °C) methanolic solutions to form stable epoxide–DETC complexes in two forms (Reaction (23)). Excess DETC is required to ensure complete binding of epoxides. Acid is then added to decompose the unreacted DETC in the solution (Reaction (24)), and the epoxide complexes are separated by reverse-phase HPLC with detection at 278 nm (Dupard-Julien et al., 2007). Phosphoric acid was employed in original procedures, but acetic acid reduces stress on HPLC columns. Epoxides are quantified by comparison of peak heights to standard curves constructed from authentic epoxides. Individual adducts can be identified by comparison with authentic adducts or by interfacing HPLC with MS detection.

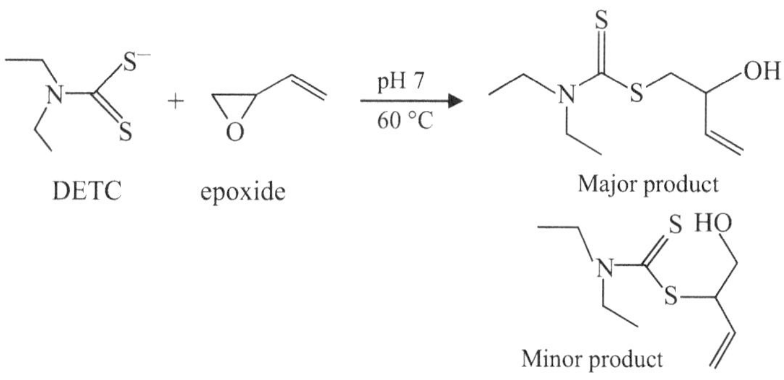

(23)

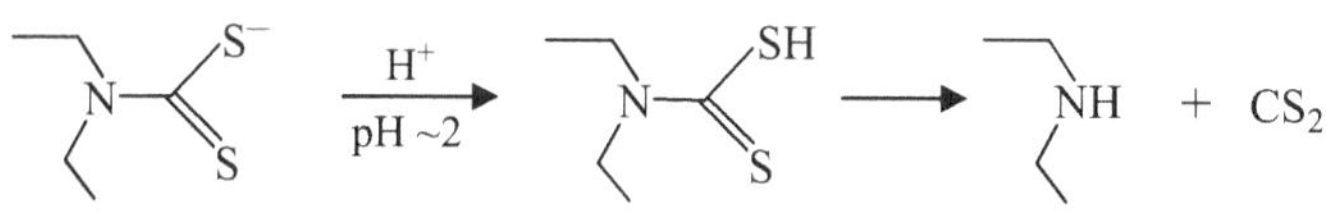

(24)

Advantages. The reaction is very sensitive (2.5 μmol/L to 0.01 mol/L), specific for epoxides (Liao, 2013), and easy to run. The DETC complex, once formed, is stable for several days if protected from light and air. Separation by HPLC provides not only quantification but in combination with MS detection can also identify the epoxides generated. This information can be very useful when trying to elucidate origins of some oxidation products. The reaction shows little influence of epoxide structure above six-carbon chain length, so a moderately long chain pure epoxide (e.g., C10 epoxide) can be used to prepare standard curves and the assay is suitable for quantitation of mixed epoxides of unknown structures (Liao, 2013). Although original assay procedures were developed for monomer epoxides, adjustment of extraction solvents and the eluting gradients makes it possible to distinguish monomer from core epoxides on phospholipids and triacylglycerols and to quantitate each fraction in a single run (Shallcross, J. and Schaich, K.M., unpublished data).

Disadvantages and precautions. DETC is a reasonably strong metal chelator (Van Damme and Oomens, 1995) so metal contaminants in fats, oils, or food extracts can interfere with the DETC assay, altering the efficiency and stoichiometry of the reaction. The reaction response is not linear with epoxide concentration, but biphasic with a break point at about 1 mmol/L. Slopes are steep in the low concentration range, then decrease at higher concentrations (Liao, 2013). In addition, high concentrations of DETC–epoxide complexes precipitate in the HPLC column. Hence, quantitative analyses are most accurate at low concentrations, and samples may need to be serially diluted to determine optimum concentrations for calculations.

Even analyses of monomer epoxides require solvent gradients to elute chain lengths longer than about six carbons in reasonable times. When oils or extracts are analyzed, more complex gradients are required, and these must be modified and optimized for the lipid classes and fatty acid chain lengths present.

Applications. This assay has been around for a long time but has seen little application in food analyses. We are using DETC regularly to monitor epoxides in oxidizing methyl linoleate and food oils (Xie, 2015) and feel it has the potential to become the main assay for lipid epoxides in foods. Full application of the DETC assay to foods will require evaluation of epoxide stability to extraction procedures and adjustment of elution gradients to accommodate the specific complex lipid components of individual extracts.

1.4.4 Carbonyl Assays

Carbonyls are secondary products of lipid oxidation, and as such, they have often been treated as if they were also of secondary importance, being only indicators that oxidation had progressed substantially into later stages. Granted, under ambient conditions, carbonyls might seem to be irrelevant because they form only at very low levels for a long time, even while hydroperoxide and epoxide formation are very active. However, carbonyl analyses must be considered critical components of any protocols for analyzing lipid oxidation because they are unique flags of changes in direction of oxidation. When hydroperoxide levels are low, carbonyls can distinguish inhibited oxidation from hydroperoxide decomposition. Unusually high concentrations of carbonyls, especially early in oxidation, signal activation of pathways other than peroxyl radical→hydroperoxide→decomposition to alkoxyl radicals→scission to carbonyls. Beyond total concentrations of carbonyls, analysis of individual carbonyl products, both volatile and nonvolatile, can provide crucial information for elucidating oxidation pathways, which in turn can guide development of more effective control of oxidation.

To encourage increased analysis of carbonyl lipid oxidation products and to provide some perspectives on limitations of current analyses, this section provides an overview of three assays for nonvolatile carbonyls – dinitrophenylhydrazine, *p*-anisidine, and thiobarbituric acid (TBA). Analysis of volatile carbonyls will be considered in Section 1.5.2 on gas chromatographic analyses.

1.4.4.1 Complexation of Carbonyls with 2,4-Dinitrophenylhydrazine (DNPH)

Principle of the assay. Carbonyls (both ketones and aldehydes) react with DNPH in acidic alcohol to form dinitrophenylhydrazones, which in the presence of base are converted to red wine–colored quinoidal ions (Reaction (25)) (Lappin and Clark, 1951).

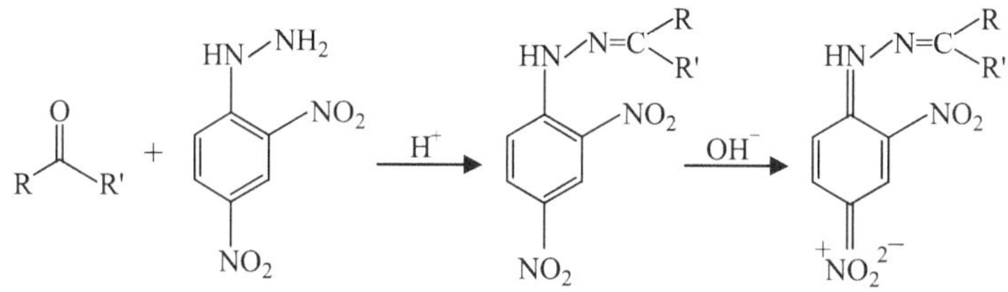

(25)

The wavelength for detection varies with the solvent and types of carbonyls. The first application of the DNPH assay to lipid carbonyls used methanol/benzene (Henick et al., 1954), in which maximum absorption was at 432 nm for saturated aldehydes and 458 nm for unsaturated aldehydes. The assay has since been modified to use less-toxic solvents, including ethanol (Yukawa et al., 1993), isopropanol (Endo et al., 2001), butanol (Schulte, 2002; Endo et al., 2003; Usuki et al., 2009), and isopropanol:acetonitrile 1:1 (Yao, 2015). In most alcohols, reactions of α,β-unsaturated carbonyls is complete, while saturated aldehydes react only about 70% (Yukawa et al., 1993) and absorption maxima are correspondingly different. In ethanol, maxima are ~435 nm (saturated aldehydes), ~460 nm (enals), and ~480 nm (dienals) and equal response (but not maximum) for all structures occurs at 425 nm (Yukawa et al., 1993). In propanol (Endo et al., 2001) and butanol (Endo et al., 2003), the corresponding maxima are 430, 450, and 457 nm with approximately equal response at 420 nm.

Early methods used heating to drive the reaction, but these procedures have largely been replaced with room temperature versions that prevent artifactual development or loss of carbonyls during reaction with DNPH.

The DNPH reaction has long been considered the classic assay for carbonyls, and it is currently being applied to measure air pollution (Lodge, 1988; Workman and Steger, 1996; Brombacher et al., 2002) and to quantitate carbonyls in proteins (Levine et al., 1990, 1994; Evans et al., 1999; Fagan et al., 1999; Shacter, 2000a,b; Armenteros et al., 2009; Luo and Wehr, 2009; Estévez, 2011). It is somewhat surprising, then, that the method is not also a routine analysis for oxidized lipids. One reason may be that traditional solution reactions were tedious and difficult to handle. The excess DNPH required for reaction as well as contaminants in reagent DNPH caused instability and interference in analysis. Three-time recrystallization of DNPH and removal of excess DNPH after reaction, such as by ion exchange chromatography (Yu et al., 1961), were time consuming and not amenable for routine analyses of many samples. Reaction solvents had to be changed depending on solubility requirements of the material being analyzed.

DNPH reactions became technically feasible and much more attractive for monitoring lipid oxidation when the need for quantitating air pollutants linked DNPH solution reactions with HPLC separation of products (Tejada, 1986; Brombacher et al., 2002; van Leeuwen et al., 2004; EPA, 2006; de M. Ochs et al., 2010). Integration of reaction and chromatography serves the dual role of isolating unreacted DNPH while separating and allowing identification of individual hydrazones, a major technical advancement that has opened the doors to new applications in analyses of carbonyls in lipid oxidation. Quantitation of total carbonyls is obtained from total peak areas minus the DNPH peaks; quantification of individual carbonyls is obtained by comparison to regression curves of authentic standards.

Advantages. DNPH-HPLC analyses provide a simple way to determine whether low hydroperoxide levels are due to low oxidation or high hydroperoxide decomposition, to identify carbonyl specific products that may be contributing to off-flavors, and to differentiate carbonyls on nonvolatile monomers versus on core structures of phospholipids and acylglycerols – information that has not been previously available (Sjövall et al., 2002; Yao, 2015). The DNPH reaction is specific for carbonyls (does not react with hydroperoxides, epoxides, or alcohols) and resulting hydrazones are sufficiently stable for groups of samples to be analyzed on autosamplers (Yao, 2015). The method is very sensitive, having detection limits on the order of 5–50 μg/L and lower quantification limits of 18–166 μg/L, with an inverse relationship to carbon number. Carbonyls are readily distinguished from ketones by MS: both aldehydes and ketones have a strong fragment at *m/z* 152 but aldehydes also have an additional fragment at *m/z* 163 (Yao, 2015). Short chain aldehydes (formaldehyde, acetaldehyde, propionaldehyde) that require special cryotrapping in GC are easy to detect by HPLC with appropriate gradients.

Capability for identification and quantitation of individual carbonyls by comparison to standards can contribute significantly to elucidating oxidation pathways while also tracking the progress of oxidation. Connecting HPLC to MS detection provides opportunities to identify products in detail not previously available. Industrial-quality control may not consider this useful immediately, but the information gained will improve our understanding of lipid oxidation, track potential sources of flavor changes, and eventually lead to more effective analytical protocols and approaches for stabilizing lipid oxidation in foods.

Disadvantages and precautions. All reagent DPNH should be recrystallized three times from solvents such as butanol or acetonitrile to eliminate reactive and optically interfering impurities. This is only a small hitch, though, since a single purification run can provide enough material for months of analyses if the DNPH is stored dry in a dark container. Perhaps the greatest limitation currently is the lack of experience and standardization of DNPH-HPLC methods for lipid carbonyls. Methods have been developed independently. DNPH reaction times, solvents and gradients vary with the source and treatment of lipids being analyzed as well as analytical goals – monomers only, specific classes, or all classes. A considerable amount of research is still needed to develop gradients that provide good peak shape and baseline resolution of all carbonyl components, particularly core carbonyls, but with advances in HPLC columns, it can be achieved.

Quantitation of carbonyls is not straightforward because each carbonyl compound has a different concentration–response curve. Even relative quantitation comparisons are not exact unless carbonyl components are matched. However, Yao (2015) has shown that using retention times to classify components by chain length, then applying average regression curves for different chain length ranges (<6 carbons, 7–10 carbons, >10 carbons) can give good approximations of total and individual carbonyl concentrations in mixtures.

Applications. Virtually limitless with proper development and refinement! The DNPH-HPLC version of the reaction has been applied to analyses of lipid carbonyls in oxidized triacylglycerols of corn and sunflower oils (Sjövall et al., 2002), in vegetable oils (Seppanen and Csallany, 2001), frying oils (Endo et al., 2001; Schulte, 2002) and heated lard (Jeong et al., 2013), meat products (Armenteros et al., 2009), and roasted and dried products (Usuki et al., 2009). It also has been used to determine acrolein levels in a wide range of oils and food extracts (Shibamoto, 2006). Currently, the major limitations are only lack of experience with the HPLC version and need for refinements in gradients to detect core carbonyls.

A number of side reactions and degradations have been identified in environmental applications of DNPH-HPLC assays (Uchiyama et al., 2011). This chemistry needs to be considered as DNPH-HPLC methods are increasingly applied to lipid analyses, especially when results are used to elucidate oxidation processes.

Applications of the DNPH reaction to analysis of protein carbonyls will be discussed in Section 1.7.5.3.

1.4.4.2 p-Anisidine Value

Principle of assay. Unsaturated aldehydes (principally 2-alkenals and 2,4-alkadienals) in animal fats and vegetable oils react with *p*-anisidine in glacial acetic acid to form a yellowish product that absorbs at 350 nm (Reaction (26)) (IUPAC, 1987; AOCS, 2011).

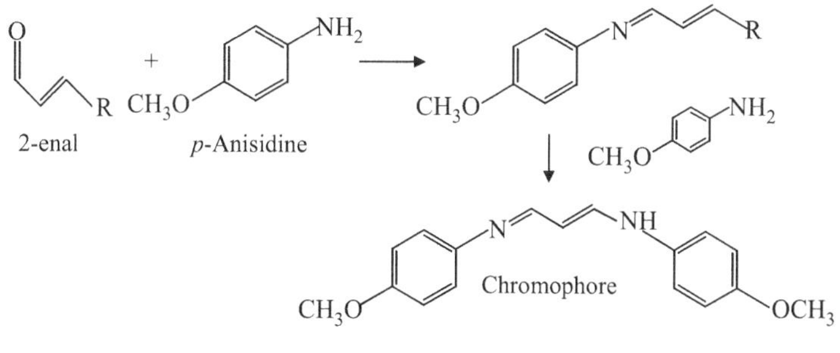

(26)

By convention, the *p*-anisidine value is defined as 100 times the change in absorbance at 350 nm (measured in a 1.0 cm cell) before and after reaction of a solution containing 1.0 g of oil in 100 mL of anisidine reagent. Combining the anisidine value (AV) with PV gives the TOTOX value, a measure of both early and secondary oxidation (Reaction (27)):

$$\text{TOTOX value} = \text{AV} + 2\text{PV} \tag{27}$$

Advantages. Standardized procedures are available for both AV (AOCS Cd 18–90, ISO 6885) and TOTOX value (AOCS Cg 3–91) (IUPAC, 1987; AOCS, 2011).

Disadvantages and precautions. Toxicity! Although not carcinogenic, *p*-anisidine is highly toxic and on the U.S. National Institutes of Occupational Safety and Health list of compounds that are Immediately Dangerous to Life or Health Concentrations (IDLH) (NIOSH, 2015). *p*-Anisidine causes blood damage upon oral ingestion, inhalation, or skin contact and, if heated strongly, may release very toxic fumes of nitrogen oxide. Considering the push to eliminate use of chlorinated hydrocarbons in lipid analyses, this reagent should also be on the discard list.

The anisidine reaction is neither quantitative nor specific. All aldehydes react but unsaturated aldehydes have higher color response than saturated, so results can give only relative, not absolute, aldehyde concentrations. Anisidine also reacts slowly with hydroperoxides so it does not clearly distinguish primary from secondary products and it should not be used to analyze oils with PVs >5 (although this seems to be a common practice). The method cannot be used with highly colored oils, particularly with carotenoids, which also absorb in the 350 nm wavelength range. To eliminate interferences, assays on lipid extracts require subtraction of a blank for the extract absorbance alone to obtain accurate values. The method also requires water-free (reactions are not fully completed in presence of water) and carbonyl-free solvents to avoid general interferences (White, 1995).

Applications. AVs and TOTOX values have been used to extensively distinguish fresh from oxidized oils, i.e., fresh oil AV <2.0 and TOTOX <4, and to track degradation in nearly every kind of oil and high-lipid food. The tendency to produce anisidine-reactive products increases with fatty acid unsaturation so the assay is most sensitive with soybean, rapeseed, flaxseed, and fish oils. Two major applications have been to monitor thermal damage in frying oils (Tompkins and Perkins, 1999) where hydroperoxides have decomposed and to evaluate refined oils where high hydroperoxides in damaged oils have decomposed to aldehydes during deodorization.

Despite extensive past applications of this assay, due to toxicity and inaccuracy this assay should now be replaced with more sensitive and accurate assays for lipid carbonyls, such as the DNPH-HPLC assay discussed in the previous section.

1.4.4.3 TBA or Thiobarbituric Acid Reactive Substances (TBARS)

Principle of the assay. Nominally, the TBA assay measures content of the secondary degradation product, malonaldehyde, which reacts with thiobarbituric acid when heated in acidic solutions to form a Schiff base adduct that is pink colored (absorbs at 532–535 nm) and fluorescent (exitation 515 nm, emission 553 nm) (Reaction (28)) (Bernheim et al., 1948; Sinnhuber et al., 1958).

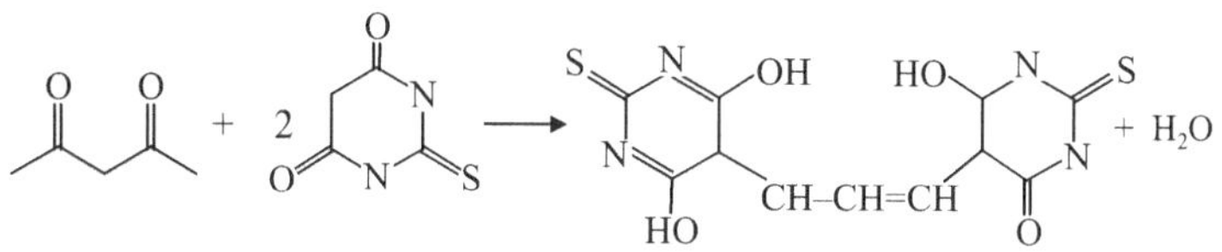

(28)

In actuality, the reaction is much more nebulous and nonspecific than this, involving alkanals, alkenals, alkadienals, and perhaps other oxidation products as well as malonaldehyde. Reactions of these alternate products can be distinguished by shifts in absorption maxima, in place of or in addition to the malonaldehyde peak. Alkanals react with TBA in a single condensation; the complexes are yellow and absorb at 450–455 nm (Tarladgis et al., 1962; Marcuse and Johansson, 1973). Acetaldehyde and other aldehydes generated during heating of reaction mixtures with acid form TBA complexes that are orange (λ_{max}=496 nm) (Kosugi et al., 1987). Alkenals absorb at primarily at 452 nm (low level at 530 nm), while 2,4-alkadienals produce the typical red chromophore that absorbs at 530 nm (Marcuse and Johansson, 1973). There is little consistency in literature reports regarding how absorption at multiple wavelengths is accounted – both peaks, only 530 nm, or only 450 nm – but it seems that the spectral variations from different aldehydes could be used diagnostically to at least tentatively identify classes of aldehydes in the sample. Such an application, although intriguing, is hampered by the interchange that occurs in aldehydes and absorption maxima depending on pH of reaction, temperature and time of heating, and relative concentrations of TBA and aldehydes. As a consequence, it is difficult to be certain of the origin of detected aldehydes – endogenous or generated in reaction. A detailed review of the complexity of aldehyde generation and reaction in TBA analyses is available for interested readers (Guillén-Sans and Guzmán-Chozas, 1998).

For the reaction, malonaldehyde and other products are harvested by four methods: (1) steam distillation (Tarladgis et al., 1960; Pikul et al., 1989; Gomes et al., 2003), (2) direct reaction of intact sample with TBA in solution (Sinnhuber et al., 1958; Witte et al., 1970; Sinnhuber and Yu, 1977), (3) preextraction of lipids before TBA reaction (Pikul et al., 1983, 1989), and (4) aqueous acid extraction of malonaldehyde before TBA reaction (Raharjo et al., 1992). Steam distillation has been extensively used with meat homogenates. Direct digestions of food samples are selected for their ease and rapidity, but they are also less specific and more prone to side reactions with nonlipids (Sørensen and Jørgensen, 1996). For example, in assays of lipid oxidation in extruded cereals, more TBA can be bound to the food matrix than react with lipid aldehydes released into solution (Schaich, K.M., unpublished data). Analysis of lipid extracts eliminates the time advantage of the method, but avoids side reactions with nonlipid reactants and provides most detailed information about oxidation in the lipid. With lipid extracts, TBA may offer some advantage over DNPH in stability, but sensitivity is lower. Numerous comparisons of the methods have been published for different foods, but with no clear decision as to method preference. Arguments for choosing a specific reaction mode for analyses in meats have been presented by Fernandez et al. (1997).

For quantitation, absorbance of samples after reaction are compared to standard curves prepared from 1,1,3,3-tetraethoxypropane (TEP), which hydrolyzes in acid to give malonaldehyde (Pikul et al., 1989). This sounds straightforward but, for accurate results, TEP recovery blanks must be run in each method to determine the percent of analytes that survives the reaction conditions, and this recovery factor is included in final TBA value calculations (Tarladgis et al., 1960; Pikul et al., 1989). Failure to include recovery factors increases variability of results, introduces error into even direct absorbance comparison, and limits ability to evaluate relative oxidation levels in different materials or different laboratories. Results of TBA analyses are usually expressed as μmol TBARS/g sample to cover all reactants, or mg malonaldehyde/kg sample if the molecular reactant is verified. It is not infrequent also to see absorbance comparisons published without calculation and normalization.

Criticisms of long reaction time, low sensitivity, and nonspecificity have engendered changes in handling, final detection, and quantitation of products. TBA adducts are still most commonly detected optically, but increasingly other methods are showing greater sensitivity and specificity, and also greater ease in handling both detection and data analysis. Calculation of third derivatives of optical spectra sharpens absorption peaks, gives more exact absorptions, and facilitates determination of different species contributing to the reaction (Fenaille et al., 2001). Use of fluorescence end points increases sensitivity and eliminates some side effects (Yagi, 1987; Williamson et al., 2003). Fourier transform technology with infrared spectroscopy (FTIR) with partial least squares chemometrics can be trained with oil samples of known TBARS values and gives equivalent direct results with oils in <5 min, but the species detected is still unidentified (Setiowaty and Che Man, 2003).

Gaining specificity in the analysis requires separation of reaction products by HPLC (Kakuda et al., 1981; Londero and Greco, 1996; Fenaille et al., 2001; Mendes et al., 2009) or GC (Fenaille et al., 2001) followed by identification of products by comparison to standards or by interfacing with mass spectrometer detectors.

Most procedures contain some heating step to increase yields of the TBA chromophore, but Kosugi et al. (1988) demonstrated the presence of artifacts from the heating process. When reacted with TBA at low temperature (5 °C), dienals formed several complexes which subsequently transformed to the typical malondiadehyde complex as the solution was heated to 100 °C (Kosugi et al., 1988). They thus proposed that actual aldehydes present were most accurately detected without heat, even if reaction times were longer. Bidder and Spika (1989) similarly recommended cold incubations to detect endogenous aldehydes and eliminate adventitious production in side reactions.

Several excellent reviews on TBA chemistry and procedures are available for interested readers (Hoyland and Taylor, 1991; Fernández et al., 1997; Guillén-Sans and Guzmán-Chozas, 1998).

Advantages. The TBA reaction has been used extensively in analyses of lipid oxidation in foods, particularly muscle foods, because it can be applied directly and does not require extraction of lipids. In general, TBA results have correlated well with sensory analyses, supporting the involvement of TBA reactants in processes that generate off-flavors and odors detected by consumers.

Disadvantages and precautions. The TBA test is a very controversial analysis, with the field of lipid oxidation being divided between those who love the TBA assay, or at least find it useful if not perfect, and those who hate the assay for its fuzzy chemistry and inappropriate applications. Although much has been learned about the species contributing to color development in the reaction, fierce arguments remain about what is actually being measured in the TBA or TBARS assays in terms of both identity and origin of reactants. For example, authentic malonaldehyde is produced primarily during the autoxidation of fatty acids with three or more double bonds (Pryor and Stanley, 1975) and is generated during the digestion stage of the assay; small amounts may also be generated in downstream oxidation of secondary lipid oxidation products such as 2-nonenal (Sinnhuber and Yu, 1977). Thus, authentic malonaldehyde of endogenous origin may actually be detected in muscle foods which are high in arachidonic acid (meats) and eicosapentaenoic and docosahexaenoic acid (fish), but products thought to be malonaldehyde from vegetable oils are more likely generated during the TBA reaction. The discussion above about the many different colored products observed in the TBA reaction also attests to its chemical complexity and opaqueness.

Another view of TBA opponents is that the TBA reaction is so nonspecific that its validity except in special applications is highly questionable. The long heating step and iron added in some versions contribute to decomposition of hydroperoxides and artifactual increases in secondary products rather than measuring compounds already present. The reaction of TBA with protein amine groups, starch and sugars, and nucleic acids leads to overestimation of oxidation when soluble nonlipid reactants are present and underestimation of oxidation when TBA binds to polymeric food components such as starches and proteins (Tarladgis et al., 1962; Baumgartner et al., 1975; Frankel, 2005a). Although some of the side reactions can be identified by appearance of peaks at different wavelengths, and these peaks can be subtracted, the overall quantitation still remains questionable unless the lipid is purified before analysis. Making the detection more specific by HPLC analyses negates time advantages of the direct digestion/optical analysis method and still does not give more or different information than the more specific HPLC-DNPH assay.

Applications. Research reports can be found citing use of the TBA assay to analyzed lipid oxidation in just about every kind of food system. Most appropriate applications have been muscle foods that have higher contents of arachidonic acid. Whether every other application is valid is arguable, but the method provides some "values" for comparison so perhaps this assay has its own niche among methods.

Still, Frankel (2005a) has stated that the TBA test is "unsuitable for complex food materials and biological systems containing nonlipid materials contributing to the color reaction" and this author concurs due to the ill-defined chemistry and high sensitivity of results to reaction conditions.

It has been argued that the oxidation artifacts caused by extraction of lipids from difficult matrices such as muscle foods and dairy and cereal products counterbalance the artifacts induced by the TBARS heating step, thereby justifying use of TBA reactions. However, pressurized solvent extractions can now extract lipids efficiently from even these difficult matrices with minimal effect on oxidation, DNPH-HPLC methods can identify and quantitate authentic malonaldehyde (Kawai et al., 1990) and differentiate other carbonyl products specifically, and both methods can be automated and handle large numbers of samples. Thus, it is time to replace the TBARS with more accurate and chemically definitive analyses. At least, if TBA is used, the method should be limited to cold digestion to trap only original products and detection only by HPLC so that individual reacting species can be identified, with only lipid products included in calculations.

1.5 PHYSICAL/INSTRUMENTAL ANALYSES OF OXIDATION PRODUCTS IN OILS AND INTACT FOODS

1.5.1 Oxygen Consumption

In one sense, oxygen consumption is an ideal measure of lipid oxidation since it is not dependent on any particular product and it provides kinetics of oxidation directly. The rate of oxygen addition to carbon radicals is diffusion controlled and almost instantaneous ($k > 10^9$ units), so oxygen consumption is essentially a measure of initiation rate. However, headspace oxygen consumption does not reflect absolute oxidation rates except perhaps with thin films, because it reflects three processes – oxidation of molecules at the sample surface, diffusion of oxygen into the sample, and oxidation of molecules in the bulk phase. In any analysis, stirring reduces effects of diffusion. All of these actions must be considered when interpreting oxygen consumption data.

Headspace oxygen is easiest to measure so is performed most frequently. Dissolved oxygen in lipid phases is more challenging because the electrode must be able to function in a hydrophobic aprotic phase or at a hydrophobic interface and also not react with oxidation products. Sensitivity is another issue since lipid oxidation occurs on a micro molecular scale – 20 mEq oxygen (10 mEq LOOH) per kg lipid in hydroperoxides makes most oils unacceptable in odor and flavor. We have not found oxygen-specific sensors that function in a lipid phase and provide the sensitivity we require, including the Foxy probe (Ocean Optics) designed for lipids. In contrast, the systems described below perform consistently and provide useful data for most research and quality control applications.

1.5.1.1 Headspace Oxygen Consumption with Oxipres™ Oxygen Bomb and Oxidograph™

Principle of analyses. The Oxipres™ (Mikrolab, Aarhus, Denmark) is an oxygen bomb (not to be confused with an oxygen bomb calorimeter) designed to monitor oxygen consumption of materials heated under controlled atmospheres. Samples of oils or foods are sealed in glass cells inside stainless steel chambers (Figure 1.3), pressurized with test gas to pressures varying from 1 to 7 bars, heated to desired test temperatures (30–199 °C), then held over time and allowed to react. Pressure in the cell is measured electronically by a sensitive pressure transducer and recorded continuously by computer while samples incubate. Cell pressure first increases as temperature equilibrates (following the universal gas law PV = nRT), then decreases as reaction ensues and oxygen consumption progresses (Figure 1.4).

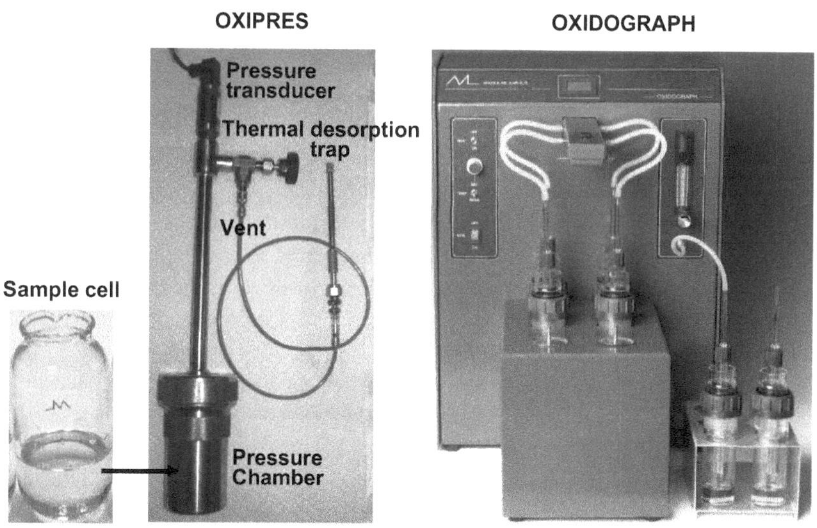

FIGURE 1.3 Sample cell configurations from Oxipres oxygen bomb and Oxidograph modification of Fira-Astell apparatus. Oxidograph picture compliments of MikroLab, Aarhus, Denmark.

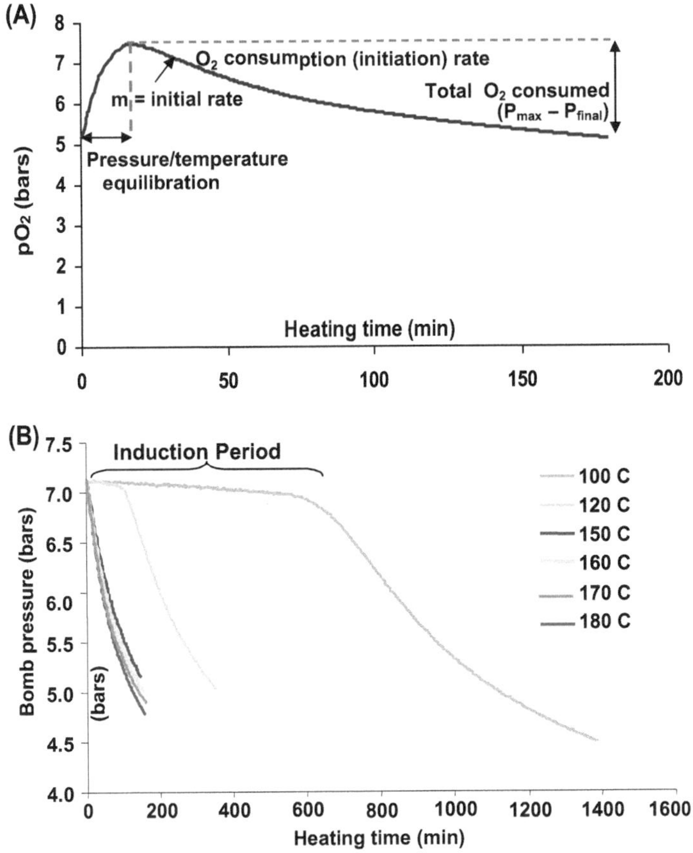

FIGURE 1.4 Typical Oxipres curves showing information derivable from them. (A) Typical curve from oil under frying model conditions (5 bars oxygen, 180 °C). (B) Curves from oil heated at different temperatures. Rapid drop of oxygen pressure indicates end of induction period. Curves shown are plotted from peak pressure time rather than start of heating so initial equilibration is not shown.

Gas and pressure should be matched to information desired. High pressures (e.g., 5 bars) of pure oxygen are generally used at moderate temperatures (e.g., 60–80 °C) when induction periods are measured to determine susceptibility of materials to oxidation or to compare antioxidant effectiveness. However, under these conditions, mechanisms and product distributions are very much skewed relative to atmospheric pressure (1 bar) while at 1 bar pressure, oxygen consumption is too low for easy visualization of pressure change (Peng, 2010). Pressurization with 2 bars purified air works well as a compromise when mechanisms and specific products are being monitored.

Information about oxidation is obtained from patterns of oxygen consumption. A marked pressure drop after slow oxygen decline indicates the end of the induction period (Figure 1.4). Induction periods are rather nebulous chemically and are used primarily for comparisons. For more definitive analyses, rates of oxidation can be determined from initial slopes (m) of the oxidation curves, and total oxygen consumption over a specified time period gives information about reaction stoichiometry (oxygen absorbed per gram of oil or sample) (Figure 1.4). When oils or pure materials are used, stoichiometry can be calculated on a molar basis and compared against measures of products to track reaction pathways (Peng, 2010; Tian, 2013).

In contrast, the Oxidograph is a refinement of the Fira-Astell apparatus for measuring oxygen consumption. It is not an oxygen bomb and handles only oils or melted fats. Samples of fats/oils in stirred sealed glass reaction vessels are warmed to desired test temperatures (range 30–150 °C) in an aluminum block then exposed to test gases at atmospheric pressure (Figure 1.3). As with the Oxipres, pressure in the cell is measured electronically by a sensitive pressure transducer and recorded continuously by computer while samples incubate. Three values can be determined:

- IP = Induction period = Time from start until rate of oxygen uptake increases (related to temp.)
- PF = Protection factor = $IP_{\text{with antioxidant}}/IP_{\text{without antioxidant}}$
- LE = Life extension = 100 × (PF − 1)%

Advantages. Both instruments detect the earliest changes in oxidation independent of actual product pathways, so give answers about oxidation faster than OSI and Rancimat (see Sections 1.5.4.1 and 1.5.4.2) (Nogala-Kalucka et al., 2005). Oxygen consumption directly reflects initiation and sensitivity to radical formation, information that is difficult to obtain in other ways. Very precise temperature and pressure control makes both systems very useful for controlled incubations as well as for shelf-life analyses and induction period determination. Analyses can be conducted on relatively small samples – a few grams or 5 mL oil. Oxygen consumption can be related to other chemical changes by removing samples at different points in oxygen consumption and measuring hydroperoxides, epoxides, aldehydes, etc. Headspace gas in both systems can be varied to test effects of atmosphere on oil oxidation. In addition, headspace in the pressurized Oxipres cell can be vented though a thermal desorption trap to collect volatile products for analysis by GC (Figure 1.3) (Repko Reader, 2013).

Interesting research possibilities have been opened by comparing induction periods determined by the Oxipres/Oxidograph (reflecting initial reactions) versus by the Rancimat or OSI (measuring secondary products). Tests of oils stabilized with various antioxidants showed shorter induction periods and larger differences between oils with the Oxipres than the Rancimat (Mikrolab, 2013). In contrast, the Oxipres could not detect differences in native winter and summer butter, but Rancimat could, and the Oxipres showed longer induction periods, especially when antioxidants were added (Gramza-Michalowska et al., 2007). The first set of results shows that the Oxipres is more sensitive to differences in initiation and early oxidation, and the oils with different fatty acid composition clearly had different susceptibilities to radical attack. In contrast, the second set of results reflects the dominance of secondary breakdown products in Rancimat analyses. The butters were basically the same fat but with endogenous differences in antioxidants and other factors. Hence, their initiation was the same (no change in the Oxipres) but the butters then diverged in secondary reactions. Coordinating results from Oxipres and Rancimat/OSI methods may help localize points of variance in oxidation under different conditions.

Disadvantages. General pressure transducers are not specific to oxygen but reflect overall changes in the headspace pressure. Volatiles released from oxidizing lipids contribute insignificantly under high oxygen pressures, but at lower pressures and lower oxygen (e.g., 1 bar air), they may add to the cell pressure and reduce apparent oxygen consumption. We have found that water vapor from high-moisture samples can also interfere with the pressure transducer. Critical moisture–temperature limits need to be determined. For example, the manufacturers have successfully monitored oxidation of mayonnaise (15% water) at 90 °C, but we have had difficulties analyzing margarines (30–40% water) at 80 °C and above. Early models of the Oxipres lacked capabilities for stirring oils, raising questions about oxygen diffusion into samples (Peng, 2010). However, current models have stirring options available. Pressurizing with pure oxygen under high pressures changes the oxidation chemistry so does not reflect reactions under ambient conditions. For induction period determination and comparison of fatty acid or antioxidant formulations, the manufacturer recommends running the Oxipres at 5 bar O_2 at 80–100 °C for greater sensitivity and to reduce oxidation time. Under these conditions, reactions run fast, making it very easy to determine sensitivity of a system or formulation to oxidation, but the oxidation processes are different from those occurring at close to ambient conditions.

Applications. Determination of induction periods, rates of oxidation, and total oxygen consumption by the Oxipres and Oxidograph have been applied to assess differences in innate susceptibility of oils with different composition or additives to oxidation (Gramza-Michalowska et al., 2007), to test and compare antioxidant effectiveness (Trojakova et al., 1999; Nogala-Kalucka et al., 2005; Bragadottir, 2007), and to study thermal stability and reactions of oils (Peng, 2010; Repko Reader, 2013; Tian, 2013) Both Oxipres and Oxidograph systems maintain very precise environmental control for investigating reaction mechanisms and effects of reaction conditions on product distributions and for coordinating oxygen consumption with products (kinetics and types) in oils.

The Oxipres also provides useful oxidation information about oxidative stability and, when combined with other product assays, oxidation processes of solid food materials within the moisture–temperature limitations noted above. Examples include extruded corn grit with different levels of sunflower oil (Ying et al., 2015), peanuts and peanut butter, potato chips and other snacks, biscuits, fish and meat meals, cookies, sponge cake, dry pet foods, milk powder, and processed cheese (Schaich, K.M., unpublished data; Mikrolabs, Aarhus, Denmark).

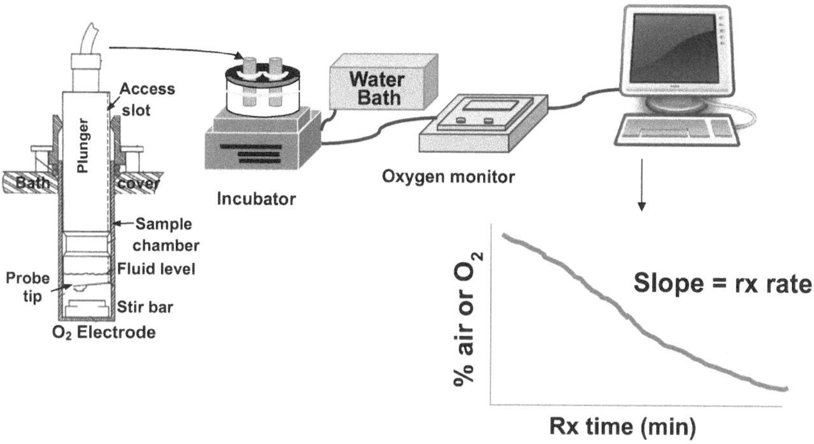

FIGURE 1.5 Diagram of YSI oxygen electrode system for recording dissolved oxygen consumption in fluid samples. *Yellow Springs Instruments, Yellow Spring, OH, USA.*

We have even used the Oxipres to test stability of yellow versus red bone marrow from different animal species (unpublished data).

Because oxygen consumption reflects initiation processes and is independent of specific product pathways, these methods merit much more attention in routine analyses. Integration of oxygen consumption with Rancimat analyses in assessing "stability," in particular, may offer an interesting means of distinguishing whether a factor acts on initiation or secondary oxidation reactions.

1.5.1.2 Dissolved Oxygen with YSI Oxygen Electrode

Principle of analyses. Sometimes it is important to determine consumption of oxygen dissolved in an oil phase or emulsion rather than in the headspace. Although an older system, the YSI oxygen electrode (Yellow Springs Instruments, Yellow Springs, OH) remains very useful for following changes in dissolved oxygen in solutions and emulsions. The system consists of a thermostatted sealed water bath containing glass sample vessels with stir bars at the bottom, and electrodes encased in epoxy sleeves (we have replaced these with Teflon sleeves for greater inertness). Samples (a few milliliters) are placed in the vessels, and the electrodes are inserted to the top of the sample. A narrow slit on the side of the electrode allows venting of all headspace and also provides a conduit for addition of catalysts or antioxidants for tests of effects on oxidation. The system has now been computerized and is available with software to control data recording and analysis (Figure 1.5).

The heart of the system is the Clark electrode that determines oxygen consumption polarographically when in contact with the oxidizing emulsion. The probe has its own microenvironment of potassium chloride encased in a thin gas-permeable Teflon membrane stretched over the end of the probe.

Oxygen diffuses from the sample through the membrane into the KCl, where it reacts at the cathode when voltage is applied:

$$O_2 + 2\,H_2O + 4\,e^- \rightarrow 4\,OH^-$$

This is an oxygen-specific reaction, and the current generated is proportional to the amount of oxygen permeating the membrane. Oxygen diffusion from emulsion or solution through the electrode membrane is driven by pressure differential. The relationship between electrode current and oxygen pressure in solution/emulsion is linear and stoichiometric. A probe current of 1 μΩ is equivalent to consumption of 8.3×10^{-11} g of oxygen per second by the probe. Thus, a typical analysis consists of establishing a baseline for the sample then starting recording with or without addition of a modifier. Current then drops at a rate proportional to their oxygen consumption. Results can be reported as rates of oxygen consumption (e.g., mol O_2/min) or stoichiometry (O_2) consumed/mol lipid in a specified time.

Advantages. This system is mechanically simple and easy to use. It is specific to oxygen and does not respond to oxidation products so gives information not readily obtainable by other methods. Relatively short analysis times (typically <30 min) are required to obtain consumption curves. The electrodes are sensitive (3–250 μL O_2/h in air-saturated solutions; 15–250 μL O_2/h in oxygen saturated solutions) and only small samples are needed (1–3 mL). Monitoring is continuous so early oxidation rates can be calculated.

Disadvantages and precautions. The system can be used only with liquids and emulsions with viscosity low enough to allow movement of the magnetic stir bar and to not impede oxygen diffusion to the electrode. Electrodes require calibration each day with oxygen or air saturated solutions, and care must be given during set up to remove all bubbles. The current assembly is not automatable and can only run three samples at a time, but this at least allows triplicate samples or one control and two test samples of the same material to be analyzed simultaneously. Because preoxidation influences oxidation rates, at least hydroperoxides in test oils or emulsions should be measured separately to establish the initial degree of oxidation. Temperature must be constant during analyses (controlled by the water bath) and when comparing different runs. A 1 °C change in temperature can cause a 4% change in the calibration and potentially more in oxygen consumption during reaction.

Applications. This technique can be used to test oxidizability of any material in aqueous solution, suspension, or dispersion. Emulsions and solutions are easy to analyze; we have also analyzed neat oils but responses are slower because oxygen diffusion is impeded by the high viscosity. Applications include comparison of source material, formulation, catalyst, and antioxidant effects on oxidation rates. Oxidation can be endogenous, initiated (e.g., with azides or Fenton reactions), or inhibited (e.g., collect baseline oxidation then add antioxidant).

1.5.2 GC of Volatile Secondary Products from Lipid Oxidation

Analysis of volatile products by GC has been a mainstay in lipid oxidation analyses, and indeed is responsible for most of our current understanding of lipid oxidation pathways. Research from, Frankel et al. (1977a,b, 1979, 1982, 1984, 1988), Frankel (1982, 1984, 1987, 1991), Grosch and Megele (1984), and Grosch (1987) in particular, identified products and their sequence of generation, and reaction schemes for decomposition of oleic, linoleic, and linolenic acids have been derived from this information.

Volatile reaction products remain a critical component of all protocols for lipid oxidation analysis because they are what consumers smell and associate with rancidity, and because the specific products detected show where scissions and other secondary reactions have occurred. Nevertheless, there are several important provisos: volatiles should not be the only analysis performed; all the volatiles should be collected, identified, and quantitated (rather than discarding all but a few) to track active pathways and insure that active pathways are monitored; and volatiles analyses should be combined with assays for other products to more fully understand how lipids are oxidizing. For example, integration of hydroperoxides, epoxide, carbonyl, and detailed volatiles analyses revealed critical contributions of peroxyl radical additions to double bonds, bypassing hydroperoxide generation (Bogusz, 2015; Xie, 2015).

That being said, the information about volatiles obtained from GC varies tremendously, both quantitatively and qualitatively, with the sampling method employed – static headspace, solid-phase microextraction (SPME), or dynamic headspace (Bogusz, 2015). Thus, analytical approaches must be tailored to instrumentation available, levels and types of products expected, and accuracy required. No single method is perfect, and overall, most information is gained by integration of multiple approaches.

1.5.2.1 Static Headspace Analysis

Principle of the analysis: Unoxidized samples are sealed in a vial and allowed to oxidize with volatiles being generated in situ, or oxidized samples are sealed in a vial and gently warmed to volatilize products already present. Headspace volumes are usually small, typically 5–20 mL, to concentrate volatiles. The analysis depends on establishment of an equilibrium between the vapor and liquid/solid phases – compounds escape into the headspace driven by their innate volatility but limited by binding by the matrix and by the degree of saturation in the vapor phase. Gas samples are withdrawn and injected into a gas chromatograph for analysis. Flame ionization detectors and internal standards provide quantitation and mass spectrometer detectors provide identification of separated volatile components (Wampler, 2004).

Advantages: Static headspace analysis has been the simplest and fastest assay available for volatile products of lipid oxidation. Basic gas chromatographs are readily available in both quality control and research laboratories.

Samples can be analyzed directly with no special sample preparation or solvent extraction, and a tremendous amount of information can be obtained from samples just by comparison of peak retention times with standards. The process is not destructive, so nonvolatile products can be measured in the same samples immediately after headspace analyses to facilitate tracking multiple pathways.

Disadvantages and precautions: Overall sensitivity is low because what is detected depends on headspace dynamics as well as volatility of analytes. Headspace equilibria limit concentrations and types of compounds that accumulate in the headspace, skewing the view of oxidation that is presented for a number of reasons. First, headspace saturation by very volatile components impedes evaporation of less volatile components. For example, rapid generation of pentane or other short alkanes from β-scission of alkoxyl radicals at the last double bond fills the headspace and represses release of longer chain products from α-scission at the same position or from other positions in the fatty acid (Bogusz and Schaich, 2013). Sensitivity is also reduced because only a small portion of the headspace, in which analytes are already diluted, is actually sampled. The low sensitivity of detection in static headspace underlies increasing current interest in SPME and purge and trap/thermal desorption as alternative sampling techniques.

A second limitation is that temperatures required to force products out of matrices into the headspace (typically 60–120 °C) can decompose or modify products, so the compounds detected may or may not be ones initially present in the sample. Although heating samples at any temperatures above 40 °C increases total volatiles collected, it also changes patterns of products. In methyl linoleate, for example, volatile C9 and C10 oxidation products barely present at 25 and 40 °C increase at higher temperatures because scissions at those positions occur more readily with higher energy available, not because they are more volatile (Bogusz, 2015; Xie, 2015).

Direct injection of the headspace sample loses short chain products due to escape, once again giving an incorrect picture of oxidation pathways. Cryogenic cooling either with a liquid nitrogen cryotrap or dry ice in the GC oven is required during injection to trap low molecular weight products of five and fewer carbons. Cryogenic cooling also focuses analytes on the initial column section and improves resolution and quantitation.

Quantitation is mostly relative, based on peak areas. Presumptive identification and approximate concentrations of major volatiles can be determined from standard curves prepared from authentic compounds. However, standards are not available for every lipid oxidation product that can be anticipated. Mass spectrometry is then required to identify unknown products. Absolute quantitation with standards is not straightforward because any standard added to the sample may alter oxidation and certainly intervenes in headspace equilibria.

Applications. Static headspace analyses are broadly applicable to analyses of most kinds of foods or model systems except those with high solvent or water contents that saturate both the headspace and the chromatography detector, or where high levels of very volatile ingredients (e.g., flavors or aromas) mask lipid oxidation products. Indeed, static headspace analyses have been one of the most extensively applied methods for monitoring lipid oxidation in all kinds of foods, with many thousands of reports published.

1.5.2.2 SPME

Principle of the analysis: SPME (not to be confused with solid-phase extraction used for sample clean-up) seeks to improve the sensitivity of headspace analyses by selectively adsorbing and concentrating volatiles onto a coated fused silica fiber attached to the end of a microsyringe plunger with a protective sheath. The fiber is inserted into the headspace vial through the septum and is given time to adsorb analytes either from the headspace above the sample or from the sample itself, if liquid (Figure 1.6). The fiber is then withdrawn back into its sheath and transferred to a gas chromatograph, where it is inserted into the hot injection port and desorbed into the chromatograph (Arthur and Pawliszyn, 1990; Zhang and Pawliszyn, 1993). An excellent introduction to SPME, its controls, and its applications has been presented by Vas and Vékey (2004).

Sensitivity and selectivity are controlled by the character and thickness of the fiber, the adsorption time and temperature, and the desorption time and temperature, all of which can be varied to maximize detection of compounds of interest (Pawliszyn, 2006). The glass fiber, usually 1–2 cm in length, is coated with solid phases applied singly or in combination to provide selectivity in adsorption of analytes. SPME analyte retention as a function of fiber polarity and cross-linking is shown graphically in Figure 1.7. The thickness of the coating largely governs quantitative retention.

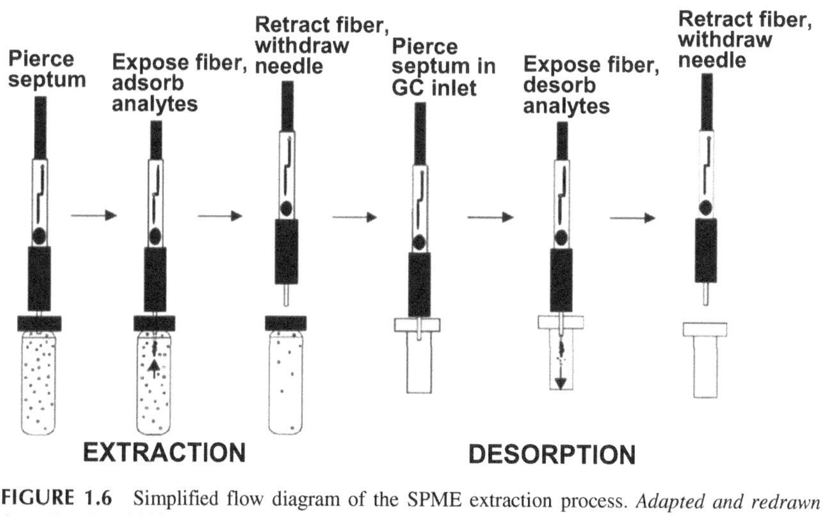

FIGURE 1.6 Simplified flow diagram of the SPME extraction process. *Adapted and redrawn from Supelco (2003); used with permission.*

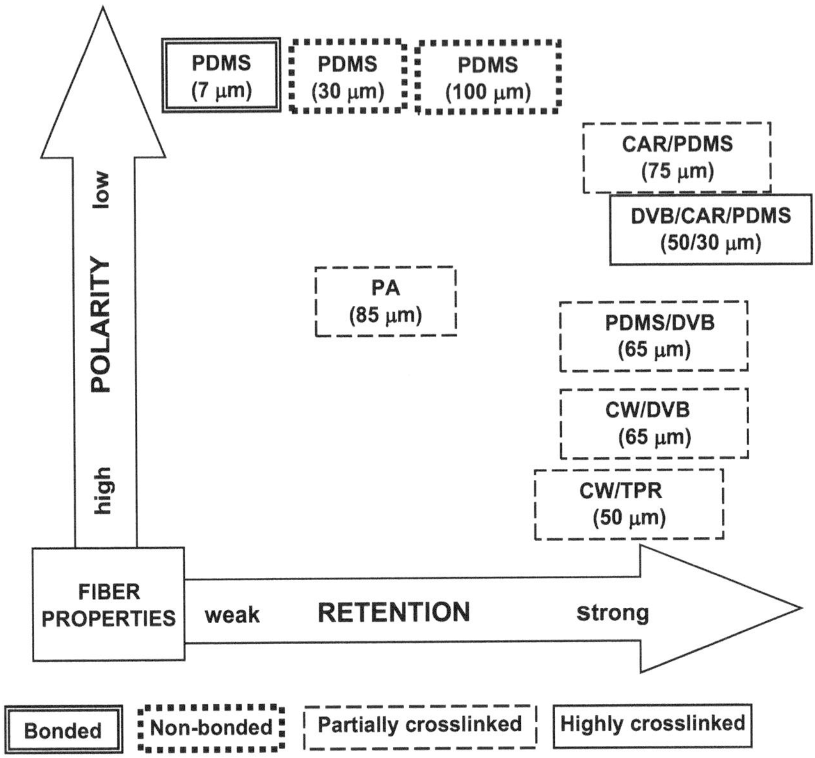

FIGURE 1.7 Relationship between molecular properties, polarity, and retention in SPME fibers. PDMS, polydimethylsiloxane; CAR, carboxen; DVB, divinylbenzene; PA, polyacrylate; CW, carbowax; TPR, templated resin. *Redrawn from Kataoka et al. (2000); used with permission.*

Polydimethylsiloxane (PDMS) is the workhorse of the method and has the broadest applications. It is a very rugged nonpolar liquid phase that withstands high injector temperatures (up to 300 °C) and works well with lipid-based materials. However, alone it is too nonpolar for lipid oxidation products. Adding divinylbenzene (DVB) to PDMS increases polarity and retention, making it a good choice for trace analyses of C6-C15 compounds. Cross-linking increases its durability.

Carboxen (CAR) is a polar particulate coating with micropores that go completely through the particles, facilitating binding and desorption of small molecules such as short chain aldehydes (Kataoka et al., 2000). By itself, CAR is not very stable to oxidizing conditions (degradation evidenced by darkening or powdering of the fiber). However, when mixed with other coatings, it tremendously increases extraction of polar compounds from nonpolar matrices, and it increases retention of short alkanes (ethane through hexane, products from β-scission of alkoxyl radicals formed at the last double bond of fatty acids) by orders of magnitude. Retention range is C2 to C12 (Mani, 1999; Shirey, 2012).

A unique three-phase fiber (DVB/CAR/PDMS) has been developed by blending carboxen with PDMS in a 30-μm inner layer on a strengthened fiber core, then coating this with a 50-μm outer layer of DVB. This DVB layer extracts longer chain and semivolatiles while the inside CAR layer adsorbs the smaller more volatile molecules (Page and Lacroix, 2000). It is this mixture of qualities that makes the DVB/CAR/PDMS fiber the most preferred fiber for samples that have a broad range of analyte polarities and chain lengths. This fiber is specifically selective for compounds in C2 to C20 range (Wardencki et al., 2007) and has turned out to be particularly useful for lipid oxidation products.

SPME sampling is much more complex than just selecting a fiber, sticking it into a vial for a few minutes, then analyzing collected compounds. Film thickness, distribution constant (binding attraction relative to the sample matrix), and binding time all interact with fiber properties in controlling what SPME detects (Mani, 1999). Fibers contain only a small amount of adsorbent (~0.5 μL on a 100-μm fiber) so their binding capacity is limited. In a mixture of analytes such as lipid oxidation products, there is an interesting dynamic in binding of the various compounds. Products do not all bind at once or at the same rate, especially when a mixed fiber is used. Unlike static headspace analysis where most volatile compounds appear first and at highest concentrations, in SPME volatiles with strong binding adsorb first and compounds with weaker binding adsorb later and less, in competition. If maximum collection of products is desired, longer times are thus needed to pick up all of the compounds. At the same time, if the film is too thin and capacity too small, with longer adsorption times paradoxically some lesser binding compounds can displace compounds originally adsorbed on the fiber surface, thus generating a different product distribution. To avoid this, capacity must be increased with a thicker coating. If, on the other hand, the analysis seeks to quantitate only one or a limited number of compounds (e.g., hexanal), thinner coatings and just enough time to maximize hexanal adsorption should be used to eliminate potential interference from other compounds. Unfortunately, SPME is not a process amenable to standardization because volatiles and release dynamics vary with the food matrix. Thus, optimum adsorption time must be determined independently for each food, fiber, and particularly the volatile target(s). Typical adsorption times are 10–30 min but can range as high as several hours (McNair and Miller, 2009).

Desorption time and temperature in the GC inlet also must be optimized for each set of conditions. Desorption times vary with the analyte characteristics but typically average about 5–20 min.

Advantages. SPME provides simple, fast sample preparation collection of volatiles with a choice of maximizing yield or selectivity based on the fiber and coating thickness selected. Orders of magnitude in sensitivity are gained over static headspace analysis because volatiles are concentrated from the full headspace, not just a limited volume, and because continual removal of volatiles by adsorption to the fiber provides a driving force moving molecules from the sample to the headspace. Thus, SPME increases detection of semivolatiles, particularly C8 to C10 aldehydes that are poorly detected in static headspace analysis. Also, SPME fibers can be used with nearly all kinds of food systems. All of these factors have contributed to a huge increase in SPME utilization in lipid oxidation analyses.

Disadvantages and precautions. While SPME detects more compounds, its selectivity also presents distinct difficulties for lipid oxidation analyses in that products are detected and quantitated based on fiber preference more than actual levels generated in reaction. Hence, product distributions are skewed relative to static headspace and may or may not be representative of actual lipid oxidation. In addition, there is always a competition among products for binding regardless of the fiber characteristics. For example, binding competition between alkanes and aldehydes shifts with adsorption time and extent of oxidation (Bogusz, 2015), and in analyses of marine phospholipids, Strecker aldehydes from Maillard reactions are more easily adsorbed than alkanals and alkenals (C. Jacobsen, personal communication). Under these conditions, either looking for specific products that have been displaced (rather than not formed) or quantitating products by total peak areas will give a totally wrong picture of oxidation.

Lipid oxidation products always have the potential to cause problems with SPME because of their reactive potential. Adsorption to the fiber may decompose some oxidation products or form new ones, such as rearrangements increase furan levels with longer extraction times and higher temperature (Adams et al., 2012). Oxidation products may also degrade the fiber coatings. Carboxen is sensitive to oxidation at elevated temperatures, becoming dark and/or powdery. It also swells in water. We have observed at least one case of fibers with carboxen fouling during analyses of oxidizing margarines. For the same reason, high-purity carrier gases are mandatory for SPME.

Degradation and fragility of fibers bring forward another issue – they often must be replaced several times during a study with many samples and extended times. Hence, because there is a significant batch to batch variation between fibers, fibers sufficient for an entire study must be obtained from the same batch to ensure consistency of analysis.

It is easy to exceed the range of the fiber capacity when using only one extraction/adsorption time, in which case correlation between volatiles generated and volatiles bound is lost. To avoid this, each system analyzed must be tested to determine binding kinetics and time for maximum binding (Jung and Ebeler, 2003; Pawliszyn, 2006; Ma et al., 2014). When saturation occurs, such as in a highly oxidizing food, smaller samples, shorter adsorption times, or thicker fiber coatings may solve the problem. Sample dilution in water or salt solution, as is recommended to reduce volatilization in general volatiles analysis, is not appropriate with lipid oxidation because these treatments will alter oxidation products.

As with static headspace, quantitation in SPME is not straightforward because it is difficult to find a standard that when added to foods with oxidizing lipids (1) will not modify the lipid oxidation, (2) will not be modified itself by lipid radicals or secondary products, (3) will not be bound by food molecules, and (4) will be released completely and bound fully by the fiber.

Applications. SPME is seeing increasing use in a wide range of lipid oxidation analyses in foods and model systems. To provide just a few examples, CAR/PDMS fibers have been used to analyze light-induced lipid oxidation in milk (30-μm coating) (Correddu et al., 2015), oxidation of linseed oil acids added to dairy beverages (Giroux et al., 2008), effect of antioxidants on oxidation of fish emulsions (Iglesias et al., 2010) and fish muscle (75 mμ) (Pazos et al., 2010), and hexanal from heme catalysis of fatty acid oxidation (75 mμ) (Lorrain et al., 2010). PDMS/DVB fibers are widely used to determine hexanal. DVB/CAR/PDMS fibers have been applied to determining effects of raspberry extracts on stability of muesli (Klensporf and Jeleń, 2008), volatiles in olive oil (Zhu et al., 2013); lipid oxidation in pork products (Estévez et al., 2003), chicken patties (Mariutti et al., 2009) and extruded pet foods (B. Bogusz and K.M. Schaich, unpublished data); as well as oxidation mechanisms in methyl linoleate (Bogusz, 2015).

1.5.2.3 Purge and Trap/Thermal Desorption

Principle of the analysis: As noted in Section 1.5.2.1, static headspace analyses are limited by disadvantageous headspace equilibria in which high volatility of some products impedes volatilization of other products. Dynamic headspace analysis was introduced in the early 1960s to overcome this problem by continuously removing volatiles from liquid or solid samples (Swinnerton et al., 1962). The technique has since developed in a number of directions to handle different types of samples. The total process involves two stages – purge and trap, followed by thermal desorption. In the purge and trap stage, the sample is heated to facilitate release of volatiles. During the process, inert gas is continuously passed over solids or through liquids in tubes or oils in flasks to carry released volatiles away from the sample and deposit them onto a tube packed with sorbent material where they are trapped and concentrated (Figure 1.8). The trap is then connected to a gas chromatograph via a special unit positioned at the inlet, and volatiles are desorbed from the tube to the column by heat under a flow of inert gas (Grimm et al., 2002) in a process called short path thermal desorption (Manura and Hartman, 1992). Volatiles are separated and analyzed by appropriate GC with detection by flame ionization, MS, or both.

As with SPME, the sorbent resins packed in the trap are the heart of the method because they determine what is retained from the gas phase and released to the GC for analysis. The two types of adsorbents most useful for lipid oxidation products are Tenax® (a porous polymer of poly (2,6-diphenyl-*p*-phenylene oxide)) and CAR (a carbon molecular sieve) (Jochman et al., 2014). Hydrophobic Tenax® (Buchem BV, Apeldoorn, Netherlands), designed specifically for trapping volatiles and semivolatiles from liquid or solid matrices, is the most broadly used resin for thermal adsorption. It binds a wide range of compound classes with chain

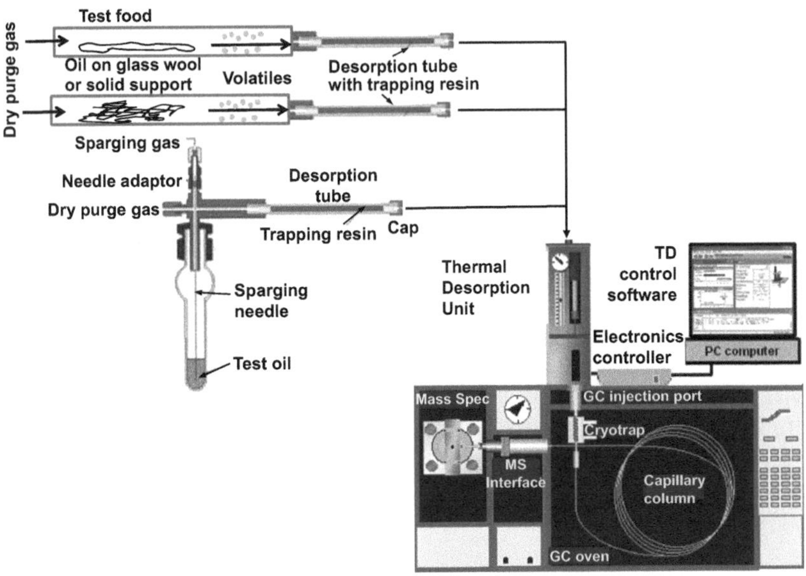

FIGURE 1.8 Typical configurations of purge and trap/thermal desorption for oils and oil-containing foods.

length greater than five carbons, and covers the chain length range expected for most volatile lipid oxidation products (up to about C11). It is not affected by water, which means it can be used to trap lipid volatiles from all kinds of food products, including emulsions. The original Tenax had problems with bleeding and impurities. This now has been modified, and a new high-purity version, Tenax®-TA, is the best general sorbent for lipid oxidation products because of its low background. Tenax® GR, a composite of Tenax TA with 23–30% graphite added to increase retention properties, has been reported to extend the analysis range to lower boiling compounds, limit sorption of water and increase the binding of aromatic compounds (Cao and Nicholas Hewitt, 1993). With lipids, its behavior is comparable to Tenax-TA but performance is very much dependent on the quality of graphite added because metals and other contaminants cause high backgrounds and degradation of analytes (T.G. Hartman, Rutgers University Mass Spectrometry Facility, personal communication). Variations in the graphite component may contribute to different results (not useful/useful) reported when Tenax GR was used with lipid oxidation products (Jochman et al., 2014; Alemán et al., 2015).

Carboxen (Supelco, Bellafonte, PA) is a molecular sieve with micropores that bind small volatiles (C2 to C5) by van der Waals interactions and water by condensation. This binding pattern complements that of Tenax, so the two resins are frequently combined in mixed bed traps. Rather than being mixed together, the resins are packed in sequence with the CAR positioned behind the Tenax to prevent breakthrough of larger compounds during binding and allow first release of the small volatiles at lower temperatures during desorption (Figure 1.9).

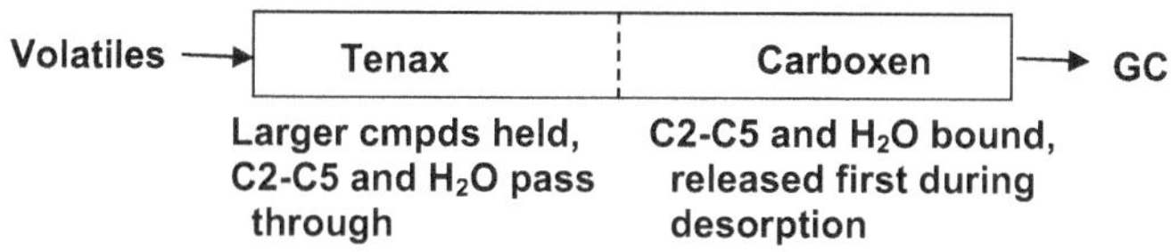

FIGURE 1.9 Configuration of mixed bed traps packed with Carboxen and Tenax.

Because the compounds trapped by CAR are so volatile, cryotrapping during desorption is mandatory when this resin is used.

Successful applications of thermal desorption require tailoring the combination of adsorption characteristics and desorption temperatures to analytes of interest to improve trapping and resolution (Manura, 1999). For example, as shown in Figure 1.10, Tenax releases short to mid-chain aldehydes and alcohols with nearly the same temperature profile as hydrocarbons up to long chain lengths (20). This means all products will be mixed in the chromatogram, making resolution and identification more difficult. Carboxen, on the other hand, releases oxygenated products at temperatures higher than hydrocarbons on Tenax. The two sorbents together, then, facilitate separation of product classes in desorption and GC – hydrocarbons desorb and elute at lower temperatures while oxidation products move only after temperature is increased.

Advantages: Often, compounds that are barely or not detected by static headspace and SPME can be trapped by thermal desorption and with fewer biases. Constant removal of volatilized compounds from the headspace provides a strong driving force to maintain continuous flow of potentially volatile compounds to the sample surface where they can evaporate, so collection of volatiles is much more complete than with static headspace or SPME methods. Concentrating volatiles on the trap increases sensitivity to parts per billion (ppb) detection (Wampler, 2004), and lack of headspace equilibrium limitations and lower binding bias improves accuracy of the analysis. The methods can be used with solid and semisolid food materials having a wider range of moisture because Tenax is not poisoned by water as are some SPME fibers, while CAR holds moisture that otherwise would create problems in static headspace analyses and in the GC.

Absolute quantitation of volatiles as μg/trap is achievable by injection of deuterated internal standards onto traps after sample adsorption (Ames et al., 2001). An internal standard cocktail consisting of 1 μL each of 10 mg/mL benzene-d6, toluene-d8, and naphthalene-d8 in methanol can provide normalization of retention times from run to run as well as quantitation by comparison of analyte peak areas to those of standards.

Disadvantages and precautions: Although simple in concept, thermal desorption must be managed carefully at each step to ensure good results (Grimm et al., 2002). Sorbent, purging times, and purging flow rates must be matched to the compounds being analyzed or some compounds may not be trapped while others may not be released. Loss of volatiles in leaks due to the

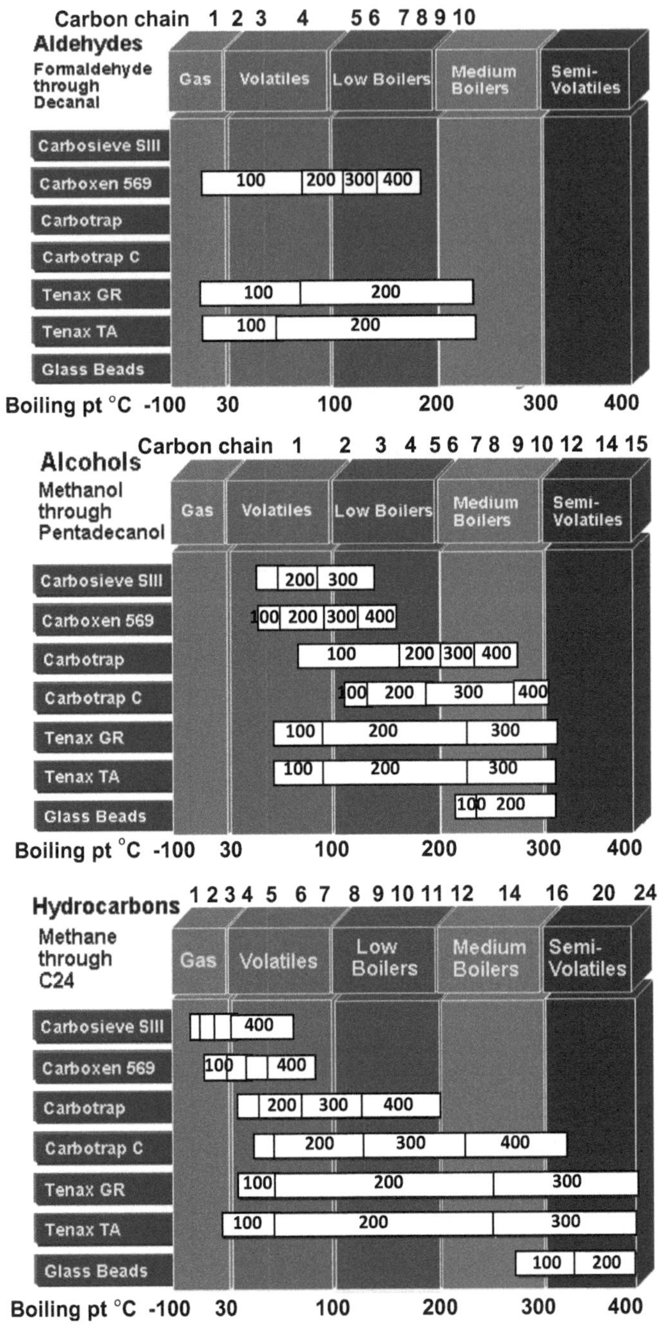

FIGURE 1.10 Properties of thermal desorption trap sorbents used to collect lipid oxidation volatiles. Numbers at the top of each chart indicate the number of carbons in straight chain compounds of each class. Numbers at the bottom of each chart give the boiling points corresponding to the chain lengths. The bars in the chart show the useful range of compounds that can be analyzed by each resin, and give the temperature required to desorb compounds of various chain lengths. *Adapted from information on the Scientific Instrument Services Website (*http://www.sisweb.com/index/referenc/bv-alc.htm; http://www.sisweb.com/index/referenc/bv-hyd.htm; http://www.sisweb.com/index/referenc/bv-aldeh.htm*). Used with permission.*

many mechanical connections and reconnections required as well as during injection is a constant threat. Desorption temperature and gas flow must be controlled carefully to avoid overheating or cold spots as well as to ensure complete desorption (Wampler, 2004). Traps must be reconditioned by gas flush and "bake out" between samples to prevent analyte carry-over. The traps, resins, and specialized instrumentation make thermal desorption substantially more expensive than static headspace and SPME, although the cost is counterbalanced by information obtained.

As with both static headspace and SPME analysis, the instability of lipid oxidation products limits some of the flexibility of thermal desorption analyses. Both purging and desorption temperatures must be restricted to the lowest temperatures sufficient for volatilization from samples and breakthrough from traps to avoid decomposing and transforming existing oxidation products as well as forming artifact products (Uhde, 2009). The combination of moisture and heat is particularly damaging to lipids, so samples may need to be dried for oxidation analyses.

Applications. Thermal desorption offers tremendous potential for following lipid oxidation in model systems in research and accelerated shelf-life (ASL) studies of a wide range of foods in quality control, as well as measuring levels and distribution of lipid oxidation products in stored foods (e.g., age and quality monitoring of processed foods in warehouses). Currently, the major impediments are availability of instrumentation, bias and fear due to unfamiliarity, and lack of development of methods for specific applications. Hopefully, these obstacles will be overcome in the near future.

In the past, short-chain lipid oxidation volatiles have been largely ignored as unimportant in food quality. Nevertheless, in keeping with the new paradigm of lipid oxidation analysis that analyzes as many products as possible to identify the presence of alternate pathways, the capability to detect these products in CAR/Tenax traps provides information critical to full understanding of lipid oxidation reactions. Including odorless short chain alkanes in analyses (n-3 fatty acids and pentane from n-6 fatty acids) can identify conditions that favor β-scissions of alkoxyl radicals at the last double bond versus those that facilitate α-scission to generate aldehydes that consumers definitely smell. Formaldehyde, acetaldehyde, and butanal signal decompositions of secondary products, contribute to off-flavors and condensations with proteins, and can explain why expected products (especially epoxides) may be missing.

1.5.3 FTIR Spectroscopy

IR spectroscopy measures absorption of electromagnetic radiation in ranges that provide energy sufficient to induce transitions between vibrational states of bonds within molecules. Three frequency (wavelength) ranges are involved – far-, mid-, and near-IR – each affecting a different type of vibrations (Table 1.5). To provide a frame of reference, the energy absorbed by molecules in mid-IR

TABLE 1.5 Frequency Ranges and Molecular Vibrations Affected in Infrared Spectroscopy

	Frequency Range		
Region	Wavelength (μm)	Wavenumber (cm^{-1})	Vibrations Excited
Near	0.75–2.5	12,500–4000	Fundamental vibrations
Middle	2.5–50	4000–400	Overtones and combinations
Far	50–1000	400–5	Backbone vibrations of large molecules

From Chong (2015); used with permission.

is between 15 kcal/mol and 1 kcal/mol (Wade, 1991). Unlike the chemical and chromatographic analyses described above which detect intact whole molecules, IR spectroscopy detects and distinguishes individual functional groups by their various vibrational modes – stretching, bending, rocking, wagging, scissoring, twisting. Each functional group absorbs energy at independent frequencies, providing a means of identification. To detect all the functional groups in a molecule or sample, IR spectrometers scan the full frequency and energy range, then plot the absorption intensity as a function of frequency (wave numbers/cm) in a spectrum. Peaks appear at frequencies where groups present in the molecule(s) absorb energy, with peak intensities proportional to the number of groups present. The peak frequencies and intensities provide a basis for identification of structure in pure molecules (the traditional application) or for following changes in materials during chemical reactions or treatments, as is most often used with lipid oxidation.

Application of IR to lipid oxidation analysis lagged for decades due to instrumental limitations. Original IR spectrometers were grating or dispersive instruments in which slits controlled the sensitivity and spectral resolution. Opening the slits increased light impinging on the sample, hence increased signal strength, but this also spread the signal and increased overlap between close peaks. Closing the slits improved resolution but also decreased sensitivity. In addition, slits focused the light source on the sample one wavelength at a time, so it took many minutes to scan a full spectrum. Lipid oxidation analyses were performed with these instruments, but the poor sensitivity and superimposition of peaks from several functional groups in products prevented large-scale application of the method.

Integration of FTIR in the early 1970s, together with the development and increasing sophistication of personal computers, provided a quantum leap in sensitivity, resolution, speed and improved data processing that now makes FTIR a very attractive analytical methodology for detecting and quantitating multiple products in intact oils, extracts, and food materials (Dufour, 2009). There are several key differences in IR introduced by Fourier transform (Subramanian and Rodriguez-Saona, 2009). First, rather than separating the IR light beam into individual wavelengths, a Michelson Interferometer around the sample chamber intercepts the impinging beam and splits it into two, each with a full frequency range. One half is directed toward a stationary mirror, and the other goes to a mirror that is continuously moving over a distance of about 2.5 μm. The beam from the stationary mirror is reflected directly back to the detector while the beam from the moving mirror which provides the wavelength variation passes through the sample on the way. The two beams recombine at the detector, producing an interference pattern set up by the absorption of discrete energies by the sample. Since all the frequencies are present and monitored simultaneously as the beam passes through the sample, the entire spectrum requires less than a few seconds to record. At this point, data are in the form of *time* of mirror movement. This value then undergoes Fourier transform, a series of complex mathematical calculations to convert it from time to frequency, using the constant velocity of the mirror. To greatly reduce signal to noise, a helium–neon laser is simultaneously directed through the Michelson interferometer as a frequency reference to align repeat scans exactly and allow digital signal averaging. This was not possible in dispersive instruments. Between improved optics and addition of signal averaging, signal-to-noise ratios increase by more than 10 times in FT instruments. Finally, computers provide fine control over mirror movement that dramatically increases spectral resolution, they digitize and store data, and open vast new capabilities for mathematical manipulation and analysis of spectra. Most systems now come with software for chemometrics, powerful statistical analyses for identifying and relating patterns in complex data. The data mining made possible by chemometrics has revolutionized IR analyses, picking out key features associated with oxidative changes rather than quantitating individual product peaks, and greatly increasing accuracy of quantitative analyses. The use of chemometrics will be described in more detail later.

Both mid- and near-IR have two sampling modes depending on how the light beam is presented to the sample material. In transmission mode, samples are presented in thin layers (<10 μm) and the IR beam passes directly through, interacting with the molecules on the way. This is the major mode for pure compounds and what most people associate with IR sampling, but most food materials are not easily accommodated, even if dehydrated. Attenuated total reflectance mode (ATR) was developed to overcome thin film handling limitations and provide an alternative for analyzing intact mixed materials directly (Sedman et al., 1999; Glassford et al., 2013). Simplistically, in reflectance mode, radiation is reflected onto the surface of a sample rather than being transmitted through it. The sample is placed in contact with a crystal of high refractive index (η), such as diamond, germanium, zinc selenide. When the IR beam passes through the crystal, some is reflected and some is transmitted, diluting

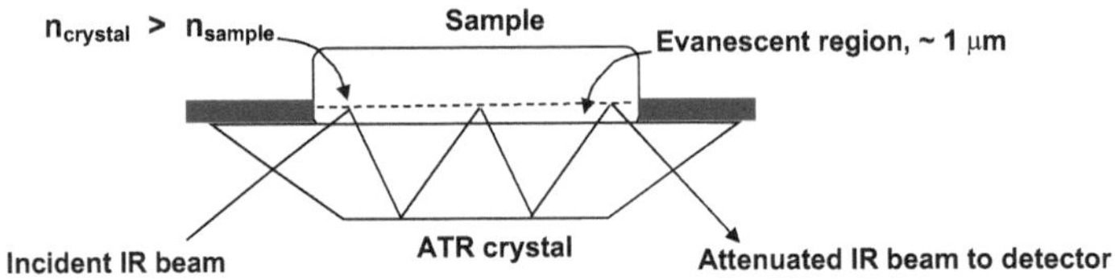

FIGURE 1.11 Diagrammatic representation of attenuated total reflectance (ATR) in IR analyses.

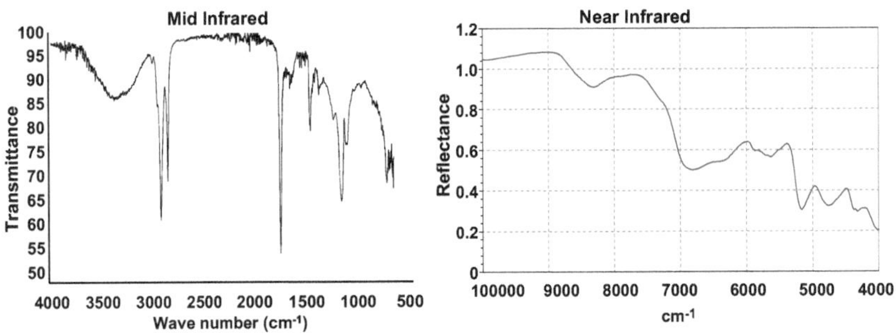

FIGURE 1.12 Comparison of infrared spectra from different frequency ranges. Mid-IR spectra have strong sharp bands, while near-IR spectra are broad and diffuse. Mid-IR, coconut oil-based margarine; near-IR, high oleic sunflower oil mixed 15:85 with rice flour.

the light beam. However, if the IR beam strikes the crystal at or above an angle called the critical angle = $\sin^{-1}(\eta_2/\eta_1)^3$, all light is reflected back into the crystal and none is transmitted. This creates an evanescent wave that projects into the sample and carries energy (Figure 1.11). As the sample absorbs energy from the evanescence, it also extracts energy from the internally reflected beam and attenuates response at the detector at wavelengths corresponding to the absorption frequencies of the sample. Comparison of beams with and without samples in place produces typical spectra by difference (Sedman et al., 1999; Glassford et al., 2013). Because of its handling ease and adaptability to a wide range of samples, attenuated total reflectance has become the most widely used IR sampling method for most materials today. Certainly, this is true with food analyses.

Difference in mid- and near-IR. The same basic spectroscopy concepts and instrument components apply to both mid- and near-IR. However, spectra from mid-IR are sharp and clear with many bands, while near-IR spectra are broad, wandering, and largely featureless to the naked eye (Figure 1.12). The defined spectra of mid-IR arise from strong fundamental vibrations of all kinds of bonds, and each group has one main identifying frequency and perhaps some weaker bands at other frequencies. In contrast, near-IR spectra arise only from vibrations of H linked to C, O, or N atoms (Stuart, 2004). Absorptions in near-IR result from overtones or combinations of the fundamental vibrations of these

3. η_2 = refractive index of the sample; η_1 = refractive index of crystal.

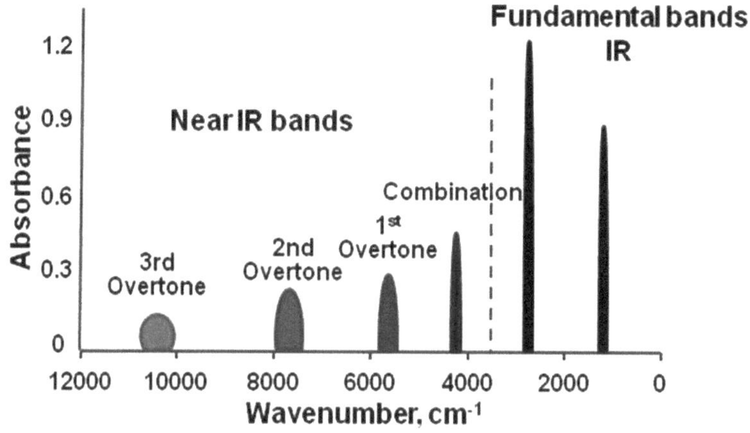

FIGURE 1.13 Relationship of mid- and near-IR bands in sharpness and intensity. *Adapted and redrawn from Matejka (2015).*

bonds detected in mid-IR, and each can have up to three overtone frequencies. Vibrational frequencies of overtones in near-IR are approximate multiples of fundamental vibrations in mid-IR, while those of combinations are the sum of fundamentals (Figure 1.13). Near IR spectra are broad and diffuse for two reasons: (1) the absorptions of overtones are weaker, and (2) each bond type has separate absorption frequencies for combinations and each of three overtones, and with many types of H-R links present in lipids or foods, substantial overlapping results.

There are trade-offs in information gained by the two spectroscopies, but they should be considered complementary rather than competitive as analytical methods.

1.5.3.1 Mid IR Analyses of Lipid Oxidation

Practical aspects. Sample mode – transmission. Penetration of mid-IR radiation into materials is very limited so samples must be analyzed in thin films of liquids in special cells, in viscous liquids or semisolids spread between potassium bromide (KBr) plates, or in solids ground with KBr and pressed into wafer-thin pellets. In the latter, particle size is critical, so pulverizing the sample–KBr mixture (1–3 mg sample/250–300 mg KBr) to a very fine dispersion is required for sharp spectra (Stuart, 2004). Water is the worst enemy of IR – it dominates the 3200–3600 cm^{-1} region of the spectrum, obscuring contributions of —OH groups from other molecules in the sample, and it erodes the halide plates. Hence, it is customary to dry samples carefully, such as under an IR lamp, before IR analyses.

Sample mode – attenuated reflectance. Special sample cells have been developed by most IR spectrometer manufacturers to replace the standard sampling unit in the sample compartment. Designs vary but share common elements of a small aperture over the crystal for deposition of the sample and a pressure screw to compress the sample firmly against the crystal for maximum contact. Particle size, packing, and contact with the crystal all have strong impacts on the detection levels, sharpness of spectra, and reproducibility so it is often recommended that samples be ground fine just before analysis. This is good practice in general, but when analyzing samples with actively oxidizing lipids grinding can cause more problems than it solves. Many scans (often 50–150) are normally run and signal averaged. However, with lipid oxidation analyses, it is best to pulverize samples only at the last minute, collect scans in small batches (e.g., 5–10 scans), then examine the progression of oxidation in batches over time. When samples are stable, successive scan batches can be combined and averaged. If samples begin to degrade, spectra can be discarded after the change point. When food products are destabilized by pulverizing and exposing new surface area, minute samples of intact food may be substituted; however, more samples will be needed to overcome the natural nonhomogeneity of the materials.

Disposable Teflon cards have been proposed as an alternative to standard sampling when studying oxidation in edible oils (Russin et al., 2003, 2004), and they have been used in some DART (direct analysis in real time)-MS analyses as well. These cards consist of a microporous polytetrafluorethylene (PTFE, Teflon) film sandwiched between two cardboard layers with apertures of 1.9 cm, and the entire assembly fits into standard IR plate holders. Three drops of oil are placed on the film and distributed evenly over the entire area, then analyzed in ATR mode. The cards simplify direct measurements on oils, but they also can be moved around with the sample remaining on them. Thus, they can be placed in incubators and removed periodically to track changes in oxidation evident in IR spectra. This makes it possible to run shelf-life studies with each sample as its own control. Precautions required to standardize use of the cards have been reviewed (Russin et al., 2003, 2004).

Advantages of mid-IR. Instrumentation designs and computerization are becoming increasingly sophisticated, so mid-IR spectra can now be obtained rapidly and easily with appropriate attention to sample handling. Only very small amounts of material (mg) are needed for analysis. The strength of mid-IR is its ability to identify molecular structures from the characteristic and well delineated absorption bands in this spectral range which can be combined to identify single molecules, analyzed separately to detect multiple components in a mixed system (e.g., lipids and proteins in a food), or integrated to follow reactions by appearance and disappearance of absorptions from different functional groups. Perhaps most importantly, mid-IR provides a means of obtaining oxidation measures on intact food samples without lipid extraction, solvents, or chemical reactions.

Intensity of mid-IR peaks follows Beer's law in being proportional to concentration of the component, so production or loss of individual functional groups or products, such as hydroperoxides, double bonds, or carbonyls can be quantitated by application of an extinction coefficient or by reference to a standard curve, both determined separately.

Disadvantages of mid-IR. Contemporary instrumentation has overcome many of the limitations of earlier IR systems. Nevertheless, for transmission mode, there still remains a substantial learning curve for sample preparation and handling to get consistent results. IR spectra are not obtained in a sealed chamber so it is common to see atmospheric water and carbon dioxide in blank spectra. Hence, most instruments now run a "background" spectrum first then automatically subtract it from every spectrum, and some instruments also purge the sample with dry nitrogen to minimize water and CO_2 interferences. Mid IR results are not quantitatively independent. Both direct quantitation, such as of hydroperoxides, and quantitative chemometric analyses depend on correlation of IR peak intensities to primary analyses determined on the same types of samples. Building and validating this database can be time consuming, but once developed the values are broadly applicable to different instruments and different types of samples.

Nonhomogeneous matrices such as foods counterbalance many of the advantages of mid-IR. Most vexing is that very small samples make it impossible to obtain "representative" examples. The small sample limitation is further exacerbated by the light beam hitting only a limited region in the material. Hence, accurate analyses for both quantitation and structure identification require averaging spectra from a large number of samples. Recommendations range from a minimum of 10 for range-finding to more than 100 for quantitation and statistical analyses. The time required for this presents a significant challenge when testing for processes that change rapidly, such as lipid oxidation.

Finally, for lipid oxidation analyses, especially in mixed food matrices, there are too many overlaps of key absorptions (—OH from alcohols, hydroperoxides, free fatty acids, amino acids, sugars; carbonyls from fatty acids and aldehydes) for easy interpretation and accurate absolute quantitation. Increasing application of chemometric statistical analyses may help distinguish shifts and changes as oxidation progresses.

Applications of mid-IR. Classically, characteristic mid-IR frequencies have been very useful for identifying multiple oxidation products in the same sample, for following oxidation (and co-oxidation) over time, and for verifying absence of the same in fresh products or incoming ingredients. Characteristic group frequencies for double bonds and expected lipid oxidation products are listed in Table 1.6. Since groups from nonlipid molecules in mixed samples are also revealed in the spectra, frequencies associated with co-oxidation of proteins are included as well.

Perhaps the most common application to lipid oxidation has been analysis of hydroperoxides and determination of PVs. Numerous reports cite PVs determined by direct comparisons of IR and standard analyses. However, applications of chemometrics have also identified other changes that accompany hydroperoxide formation and can increase the sensitivity of oxidation analyses.

TABLE 1.6 Characteristic Mid-Infrared Frequencies of Functional Groups Involved in or Affected by Lipid Oxidation: Lipid Double Bonds (Lost in Oxidation) and Oxidation Products, Protein Frequencies of Native Peptide Chain and Side Groups (Lost in Oxidation) and Some Oxidized Groups

Frequency (cm^{-1})	Functional Group	Molecular Action	Reference
Lipids			
3610–3640	Hydroperoxides, ROOH	O—H stretching	Guillén and Cabo (2000) and Borchman and Sinha (2002)
3520	Hydroperoxide-epidioxide	O—H stretching	Neff et al. (1982)
3435	Hydroperoxides, PUFA	O—H stretching	Guillén and Cabo (2000)
3444–3450	Hydroperoxides, isolated	O—H stretching	Guillén and Cabo (2000)
3470	C=O, glyceride ester	C=O overtone	Guillén and Cabo (2000)
3468	Hydroxides (alcohols), ROH	O—H stretch	Guillén and Cabo (2000)
3020–3080	Isolated C=C (*cis*)	C=C—H stretch	Borchman and Sinha (2002)
3006–3012	Conjugated dienes (*cis*)	C=C stretch	Guillén and Cabo (2000)
1746–1765	C=O in glyceryl ester	C=O stretch	Vlachos et al. (2006)
1728–1743	C=O, saturated aldehydes	C=O stretch	Guillén and Cabo (2000) and Vlachos et al. (2006)
1720	C=O in free acid	C=O stretch	Guillén and Cabo (2000) and Borchman and Sinha (2002)
1700–1730	C=O, ketones	C=O stretch	Stuart (2004)
1650–1654	C=C, conjugated	C=C stretch	Guillén and Cabo (2000) and Hayati et al. (2005)
1630	C=O, α,β-unsaturated		van de Voort et al. (1994)
950–815	epoxide (C—O)	asymmetric ring stretch	ChemistryBlog (2012)

TABLE 1.6 Characteristic Mid-Infrared Frequencies of Functional Groups Involved in or Affected by Lipid Oxidation: Lipid Double Bonds (Lost in Oxidation) and Oxidation Products, Protein Frequencies of Native Peptide Chain and Side Groups (Lost in Oxidation) and Some Oxidized Groups—cont'd

Frequency (cm^{-1})	Functional Group	Molecular Action	Reference
880–750	epoxide (C—O—C)	symmetric ring stretch	ChemistryBlog (2012)
987	*trans, trans* conjugated diene	C—H bend	
982–983	*cis, trans* conjugated diene	C—H bend	Guillén and Cabo (2000)
967–976	isolated *trans* db	C—H bend	Guillén and Cabo (2000)
914	isolated *cis* C=C	C—H bend	Guillén and Cabo (2000)
Proteins	**Backbone**[c]		Barth (2007)
3310	amide A NH	N—H stretching	
3270	amide-B H-bonded NH	N—H stretching	
1655	amide I	α-helices	
1650	amide I	C=O stretching	
1630 (1685 w)	amide I	antiparallel β-pleated sheet	
1550	amide II	NH bend + CN stretch[a]	
1490–1460	amide II′	CN stretch	
1400–1200	amide III	NH bend + CN stretch[b]	
	Side chains		
2551	SH	S—H stretching	
1716–1712	C=O in COOH, asp and glu	C=O stretching	

Continued

TABLE 1.6 Characteristic Mid-Infrared Frequencies of Functional Groups Involved in or Affected by Lipid Oxidation: Lipid Double Bonds (Lost in Oxidation) and Oxidation Products, Protein Frequencies of Native Peptide Chain and Side Groups (Lost in Oxidation) and Some Oxidized Groups—cont'd

Frequency (cm^{-1})	Functional Group	Molecular Action	Reference
1677	C=O in COOH, asn		
1695–1652/ 1663–1614	Arg $CN_3H_5{}^+$		
1688	Gln C=O		
1631	his C=C		
1626	lys NH_3^+		
1622, 1509, 1496	trp C=C, CN/NH, C—C		
	Oxidized groups		Reusch (2013)
1325, 1140	S=O sulfone		
1345	S=O sulfonic acid		
1365, 1180	S=O sulfonyl chloride		
1350–1450	S=O, sulfate		
500–540	S=S		
3550, 1665, 945	C=NOH		
960	N—O amine oxide		
1550/1530+ 1350	N=O nitroso/ nitro		

[a]*Out of phase, independent of side-chain structure.*
[b]*In phase, varies with side chains.*
[c]*Only main bands are included here; further details are available for each amino acid in Barth (2007).*

In studies of a number of oils, the most sensitive indicators of oxidation were increases in bands from *trans* double bonds (968 and 986 cm^{-1}) and a decrease in band intensity from *cis* double bonds (710 and 3006 cm^{-1}) plus a shift to lower wave numbers (Le Dreau et al., 2009a,b; Pinto et al., 2010). Loss of *cis* double bonds was proposed as a marker for lipid oxidation (Guillen and Cabo, 1999, 2000). Hydroperoxides were identified at $3500\,cm^{-1}$, along with small amounts of carbonyls ($1744\,cm^{-1}$) (Holman et al., 1958; Guillen and Cabo, 2002; Wójcicki et al., 2015).

Some examples of other approaches include use of characteristic frequencies to develop PVs, application of characteristic frequencies – partial least squares regression to quantitate hydroperoxides in virgin walnut oil (Liang et al., 2013) and low- density lipoproteins (Lam et al., 2005), application of characteristic frequencies – artificial neural networks to predict peroxide and AVs in fish oil (Klaypradit et al., 2011), and reaction of lipid hydroperoxides with triphenylphosphine and IR quantitation of the resulting triphenylphosphine oxide directly (Ma et al., 1997, 1998) and automated in flow-through cells (Yu et al., 2007).

Hydroxyl values determined in hydroxylated soybean oils using least-squares support vector machine statistical analyses of mid-IR spectra were shown to provide a much simpler yet equivalent alternative to AOCS OHV CD 13–60 (Ferrao et al., 2007).

A second area of extensive application of IR has been measurement of frying oil quality and degradation in attempt to replace the laborious multiple chemical analyses usually required. Free fatty acids, hydroperoxides, total polars, and *trans* contents determined in heated oils by FTIR have all shown good correlations with standard chemical analyses (Shen et al., 2014; Innawong et al., 2004; Tena et al., 2009, 2013; Yavari et al., 2009; Srivastava and Semwal, 2015). It may also be possible to derive measures of polymers from the fingerprint region $1500–400\,cm^{-1}$ that reflects molecular configurations.

Finally, the advent of chemometrics is now refocusing attention away from absolute values of individual products to "data mining" – identifying more complex patterns of spectral shifts that may also involve molecular configuration modifications and explain subtle quality changes associated with lipid oxidation. There are also significant efforts to improve normalization and remove variability among repeat samples, such as by calculating ratios of all bands during oxidation to normalize differences in sample preparation and thickness (Guillén and Cabo, 2000). The door remains open to many new applications of FT mid-IR to elucidate modifications in oxidizing lipids that have not been previously detectable.

1.5.3.2 Near-IR (FT-NIR)

Near-IR applications to lipid oxidation have been minimal to date. Nevertheless, the technique has the potential to become a powerful tool in lipid oxidation analysis as more experience is gained with sample handling, calibrator generation, and especially with chemometric statistical evaluations.

Practical aspects. Sampling modes used most commonly in near-IR are diffuse transmittance and diffuse reflectance, which are parallel to diffusion and ATR except that light is scattered more due to presence of particles and irregular surfaces, respectively. This makes near-IR particularly attractive for use with foods. For lipid oxidation analyses, both modes are important since oils require diffuse transmittance while solid and semisolid foods use diffuse reflectance. Oils can be analyzed by diffuse reflectance with use of a reflective adapter available from some manufacturers, in which case applications are mostly discriminatory. Transmittance, interactance, and transflectance sampling is also available for special applications.

Sample handling and computer data analysis work hand in hand for success in near-IR analyses. At least superficially, near-IR makes sample handling very easy compared to mid-IR because samples are just loaded into containers and analyzed. Regular glass can be used, such as that in petri dishes and shell vials; some manufacturers have their own sampling designs with different materials. At the same time, the nonhomogeneity of foods introduces significant variability even within samples, so that "lack of reproducibility" has been a common complaint. Newer instrument design has addressed this in reflectance mode by rotating samples or scanning them back and forth horizontally while recording repeated spectra (at least 10) to ensure representative sampling of the material. These are then averaged during data analysis.

Computer statistical analyses can pick out small differences between samples, however featureless their spectra may seem. Indeed, powerful computer analyses are what make near-IR feasible and useful on a practical level. Nevertheless, computer sensitivity to minute differences means that variability of sampling must be minimized to obtain sharp data sets with clear distinction or the within-sample variation will obscure the feature being analyzed. Because dispersion of light is affected strongly by the nature of the sample surface in diffuse transmission and especially reflectance, particle size, shape, and packing density play a critical role in the intensity and reproducibility of spectra.

Structural identification of molecules can be obtained directly from mid-IR spectra, even by visual inspection. That is not possible in near-IR due to the extensive overlapping of overtone bands. Hence, sophisticated computer analyses are necessary to sort out differences in absorption patterns and correlate them with standards (calibrators) for identification and quantitation. Unlike MS, where features of individual molecular are constant so massive libraries of fragmentation patterns can be compiled for general application, food and oil compositions are too variable so each laboratory must construct near-IR libraries for each of their materials using calibrators that are formulated exactly and analyzed independently for specific components. Basically, the computer must be trained to correlate patterns it discerns to specific qualities, quantities, and chemical identities. Data from these calibrators are then used as a basis for comparison in discriminatory and quantitative statistical analyses to answer such questions as:

- Is a particular component(s) present, and how much?
- Are these samples oxidized, and how much?
- Does this material match another material (such as same or different composition, same composition but changed?)
- Are these materials different, and how?

To answer these and other questions, chemometrics software uses multivariate statistics to correlate quality parameters or physical properties with complex instrumental data and find qualitative and quantitative information patterns. Qualitative or discriminant analyses include principal component analysis, cluster analysis, linear discriminant analysis, artificial neural networks, and Mahalanobis and Euclidean distance. Quantitative analyses include linear partial least squares regression, principal component regression, multiple linear regression, as well as nonlinear partial least squares, artificial neural networks, and support vector machines (Munck et al., 1998; Geladi, 2003; Roggo et al., 2007; Forina et al., 2009). Perhaps the greatest challenge in near-IR currently is learning how to apply these methods to glean maximum information that accurately reflects the oxidation chemistry.

Finally, application of chemometrics requires substantial "pretreatment" of the near-IR spectral data to convert the broad components to focused defined bands that can be compared with greater distinction. Thus, it is customary to subject original data at least to signal averaging of replications, baseline normalization, first and second derivatization to localize peaks, and often more manipulation. It is these manipulations that make possible computer identification of principal components that differentiate samples.

Advantages of near-IR: As with mid-IR, near-IR provides fast, sensitive, simultaneous analyses of multiple compounds in undiluted intact materials, which makes this method very attractive and useful for foods and ingredients. Nearly all forms of foods – solids, liquids, pastes, gels, powders, and intact fresh foods – can be analyzed with appropriate sampling modes. Water is transparent in the near-IR spectra region so dehydration is not a requirement. Each near-IR spectrum simultaneously provides information about most major molecular components (e.g., moisture, proteins, lipids, carbohydrates) in the sample, and this data can be mined with appropriate calibrators. However, the greatest strength of near-IR combined with chemometrics is the awesome ability to distinguish small differences between groups and to categorize materials. This makes near-IR unique in its capabilities for screening of ingredients (Fernandez-Ibanez et al., 2010), online monitoring during manufacturing and processing (Huang et al., 2008), testing identity and adulteration of oils (Yang et al., 2005), and tracking food degradation during storage.

Disadvantages of near-IR: Skills developed for interpreting mid-IR spectra cannot be transferred readily to near-IR and it is difficult to overcome old habits of looking for specific sharp bands clearly indicating the various functional groups. You don't see all atomic motions (near-IR detects only vibrations from H linked to C, O, or N atoms), the absorptions are weak, and the complex patterns of overtones make it impossible to gain more than very general information from the spectra visually. Rather, the focus in near-IR is on computer data mining. The time to obtain data from a single sample may be only a few seconds, but the time required to analyze the large number of samples needed to minimize statistical variability can take hours. Sample handling for good reproducibility requires great attention to sample details, as was noted in the introduction to this section. Mining useful information from the diffuse spectral data is laborious and the learning curve for utilizing the chemometrics is steep. The spectra all require several pretreatments (e.g., signal averaging, normalization of baselines, conversion of absorption spectra to second derivatives) to put the data into a form that can be used in statistical analyses, then multiple statistical approaches must be tested to determine which provides the best model of the data. If these factors are not recognized, near-IR may be rejected as a tool that provides too little acceptable data with too much work.

The limitation perhaps most strongly associated with near-IR is the necessity of building calibrators for each set of samples and analysis. Preparing accurate calibrators that are well-defined in composition (by formulation) and characterization (by primary chemical analyses) is time consuming but they are critical for training the computer to interpret the data. The old adage about computers "garbage in, garbage out" is certainly true here. If the calibrators do not match the samples closely enough, samples are not identified or quantitated correctly. However, for many applications, once the calibrators are developed and analyzed and the method is validated, the information can go into a library for repeated use.

Applications of near-IR. In lipid analyses, near-IR has most often been used to determine lipid content and fatty acid composition, such as in pork sausages (Ortiz-Somovilla et al., 2007) and beef (Sierra et al., 2008), respectively. Fewer applications of near-IR to lipid oxidation have been developed, probably due to the requirement of calibration at least by chemical analysis of multiple oxidation products on the same samples. In one of the earliest studies reported, hydroperoxide peaks were identified at 6849 and 4831 cm^{-1} (1460 and 2070 nm) (Holman et al., 1958). More recently, using cluster analysis and partial least squares, 0.03% free fatty acids and 0.8 mEq hydroperoxides were detected in olive oil and several vegetable oils (Armenta et al., 2007). Partial least squares and multiple linear regressions identified changes in the 2068 nm regions as most strongly associated with hydroperoxides; spectral intensity at this wavelength showed with a nearly perfect correlation with PVs determined by iodometric titration (Yildiz et al., 2002). During oxidation of olive, sunflower, and rapeseed oils, the most significant changes in near-IR spectra visually were increases at 7068 cm^{-1} and 4810 cm^{-1} (2079 and 1415 nm) assigned to hydroperoxides However, PCA revealed marked loss of double bonds (4590 and 4670 cm^{-1}, 2179 and 2141 nm) and methylene hydrogens (5870 cm^{-1} first overtone and 8600 cm^{-1} second overtone; 1704 and 1163 nm) while hydroperoxides were increasing (4804 and 6950 cm^{-1}, 2082 and 1439 nm) (Wójcicki et al., 2015). Loss of double bonds thus might be an even better marker of oxidation than hydroperoxides because the change is continuous in one direction and does not involve secondary changes. Such detailed concurrent chemical changes are very difficult, if not impossible, to observe by chemical reactions where multiple simultaneous analyses are required and detection limits are not the same on each assay. These results demonstrate the power of chemometrics for identifying key information in near-IR spectra as well as the value of looking beyond set traditional products to obtain useful information about lipid oxidation.

1.5.4 Evaluation of Stability or Resistance to Oxidation

There is a running controversy over whether tests of early or secondary products provide the most accurate and useful prediction of oxidative stability, or resistance to oxidation. For basic research, following early products and their progressive transformation from induction period to active oxidation is most important because details of the entire process are important. The Oxidograph already discussed in Section 1.5.1.1 was indeed designed to test stability at the very beginning of oxidation. In contrast, for the food industry, it is often more important to know how fast secondary products that account for off-odors and flavors detected by consumers develop under different conditions. At the same time, industry does not have the luxury of waiting for months to a year to observe degradation at rates under ambient conditions. Thus, three analyses – OSI, Rancimat, and Schaal oven test (now largely replaced by ASL) – have been extensively applied to rapidly assess rates of oxidation progression to unacceptable points in secondary stages. All of these assays use substantially elevated temperatures to accelerate oxidation so that processing or formulation, such as fatty acid composition or antioxidant content, can be evaluated rapidly. The Schaal oven test temperature is 63 °C (corresponding to 6–10 days at 21 °C) (Pomeranz and Meloan, 2000), accelerated shelf-life studies are usually conducted at 40 or 60 °C, and the Rancimat and OSI analyses usually are run at 100–130 °C (Wrolstad et al., 2005).

Of the machine tests, OSI and Rancimat are usually justified as pushing the entire oxidation process through faster based on Arrhenius kinetics, and to prove this they are traditionally correlated with sensory evaluations, PVs, or other measures of lipid oxidation. However, both methods reflect mostly breakdown of initial products and oxidation of aldehydes to fatty acids that are highly conductive. Catalytic factors do not act all at one place or the same place in lipid oxidation (Schaich, 2005). Thus, as noted previously, rather than using these tests alone, it might prove highly useful to combine these assays of secondary products with Oxidigraph analyses of oxygen uptake to differentiate which stage of oxidation – initiation or progression – is being affected by specific processing, treatment, or formulation.

ASL testing has also assumed Arrhenius kinetics but, not being tied to a particular instrument, provides considerably more flexibility in analyzing different stages of oxidation, singly or (more effectively) in combination.

1.5.4.1 Rancimat and OSI

These are fundamentally the same procedure but the instruments are from different manufacturers – Rancimat (Metrohm) and OSI (Omnion).

Principle of the analysis: Historically, this assay is based on observations that oxidation products, being polar, impart conductivity to solutions (Pardun and Kroll, 1972). Fat or oil is heated to high temperatures (≥100 °C) and oxidized by bubbling air through the sample. Volatile secondary products from lipid oxidation are carried by the air into a tube with distilled water where conductivity is constantly monitored between two platinum electrodes, recorded, and plotted against time to produce oxidation curves. Conductivity slowly increases as oxygenated products are generated until sudden marked increases in conductivity indicate the end of induction period. The point of inflection in the curve denotes the stability time. This point is calculated in the Rancimat methods by the intersection of the slopes before and after the rate change, while in the OSI method it is determined by second derivative of the continuous curve. Results are expressed as stability time, time to active oxidation, or time to reach a specified increase in conductivity in the Rancimat, or as % Extension over reference = 100 x [OSI(sample) − OSI(blank)/OSI(blank)] for the OSI system.

Advantages: Both systems are easy to set up with samples, runs are fully automated and handle multiple samples simultaneously (8 for Rancimat and 24 for OSI), and data are acquired continuously so end points are clear (no interpolation needed). National and international standardized procedures are available: American Oil Chemists' Society Cd 12b-92, Sampling and analysis of commercial fats and oils: OSI; International Organization for Standardization ISO 6886:2006, Animal and vegetable fats and oils – determination of oxidation stability (accelerated oxidation test); Japan Oil Chemists' Society 2.4.28.2–93 fat stability test on autoxidation, Conductometric Determination Method 2.4.28.2–93; Swiss Food Manual (Schweizerisches Lebensmittelbuch, SLB), Section 1.7.5.4. Perhaps most usefully, correlations have been established to facilitate estimation of shelf-life (Table 1.7).

Disadvantages and precautions: The chemistry measured is a "black box" based on tertiary oxidation products that require a long progression in oxidation before they are generated – formation of LOOH, decomposition to secondary products, oxidation of aldehydes to short-chain acids (formic, acetic, propionic) that are volatile and contribute most to conductivity. Formic acid contributes most, acetic much less, others small amounts. These are not products typically associated with lipid oxidation, or even measured with chemical assays.

TABLE 1.7 Predictions of Shelf Life From OSI and Rancimat Induction Times for Oils With PV <2 mEq/kg

OSI/Rancimat Stability Time (h)	Predicted Shelf Life
0–5	No guarantee
6–12	6 months
12–20	1 year
>20	≥2 years

From Floratech (2015); used with permission.

Any factors that alter oxidation the direction of oxidation, i.e., through hydroperoxides versus addition or rearrangement, will also alter Rancimat/OSI results that depend on specific products. The acids detected occur downstream of aldehydes and other products that have distinctive odors and flavors, so actually overestimate stability. The method detects only volatiles compounds while nonvolatiles that are just as important in maintaining oxidation do not register. Thus, stability determined by Rancimat/OSI should be combined with other assays, both sensory and chemical, for verification.

Stability indexes determined by OSI and Rancimat methods are completely dependent on the conditions used (temperature and oxygen flow), and these must be specified for any comparison between runs or laboratories. Temperatures normally used are determined by the characteristics of the oils being tested. Saturated fats require higher temperatures to maintain liquidity and to increase oxidation (120 °C), standard vegetable oils degrade more rapidly so are tested at 100–110 °C, and fish oils or other very sensitive oils are analyzed at 68 °C. Temperatures higher than 120 °C should not be used because they put the oil into thermal degradation rather than oxidation mode (Tian, 2013), thus giving induction periods too short to measure accurately, and they greatly increase loss of water in the detector cell. Demonstration of strong linear inverse relationships between induction periods and temperature in vegetable oils (Hasenhuettl and Wan, 1992) and fish oils (Méndez et al., 1997) did verify that reactions contributing products with conductivity followed Arrhenius kinetics under the conditions of the analyses (decreasing induction periods by a factor of about 2 for every 10 °C increase in operating temperature) but at the same time demonstrated that small differences in conditions have a large impact on results.

Less attention has been given to the role of oxygen in Rancimat and OSI analyses, but control of oxygen flow rates is also critical since oxygen drives the reactions generating products. The main concern is maintaining sufficient flow to avoid oxygen depletion while not blowing products through the collecting water. However, oxidation products shift under different levels of oxygen, and effects of oxygen on types and levels of products formed become more pronounced as temperatures increase (Peng, 2010; Tian, 2013). At the same time, samples in real life are almost never exposed to full oxygen. Thus, oxygen flow must be balanced against temperature in determining optimum conditions for each test oil.

Applications: Rancimat and OSI are applicable only to oils, although application to low density foods with open surfaces (e.g., potato chips) have also been reported (Barrera-Arellano and Esteves, 1992). These two methods have been extensively used to evaluate efficiency of oil refining, compare stability of different oils, test effects of genetic modifications on oxidative stability of oils, compare and project long-term effects of antioxidants, and more.

Detailed protocols for Rancimat/OSI analyses are available in (Wrolstad et al., 2005).

1.5.4.2 Schaal Oven Test (Oven Storage Test, AOCS Method Cg 5–97)

Principle of the analysis: A sample is placed in an oven with forced air flow at elevated temperature between ambient and 80 °C; near 60 °C is the most common temperature. The initial Schaal test determined the end point from weight gain during oxidation and from appearance of rancid odors, which required trained evaluators (Rossell, 1994). Although this assay has largely been abandoned in its original form, the basic concepts nevertheless still provide the basis for ASL testing.

Advantages: Originally developed to test crackers, the assay is very simple, requires no special equipment, and can be used on oils or food samples with various matrices.

Disadvantages and precautions: Even though it sounds simple, the Schaal oven test is highly variable, tedious, and not practical for routine analyses. Reaction conditions are difficult to control through all the repeated handling, oxidations are highly sensitive to trace contaminations in glassware, it cannot be automated, and it is usually only semiquantitative at best. Incubations require long times and do not allow for incubating multiple samples open in same chamber if sensory tests are to be performed.

Appropriate applications: This old assay has been largely replaced by Oxipres and ASL analyses for greater accuracy. The oven test should be reserved for situations in which no other options are available and only general estimates of oxidative stability are needed.

1.5.4.3 ASL Tests

ASL studies these days provide the basic structure of what research and quality control laboratories alike do all the time – follow the lipid oxidation – only incubations are conducted with much more sophistication and control than the early Schaal oven tests. Typically, samples are weighed into matched containers and incubated closed or open in incubators with precisely controlled temperatures, shaking, and often specific atmospheres. Light with exact wavelengths may be supplied, or samples may be held in the dark. Samples are withdrawn periodically for analysis not by weighing and tasting but by a wide range of chemical and instrumental analyses to determine progression of oxidation. The same basic approach may be used to study fundamental oxidation reactions, compare stability of different oils or foods, compare processing and formulation of foods, test catalysts and antioxidants, guide genetic modifications, determine "sell by" dates, or answer any number of other research questions, practical or fundamental.

ASL studies serve an important purpose but not without detriments. Like the Schaal oven Test, ASL is based on the assumption that oxidation mechanisms, catalyses, and antioxidant effects at moderately elevated temperatures, usually 40 and/or 60 °C, are the same as at room temperature, just faster, and thus projections of oxidation that will occur can be obtained in much shorter times. This assumption has some shortcomings because, in actuality, lipid oxidation is a series of reactions, not just one, and every step active in forming, transforming, or decomposing lipid oxidation products has a different activation energy. Similarly, all catalysts found in stored fats, oil, or foods – heat, lipoxygenase, metals, irradiation, high oxygen, etc. – promote specific pathways that differ from fundamental autoxidations, and each of these also has its own activation energy as well as specific products enhanced. This means that the presence and balance between different pathways and catalysts must change with increasing temperature as some pathways become facilitated and others saturate, so the same lipid oxidation products should not be expected at all temperatures. Results of ASL testing may reveal how oxidation changes in rates and product distributions under the catalyzed conditions (how oils or foods handle test stresses) but should be applied with great caution, and perhaps even skepticism, to predict oxidation under milder conditions.

Of course, the corollary of varying reactions is that the picture of oxidation generated is directed by the product(s) analyzed, by the frequency of analyses, and whether the results are interpreted broadly or narrowly. The judgment about shelf-life is different if only hydroperoxides are measured and they peak at 10 days then degrade versus if only hexanal is measured and it is still accumulating after 20 days. Similarly, if only hexanal is measured but at elevated temperatures, more C9 products are generated (e.g., nonenal and decadienal), oxidation levels are tremendously underestimated and shelf-life may be substantially overestimated because oxidation pathways have shifted and not been detected. These are all examples we have observed in our laboratory (Bogusz, 2015; Xie, 2015). To catch all that is going on, you need to cast your analytical nets broadly and measure as many products as can be handled accurately. Moreover, whenever possible, analyses should include differentiation of individual products, not just total classes. In both volatile and nonvolatile products, we have observed notable changes in distributions of aldehydes and alkanes even when variations in total aldehyde levels are small. The product distributions reveal shifts in pathways that we would have missed if only total carbonyls were quantitated. Finally, although ASL is intended to model long-term storage, samples should be analyzed as frequently as feasible, not just once a week or month, to observe detailed patterns of change.

1.6 SENSORY ANALYSES TO ESTABLISH CORRELATIONS OF CHEMICAL AND PHYSICAL ASSAYS

Sensory analyses of lipid oxidation are critical components of any evaluation protocol, particularly when actual foods are involved, because they provide grounding to the chemical assays and answer the question, "What do these chemical measures mean in terms of perceived product quality?" The general correlations of PVs that define fresh (<1), questionable (5), and unacceptable

(10) were determined by integration of sensory evaluations with peroxide assays. Similarly, taste thresholds and odor characteristics of specific aldehydes and other oxidation products were determined by sensory panels (cf Frankel, 2005b, and references therein).

Clearly, sensory analyses of lipid oxidation are invaluable but should probably be reserved for answering specific questions for several reasons. Regulatory hurdles for sensory analyses of oils or foods with oxidation these days are steep because of the requirement to demonstrate safety and lack of toxicity before allowing consumption. Sensory panels are expensive and need to be carefully screened and trained for lipid oxidation analyses. Also, the large numbers of food samples required may be readily available in industry, but are often beyond the economic reach of basic research.

A full consideration of sensory analyses of lipid oxidation is beyond the scope of the chapter. Interested readers are directed to Chapters 3 to 5 in Warner and Eskin (1995) and Civille and Dus (1992) for detailed information.

1.7 ANALYSES FOR CO-OXIDATION OF PROTEINS

As noted in the Introduction, this chapter argues for a new paradigm in lipid oxidation analysis. Previous sections of this chapter described analyses of multiple products that must be monitored to track lipid oxidation accurately. The second part of this new paradigm adds analyses of protein co-oxidation to follow the footprints of lipid oxidation and determine the broader extent of oxidative degradation in foods. Detailed protocols are beyond the scope of this book and are available in too many molecular biology and biochemistry texts to list here. Rather, analyzing co-oxidized proteins requires different thinking than when analyzing native proteins, even when the same assays are used. This chapter will focus on approaches that can provide information about the nature and extent of protein co-oxidation from lipids, and how assays and interpretation of results must be modified for damaged proteins.

Co-oxidation of proteins has been largely ignored as a practical problem in foods until recently, despite research evidence that it occurs (Schaich, 2008 and references therein). Protein oxidation is intertwined with lipid oxidation intimately and mechanistically because lipids and proteins are often closely associated in food structures and in membranes, and because literally every intermediate and product of lipid oxidation reacts with proteins. Details of mechanisms involved in protein co-oxidation may be found in Schaich (2008) and references therein. These interactions have antioxidant effects on the lipid radical chain reactions while redirecting and broadcasting oxidation to proteins and other biomolecules, and also mediating degradation of food texture, flavor, color, and nutrition in ways that may not be obviously connected to lipid oxidation. For example, peanut butter hardens without dehydration as it oxidizes in sealed laminate packages, tortilla chips become tough and brittle, and flour loses bread-making ability over time without development of typical rancid odors and flavors (Schaich, 2014). These changes all result from protein co-oxidation. Warmed-over flavors in meats, which have never been successfully attributed to specific lipid oxidation products, are probably also from protein co-oxidation. Thus, when measures of lipid oxidation are low but quality loss is evident in complex foods, protein co-oxidation is always a likely explanation.

The task of analyzing protein co-oxidation is not as straightforward as analyzing native proteins in natural materials, whether plant or animal. First, co-oxidized proteins are, by definition, modified, so they should not be expected to behave "normally" in any analysis, as will be shown in discussions of methods below. In addition, one particular issue that makes interpreting protein changes in foods more challenging than in biological tissues is that processing damage is imposed on or occurs parallel to oxidation damage, and the two processes may be independent or interactive. There are no foolproof methods for separating forces, but, at a minimum, base levels of protein chemical properties should be established for each food product before processing. Comparisons of these with values determined immediately after processing give an estimate of mostly processing-induced damage, while changes in properties during or after storage reflect oxidation damage.

This section describes a number of analyses that are used to characterize oxidized proteins. Some are general and can easily be used in industrial quality control. Others provide more molecular detail or are specialized products. These are probably more applicable to research on protein oxidation but can also give industry information about specific chemistry to counteract in developing stabilization approaches. Which assays are used in a given application will depend on the proteins involved, the information most needed, and the instrumentation and manpower available.

Analysis of protein oxidation in foods is still relatively limited due to lack of awareness of its importance and unfamiliarity with detection methods. Currently, analyses of protein oxidation are mostly comparative, designed to identify which conditions cause protein oxidation in foods, but at least this is a start. More widespread application of the methods described below are needed to correlate levels of specific protein oxidation products to other degradative chemical and property changes in foods in order to determine "acceptable" and "reject" levels for different foods.

1.7.1 Extraction of Co-oxidized Proteins

Every protein has different structural organization and solubility characteristics that impart unique extraction requirements. It is not possible for this chapter to cover procedures for all proteins, but a few general guidelines are offered here, with particular reference to modifications required for oxidized proteins. One cardinal rule to keep in mind is that inability to extract proteins by normal methods merely reflects the protein modifications which are the target of investigation. Proteins that are not extracted cannot be analyzed. Therefore, be prepared to "think chemistry" and be creative in designing alternate extraction procedures that accommodate protein changes. Some typical modifications include the following:

- Because aggregated or cross-linked proteins react poorly with most protein reagents, a denaturing solvent such as SDS or dimethylsulfoxide is often needed to extract co-oxidized proteins.
- For very polar or charged proteins where surface residues are modified by lipids, a change in buffer, pH, or salt strength may be needed to improve solubilization.
- For proteins with reasonable content of sulfur amino acids, extraction both with and without a reducing agent such as 2-mercaptoethanol to disrupt disulfide bonds is required. This provides a more accurate picture of protein content, and an increase in solubility with reducing agent also reveals involvement of disulfide cross-links.
- In foods with higher fat content, protein yields may sometimes be improved by lipid extraction before protein solubilization.
- In processed foods where proteins become closely associated with gelatinized starches in the food matrix (e.g., baked products or extruded snacks or pet foods), predigestion with mixtures of amylases may be necessary for accurate recovery of proteins (Strange and Schaich, 2000; Wang, 2000).

1.7.2 Loss of Protein Solubility with and without Disulfide Reduction

One of the earliest indicators of protein modification is loss of solubility or ability to extract proteins in given solvents. Unfortunately, all chemical assays for protein content in solution depend on reactions of some reagent with amino acid residues accessible on the protein surface. Since all proteins have different amino acid compositions, they all react differently in the assays, making individual standard curves necessary. This problem is compounded for co-oxidized proteins because the amino acids that react in the assays are same ones most modified by oxidation. Thus, it is common to have oxidized proteins show *apparent* decreased solubility while they remain in suspension. Inaccurate estimates of protein in solution is particularly a detriment when other assays are normalized to "measured" protein content because it causes overestimation of individual amino acids and overloading of electrophoresis gels.

Since there is no perfect assay for protein solubility, the best approach is to keep multiple assays available in the laboratory and select the one that best fits the amino acid composition of the proteins involved, the conditions of extraction, and the sensitivity needed. Indeed, it is often useful to combine more than one assay with different residue specificity to generate alternate views of solubility and derive information about amino acids involved in modifications. The information in the following sections will provide some guidelines for mixing and matching protein concentration assays when analyzing protein oxidation. We have included discussion of only the major assays for which we have experience in applications to oxidized proteins. As will be shown, all have limitations when oxidation modifies amino acid binding sites. We have avoided using fluorescence-based assays such as fluorescamine and *o*-phthalaldehyde because they react with lysine and glutamine, major targets of oxidation, and because fluorescence yields are so sensitive to environment and other factors. Other methods such as colloidal gold (Harrison et al., 2008), and polyhydroxybenzenesulfonephthalein dye–metal complexes (Antharavally et al., 2009) are being investigated but little information about their binding specificities is yet available. Lack of inclusion of other methods does not imply that they are not effective or useful, just not tested and validated yet with oxidized proteins.

Details of assay protocols are available in numerous texts and online; see, for example, (Noble and Bailey, 2009; Walker, 2009; ThermoScientific, 2010; Noble, 2014) and the Methods in Molecular Biology series. The Noble and Bailey (2009) chapter and Pierce Protein Assay Selection Guide (Pierce, 2014), in particular, are highly recommended for their excellent evaluations and comparisons of the various protein assays, including details of compatible and interfering substances.

1.7.2.1 Lowry Assay

Principles of the analysis. Proteins first react with alkaline cupric sulfate in the presence of sodium potassium tartrate to form a tetradentate cuprous complex (four peptide bonds:one atom of copper). The Cu^+ in the complex then reduces phosphomolybdic-phosphotungstic acid (Folin–Ciocalteu phenol reagent) to generate an intense blue color (Reaction (29)). Although the peak absorbance is rather broad, the reaction is usually monitored at 750 nm where few other substances interfere (Lowry et al., 1951; Noble and Bailey, 2009).

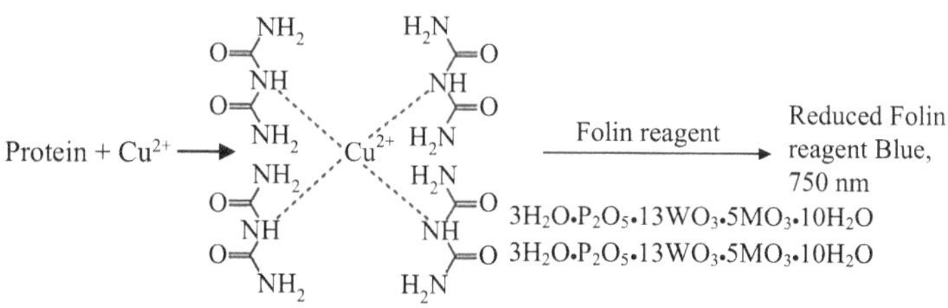

(29)

Advantages: The Folin reagent reaction with intact proteins is relatively constant from protein to protein (Dorsey et al., 1977) and it does not react with free amino acids except tyrosine and tryptophan. The assay tolerates up to 1% SDS in the sample, an important consideration considering the extensive use of SDS in extracting proteins.

Disadvantages and precautions: The Lowry assay has the lowest sensitivity of all those discussed here (5–100 μg) (Noble and Bailey, 2009). The reaction is relatively slow, and assay response is dependent on protein size and amino acid composition. Color formation increases with peptide size and with content of tyrosine, tryptophan, cysteine, histidine, and asparagine which contribute most of the reducing equivalents. Since these residues are readily attacked by oxidizing lipid radicals and products, this assay often exhibits very low reaction with oxidized proteins. High proline contents in the protein (e.g., gelatin) also interfere with protein–copper complex formation and color development. Handling is very awkward since Folin reagent must be added to each tube after precisely 10-minute reaction between protein and copper while simultaneously mixing each tube to avoid precipitation of the Folin reagent. This limits the number of samples that can be analyzed at the same time. The assay is subject to many detractions and interferences including reducing substances and free thiols, chelating agents, strong acids or strong bases, and detergents and potassium ions in the sample (Waterborg, 2009).

Appropriate applications: This traditional assay has been largely replaced with other assays that are more sensitive and less subject to interferences.

1.7.2.2 Bicinchoninic Acid (BCA) Assay

Principles of the analysis. Proteins form blue complexes with cupric ions in alkaline solution containing sodium potassium tartrate, reducing the copper in the process (Smith et al., 1985). Binding sites are peptide bonds plus cysteine or cystine, tyrosine or tryptophan residues, four to six per molecule of copper (Weichelman et al., 1988). With addition of BCA, which has stronger metal binding constants, the cuprous ions are released and transferred to the BCA to form intensely colored purple chelates (2 BCA:1 Cu^{1+}) that absorb at 562 nm (Smith et al., 1985) (Reaction (30a,b)):

$$\text{Protein} + Cu^{2+} \rightarrow Cu^{1+} - \text{Protein (blue)} \quad (30a)$$

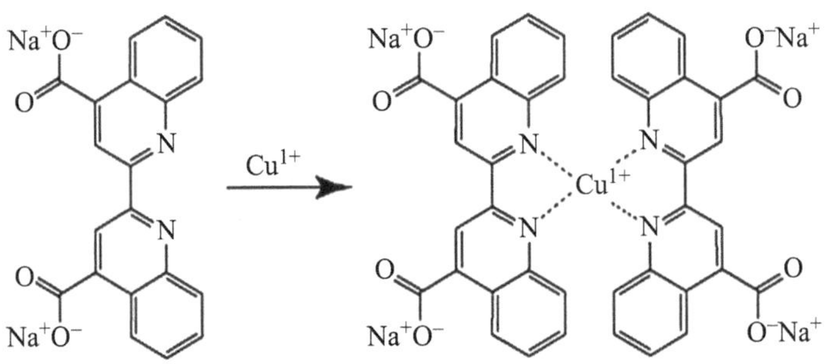

(30b)

BCA, Na salt

BCA-Copper complex, purple

Protocols for the reaction use either 37 or 60 °C as the reaction temperature. Higher temperatures increase reaction response due to partial protein denaturation with exposure of reactive aromatic residues and more complete reaction of the amino acids with the copper (Weichelman et al., 1988). However, higher reaction temperatures can only be used with proteins that do not aggregate at 60 °C. Since behavior of damaged proteins often differs from that of their native counterparts, both temperatures must be tested when using the BCA assay with oxidized or otherwise modified proteins. Similarly, the reaction is normally allowed to proceed for 30 min, but damaged proteins may need longer incubation times for full reaction. Protein concentrations are determined and reported with reference to a standard curve prepared from serial dilutions of a common protein such as bovine serum albumin (BSA) or a protein that closely matches the amino acid content (type and number) of the protein to be analyzed (Noble and Bailey, 2009).

Because oxidized and processed proteins often lose solubility, a solid-phase version of the BCA assay was developed for direct assay of protein in intact food materials such as cereals (Chan and Wasserman, 1993a). Here, the food material is suspended in the BCA reagent, incubated with sonication, and the reacted BCA released into solution is analyzed at 562 nm after centrifugation to remove solids. We have found that this modification works very well as long as the component cysteines are not damaged.

Advantages: This is the most sensitive protein assay. The standard BCA assay detects protein concentrations from 20 to 2000 μg/mL; the micro-BCA assay that uses diluted samples is still more sensitive and has a narrower dynamic range of 0.1–25 μg/mL. Standard cuvette methods are easily adapted to manual or auto-dispensing microplates (Redinbaugh and Turley, 1986), and it is compatible with up to 5% detergents so can be used with most SDS extracts. Only tripeptides and larger polypeptides or proteins reduce enough copper for the BCA reaction, so single amines and dipeptides, such as from hydrolyses during processing, do not interfere.

Disadvantages and precautions: Despite its sensitivity and wide use, two factors limit the usefulness of the BCA assay with oxidized proteins. First, BCA–copper complex formation is sensitive to the presence of reducing substances, chelating agents and strong acids or strong bases in the sample, particularly in the micro-BCA assay, so it cannot be used when protein extractions are performed with 2-mercaptoethanol or dithiothreitol unless these are removed by dialysis or proteins are separated by precipitation in ice-cold acetone (Gates, 1991) or deoxycholate and trichloroacetic acid (Brown et al., 1989), followed by washing. Two interferences are potentially critical for lipid co-oxidations of proteins. Lipids associated with proteins also react with copper and lead to overestimation of protein content (Kessler and Fanestil, 1986); this interference can be eliminated by 2% SDS (Morton and Evans, 1992), which is a common component of solvents for protein extraction from foods.

Even more importantly, the four amino acids that bind and reduce the copper in the BCA assay – cysteine, cystine, tryptophan, and tyrosine – are major targets for damage during processing and for reaction with oxidizing lipids, particularly early radicals. Consequently, proteins that are accurately quantitated in their native form are significantly underestimated when they are oxidized. This leads to significant overestimation of other chemical changes whose calculations are based on protein content, and it causes particular problems in applying consistent accurate quantities of protein to electrophoresis gels.

At the same time, BCA has recently been shown to react with hydroxyl ketones as well as peptide bonds (Weiser et al., 2012). Similar structures may be produced in oxidized proteins, leading to aberrantly high apparent protein concentrations. Thus, the BCA assay presents some very serious limitations for quantitating concentrations of oxidized proteins in solution. At a minimum, caution is warranted when interpreting data from this assay applied to oxidized proteins.

Finally, in terms of handling, the BCA reaction does not have a true end point; like the Lowry method, the final color continues to develop over time. Thus, reaction timing must be precisely controlled. When more than one sample is analyzed simultaneously, starting times for samples must be staggered to obtain accurate results.

Appropriate applications: The assay is applicable to all native proteins. However, when proteins being analyzed have high cysteine contents that may be oxidized (e.g., to sulfoxides), an assay with different amino acid binding sites should be used or apparent loss of protein solubility should be verified by a method independent of amino acids, such as Kjeldahl or elemental analysis.

1.7.2.3 Bradford Assay

Principles of the analysis. The Bradford assay eliminates the copper intermediate and the interferences associated with it in the Lowry and BCA assays by binding Coomassie Brilliant Blue (CBB) dye directly to the protein (Bradford, 1976).

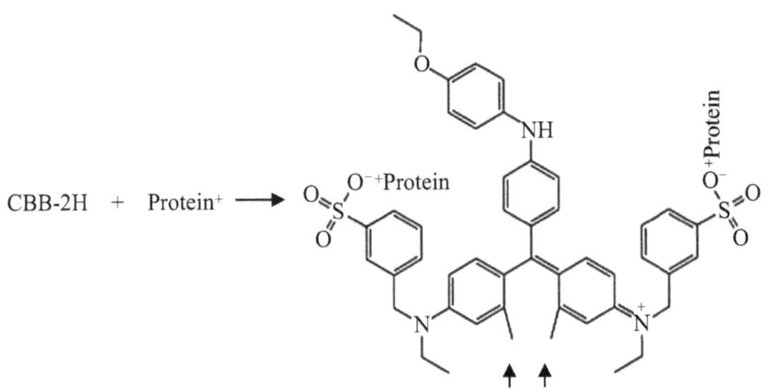

(31)

Free Coomassie Blue exists in three main ionic forms with different colors depending on pH. For Coomassie Blue G-250, the forms are at pH<0.3 – double cation, red (465–470 nm); pH 1.3 – neutral, green (650 nm); and pH>1.3 – anion, blue (590 nm) (Chial et al., 1993). The blue form is thought to be the active form, binding to proteins initially via charge complexes between its sulfonic acid groups and positively charged amino acids side-chains (arginine, histidine, lysine) on the protein (Tal et al., 1985; de Moreno et al., 1986) (Reaction (31)). The charge complexes disrupt the structure of the protein and cause a partial unwinding, exposing hydrophobic pockets. CBB then binds in these pockets by van der Waals interactions, further stabilizing the complexes (Katrahalli et al., 2010). When bound to protein, Coomassie G-250 is unprotonated and blue, with absorbance at 620 nm. However, at this wavelength there is too much overlap from the green form absorbance, so the assay is normally run at 595 nm (Bradford, 1976; Noble and Bailey, 2009; Noble, 2014). Since this wavelength is so close to the unreacted blue form, it is critical to run the reaction at pH close to 1.3 to provide sufficient blue form for reaction while minimizing background from unbound dye (Chial et al., 1993).

At this point, some further explanations of Coomassie Blue dyes are in order. Originally developed in Ghana as wool dyes, the two main forms used in protein chemistry are the original R-250 and colloidal G-250. 250 denotes the purity of the dye. Structurally, the two forms differ only in the absence (R) or presence (G) of two methyl adducts on the benzene rings, as noted by the arrow positions in Reaction (31), and their color in acid solution (reddish vs greenish, respectively) (Merril, 1990). The scientific literature is replete with contradictory reports of relative sensitivity and preference of each form, some of which may arise from differences in the purity and concentration of the dye in a given commercial product. For the purposes of this discussion, suffice it to say that G-250 is the current preferred form for the Bradford assay because its ionization states are more advantageous for charge interactions with proteins, the added two methyl groups strengthen the van der Waals interactions, and it is easier to handle.

Quantitation of protein comes from Beer's law relationships between protein concentration and the absorbance of protein–dye complexes in solution. Protein concentrations are thus determined by comparison of experimental absorbances with standard curves prepared from a purified protein the same as being analyzed when possible, or having comparable amino acid composition. Purified BSA or immunoglobulin G are the conventional standards used when protein compositions are unknown or pure proteins are not available. Usually absorbance values at 595 nm are used to generate the standard curve, but sensitivity and accuracy at low concentrations can be increased approximately 10 times by plotting the ratio of absorbance readings at 595 and 450 nm (Zor and Selinger, 1996).

Advantages: The Bradford method is very sensitive, with lower detection limit of 0–50 μg in standard methods (Noble and Bailey, 2009) and nanograms detectable in recent modified versions (Grintzalis et al., 2015, 2009). It requires the smallest samples and shortest analytical times and has low unit cost (Chutipongtanate et al., 2012). It will tolerate normal levels of most salts, solvents, buffers, thiols, reducing substances and metal chelating agents encountered in protein samples (Noble and Bailey, 2009), and is adaptable to both microplate and cuvette protocols.

Disadvantages and precautions. The Bradford protein assay is inhibited by the presence of detergents, strong acids, or bases, so extracts containing these may need to be dialyzed or neutralized before analysis, or the proteins can be precipitated in trichloroacetic acid or acetone then resuspended in neutral buffer. The linear range is rather short, so samples should be tested with at least one dilution to make sure concentrations are below the plateau region. Most problematic for oxidized proteins, acids (arginine, histidine, and lysine) that bind the dye are also major targets for reactions with radicals and products from oxidizing lipids. Loss of these binding sites reduces reactivity and apparent solubility in solution and impedes staining on polyacrylamide gels.

Appropriate applications. Like the BCA assay, the Bradford method is applicable to all native proteins. However, when binding inhibition is suspected in oxidized proteins (i.e., when lysine, arginine, and histidine are damaged and no longer can establish charge complexes), apparent loss of protein solubility should be verified by a method independent of amino acids, such as Kjeldahl or elemental analysis.

1.7.3 Sulfhydryl/Disulfide and Sulfur Oxide Analysis

1.7.3.1 Sulfhydryl/Disulfide

Principle of the analysis: The prevailing method for SH/SS determination is by reaction with Ellman's reagent, 5,5'-dithiobis-(2-nitrobenzoic acid) (DTNB) (Ellman, 1958, 1959). Since DTNB is a disulfide, it undergoes thiol-disulfide interchange with thiol groups in free compounds or on proteins (Reaction (32)).

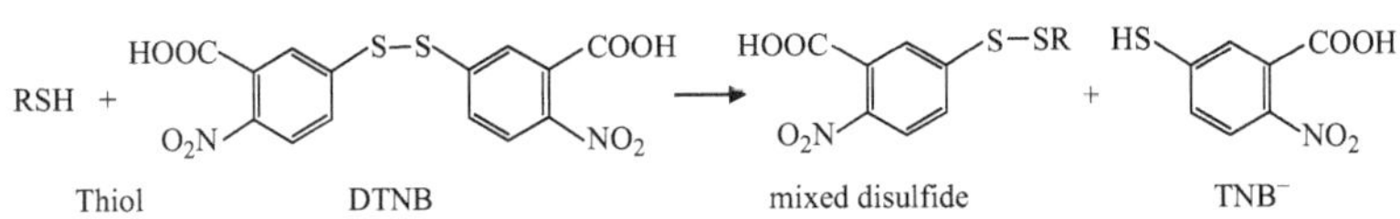

(32)

The fundamental reaction for detecting free thiols has been well characterized. Each thiol reacting with DTNB adds across the disulfide bond to form a mixed disulfide product and release one 2-nitro-5-thiobenzoate anion (TNB^-), which ionizes to the TNB^{2-} dianion under mild alkaline conditions. TNB^{2-} is yellow and can be detected optically at 412 nm. Thiol concentrations are then quantitated from Beer's law using an extinction coefficient if standard conditions are used or by comparison to a standard curve prepared from a pure thiol such as cysteine if reaction conditions have been modified. The extinction coefficient of the dianion varies with the solvent used (Table 1.8), so the value of 13,600 mol/(L s) originally cited should not be applied automatically. Lower detection limit is one thiol group per protein in 2 nmol of protein.

Disulfides are determined by the now classical method of Thannhauser et al. (1984, 1987), which creates a thiol-disulfide interchange reaction in reverse. With Ellman's reagent, free thiols add to the DTNB disulfide bond, but disulfides cannot do this. If S–S are reduced in situ, DTNB is reduced as well and the mixed disulfide does not form. Thannhauser's method first reduces DTNB asymmetrically with sodium sulfite to generate $NTSB^{2-}$ (disodium 2-nitro,5-thiosulfo benzoic acid) and NTB nitrothiobenzoic acid at pH 9.5. Protein is dissolved in guanidine HCl or thiocyanate to unwind the peptide chain and expose all sulfur amino acids. For reaction, each disulfide is reduced with sodium sulfite to a thiosulfonate and a thiyl anion (Reaction (33)), the latter of which adds

TABLE 1.8 Molar Extinction Coefficients of Dithio-bis-Nitrobenzoic Acid (DTNB), Ellman's Reagent, Used in Thiol/Disulfide Assays

Solvent	ε (mol/(L s)) at 412 nm
2% SDS	12,500
0.1 mol/L Phosphate, pH 8.0, 1 mmol/L EDTA	14,150
Buffered 6 mol/L guanidine hydrochloride	13,700
6 mol/L Guanidine hydrochloride	13,880
Guanidine thiocyanate	13,600 (original)
8 mol/L Urea	14,290

Excerpted from ThermoScientific (2011); used with permission.

across the S—S bond of NTSB to give the standard Ellman's reagent mixed disulfide (Reaction (34)).

$$RSSR' + SO_3^{2-} \longleftrightarrow RSSO_3^- + R'S^- \qquad (33)$$

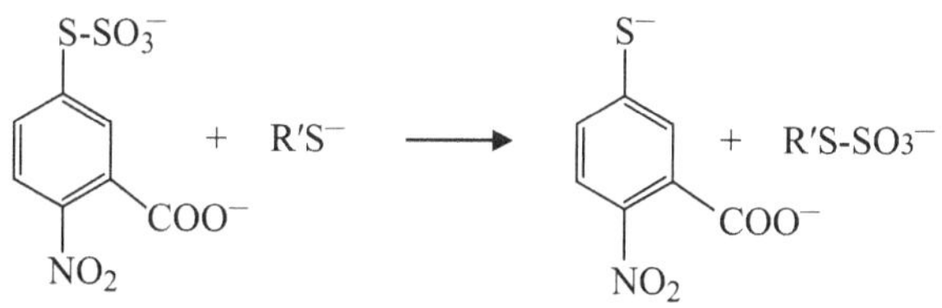

(34)

This product is detected at 412 nm and quantitated by comparison to a standard curve prepared from cystine. An extinction coefficient of 13,600 mol/(L cm) per thiol or disulfide is normally used, but this can be modified with specific proteins. The thiol value determined in this assay actually accounts for total thiols from both thiols and disulfides, so the free thiol content determined in unreduced proteins by the standard DTNB reaction must subtracted from the total thiol content to determine S—S content by difference. The assay is rapid, requiring 3–5 min for oligopeptides and 20 min for proteins, and detects as little as 10^{-8} mol of disulfide bonds, with an error of ±3%.

This basic Ellman's reaction has been modified in numerous ways to achieve different goals. DTNB reacts only with thiols available on the surface of proteins, the loss of which during processing and storage can suggest protein aggregation, disulfide cross-linking, or thiol oxidation to various sulfur oxides. Gain of surface thiols under the same circumstances indicates either unwinding of native structure to expose additional cysteines or scission of native disulfides. Determination of total protein thiols requires denaturation before assay to expose thiols that are buried in native configuration or by thermal or oxidative aggregation. The difference between this value and surface thiols provides a measure of protein folding or aggregation. 6 mol/L guanidinium salts are preferred for protein denaturation in this assay because they are stable and provide no interference. Urea must be prepared fresh and used with caution because it readily degrades to form cyanates which will react with thiol groups. Adding the disulfide assay to quantitate total thiols and distinguish disulfides provides a mass balance for thiols. When thiols oxidize only to S—S cross-links, total thiols remain constant. When total thiols decrease, other mechanisms for loss, such as oxidation to sulfur oxides by lipid radicals and complexation with lipid oxidation aldehydes, are probably active.

The Ellman assay detects free thiols only in solution, but oxidized proteins often lose solubility and become difficult to extract, and reducing agents required to solubilize some proteins destroy existing disulfide bonds. Taking advantage of the fact that NTB^{2-}, the color-producing agent in both Ellman's SH assay and Thannhauser's S—S assay, does not remain bound to protein but is released into solution, Chan and Wasserman (1993b) developed a solid-phase assay to detect both soluble and particulate thiol and disulfide groups in combined simplified procedures. The procedure was designed for extruded cereal proteins but we have used it with other types of foods as well. Briefly, for free thiols, dried ground food samples (10–30 mg) are suspended in 8 mol/L urea (freshly prepared), 10 mmol/L DTNB, 3 mmol/L ethylenediaminetetraacetic acid (EDTA), 1% SDS, and 0.2 mol/L Tris–HCl, pH 8, and allowed to react (required time determined by the material), then the samples are centrifuged to remove solids, and the absorbance of the supernatant is read at 412 nm (modified Ellman's assay). Procedures are basically the same for disulfide and total thiol except that Tris–HCl, pH 9.5, and $NTSB^{2-}$ are substituted (modified Thannhauser assay). The reaction is run without sulfite to determine disulfides and with sulfite reduction to determine total thiols. Disulfide content is calculated from the difference between total thiols after and before reduction of disulfide bonds with sulfite. Total cysteine is calculated as $[-SH]+2[-S-S-]$.

Advantages. Both Ellman and Thannhauser reactions are fast, simple, and stoichiometrically quantitative; they are very sensitive and can detect small changes in oxidized proteins. Together, the assays provide considerable flexibility for investigating roles of SH/SS in protein modifications by analyzing thiols with and without denaturation and with and without reduction of disulfides. The solid-phase version of the assay (Chan and Wasserman, 1993b) provides a rapid, convenient method for quantifying thiol and disulfide contents of oxidized proteins without prior extraction. It overcomes drawbacks of incomplete extraction or solubilization of damaged proteins, and it greatly reduces sulfhydryl oxidation and protein loss during handling and reactions.

Disadvantages and precautions: Basic thiol detection by Ellman's reagent requires removal of reducing agents from extraction media either by dialysis or by protein precipitation and resuspension before analysis. Reagent DTNB must be recrystallized from ethanol to remove impurities that interfere with the reactions. DTNB supplied in kits presumably has already had this done. Thiols are sensitive to oxidation and both TNB^- and $NTSB^{2-}$ anions degrade in light (Walmsley et al., 1987), so most accurate results are obtained when reactions are run in dim light under anoxic conditions, such as argon-sparged solutions and headspace maintained under argon or reaction run in a glove box.

Applications. The solution and solid-phase methods described here are applicable to determining native cysteine and cystine contents of all food proteins, as well as SH/SS modifications during processing and storage.

1.7.3.2 Sulfur Oxides

Sulfur oxides are almost never analyzed in food proteins, probably because it is assumed that oxidized cysteines only form disulfide cross-links. However, formation of methionine sulfoxide is well recognized so there is every reason to expect parallel reactions of cysteine to form cysteine sulfoxides and sulfinic, sulfenic, and sulfonic acids. We have recently observed almost complete loss of detected thiols and disulfides in aged flours, accompanied by reduction in polymerization of high molecular weight glutenins and are currently testing to determine whether surface oxidation of cysteine residues to sulfoxides is responsible. Sulfur oxides may at least be detectable by mid-IR (Reusch, 2013), even if they are not quantifiable. Rudyk and Eaton (2014) have recently reviewed methods for measuring oxidation states of sulfur in proteins, but quantitative methods for analyzing sulfur oxides in food proteins remain to be developed.

1.7.4 Polymerization and Fragmentation (SDS-PAGE)

Oxidation often causes cross-linking of proteins by radicals, disulfide bonds, and Schiff base or Michael bridges, and formation of dicarbonyls in the protein backbone leads to scission of peptide chains (Schaich, 2008). Following these molecular weight changes is one of the easiest ways to detect protein co-oxidation. The most straightforward method to determine either molecular weight distributions in mixtures of proteins or changes in molecular weights of modified proteins is separation by polyacrylamide gel electrophoresis with SDS.

Principles of the analysis. The gel-forming system consists of four components: base polymer acrylamide ($CH_2{=}CH{-}CO{-}NH_2$), cross-linker bisacrylamide ($CH{=}CH{-}CO{-}NH{-}CH_2{-}NH{-}CO{-}CH{=}CH_2$), ammonium persulfate (radical initiator), and accelerator TEMED (tetramethylethylenediamine). TEMED forms free radicals in the ammonium persulfate, which in turn initiate free radical chain reactions between acrylamide and bisacrylamide, and copolymerization between these two agents forms a gel. The gel is an open network with pores of sizes that vary with the concentrations of the acrylamide components. The porous structure of the gel acts as a kind of sieve, permitting or retarding movement through it depending on molecular size. Higher total acrylamide creates smaller pores and separates smaller proteins while excluding large proteins.

Low acrylamide concentrations do the opposite, separating high molecular weight proteins while not retaining very small proteins. Similarly, at a given percent acrylamide, low bis concentrations open up the matrix and expand the protein size accepted while higher bis has the effect of increasing the mesh size of the gel (smaller pores) and retaining more small peptides. Pore sizes are smallest at 5% bisacrylamide and increase at both lower and higher values.

In routine analyses, total acrylamide levels (%T) typically range from 7–20%. To provide a frame of reference, some examples of molecular masses separated by various acrylamide concentrations are listed in Table 1.9. These should be used as general guidelines only since the limits of protein mass, low and high, that can be handled by a give percent acrylamide vary with the percent cross-linker added as well as the gel buffer system, such as Tris–HCl, Tris–glycine, Tris–acetate, etc.

As shown in Table 1.9, gels can be single (isocratic) or a range of (gradient) acrylamide concentrations. The choice of conditions depends on the molecular weight distribution of proteins in the extracts, and acrylamide concentrations of the gels must be tailored to each type or mixture of proteins analyzed. When peptides to be analyzed have a very wide molecular weight range or have many peptide bands (native or modified) very similar in weight, gradient gels can

TABLE 1.9 Protein Sizes Separated by Different Total Acrylamide (%T) Concentrations in Isocratic and Gradient Gels. Tris–glycine Gels, 5% Cross-linker Is Assumed

%T	Protein Mol Mass (kDa)
7	45–200
7.5	50–200
8	40–200
10	25–200
12	20–100
15	3–100
20	1–50
4–20	5–200
10–20	5–150
4–12	25–250
8–16	15–200

Adapted from Lonza (2015); used with permission.

effectively extend the working range and improve separation of close bands. Most foods, for example, have many peptides with molecular weights that cluster in certain regions. In such cases, isocratic gels can compress these regions and reduce resolution of individual bands. Cross-linking, fragmentation, and disruption of normal quaternary organization in oxidized proteins can both subtly and dramatically shift molecular weight patterns, so the analyst must be willing to explore isocratic acrylamide concentrations that are different than the native proteins or addition of a gradient in order to track all protein fractions. Finding the optimum conditions for different protein extracts of native and modified proteins can require a lot of trial and error.

Today, slab gels are used almost exclusively for protein separation. A thin acrylamide–bisacrylamide solution is prepared in buffer and poured between two glass plates to gel; this is the resolving or separating gel. A special comb placed in the liquid acrylamide at the top creates sample wells as the gel solidifies. To separate proteins, the gel assembly with protein loaded into the wells is placed between two solvent reservoirs containing buffers. Each reservoir has an electrode, usually with the negative electrode at the top and the positive electrode at the bottom, but this can be reversed. When an electrical field is applied to the system, charges on the protein begin driving each peptide toward the electrode with opposite charge (Garfin, 1990).

If proteins are loaded directly, they diffuse slowly into the resolving gel at different rates, and this leads to fuzzing and reduced resolution of separated peptide bands. To provide cleaner patterns and tighter peptide bands, electrophoresis is routinely run with a narrow stacking gel (4–5% acrylamide) poured at the top of the resolving gel. Stacking gels have large pores that allow proteins up to about 250kDa to enter and become concentrated. Now add to this a discontinuous buffer system in which the gels are prepared most commonly with Tris–HCl buffers (pH 6.8 for stacking gel, pH 8.8 for resolving gel) and the reservoir buffer is Tris-glycine pH 8.3. When current is applied, chloride ions (−) from the two gels migrate in the front while glycine ions (+) from the reservoir buffer trail behind and proteins are stacked or focused between these two fronts in tight bands. Proteins remain stacked until glycine ions finally enter the resolving gel at higher pH, become deprotonated, and assume a net (−) charge. At this point, all molecules begin to migrate and separate according to molecular mass.

Electrophoresis has two major modes of use in studying oxidized proteins: native gels and SDS gels. Both can be run with or without stacking gels. Native gels separate proteins primarily by charge balance, and can provide important insights into surface modifications (see Section 1.7.5.1). In contrast, SDS-PAGE, otherwise known as the Laemmli system, includes 1% SDS and 0.1 mol/L mercaptoethanol in the electrophoresis buffer to separate proteins by molecular mass (Laemmli, 1970). SDS is an anionic surfactant that denatures proteins, exposing buried amino acids, then binds to proteins at a constant level of about 1.4mg SDS/mg of protein, while mercaptoethanol reduces disulfide cross-links and facilitates complete unwinding of proteins. The bound SDS imparts a negative charge and hydrodynamic volume to the protein that is proportional to the molecular mass. Since the charge-to-mass ratio is the same for all proteins, in an electrical field proteins are move at a rate determined strictly by their size, i.e., total charge and ability to move through the pores of the gel. Smaller proteins have less charge, but they can pass through the gel matrix while larger proteins are impeded. Thus, proteins distribute in an SDS gel with smallest proteins at the bottom (+ electrode) and largest proteins at the top (− electrode). SDS-PAGE applications are critical for detecting cross-linking, fragmentation, and rearrangements of oxidized proteins. SDS-PAGE makes it easy to determine which peptide fractions have been most modified (bands are blurred, lessened, or lost altogether), and it reveals formation of new bands. However, these results must be combined with other method such as specific antibodies (Partridge et al., 2003) or isolation of bands followed by mass spectrometric or other analysis reference to track the source of new bands.

Finally, gels must be stained in order to visualize separated protein bands. Most traditional stains are positive, i.e., the proteins are dyed and stand out against a clear background. A number of new negative stains are also now available that bind to the gel matrix exclusively so the proteins stand out as clear bands. Since oxidized proteins have their own uniquc issucs with staining, this discussion will be limited to Coomassie Blue and silver stain. Information about additional stains may be found in (Steinberg, 2009).

Coomassie Blue R-250 (in contrast to G-250) used in the Bradford protein assay has been the standard stain for PAGE gels almost since the development of electrophoresis. As noted in the Bradford protein assay section, Coomassie blue associates with positively charged amino acids on the protein surface (arginine, histidine, lysine) by charge interactions and with aromatic amino acids and peptide regions by hydrophobic or van der Waals interactions (Roth, 2010). Eluted gels are removed from the plates, soaked in water to remove SDS then in Coomassie Blue R-250 (0.1% in methanol:glacial acetic acid:water 4:1:5) to fix the proteins and allow time for the dye to associate with the protein target sites. R-250 stains the entire gel, so destaining in the same solvent without dye is required to remove the nonspecifically bound dye. Proteins then appear as blue bands with intensity proportional to the number of reactive residues in the peptide. As little as 0.1–1.0 μg protein per band can be detected (Brunelle and Green, 2014). Coomassie R-250 works very well with native proteins, but we have encountered significant loss of dye binding in proteins that have been modified by processing and oxidation, where the three target amino acids are frequently damaged. In such cases, bands appear faded and can even be lost. Increasing protein concentrations on the gel does not correct the problem.

There are many statements in the literature that Coomassie Blue G-250 will not work for staining PAGE gels. Nevertheless, procedures for use of the colloidal dye have been developed and shown to be preferable to R-250 in some cases. Staining solvents for G-250 are ammonium sulfate–phosphoric acid–methanol mixtures and the colloidal form of the dye does not become entrapped in the gel matrix. Thus, unbound dye can easily be washed out with water, giving a clear background on the gel (Neuhoff et al., 1985, 1988; Kavran and Leahy, 2014). CBB G-250 has an extra two methyl groups that increase binding of the dye to proteins by van der Waals interactions, replacing potentially lost charge associations. We are in the process of evaluating whether use of this dye can improve staining of oxidized proteins.

Protocols for staining gels with Coomassie blue are available extensively in many Protocols series. Some specific examples may be found in (Grintzalis et al., 2009; Dyballa and Metzger, 2012; Brunelle and Green, 2014).

Silver staining is 100 time more sensitive that Coomassie Blue, detecting 10–100 ng per peptide band (Merril et al., 1983, 1984; Merril, 1990). Silver ions bind to sulfhydryl and carboxyl groups of proteins and become reduced to free metallic silver, which then deposits on the peptide backbones close to where it was formed (Rabilloud, 1990). This is opposite the Coomassie binding character. Because most proteins have both basic and acidic amino acid components, peptide patterns are usually the same whether stained by Coomassie Blue or silver. However, when proteins are predominantly one or the other – basic or acidic, their gels can look very different with the two dyes.

Silver staining procedures are complex and very easy to overdevelop. All of the many variations published in the literature have in common the following same basic steps: (1) fix proteins and remove interfering compounds in gels; (2) sensitize with thiosulfate to increase the sensitivity and contrast of the staining; (3) impregnate gel with either a silver nitrate solution or a silver-ammonia complex solution, bind to proteins by charge complexes; (4) rinse unbound silver out of gel then develop with formaldehyde or other agent to build up the silver metal image; and (5) stop reaction and rinse to end development prior to excessive background formation and to remove excess silver ion and other chemicals prior to further processing (Chevallet et al., 2006). Resulting band color varies from dark brown with short exposure times to a range of rusty to orange to dark yellow colors with extended exposure times (Dunn, 1996). Gel staining with silver is very temperature-dependent, so the laboratory or work area should be maintained at about 25 °C during the entire staining process (Chevallet et al., 2006).

Examples of protocols with detailed procedures and helpful hints may be found in (Patel et al., 1988; Dunn, 1996; Chevallet et al., 2006; Gromova and Celis, 2006; Kavran and Leahy, 2014).

An important note: Peptide patterns are usually the same whether stained by Coomassie Blue or silver because most proteins have both basic and acidic amino acid components. However, when proteins are predominantly one or the other – basic or acidic, their gels can look very different with the two dyes. Similarly, modification of binding residues in oxidized proteins changes staining patterns with Coomassie Blue and silver differently. Thus, comparing peptide patterns obtained from the two stains can provide some very useful clues about sites of damage that can aid in interpreting PAGE results as well as guide subsequent detailed analyses for specific products.

Advantages. Electrophoresis offers considerable flexibility in tailoring protein separations to different molecular weight distributions and protein properties by adjusting acrylamide concentrations in isocratic and gradient gels, altering running buffer strengths and pH, using SDS versus native gels, and by varying the dye used for visualization of peptide bands. Only very small amounts of proteins are needed, typically 1–100 ng depending on the stain. Multiple samples can be analyzed on the same gel (typically 12 in 5 cm gels), allowing controls and replicates to be run simultaneously, or many variants to be compared with room left for molecular weight standards.

Disadvantages and precautions. Acrylamide is a potent neurotoxin so must be handled in a hood with gloves, taking special care not to inhale any particles. Gels are fragile. Learning to prepare them consistently and handle them without damage requires a lot of practice and constant attention to detail. Commercial pre-cast gels are available, but at high cost. Because acrylamide cross-linking is greatly accelerated by oxygen, solutions must be carefully deaerated or the acrylamide solidifies before the gel can be cast. Keeping the acrylamide warm facilitates casting. Electrophoresis is not a fast process. Much time (at least two days or longer) is required to pour gels, run separations, fix-stain-detain gels, and scan the final stained gels.

Salts, detergents, denaturants, traces of organic solvents, and carbohydrate contaminants in protein extracts all interfere with SDS-PAGE, changing migration and reducing band focusing. Thus, it is customary with native protein extracts and even more critical with oxidized proteins, to clean up the extracts before subjecting the proteins to PAGE. A number of options are available depending on the nature of the contaminant, the amount of protein available, and the susceptibility of the test peptides to further oxidation (Evans et al., 2009). Dialysis is one option, but requires long times and large amounts of protein; dialysis slide cassettes are an option for limited samples. Acid precipitation (e.g., trichloroacetic acid) is a standard method with native proteins, but runs the risk of modifying oxidation sites in oxidized proteins. Precipitation in ice-cold acetone is effective, can be handled on microscale, does not damage modified proteins, and has the additional advantage of dissolving any unbound lipids that have been carried along in the extract. We have found this method very useful with a number of modified protein applications (Partridge et al., 2003). Traces of organic solvents can easily be removed by vacuum freeze evaporation/concentration while still protecting protein.

Finally, a word to the wise regarding electrophoresis of oxidized proteins: ***Do not*** expect to get nice-looking gels with clean banding patterns! Damage to surface groups as well as random fragmentation and cross-linking all lead to modifications that can result in considerable smearing of bands. Also, loss of dye binding should be expected, especially with Coomassie blue, since lysine and arginine are among the amino acids most sensitive to oxidation.

Silver staining is less susceptible but still shows aberrations if acidic groups are lost, such as by decarboxylation (Schaich, 2014). Covalent complexation of proteins with starches and lipids in foods further complicates migration patterns and dye binding. Interpreting electrophoresis band patterns of oxidized proteins from foods is most often not a quandary of "what changes can we see?" but rather, "what chemistry could have happened to make such dramatic changes?'

Applications. Electrophoresis ranks with solubility as the analyses most used as indicators of protein oxidation and modification. SDS gels for detection of cross-linking and fragmentation are the most common application, but more extensive use of native gels and differential staining can provide even more insights into other molecular changes that have occurred. We have used SDS gels to study protein oxidation in baked and fried tortilla chips, laminate-packaged peanut butter, and flour stored in different packaging (Schaich, 2014). Wu et al. (2009) used electrophoresis to determine that azide-generated peroxyl radicals induced disulfide cross-linking specifically in the β-conglycinins.

1.7.5 Surface Modifications

Lipid radicals, hydroperoxides, aldehydes, and epoxides all react with amino acid side chains to modify protein surfaces in ways that interfere with normal protein function and degrade protein structure in foods. Surface modifications include oxidation of amines, alcohols, and thiols to oxides and carbonyls; deamination and decarboxylation; and complexation of lipid oxidation products through Michael addition and Schiff base condensations. Unfortunately, there is no way to test for all of these modifications in a single assay; each modification must be analyzed independently. How many and which changes should be followed as markers of oxidation depends on the protein oxidized, the chemistry suspected to underlie observed macroscopic or functional modifications, and the loss of function or altered behavior that needs to be explained.

1.7.5.1 Native PAGE

Principles of the analysis. General principles of polyacrylamide gel electrophoresis were described above. In contrast to SDS-PAGE which seeks to determine molecular weights, native gels contain no denaturing agent so proteins retain their native (or modified) configuration and protein migration is governed by molecular properties – surface charge mainly, followed by shape, and size. Protein charge is determined by the relative composition of NH_3^+ and COO^- groups on the peptide surface. In general, the higher the charge density, the faster and farther proteins will move in the gel, with acidic proteins moving toward the cathode at the bottom and basic proteins staying closer to the anode at the top. At the same time, the gel matrix imposes a secondary sieving effect impeding protein movement according to hydrodynamic size and shape. Because charge driving forces are sometimes opposite the mass impeding effects, proteins do not distribute by size on the gels, and peptide banding patterns are usually very different than in SDS gels for the same proteins. Position of peptides on the gels is not easily deduced by logic. Hence, pure controls must be run to identify positions of different proteins on the gels.

Advantages. This is the fastest, most straightforward way to assess whether surface modifications have occurred on oxidized proteins. Because no denaturants are used in native PAGE, subunit interactions within a multimeric protein are generally retained and information can be gained about the quaternary structure. Furthermore, by direction of migration shifts and changes in Coomassie Blue versus silver staining, deductions can be made about the potential for deamination and decarboxylation or blocking of the respective amine and acid groups.

Disadvantages. Running these gels generates more frictional heat than SDS gels so it is critical to keep the apparatus cool to minimize the effects of denaturation and proteolysis and to prevent band blurring. Most gel systems have a cooling jacket around the gel; this needs to be filled with very cold water or ice slurries. Interpreting band patterns and tracking oxidation changes of a specific protein in a mixture is difficult due to mixed-mode migration. Controls of individual unmodified proteins or extracts from native materials before processing should be run on each gel for comparison. Reproducibility is a problem; native gels are much more sensitive to variations in electrical current from the source or across the gel as well as to microscopic variations in each gel. Pre-cast commercial gels increase reproducibility but at high cost.

Applications. Native gels are used when information other than molecular weight is needed. For example, in protein groups such as gliadins where many peptides have similar molecular weights but different properties, gliadins run very closely together or even overlap in SDS gels but can be completely separated in native gels (Partridge et al., 2003). Native gels are also useful for detecting proteins with many charged groups that may have been decarboxylated or deaminated, such as peanut proteins (Wan Ibadullah, 2013) and meat myofibrillar proteins, as well as proteins in which charge associations between quaternary structures have been disrupted, such as peanut arachins and conarachins (Wan Ibadullah, 2013).

1.7.5.2 Fluorescence

1.7.5.2.1 Loss of Tryptophan

All aromatic amino acids exhibit fluorescence in solution and in proteins (Table 1.10). Tryptophan exhibits the strongest intrinsic fluorescence, with lesser contributions from tyrosine, and the least from phenylalanine (Teale and Weber, 1957). Interestingly, this is also the order of susceptibility to oxidation by lipids and other reactive oxygen species; tryptophan, in particular, is very susceptible to loss during processing and storage. Fluorescence of the aromatic amino acids shifts in wavelengths with changes in protein structure and amino acid environment. Tryptophan residues that are totally buried in folded proteins

TABLE 1.10 Fluorescence Characteristics of the Aromatic Amino Acids

	Absorption		Fluorescence	
Amino Acid	Wavelength (nm)	Absorptivity	Wavelength (nm)	Quantum Yield
Tryptophan	280	5600	348	0.20
Tyrosine	274	1400	303	0.14
Phenylalanine	257	200	282	0.04

Held (2003), used with permission.

fluoresce at about 330 nm whereas residues exposed to water after denaturation have maximal fluorescence at a wavelength of about 340–350 nm (Lakowicz, 1983). Fluorescence decreases in intensity or is lost altogether when the residues are oxidized or otherwise chemically modified. Thus, both intensity and wavelength of tryptophan emission can serve as rapid, simple markers for protein oxidation.

Intrinsic fluorescence from aromatic amino acids (primarily tryptophan) is determined in protein extracts (in extraction buffer) by recording fluorescence emission spectra from 300 to 400 nm, with excitation at 280 nm to detect all aromatic amino acids and at 295 nm to detect tryptophan separately. The emission maximum wavelength for tryptophan is diagnostic for native versus modified protein configuration and residue environment. Decreased intensity is diagnostic for residue modification. Unfortunately, tryptophan is damaged extensively by heat as well as oxidation, so loss measured in isolated samples cannot unequivocally be attributed to lipid oxidation. However, processing damage can be separated from oxidation at least presumptively by analyzing spectra immediately after processing (thermal or other physical stress) and after incubation or storage where oxidation occurs.

Fluorescence intensity varies with temperature, for most proteins decreasing at different rates as temperature increases. Thus, assays for tryptophan fluorescence require tightly controlled constant temperature (Gally and Edelman, 1962).

Utrera and Estévez (2012) used intrinsic fluorescence to follow oxidation of tryptophan in solution by Fenton reactions as well as protection by various phenolic compounds. We use shifts and loss of tryptophan fluorescence in all our studies of protein oxidation in foods.

1.7.5.2.2 Appearance of Fluorescent Products

This phenomenon was largely discovered by A.L.Tappel and his colleagues (Chio and Tappel, 1969; Fletcher and Tappel, 1970; Fletcher et al., 1973; Malshet and Tappel, 1973; Dillard and Tappel, 1984), and has been used frequently as a marker of lipid carbonyl–protein interactions. A general misperception has been that the fluorescence arises from Schiff base products of any lipid carbonyl reaction with protein amino groups (Schaich, 2008). However, detailed model system studies have shown that while simple Schiff base adducts ($-N{=}CH-CH_2-CHO$ or $-NH-CH{=}CH-CHO$) contribute to browning (Zamora et al., 2000), not all are fluorescent (Itakura and Uchida, 2001). Rather, pyridinium and pyrrole products resulting from reactions of proteins with lipid hydroperoxides as well as carbonyl secondary products are the main sources of strong fluorescence detected in foods [(Schaich, 2008) and references therein], and the structural feature responsible for the fluorescence is the imino-ene conjugation ($R-N{=}CH-CH{=}CH-NH-R$) originally identifiedby Malshet and Tappel (1973). Fluorescent products exist in both monomer forms and cross-links, so contribute both to surface modifications and to cross-linking in co-oxidized proteins, respectively.

Fluorescent co-oxidation products require elevated temperatures to form; only trace levels, if any, are observed at refrigeration or room temperatures. Little to no fluorescence develops in reactions at refrigerated or room temperatures, but at ASL testing temperatures of 40 and 60 °C and higher food-processing temperatures, fluorescent compounds become major products that contribute to characteristic browning and flavor production (Hidalgo et al., 1999, 2005).

Spectral characteristics of fluorescent products vary tremendously depending on the originating lipid oxidation products and proteins, the numbers and types of multiple products formed, reaction conditions, and solvent. Generally, fluorescence ranges are 350–395 nm excitation and 425–470 nm emission (Schaich, 2008). Linear and cyclic products are often both present and have different spectral ranges, with emission maxima near 430 nm for linear products (Nadkarni and Sayre, 1995) and 470 nm for pyrrole (Hidalgo and Zamora, 1993) or pyridinium (Kikugawa and Ido, 1984; Kikugawa and Beppu, 1987) cyclic products. These differences can be used diagnostically in analyzing oxidized proteins.

Procedures for measuring carbonyl-amine fluorescence are rather ill-defined, with each laboratory using its own method, for the most part. Protein fluorescence has been analyzed in buffer solutions (Viljanen et al., 2004), in aqueous extracting media, and in solvent extracts, usually chloroform:methanol 1:1 (Chio and Tappel, 1969; Fletcher and Tappel, 1970; Dillard and Tappel, 1971, 1973, 1984; Fletcher et al., 1973; Malshet and Tappel, 1973). Fluorescence characteristics vary with the solvent, ex/em maxima and emission intensity both increasing increase with solvent polarity (Kikugawa and Ido, 1984; Kikugawa and Beppu, 1987). To determine excitation and emission maxima for each system, first identify the wavelength of maximum excitation by recording the UV/VIS absorption of the protein solution from 300 to 420 nm. Then use this value as the excitation set point in the fluorometer and scan emissions from 400 to 500 nm. Since there is often shifting in the fluorescence maximum as oxidation progresses, it is often advantageous to scan emissions over a broad range.

For quantitation, spectral wavelengths and intensity are calibrated with quinine sulfate and reported as photons/mg protein.

Careful sample handling is critical to avoid artifacts in fluorescence analyses. Contaminants in extracts catalyze both secondary photoreactions and fluorescence quenching occurs, especially in aqueous solvents. Hence, extracts should be prepared and analyzed as rapidly as possible and protected from light to avoid loss by decay and alteration of fluorescence.

Finally, while protein fluorescence can serve as a general marker for oxidative degradation, great care must be exercised in interpreting the data. As introduced briefly in the discussion above, many factors affect the intensity and the wavelengths of emission, and emissions from complex systems such as foods probably reflect many fluorescent species. Thus, assignment of fluorescent structures based solely on excitation and emission maxima is limited to presumptive distinction between linear and ring fluorescent complexes. However, even this information can support other analytical results (e.g., electrophoresis) and guide decisions about confirmatory and auxiliary testing.

1.7.5.3 Protein Carbonyls

Development of methods for analyzing protein carbonyls has been largely driven by the need to understand roles of these oxidation products in pathology (Shacter, 2000a,b). Only recently have methods been applied to understand effects of protein carbonyls in foods, particularly meats (Estévez, 2011). Protein carbonyls, developed from α- and β-scission of alkoxyl radicals formed on the peptide backbone as well as from oxidation of amino acid side chains, are more stable than protein and lipid hydroperoxides so provide very sensitive markers for oxidation. Most protein preparations contain about 1 nmol carbonyl/mg protein or ~0.05 mol carbonyl/mol protein, and these levels increase by orders of magnitude during oxidation in foods. Kits now available from multiple manufacturers for all the assays described below are expensive, but they often provide greater sensitivity and reproducibility than running the assays "from scratch" in the laboratory, particularly when reagent purity is critical. Guidelines are provided by each manufacturer, but optimal dilutions/concentrations should be determined for every application by the end user.

Currently, protein carbonyl values are mostly comparative. Increases indicate that protein oxidation has occurred but give no hint about origin, reactions involved, or impact on food quality. More studies are needed to correlate specific protein carbonyl levels to other degradative changes in proteins, lipids, nutritional qualities, and food properties to learn how best to interpret and use results of the analyses. Oxidation studies also need to be coordinated with metabolism and toxicity studies to determine levels at which oxidized proteins may begin to cause physiological damage.

1.7.5.3.1 Total Carbonyls

Total protein carbonyls, like lipid carbonyls, are determined by reaction with recrystallized 2,4-dinitrophenylhydrazine (DNPH). The basic approach was developed by Levine et al. (1990, 1994) and analytical versions are now available in solution (spectrophotometric analysis), by enzyme-linked immunosorbent assay (ELISA), by electrophoresis followed by Western blotting and specific antibody reaction, or by HPLC. An excellent review has compared and evaluated these various procedures (Dalle-Donne et al., 2003).

1.7.5.3.1.1 Solution Reactions. The basic reaction mechanism is the same as that described above for lipid carbonyls, although reaction conditions are more complex. Solutions of proteins or protein extracts are reacted with DNPH in acid at room temperature in the dark for 1 h, derivatized proteins are precipitated by trichloroacetic acid then separated from unreacted DNPH and other contaminants by centrifugation, the protein pellet is washed 3x with ethanol:ethyl acetate 1:1, centrifuged, pellet resuspended in 6 mol/L guanidine- HCl, then optical absorbance is read at 276 nm to determine protein concentration ($\varepsilon = 9460\,M^{-1}\,cm^{-1}$) and at 370 nm to determine protein hydrazone concentrations ($\varepsilon = 22{,}000\,M^{-1}\,cm^{-1}$) (Levine et al., 1990; Shacter et al., 1994b). Protein carbonyls are expressed as nmol carbonyl/mg protein.

Use of this assay has been reported in numerous studies, with almost as many variations as reports. Some procedures are conducted with protein extracts and some with food homogenates to save time or when protein extractions are difficult. However, getting useful results from this reaction depends critically on removing two key interferences – nonprotein contaminants in the extracts and unreacted DNPH after the reaction. Any procedures omitting these clean-up steps are highly questionable. In food homogenates, in particular, DNPH reacts with any lipid or other carbonyls present, and these will all be accounted for if not removed. The final protein pellet wash in the procedure described above was designed to remove both interferences, but protein extracts from tissues and foods often need additional steps. Precipitating proteins from initial extracts in 10% trichloroacetic acid as a first step removes nonprotein DNPH-reactive materials and gives cleaner reactions, and adding a 20% TCA wash after DNPH reaction plus increasing the volume of ethyl acetate more effectively removes all traces of unreacted DNPH (Fagan et al., 1999).

Advantages. The spectrophotometric assay requires no special equipment or reagents and can be handled in any laboratory. Protein concentrations in the test solution can be determined by any acceptable method for normalization of carbonyl contents.

Disadvantages. This has the lowest sensitivity of all the versions of the assay, relatively large amounts of protein are required, the handling required is extensive and tedious, and many blanks must be included in the reaction. The solution assay is prone to interferences from contaminants and inaccuracies due to loss of solubility of some DNPH-derivatized proteins. This is not an assay easily applied to large numbers of samples, so it is not surprising that most applications have been in biological studies rather than with food proteins where many treatments need to be compared. From a fundamental perspective, this is a "black box" assay where all carbonyls are counted but the source of the carbonyls is not identified.

Applications. Despite its limitations, the solution DNPH assay is considered the gold standard against which accuracy of other protein carbonyl assays are measured, and it is used to quantitate standards for ELISA and immunoblot methods. When handled correctly, spectrophotometric detection of protein hydrazones provides a value for comparison of protein susceptibility to oxidation, such as between different proteins in a formulation and different processing methods of a food material. In foods, this basic DNPH reaction has been used to detect protein carbonyls in meat muscle (Armenteros et al., 2009) and fish muscle and cheese slices (Fernandez et al., 2014).

1.7.5.3.1.2 Enzyme-Linked Immunosorbent Assay (ELISA). *Principle of the assay*. Antibody versions of the DNPH assay were developed to eliminate interferences of the solution reaction and to increase sensitivity by specifically detecting DNPH labeled proteins. ELISA makes the analyses quantitative by serially binding DNPH-labeled proteins to the surface of wells in polystyrene microplates (typically 96 well), then a primary anti-DNPH antibody, a secondary antibody carrying an enzyme such as horse radish peroxidase (HRP) that will provide a marker reaction, and finally an agent that will react with HRP to produce color for quantitation. Unreacted reagents and contaminating materials are washed away between each step so that very low background and high specificity results. Because this process is designed to be handled on 96 well plates and special equipment is available for both the reagent deposits and the washes, the total procedure is actually easier and faster than it seems from this description.

As with any complex procedures, controversy over optimum procedures is bound to develop. With ELISA of protein carbonyls, points of disagreement focus mainly on the protein binding step, particularly whether to bind native or derivatized protein and how much to bind. Initial procedures used saturating concentrations of 1 μg oxidized proteins in each well and high concentrations of DTNB (Buss et al., 1997; Buss and Winterbourn, 2002). Other studies claim that more consistent results with lower backgrounds are obtained by binding less protein than binding capacity of polystyrene (0.25–0.3 ug/well), reacting DTNB with protein *after* binding, and using lower concentrations of DTNB (e.g., 0.05 mmol/L) (Alamdari et al., 2005; Uehara and Rao, 2015). Both procedures have their proponents, and commercial kits are available with each approach. The bottom line is that, especially for oxidized food proteins where there is little experience base and oxidation levels should be much higher than in biological tissues, alternate ELISA protocols must be tested and optimized for each new material analyzed.

Advantages. ELISA is very sensitive. As little as 5 μg/mL or one ug/well DNPH-conjugated protein can be detected in a standard microplate reader (Alamdari et al., 2005) so the method is very useful when amounts of isolated protein are limited. It increases specificity and sensitivity over the standard spectrophotometric DNPH assay and provides quantitative results. The method easily handles a large numbers of samples in parallel, and no concentration or precipitation steps contribute to sample loss.

Disadvantages and precautions. Quantitation is totally dependent on complete binding of the protein to the well surface by hydrophobic interactions and other reagents in most extracts of food proteins. Protein binding is extremely difficult to determine and is inhibited by detergents (Uehara and Rao, 2015). Thus, proteins in food extracts must be isolated by dialysis or precipitation and washing before analyzing in ELISA. Careful handling is required to avoid loss of variable amounts of protein in washing steps. Oxidized proteins for development of standard curves must be synthesized and standardized by the solution DNPH assay. Procedures are still not well-developed but must be tested and optimized for each system. Assays are not as applicable to highly oxidized proteins where aggregation blocks reaction sites and secondary reactions transform carbonyl adducts (Augustyniak et al., 2015). Finally, a major impediment is cost. ELISA equipment, kits, and antibodies are all quite expensive.

Applications. ELISA assays of DNPH-labeled proteins is suitable for plasma, serum, cell lysates, extracts, or purified proteins. The capability to handle large numbers of samples simultaneously in a single plate and its increase in sensitivity over the solution assay has made this a very popular technique for testing protein oxidation in biological samples where proteins are easily isolated (Dalle-Donne et al., 2003; Khatoon et al., 2012). However, parallel applications have not yet been developed for detecting oxidation of food proteins because the importance of protein oxidation in foods has only recently attracted industry attention, the requirements for protein isolation add significantly to handling time and labor, and the cost per analysis is high relative to other options.

1.7.5.3.1.3 Carbonyls in Individual Proteins by Immunoblotting. As has been noted, all of the DNPH methods described so far measure total carbonyls and do not distinguish which peptides are affected. Are all oxidized equally or are some particularly sensitive? Because this information can be very important in foods where individual proteins have unique functions (e.g., structure, gelation, foam formation, hydration, emulsification, water binding), determining which proteins have been oxidized in an extract by immunoblotting can be a very useful adjunct to quantitative carbonyl analyses.

Principle of the assay. Immunoblotting, or Western blotting, was originally developed to make separated proteins available to antibodies on an inert matrix for detection of specific properties (Towbin et al., 1979). The process involves six steps similar to those of ELISA: (1) separation of protein mixtures by gel electrophoresis, either SDS or native; (2) transfer of unstained proteins in the same separation pattern from the gel to blotting paper by electrical current in a special "sandwich" unit; (3) incubation of the membrane with a generic protein (usually bovine serum albumin) to bind to open sites on the membrane; (4) binding of an anti-DNPH antibody, washing to remove unbound antibody; (5) binding of a secondary antibody conjugated to a reporter enzyme;

(6) reaction of the enzyme with a substrate to create a colored marker for detection of antibody–protein complexes, washing to remove unreacted substrate. A number of different enzymes and substrates are available allowing for detection by different methods (color, fluorescence, chemiluminescence, etc.) but most common is horseradish peroxidase with tetramethylbenzidine for color detection or Luminata Forte for chemiluminescence detection. A parallel gel run simultaneously with the same proteins but stained normally with silver or Coomassie blue provides a normal PAGE pattern for peptide identification (Figure 1.14). Detailed procedures are available on many websites and in many molecular biology textbooks and compendia of protocols.

Blotting membranes were originally made of nitrocellulose. However, the dominant membranes in use now are of PVDF (polyvinyl difluoride) or nylon which are sturdier materials, have better retention of adsorbed proteins, more resistance to discoloration, lower background levels and higher sensitivity than nitrocellulose membranes, particularly when fluorescence or chemiluminescence detection is used (Shewry and Fido, 1998).

The traditional immunoblotting DNPH assay starts with reacting the test proteins with DNPH, then separates the complexes by PAGE before transfer to a membrane and reaction with antibodies specifically raised against DNPH-proteins. However, there is some controversy over whether the DNPH derivatization of oxidized proteins should be performed before or after transfer from PAGE gel to membrane. A key argument supporting DNPH reaction after blotting is that derivatization of proteins before PAGE, as in the original assay, requires excessive handling and leads to inaccuracies and reduced sensitivity

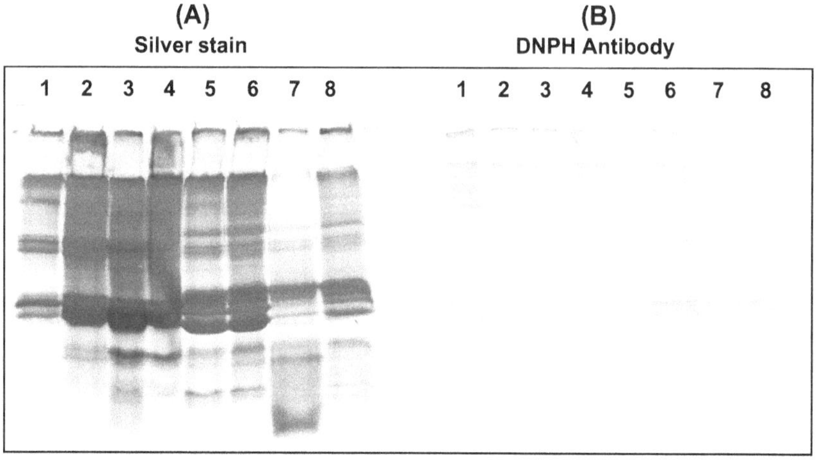

FIGURE 1.14 Detection of protein carbonyls in extracts from cornmeal processed and aged in different ways. (A) PAGE gel with silver staining to establish peptide pattern. (B) Western blot of the same proteins reacted with DNPH antibodies. Colored regions denote presence of protein carbonyls.

due to degradation of derivatized proteins during electrophoresis. DNPH binding also alters protein charges and electrophoretic migration, so it is not possible to directly compare the patterns from oxidized proteins on Western blots with total proteins on gels (Robinson et al., 1999). Thus, available kits from different companies are split between the two methods (DNPH reaction before electrophoresis versus after blotting). Each laboratory needs to test the two approaches with their particular proteins and technical skills to determine which method is best for their application.

Advantages. The assay is very sensitive, detecting ≤1 pmol of protein carbonyl (Shacter et al., 1994a), and it determines which individual proteins are oxidized (Figure 1.14). Identification of oxidized fractions provides important supplementary data for integrating with other oxidation analyses. In highly oxidized fractions, antibodies often detect modified fractions that are poorly detected by electrophoresis dyes and scanning. With the use of computerized gel scanners, results can be made at least semi-quantitative. Kits are available from many commercial sources.

Disadvantages. Western blots are time-consuming and have a steep learning curve to develop the expertise required to obtain reproducible results with low background. Experimental conditions must be optimized for each type of protein, antibody, and modification target. Without a scanner, results are only diagnostic, but even that can be useful.

Applications. Immunoblots have seen a number of applications for detecting oxidized proteins in foods. In studies of frozen fish, spectrophotometric assays for quantitation were combined with immunoblots to identify fish protein classes affected (Baron et al., 2007). DNPH reaction with PAGE and immunoblotting showed that in beef steaks in high-oxygen modified atmosphere packing, postmortem degradation of troponin-T and desmin were unaffected but the myosin heavy chain specifically became highly oxidized and cross-linked (Kim et al., 2010). We have used immunoblots to detect protein oxidation in extruded proteins, baked and fried tortillas (Dong, 2011), and laminate-packaged peanut butter (Wan Ibadullah, 2013).

1.7.5.3.2 Individual Carbonyls

As evidence accumulates demonstrating the presence of carbonyls in co-oxidized proteins, interest has also intensified in going beyond undifferentiated analysis of total protein carbonyls to determining oxidation of specific amino acids. Lysine and arginine are particularly susceptible to oxidation by lipids, forming α-aminoadipic semialdehyde (AAS) and the γ-glutamic semialdehyde (GGS), respectively, and these products are proving to be useful oxidation markers in both physiological and food proteins. Early assays for these modified amino acids were complicated and difficult to handle, and not amenable to analyzing many samples. Recently, simpler procedures for detection and quantitation of these two protein carbonyls have been developed using reductive amination by aminobenzoic acid, acid hydrolysis to release the products, then

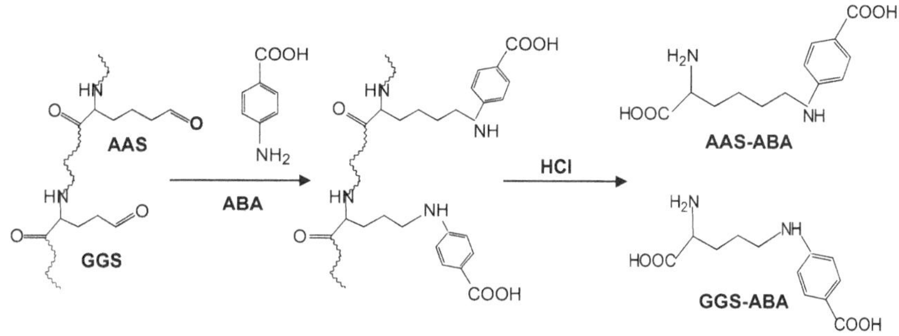

FIGURE 1.15 Aminobenzoic acid (ABA) reaction for detection of α-aminoadipic semialdehyde (AAS) and the γ-glutamic semialdehyde (GGS).

reverse-phase HPLC analysis of the products (Figure 1.15) (Akagawa et al., 2006, 2009). Modifications of these procedures are being applied to determining protein oxidation in meats (Estévez et al., 2009; Utrera et al., 2011; Utrera and Estévez, 2013).

In the basic reaction, extracted proteins isolated by dialysis or trichloroacetic acid precipitation are dissolved in 2-(*N*-morpholino)ethanesulfonic acid buffer, then reacted with aminobenzoic acid and sodium cyanoborohydride in the dark. The derivatized protein is precipitated in trichloroacetic acid (TCA), centrifuged, and washed with TCA then ethanol. The pellet is then hydrolyzed in 6N HCl, evaporated to dryness under vacuum, reconstituted in buffer, and analyzed by HPLC with fluorescence detection at 283 nm excitation and 350 nm emission. Lower detection limits are picomoles per mg protein.

In comparisons of methods for determining protein oxidation in meats, protein carbonyl determination by both DNPH-HPLC and tryptophan fluorescence assays was affected by the composition or the structure of the food matrix, while α-aminoadipic semialdehyde (AAS) and the γ-glutamic (GGS) semialdehyde detection was independent (Armenteros et al., 2009), perhaps due to the additional protein processing that was involved. Thus, it was proposed that these semialdehydes, particularly GGS, could be used as indicators of protein oxidation in meat products, replacing TBARS numbers. More extensive application of these analyses should begin to identify which proteins are particularly prone to AAS and GGS formation, and what conditions facilitate the oxidation.

1.7.5.4 IR Spectroscopy

Table 1.6 listed IR frequencies of key lipid oxidation products and included also key frequencies for peptides bonds, amino acids likely to be oxidized, and some oxidation products of sulfur and nitrogen. Protein frequencies were included because intact food samples are being analyzed increasingly by mid-IR, especially by attenuated reflectance, so components of both lipids and proteins will be present in the spectra. In the past, there was so much overlap of bands in the amide and carbonyl regions that definitive information was difficult to obtain. However, chemometrics is changing the scene drastically, so it should now become possible to distinguish loss of native group frequencies as well as appearance of oxidation products from both lipids and proteins in the same sample. This offers one interesting technique for monitoring lipid oxidation while at the same following its footprints into co-oxidations. The methods certainly merit more extensive application in both quality control and research.

1.8 SUMMARY AND CONCLUSIONS

Not very long ago, the options for routine monitoring of lipid oxidation were rather limited. CDs and hydroperoxides were measured in test tube assays and hexanal and other volatiles were detected by GC. Because these limited analyses alone do not present fully or accurately portray the extent of lipid oxidation, this chapter has argued for a new paradigm in analysis of lipid oxidation – one that monitors multiple lipid products from alternate pathways, looks for quantitative and stoichiometric amounts of products rather than relative changes, integrates chemical and instrumental analyses of extracts and intact foods, and looks beyond lipid oxidation alone to track its footprints in co-oxidations of other molecules. Co-oxidation targets covered in this chapter were proteins.

Material in the chapter was organized as a mini course and attempted to include important details and considerations that are seldom found in general reviews. An obvious major goal was to provide guides for choice and handling of analyses. A less obvious but no less important goal was also to stimulate students, technicians, researchers, and professors to think critically about the methods, to pay close attention to all aspects of the analytical environment, not just copy procedures blindly from a publication, and overall to "Think Chemistry" at every step of each procedure. There is much still to be learned about mechanisms by which lipids and proteins oxidize and how their oxidations affect food properties and qualities. Hopefully, this chapter will stimulate more interest in the analytical challenges of lipid and protein oxidation and will encourage more detailed research.

REFERENCES

Adams, A., Van Lancker, F., De Meulenaer, B., Owczarek-Fendor, A., De Kimpe, N., 2012. On-fiber furan formation from volatile precursors: a critical example of artefact formation during Solid-Phase Microextraction. J. Chromatogr. B 897, 37–41.

Akagawa, M., Sasaki, D., Ishii, Y., Kurota, Y., Yotsu-Yamashita, M., Uchida, K., Suyama, K., 2006. New method for the quantitative determination of major protein carbonyls, α-aminoadipic and γ-glutamic semialdehydes: investigation of the formation mechanism and chemical nature in vitro and in vivo. Chem. Res. Toxicol. 19, 1059–1065.

Akagawa, M., Suyama, K., Uchida, K., 2009. Fluorescent detection of α-aminoadipic and γ-glutamic semialdehydes in oxidized proteins. Free Radic. Biol. Med. 46, 701–706.

Alamdari, D.H., Kostidou, E., Paletas, K., Sarigianni, M., Konstas, A.G., Karapiperidou, A., Koliakos, G., 2005. High sensitivity enzyme-linked immunosorbent assay (ELISA) method for measuring protein carbonyl in samples with low amounts of protein. Free Radic. Biol. Med. 39, 1362–1367.

Alemán, M., Bou, R., Guardiola, F., Durand, E., Villeneuve, P., Jacobsen, C., Sørensen, A.D., 2015. Antioxidative effect of lipophilized caffeic acid in fish oil enriched mayonnaise and milk. Food Chem. 167, 236–244.

Ames, J.M., Duckham, S.C., Bakker, J., 2001. Choice and use of standards for dynamic headspace trapping and application to the analysis of the volatiles of baked potato. Adv. Exp. Med. Biol. 488, 133–141.

Anonymous, 1998. Evidence on the carcinogenicity of methyl tertiary butyl ether (MBTE). California Environmental Protection Agency. http://oehha.ca.gov/prop65/pdf/dhidmt1.pdf (accessed 05.07.15.).

Antharavally, B.S., Mallia, K.A., Rangaraj, P., Haney, P., Bell, P.A., 2009. Quantitation of proteins using a dye-metal-based colorimetric protein assay. Anal. Biochem. 385, 342–345.

AOCS, 2011. *p*-Anisidine value: AOCS Official Method Cd 18–90(11). In: Firestone, D. (Ed.), Official Methods and Recommended Practices of the American Oil Chemists' Society, sixth ed. AOCS Press (second printing), Champaign, IL.

APJCN-NHRI, 2015. Important Basics Food Charts. FAT. http://apjcn.nhri.org.tw/server/info/books-phds/books/foodfacts/html/data/data2g.html (accessed 05.07.15.).

Armenta, S., Garrigues, S., de la Guardia, M., 2007. Determination of edible oil parameters by near infrared spectrometry. Anal. Chim. Acta 596, 330–337.

Armenteros, M., Heinonen, M., Ollilainen, V., Toldrá, F., Estévez, M., 2009. Analysis of protein carbonyls in meat products by using the DNPH-method, fluorescence spectroscopy and liquid chromatography-electrospray ionisation-mass spectrometry (LC-ESI-MS). Meat Sci. 83, 104–112.

Arthur, C.L., Pawliszyn, J., 1990. Solid phase microextraction with thermal desorption using fused silica optical fibers. Anal. Chem. 62, 2145–2148.

ASTDR, 2014. Priority List of Hazardous Substances. http://www.atsdr.cdc.gov/spl/ (accessed 14.05.15.).

Augustyniak, E., Adam, A., Wojdyla, K., Rogowska-Wrzesinska, A., Willetts, R., Korkmaz, A., Atalay, M., Weber, D., Grune, T., Borsa, C., Gradinaru, D., Chand Bollineni, R., Fedorova, M., Griffiths, H.R., 2015. Validation of protein carbonyl measurement: a multi-centre study. Redox Biol. 4, 149–157.

Banerjee, D., Jacob, J., Kunjamma, G., Madhusoodanan, U.K., Ghosh, S., 2004. Measurement of urinary hydrogen peroxide by FOX-1 method in conjunction with catalase in diabetes mellitus—a sensitive and specific approach. Clin. Chim. Acta 350, 233–236.

Baron, C.P., Bro, R., Skibsted, L.H., Andersen, H.J., 1997. Direct measurement of lipid peroxidation in oil-in-water emulsions using multiwavelength derivative UV-Spectroscopy. J. Agric. Food Chem. 45, 1741–1745.

Baron, C.P., Kjaersgard, I.V.H., Jessen, F., Jacobsen, C., 2007. Protein and lipid oxidation during frozen storage of rainbow trout (*Oncorhynchus mykiss*). J. Agric. Food Chem. 55, 8118–8125.

Barrera-Arellano, D., Esteves, W., 1992. Oxidative stability of potato chips determined by Rancimat. J. Am. Oil Chem. Soc. 69, 335–337.

Barth, A., 2007. Infrared spectroscopy of proteins. Biochim. Biophys. Acta Bioenerg. 1767, 1073–1101.

Battino, R., Clever, H.L., 1966. The solubility of gases in liquids. Chem. Rev. 66, 395–463.

Battino, R., Rettich, T.R., Tominaga, R., 1983. The solubility of oxygen and ozone in liquids. J. Phys. Chem. Ref. Data 12, 163–178.

Baumgartner, W.A., Baker, N., Hill, V.A., Wright, E.T., 1975. Novel interference in thiobarbituric acid assay for lipid peroxidation. Lipids 10, 309–311.

Belhaj, N., Arab-Tehrany, E., Linder, M., 2010. Oxidative kinetics of salmon oil in bulk and in nanoemulsion stabilized by marine lecithin. Process Biochem. 45, 187–195.

Bernheim, F., Bernheim, M.L.C., Wilbur, K.M., 1948. The reaction between thiobarbituric acid and the oxidation products of certain lipids. J. Biol. Chem. 174, 257–264.

Bidder, T.G., Sipka, B., 1989. A modification of the thiobarbituric acid reaction. Lipids 24, 656–658.

Bligh, E.G., Dyer, W.J., 1959. A rapid method of total lipid extraction and purification. Can. J. Biochem. Physiol. 37, 911–917.

Bogusz, B., Schaich, K.M., 2013. Monitoring volatile product formation to track lipid oxidation pathways in methyl linoleate. In: Symposium on Lipid Oxidation: Reconsideration of Mechanisms, American Chemical Society National Meeting, Indianapolis, IN.

Bogusz, B.A., 2015. Assessing Presence of Alternate Lipid Oxidation Pathways from Volatile Products Detected by Gas Chromatography. Ph.D. Food Science, Rutgers University, New Brunswick, NJ.

Borchman, D., Sinha, S., 2002. Determination of products of lipid oxidation by infrared spectroscopy. In: Armstrong, D. (Ed.), Determination of Products of Lipid Oxidation by Infrared Spectroscopy. Humana Press, New York, pp. 21–28.

Bou, R., Codony, R., Tres, A., Decker, E.A., Guardiola, F., 2008. Determination of hydroperoxides in foods and biological samples by the ferrous oxidation-xylenol orange method: a review of the factors that influence the method's performance. Anal. Biochem. 377, 1–15.

Bradford, M.M., 1976. A rapid and sensitive method for the quantitation of microgram quantities of protein utilizing the principle of protein-dye binding. Anal. Biochem. 72, 248–254.

Bradley, D., Williams, G., Lawton, M., 2010. Drying of organic solvents: quantitative evaluation of the efficiency of several desiccants. J. Org. Chem. 75, 8351–8354.

Bragadottir, M., 2007. Methods for measuring lipid oxidation and their application to fish meal. J. Aquat. Food Prod. Technol. 16, 115–136.

Brombacher, S., Oehme, M., Dye, C., 2002. Qualitative and quantitative analysis of carbonyl compounds in ambient air samples by use of an HPLC-MS(n) method. Anal. Bioanal. Chem. 372, 622–629.

Brown, R.E., Jarvis, K.L., Hyland, K.J., 1989. Protein measurement using bicinchoninic acid: elimination of interfering substances. Anal. Biochem. 180, 136–139.

Brunelle, J.L., Green, R., 2014. Coomassie blue staining. Methods Enzymol. 541, 161–167.

Burfield, D.R., Lee, K.-H., Smither, R.H., 1977. Desiccant efficiency in solvent drying. A reappraisal by application of a novel method for solvent water assay. J. Org. Chem. 42, 3060–3065.

Buss, H., Chan, T.P., Sluis, K.B., Domigan, N.M., Winterbourn, C.C., 1997. Protein carbonyl measurement by a sensitive ELISA method. Free Radical Biol. Med. 23, 361–366.

Buss, I.H., Winterbourn, C.C., 2002. Protein carbonyl measurement by ELISA. Methods Mol. Biol. 186, 123–128.

Cao, X.-L., Nicholas Hewitt, C., 1993. Evaluation of tenax-GR adsorbent for the passive sampling of volatile organic compounds at low concentrations. Atmos. Environ., Part A 27, 1865–1872.

Chan, H.W.-S., Coxon, D.T., 1987. Autoxidation of Unsaturated Lipids. Academic Press, London.

Chan, K.-Y., Wasserman, B.P., 1993a. Rapid solid-phase determination of cereal protein using bicinchoninic acid. Cereal Chem. 70, 27–28.

Chan, K., Wasserman, B.P., 1993b. Direct colorimetric assay of free thiol groups and disulfide bonds in suspensions of solubilized and particulate cereal proteins. Cereal Chem. 70, 22–26.

ChemistryBlog, 2012. Epoxide Infrared Spectra. http://chemistrytextbookcrawl.blogspot.com/2012/12/epoxide-infrared-spectra.html (accessed 10.07.15.).

Chevallet, M., Luche, S., Rabilloud, T., 2006. Silver staining of proteins in polyacrylamide gels. Nat. Protoc. 1, 1852–1858.

Chial, H.J., Thompson, H.B., Splittgerber, A.G., 1993. A spectral study of the charge forms of Coomassie blue G. Anal. Biochem. 209, 258–266.

Chio, K.S., Tappel, A.L., 1969. Synthesis and characterization of the fluorescent products derived from malondialdehyde and amino acids. Biochemistry 8, 2821–2827.

Chong, N., 2015. Principles of Spectroscopy. Infrared Spectral Ranges. Slide 39. http://mtweb.mtsu.edu/nchong/Spectroscopy-CHEM6230.pdf (accessed 15.07.15.).

Christie, W.W., 1976. Lipid Analysis. Pergamon Press, New York.

Chutipongtanate, S., Watcharatanyatip, K., Homvises, T., Jaturongkakul, K., Thongboonkerd, V., 2012. Systematic comparisons of various spectrophotometric and colorimetric methods to measure concentrations of protein, peptide and amino acid: detectable limits, linear dynamic ranges, interferences, practicality and unit costs. Talanta 98, 123–129.

Civille, G.V., Dus, C.A., 1992. Sensory evaluation of lipid oxidation in foods. In: Sensory Evaluation of Lipid Oxidation in Foods, vol. 500. American Chemical Society, Washington, D.C., pp. 279–289.

Clark, D.L., 2001. Peroxides and peroxide forming compounds. Chem. Health Saf. September/October, 12–22.

Clayden, J., Greeves, N., Warren, S., 2006. Organic Chemistry. Oxford University Press, Oxford, UK.

Connor, H.D., Gao, W., Mason, R.P., Thurman, R.G., 1994. New reactive oxidizing species causes formation of carbon-centered radical adducts in organic extracts of blood following liver transplantation. Free Radic. Biol. Med. 16, 871–875.

Corongiu, F.P., Milia, A., 1983. An improved and simple method for determining diene conjugation in atuoxidized poyunsaturated fatty acids. Chem.-Biol. Interact. 44.

Correddu, F., Nudda, A., Manca, M.G., Pulina, G., Dalsgaard, T.K., 2015. Light-induced lipid oxidation in sheep milk: effects of dietary grape seed and linseed, alone or in combination, on milk oxidative stability. J. Agric. Food Chem. 63, 3980–3986.

Crowe, T., White, P., 2001. Adaptation of the AOCS official method for measuring hydroperoxides from small-scale oil samples. J. Am. Oil Chem. Soc. 78, 1267–1269.

Cunningham, M.L., Chang, S.Y., Sipes, I.G., 1985. Covalent adduct formation and chloroform production after free radical attack on fatty acids by carbon tetrachloride reactive intermediates. Toxicology 37, 297–305.

Cyberlipid, 2014. Lipid Extraction: Solvents. http://www.cyberlipid.org/cyberlip/home0001.htm. (accessed 28.04.15.).

Dalle-Donne, I., Rossi, R., Giustarini, D., Milzani, A., Colombo, R., 2003. Protein carbonyl groups as biomarkers of oxidative stress. Clin. Chim. Acta 329, 23–38.

de M. Ochs, S., Fasciotti, M., Barreto, R.P., de Figueiredo, N.G., Albuquerque, F.C., Massa, M.C., Gabardo, I., Pereira Netto, A.D., 2010. Optimization and comparison of HPLC and RRLC conditions for the analysis of carbonyl-DNPH derivatives. Talanta 81, 521–529.

de Moreno, M.R., Smith, J.F., Smith, R.V., 1986. Mechanisms studies of coomassie blue and silver staining of proteins. J. Pharm. Sci. 75, 907–911.

DeLloyd, D., 2006. Solvents drying and drying agents. Delloyd's Lab Tech Resources, Reagents, and Solutions, St. Augustine, West Indies. http://www.cyberlipid.org/cyberlip/home0001.html (accessed 04.05.15.).

DeLong, M., Prange, R.K., Hodges, D.M., Forney, C.F., Bishop, M.C., Quilliam, M., 2002. Using a modified ferrous oxidation-xylenol orange (FOX) assay for detection of lipid hydroperoxides in plant tissue. J. Agric. Food Chem. 50, 248–254.

Dillard, C.J., Tappel, A.L., 1971. Fluorescent products of lipid peroxidation of mitochondria and microsomes. Lipids 6, 715–721.

Dillard, C.J., Tappel, A.L., 1973. Fluorescent products from reaction of peroxidizing polyunsaturated fatty acids with phosphatidyl ethanolamine and phenylalanine. Lipids 8, 183–189.

Dillard, C.J., Tappel, A.L., 1984. Fluorescent damage products of lipid peroxidation. Methods Enzymol. 105, 337–341.

Dobarganes, M.C., Velasco, J., 2002. Analysis of lipid hydroperoxides. Eur. J. Lipid Sci. Technol. 104, 420–428.

Dodds, E., McCoy, M., Geldenhuys, A., Rea, L., Kennish, J., 2004. Microscale recovery of total lipids from fish tissue by accelerated solvent extraction. J. Am. Oil Chem. Soc. 81, 835–840.

Dong, Y., 2011. Protein Modifications in Baked versus Fried Tortilla Chips. M.S. Food Science, Rutgers University, New Brunswick, NJ.

Dorsey, T.E., McDonald, P.W., Roels, O.A., 1977. A heated Biuret-Folin protein assay which gives equal absorbance with different proteins. Anal. Biochem. 78.

Dufour, E., 2009. Principles of infrared spectroscopy. In: Sun, D.-W. (Ed.), Principles of Infrared Spectroscopy. Academic Press, Burlington, MA, pp. 3–28.

Dunn, M.J., 1996. Detection of proteins in polyacrylamide gels by silver staining. In: Walker, J.M. (Ed.), Detection of Proteins in Polyacrylamide Gels by Silver Staining. Humana Press, Totowa, NJ, pp. 229–234.

Dupard-Julien, C., Kandlakunta, B., Uppu, R., 2007. Determination of epoxides by high-performance liquid chromatography following derivatization with N,N-diethyldithiocarbamate. Anal. Bioanalyt. Chem. 387, 1027–1032.

Durbetaki, A.J., 1956. Direct titration of oxirane oxygen with hydrogen bromide in acetic acid. Anal. Chem. 28, 2000–2001.

Dyballa, N., Metzger, S., 2012. Fast and sensitive coomassie staining in quantitative proteomics. In: Marcus, K. (Ed.), Fast and Sensitive Coomassie Staining in Quantitative Proteomics. Springer Science+Business Media, New York, pp. 47–59.

Ellman, G.L., 1958. A colorimetric method for determining low concentrations of mercaptans. Arch. Biochem. Biophys. 74, 443–450.

Ellman, G.L., 1959. Tissue sulfhydryl groups. Arch. Biochem. Biophys. 82, 70–77.

Endo, Y., Li, C.-M., Tagiri-Endo, M., Fujimoto, K., 2001. A modified method for the estimation of total carbonyl compounds in heated and frying oils using 2-propanol as a solvent. J. Am. Oil Chem. Soc. 78, 1021–1024.

Endo, Y., Tominaga, M., Tagiri-Endo, M., Kumozaki, K., Kouzi, H., Shiramasa, H., Miyakoshi, K., 2003. A modified method to estimate total carbonyl compounds in frying oils using butanol as a solvent. J. Oleo Sci. 52, 353–358.

EPA, California., 2006. Standard operating procedures for determination of aldehyde and ketone compounds in automotive source samples by high performance liquid chromatography. SOP MLD 104. http://www.arb.ca.gov/testmeth/slb/sop104v3.pdf (accessed 10.07.15.).

Estévez, M., 2011. Protein carbonyls in meat systems: a review. Meat Sci. 89, 259–279.

Estévez, M., Morcuende, D., Ventanas, S., Cava, R., 2003. Analysis of volatiles in meat from Iberian pigs and lean pigs after refrigeration and cooking by using SPME-GC-MS. J. Agric. Food Chem. 51, 3429–3435.

Estévez, M., Ollilainen, V., Heinonen, M., 2009. Analysis of protein oxidation markers α-aminoadipic and γ-glutamic semialdehydes in food proteins using liquid chromatography (LC)–Electrospray ionization (ESI)–multistage tandem mass spectrometry (MS). J. Agric. Food Chem. 57, 3901–3910.

Evans, D.R.H., Romero, J.K., Westoby, M., 2009. Concentration of proteins and removal of solutes. Methods Enzymol. 463, 97–120.

Evans, P., Lyras, L., Halliwell, B., 1999. Measurement of protein carbonyls in human brain tissue. Methods Enzymol. 300, 145–156.

Fagan, J.M., Sleczka, B.G., Sohar, I., 1999. Quantitation of oxidative damage to tissue proteins. Int. J. Biochem. Cell Biol. 31, 751–757.

Fajardo, A.R., Cerdán, L.E., Medina, A.R., Fernández, F.G.A., Moreno, P.A.G., Grima, E.M., 2007. Lipid extraction from the microalga *Phaeodactylum tricornutum*. Eur. J. Lipid Sci. Technol. 109, 120–126.

Fenaille, F., Mottier, P., Turesky, R.J., Ali, S., Guy, P.A., 2001. Comparison of analytical techniques to quantify malondialdehyde in milk powders. J. Chromatogr. A 921, 237–245.

Fernandez-Ibanez, V., Fearn, T., Soldado, A., de la Roza-Delgado, B., 2010. Development and validation of near infrared microscopy spectral libraries of ingredients in animal feed as a first step to adopting traceability and authenticity as guarantors of food safety. Food Chem. 121, 871–877.

Fernández, J., Pérez-Álvarez, J.A., Fernández-López, J.A., 1997. Thiobarbituric acid test for monitoring lipid oxidation in meat. Food Chem. 59, 345–353.

Fernandez, M., Ganan, M., Guerra, C., Hierro, E., 2014. Protein oxidation in processed cheese slices treated with pulsed light technology. Food Chem. 159, 388–390.

Ferrao, M.F., Godoy, S.C., Gerbase, A.E., Mello, C., Furtado, J.C., Petzhold, C.L., Poppi, R.J., 2007. Non-destructive method for determination of hydroxyl value of soybean polyol by LS-SVM using HATR/FT-IR. Anal. Chim. Acta 595, 114–119.

Fishwick, M.J., Swoboda, P.A.T., 1977. Measurement of oxidation of polyunsaturated fatty acids by spectrophotometric assay of conjugated derivatives. J. Sci. Food Agric. 28, 387–393.

Fletcher, B.L., Dillard, C.J., Tappel, A.L., 1973. Measurement of fluorescent lipid peroxidation products in biological systems and tissues. Anal. Biochem. 52, 1–9.

Fletcher, B.L., Tappel, A.L., 1970. Fluorescent modification to serum albumin by lipid peroxidation. Lipids 6, 172–175.

Floratech, 2015. Oxidative Stability Determination. Floratech Technical Bulletin MKT 11. http://www.floratech.com/fileMgr/upload/files/ARTICLES/REPRINTS/M013.pdf (accessed 05.05.15.).

Folch, J., Ascoli, I., Lees, M., Meath, J.A., LeBaron, F.N., 1951. Preparation of lipide extracts from brain tissue. J. Biol. Chem. 191, 833–841.

Folch, J., Lees, M., Stanley, G.H.S., 1957. A simple method for the isolation and purification of total lipids from animal tissues. J. Biol. Chem. 226, 497–509.

Forina, M., Casale, M., Oliveri, P., 2009. Application of chemometrics to food chemistry. In: Brown, S.D., Tauler, R., Walczak, B. (Eds.), Comprehensive Chemometrics. Elsevier, Oxford, pp. 75–128.

Frankel, E.N., 1982. Volatile lipid oxidation products. Prog. Lipid Res. 22, 1–33.

Frankel, E.N., 1984. Lipid oxidation: mechanisms, products, and biological significance. J. Am. Oil Chem. Soc. 61, 1908–1917.

Frankel, E.N., 1987. Secondary products of lipid oxidation. Chem. Phys. Lipids 44, 73–85.

Frankel, E.N., 1991. Recent advances in lipid oxidation. J. Sci. Food Agric. 54, 495–511.

Frankel, E.N., 2005a. 2-Thiobarbituric acid (TBA) value. In: Lipid Oxidation, second ed. The Oily Press, Bridgewater, UK, pp. 108–110.

Frankel, E.N., 2005b. Lipid Oxidation, second ed. The Oily Press, Bridgewater, UK.

Frankel, E.N., Neff, W.E., Bessler, T.R., 1979. Analysis of autoxidized fats by gas chromatography-mass spectrometry: V. Photosensitized oxidation. Lipids 14, 961–967.

Frankel, E.N., Neff, W.E., Rodwedder, W.K., Khambay, B.P.S., Garwood, R.F., Weedon, B.C.L., 1977a. Analysis of autoxidized fats by gas chromatography-mass spectrometry: III. Methyl linolenate. Lipids 12, 1055–1061.

Frankel, E.N., Neff, W.E., Rohwedder, W.K., Khambay, B.P.S., Garwood, R.F., Weedon, B.C.L., 1977b. Analysis of autoxidized fats by gas chromatography-mass spectrometry: I. Methyl oleate. Lipids 12, 901–907.

Frankel, E.N., Neff, W.E., Selke, E., 1984. Analysis of autoxidized fats by gas chromatography-mass spectrometry. IX. Homolytic vs heterolytic cleavage of primary and secondary oxidation products. Lipids 19, 790–800.

Frankel, E.N., Neff, W.E., Selke, E., Brooks, D.D., 1988. Analysis of autoxidized fats by gas chromatography-mass spectrometry: X. Volatile thermal decomposition products of methyl linolenate dimers. Lipids 23, 295–298.

Frankel, E.N., Neff, W.E., Selke, E., Weisleder, D., 1982. Photosensitized oxidation of methyl linoleate: secondary and volatile thermal decomposition products. Lipids 17, 11–18.

Fukuzawa, K., Fujisaki, A., Akai, K., Tokumura, A., Terao, J., Gebicki, J.M., 2006. Measurement of phosphatidylcholine hydroperoxides in solution and in intact membranes by the ferric–xylenol orange assay. Anal. Biochem. 359, 18–25.

Gally, J.A., Edelman, G.M., 1962. The effect of temperature on the fluorescence of some aromatic amino acids and proteins. Biochim. Biophys. Acta 60, 499–509.

Gardner, H.W., Plattner, R.D., Weisleder, D., 1985. The epoxyallylic radical from homolysis and rearrangement of methyl linoleate hydroperoxide combines with the thiyl radical of *N*-acetylcysteine. Biochim. Biophys. Acta 834, 65–74.

Garfin, D.E., 1990. One dimensional gel electrophoresis. Methods Enzymol. 182, 425–441.

Gates, R.E., 1991. Elimination of interfering substances in the presence of detergent in the bicinchoninic acid protein assay. Anal. Biochem. 196, 290–295.

Gay, C., Collins, J., Gebicki, J.M., 1999a. Determination of iron in solutions with the ferric-xylenol orange complex. Anal. Biochem. 273, 143–148.

Gay, C., Collins, J., Gebicki, J.M., 1999b. Hydroperoxide assay with the ferric-xylenol orange complex. Anal. Biochem. 273, 149–155.

Geladi, P., 2003. Chemometrics in spectroscopy. Part 1. Classical chemometrics. Spectrochim. Acta, Part B 58, 767–782.

Giroux, H.J., Acteau, G., Sabik, H., Britten, M., 2008. Influence of dissolved gases and heat treatments on the oxidative degradation of polyunsaturated fatty acids enriched dairy beverage. J. Agric. Food Chem. 56, 5710–5716.

Glassford, S.E., Byrne, B., Kazarian, S.G., 2013. Recent applications of ATR FTIR spectroscopy and imaging to proteins. Biochim. Biophys. Acta 1934, 2849–2858.

de Azevedo Gomes, H., da Silva, E.N., do Nascimento, M.R.L., Fukuma, H.T., 2003. Evaluation of the 2-thiobarbituric acid method for the measurement of lipid oxidation in mechanically deboned gamma irradiated chicken meat. Food Chem. 80, 433–437.

Gramza-Michalowska, A., Korczak, J., Regula, J., 2007. Use of plant extracts in summer and winter season butter oxidative stability improvement. Asia Pac. J. Clin. Nutr. 16 (Suppl. 1), 85–88.

Grau, A., Codony, R., Rafecas, M., Barroeta, A.C., Guardiola, F., 2000. Lipid hydroperoxide determination in dark chicken meat through a ferrous oxidation-xylenol orange method. J. Agric. Food Chem. 48, 4136–4143.

Grimm, C., Lloyd, S., Miller, J., Spanier, A.M., 2002. The analysis of food volatiles using direct thermal desorption. In: Marsili, R. (Ed.), The Analysis of Food Volatiles Using Direct Thermal Desorption. Marcel Dekker, New York, pp. 55–75.

Grintzalis, K., Georgiou, C.D., Schneider, Y.-J., 2015. An accurate and sensitive coomassie brilliant blue G-250-based assay for protein determination. Anal. Biochem. 480, 28–30.

Grintzalis, K., Papapostolou, I., Georgiou, C., 2009. Protocol for the quantification of protein ng quantities by a coomassie brilliant blue G-250-based hydrophobic assay. Nat. Protoc. Exch. http://www.sciencedirect.com/science/article/pii/S0003269715001323 (accessed 15.07.15.).

Gromova, I., Celis, J.E., 2006. Protein detection in gels by silver staining: a procedure compatible with mass spectrometry. In: Celis, J.E. (Ed.), Cell Biology: A Laboratory Handbook, vol. 4. Elsevier, Amsterdam, pp. 219–223.

Grosch, W., 1987. Reactions of hydroperoxides – products of low molecular weight. In: Chan, H.W.-S. (Ed.), Autoxidation of Unsaturated Lipids. Academic Press, London, pp. 95–139.

Grosch, W., Megele, J., 1984. Preparation, isomerization, and breakdown of hydroperoxides from methyl oleate. In: Bors, W., Saran, M., Tait, D. (Eds.), Oxygen Radicals in Chemistry and Biology. Walter de Gruyter & Co., Berlin, pp. 249–256.

Gruszka, J., Pawlak, A., Kruk, J., 2008. Tocochromanols, plastoquinol, and other biological prenyl-lipids as singlet oxygen quenchers – determination of singlet oxygen quenching rate constants and oxidation products. Free Radic. Biol. Med. 45, 920–928.

Guillén-Sans, R., Guzmán-Chozas, M., 1998. The thiobarbituric acid (TBA) reaction in foods: a review. Crit. Rev. Food Sci. Nutr. 38, 315–330.

Guillen, M.D., Cabo, N., 1999. Usefulness of the frequency data of the Fourier transform infrared spectra to evaluate the degree of oxidation of edible oils. J. Agric. Food Chem. 47, 709–719.

Guillen, M.D., Cabo, N., 2002. Fourier transform infrared spectra data versus peroxide and anisidine values to determine oxidative stability of edible oils. Food Chem. 77, 503–510.

Guillén, M.D., Cabo, N., 2000. Some of the most significant changes in the Fourier transform infrared spectra of edible oils under oxidative conditions. J. Sci. Food Agric. 80, 2028–2036.

Gunstone, F.D., Padley, F.B., 1997. Lipid Technologies and Applications. Marcel Dekker, New York.

Hara, A., Radin, N.S., 1978. Lipid extraction of tissues with a low-toxicity solvent. Anal. Biochem. 90, 420–426.

Hara, S., Totani, Y., 1988. A highly sensitive method for the micro-determination of lipid hydroperoxides by potentiometry. J. Am. Oil Chem. Soc. 65, 1948–1950.

Harrison, G., Haffey, P., Golub, E.E., 2008. A nanogram-level colloidal gold single reagent quantitative protein assay. Anal. Biochem. 380, 1–4.

Hasenhuettl, G., Wan, P., 1992. Temperature effects on the determination of oxidative stability with the metrohm rancimat. J. Am. Oil Chem. Soc. 69, 525–527.

Hayati, I.N., Man, Y.B.C., Tan, C.P., Aini, I.N., 2005. Monitoring peroxide value in oxidized emulsions by Fourier transform infrared spectroscopy. Eur. J. Lipid Sci. Technol. 107, 886–895.

Haynes, R.K., Vonwiller, S.C., 1990. Iron(III) and copper(II) catalyzed transformation of fatty acid hydroperoxides: efficient generation of peroxy radicals with Cu(II) trifluoromethane sulfonate. J. Chem. Soc. Chem. Commun. 1102–1104.

Held, P., 2003. Peptide and amino acid quantification using UV fluorescence in synergy HT multi-mode microplate reader. BioTek. http://www.biotek.com/resources/articles/peptides-amino-acids-fluorescence.html (accessed 25.05.15.).

Henick, A.S., Benca, M.F., Mitchell, J.H.J., 1954. Estimating carbonyl compounds in rancid fats and foods. J. Am. Oil Chem. Soc. 31, 88–91.

Hermes-Lima, M., Willmore, W.G., Storey, K.B., 1995. Quantification of lipid peroxidation in tissue extracts based on Fe(III)xylenol orange complex formation. Free Radic. Biol. Med. 19, 271–280.

Herraro, M., Castro-Puyana, M., Mendiola, J.A., Ibanez, E., 2013. Compressed fluids for the extraction of bioactive compounds. Trends Anal. Chem. 43, 67–83.

Hidalgo, F.J., Alaiz, M., Zamora, R., 1999. Effect of pH and temperature on comparative nonenzymatic browning of proteins produced by oxidized lipids and carbohydrates. J. Agric. Food Chem. 47, 742–747.

Hidalgo, F.J., Gallardo, E., Zamora, R., 2005. Strecker type degradation of phenylalanine by 4-hydroxy-2-nonenal in model systems. J. Agric. Food Chem. 53, 10254–10259.

Hidalgo, F.J., Zamora, R., 1993. Fluorescent pyrrole products from carbonyl-amine reactions. J. Biol. Chem. 268, 16190–16197.

Holman, R., Nickell, C., Privett, O., Edmondson, P., 1958. Detection and measurement of hydroperoxides by near infrared spectrophotometry. J. Am. Oil Chem. Soc. 35, 422–425.

Hornero-Méndez, D., Pérez-Gálvez, A., Mínguez-Mosquera, M., 2001. A rapid spectrophotometric method for the determination of peroxide value in food lipids with high carotenoid content. J. Am. Oil Chem. Soc. 78, 1151–1155.

Hoyland, D.V., Taylor, A.J., 1991. A review of the methodology of the 2-thiobarbituric acid test. Food Chem. 40, 271–291.

Huang, H., Yu, H., Xu, H., Ying, Y., 2008. Near infrared spectroscopy for on/in-line monitoring of quality in foods and beverages: a review. J. Food Eng. 87, 303–313.

Iglesias, J., Pazos, M., Lois, S., Medina, I., 2010. Contribution of galloylation and polymerization to the antioxidant activity of polyphenols in fish lipid systems. J. Agric. Food Chem. 58, 7423–7431.

Innawong, B., Mallikarjunan, P., Irudayaraj, J., Marcy, J.E., 2004. The determination of frying oil quality using Fourier transform infrared attenuated total reflectance. LWT - Food Sci. Technol. 37, 23–28.

International Dairy Federation, 1991. IDF Standard 74A. Anhydrous milkfat: determination of the peroxide value. In: International IDF Standards. International Dairy Federation, Brussels, Belgium.

Itakura, K., Uchida, K., 2001. Evidence that malondialdehyde-derived aminoenimine is not a fluorescent age pigment. Chem. Res. Toxicol. 14, 473–475.

IUPAC, 1987. Determination of the *p*-anisidine value, method 2.504. In: IUPAC Standard Methods for the Analysis of Oils, Fats and Derivatives, seventh ed. Alden Press, Oxford, pp. 210–211.

Iverson, S., Lang, S.C., Cooper, M., 2001. Comparison of the Bligh and Dyer and Folch methods for total lipid determination in a broad range of marine tissue. Lipids 36, 1283–1287.

Jeong, M., Yeo, J., Jang, E., Lee, J., 2013. Monitoring of radical scavenging compounds from oxidized lipids (RSOL) by 2,4-dinitrophenylhydrazine (DNPH) derivatization in thermally oxidized linoleic acid and lard systems. Food Sci. Biotechnol. 22, 161–165.

Jiang, Z., Woollard, A., Wolff, S., 1991. Lipid hydroperoxide measurement by oxidation of Fe^{2+} in the presence of xylenol orange. comparison with the TBA assay and an iodometric method. Lipids 26, 853–856.

Jochman, M.A., Laaks, J., Schmidt, T.C., 2014. Solvent-free extraction and injection techniques. In: Dettmer-Wilde, Engewald, W. (Eds.), Practical Gas Chromatography. A Comprehensive Reference. Springer, Heidelberg, pp. 371–412.

Jolley, J.E., Hildebrand, J.H., 1958. Solubility, entropy and partial molal volumes in solutions of gases in non-polar solvents. J. Am. Chem. Soc. 80, 1050–1054.

Jung, D.-M., Ebeler, S.E., 2003. Headspace solid-phase microextraction method for the study of the volatility of selected flavor compounds. J. Agric. Food Chem. 51, 200–205.

Kakuda, Y., Stanley, D.W., Van De Voort, F.R., 1981. Determination of TBA number by high performance liquid chromatography. J. Am. Oil Chem. Soc. 58, 773–775.

Kataoka, H., Lord, H.L., Pawliszyn, J., 2000. Applications of solid-phase microextraction in food analysis. J. Chromatogr. A 880, 35–62.

Katrahalli, U., Kalanur, S.S., Seetharamappa, J., 2010. Interaction of bioactive coomassie brilliant blue G with protein: insights from spectroscopic methods. Sci. Pharm. 78, 869–880.

Kavran, J.M., Leahy, D.J., 2014. Silver staining of SDS-polyacrylamide gel. Methods Enzymol. 541, 169–176.

Kawai, S., Fuchiwaki, T., Higashi, T., Tomita, M., 1990. High-performance liquid chromatographic determination of malonaldehyde using p-nitrophenylhydrazine as a derivatizing reagent. J. Chromatogr. A 514, 29–35.

Kessler, R.J., Fanestil, D.D., 1986. Interference by lipids in the determination of protein using bicinchoninic acid. Anal. Biochem. 159, 138–142.

Khatoon, F., Moinuddin, Alam, K., Ali, A., 2012. Physicochemical and immunological studies on 4-hydroxynonenal modified HSA: implications of protein damage by lipid peroxidation products in the etiopathogenesis of SLE. Hum. Immunol. 73, 1132–1139.

Khor, H.T., Chan, S.L., 1985. Comparative studies of three solvent mixtures for the extraction of soybean lipids. J. Am. Oil Chem. Soc. 62, 98–99.

Kikugawa, K., Beppu, M., 1987. Involvement of lipid oxidation products in the formation of fluorescent and cross-linked proteins. Chem. Phys. Lipids 44, 277–296.

Kikugawa, K., Ido, Y., 1984. Studies on peroxidized lipids. V. Formation and characterization of 1,4-dihydropyridine-3,5-dicarbaldehydes as model of fluorescent components in lipofuschin. Lipids 19, 600–608.

Kim, Y.H., Huff-Lonergan, E., Sebranek, J.G., Lonergan, S.M., 2010. High-oxygen modified atmosphere packaging system induces lipid and myoglobin oxidation and protein polymerization. Meat Sci. 85, 759–767.

Kislik, V.S., 2012. Conventional (classical) principles and practices of solvent extraction. In: Solvent Extraction: Classical and Novel Approaches. Elsevier, Oxford, UK, pp. 3–68.

Klaypradit, W., Kerdpiboon, S., Singh, R., 2011. Application of artificial neural networks to predict the oxidation of menhaden fish oil obtained from Fourier transform infrared spectroscopy method. Food Bioprocess Technol. 4, 475–480.

Klein, R.A., 1970. The detection of oxidation in liposome preparations. Biochim. Biophys. Acta 210, 486–489.

Klensporf, D., Jeleń, H.H., 2008. Influence of the addition of raspberry seed extract on changes in the volatile pattern of stored model breakfast cereal. J. Agric. Food Chem. 56, 3268–3272.

Kolthoff, I.M., Medalia, A.I., 1949. The reaction between ferrous iron and peroxides. III. Reaction with cumene hydroperoxide, in aqueous solution. J. Am. Chem. Soc. 71, 3789–3792.

Kolthoff, I.M., Medalia, A.I., 1951. Determination of organic peroxides by reaction with ferrous iron. Anal. Chem. 23, 595–603.

Kosugi, H., Kato, T., Kikugawa, K., 1987. Formation of yellow, orange, and red pigments in the reaction of alk-2-enals with 2-thiobarbituric acid. Anal. Biochem. 165, 456.

Kosugi, H., Kato, T., Kikugawa, K., 1988. Formation of red pigment by a two-step 2-thiobarbituric acid reaction of alka-2,4-dienals. Potential products of lipid oxidation. Lipids 23, 1024–1031.

Kretschmer, C.B., Nowakowsici, J., Wiebe, R., 1946. Solubility of oxygen and nitrogen in organic solvents from -25° to 50°C. Ind. Eng. Chem. 39, 506–509.

Laemmli, U.K., 1970. Cleavage of structural proteins during the assembly of the head of bacteriophage T4. Nature 227, 680–685.

Lakowicz, J.R., 1983. Protein fluorescence. In: Principles of Fluorescence Spectroscopy. Plenum Press, New York, pp. 342–385.

Lam, H., Proctor, A., Nyalala, J., Morris, M., Smith, W., 2005. Fourier transform infrared spectroscopy evaluation of low density lipoprotein oxidation in the presence of quercetin, catechin, and α-tocopherol. Lipids 40, 569–574.

Lappin, G.R., Clark, L.C., 1951. Colorimetric method for determination of traces of carbonyl compounds. Anal. Chem. 23, 541–543.

Le Dreau, Y., Dupuy, N., Artaud, J., Olliver, D., Kister, J., 2009a. Infrared study of aging of edible oils by oxidative spectroscopic index and MLS-ALS chemometric method. Talanta 77, 1748–1756.

Le Dreau, Y., Dupuy, N., Gaydou, V., Joachim, J., Kister, J., 2009b. Study of jojoba oil aging by FTIR. Anal. Chim. Acta 642, 163–170.

Levine, R.L., Garland, D., Oliver, C.N., Amici, A., Climent, I., Lentz, A., Ahn, B., Shaltiel, S., Stadtman, E.R., 1990. Determination of carbonyl content in oxidatively modified proteins. Methods Enzymol. 186, 464–478.

Levine, R.L., Williams, J.A., Stadtman, E.P., Shacter, E., 1994. Carbonyl assays for determination of oxidatively modified proteins. Methods Enzymol. 233, 346–357.

Liang, K., Zhang, Q., Wei Cong, W., 2012. Enzyme-assisted aqueous extraction of lipid from microalgae. J. Agric. Food Chem. 60, 11771–11776.

Liang, P., Chemn, C., Zhao, S., Ge, F., Liu, D., Liu, B., Fan, Q., Han, B., Xiong, X., 2013. Application of Fourier transform infrared spectroscopy for the oxidation and peroxide value evaluation in virgin walnut oil. J. Spectrosc. Article ID 138728. http://www.oalib.com/paper/3078579# (accessed 25.05.15.).

Liao, C.-H., 2013. Comparison of Chemical Assays Used to Determine Epoxides in Oxidized Lipids. M.S. Food Science, Rutgers University, New Brunswick, NJ.

Lodge Jr., J.P., 1988. Determination of C1–C5 aldehydes in ambient air and source emissions as 2,4-dinitrophenylhydrazones by HPLC. In: Methods of Air Sampling and Analysis, third ed. CRC Press, Boca Raton, FL, pp. 293–295.

Logan, A., Nienaber, U., Pan, X., 2013. Lipid Oxidation: Challenges in Food Systems. AOCS Press, Urbana, IL.

Londero, D., Greco, P.L., 1996. Automated high-performance liquid chromatographic separation with spectrofluorometric detection of a malondialdehyde-thiobarbituric acid adduct in plasma. J. Chromatogr. A 729, 207–210.

Lonza, 2015. Section X. Protein Separation in Polyacrylamide Gels. http://www.genehk.com/news/doc/Lonza_SourceBook_Protein_Separation.pdf (accessed 15.07.15.).

Lorrain, B., Dangles, O., Genot, C., Dufour, C., 2010. Chemical modeling of heme-induced lipid oxidation in gastric conditions and inhibition by dietary polyphenols. J. Agric. Food Chem. 58, 676–683.

Lowry, O.H., Rosebrough, N.J., Farr, A.L., Randall, R.J., 1951. Protein measurement with the Folin phenol reagent. J. Biol. Chem. 193, 265–275.

Lundberg, W.O. (Ed.), 1961. Autoxidation and Antioxidants, vol. 1. Interscience, New York.

Luo, S., Wehr, N.B., 2009. Protein carbonylation: avoiding pitfalls in the 2,4-dinitrophenylhydrazine assay. Redox Rep. 14, 159–166.

Ma, C., Ji, J., Tan, C., Chen, D., Luo, F., Wang, Y., Chen, X., 2014. Headspace solid-phase microextraction coupled to gas chromatography for the analysis of aldehydes in edible oils. Talanta 120, 94–99.

Ma, K., van de Voort, F., Ismail, A., Sedman, J., 1998. Quantitative determination of hydroperoxides by Fourier transform infrared spectroscopy with a disposable infrared card. J. Am. Oil Chem. Soc. 75, 1095–1101.

Ma, K., van de Voort, F., Sedman, J., Ismail, A., 1997. Stoichiometric determination of hydroperoxides in fats and oils by Fourier transform infrared spectroscopy. J. Am. Oil Chem. Soc. 74, 897–906.

Malshet, M.G., Tappel, A.L., 1973. Fluorescent products of lipid peroxidation. I. Structural requirement for fluorescence in conjugated schiff bases. Lipids 8, 194–198.

Mani, V., 1999. Properties of commercial SPME coatings. In: Pawliszyn, J. (Ed.), Applications of Solid Phase Microextraction. The Royal Society of Chemistry, Cambridge, UK, pp. 57–72.

Manura, J.J., 1999. Selection and use of adsorbent resins for purge and trap thermal desorption applications. SIS Application Note #32. http://www.sisweb.com/referenc/applnote/app-32.htm (accesses 10.07.15.).

Manura, J.J., Hartman, T.G., 1992. Applications of a short-path thermal desorption GC accessory. Am. Lab. 24, 46–52.

Marcuse, R., Johansson, L., 1973. TBA (thiobarbituric acid) test for rancidity grading. II. TBA reactivity of different aldehyde classes. J. Am. Oil Chem. Soc. 50, 387–391.

Mariutti, L.R.B., Nogueira, G.C., Bragagnolo, N., 2009. Solid phase microextraction - gas chromatography for the evaluation of secondary lipid oxidation products in chicken patties during long-term storage. J. Braz. Chem. Soc. 20.

Massey, L.K., 2003. Permeability Properties of Plastics and Elastomers: A Guide to Packaging and Barrier Materials. William Andrew Publishing, Norwich, NY

Matejka, P., 2015. NIR Spectrometry. Vysoka Skola Chemicko-Technologicka v Praze. http://old.vscht.cz/anl/vibspec/NIR%20spectrometry.pdf (accessed 15.07.15.).

Matyash, V., Liebisch, G., Kurzchalia, T.V., Shevchenko, A., Schwudke, D., 2008. Lipid extraction by methyl-tert-butyl ether for high-throughput lipidomics. J. Lipid Res. 49, 1137–1146.

McCormick, J.M., 2006. The Grasshopper's Guide to Cleaning Glassware. Truman State University, Truman, Kirksville, MO. http://chemlab.truman.edu/Miscellaneous_files/Cleaning.htm (accessed 23.04.15.).

McNair, H.M., Miller, J.M., 2009. Basic Gas Chromatography, second ed. John Wiley & Sons, Hoboken, New Jersey.

Mehlenbacher, V.C., 1960. The Analysis of Fats and Oils. Garrard Press, Champaign, IL.

Mendes, R., Cardoso, C., Pestana, C., 2009. Measurement of malondialdehyde in fish: a comparison study between HPLC methods and the traditional spectrophotometric test. Food Chem. 112, 1038–1045.

Méndez, E., Sanhueza, J., Speisky, H., Valenzuela, A., 1997. Comparison of Rancimat evaluation modes to assess oxidative stability of fish oils. J. Am. Oil Chem. Soc. 74, 331–332.

Mendiola, J.A., Herrero, M., Cifuentes, A., Ibanez, E., 2007. Use of compressed fluids for sample preparation: food applications. J. Chromatogr. A 1152, 234–246.

Merril, C.R., 1990. Gel-staining techniques. Methods Enzymol. 182, 477–488.

Merril, C.R., Goldman, D., Van Keuren, M.L., 1983. Silver staining methods for polyacrylamide gel electrophoresis. Methods Enzymol. 96, 230–239.

Merril, C.R., Goldman, D., Van Keuren, M.L., 1984. Gel protein stains: silver stain. Methods Enzymol. 104, 441–447.

Mihaljevic, B., Katusin-Razem, B., Razem, D., 1996. The reevaluation of the ferric thiocyanate assay for the lipid hydroperoxides with special consideration of the mechanistic aspects of the response. Free Radic. Biol. Med. 21, 53–63.

Mikrolab, M.L., 2013. Oxipres and Oxidograph. http://www.mikrolab.dk/Files/filer/Diverse/MLOX2013.PPT (accessed 23.04.15.).

Min, D.B., Ellefson, W.C., 2010. Fat analysis. In: Nielsen, S.S. (Ed.), Food Analysis, fourth ed. Springer, New York, pp. 117–132.

Moreau, R., Powell, M., Singh, V., 2003. Pressurized liquid extraction of polar and nonpolar lipids in corn and oats with hexane, methylene chloride, isopropanol, and ethanol. J. Am. Oil Chem. Soc. 80, 1063–1067.

Morton, R.E., Evans, T.A., 1992. Modification of the bicinchoninic acid protein assay to eliminate lipid interference in determining lipoprotein protein content. Anal. Biochem. 204, 332–334.

Munck, L., Norgaard, L., Engelsen, S.B., Bro, R., Andersson, C.A., 1998. Chemometrics in food science – a demonstration of the feasibility of a highly exploratory, inductive evaluation strategy of fundamental scientific significance. Chemom. Intell. Lab. Syst. 44, 31–60.

Nadkarni, D.V., Sayre, L.M., 1995. Structural definition of early lysine and histidine adduction chemistry of 4-hydroxynonenal. Chem. Res. Toxicol. 8, 284–291.

Navas, J.A., Tres, A., Codony, R., Boatella, J., Bou, R., Guardiola, F., 2004. Modified ferrous oxidation-xylenol orange method to determine lipid hydroperoxides in fried snacks. Eur. J. Lipid Sci. Technol. 106, 688–696.

Neff, W.E., Frankel, E.N., Weisleder, D., 1982. Photosensitized oxidation of methyl linolenate. Secondary products. Lipids 17, 780–790.

Neuhoff, V., Arnold, N., Taube, D., Ehrhardt, W., 1988. Improved staining in polyacrylamide gels including isoelectric focusing gels with clear background at nanogram sensitivity using Coomassie Brilliant Blue G-250 and R-250. Electrophoresis 9, 255–262.

Neuhoff, V., Stamm, R., Hansjorg, E., 1985. Clear background and highly sensitive protein staining with Coomassie blue dyes in polyacrylamide gels: a systematic analysis. Electrophoresis 6, 427–488.

NIOSH, 2015. Chemical Listing and Documentation of Revised IDLH Values (as of 3/1/95): *p*-Anisidine. http://www.cdc.gov/niosh/idlh/104949.html (accessed 15.07.2015).

Noble, J.E., 2014. Quantification of protein concentration using UV absorbance and Coomassie dyes. Methods Enzymol. 536, 17–26.

Noble, J.E., Bailey, M.J.A., 2009. Quantitation of protein. Methods Enzymol. 463, 73–95.

Nogala-Kalucka, M., Korczak, J., Dratwia, M., Lampart-Szczapa, E., Siger, A., Buchowski, M., 2005. Changes in antioxidant activity and free radical scavenging potential of rosemary extract and tocopherols in isolated rapeseed oil triacylglycerols during accelerated tests. Food Chem. 93, 227–235.

Nourooz-Zadeh, J., Tajaddini-Sarmadi, J., Wolff, S.P., 1995. Measurement of hydroperoxides in edible oils using the ferrous oxidation in xylenol orange assay. J. Agric. Food Chem. 43, 17–21.

Oette, K., Peterson, M.L., McAuley, R.L., 1963. A highly sensitive method for measurement of lipid hydroperoxides by iodimetry and amperometric endpoint. J. Lipid Res. 4, 212–215.

Oishi, M., Onishi, K., Nishijima, M., Nakagomi, K., Nakazawa, H., Uchiyama, S., Suzuki, S., 1992. Rapid and simple coulometric measurements of peroxide value in edible oils and fats. J. Assoc. Off. Anal. Chem. 75, 507–510.

Ortiz-Somovilla, V., Espana-Espana, F., Gaitan-Jurado, A.J., Perez-Aparicio, J., De Pedro-Sanz, E.J., 2007. Proximate analysis of homogenized and minced mass of pork sausages by NIRS. Food Chem. 101, 1031–1040.

Osawa, C.C., Goncalves, L.A.G., Ragazzi, S., 2007. Determination of hydroperoxides in oils and fats using kits. J. Sci. Food Agric. 87, 1659–1666.

Page, B.D., Lacroix, G., 2000. Analysis of volatile contaminants in vegetable oils by headspace solid-phase microextraction with carboxen-based fibres. J. Chromatogr. A 873, 79–94.

Pardun, H., Kroll, E., 1972. Bestimmung der Oxydationsstabilität von Ölen und Fetten mit Hilfe einer automatischen Version des SWIFT-Testes. Fette-Seifen-Anstrichmittel 74, 366–375.

Parr, L.J., Swoboda, P.A.T., 1976. The assay of conjugable oxidation products applied to lipid deterioration in stored foods. Int. J. Food Sci. Technol. 11, 1–12.

Partridge, M.A.K., Jiang, Y., Skerritt, J.H., Schaich, K.M., 2003. Immunochemical and electrophoretic analysis of soluble modified proteins in extruded wheat flour. Cereal Chem. 80, 791–798.

Patel, K., Easty, D.J., Dunn, M.J., 1988. Detection of proteins in polyacrylamide gels using an ultrasensitive silver staining technique. In: Walker, J.M. (Ed.), Methods in Molecular Biology. In: New Protein Techniques, vol. 3. Humana Press, Clifton, NJ, pp. 159–168.

Pawliszyn, J., 2006. Quantitative aspects of SPME. In: Pawliszyn, J. (Ed.), Quantitative Aspects of SPME. The Royal Society of Chemistry, pp. 3–21.

Pazos, M., Iglesias, J., Maestre, R., Medina, I., 2010. Structure–activity relationships of polyphenols to prevent lipid oxidation in pelagic fish muscle. J. Agric. Food Chem. 58, 11067–11074.

Peng, T., 2010. Evaluation of Oxygen Bomb Methodology for Studying Frying Chemistry. M.S. Food Science, Rutgers University, New Brunswick, NJ.

Pierce, 2014. Thermo Scientific Pierce Protein Assay Selection Guide. https://tools.lifetechnologies.com/content/sfs/brochures/1602789-Protein-Assay-Selection-Poster.pdf (accessed 20.07.15.).

Pikul, J., Leszczynski, D.E., Kummerow, F.A., 1983. Elimination of sample autoxidation by butylated hydroxytoluene additions before thiobarbituric acid assay for malonaldehyde in fat from chicken meat. J. Agric. Food Chem. 31, 1338–1342.

Pikul, J., Leszczynski, D.E., Kummerow, F.A., 1989. Evaluation of three modified TBA methods for measuring lipid oxidation in chicken meat. J. Agric. Food Chem. 37, 1309–1313.

Pinto, R.C., Locquet, N., Eveleigh, L., Rutledge, D.N., 2010. Preliminary studies on the mid-infrared analysis of edible oils by direct heating on an ATR diamond crystal. Food Chem. 120, 1170–1177.

Pokorný, J., 2005. Volumetric analysis of oxidized lipids. In: Kamal-Eldin, A., Pokorný, J. (Eds.), Analysis of Lipid Oxidation, AOCS Press, Champaign, IL, pp. 8–16.

Pomeranz, Y., Meloan, C.E., 2000. Lipids. In: Food Analysis: Theory and Practice, third ed. Aspen Publishers, Gaithersburg, MD, pp. 678–732.

Princeton University, 2014. Aqua Regia. Princeton University Environmental Health and Safety, Princeton, NJ. http://web.princeton.edu/sites/ehs/labsafetymanual/cheminfo/aquaregia.htm (accessed 23.04.15.).

Pryor, W.A., Stanley, J.P., 1975. A suggested mechanism for the production of malonaldehyde during the autoxidation of polyunsaturated fatty acids. Nonenzymatic production of prostaglandin endoperoxides during autoxidation. J. Org. Chem. 40, 3615–3617.

Qin, X., 2011. Thermal Desorption Studies of Lipid Degradation at Elevated Temperatures. M.S. Food Science, Rutgers University, New Brunswick, NJ.

Rabilloud, T., 1990. Mechanisms of protein silver staining in polyacrylamide gels: a 10-year synthesis. Electrophoresis 11, 785–794.

Raharjo, S., Sofos, S., Schmidt, G.R., 1992. Improved speed, specificity, and limit of determination of an aqueous acid extraction thiobarbituric acid-C_{18} method for measuring lipid peroxidation in beef. J. Agric. Food Chem. 40, 2181–2185.

Ramluckan, K., Moodley, K.G., Bux, F., 2014. An evaluation of the efficacy of using selected solvents for the extraction of lipids from algal biomass by the soxhlet extraction method. Fuel 116, 103–108.

Recknagel, R.O., Glende Jr., E.A., 1984. Spectrophotometric detection of lipid conjugated dienes. Methods Enzymol. 105, 331–337.

Redemann, C.E., 1942. Peroxides in isopropanol. J. Am. Chem. Soc. 64, 3049–3050.

Redinbaugh, M.G., Turley, R.B., 1986. Adaptation of the bicinchoninic acid protein assay for use with microtiter plates and sucrose gradient fractions. Anal. Biochem. 153, 267–271.

Reichardt, C., 2003. Solvents and Solvent Effects in Organic Chemistry, third ed. Wiley-VCH Publishers, Weinheim.

Reis, A., Rudnitskaya, A., Blackburn, G.J., Fauzi, N.M., Pitt, A.R., Spickett, C.M., 2013. A comparison of five lipid extraction solvent systems for lipidomic studies of human LDL. J. Lipid Res. 54, 1812–1824.

Repko Reader, I., 2013. Thermal Desorption Studies of Volatiles Released during Heating of Food Oils. M.S. Food Science, Rutgers University, New Brunswick, NJ.

Reusch, W., 2013. Infrared Spectroscopy. https://www2.chemistry.msu.edu/faculty/reusch/VirtTxtJml/Spectrpy/InfraRed/infrared.htm (accessed 15.07.15.).

Richards, M.P., Feng, Y., 2000. Choosing the proper organic solvent for the determination of lipid hydroperoxides by the ferric thiocyanate assay. Anal. Biochem. 278, 232–235.

Robinson, C.E., Keshavarzian, A., Pasco, D.S., Frommel, T.O., Winship, D.H., Holmes, E.W., 1999. Determination of protein carbonyl groups by immunoblotting. Anal. Biochem. 266, 48–57.

Rockenbauer, A., Tudos, F., 1979. Solvent effects of ESR spectra of neutral nitroxide free radicals: I. Localized solvent-radical adducts in mixtures. Adv. Mol. Relax. Interact. Processes 15, 207–212.

Roggo, Y., Chalus, P., Maurer, L., Lema-Martinez, C., Edmond, A., Jent, N., 2007. A review of near infrared spectroscopy and chemometrics in pharmaceutical technologies. J. Pharm. Biomed. Anal. 44, 683–700.

Rollie, M.E., Patonay, G., Warner, I.M., 1987. Deoxygenation of solutions and its analytical applications. Ind. Eng. Chem. Res. 26, 1–6.

Rose, H.G., Oklander, M., 1965. Improved procedure for the extraction of lipids from human erythrocytes. J. Lipid Res. 6, 428–431.

Rosen, G.T., Britigan, B.E., Halpern, H.J., Pou, S., 1999. Free Radicals: Biology and Detection by Spin Trapping. Oxford University Press, New York.

Rossell, J.B., 1994. Measurement of rancidity. In: Allen, J.C., Hamilton, R.J. (Eds.), Rancidity in Foods, third ed. Blackie Academic & Professional, London, UK, pp. 22–53.

Roth, C., 2010. Technical Info: Coomassie Brilliant Blue and Mechanisms of Protein Staining and Bradford-Assay. https://www.carlroth.com/website/de-de/pdf/Coomassie_E.pdf (accessed 20.07.15.).

Rudyk, O., Eaton, P., 2014. Biochemical methods for monitoring protein thiol redox states in biological systems. Redox Biol. 2, 803–813.

Russin, T.A., van de Voort, F.R., Sedman, J., 2003. Novel method for rapid monitoring of lipid oxidation by FTIR spectroscopy using disposable IR cards. J. Am. Oil Chem. Soc. 80, 635–641.

Russin, T.A., van de Voort, F.R., Sedman, J., 2004. Rapid determination of oxidative stability of edible oils by FTIR spectroscopy using disposable IR cards. J. Am. Oil Chem. Soc. 81, 111–116.

Schäfer, K., 1998. Accelerated solvent extraction of lipids for determining the fatty acid composition of biological material. Anal. Chim. Acta 358, 69–77.

Schaich, K.M., 2002. Free radical generation during extrusion: a critical contributor to texturization. In: Ho, C.T. (Ed.), Free Radical Generation during Extrusion: A Critical Contributor to Texturization Free Radicals in Foods: Chemistry, Nutrition, and Health. American Chemical Society, Washington, DC, pp. 35–48.

Schaich, K.M., 2005. Lipid oxidation in fats and oils: theoretical aspects. In: Shahidi, F. (Ed.), Bailey's Industrial Fats and Oils, sixth ed. John Wiley, New York, pp. 2681–2767.

Schaich, K.M., 2008. Co-oxidations of oxidizing lipids: reactions with proteins. In: Kamal-Eldin, A., Min, D.B. (Eds.), Lipid Oxidation Pathways, vol. 2. CRC Press, Boca Raton, FL, pp. 183–274.

Schaich, K.M., 2012. Thinking outside the classical chain reaction box of lipid oxidation. Lipid Technol. 24, 55–58.

Schaich, K.M., 2013a. Challenges in elucidating lipid oxidation mechanisms: when, where, and how do products arise? In: Nienaber, U., Logan, A., Pan, X. (Eds.), Lipid Oxidation: Challenges in Food Systems AOCS Press, Champaign, IL, pp. 1–52.

Schaich, K.M., 2013b. Challenges in analyzing lipid oxidation: are one product and one sample concentration enough? In: Logan, A., Nienaber, U., Pan, X. (Eds.), Lipid Oxidation: Challenges in Food Systems. American Oil Chemists' Society, Champaign, IL, pp. 53–128.

Schaich, K.M., 2014. Lipid co-oxidation of proteins: one size does not fit all. Inform 25, 134–139.

Schaich, K.M., Fisher, C., King, R.T., 1993. Formation and reactivity of free radicals in curcuminoids: an EPR study. In: Ho, C.T., Huang, M.T., Rosen, R.T. (Eds.), Food Phytochemicals for Cancer Prevention II: Teas, Spices, and Herbs, ACS Symposium Series 547. American Chemical Society, Washington, DC, pp. 204–221.

Schmid, P., 1973. Extraction and purification of lipids. II. Why is chloroform-methanol such a good lipid solvent? Physiol. Chem. Phys. 5, 141–150.

Schmid, P., Hunter, E., 1971. Extraction and purification of lipids. I. Solubility of lipids in biologically important solvents. Physiol. Chem. Phys. 3, 98–102.

Schulte, E., 2002. Determination of higher carbonyl compounds in used frying fats by HPLC of DNPH derivatives. Anal. Bioanal. Chem. 372, 644–648.

Scott, G., 1965. Atmospheric Oxidation and Antioxidants. Elsevier, London.

Sedman, J., van de Voort, F.R., Ismail, A.A., 1999. Attenuated total reflectance spectroscopy: principles and applications in infrared analysis of food. In: Mossoba, M.M. (Ed.), Spectral Methods In Food Analysis. Marcel Dekker, New York, pp. 397–425.

Seppanen, C.M., Csallany, A.S., 2001. Simultaneous determination of lipophilic aldehydes by high-performance liquid chromatography in vegetable oil. J. Am. Oil Chem. Soc. 78, 1253–1260.

Setiowaty, G., Che Man, Y.B., 2003. A rapid Fourier transform infrared spectroscopic method for the determination of 2-TBARS in palm olein. Food Chem. 81, 147–154.

Shacter, E., 2000a. Protein oxidative damage. Methods Enzymol. 319, 428–436.

Shacter, E., 2000b. Quantification and significance of protein oxidation in biological samples. Drug Metab. Rev. 32, 307–326.

Shacter, E., William, J.A., Lim, M., Levine, R.L., 1994a. Differential susceptibility of plasma proteins to oxidative modification by western blot immunoassay. Free Radic. Biol. Med. 17, 429–437.

Shacter, E., Williams, J.A., Stadtman, E.R., Levine, R.L., 1994b. Determination of carbonyl groups in oxidized proteins. In: Punchard, N., Kellym, F. (Eds.), Free Radicals: A Practical Approach. Oxford University Press, Oxford, UK, pp. 159–170.

Shantha, N.C., Decker, E.A., 1994. Rapid, iron-based spectrophotometric methods for determination of peroxide values of food lipids. J. AOAC Int. 77, 421–424.

Shen, Y., Chen, S., Du, R., Xiao, Z., Huang, Y., Rasco, B., Lai, K., 2014. Rapid assessment of the quality of deep frying oils used by street vendors with Fourier transform infrared spectroscopy. J. Food Meas. Charac. 8, 336–342.

Shewry, P.R., Fido, R.J., 1998. Protein blotting: principles and applications. In: Rapley, R., Walker, J.M. (Eds.), Molecular Biomethods Handbook. Humana Press, Totowa, NJ, pp. 435–444.

Shibamoto, T., 2006. Analytical methods for trace levels of reactive carbonyl compounds formed in lipid peroxidation systems. J. Pharm. Biomed. Anal. 41, 12–25.

Shirey, R.E., 2012. SPME commercial devices and fibre coatings. In: Pawliszyn, J. (Ed.), Handbook of Solid Phase Microextraction. Elsevier, London, UK, pp. 99–133.

Shirono, K., Morimatsu, T., Takemura, F., 2008. Gas solubilities (CO_2, O_2, Ar, N_2, H_2, and He) in liquid chlorinated methanes. J. Chem. Eng. Data 53, 1867–1871.

Sierra, V., Aldai, N., Castro, P., Osoro, K., Coto-Montes, A., Olivan, M., 2008. Prediction of the fatty acid composition of beef by near infrared transmittance spectroscopy. Meat Sci. 78, 248–255.

Sigma-Aldrich, 2012. Solvent Stabilizer Systems. https://www.sigmaaldrich.com/chemistry/solvents/learning-center/stabilizer-systems.html (accessed 15.07.15.).

Singleton, J., Stikeleather, L., 1999. A solvent extractor system for the rapid extraction of lipids and trace bioactive micronutrients in oilseeds. J. Am. Oil Chem. Soc. 76, 1461–1466.

Sinnhuber, R., Yu, T.C., 1977. The 2-thiobarbituric acid reaction, an objective measure of oxidative deterioration occurring in fats and oils. J. Jpn. Soc. Fish. Sci. 26, 259–267.

Sinnhuber, R.O., Yu, T.C., Yu, T.C., 1958. Characterization of the red pigment formed in the 2-thiobarbitruic acid determination of oxidative rancidity. Food Res. 23, 626–634.

Sjövall, O., Kuksis, A., Kallio, H., 2002. Formation of triacylglycerol core aldehydes during rapid oxidation of corn and sunflower oils with tert-butyl hydroperoxide/Fe^{2+}. Lipids 37, 81–94.

Skoog, D.A., West, D.M., Holler, F.J., 1996. Fundamentals of Analytical Chemistry, seventh ed. Saunders College Publishing, Philadelphia, PA.

Smedes, F., 1999. Determination of total lipid using non-chlorinated solvents. Analyst 124, 1711–1718.

Smith, P.K., Krohn, R.I., Hermanson, G.T., Mallia, A.K., G, F.H., Provenzano, M.D., Fujimoto, E.K., Goeke, N.M., Olson, B.J., Klenk, D.C., 1985. Measurement of protein using bicinchoninic acid. Anal. Biochem. 150, 76–85.

Södergren, E., Nourooz-Zadeh, J., Berglund, L., Vessby, B., 1998. Re-evaluation of the ferrous oxidation in xylenol orange assay for the measurement of plasma lipid hydroperoxides. J. Biochem. Biophys. Methods 37, 137–146.

Sørensen, G., Jørgensen, S., 1996. A critical examination of some experimental variables in the 2-thiobarbituric acid (TBA) test for lipid oxidation in meat products. Z. Lebensm. Unters. Forch. 202, 205–210.

Srivastava, Y., Semwal, A., 2015. A study on monitoring of frying performance and oxidative stability of virgin coconut oil (VCO) during continuous/prolonged deep fat frying process using chemical and FTIR spectroscopy. J. Food Sci. Technol. 52, 984–991.

Steinberg, T.H., 2009. Protein gel staining methods: an introduction and overview. Methods Enzymol. 463, 541–562.

Steltzer, E., 2012. Evaluation of Chemical Assays for Determining Hydroperoxides Levels in Oxidized Lipids. M.S. Food Science, Rutgers University, New Brunswick, NJ.

Strange, D.E., Schaich, K.M., 2000. Extraction of lipids from extruded corn-soy blends. J. Food Lipids 7, 217–224.

Stuart, B.H., 2004. Infrared Spectroscopy-Fundamentals and Applications. Wiley Analytical Techniques in the Sciences (AnTs), Hoboken, NJ.

Subramanian, A., Rodriguez-Saona, L., 2009. Fourier transform infrared (FTIR) spectroscopy. In: Sun, D.-W. (Ed.), Infrared Spectroscopy for Food Quality Analysis and Control. John Wiley & Sons, Hoboken, NJ, pp. 145–178.

Supelco, 2003. Capillary GC Column Choices for Residual Solvent Analyses Using Direct Injection or Solid Phase Microextraction* (SPME). Bulletin 933. Bellefonte, PA. https://www.sigmaaldrich.com/content/dam/sigma-aldrich/docs/Supelco/Bulletin/11685.pdf (accessed 15.07.15.).

Swinnerton, J.W., Linnenbom, V.J., Cheek, C.H., 1962. Determination of dissolved Gases in aqueous solutions by gas chromatography. Anal. Chem. 34, 483–485.

Tal, M., Silberstein, A., Nusser, E., 1985. Why does Coomassie Brilliant Blue R interact differently with different proteins? J. Biol. Chem. 260, 9976–9980.

Tarladgis, B., Pearson, A.M., Dugan, L.R.J., 1962. The chemistry of the 2-thiobarbituric acid test for the determination of oxidative rancidity in foods. I. Some important side reactions. J. Am. Oil Chem. Soc. 39, 34–39.

Tarladgis, B., Watts, B.M., Younathan, M.T., Dugan, L.R.J., 1960. A distillation method for the quantitative determination of malonaldehyde in rancid foods. J. Am. Oil Chem. Soc. 37, 44–48.

Teale, F.W.J., Weber, G., 1957. Ultraviolet fluorescence of the aromatic amino acids. Biochem. J. 65, 476–482.

Tejada, S.B., 1986. Evaluation of silica gel cartridges coated *in situ* with acidified 2,4-dinitrophenylhydrazine for sampling aldehydes and ketones in air. Int. J. Environ. Anal. Chem. 26, 167–185.

Tena, N., Aparicio, R., Garcia-Gonzalez, D.L., 2013. Time course analysis of fractionated thermoxidized virgin olive oil by FTIR spectroscopy. J. Agric. Food Chem. 61, 3212–3218.

Tena, N., Aparico, R., Garcia-Gonzalez, D.L., 2009. Thermal deterioration of virgin olive oil monitored by STR-FTIR analysis of trans content. J. Agric. Food Chem. 57, 9997–10003.

Thannhauser, T.W., Knishi, Y., Scheraga, H.A., 1987. Analysis for disulfide bonds in peptides and proteins. Methods Enzymol. 141, 115–119.

Thannhauser, T.W., Konishi, Y., Scheraga, H.A., 1984. Sensitive quantitative analysis of disulfide bonds in polypeptides and proteins. Anal. Biochem. 138, 181–188.

ThermoScientific, 2010. Thermo Scientific Pierce Protein Assay Technical Handbook. https://tools.lifetechnologies.com/content/sfs/brochures/1602063-Protein-Assay-Handbook.pdf (accessed 29.04.15.).

ThermoScientific, 2011. Ellman's Reagent. https://tools.lifetechnologies.com/content/sfs/manuals/MAN0011216_Ellmans_Reag_UG.pdf (accessed 20.05.15.).

Tian, X., 2013. Chemical Factors Affecting Degradation Processes of Vegetable Oils during Frying. Ph.D. Food Science, Rutgers University, New Brunswick, NJ.

Tompkins, C., Perkins, E.G., 1999. The evaluation of frying oils with the *p*-anisidine value. J. Am. Oil Chem. Soc. 76, 945–947.

Toschi, T.G., Bendini, A., Ricci, A., Lercker, G., 2003. Pressurized solvent extraction of total lipids in poultry meat. Food Chem. 83, 551–555.

Towbin, H., Staehelin, T., Gordon, J., 1979. Electrophoretic transfer of proteins from polyacrylamide gels to nitrocellulose sheets: procedures and some applications. Proc. Natl. Acad. Sci. U.S.A. 76, 4350–4354.

Trojakova, L., Reblova, Z., Pokorny, J., 1999. Determination of the oxidation stability of fats and oils using the Oxipres apparatus. Czech. J. Food Sci. 17, 68–72.

Turk, E., 1998. Phosgene from chloroform. Chem. Eng. News 76, 6.

Uchiyama, S., Inaba, Y., Kunugita, N., 2011. Derivatization of carbonyl compounds with 2,4-dinitrophenylhydrazine and their subsequent determination by high-performance liquid chromatography. J. Chromatogr. B 879, 1282–1289.

Uehara, H., Rao, V.A., 2015. Metal-mediated protein oxidation: applications of a modified ELISA-based carbonyl detection assay for complex proteins. Pharm. Res. 32, 691–701.

Uhde, E., 2009. Application of solid sorbents for the sampling of volatile organic compounds in indoor air. In: Salthammer, T., Uhde, E. (Eds.), Organic Indoor Air Pollutants, second ed. Wiley-VCH Verlag GmbH & Co., Weinheim, pp. 3–18.

Undeland, I., Härröd, M., Lingnert, H., 1998. Comparison between methods using low-toxicity solvents for the extraction of lipids from herring (*Clupea harengus*). Food Chem. 61, 355–365.

University of Illinois, 2015. Aqua Regia. University of Illinois, Division of Research Safety, Urbana, IL. https://www.drs.illinois.edu/SafetyLibrary/AquaRegia (accessed 23.04.14.).

U. N. Environment. Programme, 1998. Methyl Tertiary-butyl Ether. Switzerland, Geneva. http://www.inchem.org/documents/ehc/ehc/ehc206.htm (accessed 23.04.15.).

Usuki, R., Shibata, Y., Tagiri-Endo, M., Endo, Y., 2009. Estimation of total carbonyl compounds in roasted and dried foods by the tentative method using 1-butanol as a solvent. J. Oleo Sci. 58, 23–25.

Utrera, M., Estévez, M., 2012. Analysis of tryptophan oxidation by fluorescence spectroscopy: effect of metal-catalyzed oxidation and selected phenolic compounds. Food Chem. 135, 88–93.

Utrera, M., Estévez, M., 2013. Oxidative damage to poultry, pork, and beef during frozen storage through the analysis of novel protein oxidation markers. J. Agric. Food Chem. 61, 7987–7993.

Utrera, M., Morcuende, D., Rodríguez-Carpena, J.-G., Estévez, M., 2011. Fluorescent HPLC for the detection of specific protein oxidation carbonyls – α-aminoadipic and γ-glutamic semialdehydes – in meat systems. Meat Sci. 89, 500–506.

Van Damme, F., Oomens, A.C., 1995. Determination of residual free epoxide in polyether polyols by derivatization with diethylammonium N,N-diethyldithiocarbamate and liquid chromatography. J. Chromatogr. A 696, 41–47.

Van de Voort, F.R., Ismail, A.A., Sedman, J., Dubois, J., Nicodermo, T., 1994. The determination of peroxide value by Fourier transform infrared spectroscopy. J. Am. Oil Chem. Soc. 71, 921–926.

van Leeuwen, S.M., Hendriksen, L., Karst, U., 2004. Determination of aldehydes and ketones using derivatization with 2,4-dinitrophenylhydrazine and liquid chromatography–atmospheric pressure photoionization-mass spectrometry. J. Chromatogr. A 1058, 107–112.

Vas, G., Vékey, K., 2004. Solid-phase microextraction: a powerful sample preparation tool prior to mass spectrometric analysis. J. Mass Spectrom. 39, 233–254.

Viljanen, K., Kivikari, R., Heinonen, M., 2004. Protein-lipid interactions during liposome oxidation with added anthocyanins and other phenolic compounds. J. Agric. Food Chem. 52, 1104–1111.

Vlachos, N., Skopelitis, Y., Psaroudaki, M., Konstantinidou, V., Chatzilazarou, A., Tegou, E., 2006. Applications of Fourier transform-infrared spectroscopy to edible oils. Anal. Chim. Acta 573, 459–465.

Wade Jr., L.G., 1991. Organic Chemistry, second ed. Prentice-Hall, Englewood Cliffs, NJ.

Walker, J.M., 2009. The Protein Protocols Handbook, third ed. Humana Press, New York.

Walmsley, T.A., Abernathy, M.H., Fitzgerald, H.P., 1987. Effect of daylight on the reaction of thiols with Ellman's reagent, 5,5'-dithiobis(2-nitrobenzoic acid). Clin. Chem. 33, 1928–1931.

Wampler, T., 2004. Analysis of food volatiles using headspace-gas chromatographic techniques. In: Marsili, R.T. (Ed.), Flavor, Fragrance, and Odor Analysis. Marcel Dekker, New York, pp. 25–54.

Wan Ibadullah, W.Z., 2013. Lipid-Protein Interactions in Peanut Butter. Ph.D. Food Science, Rutgers University, New Brunswick, NJ.

Wang, S., 2000. Protein-Starch Interactions during Wheat Flour Extrusion: Formation and Effects on Textural Properties of Extrudates. Ph.D. Food Science, Rutgers University, New Brunswick, NJ.

Wardencki, W., Curyło, J., Namieśnik, J., 2007. Trends in solventless sample preparation techniques for environmental analysis. J. Biochem. Biophys. Methods 70, 275–288.

Warner, K., Eskin, N.A.M., 1995. Methods to Assess Quality and Stability of Oils and Fat-Containing Foods. AOCS Press, Champaign, IL.

Waterborg, J.H., 2009. The Lowry method for protein quantitation. In: Walker, J.M. (Ed.), The Protein Protocols Handbook, third ed. Human Press, New York, pp. 7–10.

Weichelman, K.J., Braun, R.D., Fitzpatrick, J.D., 1988. Investigation of the bicinchoninic acid protein assay: identification of the groups responsible for color formation. Anal. Biochem. 175, 231–237.

Weiser, J.R., Ricapito, N.G., Yueh, A., Weiser, E.L., Putnam, D., 2012. A mechanistic analysis of the quantitation of α-hydroxy ketones by the bicinchoninic acid assay. Anal. Biochem. 430, 116–122.

Wells, M.A., Dittmer, J.C., 1963. The use of sephadex for the removal of nonlipid contaminants from lipid extracts. Biochemistry 2, 1259–1263.

White, P.J., 1995. Conjugated diene, anisidine value, and carbonyl value analyses. In: Warner, K., Eskin, N.A.M. (Eds.), Methods to Assess Quality and Stability of Oils and Fat-Containing Foods. AOCS Press, Champaign, IL, pp. 159–178.

Williamson, K., Hensley, K., Floyd, R., 2003. Fluorometric and colorimetric assessment of thiobarbituric acid-reactive lipid aldehydes in biological matrices. In: Hensley, K., Floyd, R. (Eds.), Methods in Biological Oxidative Stress. Humana Press, pp. 57–65.

Witte, V.C., Krause, G.F., Bailey, M.E., 1970. A new extraction method for determining 2-thiobarbituric acid values of pork and beef during storage. J. Food Sci. 35, 582–585.

Wójcicki, K., Khmelinskii, I., Sikorski, M., Sikorska, E., 2015. Near and mid infrared spectroscopy and multivariate data analysis in studies of oxidation of edible oils. Food Chem. 187, 416–423.

Wolff, S.P., 1994. Ferrous ion oxidation in presence of ferric ion indicator xylenol orange for measurement of hydroperoxides. Methods Enzymol. 233, 182–189.

Workman Jr., G.S., Steger, J., 1996. Field Validation of the DNPH Method for Aldehydes and Ketones. U.S. Environmental Protection Agency Report No. EPA/600/SR-96/050, Research Triangle Park, NC.

Wrolstad, R.E., Acree, T.E., Penner, M.H., Reid, D.S., Schwarts, S.J., Shoemaker, C.F., Smith, D.M., Sporns, P., 2005. Lipid oxidation/stability. In: Handbook of Food Analytical Chemistry, Water, Proteins, Lipids, and Carbohydrates, vol. 1. John Wiley & Sons, Hoboken, NJ, pp. 513–564.

Wu, W., Zhang, C., Kong, X., Hua, Y., 2009. Oxidative modification of soy protein by peroxyl radicals. Food Chem. 116, 295–301.

Wu, Z., Ahmad, R., Armstrong, D.A., 1977. Formation of lipoamide anion radicals by hydroxyl, formate and alcohol radicals at pH 6–9. Radiat. Phys. Chem. 23, 251–257.

Xie, J., 2015. Evidence for Multiple Oxidation Pathways from Non-volatile Products of Methyl Linoleate. PhD. Food Science, Rutgers University, New Brunswick, NJ.

XploSafe, 2012. XPell CE Indicating Pellets for Peroxides Prevention in Organic Solvents. http://www.xplosafe.com/products/xpell-ce (accessed 23.04.15.).

Yagi, K., 1987. Lipid peroxides and human disease. Chem. Phys. Lipids 45, 337–351.

Yang, H., Irudayaraj, J., Paradkar, M.M., 2005. Discriminant analysis of edible oils and fats by FTIR, FT-NIR and FT-Raman spectroscopy. Food Chem. 93, 25–32.

Yao, L., 2015. Development of New Methods for Analyzing Lipid Oxidation: Accelerated Solvent Extraction and HPLC-dnph Analyses of Carbonyl Oxidation Products. Ph.D. Food Science, Rutgers University, New Brunswick, NJ.

Yao, L., Schaich, K.M., 2014. Accelerated solvent extraction improves efficiency of lipid removal from dry pet food while limiting lipid oxidation. J. Am. Oil Chem. Soc. 92, 141–151.

Yavari, A.A., Hamedi, M., Haghbin, S., 2009. Investigation of frying oil quality using VIS/NIR hyperspectral analysis. J. Am. Oil Chem. Soc. 86, 941–947.

Yildiz, G., Wehling, R., Cuppett, S., 2002. Monitoring PV in corn and soybean oils by NIR spectroscopy. J. Am. Oil Chem. Soc. 79, 1085–1089.

Yildiz, G., Wehling, R.L., Cuppett, S.L., 2003. Comparison of four analytical methods for the determination of peroxide value in oxidized soybean oils. J. Am. Oil Chem. Soc. 80.

Yin, H., Porter, N.A., 2003. Specificity of the ferrous oxidation of xylenol orange assay: analysis of autoxidation products of cholesteryl arachidonate. Anal. Biochem. 313, 319–326.

Ying, D., Edin, L., Cheng, L., Sanguansri, L., Augustin, M.A., 2015. Enhanced oxidative stability of extruded product containing polyunsaturated oils. LWT - Food Sci. Technol. 62, 1105–1111.

Yu, T.C., Day, E.A., Suinnhuber, R.O., 1961. Autoxidation of fish oils. I. Identification of volatile monocarbonyl compounds from autoxidized salmon oil. J. Food Sci. 26, 192–197.

Yu, X., van de Voort, F.R., Sedman, J., 2007. Determination of peroxide value of edible oils by FTIR spectroscopy with the use of the spectral reconstitution technique. Talanta 74, 241–246.

Yukawa, N., Takamura, H., Matoba, T., 1993. Determination of total carbonyl compounds in aqueous media. J. Am. Oil Chem. Soc. 70, 881–884.

Zamora, R., Alaiz, M., Hidalgo, F.J., 2000. Contribution of pyrrole formation and polymerization to the nonenzymatic browning produced by amino-carbonyl reactions. J. Agric. Food Chem. 48, 3152–3158.

Zhang, Z., Pawliszyn, J., 1993. Headspace solid-phase microextraction. Anal. Chem. 65, 1843–1852.

Zhu, H., Li, X., Shoemaker, C.F., Wang, S.C., 2013. Ultrahigh performance liquid chromatography analysis of volatile carbonyl compounds in virgin olive oils. J. Agric. Food Chem. 61, 12253–12259.

Zor, T., Selinger, Z., 1996. Linearization of the Bradford protein assay increases its sensitivity: theoretical and experimental studies. Anal. Biochem. 236, 302–308.

Chapter 2

Determination and Prediction of Shelf Life of Oils/Fats and Oil/Fat–Based Foods

Lara Manzocco, Sonia Calligaris, Monica Anese and Maria C. Nicoli
Dipartimento di Scienze degli Alimenti, Università di Udine, Udine, Italy

2.1 SHELF-LIFE ASSESSMENT ISSUES

Food quality is actually well known to be a dynamic state continuously moving to reduced levels until reaching the quality level that discriminates products that are still acceptable for consumption from those that are no longer acceptable (Kramer and Twigg, 1968). This quality level is generally defined as the acceptability limit beyond which the product becomes unsuitable for consumption due to legal, nutritional, or sensory criticisms. Based on these considerations, shelf life is the finite length of time after food production and packaging needed to reach this acceptability level under well-defined storage conditions (Nicoli, 2012). On the basis of this definition, attention should be paid to the difference existing between a shelf-life study aiming to correctly estimate the product shelf life and a stability study addressed to measure quality depletion kinetics. Although a wide number of studies dealing with food stability are available in the literature, they often do not allow shelf-life data to be obtained due to lack of a clearly defined acceptability limit.

Shelf-life assessment of foods not only provides exciting methodological challenges to food scientists but is a vital process for food companies willing to maintain their brand reputation in the market. According to the legislation of many countries, food companies are required to attribute a shelf life to their products (Robertson, 2009). Food dating is not compulsory in the United States on a national level, with the only exception of infant formula and baby foods, but many U.S. states require dating of some food categories (Nicoli, 2012). The EC Regulation 1169/2011 (EC, 2011) defines *shelf life* as the date until which the food retains its specific properties when properly stored. According to this definition, food dating responsibility completely relies on the food companies. The companies are required to assess the date, preferably on the basis of appropriate shelf-life studies. Food Safety International

Oxidative Stability and Shelf Life of Foods Containing Oils and Fats
http://dx.doi.org/10.1016/B978-1-63067-056-6.00002-1

Standards have produced even more restrictive requirements to guarantee food safety and quality (BRC Global Standard for Food Safety, ver. 6, 2011; IFS Food, ver. 6, 2012; ISO 22000:2005 Food safety management system, ISO, 2005). These standards require documented protocols for shelf-life definition to be adopted when developing new products as well implementation of an in-place system of ongoing shelf-life assessment. In addition, records and results from shelf-life tests will be used to monitor the market performances of consolidated productions. Further situations, such as changes in formulation, processing, packaging, and storage, can also require shelf-life studies to provide essential information to support the company decision process.

As explained in the BRC guidelines (2006), "A proper shelf-life determination is vital to verify how long the product will last before it deteriorates to the point where it is no longer salable. Shelf-life testing may be an arduous process depending on the product, but it is certainly cheaper than the product recall and will maintain your company's reputation." This statement is particularly true when considering oils/fats and oil/fat–based foods. They are mainly microbiologically stable foods, stored at ambient or frozen conditions and experiencing oxidative reactions during storage that lead to product unacceptability. As extensively reviewed in the literature, shelf life depends on the complex interplay among several compositional, environmental, and packaging factors. In Table 2.1 the main variables affecting lipid oxidation, and thus shelf life, during storage are reported.

In this chapter, shelf-life assessment strategies of oils/fats and oil/fat–based foods will be discussed. The definition of *critical indicators* and the choice of the acceptability limit, as well as possible methodologies for shelf-life testing,

TABLE 2.1 Intrinsic and Environmental Variables Affecting Shelf Life of Oils/Fats and Oil/Fat–Based Foods

Food Intrinsic Variables	Package Intrinsic Variables	Environmental Variables
Features of the oxidizing molecule (e.g., fatty acid unsaturation degree)	Barrier properties to moisture, light, and gases	Temperature
Concentration of antioxidant and prooxidant molecules	Atmosphere composition inside the package	Intensity and wavelength of light radiation
Activity of oxidative enzymes	Active packaging	Relative humidity
Physical and physicochemical properties (e.g., physical state, presence of interfaces)	Temperature dependence of package performance	

will be presented. The focus is on methods of shelf-life evaluation under actual storage conditions and methods of shelf-life prediction based on exploitation of proper accelerating environmental factors. The methodology will be illustrated with reference to case studies showing uncertainties and possible risks arising during shelf-life assessment.

2.2 SHELF-LIFE ASSESSMENT STRATEGIES

A shelf-life study of food is articulated into three fundamental steps. Preliminarily, the most critical event leading to product quality depletion should be identified and a relevant acceptability limit chosen. It is then necessary to evaluate the changes in the selected quality indicator as a function of time under storage conditions that mimic the expected storage conditions (real-time shelf-life testing) or under environmental conditions able to accelerate quality depletion (accelerated shelf-life testing (ASLT)). Finally, data are modeled to obtain a shelf-life estimation or prediction, respectively. The shelf-life assessment strategies depicted in Figure 2.1 can be applied to oils/fats and oil/fat–based foods, for which oxidation is well known to be the main critical event leading to product unacceptability.

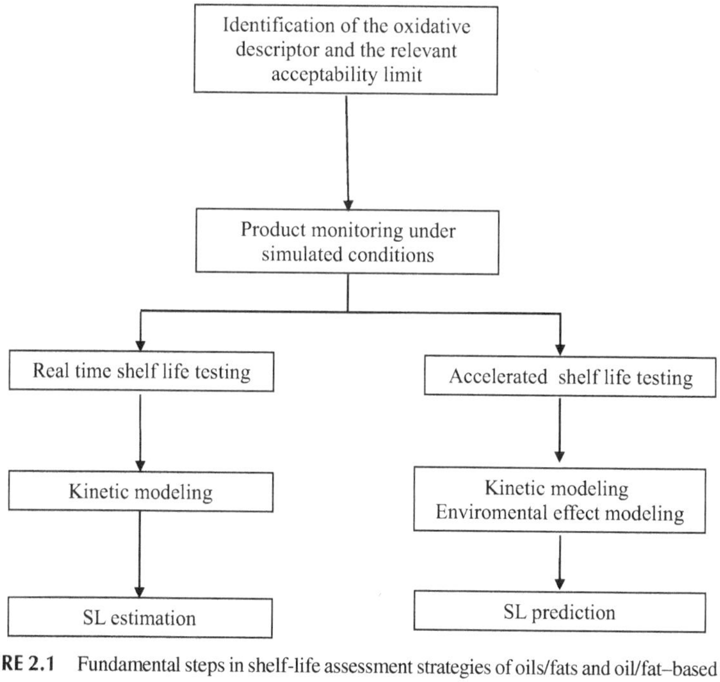

FIGURE 2.1 Fundamental steps in shelf-life assessment strategies of oils/fats and oil/fat–based foods.

2.3 DEFINITION OF THE ACCEPTABILITY LIMIT AND RELEVANT CRITICAL INDICATOR OF OXIDATION

Lipid oxidation during food storage can be monitored by using different indicators. In most cases, due to the complexity of the overall oxidative reaction, a concomitant change in a number of different oxidative indices is generally observed.

Table 2.2 shows some recent literature examples of applied indicators to monitor oxidation during storage of oil/fat–containing foods. Due to the huge number of literature data available, Table 2.2 should not be considered exhaustive. It should be noted that more than one index is generally used to follow

TABLE 2.2 Some Recent Examples of Analytical Index Used to Monitor the Evolution of Oxidation during Storage by Different Authors in Some Oil/Fat–Based Foods

Product	Analytical Index	References
Perilla oil	Peroxide value, anisidine value	Shim and Lee (2011) and Jung at al. (2012)
Walnut oil	Peroxide value, acid value, conjugates dienes, tocopherol content, descriptive sensory analysis	Martinez et al. (2011)
Extra virgin olive oil	Peroxide value, conjugated dienes, FTIR absorbance	Rizzo et al. (2014) and Nenadis et al. (2013)
California almonds	Peroxide value	Lin et al. (2008)
Shrimp	Peroxide value, thiobarbituric acid, anisidine value, volatile, rancid odor	Takeungwongtrakul and Benjakul (2014)
Seabass	Peroxide value, acidity, thiobarbituric acid, sensory profile	Sae–leaw and Benjakul (2014)
Potato chips	Peroxide value, descriptive sensory analysis	Rababah et al. (2012)
Oat biscuits	Peroxide, anisidine value, fatty acid profile, descriptive sensory analysis	Cognat et al. (2014)
Roasted coffee	Free fatty acid and triacylglycerol	Toci et al. (2013)
Sausage	Conjugate dienes, thiobarbituric acid	Karwowska et al. (2014)
Dairy beverage formulation	Conjugate dienes, dissolved oxygen, volatiles	Boroski et al. (2012)

oxidation during storage due to the complex interplay of oxidative reactions. It is convenient to monitor at least one primary and one secondary oxidation index to have a clearer picture of the ongoing oxidation situation. Instrumental analysis is widely used, but sensory analysis of descriptive indicators is often applied. This is due to the fact that oxidation is generally associated to the formation of off-odors, off-flavors, or color modification that is easily appreciable by the consumer's senses. Although providing information about the level of oxidation, data about the evolution of these oxidative indices could have no relation with the consumer decision to eat/buy the product. In other words, the changes in these indices do not allow any sound suggestion about the consumer sensory satisfaction to be achieved.

Although it is evident that the critical indicator of oxidation to be used in shelf-life studies should be chosen among them, two basic questions arise: Which is the best one for the shelf-life study? Which is the acceptability limit to be used to estimate shelf life?

The simplest criterion of choice is based on the selection of the indicator that first changes during storage. However, the questions are more complex and require additional considerations about the links between the critical indicator and the acceptability limit. In fact, the definition of the acceptability limit may result from the application of different criteria discriminating the acceptable product from the no longer acceptable one. To this regard, the criteria for the definition of the shelf-life acceptability limits for food undergoing oxidation are based on (1) adherence to legal requirements or (2) consumer sensory satisfaction.

2.3.1 Adherence to Legal Requirements

When the national or international regulations provide specific requirements for the product to be commercialized, the acceptability limit is compulsory and the critical indicator derived thereby. For example, the unique compulsory indication relevant to the development of oxidation in oils/fats is relevant to olive oils (European Regulation 1531/2001). The latter are subdivided into different categories according to quality requirements, including oxidation indices (e.g., peroxide value, acidity, conjugated dienes) and sensory attributes. For instance, the European Commission defines the limit concentration of peroxides in extra virgin olive oil and olive-pomace oil to be 20 and 15 mEq/kg_{oil}, respectively. It should be noted that these values are relevant to just produced oils. However, considering that conformity to regulatory indications should be guaranteed throughout the shelf life, such values can be considered as acceptability limits to estimate shelf life. Noncompulsory recommendations on the oxidation level for other fats and oils are also available in the relevant Official Codex Standards (Codex Alimentarious, 1999). However, these indications only refer to bulk lipids; no advice for oil/fat–based foods is provided.

Compulsory acceptability limits can be derived from voluntary label claims. In fact, according to the regulation, producers must guarantee product conformity to any claim reported on the label. In this context, the concentration of a bioactive molecule claimed on the label can be regarded as shelf-life acceptability limit. Additionally, the regulation on packaging contact materials should also be taken into account to assess if the contaminant concentration in the oils/fats or oil/fat–based foods is below the acceptable migration level. Whenever migration is the prevalent event occurring during storage, the maximum allowed value of the migrant in the product could be used as acceptability limit to estimate the shelf life.

2.3.2 Consumer Sensory Satisfaction

When no indication about the acceptability limit is provided by the regulation or can be derived from it, the acceptability limit and the relevant critical indicator can be freely chosen according to quality industry policy. This choice may have a dramatic effect on the brand reputation and product positioning on the market. In fact, based on their level of sensory satisfaction, consumers make their own evaluation and decide whether the producer should be "punished" or "awarded" with their fidelity. The producer should thus define the maximum level of oxidation and then the minimum level of consumer sensory satisfaction the product has at shelf life. This definition is generally the result of a decision process that merges the different consumer, production, delivery structure, and business needs. In any case, the decision maker is the producer who chooses the acceptability limit by defining a set of requirements for the commercialization of its own product.

The identification of the oxidation limit can be rationally made by using data about consumer satisfaction acquired from previous company records, literature information, and competitor data. Unfortunately, when dealing with new foods, information about the historical performance of the product on the market is not available. In these cases, the acceptability limit choice is fraught with the risk of critical overestimation or disadvantageous underestimation of the shelf life.

In order to evaluate consumer sensory satisfaction, sensory analysis performed on consumers should be applied. Some companies perform affective analysis by asking consumers for their level of satisfaction on different hedonic scales. Muñoz et al. (1992) suggested a value of 6 on a 9-point hedonic scale to be used as acceptability limit for quality control specification. This acceptability limit is highly subjected to a wide individual discretion. In order to use the consumer satisfaction level expressed on hedonic scale as an indicator of consumer sensory satisfaction, the relation between the two should be known.

Consumer sensory satisfaction can be assessed using survival analysis methodology (Hough and Garitta, 2012). In this case, the product is analyzed during storage by asking the consumers to provide a response regarding acceptability/unacceptability. Data are elaborated by survival analysis and can be used to directly estimate the shelf life of the product. However, such methodology can also be applied to support the choice of the oxidation limit to be subsequently used for shelf-life computation. In practical terms, once data on food sensory acceptability/unacceptability are gathered, the function of consumer satisfaction of the food as a function of storage time is produced. Figure 2.2 shows an example of consumer satisfaction functions obtained through survival analysis of acceptability data of all bran crackers and biscuits stored at room temperature. When these functions are available, the food company can choose to be exposed to more or less risk by selecting, as acceptability limit, a certain level of consumer satisfaction. In other words, the acceptability limit becomes the minimum percentage of consumers that the company requires to satisfy when eating the product stored to the end of shelf life. Some literature examples about the use of survival analysis to estimate consumer satisfaction during storage of different oils/fats or oil/fat–based foods are shown in Table 2.3. A 50% consumer satisfaction is generally chosen as an acceptability limit to estimate product shelf life, but higher values are also reported in the literature (Table 2.3). Actually, there are also companies preferring to apply 100% consumer satisfaction as the acceptability limit. This means that all consumers eating the product at the end of shelf life should find it acceptable.

It is noteworthy that the development of consumer satisfaction functions requires a high sample size, assembling large consumer panels as well as the application of appropriate statistical techniques. These conditions make survival analysis, despite being powerful and accurate, difficulty applicable by company operators on a routine basis. To save time and meet industrial needs, instrumental or sensory oxidative indices whose evolution is correlated to consumer sensory satisfaction can be identified. The possible procedural steps to identify oxidation indices that are easily detectable and related to consumer sensory satisfaction are shown in Table 2.4. The table also shows an example of possible output of the different procedural steps with reference to potato chips stored at

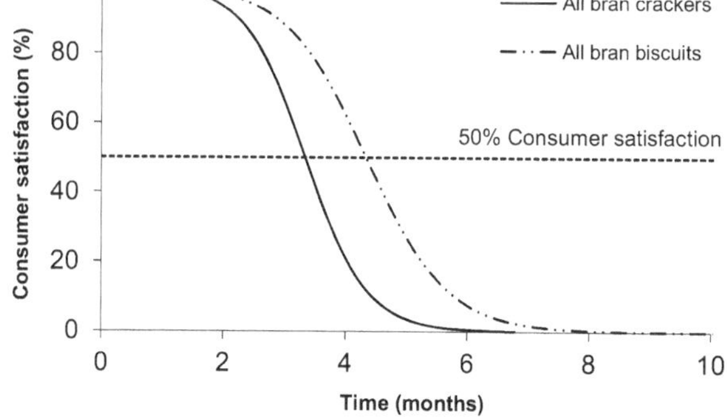

FIGURE 2.2 Consumer satisfaction (%) for all bran crackers and biscuits stored at room temperature.

TABLE 2.3 Literature Examples on the Use of Survival Analysis to Estimate Consumer Satisfaction During Storage of Different Oils/Fats or Oil/Fat–Based Foods and the Acceptability Limit Used to Calculate Shelf Life

Food	Acceptability Limit (% Consumer Satisfaction)	References
Powdered milk	30	Hough et al. (2002)
Minced meat	50	Hough et al. (2006)
Roasted and ground coffee	50	Cardelli and Labuza (2001)
Biscuits	50	Calligaris et al. (2007a)
Breadsticks	50	Calligaris et al. (2008)
White pan bread	50 or 75	Gambaro et al. (2004)
Muffins	50 or 80	Baixauli et al. (2008)
Brown pan bread	75	Gimenez et al. (2007)
Fruit-filled snack bars	50	Corrigan et al. (2012)

TABLE 2.4 Procedural Steps to Define the Acceptability Limit and the Relevant Critical Indicator of Oxidation with Examples of Possible Output of the Different Procedural Steps with Reference to Potato Chips

Procedure	Output
1. Assessment of consumer sensory satisfaction during storage	Function of consumer satisfaction vs time
2. Evaluation of sensory or instrumental oxidative indices during storage	Function of off-flavor and peroxides vs time
3. Correlation analysis between consumer satisfaction and off-flavor, off-odor, peroxides	Regression coefficients
4. Identification of the oxidative indicator best correlating with consumer satisfaction	Selection of peroxide value based on the highest regression coefficient
5. Choice of the acceptability limit based on the minimum tolerable consumer satisfaction at shelf life	Selection of 50% consumer satisfaction
6. Modeling of the relations between the selected oxidative indicator and consumer satisfaction	Function of peroxide value vs consumer satisfaction
7. Computation of the value of the oxidative index corresponding to the chosen acceptability limit	Value of peroxide value accounting for 50% consumer satisfaction

ambient temperature. This procedure is based on the concomitant assessment of the consumer sensory satisfaction function and easily detectable oxidative indices during product storage. In the case of potato chips, these steps produced the consumer satisfaction function and the evolution of off-flavor and peroxides during storage. Correlation analysis is then used to identify the oxidative index best correlated to consumer satisfaction. By comparing regression coefficients, the oxidative indicator best related with the consumer acceptability resulted to be peroxide value. The relation between the selected oxidative indicator and consumer satisfaction is then modeled. This function is used to estimate the value of peroxide value corresponding to the chosen acceptability limit of consumer satisfaction (50%). This value may be routinely applied to evaluate the shelf life of the product in the industry quality control programs.

The advantages of the exploitation of easily detectable oxidative indices, compared with consumer satisfaction, are undoubted. Once the oxidative indicator correlating with consumer satisfaction has been identified, further consumer tests can be skipped, yet allowing reliable shelf-life data to be obtained. It should be noted that not always simple indicators, such as peroxide value, are well correlated with consumer satisfaction. In some cases, more advanced analyses, including those of volatiles, are necessary to identify the best index accounting for consumer satisfaction.

2.4 TESTING AND DATA MODELING

Once the appropriate oxidative indicator and the relevant acceptability limit are set, it is necessary to estimate the length of time needed to reach such a critical value. This step, generally defined as shelf-life testing, implies the monitoring of the evolution of the oxidative indicator during food storage under conditions that reasonably foresee the situation expected on the market shelf. Data obtained are then modeled to obtain the parameters describing/predicting the oxidation kinetics.

To get shelf-life data, two different strategies can be pursued: actual shelf-life testing and accelerated shelf-life testing (Figure 2.1). The former is theoretically applicable for shelf-life estimation of any food category. However, it becomes profitable in case of perishable foods with fast quality decay kinetics. By contrast, the prediction of long shelf-lives is traditionally obtained by accelerating shelf-life experiments performed under environmental conditions able to speed up deterioration kinetics. This is the case of many ambient stable and frozen products in which quality depletion during storage is in most cases attributable to oxidative reactions.

2.4.1 Actual Shelf-Life Testing

The basic requirement to get reliable shelf-life data by actual testing is that the environmental factors (e.g., temperature, humidity, and light) during storage are kept constant and analogous as much as possible to those experienced by the product on the market shelf. Because temperature oscillations are very frequent during food storage, it could be useful to perform the shelf-life test under the worst situation expected on the market shelf. If the food is photosensitive and packed in a see-through container, the light exposure suffered by the product could become critical. In these cases, the test should be performed under the light intensity commonly found on the market shelves (600–800 lux).

Data describing the evolution of the oxidative indicator as a function of time are then submitted to modeling according to the fundamental kinetic principles. The rate of changes of an oxidative indicator (*I*) can be calculated by integrating the general kinetic equation:

$$\int_{I_0}^{I} \frac{dI}{I^n} = \int_{0}^{t} k\,dt \tag{1}$$

where *k* is the rate constant and *n* is the reaction order. The general rate law can be integrated to obtain the equations of the pseudo zero, first, second, or *n* order. As previously noted, oxidation reactions show complex pathways and their rate is often dependent on a huge number of factors (Table 2.1). Indeed, their evolution can be frequently the result of different reactions taking place simultaneously or consecutively. Thus, the reaction order *n* does not give any indications on the true reaction mechanisms involved, and *k* is an "apparent" rate constant.

Reports dealing with this well-recognized procedure have produced huge amounts of data on the oxidation rates of several foods undergoing oxidation. Zero order was used to model peroxide changes and K232 dienes in both oils/fats (e.g., extra virgin olive oil, salmon oil, perilla oil, walnut oil) (Calligaris et al., 2006; Huang and Sathivel, 2008; Shim and Lee, 2011) and oil/fat–based foods (potato chips, biscuits, almond-based products) (Houhoula and Oreopolou, 2004; Calligaris et al., 2007a; Tazi et al., 2009). Similarly, the first-order kinetic equation was used in the case of oxygen consumption of soybean oil (Colakoglu, 2007) or K270 dienes in extra virgin olive oil (Gutierrez and Fernandez, 2002; Mascebo-Campos et al., 2008). Second-order reaction was less frequently applied as in the case of pigment oxidation (Ghidouche et al., 2013).

Once the reaction rate constant has been calculated, the shelf life can be computed by solving the integrated forms of Eqn (1) as a function of time:

$$\mathrm{SL} = \frac{1}{k} \int_{I_0}^{I_{\mathrm{lim}}} \frac{dI}{I^n} \tag{2}$$

where I_0 is the value of the oxidative indicator just after food production, I_{lim} is the oxidative value corresponding to the previously defined acceptability limit. Table 2.5 shows the shelf-life equations based on zero-, first-, second-, and *n*-order reaction kinetics.

Unfortunately, more complex situations can be observed for oxidative indexes leading to inapplicability of the classic kinetics approach.

TABLE 2.5 Shelf-Life Equations Based on Zero-, First-, Second-, and *n*-Order Oxidative Reaction Kinetics

Reaction Order	Shelf-Life Equation
$n=0$	$SL=\dfrac{I_{\lim}-I_o}{k}$
$n=1$	$SL=\dfrac{\ln I_{\lim}-\ln I_o}{k}$
$n=2$	$SL=\dfrac{\dfrac{1}{I_{\lim}}-\dfrac{1}{I_o}}{k}$
$n\neq1$	$SL=\dfrac{\dfrac{1}{I_{\lim}^{n-1}}-\dfrac{1}{I_o^{n-1}}}{k}$

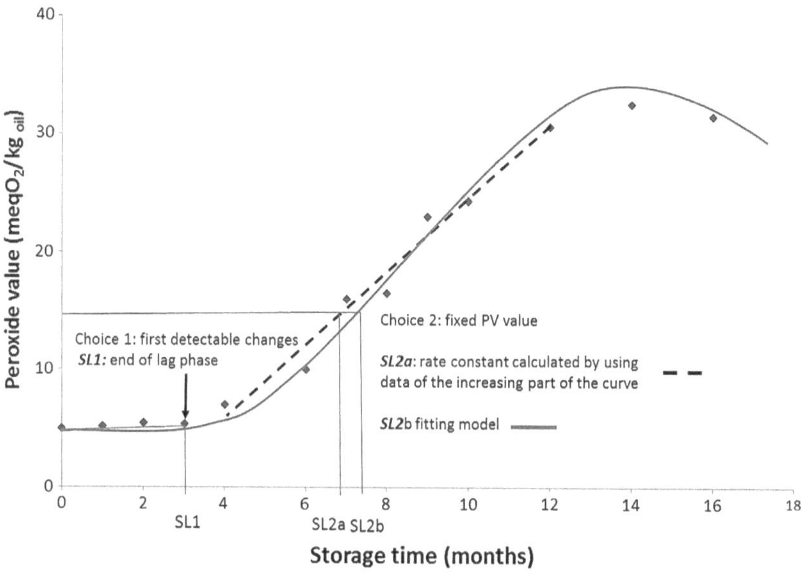

FIGURE 2.3 Schematic representation of the evolution of peroxide value as a function of food storage time and possible methodologies to identify shelf life.

A case in point is that relevant to the evolution of peroxide index as a function of storage time that is well known to follow the typical bell-shaped curve. After the induction period, the progressive increase of peroxides occurs until a maximum is reached, indicating that the oxidative reaction approaches termination (Figure 2.3). Two strategies could be applied to assess shelf life, depending on the level of peroxide value chosen as acceptability limit:

- Choice 1: first detectable changes approach that is when the shelf life is defined as the time at which an increase of peroxide value is measured. In this case, the induction period can be regarded as the product shelf life (see *SL1* in Figure 2.3).
- Choice 2: fixed change level approach that is when a selected value of peroxide is chosen as acceptability limit. In this case, data relevant to the exponential increase phase should be modeled by considering this part of the curve separately from that of the induction period (see *SL2a* in Figure 2.3). Alternatively, the entire evolution of the peroxide value during storage can be modeled by identifying an empirical mathematical model fitting the curve (see *SL2b* in Figure 2.3). Different mathematical models have been applied in literature (Aragao et al., 2009; Özilgen and Özilgen, 1990; Cunha et al., 1998; Corradini and Peleg, 2007; Calligaris et al., 2008; Imai et al., 2008; Odriozola-Serrano et al., 2009). In these cases, the fitting curve parameters should then be integrated in the shelf-life equation for its estimation.

2.4.2 Accelerated Shelf-Life Testing

ASLT is exploited when oxidation slowly advances under actual storage conditions. In these cases, shelf-life testing can be performed under environmental conditions that speed up oxidative reactions and data are used to extrapolate kinetics at the conditions usually experienced by the product. When performing ASLT, a proper accelerating factor to be used in the test should be chosen. The absolute necessary requirement to apply ASLT is the knowledge of the relationship between the accelerating factor and oxidation rate. Only if this requisite is complied, the extrapolation to actual conditions could be used to predict shelf life.

The main environmental factors exploited to perform ASLT of oil/fats and oil/fat–based products are temperature and light. Temperature is certainly the most widely used accelerating factor in ASLT due to its critical role in affecting oxidative kinetics as well as the availability of a theoretical mathematical equation describing the temperature sensitivity of reaction rates: Arrhenius equation (3) (Arrhenius, 1901):

$$k = k_o \cdot e^{-\frac{E_a}{RT}} \tag{3}$$

in which k is the reaction rate constant, R is the molar gas constant (8.31 J/K mol), T is the absolute temperature (K), E_a is the apparent activation energy (J/mol), and k_o is the so-called pre-exponential factor. When fulfilled, the Arrhenius equation allows the reaction rate at a desired temperature to be extrapolated by measuring the oxidation rate at least at three different temperatures.

The Arrhenius equation has been successfully used to estimate the temperature dependence of oxidation rate for oils/fats and oil/fat–based foods.

A number of data can be found in the literature relevant to the application of the Arrhenius equation to describe the temperature dependence of oxidative reactions by using different oxidation indices. E_a values for the different products is reported to greatly vary from 9 to about 200 kJ/mol (Manzocco et al., 2010).

In the temperature range where Arrhenius is fulfilled, a direct and simple index of how the reaction rate or shelf life changes with temperature can be computed. This index is called the Q_{10} value (Labuza and Schmidl, 2005):

$$Q_{10} = \frac{k_{T+10}}{k_T} = \frac{\mathrm{SL}_T}{\mathrm{SL}_{T+10}} \tag{4}$$

where k_T or SL_T is the rate constant/shelf-life at a certain temperature, and $k_{T+10}/\mathrm{SL}_{T+10}$ is the rate constant/shelf-life at a temperature higher by 10 °C. This index allows a rapid indication of the increase of reaction rate/decrease of shelf-life as a consequence of a 10 °C temperature increase to be obtained. The Q_{10} value is not constant but is related to E_a by the following equation:

$$Q_{10} = e^{\frac{10E_a}{RT(T+10)}} \tag{5}$$

Following the same approach of Q_{10} index, the shelf-life factor can be calculated. It is defined as the ratio between the shelf-life value at a given temperature T and the shelf-life at a reference temperature T_{ref} (Anese et al., 2012):

$$\text{Shelf-life factor} = \frac{\mathrm{SL}_T}{\mathrm{SL}_{T_{\mathrm{ref}}}} \tag{6}$$

Similarly to Q_{10}, this index can quickly provide an indication of the decrease in shelf-life as a consequence of a given temperature increase.

Both indices are very easily understandable and applicable in the common practice. However, it should be taken in mind that to use this index in the common practice, there is a need to know the value of the activation energy as well as the temperature interval of Arrhenius model validity. Only after this preliminary step is it possible to choose the maximum temperature at which the test can be performed to obtain the estimated shelf-life in the shortest possible time.

The successful application of the Arrhenius model or Q_{10} approach requires that food is able to withstand the temperature increase without inducing the development of phenomena other than oxidation. Although the Arrhenius equation is frequently applied, unfortunately, in many cases, deviations from the Arrhenius equation are observed, impairing shelf-life prediction. Figure 2.4 shows an example of positive deviation from the Arrhenius behavior of zero-order rate constants of peroxide and hexanal formation in palm oil stored at increasing temperatures from 15 to 60 °C (Calligaris, 2003). It can be noted that reaction rate of both oxidative indices abruptly changed when palm oil was stored at 15 °C. In particular, oxidation rates at this temperature were higher than those expected on the basis of the Arrhenius equation, leading to a positive deviation from linearity.

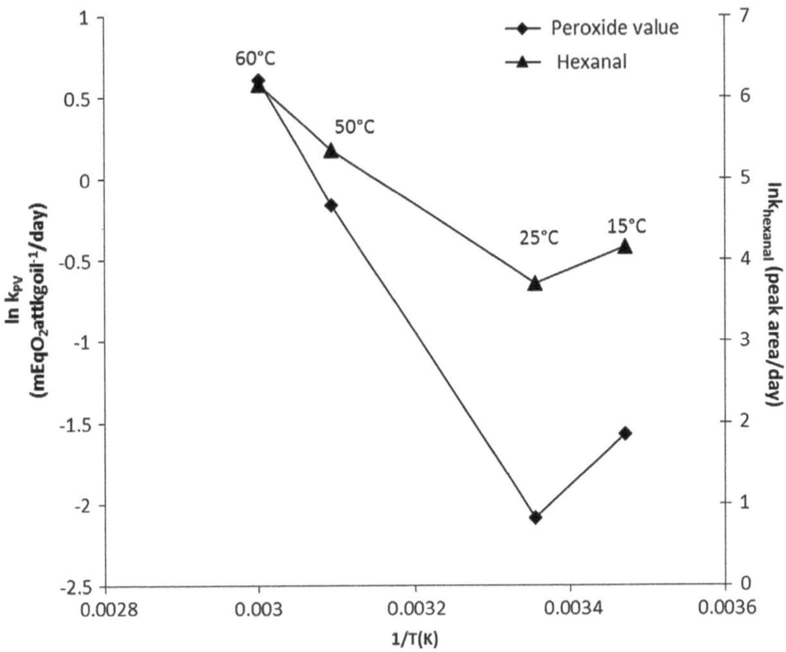

FIGURE 2.4 Arrhenius plot of zero-order rate constants of peroxide (k_{PV}) and hexanal ($k_{hexanal}$) formation in palm oil stored at increasing temperatures from 15 to 60 °C. *Modified from Calligaris (2003).*

TABLE 2.6 Factors Potentially Causing Deviations from the Arrhenius Behavior of Oxidation Rate

Factor	Effect of Temperature Increase
Reaction pathway	Alteration of the rate-determining step and thus the temperature dependence of oxidation
Oxygen concentration	Modification of gas solubility
Prooxidant and antioxidant concentrations	Degradation/formation of molecules modulating oxidation rate
Physical state of oils/fats	Modification of chemical and physical properties of the oxidizing matrix
Physical structure of oil/fat–based foods	Modification of reactant concentration, mobility and compartmentalization

The temperature interval adopted to perform ASLT is critical to avoid dangerous deviations from linearity of the Arrhenius equation. Thus, it is advisable to verify the linearity of the Arrhenius equation before its use as predictive tool.

Different factors could potentially cause deviations from the Arrhenius behavior. Table 2.6 shows a summary of the main causes of Arrhenius deviations.

The overall oxidation rate depends on the slowest, and thus rate-determining, step. As temperature changes, these steps can have different E_a and k_o values, leading to non-Arrhenius behavior. This is even more important considering that a specific pathway can dominate at lower temperatures while another one takes over at higher temperature. For instance, ASLT performed at temperatures higher than 100 °C should be avoided due to the development of excessive rancidity, certainly not relevant to the product performance under conventional storage conditions (Frankel, 2005). In addition, the more polyunsaturated the oils, the lower are the temperatures that should be used to test their oxidative stability; for instance, vegetable oils should be tested at temperatures lower than 60 °C, while fish oils should be tested only below 40 °C (Frankel, 2005).

Oxygen solubility in food decreases with increasing temperature. When oxidation rate depends on oxygen availability, deviations from the Arrhenius equation may occur. To this regard, oxidation rate in oil-in-water emulsions stored at −30 °C is higher than that predicted by the Arrhenius equation (Calligaris et al., 2007b) due to the limiting concentration of oxygen at this temperature. Oxygen concentration was also found to be critical in determining Arrhenius deviation of carotenoid oxidation in tomato derivatives stored below 0 °C (Manzocco et al., 2006). The increase in temperature may promote thermal degradation of minor prooxidant or antioxidant compounds that are expected to steer the overall oxidation rate. In addition, in complex foods containing not only lipids but also carbohydrates and proteins, novel compounds such as Maillard reaction products with prooxidant or antioxidant capacity could be generated (Zamora and Hidalgo, 2005; Echavarria et al., 2012).

Temperature increase may modify the physical state of the oils/fats as well as the physical structure of oil/fat–based foods potentially affecting the physicochemical properties of the oxidizing matrix. Relevant changes in reactant concentration and solubility, pH, ionic strength, water activity, and viscosity could thus lead to significant deviations from the Arrhenius plot linearity (Parker and Ring, 1995; Calligaris et al., 2004, 2006, 2007b, 2008; Manzocco et al., 2006). The Arrhenius deviations shown in Figure 2.4 are actually due to the crystallization of the highly saturated fraction of palm oil and the concentration of unsaturated triglycerides and solubilized oxygen in the liquid phase. Similarly, the modification of temperature can show dramatic effects on the physical state and structural architecture of oil/fat–based foods.

For instance, melting of ice crystals on temperature change during storage of lipid containing frozen foods can significantly modify oxidation rate due to both changes in reactant concentration and mobility. Moreover, if powdered foods progressively adsorb water during storage, the rate of oxidative reactions could unexpectedly change as well described by the well-known inverse bell-shaped curve of oxidation rate as a function of water activity. At least but not last, when emulsion physical properties changes during storage (i.e., increasing oil droplet particle size), the different surface properties and distribution of reactants could also greatly influence the sample oxidability.

When the Arrhenius equation is not applicable, it requires adaptation to the specific circumstances of the product being considered. In other words, starting from the understanding of the physicochemical phenomena leading to the Arrhenius deviation, other models could be identified. For instance, a modified Arrhenius equation, accounting for phase transitions of lipids or water, was successfully applied to describe the temperature dependence of oxidation in different food systems:

$$k = k_0 \cdot \Delta k \cdot e^{-\frac{E_a}{RT}} \tag{7}$$

where Δk is a corrective factor, defined as the ratio between forces and resistances responsible for positive and negative deviations from the Arrhenius behavior, respectively. Table 2.7 shows examples of forces and resistances used to compute Δk in different systems. By inserting the corrective factor Δk in the Arrhenius equation, a straight line in the entire temperature range of observation, including the one associated to deviations from the conventional Arrhenius equation, was observed. This provides the undoubted advantage of allowing ASLT even in the temperature range where deviations occur. Once defined the new equation describing the temperature dependence of oxidation rate, Q_{10} value could be also computed as previously described.

Finally, it is noteworthy that when the value of the activation energy is very low, temperature is definitely not the proper accelerating factor for ASLT. In these cases, the increase in temperature during food storage does not allow one to significantly

TABLE 2.7 Possible Forces and Resistance Responsible for Arrhenius Deviations, Used to Estimate the Corrective Factor Δk (Calligaris et al., 2012)

Sample	Forces (Positive Deviations)	Resistances (Negative Deviations)
Sunflower oil	Liquid fraction of oil	Viscosity
Extra virgin olive oil	Liquid fraction of oil	Polyphenol concentration in the liquid phase
Oil-in-water emulsion	Liquid fraction of oil Solute concentration upon ice formation Oxygen solubility	Viscosity
Tomato derivatives	Solute concentration upon ice formation Oxygen solubility	
Biscuits	Liquid fraction of oil	
Breadsticks	Liquid fraction of oil	

reduce the time necessary to appreciate the rate of oxidation. For instance, the percentage of time that can be saved when the temperature of the shelf-life test is increased by 10 °C is 60% if the activation energy is about 80 kJ/mol and only 10% when activation energy is about 10 kJ/mol (Calligaris et al., 2012).

When no temperature dependence of oxidation (E_a approaching 0) is observed, the use of ASLT based on temperature as accelerating factor is precluded. To this regard, light-induced oxidative reactions often show low values of activation energies. In these conditions, light may be the acceleration factor to be used instead of temperature in ASLT. The latter can be performed by exposing oils/fats or oil/fat–based foods to light with increasing intensity. To our knowledge, there are few examples of the use of light as accelerating factor in ASLT. Manzocco et al. (2012) developed a mathematical model predicting the effect of light intensity on oxidation rate of soybean and sunflower oil (Figure 2.5). The light dependence of peroxide formation rate constants was well described by the following equation:

$$k = k_d + E_1 \cdot L^{E_2} \tag{8}$$

where L is light intensity (lx), k_d is the reaction rate under dark, and E_1 and E_2 are the experimental parameters of the model, namely the electromagnetic energy required to activate the reaction. This equation represents actually a model allowing prediction of reaction rate as a function of the light intensity at a given temperature T. A similar approach was also efficaciously used to predict oxidation rate of carotenoid-containing beverages (Manzocco et al., 2008).

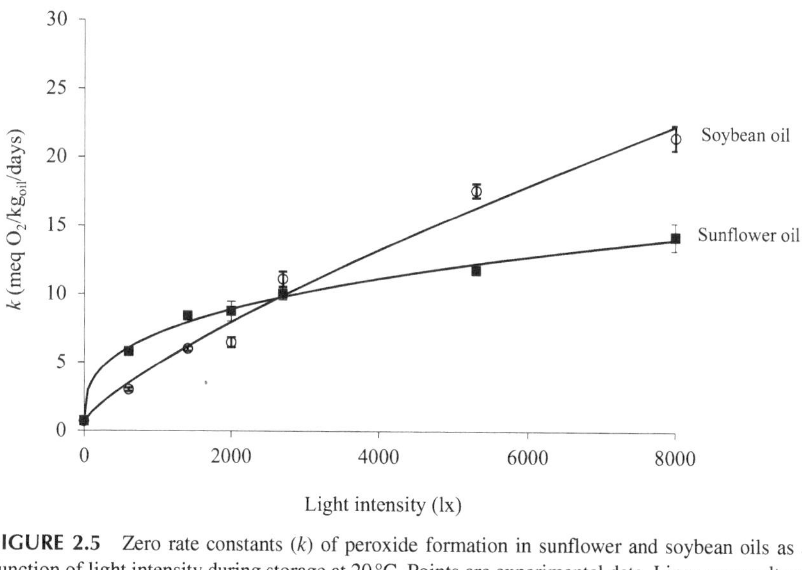

FIGURE 2.5 Zero rate constants (k) of peroxide formation in sunflower and soybean oils as a function of light intensity during storage at 20 °C. Points are experimental data. Lines are results of best-fitting analysis. *Modified from Manzocco et al. (2012).*

The exploitation of light as an accelerating factor may allow solving the difficult task of predicting shelf life of photosensitive food usually marketed in the presence of light. Additional advantages could be the possibility to exploit light instead of temperature to quickly predict shelf life of foods not withstanding high temperature.

2.4.3 Case Study of Shelf-Life Testing under Actual and Accelerated Conditions

The procedural steps needed to make rational decisions for reliable shelf-life dating previously described are illustrated in detail by taking into consideration an example relevant to a ready meal consisting of frozen smoky bacon and tomato pasta. The choice of a multicomponent product as the study case allows us the critical discussion of the different phases of a shelf-life testing protocol when the acceptability limit is unknown. Moreover, the storage at frozen temperature is an interesting case since the acceleration of quality depletion by temperature increase cannot be extended in a wide range of temperatures. In fact, the thawing of the product is undesired during ASLT since this could cause changes in food composition and physical structure affecting the Arrhenius behavior of alterative reactions.

Frozen pasta is a valuable commercial product due to its convenience, competitive price, and extended shelf life. Precooked pasta is presented separately from the sauce pellet pieces or mixed with the sauce. It generally comes in an oriented polypropylene package and is maintained at a nominal temperature of −18 °C (European market) or −12 °C (U.S. market) during transportation, storage, and distribution.

To find the proper quality index describing quality depletion during storage, the main chemical and physical events have to be evaluated during storage for each food component (i.e., pasta, tomato sauce, and bacon pieces). Quality depletion of pasta can be attributed to surface dehydration, analyzed by weight loss. Carotenoid and lipid oxidation can be regarded as the main alterative event in tomato sauce and can be described by color bleaching based on tristimulus colorimeter and peroxide value by iodometric titration method, respectively. In addition, the bacon component of the sauce could develop an undesired rancid flavor, which can be monitored by descriptive sensory analysis. These phenomena were monitored using the analytical descriptors of weight loss, color, peroxides, and off-flavor. In order to identify the earliest quality decay event, the product was submitted to a stability test at −7 °C. Figure 2.6 shows the evolution of the quality depletion indicators, expressed as percentage of their normalized intensity. It is evident that weight loss and peroxides changed very slowly and were thus excluded as shelf-life indicators. The choice between carotenoid bleaching and bacon off-flavor could be not taken for granted. Instinctively, one would choose bleaching of tomato sauce as critical indicator due to its early change. However, it may not be linked to consumer acceptability. A rational

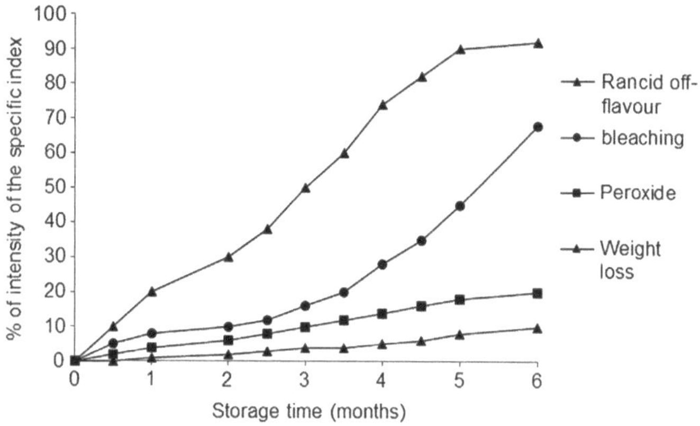

FIGURE 2.6 Changes in the specific index as a function of storage time at −7 °C of frozen smoky bacon and tomato pasta.

TABLE 2.8 Score of Bacon Off-flavor Corresponding to Given Percentages of Consumers Satisfaction

Consumer Satisfaction (%)	Rancid Off-flavor Score (1–9 Scale)
90	1.2
75	2.6
50	3.8
25	4.5
10	5.9

decision of the most suitable indicator could only be derived by analyzing the consumer sensory satisfaction by survival analysis (data not shown) and performing regression analysis between sensory satisfaction and the oxidative indicators. Regression coefficient highlighted that the development of bacon off-flavor ($R=0.95$, $p<0.05$) was better correlated to the decrease of consumer satisfaction compared with tomato bleaching ($R=0.84$, $p<0.05$). Thus, bacon off-flavor was selected as the critical indicator accounting for consumer satisfaction.

Once we defined the analytical indicator best related with the consumer satisfaction, its value in correspondence of the acceptability limit defined on the basis of consumer satisfaction was computed. Table 2.8 shows the estimated values of bacon off-flavor corresponding to given percentages of consumer satisfaction. The off-flavor score of 3.8, corresponding to 50% consumer satisfaction, was selected as the acceptability limit to evaluate the shelf life.

In order to develop a mathematical tool to be applied in the shelf-life prediction of the product, off-flavor versus time was monitored on samples stored at −12, −18, and −25 °C. As stressed before, ASLT can replace the normal storage test only when the Arrhenius fulfillment is experimentally proven. Figure 2.7 shows the off-flavor score as a function of storage time of the tomato sauce during storage at different temperatures. In the figure, the zero-order rate constants computed by means of linear regression analysis are also reported. To describe the temperature dependence of off-flavor changes, the Arrhenius equation can be used (Figure 2.8). A good Arrhenius behavior was obtained and the off-flavor

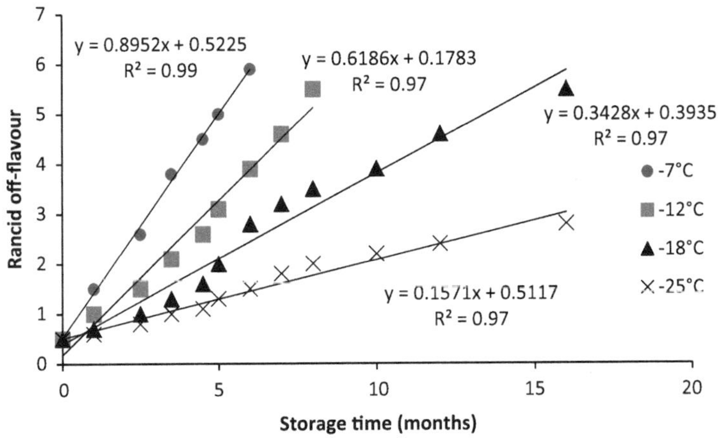

FIGURE 2.7 Changes in rancid off-flavor score of frozen smoky bacon and tomato pasta as a function of storage time at −7, −12, −18, and −25 °C.

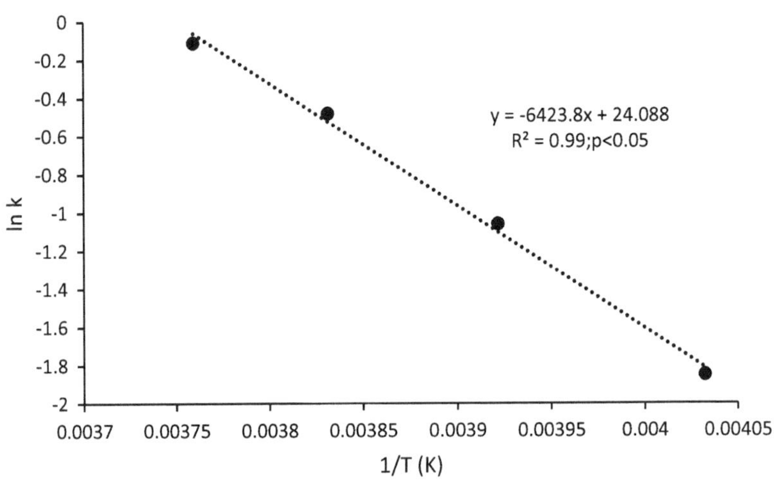

FIGURE 2.8 Arrhenius plot of rancid off-flavor formation in frozen smoky bacon and tomato pasta.

formation showed an E_a value of 53.37 kJ/mol, as calculated from the regression coefficient. This model can be used to compute the rate constant of oxidation at any temperature within the −25/−7 °C temperature range. Then, these k values can be used to estimate shelf life by using the defined acceptability limit and the following shelf-life equation:

$$SL = \frac{Score_{lim} - Score_0}{k_T} \tag{9}$$

where $Score_{lim}$ is the acceptability limit (3.8), $Score_0$ is the off-flavor score value at time 0, and k_T is the rate constant at temperature T. To convert these results in an easy instrument to be applied by the food industry, the shelf-life factor, defined as the ratio between the shelf-life value at a given temperature T and the shelf life at a reference temperature T_{ref}, can be calculated (Anese et al., 2012):

$$\text{Shelf-life factor} = \frac{SL_T}{SL_{T_{ref}}} \tag{10}$$

Considering $T_{ref} = -7$ °C, the shelf-life factor for −12 and −18 °C is in the presented example 1.44 and 2.61, respectively. These values can be used in the ongoing shelf-life validation during the quality management system activities of the company. In fact, it is possible to directly perform test at −7 °C to estimate with the calculated shelf-life factor the product shelf life at −12 or −18 °C, depending on the reference market.

2.5 CONCLUSION

Shelf life of oils/fats and oil/fat–containing foods can be regarded as the result of two interacting aspects: (1) the evolution of oxidative processes and (2) the effects of the development of oxidation on product adherence to legal requirements and, much more often, on consumer acceptability. The determination of shelf-life thus requires the ability to monitor, model, and combine information relevant to quality evolution and consumer perception of the product. The generation of shelf-life models for products experiencing oxidative reactions could be an arduous task due to the huge number of environmental and compositional factors affecting the oxidation rate. Open issues of shelf-life assessment of oils/fats and oil/fat–containing foods are mainly related to (1) the identification of strategies to adequately set the acceptability/oxidative limit and (2) the possibility to speed up shelf-life testing without reducing its predictive capability.

Finally, it is noteworthy that caution should be paid to any shelf-life estimation/prediction carried out under laboratory conditions. The discrepancy between laboratory and market shelf-life is recognized as a critical hazard. It is surprising that there is only circumstantial information about the possible error deriving from attributing a product a shelf life based only on laboratory tests. The latter do not take into account any changes in environmental conditions (e.g., temperature, light) during distribution and storage and in batch variability. For this reason, discussion about possible procedure to quickly obtain reliable data on market shelf life should not be neglected.

REFERENCES

Anese, M., Lanciotti, R., Gardini, F., Lagazio, C., 2012. Case studies. In: Nicoli, M.C. (Ed.), Shelf Life Assessment of Food. CRC Press, Boca Raton, pp. 247–294.

Aragao, G.M.F., Corradini, M.G., Peleg, M., 2009. A phenomenological model of the peroxide value's rise and fall during lipid oxidation. J. Am. Oil Chem. Soc. 85, 1143–1153.

Arrhenius, S.A., 1901. Larobok I Teoretisk Elektrokeni. Quando & Handel, Leipzig.

Boroski, M., Giroux, E.J., Sabik, H., Petit, E.V., Visentainer, J.V., Matumoto-Pintro, P.T., Britten, M., 2012. Use of oregano extract and oregano essential oil as antioxidants in functional dairy beverage formulations. LWT-Food Sci. Technol. 47, 167–174.

Baixauli, R., Salvador, A., Fiszman, S.M., 2008. Textural and colour changes during storage and sensory shelf life of muffins containing resistant starch. Eur. Food Res. Technol. 226 (3), 523–530.

BRC, 2006. BRC Global Standard Guidelines: Shelf Life Determination. British Retail Consortium, London.

BRC, 2011. BRC Global Standard for Food Safety. Issue 6. British Retail Consortium, London.

Calligaris, S., 2003. Effect of Compositional and Structural Conditions on Lipis Stability in Food Systems (Ph.D. thesis). University of Udine.

Calligaris, S., Manzocco, L., Lagazio, C., 2012. Modeling chemical, physical and sensory indicators. In: Nicoli, M.C. (Ed.), Shelf Life Assessment of Food. CRC Press, Boca Raton, pp. 75–126.

Calligaris, S., Da Pieve, S., Kravina, G., Manzocco, L., Nicoli, M.C., 2008. Shelf-life prediction of bread sticks by using oxidation indices: a validation study. J. Food Sci. 73 (2), E51–E56.

Calligaris, S., Manzocco, L., Nicoli, M.C., 2007b. Modelling the temperature dependence of oxidation rate in water-in-oil emulsions stored at sub-zero temperatures. Food Chem. 101, 1019–1024.

Calligaris, S., Manzocco, L., Conte, L.S., Nicoli, M.C., 2004. Application of a modified Arrhenius equation for the evaluation of oxidation rate of sunflower oil at subzero temperatures. J. Food Sci. 69, 361–366.

Calligaris, S., Manzocco, L., Kravina, G., Nicoli, M.C., 2007a. Shelf-life modeling of bakery products by using oxidation indexes. J. Agric. Food Chem. 55, 2004–2009.

Calligaris, S., Sovrano, S., Manzocco, L., Nicoli, M.C., 2006. Influence of crystallisation on the oxidative stability of extra virgin olive oil. J. Agric. Food Chem. 54 (2), 529–535.

Cardelli, C., Labuza, T.P., 2001. Application of Weibull Hazard Analysis to the determination of the shelf-life of roasted ground coffee. LWT-Food Sci. Technol. 34 (5), 273–278.

Codex Alimentarious, 1999. Codex Standard for Named Vegetable Oils 210–1999.

Cognat, C., Shepherd, T., Verrall, S.R., Stewart, D., 2014. Relationship between volatile profile and sensory development of an oat-based biscuit. Food Chem. 160, 72–81.

Colakoglu, A.S., 2007. Oxidation kinetics of soybean oil in the presence of monoleion, stearic acid and iron. Food Chem. 101, 724–728.

Corradini, M.G., Peleg, M., 2007. Shelf life estimation from accelerated data. Trends Food Sci. Technol. 18, 37–47.

Corrigan, V., Hedderley, D., Harvey, W., 2012. Modeling the shelf life of fruit-filled snack bars using survival analysis and sensory profiling techniques. J. Sens. Stud. 27, 403–416.

Cunha, L.M., Oliveira, F.A.R., Oliveira, J.C., 1998. Optimal experimental design for estimating the kinetic parameters of processes described by the Weibull probability distribution function. J. Food Eng. 37, 175–191.

EC (European Commission), November 22, 2011. Regulation No. 1169/2011 of the European Parliament and of the Council of 28 January 2002. Off. J. Eur. Communities L304/18, 1–46.

Echavarria, A.P., Pagan, J., Ibarz, A., 2012. Melanoidins formed by Maillard Reaction in food and their biological activity. Food Eng. Rev. 4, 203–223.

Frankel, E.N., 2005. Stability methods. In: Frankel, E.N. (Ed.), Lipid Oxidation. The Oil Press, Davis, pp. 165–186.

Gambaro, A., Fiszman, S.M., Gimenez, A., Varela, P., Salvador, A., 2004. Consumer acceptability compared with sensory and instrumental measures of white pan bread: sensory shelf-life estimation by survival analysis. J. Food Sci. 69 (9), S401–S405.

Ghidouche, S., Rey, B., Michel, M., Galaffu, N., 2013. A Rapid tool for the stability assessment of natural food colours. Food Chem. 139, 978–985.

Gimenez, A., Varela, P., Salvador, A., Ares, G., Fiszman, S., Garitta, L., 2007. Shelf life estimation of brown pan bread: a consumer approach. Food Qual. Prefer. 18, 196–204.

Gutierrez, F., Fernandez, J.L., 2002. Determinant parameters and components in the storage of virgin olive oil. Prediction of storage time beyond which the oil is no longer of "Extra" quality. J. Agric. Food Chem. 50, 571–577.

Hough, G., Sanchez, R.H., Garbarini de Pablo, G., Sanchez, R.G., Calderon-Villaplana, S., Gimenez, A.M., Gambaro, A., 2002. Consumer acceptability versus trained sensory panel scores of powdered milk shelf-life defect. J. Dairy Sci. 85, 2075–2080.

Hough, G., Garitta, L., Gómez, G., 2006. Sensory shelf-life predictions by survival analysis accelerated storage models. Food Qual. Prefer. 17, 468–473.

Hough, G., Garitta, L., 2012. Methodology for sensory shelf-life estimation: a review. J. Sens. Stud. 27 (3), 127–137.

Houhoula, D.P., Oreopolou, V., 2004. Predictive study for the extent of deterioration of potato chips during storage. J. Food Eng. 65, 427–432.

Huang, J., Sathivel, S., 2008. Thermal and rheological properties and effects of temperature on the viscosity and oxidation rate of unpurified salmon oil. J. Food Eng. 89, 105–111.

Imai, H., Maeda, T., Shima, M., Adachi, S., 2008. Oxidation of methyl linoleate in oil-in-water micro- and nanoemulsion systems. J. Am. Oil Chem. Soc. 85, 809–815.

IFS Food. Version 6, 2012. IFS Management GmbH, Berlin.

ISO 22000, 2005. Food Safety Management Systems. Requirements for Any Organization in the Food Chain.

Jung, D.M., Yoon, S.H., Jung, M.Y., 2012. Chemical properties and oxidative stability of perilla oils obtained from roasted perilla seeds as affected by extraction methods. J. Food Sci. 77 (12), C1249–C1255.

Karwowska, M., Wojciak, K., Dolatowski, Z.J., 2014. Comparative studies on lipid oxidation of organic model sausage without nitrite produced with the addition of native or autoclaved mustard seed and acid whey. Int. J. Food Sci. Technol. 49, 2563–2570.

Kramer, A., Twigg, B., 1968. Measure of frozen food quality and quality changes. In: Tressler, D.K. (Ed.), The Freezing Preservation of Foods, fourth ed. vol. 2. AVI Publ. Co., Westport.

Labuza, T.P., Schmidl, M.K., 2005. Accelerated shelf life testing of foods. Food Technol. 39, 57–64.

Lin, X.Y., Wu, J., Zhu, R.B., Chen, P., Huang, G.W., Li, Y., Ye, N.H., Huang, B.H., Lai, Y.P., Zhang, H., Lin, W.Y., Lin, J.J., Wang, Z.B., Zhang, H., Ruan, R.G., 2008. California almond shelf life: lipid deterioration during storage. J. Food Sci. 77 (6), C583–C593.

Mancebo-Campos, V., Fregapane, G., Desamparados-Salvador, M., 2008. Kinetic study for the development of an accelerated oxidative stability test to estimate virgin olive oil potential shelf life. Eur. J. Lipid Sci. Technol. 110 (10), 967–976.

Manzocco, L., Panozzo, A., Calligaris, S., 2012. Accelerated shelf life testing (ASLT) of oils by light and temperature exploitation. J. Am. Oil Chem. Soc. 89 (4), 577–583.

Manzocco, L., Calligaris, S., Nicoli, M.C., 2010. Method of shelf-life determination and prediction. In: Decker, E., McClements, J., Elias, R. (Eds.), Oxidation in Foods and Beverages and Antioxidant Applications. Woodhead Publishing Limited, Cambridge, pp. 196–219.

Manzocco, L., Calligaris, S., Nicoli, M.C., 2006. Modeling bleaching of tomato derivatives at subzero temperatures. J. Agric. Food Chem. 54 (4), 1302–1308.

Manzocco, L., Kravina, G., Calligaris, S., Nicoli, M.C., 2008. Shelf life modelling of photosensitive food: the case of coloured beverages. J. Agric. Food Chem. 56, 5158–5164.

Martınez, M., Barrionuevo, G., Nepote, V., Grosso, N., Maestri, D., 2011. Sensory characterisation and oxidative stability of walnut oil. Int. J. Food Sci. Technol. 46, 1276–1281.

Muñoz, M., Civille, V.G., Carr, B.T., 1992. Sensory Evaluation in Quality Control. Van Mostrand Reinhold, New York.

Nenadis, N., Tsikouras, I., Xenikakis, P., Tsimidou, M.Z., 2013. Fourier transform mid-infrared spectroscopy evaluation of early stages of virgin olive oil autoxidation. Eur. J. Lipid Sci. Technol. 115, 526–534.

Nicoli, M.C., 2012. The shelf life assessment process. In: Nicoli, M.C. (Ed.), Shelf Life Assessment of Food. CRC Press, Boca Raton, pp. 17–36.

Odriozola-Serrano, I., Soliva-Fortuny, R., Martin-Belloso, O., 2009. Influence of storage temperature on the kinetics of the changes in anthocyanins, vitamin C, and antioxidant capacity in fresh-cut strawberries stored under high-oxygen atmospheres. J. Food Sci. 74, C184–C191.

Özilgen, S., Özilgen, M., 1990. Kinetic model of lipid oxidation in foods. J. Food Sci. 55, 498–501.

Parker, R., Ring, S.G., 1995. A theoretical analysis of diffusion-controlled reactions in frozen solution. Cryo-Letters 16, 197–208.

Rababah, T.M., Feng, H., Yang, W., Yucel, S., 2012. Fortification of potato chips with natural plant extracts to enhance their sensory properties and storage stability. J. Am. Oil Chem. Soc. 89, 1419–1425.

Rizzo, V., Torri, L., Licciardello, F., Piergiovanni, L., Muratore, G., 2014. Quality changes of extra virgin olive oil packaged in coloured polyethylene terephthalate bottles stored under different lighting conditions. Packag. Technol. Sci. 27, 437–448.

Roberson, G.L., 2009. Food packaging and shelf life. In: Robertson, G.L. (Ed.), Food Packaging and Shelf Life. A Practical Guide. Taylor & Francis Group, Boca Raton, pp. 1–16.

Sae–leaw, T., Benjakul, T., 2014. Fatty acid composition, lipid oxidation, and fishy odour development in seabass (*Lates calcarifer*) skin during iced storage. Eur. J. Lipid Sci. Technol. 116, 855–894.

Shim, S.D., Lee, S.J., 2011. Shelf-life prediction of perilla oil by considering the induction period of lipid oxidation. Eur. J. Lipid Sci. Technol. 113 (7), 904–909.

Takeungwongtrakul, S., Benjakul, S., 2014. Oxidative stability of lipids from hepatopancreas of Pacific white shrimp (*Litopenaeus vannamei*) as affected by essentialoils incorporation. Eur. J. Lipid Sci. Technol. 116, 987–995.

Tazi, S., Plantevin, F., Di Falco, C., Puigserver, A., Ajandouz, E.H., 2009. Effect of light, temperature and water activity on kinetics of lipoxidation in almond-based products. Food Chem. 15, 958–964.

Toci, A.T., Neto, V.J.M.F., Torres, A.G., Farah, A., 2013. Changes in triacylglycerols and free fatty acids composition during storage of roasted coffee. LWT-Food Sci. Technol. 50, 581–590.

Zamora, R., Hidalgo, F., 2005. Coordinate contribution of lipid oxidation and maillard reaction to the nonenzymatic food browning. Crit. Rev. Food Sci. Nutr. 45, 49–59.

Chapter 3

Sensory Evaluation of Oils/Fats and Oil/Fat–Based Foods

Xiaoqing Yang and Robin A. Boyle
Kalsec®, Inc., Kalamazoo, MI, USA

3.1 INTRODUCTION

Sensory evaluation is a key method to assess the flavor quality of foods because it measures what consumers really perceive; however, it is a subjective method. For example, one consumer may describe a sample as unpalatable while another consumer may consider the same sample acceptable. These differences are common in sensory evaluation and can be explained by nationality, culture, individual variation, etc. (Bryhni et al., 2002). To obtain more objectiveness through sensory evaluation, a reasonable number of panelists must be trained. Training protocols and parameters need to be carefully designed based on the intrinsic characteristics of the tested food matrix and any possible variation among different samples caused by food formulation, processing, packaging, storage conditions, etc. Therefore, sensory evaluation can be very time-consuming and costly.

To overcome the disadvantages of sensory evaluation, numerous analytical methods to detect hydrolysis and oxidation molecules in foods are routinely used in the food industry and academia to measure the stability and quality of foods. Although not directly associated with the sensory quality of foods, these methods provide objective measurements and possible insight into the origin of flavor quality deterioration. However, there is no universal analytical method that can explain all of the changes in every food system. Most analytical methods require sample preparation, such as proper heating to measure volatile compounds in food samples by gas chromatography and solvent extraction to measure peroxides in snacks. Therefore, it is necessary to select an appropriate analytical method as well as sample preparation method for a particular food matrix. Details of numerous analytical methods to measure hydrolysis and oxidation molecules in various food matrices have already been discussed in other chapters of this book.

Sensory science is "a scientific discipline used to evoke, measure, analyze, and interpret reactions to those characteristics of food and other materials as they are perceived by the senses of sight, smell, touch, taste, and hearing"

Oxidative Stability and Shelf Life of Foods Containing Oils and Fats
http://dx.doi.org/10.1016/B978-1-63067-056-6.00003-3

(Stone and Sidel, 2004). Sensory science plays a major role in the consumer products industry. In addition to the evaluation of food and beverages, sensory science is used for testing personal hygiene products, household products, and environmental odors (Meilgaard et al., 2006; Stone and Sidel, 2004). While many aspects of food and beverages can be measured using analytical techniques, sensory science, using human assessors, provides unique insight into the sensory perception of those foods and beverages. Sensory evaluation provides support to research and development, marketing, and quality control. New product development and existing product improvement use sensory science to acquire information about product differences, consumer preferences, and attribute intensities. Marketing departments rely on the information gathered from sensory evaluation to identify consumer segments, support advertising claims, and monitor competition. Existing product maintenance can be monitored by a sensory quality control department to ensure that finished products are produced consistently. The effects of processing or ingredient changes, plant-to-plant variation, pilot-plant scale-up, cost reductions, and packaging changes can all be measured using sensory evaluation. In some cases, it may be possible to correlate certain sensory properties with analytical or instrumental analyses. Ultimately, sensory evaluation should aid in making sound business decisions using the best information available (Meilgaard et al., 2006; Sidel and Stone, 1993).

3.2 PROPER TEST DESIGN/TEST CONTROLS

To accurately measure product differences, all features of a sensory test must be controlled or measured. These features are grouped into three major categories: test, product, and panel controls. While the exact testing protocol will be dictated by the type of test being used, certain test controls are essential for product evaluation.

The physical nature of where the test is taking place will have an effect on the outcome of the test. Testing rooms should be comfortable, temperature/humidity controlled, neutral locations, free of loud noises, distractions, and odors. Sensory tests are expensive and time-consuming, and a properly designed testing room will minimize panelists' bias and eliminate the need to measure variation not resulting from the products themselves. A preparation area separate from but near to the testing room allows technicians to efficiently prepare and serve test samples and clean-up after tests are complete. Test samples should be prepared and presented under controlled conditions. Cooking time, serving temperature, and the amount and appearance of sample presented should be standardized to minimize sample-to-sample variation. Samples should be served in containers appropriate for the type of sample being tested. Samples should be presented blind-coded, preferably with a three-digit number, in a randomized and balanced order. Randomized, balanced serving order will minimize order bias. The number of samples presented should take into consideration the type of sample being tested. Trained panelists can evaluate up to 30 samples if measuring appearance only (Meilgaard et al., 2006). If samples are highly fatiguing or have a large carry-over effect, one or two samples may be the maximum number that can be evaluated during a single session. Panelists, consumers or highly trained, need detailed and clear instructions on how to evaluate the samples presented to them. The amount of sample to evaluate, the delivery of the sample, how to fill out the ballot, and how to dispose of the test sample must be conveyed before testing begins. A sensory panel is an essential tool for measuring differences among products and calibration- and standardization of a sensory panel will minimize variation among panelists, resulting on more powerful results (Meilgaard et al., 2006).

3.3 TYPES OF SENSORY TESTS

3.3.1 Affective Tests

Affective tests are divided into two main categories: qualitative and quantitative. Qualitative consumer research includes one-on-one interviews, focus groups, and consumer panels. Qualitative research is used in conjunction with and to help interpret the results of quantitative consumer studies and/or discriminative tests using trained panelists. Quantitative consumer studies measure preference and acceptance (the degree of liking) among products. The 9-point hedonic scale is the most widely used scale for measuring acceptance/liking. Consumers are presented with the products in question and are asked to score their liking on a scale of one to nine, where 1 = dislike extremely, 9 = like extremely, and 5 = neither like nor dislike.

3.3.2 Discrimination Tests

Discrimination testing is an important way to test whether changes in ingredients and processes, for cost reduction purposes or otherwise, affect the product in question. They compare the qualitative and/or qualitative variation among products with no indication of preference or acceptance. Discrimination testing also supports quality control practices and aids in compliance with health initiatives (Ennis, 2012). Overall difference tests, including triangle, tetrad, and duo-trio tests, ask whether a difference exists between two samples. Attribute difference tests, including paired comparison, ranking, and category scale tests, asks if a difference exists between two samples based on a particular attribute.

Triangle testing is one of the most popular types of discrimination tests where judges are presented with three samples, two identical and one odd, and are asked to identify the odd sample (Ennis, 2012; Ennis and Rousseau, 2012). Triangle tests are useful discrimination tests because judges only need to be familiar with the testing format and the nature of the difference, if detected, does not need to be specified. A triangle test is used to determine if an overall difference exists between two products. While triangle tests are statistically more powerful than other difference tests, products that cause sensory fatigue or have high carryover effects should not be tested. Triangle tests require large sample sizes to be effective, 20–40 panelists are typically used (Ennis and Rousseau, 2012). Samples should be presented in a random and balanced order based on the following six possible combinations (AAB, ABB, ABA, BAB, BAA, and BBA). Statistical tables, like those found in Meilgaard et al. (2006), can be referenced to determine significance based on the total number of responses, number of correct responses, and α-level.

Tetrad tests are increasing in popularity (Masuoka et al., 1995; Ennis and Jesionka, 2011; Garcia et al., 2012) and in some cases have proven to be a more sensitive method for overall difference testing than triangle tests (Garcia et al., 2012). During a tetrad test, judges are presented with four samples and are asked to group the samples into two groups of two based on similarity (Ennis, 2012). The benefit of the tetrad test is that it has the ability to detect differences more often and it requires a smaller sample size; as long as the effect size (size of the difference between two groups) does not decrease by more than one third the effect size of the triangle test, the tetrad test is a more powerful discrimination test (Ennis, 2012; Ennis and Rousseau, 2012). Tetrad tests, compared with triangle tests, are more stable in the presence of sample and panelist variability. Despite these benefits, tetrad tests require a fourth sample to be presented to the judges, and a fourth sample may reduce the effect size (magnitude of the treatment effect) to a point where the theoretical advantage of the tetrad test is lost (Ennis, 2012; Ennis and Rousseau, 2012). Sensory products that cause fatigue (i.e., pungent products) should not be tested using a tetrad test (Garcia et al., 2012). Samples should be presented in a randomized and balanced order based on the following six possible combinations (BAAB, AABB, BABA, ABAB, BBAA, and ABBA). The same statistical tables that are used to determine significance for a triangle test are used to determine significance in a tetrad test. Despite to the recent increase in the popularity of the tetrad test, the literature is lacking published research using tetrad testing for measuring oxidized aromas and flavors in foods.

The duo-trio test is used to determine if a difference exists between two products. In a duo-trio, the panelists are presented with three samples: a known reference sample and two blind-coded samples, and are asked which blind coded sample matches the reference sample. This test is simple and easily understood and can be used with trained or untrained panelists. When using trained panelists, the constant reference mode should be used in which the reference sample is the same every time. When using untrained panelists or when both coded samples are unknown, the balanced reference mode is used in which either sample being compared can be used as the reference sample (Meilgaard et al., 2006).

Difference-from-control tests are used to determine if differences exist between one or more samples compared with a control and to measure the size of the difference. Difference-from-control tests are useful when differences are known to exist and knowing the size of the difference affects whether the test objective has been met. They are also useful for evaluating products that are not homogeneous, like meats and baked goods. Trained or untrained panelists are given one or more blind-coded samples and are asked how different the sample is compared with a labeled control sample. The size of the difference is measured by the panelist using a verbal or numerical category scale. Panelists should be familiar with the test format and the scale being used and should know that blind controls may be included as part of the sample set (Meilgaard et al., 2006).

Paired comparison tests, also known as directional difference and two-alternative forces choice tests, measure the way in which a particular sensory attribute differs between two samples. Two-sided or bilateral paired comparison tests do not have a "correct" answer because developers are interested in how the attribute being tested is perceived in both samples. One-sided or unilateral paired comparison tests do have a "correct" answer and developers have a degree of certainty about which sample has a stronger sensory perception with respect to the attribute in question. Paired comparison tests are simple to conduct and use either untrained or trained panelists. The minimum number of panelists used is determined based on the type of test, one- or two-sided, and on the test-sensitivity parameters. Tables, like those found in Meilgaard et al. (2006), can be referenced to determine the minimum number of panelists needed and to determine the minimum number of correct, or directional responses required for significance.

3.3.3 Threshold Testing

Threshold testing measures the limits of sensory perception of individuals or groups of individuals (ASTM, 2004a,b). Developers are often interested in knowing the point at which certain substances affect the acceptability of food. The American Society for Testing and Materials (ASTM, 2004a,b) and the International Organization for Standardization (ISO, 2002) support the use of the three-alternative force choice (3-AFC) method for determining threshold. During a 3-AFC threshold test, panelists are presented with multiple sets of three samples where each set contains two controls and a test sample. Sets of samples are presented in a series of ascending concentration; group threshold can be calculated based on individual threshold values. Because methods for determining threshold vary depending on the test objective and because the methods used can greatly influence results (Doty et al., 1986; Hangartner and Paduch, 1998), careful consideration should be used to ensure alignment of test objectives and methods to ensure correct interpretation of results.

3.3.4 Descriptive Analysis

Descriptive analysis uses panelists that are trained to detect and describe differences among products. The panelists must be able to indicate which sensory attributes are present in a product and be able to measure the intensity of those attributes. Descriptive analysis techniques allow the sensory scientist to obtain detailed information about products including subtle differences in important sensory attributes. The most popular and widely used descriptive analysis methods include quantitative descriptive analysis, the Spectrum™ method, and time intensity but are not inclusive of all descriptive techniques available for product evaluation.

Quantitative Descriptive Analysis (QDA®) was developed to advance the field of sensory science by using human assessors as measuring instruments. Panelists are selected based on their ability to discriminate between products and are trained using product and attribute references. Intensities are scored on a 15-cm-line scale; data are statistically analyzed and often visually represented in the form of spider diagrams. While QDA® was a significant step forward with respect to formal sensory evaluation, several drawbacks remain. Panelists are not formally trained, which may lead to the development of erroneous attributes and the use of different parts of the scale. Panel leaders act as facilitators, and the lack of direction could lead to some panelists dominating lexicon development.

The Spectrum™ descriptive analysis method solves many of the issues with QDA® and was developed to provide reproducible sensory results for a variety of products (Munoz et al., 1992; Munoz and Civille, 1998). Panelists are trained using a standard set of attribute references that correspond with set intensities scored on a 15-cm-line scale (Meilgaard et al., 2006). The panel leader provides authority and structure and works with panelists to develop lexicons based on common terminology.

Time intensity is a method for tracking changes in sensory attributes over time. In foods and beverages, time intensity testing is useful for measuring attributes that linger for long periods of time, like pungency and bitterness. An attribute's time-intensity profile may be the most important aspect in explaining differences among products (Overbosch et al., 1986).

3.4 EXAMPLES FROM THE LITERATURE

Xu et al. (1999) measured the acceptability of six frying oils based on the liking ratings of five sensory attributes: color, taste, overall quality, greasiness, and crispiness. Their objective was to determine the frying performance of three high-oleic canola oils compared with a partially hydrogenated canola oil, medium-high-oleic sunflower oil, and palm olein. Chemical analysis, including fatty acid analysis, iodine value (IV), color index (CI), dielectric constant, free fatty acid content (FFA), and total polar compounds (TPC), was also conducted. Chips were fried in the oils at 190±2 °C for 3.5 min every 10 min on sensory panel days. Sensory analysis was conducted by 36 panelists on days 1, 3, 6, 8, and 10. Color acceptability significantly decreased as frying time increased and was highly correlated with TPC and FFA present in the oils ($r=-0.665$

and −0.661, $P<0.001$). The degree of greasiness of all the oils significantly increased with frying time and was also highly correlated with TPC ($r=0.753$, $P<0.001$). Partially hydrogenated canola oil had the greatest degree of unpleasant off-flavor and, not surprisingly, the lowest scores for taste acceptability and overall quality. The authors were able to use sensory data from consumer liking scores and analytical measurements to explain the sensory perception of different types of oils.

Moore et al. (2012) studied the aroma effects of omega-3 fatty acid–fortified dairy-based beverages compared with commercially available ultra-high temperature (UHT) processed milk. Triangle tests with untrained panelists were used to determine if differences could be detected between dairy beverages containing fish oil and standard UHT processed milk. Sample were stored for 1 week and then presented to the panelists. No significant differences in aroma were detected between the control and the fish oil–fortified dairy beverage, suggesting that fish oil does not have a major impact on the aroma profile of dairy-based beverages.

Petersen et al. (2012) used duo-trio, paired comparison tests and analytical methods to evaluate oxidation in conventional and high-oleic rapeseed oil. The constant reference method duo-trio test using trained panelists was conducted on aged samples (samples taken every other day for 26 days stored at 40 °C) of conventional and high-oleic rapeseed oil compared with a fresh reference for each. Panelists were asked to identify the sample that was different from the control; selection of the aged sample would qualify as a correct answer. For paired comparison tests, trained panelists were presented with two samples of either conventional or high-oleic rapeseed oil and were asked which sample was more rancid with regard to aroma and flavor. Duo-trio results showed a higher stability for high-oleic rapeseed oil (26 days) compared with conventional rapeseed oil (22 days). Paired comparison tests showed a difference in conventional rapeseed oil after 14 days of storage compared with fresh conventional rapeseed oil. High-oleic rapeseed oil was distinguishable from fresh oil after 14 and 20 days for flavor and aroma, respectively; high-oleic rapeseed oil was different from all other aged high-oleic samples after 26 days of storage. Analytical methods (HS-SPME for propanal, E,E-2,4-hexadienal, and E-2-heptenal) detected differences between samples of conventional rapeseed oil aged for 12 days compared with the fresh reference sample; differences in aged high-oleic oil occurred at 16 days' storage.

Rognlien et al. (2012) evaluated the sensory characteristics of chile-lime yogurt enriched with butter oil and fresh and oxidized fish oil. Preliminary experiments, using triangle testing, were conducted to determine if untrained panelists could detect fresh and oxidized fish oil in plain yogurt samples. Significant differences were detected for fresh versus oxidized fish oil ($P=0.0111$) and butter versus oxidized fish oil ($P=0.0004$); however, differences were less evident when panelists were presented with butter oil and fresh fish oil ($P=0.0589$). Triangle testing was also used to determine flavor concentrations for consumer and trained panel evaluations. Trained panelists evaluated six yogurt samples for five flavor attributes: lime, sweet, heat, acid, and oxidized following methods comparable to the Spectrum™ method (Meilgaard et al., 2006). No significant differences were detected between the samples made with butter oil. Samples made with high levels of oxidized fish oil were rated significantly higher in oxidized flavor and significantly lower in lime and acid ($P < 0.05$). Consumers evaluated the same yogurt samples for flavor and overall acceptance using a 9-point hedonic scale. Overall and flavor acceptability was bimodally distributed, which may have been caused by some of the consumers liking the savory chile flavor and others not. The addition of fish oil lowered flavor acceptability compared with the yogurt samples with butter oil (4.82 versus 5.40, respectively). The authors conclude that the overall low flavor and overall acceptability scores may be due to consumers' preconceived notions that sweet, fruit-flavored yogurt is the norm; these yogurt samples lacked additional sweeteners and contained savory flavors.

Gómez-Alonso et al. (2004) evaluated the recognition threshold, the level at which a certain stimulus can be recognized and identified, of rancid aroma in aged olive oil. Twelve olive oil samples were stored in the dark at different temperatures (25, 40, 50, 60, and 75 °C) and evaluated by a sensory panel at 12 stages of oxidation. Panelists were instructed to mark samples of olive oil based on whether a rancid defect was present compared with a fresh sample of olive oil. The authors found an inverse relationship between storage temperature and time; samples stored at high temperatures required less time for recognition threshold levels of rancid aroma to develop.

Emrick et al. (2005) studied the heat and warmed-over flavor (WOF) intensity of precooked chicken using time-intensity methods and descriptive analysis with a trained sensory panel. Cooked meats that are refrigerated or frozen then reheated before consumption can produce off notes, typically described as WOF. WOF can be described as cardboard, painty, metallic, and rancid. The study looked at the heat (pungency) intensity and WOF of precooked, chopped, and formed chicken patties at three fat levels (5%, 7%, and 9%) and three concentrations of red pepper (0%, 0.2%, and 0.4%). The samples were stored for 9 weeks and evaluated by the sensory panel every other week. Panelists were trained in time intensity using reference water solutions dosed with varying levels of capsaicin. Test samples were rated for heat intensity every 15 s over a period of 3 min on a scale of 0–150 with threshold (12), slight (50), and moderate (100) anchors acting as reference points on the scale. Panelists were trained in WOF using fresh and cooked, refrigerated, then reheated samples of chicken patties. Panelists evaluated WOF on a 0–15 point line scale. Contrary to other studies (Carden et al., 1999), the authors found that the chicken patties with higher fat content (7% and 9%) had a higher total heat intensity (measured as area under the curve) and maximum heat intensity (maximum point on the heat intensity curve) and speculated that the fat is a carrier for the red pepper due to the nature of the food system. As the concentration of red pepper increased, the perceived

WOF intensity decreased ($P<0.005$) and, not surprisingly, the chicken patties with the highest fat content (9%) had the highest WOF intensity. The authors theorize that the red pepper may be masking the perception of WOF or the red pepper may be acting as an inhibitor of oxidation. Other studies (Chang and Chen, 1998; Larson, 1988) have suggested that chili peppers may have the potential to be natural antioxidants, and with recent shifts in consumer perspectives of food additives, antioxidants from plant sources are gaining popularity.

Villarino et al. (2007) evaluated virgin coconut oil (VCO) and refined, bleached, and deodorized (RBD) coconut oil to characterize the sensory characteristics unique to each type of oil and to describe the differences between them. Trained panelists evaluated four VCO and one RBD samples using a descriptive analysis method combining techniques from both QDA® and the Spectrum™ method. The VCO samples were produced using the following processes: fermentation, fermentation with heat, and centrifugation with heat. Panelists were trained using reference standards and generated attributes for aroma, flavor, and appearance, which were scored on a 150-mm line scale. Through descriptive analysis, the authors found a significant difference in appearance among the samples; all VCO samples were nearly colorless while the one RBD sample had a distinct yellow color possibly resulting from mechanical, chemical, and thermal processing. Processing differences also had an effect on aroma and flavor; the RBD sample had no perceptible aroma or flavor, while the virgin samples were described as nutty, cocojam, latik (sweet cooked coconut notes), and rancid. The VCO sample made with fermented coconut milk and water (no heat) was described as the most rancid and significantly lower in cocojam and latik notes. By using a trained panel and descriptive methods, the authors were able to determine how different processing methods affect the sensory perception of coconut oils.

Martinez et al. (2002) combined consumer preference data and QDA® to produce an extended preference map describing the sensory properties that drive consumer liking and preference of cracker-type biscuits. Trained panelists scored aroma, flavor, and texture attributes for nine commercial brands of biscuits using a 100-mm unstructured line scale. Consumers were asked to rate the acceptability of samples based on appearance, flavor, texture, and overall acceptability. Consumers also used a 100-mm unstructured line scale with "dislike very much" at one extreme and "like very much" at the other extreme. Not surprisingly, high consumer acceptability scores for appearance, flavor, and texture were predictors of overall acceptance. Of the nine samples, the sample made with bovine fat, yeast, glucose syrup, and flavoring agents was rated lowest for texture, flavor, and overall acceptability and was rated highest by trained panelists for greasiness and rancid and margarine flavor (55, 64, and 20 on a 0–100 scale, respectively). The sample made with hydrogenated vegetable oil, malt extract, high-fructose corn syrup, sodium bicarbonate, lecithin, and powered skim milk was rated highest for texture and overall acceptance by consumers and trained panelists gave the sample high scores for noise and crispness

(79 and 78) and low for rough surface, rancid, and fermented flavor (44, 1, and 4). By combining trained panel and consumer data, the authors learned which sensory descriptors were driving consumer liking, information that could be used for future research and market development of cracker-type biscuits.

Ramirez et al. (2001) combined consumer acceptability and trained panel data scored for oxidized flavor to determine the effects of temperature and light on the shelf life of refined sunflower oil. Trained panelists were presented with five samples of sunflower oil including fresh oil and oil stored for 20 and 35 days at 60°C with and without light exposure. Oxidized flavor was measured on a 0–10 line scale following AOCS (1989) guidelines. Consumers were presented with cubed potatoes mixed with the different oil samples described earlier. Consumers rated acceptability of the oil on a 9-point hedonic scale. To establish sensory failure criteria, trained panelists and consumers evaluated sunflower oil samples that had been stored at 60 °C for 0, 15, 30, 45, 60, and 75 days. Shelf life was measured by comparing control (no oxidation) sunflower oil to oil samples stored at different temperatures (35, 45, and 60 °C) in the dark and exposed to 12 h of light per day. Nine of 60 consumers liked the more oxidized samples and were removed from the analysis for estimating sensory failure. Sensory failure was calculated by subtracting the Fisher's LSD (calculated from a two-way ANOVA) from the average consumer acceptability score of the control sample. The resulting sensory failure point of 5.8 (on the 9-point hedonic scale) corresponded to 3.6 on the 0–10 scale for oxidized flavor. Using the sensory failure estimation of 3.6, shelf life was estimated using linear regression. Light exposure significantly reduced shelf life; samples stored at 20 °C had an estimated shelf life of 281 days, while samples stored at 60 °C had a shelf life of 17 days. Shelf life of samples stored in the dark ranged from 24 days for samples at 60 °C and an estimated 1140 days for samples at 20°C (rough estimate based on extrapolation). Sunflower oil makes up 12% of the yearly per-capita consumption of vegetable oil, and understanding the relationship between storage conditions, oxidized flavor scores, and consumer acceptability is critical for delivering high-quality products.

3.5 CORRELATION OF SENSORY DATA WITH CHEMICAL ANALYSIS DATA

Flavor is one of the most important qualities of foods. The generation of off-odors and off-flavors in foods especially in oils/fats and oils/fats–based foods is usually associated with oxidative and/or hydrolytic degradations of the lipid components, and these major degradation reactions are summarized in Figure 3.1 (Robards et al., 1988). Factors affecting lipid oxidation include the degree of unsaturation of lipids; exposure to light, oxygen, or certain enzymes (lipoxygenases); presence of transition metal ions, pro-oxidants, or antioxidants, and so on, while the lipid hydrolysis pathway usually involves hydrothermal or enzymatic (lipase) reactions. Details of mechanisms involved in each pathway will not be

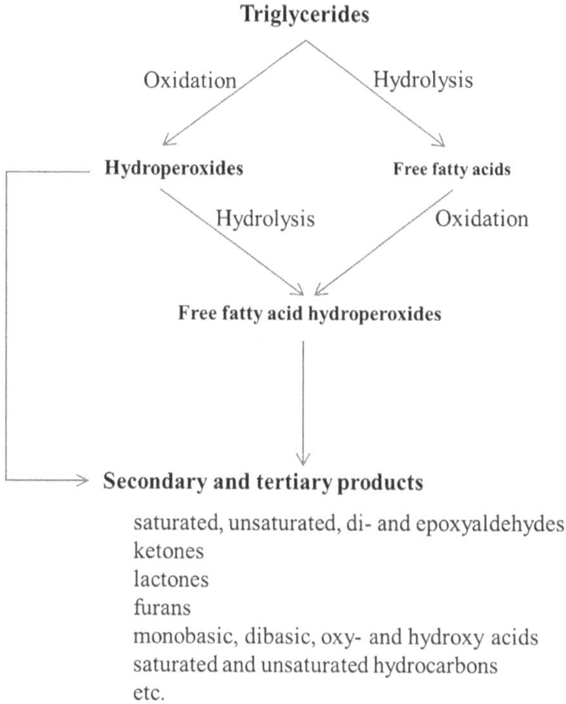

FIGURE 3.1 Oxidative and hydrolytic degradation reaction pathways. *Adapted from Robards et al. (1988) with permission of The Royal Society of Chemistry.*

covered in this chapter because they have been extensively reviewed (Robards et al., 1988; Talbot, 2011) and have been described in other chapters of this book.

The aim of this section is to establish the effectiveness of chemical analysis data in measuring food sensory quality, by examining the correlation between sensory data and analytical measurements. Since a vast number of analytical methods are available to evaluate hydrolysis and oxidation in oils/fats and oil/fat–based foods, and a great quantity of chemical compounds have been identified as markers responsible for off-odors and off-flavors in various food applications, this section will focus primarily on the most commonly used analytical methods and the key chemical markers. The correlation may provide useful information to conduct simple, fast, and cost-effective studies to predict the sensory quality and shelf life of oils/fats and oil/fat–based foods.

3.5.1 Free Fatty Acids

Free fatty acid (FFA) measurement is the simplest method to monitor hydrolytic rancidity in lipid-containing foods. They contribute to rancidity either directly or by generating subsequent oxidation compounds as shown in Figure 3.1. However, it cannot be neglected that the flavor influence of FFAs can be positive.

For example, certain short-chain FFAs were identified as important cheese flavor indicators to assess quality of Swiss cheese (Koca et al., 2007).

In most cases, FFAs are still the culprits for rancidity because they are easier to undergo auto-oxidation than intact triglycerides, especially in oils rich in short-chain FFAs such as lauric acid (e.g., coconut and palm kernel oils) because short-chain FFAs are more soluble in water than are long-chain FFAs (Gan et al., 2005a; Villarino et al., 2007), which means water in foods is more easily accessible to short-chain FFAs (Nawar, 1969). In a study to correlate FFA concentrations with rancidity score in butter, rancid and goat-like flavors were found to be principally caused by C_4 and C_6 FFAs (rancidity score = 1.93163 + 0.01754 butyric acid + 0.00054 C_{18} FFA congeners) and soapy and bitter flavors by C_{10} and C_{12} FFAs (Woo and Lindsay, 1983).

Commodity oils (e.g., soybean and sunflower oils) are predominantly sources of relatively long-chain (C_{16}–C_{18}) FFAs, which do not hydrolyze easily compared with short-chain FFAs; however, when these oils are used as frying media, the combination of high temperature and water introduced by foods leads to rapid generation of FFAs. For example, the increase of FFAs in frying oil as well as fried food products is strongly correlated with the quality of both the oil and the fried products (Choe and Min, 2007; Chung et al., 2004). In other similar research, the storage stability of potato chips fried in four types of canola oils (regular, low-linoleic acid, high-oleic, and hydrogenated canola oils) was studied (Petukhov et al., 1999), and the results showed that oil types had difference influences on the sensory quality of potato chips. The combination of chemical analysis of both the chips and the oils extracted from chips and the sensory evaluation results of the stored chips showed that both hydrolytic and oxidative products contributed to the generation of off-flavors in chips. Therefore, it is worth examining both FFAs and oxidative compounds to assess the sensory quality of oils/fats and oil/fat–based foods.

3.5.2 Peroxide Value

Peroxides as primary oxidation compounds are formed in the early stages of lipid oxidation, and they may be subject to subsequent oxidation reactions to generate a variety of nonvolatile and volatile secondary products (Dobarganes and Velasco, 2002). Peroxide value (PV) is one of the most widely used quality indicators in food industry, and a number of analysis methods have been developed to measure PV in oils/fats and oil/fat–based foods (Dobarganes and Velasco, 2002). However, since peroxides are susceptible to break down with time, measuring the PV trend as a function of time is recommended to give a complete picture of lipid oxidation status (Shahidi and Wanasundara, 2002).

Although widely used, the relationship between PV and the sensory quality of oils/fats or oil/fat–based foods is yet inconclusive. For example, in one study, PV was correlated with the painty odor development in regular and low-linolenic canola oils (Malcolmson et al., 1996), while in another study to evaluate red palm oil and refined, bleached, and deodorized palm oils, poor correlation coefficients were found (Idris et al., 1992). Another example is that PV showed excellent correlation with the rancidity rating of salmon (Aubourg et al., 2005), while no significant correlation was found in precooked beef patties (Thongtan et al., 2005). The problems may rise from interference caused by other lipid-soluble components, lack of sensitivity of the selected method to measure PV, or additional oxidation caused by solvent extraction of the oils/fats from foods. Other factors, such as the lipid composition and flavor profile of the tested matrices, also play an important role. Therefore, the correlation between PV and sensory data needs to be examined on a case-by-case basis.

3.5.3 Conjugated Dienes and Trienes

The formation of hydroperoxides generated from unsaturated fatty acids such as linoleic and linolenic acids is usually also accompanied by the generation of conjugated dienes (CD) due to the rearrangement of the double bonds, which absorb ultraviolet (UV) light at 230–235 nm. Conjugated trienes (CT) will also be produced, which give rise to a UV absorption peak at 268 nm (Shahidi and Zhong, 2005). CD and CT measurements are simple and fast; however, the measurements have less specificity and sensitivity than PV measurement (Gordon, 2001; Antolovich et al., 2002), and the result could be interfered by coexisting or co-extracted compounds such as carotenoids (Wheatley, 2000; Shahidi and Wanasundara, 2002).

Positive correlations between the sensory and CD/CT data have been reported in several foods, such as salami (Larrauri et al., 2013) and peanuts (Nepote et al., 2009; Riveros et al., 2010, 2013). However, using CD/CT as chemical indicators to evaluate the sensory quality of foods remains questionable. For example, poor correlations were found in frozen whole mackerel (Stodolnik et al., 2005) and in rapeseed and soybean oils, although both oils contain high levels of linoleic and linolenic acids (López-Aguilar et al., 2006; Pawłowicz et al., 2013). In a frying study, only fine-tuning of the frying oils by adding different rosemary extracts as antioxidants greatly influenced the correlations between the CD of the oils and the flavor quality of fried product (Reblova et al., 1999). Therefore, factors influencing the generation of conjugated double bonds, such as CD and CT, need to be carefully studied. And the use of two or more other methods together with CD/CT measurements to evaluate oxidation in foods is highly recommended.

3.5.4 Anisidine Value

The anisidine value (AV) method measures the secondary oxidation compounds, primarily 2-alkenals and 2,4-alkadienals generated due to hydroperoxide decomposition, and it is more sensitive to unsaturated aldehydes (Gordon, 2001). Since AV represents the content of secondary oxidation products, it is used instead of, or together with, PV to evaluate fats and oils. For example, AV is considered to be a more reliable indicator of oxidative rancidity than PV in a long-term storage stability study of palm-olein oil (van der Merwe et al., 2004).

However, this does not mean AV is a better chemical marker than PV. For example, in a study to evaluate the oxidative stability of omega-3–enriched fat spreads (oil-in-water-in-oil emulsions), PV showed fair correlations with the sensory quality of the samples while AV showed no correlations (O'Dwyer et al., 2013). In another study to evaluate the shelf life of oatcakes, good correlations were found between PV and several sensory parameters, such as aroma, flavor, and aftertaste of the oatcakes (0.78, 0.75, and 0.72, respectively), while AV did not correlate with any of these sensory parameters (Cognat et al., 2014). One hypothesis to explain this was that nonvolatile aldehydes measured by the AV method were converted into volatiles through the production of additional peroxides; it was also proposed that the n-3 polyunsaturated fatty acid content in oatcake is very low (on average, 4.3 mg/g of oil extracted); therefore, the aldehyde level that can be measured by the AV method was not sufficient to reflect the rancidity of oatcakes (Cognat et al., 2014).

Storage condition (temperature) also tends to have an influence on using AV as a sensory quality indicator. When determining the effects of various oils on the chemical and sensory stability of tortilla chips under a Schaal oven and practical storage conditions, the PV and AV of oils extracted from chips tended to support sensory data of Schaal oven–stored chips; however, they did not support that of the practical stored chips (Cognat et al., 2014). This difference is probably due to the altered oxidation pathways or reaction rates of lipids under different storage conditions, but the mechanism remains unknown.

These observations suggested that although AV measures the secondary oxidation products, its low specificity and tendency to be influenced by other factors, such as the physical states of lipids, fatty acid compositions, and storage conditions, greatly limit its application as a good chemical indicator to evaluate the sensory quality of foods.

3.5.5 Thiobarbituric Acid

The thiobarbituric acid (TBA) test measures malonaldehyde (MDA) produced due to the oxidation of fatty acids with three or more double bonds, and it measures other TBA reactive substances such as 2-alkenals and 2,4-alkadienals. Therefore, TBA is also referred to as TBARs (TBA reactive substances). So far, no data have been reported to investigate the correlation between TBA and sensory quality of snack foods, but TBA was found satisfactory to assess the quality of fats and oils, such as lard, cooking fat, and soybean, sunflower, and rapeseed oils, in the early stages of rancidity (Pokorný et al., 1985). TBA is also commonly used as a quality marker in fish and meat industry, but regarding its correlation with sensory assessment of fish and meat, various results have been reported. For example, positive correlation between TBA and rancidity rating was found in Coho salmon (Aubourg et al., 2005), while the increase of TBA in frozen Baltic cod and herring, mackerel, horse mackerel, and hake was not accompanied by the development of rancid odor (Kolakowska and Deutry, 1983). In the first study, lipid hydrolysis showed better correlation than oxidation with rancidity development, and in the second study, it was proposed that the presence of interfering substances such as protein may affect the TBA measurement. For the same fish species, environmental conditions showed influence on the fish shelf life determined by the sensory quality and chemical marker levels including the TBA of fish (Bogdanović et al., 2012) mostly due to different moisture, protein, and fat contents in the fish.

Sensory quality deterioration of fish can also be caused by microorganism spoilage (Ruiz-Capillas and Moral, 2001; Bogdanović et al., 2012), which can be measured by total volatile base nitrogen (TVB-N) and trimethylamine nitrogen (TMA-N). Thus, when trying to establish a correlation between TBA and sensory quality of fish, researchers also need to consider the influence from environmental conditions and spoilage microflora. Contrary correlations were also reported in meat. For example, a positive correlation was observed in a study to evaluate the antioxidant effects of several spices in ground beef during refrigerated storage despite the distinctive flavors imparted by different spices (Vasavada et al., 2006), while TBA showed no correlation with the odor, flavor, color, appearance, and texture in a similar study to evaluate oxidation in frozen, precooked beef patties (Thongtan et al., 2005).

For TBA measurement, there are a number of interfering factors influencing the TBA results in fish and meat as discussed earlier. For example, acids, esters, sugars, imides and amides, amino acids, and oxidized proteins can also react with TBA (Guillén-Sans and Guzmán-Chozas, 1998). Different sample preparation methods, including a test on the whole sample, a test on the extraction of the sample, a test on steam distillation of the sample, and a test on extracted lipid from the sample, also influence the accuracy and reproducibility of the measurements (Hoyland and Taylor, 1991). These studies were conducted using different materials under different processing and storage conditions, so they cannot be compared directly. However, examining these studies is necessary for researchers to improve the study methodologies, including sample preparation methods, measurement methods, sensory evaluation, and data analysis methods, to establish better correlation between TBA and sensory quality of fish and meat.

3.5.6 Volatiles Measured by Gas Chromatography and Related Instrumental Analysis Methods

During later stages of lipid oxidation, a wide range of volatile compounds can be generated in oils/fats and oil/fat–based foods. These volatile compounds include alcohols, aldehydes, ketones, alkanes, and organic acids. Many of them are directly responsible for off-odors and off-flavors. Gas chromatography

(GC), usually coupled with mass spectrometry (MS), has been widely used to identify and quantify these volatile molecules as it generates an accurate profile of volatile components with high sensitivity. Moreover, this technique also reduces the sample preparation work. For example, a whole food sample can be directly measured without any extraction procedure. Table 3.1 summarizes the most commonly used sampling methods, which include static headspace, dynamic headspace (purge and trap), solid phase microextraction (SPME) and stir bar sorptive extraction (SBSE) methods.

Correlations between total volatiles or specific volatile compound(s) and the rancidity odor/flavor of foods by sensory evaluation have been reported extensively. For example, 17 of 55 volatile compounds measured by SPME headspace GC were selected to assess the stability of conventional rapeseed and high-oleic rapeseed oils. Multivariate statistical analysis of the GC and sensory results showed that propanal, E,E–2,4-hexadienal, and E–2-heptenal were good

TABLE 3.1 Summary of Main Sampling Methods for Evaluation of Volatile Compounds by GC

Sampling Technique	Advantages	Limitations
Static headspace	Directly measures volatile compounds representative of the food sample itself	Sample-size limitation; low sensitivity
Dynamic headspace	Much better sensitivity than static headspace method with ample sample size allowed for analysis	Volatile compounds loss due to purge and trap process and the use of splits
Solid phase micro-extraction (SPME)	Better sensitivity than static headspace; faster and simpler than dynamic headspace	Sampling parameters (time, temperature, fiber selection, etc.), partition coefficient between sample and headspace and between the headspace and fiber coating, and possible competition for absorption of different volatiles on the fiber coating and so on may influence the analysis results
Stir bar sorptive extraction (SBSE)	Much higher sensitivity than SPME	Difficult for automated sampling; sampling parameters (stirring time, temperature, coating material of the stir bar, etc.) also need to be carefully selected

Summarized from Irwin and Hedges (2004)

chemical markers to differentiate conventional oil, while 3-octanone and octanal were more suitable to differentiate high-oleic oil regarding the different oxidative qualities of the oils (Petersen et al., 2012). A total of 106 volatile products were detected in three fish oils using SPME GC–MS, and eight of them correlated with sensory characteristics of all the oils in either a positive or a negative way (Sullivan Ritter and Budge, 2012). In fish oil–enriched mayonnaise evaluated by dynamic headspace GC–MS, 3,5-octadien-2-one was found to be strongly correlated with oxidation (Hartvigsen et al., 2000; Jacobsen et al., 2000). Volatile compounds are also good markers to assess the quality of fish and meat. For example, total volatile aldehydes rather than any individual aldehyde was highly correlated ($r=0.83$) with oxidized flavor in channel catfish (Brannan and Erickson, 1996). Similar correlations were also found between volatile compounds and consumer acceptability in sausages (Olivares et al., 2011).

Considerable amount of researches have been conducted to prove the effectiveness of using GC and GC–MS to assess the sensory quality of foods; however, this requires analyzing the development trends of a number of molecules, and the work can be very tedious. That is because flavor chemistry has identified more than 7000 flavor compounds during the past decade (Rowe, 2005), including both good and bad flavor indicators in different food applications. Often, different foods generate different groups of volatiles during storage/processing, which further poses challenges to food scientists to target a specific chemical marker. Recent studies proved that many of these lipid oxidation–related volatiles also correlated with the production of hexanal; therefore, hexanal has been one of the usual reported oxidation markers for evaluation of off-odors and off-flavors in foods. Examples of hexanal as a marker include fats and oils (Arvanitoyannis and Vlachos, 2007; Jeleń et al., 2007), frying oils and fried foods (Brewer et al., 1999; Warner, 2009), fish and meat (Fereidoon and Ronald, 1994; Tikk et al., 2008; Miyasaki et al., 2011), nuts (Lee et al., 2002a; Zajdenwerg et al., 2011), and low-fat foods such as various cereals (Fritsch and Gale, 1977). Hexanal, however, does not contribute directly to off-flavors and is closely correlated to the development of rancidity. Similarly, other aldehydes, such as pentanal, nonanal, and pentane, have also been proved to be used as effective rancidity indices. This does not mean that hexanal or any other individual aldehyde can be used as a universal chemical marker to evaluate rancidity development that applies to all the food applications. In some cases, poor correlation between hexanal level and sensory evaluation has also been noted (Crackel et al., 1988; Neilson et al., 2006).

GC-olfactometry (GC-O), also known as "GC-sniff," represents a combination of both instrumental and sensorial analysis methods. It is also an effective tool to assess the odor-active compounds concentrations and odor-intensity relationships, because not all molecules contribute equally to the whole flavor profile of the tested sample (d'Acampora Zellner et al., 2008). After its introduction, GC-O has been widely used to identify important food flavor/aroma compounds, including both the positive and negative ones, to assess the sensory quality of various foods. For example, 78 volatiles were correlated with fishy notes and 27 volatiles were correlated with rancid off-flavors in fish oil–enriched mayonnaise measured by GC-O (Hartvigsen et al., 2000). Other examples, including the potent odorants responsible for various flavors in dairy and meat products, have been extensively reviewed (d'Acampora Zellner et al., 2008). Overall, the introduction of GC-O was a breakthrough for flavor analysis to screen odor-active compounds in foods, which is promising to establish reliable correlations between sensory and analytical analysis data.

3.5.7 Other Techniques

Apart from the most commonly used chemical analysis methods and chemical markers described here, interest in using some other existing techniques as well as relatively new techniques to analyze food quality has also increased in recent years.

For instance, fluorescence spectroscopy has been used as a rapid and sensitive method to evaluate oxidation in fats and oils (Tan et al., 1995; Xu, 2000). Studies also showed strong correlations between fluorescence data and sensory quality of various food applications (Jensen et al., 2004; Veberg et al., 2006a; Veberg et al., 2007). Because samples can be measured nondestructively by fluorescence spectroscopy, this technique is promising for online real-time evaluation of oxidation status in food samples (Veberg et al., 2006b).

Another two examples are electronic nose (e-nose) and electronic tongue (e-tongue). E-nose contains an array of gas sensors, and a signal-collecting unit has been shown to have high sensitivity toward various flavor compounds in foods. The principle of e-tongue is similar to that of the e-nose, except that the sensors of e-tongue are designed for liquids; therefore, they react to nonvolatile compounds in liquid samples. It has been reported that the liquid sensors often possess higher selectivity and lower detection limits than the gas sensors used in e-nose (Baldwin et al., 2011). The applications of e-nose and e-tongue include inspection of ingredients quality, detection of flavor changes of foodstuffs, and possibly prediction of products shelf life (Röck et al., 2008; Baldwin et al., 2011). Regarding the shelf-life prediction, positive correlations between the signals from e-nose/e-tongue, sensory data obtained from sensory panel, and various lipid degradation compounds measured by GC–MS have been established. Examples include fats and oils (Gan et al., 2005b; Cosio et al., 2007; Mildner-Szkudlarz et al., 2008), meat and poultry (Haugen et al., 2006; Tikk et al., 2008), fish (Haugen and Undeland, 2002; Olafsdottir et al., 2005; Limbo et al., 2009), and nuts (Pastorelli et al., 2007). These results indicated that the e-nose and e-tongue techniques could be used as rapid, nondestructive, and economically feasible methods and could offer valid alternatives to traditional analytical and sensory evaluation methods. Most studies in this area focus more on the application of e-nose rather than of e-tongue, because e-tongue mainly detects and predicts the taste attributes of foods, such as bitter, sweet, sour, and salty. More studies will be needed in the future to relate these taste attributes with the sensory quality deterioration of oils/fats and oil/fat–based foods.

Relatively new analytical methods that can be used to measure oxidation in foods have also been developed recently. For example, electron spin resonance (ESR) spectroscopy measures free radicals due to the paramagnetic properties of the unpaired electrons in radicals, which allows detection of the onset of oxidation in foods (Andersen and Skibsted, 2002). Another example is nuclear magnetic resonance (NMR), which could qualitatively and quantitatively detect both primary and secondary oxidation compounds as well as the degradation of food components; therefore, NMR may help generate a relatively more complete picture of food stuffs degradation in a variety of applications (Martínez-Yusta et al., 2014). Although the correlations between these measurement results and sensory data are still missing, these relatively new methods may become more common in the future to predict the sensory quality and possibly shelf life of foods. Further studies are needed to investigate the reliability, as well as the advantages and disadvantages related to these methods.

3.6 DATA ANALYSIS

When establishing correlations between two different data sets, such as analytical and sensory data described in this chapter, it is important to adopt an appropriate data analysis tool. A simple method is to select a single parameter from both sets of data. For example, in the study to correlate sensory data and levels of chemical markers produced in Brazil nuts during storage, positive correlation coefficients were obtained by linear regression analysis between the sensory ratings and several chemical compounds, including PV, AV and hexanal (Zajdenwerg et al., 2011). In another study to evaluate the shelf life of olive oil under different storage conditions (Kanavouras et al., 2004), a mathematical model for calculating the probability that the olive oil would not have reached the end of its shelf life was established. The model was based on the concentration of an individual oxidation compound, such as hexanal and 2-pentyl furan. A limitation of the study was, considering that the sensory quality deterioration of food is a more complex process, the accuracy of the model to predict the real shelf life of olive oil remains questionable.

More advanced statistical analysis methods, such as multivariate analysis methods, can be used. Multivariate analysis methods represent a set of statistical analysis tools for shelf-life studies which allows all quality factors associated with shelf life to be involved in its prediction. One of the most applied multivariate analysis methods is principal component analysis (PCA) (Pedro and Ferreira, 2006). For example, PCA can be used to identify the most important variables responsible for causing different sensory attributes in the same samples during storage or among different samples (Nielsen et al., 1997; Anderson and Jones, 1999; Arvanitoyannis and Vlachos, 2007). It can also be used to study the relationship between sensory attributes and instrumental analysis results (Lee et al., 2002b; Trihaas et al., 2005). One more important application of PCA is to determine the shelf life of foods (Marsili, 2000; Pedro and Ferreira, 2006; Limbo et al., 2010). Multivariate analysis methods are more appropriate to establish correlations between sensory and chemical analysis results, because food-quality changes are usually nonlinear and multivariate (Martins et al., 2008). A great number of factors could exert influence on the sensory quality and the development of off-flavor compounds. Such factors include but are not limited to oxygen and water content in the local environment, surface area exposed to oxygen and light, and various processing and storage conditions. Therefore, choosing the most relevant factors to correlate with sensory data is of great importance. Multivariate analysis, such as PCA and other data-reduction techniques, offers an advantageous data processing tool to evaluate the sensory quality and to predict the shelf life of foods through instrumental analysis.

Furthermore, it is important to note that no matter what data analysis method is used, a model or equation that applies to one food application is highly unlikely to be applied to another. Even for the same food application, any changes such as environmental parameters changes may affect the accuracy of the model/equation because the chemical reaction pathways may not be the same. Therefore, the reliability of each model/equation needs to be examined with caution.

3.7 CONCLUSIONS

Foods are extremely complex mixtures consisting of many organic and inorganic components, such as lipids, proteins, minerals, water, and so on. Thus, the sensory quality deterioration of foods means a wide range of possible reaction products can be produced, and the type and concentrations of those reaction products depend on a number of factors. The major ones include:

1. Different formulations/compositions of foods
2. Possible interactions between different food ingredients, which may interfere with the food ingredients degradation pathways
3. Various processing conditions, such as thermal treatment and homogenization
4. Various storage conditions, such as packaging materials, and exposure to oxygen and light
5. Intrinsic and/or added flavors that may affect the sensory evaluation results
6. Presence of endogenous antioxidants/pro-oxidations/oxidation catalysts

Therefore, it is of great difficulty to establish a generic chemical marker such as PV or a group of markers such as total volatile aldehydes for all food applications to completely replace sensory evaluation. Although positive correlations between sensory and chemical analysis results have been widely studied and reported as described in this chapter, failures are also not uncommon – one marker may be a good sensory quality indicator in one study but poor relationship may be observed in another. Cases like this frequently appear in published literature due to the complexity of foods, different analytical instruments and methods used, subjectivity of sensory panels, different data processing methods applied, and so on. There is unlikely to be any generic solution that applies to all food systems, and each case needs to be treated as a separate problem (Irwin and Hedges, 2004).

For oils/fats and oil/fat–based foods, volatiles measured by GC–MS tend to give a more accurate picture of what is happening at the molecular level than the other chemical analysis methods, such as PV and AV. The volatiles profile has been broadly related to the key sensory attributes, especially the rancidity note that leads to the rejection of the sample by the sensory panel. Compared with other lipid hydrolysis and oxidation-resulting chemicals, such as FFA and peroxides, the volatile aldehydes are mainly responsible for potent off-flavors due to their low odor threshold values. Other volatile oxidation products such as furan derivatives, vinyl alcohols, ketones, and alkynes also contribute to the formation of undesirable flavor notes to a varying extent (Kochhar, 1996). Substantial evidence published in the literature suggests good correlations between key volatile compounds and certain off-flavor notes. However, researchers still need to be cautious because different volatiles are produced through different pathways affected by the complexity of foods and other factors discussed above. Case-by-case studies are needed to choose the most appropriate volatile compounds as sensory quality indicators. Also, the flavor/odor threshold values of different volatiles and even the same volatile compound, such as hexanal, differ greatly among various food systems. Therefore, the determination of the cutoff criterion of the chosen compound(s) in different food systems needs separate studies. The odor threshold concentrations of several selected volatiles in different matrices are listed in Table 3.2.

Overall consumers are still the most appropriate tool to determine when a food product is acceptable. Sensory evaluation is likely to be the gold standard when assessing the quality and shelf life of foods. Despite the advantages and disadvantages associated with both sensory and chemical analysis methods, the correlation studies provide valuable insight into the sensory quality deterioration process of various foods. This helps researchers to better understand the chemistry of off-flavor generation. And from there, strategies to eliminate the flavor quality changes can be developed, such as adding free radical scavengers to inhibit lipid auto-oxidation in foods, improving the packaging materials, and adjusting the storage and transportation conditions to slow down photo-oxidation and/or other unfavorable degradation reactions. These measures could ultimately provide cost-effective ways leading to food quality improvements. Another benefit is that, although in most cases the chemical markers are only comparable within the same food matrix, utilization of these markers makes food stability studies much easier. Any stability improvement strategy could be evaluated through chemical analysis without conducting sensory assessments repeatedly for all the treatments made. Generally, the efforts and studies that have been made to correlate sensory and chemical analysis greatly help to develop more rapid, efficient, and reliable ways for researchers to gain deeper knowledge regarding food science and analytical chemistry. This will also lead to improvements of food product qualities for the food industry and consumers.

TABLE 3.2 Odor Thresholds of Selected Volatiles

Volatile Compound	Odor Threshold (ppm)	Matrix	Odor Descriptors	References
2,4-Decadienal	0.135 0.47	High oleic safflower oil Lean ground beef	– – –	Buttery et al. (1973) Brewer and Vega (1995)
Heptanal	0.25 0.23	High oleic safflower oil Lean ground beef	– – –	Buttery et al. (1973) Brewer and Vega (1995)
2,4-Heptadienal	0.36	Virgin olive oil	Fatty, nutty	Morales et al. (2005)
Hexanal	0.12 5.87 0.08 0.075	High oleic safflower oil Lean ground beef Virgin olive oil Refined vegetable oil	– – Green apple, grassy Green, sweet	Buttery et al. (1973) Brewer and Vega (1995) Morales et al. (2005) Aparicio and Luna (2002)
2-Hexanal	1.125 7.87 0.42	Refined sunflower oil Lean ground beef Virgin olive oil	Green, astringent – Bitter almonds, green	Guth and Grosch (1993) Brewer and Vega (1995) Morales et al. (2005)
3-Hexanal	0.003	Refined vegetable oil	Green	Aparicio and Luna (2002)
Hexanol	0.4	Refined vegetable oil	Undesirable	Aparicio and Luna 2002

Nonanal	1.0 13.5	High oleic safflower oil Refined vegetable oil	– Rancid	Buttery et al. (1973) Aparicio and Luna (2002)
1-Octen-3-one	0.01	Virgin olive oil	Mushroom, mold, pungent	Morales et al. (2005)
Pentanal	2.67	Lean ground beef	–	Brewer and Vega (1995)
Pentanol	0.47	Refined vegetable oil	Fruity	Aparicio and Luna (2002)
1-Penten-3-one	0.0055 0.05	High oleic safflower oil Refined vegetable oil	– Green	Buttery et al. (1973) Aparicio and Luna (2002)

REFERENCES

AOCS, 1989. Flavor Panel Evaluation of Vegetable Oils. AOCS recommended practice Cg 2-83. AOCS, Champaign, Illinois.

ASTM, 2004a. Standard practice for determination of odor and taste thresholds by a forced-choice ascending concentration series method of limits. ASTM standard practice E 679–04. ASTM, West Conshohocken, Pennsylvania.

ASTM, 2004b. Standard practice for calculating individual and group sensory thresholds from forced-choice data sets of intermediate size. ASTM standard practice E 1432–04. ASTM, West Conshohocken, Pennsylvania.

Andersen, M.L., Skibsted, L.H., 2002. Detection of early events in lipid oxidation by electron spin resonance spectroscopy. Eur. J. Lipid Sci. Technol. 104 (1), 65–68.

Anderson, J.C., Jones, B.D., 1999. Principal factor analysis of extruded sorghum and peanut bar changes during accelerated shelf life studies. J. Food Sci. 64 (6), 1059–1063.

Antolovich, M., Prenzler, P.D., Patsalides, E., McDonald, S., Robards, K., 2002. Methods for testing antioxidant activity. Analyst 127 (1), 183–198.

Aparicio, R., Luna, G., 2002. Characterisation of monovarietal virgin olive oils. Eur. J. Lipid Sci. Technol. 104 (9–10), 614–627.

Arvanitoyannis, I.S., Vlachos, A., 2007. Implementation of physicochemical and sensory analysis in conjunction with multivariate analysis towards assessing olive oil authentication/adulteration. Crit. Rev. Food Sci. Nutr. 47 (5), 441–498.

Aubourg, S.P., Vinagre, J., Rodríguez, A., Losada, V., Larraín, M.A., Quitral, V., Gómez, J., Maier, L., Wittig, E., 2005. Rancidity development during the chilled storage of farmed coho salmon (*Oncorhynchus kisutch*). Eur. J. Lipid Sci. Technol. 107 (6), 411–417.

Baldwin, E.A., Bai, J., Plotto, A., Dea, S., 2011. Electronic noses and tongues: applications for the food and pharmaceutical industries. Sensors 11 (5), 4744–4766.

Bogdanović, T., Šimat, V., Frka-Roić, A., Marković, K., 2012. Development and application of quality Index method scheme in a shelf life study of wild and fish farm affected bogue (*Boops boops*, L.). J. Food Sci. 77 (2), S99–S106.

Brannan, R.G., Erickson, M.C., 1996. Sensory assessment of frozen stored channel catfish in relation to lipid oxidation. J. Aquat. Food Prod. Technol. 5 (1), 67–80.

Brewer, M.S., Vega, J.D., 1995. Detectable odor thresholds of selected lipid oxidation compounds in a meat model system. J. Food Sci. 60 (3), 592–595.

Brewer, M.S., Vega, J.D., Perkins, E.G., 1999. Volatile compounds and sensory characteristics of frying fats. J. Food Lipids 6 (1), 47–61.

Bryhni, E.A., Byrne, D.V., Rødbotten, M., Claudi-Magnussen, C., Agerhem, H., Johansson, M., Lea, P., Martens, M., 2002. Consumer perceptions of pork in Denmark, Norway and Sweden. Food Qual. Preference 13 (5), 257–266.

Buttery, R.G., Guadagni, D.G., Ling, L.C., 1973. Flavor compounds. Volatilities in vegetable oil and oil-water mixtures. Estimation of odor thresholds. J. Agric. Food Chem. 21 (2), 198–201.

Carden, L.A., Penfield, M.P., Saxton, A.M., 1999. Perception of heat in cheese sauces as affected by capsaicin concentration, fat level, fat mimetic and time. J. Food Sci. 64 (1), 175–179.

Chang, M., Chen, T., 1998. "Hotness" stability of chicken hot-wing products as affected by preparation methods and storage. Poult. Sci. 77 (4), 627–631.

Choe, E., Min, D.B., 2007. Chemistry of deep-fat frying oils. J. Food Sci. 72 (5), R77–R86.

Chung, J., Lee, J., Choe, E., 2004. Oxidative stability of soybean and sesame oil mixture during frying of flour dough. J. Food Sci. 69 (7), 574–578.

Cognat, C., Shepherd, T., Verrall, S.R., Stewart, D., 2014. Relationship between volatile profile and sensory development of an oat-based biscuit. Food Chem. 160 (0), 72–81.

Cosio, M.S., Ballabio, D., Benedetti, S., Gigliotti, C., 2007. Evaluation of different storage conditions of extra virgin olive oils with an innovative recognition tool built by means of electronic nose and electronic tongue. Food Chem. 101 (2), 485–491.

Crackel, R.H.L., Gray, J.I., Booren, A.M., Pearson, A.M., Buckley, D.J., 1988. Effect of antioxidants on lipid stability in restructured beef steaks. J. Food Sci. 53 (2), 656–657.

d'Acampora Zellner, B., Dugo, P., Dugo, G., Mondello, L., 2008. Gas chromatography–olfactometry in food flavour analysis. J. Chromatogr. A 1186 (1–2), 123–143.

Dobarganes, M.C., Velasco, J., 2002. Analysis of lipid hydroperoxides. Eur. J. Lipid Sci. Technol. 104 (7), 420–428.

Doty, R.L., Gregor, T.P., Settle, R.G., 1986. Influence of intertrial interval and sniff-bottle volume on phenyl ethyl alcohol odor detection thresholds. Chem. Sens 11 (2), 259–264.

Emrick, M.E., Penfield, M.P., Bacon, C.D., Van L aack, R.V.L., Brekke, C.J., 2005. Heat Intensity and Warmed-over Flavor in Precooked Chicken Patties Formulated at 3 Fat Levels and 3 Pepper Levels. J. Food Sci. 70 (9), S600–S604.

Ennis, J.M., 2012. Guiding the switch from triangle testing to tetrad testing. J. Sens. Stud. 27 (4), 223–231.

Ennis, J.M., Jesionka, V., 2011. The power of sensory discrimination testing methods revisited. J. Sens. Stud. 26 (5), 371–382.

Ennis, J.M., Rousseau, B., 2012. Reducing costs with tetrad testing. Inst. Percept. 15, 3–4.

Fereidoon, S., Ronald, B.P., 1994. Hexanal as an indicator of the flavor deterioration of meat and meat products. In lipids in food flavors. Am. Chem. Soc. 558, 256–279.

Fritsch, C.W., Gale, J.A., 1977. Hexanal as a measure of rancidity in low fat foods. J. Am. Oil Chem. Soc. 54 (6), 225–228.

Gan, H.L., Man, Y.B.C., Tan, C.P., NorAini, I., Nazimah, S.A.H., 2005a. Characterisation of vegetable oils by surface acoustic wave sensing electronic nose. Food Chem. 89 (4), 507–518.

Gan, H.L., Tan, C.P., Che Man, Y.B., NorAini, I., Nazimah, S.A.H., 2005b. Monitoring the storage stability of RBD palm olein using the electronic nose. Food Chem. 89 (2), 271–282.

Garcia, K., Ennis, J.M., Prinyawiwatkul, W., 2012. A large-scale experimental comparison of the tetrad and triangle tests in children. J. Sens. Stud 27 (4), 217–222.

Gómez-Alonso, S., Mancebo-Campos, V., Salvador, M.D., Fregapane, G., 2004. Oxidation kinetics in olive oil triacylglycerols under accelerated shelf life testing (25–75 °C). Eur. J. Lipid Sci. Technol. 106 (6), 369–375.

Gordon, M.H., 2001. 4-Measuring antioxidant activity. In: Pokorny, J., Yanishlieva, N., Gordon, M. (Eds.), Antioxidants in Food. Woodhead Publishing, pp. 71–84.

Guillén-Sans, R., Guzmán-Chozas, M., 1998. The thiobarbituric acid (TBA) reaction in foods: a review. Crit. Rev. Food Sci. Nutr. 38 (4), 315–350.

Guth, H., Grosch, W., 1993. Quantitation of potent odorants of virgin olive oil by stable-isotope dilution assays. J. Am. Oil Chem. Soc. 70 (5), 513–518.

Hangartner, M., Paduch, M., 1988. Interface human nose-olfactometer: measurement of odor emissions. Proceedings of Workshop, Annex V, 53, Commission of the EEC.

Hartvigsen, K., Lund, P., Hansen, L.F., Hølmer, G., 2000. Dynamic headspace gas chromatography/mass spectrometry characterization of volatiles produced in fish oil enriched mayonnaise during storage. J. Agric. Food Chem. 48 (10), 4858–4867.

Haugen, J.-E., Lundby, F., Wold, J.P., Veberg, A., 2006. Detection of rancidity in freeze stored Turkey meat using a commercial gas-sensor array system. Sens. Actuators B Chem. 116 (1–2), 78–84.

Haugen, J.-E., Undeland, I., 2002. Lipid oxidation in herring fillets (*Clupea harengus*) during ice storage measured by a commercial hybrid gas-sensor array system. J. Agric. Food Chem. 51 (3), 752–759.

Hoyland, D.V., Taylor, A.J., 1991. A review of the methodology of the 2-thiobarbituric acid test. Food Chem. 40 (3), 271–291.

Idris, N., Abdullah, A., Halim, A.H., 1992. Evaluation of palm oil quality: correlating sensory with chemical analyses. J. Am. Oil Chem. Soc. 69 (3), 272–275.

Irwin, J.W., Hedges, N., 2004. Measuring lipid oxidation. In: Steele, R. (Ed.), Understanding and Measuring the Shelf Life of Food. Woodhead Publishing, pp. 289–316.

ISO, 2002. General Guidance for Measuring Odour, Flavour and Taste Detection Thresholds by a Three-Alternative Forced-Choice Procedure. ISO Standard 13301, ISO, Geneva, Switzerland.

Jacobsen, C., Hartvigsen, K., Lund, P., Adler-Nissen, J., Hølmer, G., Meyer, A.S., 2000. Oxidation in fish-oil-enriched mayonnaise—assessment of the efficacy of different tocopherol antioxidant systems by discriminant partial least squares regression analysis. Eur. Food Res. Technol. 210 (4), 242–257.

Jeleń, H.H., Mildner-Szkudlarz, S., Jasińska, I., Wąsowicz, E., 2007. A headspace—SPME—MS method for monitoring rapeseed oil autoxidation. J. Am. Oil Chem. Soc. 84 (6), 509–517.

Jensen, P., Christensen, J., Engelsen, S., 2004. Oxidative changes in pork scratchings, peanuts, oatmeal and muesli viewed by fluorescence, near-infrared and infrared spectroscopy. Eur. Food. Res. Technol. 219 (3), 294–304.

Kanavouras, A., Hernandez-Münoz, P., Coutelieris, F.A., 2004. Shelf life predictions for packaged olive oil using flavor compounds as markers. Eur. Food Res. Technol. 219 (2), 190–198.

Koca, N., Rodriguez-Saona, L.E., Harper, W.J., Alvarez, V.B., 2007. Application of Fourier transform infrared spectroscopy for monitoring short-chain free fatty acids in Swiss cheese. J. Dairy Sci. 90 (8), 3596–3603.

Kochhar, S.P., 1996. Oxidative pathways to the formation of off-flavours. In: Saxby, M.J. (Ed.), Food Taints and Off-flavours. Springer, US, pp. 168–225.

Kolakowska, A., Deutry, J., 1983. Some comments on the usefulness of 2-thiobarbituric acid (TBA) test for the evaluation of rancidity in frozen fish. Food Nahrung 27 (5), 513–518.

Larrauri, M., Barrionuevo, M.G., Riveros, C., Mestrallet, M.G., Zunino, M.P., Zygadlo, J.A., Grosso, N.R., Nepote, V., 2013. Effect of peanut skin extract on chemical stability and sensory properties of salami during storage. J. Sci. Food Agric. 93 (7), 1751–1757.

Larson, R.A., 1988. The antioxidants of higher plants. Phytochemistry 27 (4), 969–978.

Lee, S.Y., Guinard, J.X., Krochta, J.M., 2002a. Relating sensory and instrumental data to conduct an accelerated shelf life testing of whey-protein-coated peanuts. In freshness and shelf life of foods. Am. Chem. Soc. 836, 175–187.

Lee, S.Y., Trezza, T.A., Guinard, J.X., Krochta, J.M., 2002b. Whey-protein-coated peanuts assessed by sensory evaluation and static headspace gas chromatography. J. Food Sci. 67 (3), 1212–1218.

Limbo, S., Sinelli, N., Torri, L., Riva, M., 2009. Freshness decay and shelf life predictive modelling of European sea bass (*Dicentrarchus labrax*) applying chemical methods and electronic nose. LWT Food Sci. Technol. 42 (5), 977–984.

Limbo, S., Torri, L., Sinelli, N., Franzetti, L., Casiraghi, E., 2010. Evaluation and predictive modeling of shelf life of minced beef stored in high-oxygen modified atmosphere packaging at different temperatures. Meat Sci. 84 (1), 129–136.

López-Aguilar, J., Valerio-Alfaro, G., Monroy-Rivera, J., Medina-Juárez, L., O'mahony, M., Angulo, O., 2006. Evaluation of a simple and sensitive sensory method for measuring rancidity in soybean oils. Grasas y Aceites 57 (2), 149–154.

Malcolmson, L.J., Vaisev-Genser, M., Przybylski, R., Ryland, D., Eskin, N.A.M., Armstrong, L., 1996. Characterization of stored regular and low-linolenic canola oils at different levels of consumer acceptance. J. Am. Oil Chem. Soc. 73 (9), 1153–1160.

Marsili, R.T., 2000. Shelf life prediction of processed milk by solid-phase microextraction, mass spectrometry, and multivariate analysis. J. Agric. Food Chem. 48 (8), 3470–3475.

Martinez, C., Santa Cruz, M.J., Hough, G., Vega, M.J., 2002. Preference mapping of cracker type biscuits. Food Qual. Preference 13 (7–8), 535–544.

Martínez-Yusta, A., Goicoechea, E., Guillén, M.D., 2014. A review of thermo-oxidative degradation of food lipids studied by 1 h NMR spectroscopy: influence of degradative conditions and food lipid nature. Compr. Rev. Food Sci. Food Saf. 13 (5), 838–859.

Martins, R.C., Lopes, V.V., Vicente, A.A., Teixeira, J.A., 2008. Computational shelf life dating: complex systems approaches to food quality and safety. Food Bioprocess Technol. 1 (3), 207–222.

Masuoka, S., Hatjopoulos, D., O'Mahony, M., 1995. beer bitterness detection: testing thurstonian and sequential sensitivity analysis models for triad and tetrad methods. J. Sens. Stud. 10 (3), 295–306.

Meilgaard, M.C., Carr, B.T., Civille, G.V., 2006. Sensory Evaluation Techniques, fourth ed. CRC Press, FL.

van der Merwe, G.H., du Plessis, L.M., Taylor, J.R.N., 2004. Changes in chemical quality indices during long-term storage of palm-olein oil under heated storage and transport-type conditions. J. Sci. Food Agric. 84 (1), 52–58.

Mildner-Szkudlarz, S., Jeleń, H.H., Zawirska-Wojtasiak, R., 2008. The use of electronic and human nose for monitoring rapeseed oil autoxidation. Eur. J. Lipid Sci. Technol. 110 (1), 61–72.

Miyasaki, T., Hamaguchi, M., Yokoyama, S., 2011. Change of volatile compounds in fresh fish meat during ice storage. J. Food Sci. 76 (9), C1319–C1325.

Moore, R.L., Duncan, S.E., Rasor, A.S., Eigel, W.N., O'Keefe, S.F., 2012. Oxidative stability of an extended shelf life dairy-based beverage system designed to contribute to heart health. J. Dairy Sci. 95 (11), 6242–6251.

Morales, M.T., Luna, G., Aparicio, R., 2005. Comparative study of virgin olive oil sensory defects. Food Chem. 91 (2), 293–301.

Munoz, A.M., Civille, G.V., 1998. Universal, product, and attribute scaling and the development of common lexicons in descriptive analysis. J. Sens. Stud. 13 (1), 57–75.

Munoz, A.M., Civille, G.V., Carr, B.T., 1992. Sensory Evaluation in Quality Control. Chapman and Hill, New York.

Nawar, W.W., 1969. Thermal degradation of lipids. J. Agric. Food Chem. 17 (1), 18–21.

Neilson, A.P., Pahulu, H.F., Ogden, L.V., Pike, O.A., 2006. Sensory and nutritional quality of dehydrated potato flakes in long-term storage. J. Food Sci. 71 (6), S461–S466.

Nepote, V., Olmedo, R.H., Mestrallet, M.G., Grosso, N.R., 2009. A study of the relationships among consumer acceptance, oxidation chemical indicators, and sensory attributes in high-oleic and normal peanuts. J. Food Sci. 74 (1), S1–S8.

Nielsen, B.R., Stapelfeldt, H., Skibsted, L.H., 1997. Early prediction of the shelf life of medium-heat whole milk powders using stepwise multiple regression and principal component analysis. Int. Dairy J. 7 (5), 341–348.

O'Dwyer, S., O'Beirne, D., Ní Eidhin, D., Hannon, J., O'Kennedy, B., 2013. Oxidative stability of tuna fat spreads (O/W/O emulsions) using conventional lipid oxidation methods, SPME-GC/MS and sensory analysis. Eur. Food Res. Technol. 237 (3), 385–398.

Olafsdottir, G., Jonsdottir, R., Lauzon, H.L., Luten, J., Kristbergsson, K., 2005. Characterization of volatile compounds in chilled cod (*Gadus morhua*) fillets by gas chromatography and detection of quality indicators by an electronic nose. J. Agric. Food Chem. 53 (26), 10140–10147.

Olivares, A., Navarro, J.L., Flores, M., 2011. Effect of fat content on aroma generation during processing of dry fermented sausages. Meat Sci. 87 (3), 264–273.

Overbosch, P., Enden, J.C.v. d, Keur, B.M., 1986. An improved method for measuring perceived intensity/time relationships in human taste and smell. Chem. Sens. 11 (3), 331–338.

Pastorelli, S., Torri, L., Rodriguez, A., Valzacchi, S., Limbo, S., Simoneau, C., 2007. Solid-phase micro-extraction (SPME-GC) and sensors as rapid methods for monitoring lipid oxidation in nuts. Food Addit. Contam. 24 (11), 1219–1225.

Pawłowicz, R., Gromadzka, J., Tynek, M., Tylingo, R., Wardencki, W., Karlovits, G., 2013. The influence of the UV irradiation on degradation of virgin rapeseed oils. Eur. J. Lipid Sci. Technol. 115 (6), 648–658.

Pedro, A.M.K., Ferreira, M.M.C., 2006. Multivariate accelerated shelf life testing: a novel approach for determining the shelf life of foods. J. Chemom. 20 (1–2), 76–83.

Petersen, K.D., Kleeberg, K.K., Jahreis, G., Busch-Stockfisch, M., Fritsche, J., 2012. Comparison of analytical and sensory lipid oxidation parameters in conventional and high-oleic rapeseed oil. Eur. J. Lipid Sci. Technol. 114 (10), 1193–1203.

Petukhov, I., Malcolmson, L.J., Przybylski, R., Armstrong, L., 1999. Storage stability of potato chips fried in genetically modified canola oils. J. Am. Oil Chem. Soc. 76 (8), 889–896.

Pokorný, J., Valentová, H., Davídek, J., 1985. Modified determination of 2-thiobarbituric acid value in fats and oils. Food Nahrung 29 (1), 31–38.

Ramirez, G., Hough, G., Contarini, A., 2001. Influence of temperature and light exposure on sensory shelf life of commercial sunflower oil. J. Food Qual. 24 (3), 195–204.

Reblova, Z., Kudrnova, J., Trojakova, L., Pokornya, J.A.N., 1999. Effect of rosemary extracts on the stabilization of frying oil during deep fat frying. J. Food Lipids 6 (1), 13–23.

Riveros, C.G., Mestrallet, M.G., Gayol, M.F., Quiroga, P.R., Ncpote, V., Grosso, N.R., 2010. Effect of storage on chemical and sensory profiles of peanut pastes prepared with high-oleic and Normal peanuts. J. Sci. Food Agric. 90 (15), 2694–2699.

Riveros, C.G., Mestrallet, M.G., Quiroga, P.R., Nepote, V., Grosso, N.R., 2013. Preserving sensory attributes of roasted peanuts using edible coatings. Int. J. Food Sci. Technol. 48 (4), 850–859.

Robards, K., Kerr, A.F., Patsalides, E., 1988. Rancidity and its measurement in edible oils and snack foods. A review. Analyst 113 (2), 213–224.

Röck, F., Barsan, N., Weimar, U., 2008. Electronic nose: current status and future trends. Chem. Rev. 108 (2), 705–725.

Rognlien, M., Duncan, S.E., O'Keefe, S.F., Eigel, W.N., 2012. Consumer perception and sensory effect of oxidation in savory-flavored yogurt enriched with N-3 lipids. J. Dairy Sci. 95 (4), 1690–1698.

Rowe, D.J., 2005. Chemistry and Technology of Flavors and Fragrances. CRC Press, Boca Raton, FL.

Ruiz-Capillas, C., Moral, A., 2001. Correlation between biochemical and sensory quality indices in hake stored in ice. Food Res. Int. 34 (5), 441–447.

Shahidi, F., Wanasundara, U.N., 2002. Methods for measuring oxidative rancidity in fats and oils. In: Akoh, C.C., Min, D.B. (Eds.), Food Lipids: Chemistry, Nutrition, and Biotechnology. Marcel Dekker, New York.

Shahidi, F., Zhong, Y., 2005. Lipid Oxidation: Measurement Methods. In Bailey's Industrial Oil and Fat Products. John Wiley and Sons, Inc.

Sidel, J.L., Stone, H., 1993. The role of sensory evaluation in the food industry. Food Qual. Preference 4 (1–2), 65–73.

Stodolnik, L., Stawicka, A., Szczepanik, G., Aubourg, S.P., 2005. Rancidity inhibition study in frozen whole mackerel (*Scomber scombrus*) following flaxseed (*Linum usitatissimum*) extract treatment. Grasas y Aceites 56 (3), 198–204.

Stone, H., Sidel, J.L., 2004. Sensory Evaluation Practices, third ed. Elsevier Academic Press, San Diego.

Sullivan Ritter, J., Budge, S., 2012. Key lipid oxidation products can be used to predict sensory quality of fish oils with different levels of EPA and DHA. Lipids 47 (12), 1169–1179.

Talbot, G., 2011. The stability and shelf life of fats and oils. In: Kilcast, D., Subramaniam, P. (Eds.), Food and Beverage Stability and Shelf Life. Woodhead Publishing, pp. 683–715.

Tan, Y.A., Chong, C.L., Low, K.S., 1995. Relationship between laser-induced fluorescence intensity and crude palm oil quality. J. Sci. Food Agric. 67 (3), 375–379.

Thongtan, K., Toma, R.B., Reiboldt, W., Daoud, A.Z., 2005. Effect of rosemary extract on lipid oxidation and sensory evaluation of frozen, precooked beef patties. Foodserv. Res. Int. 16 (3–4), 93–104.

Tikk, K., Haugen, J.-E., Andersen, H.J., Aaslyng, M.D., 2008. Monitoring of warmed-over flavour in pork using the electronic nose—correlation to sensory attributes and secondary lipid oxidation products. Meat Sci. 80 (4), 1254–1263.

Trihaas, J., Vognsen, L., Nielsen, P.V., 2005. Electronic nose: new tool in modelling the ripening of Danish blue cheese. Int. Dairy J. 15 (6–9), 679–691.

Vasavada, M.N., Dwivedi, S., Cornforth, D., 2006. Evaluation of garam masala spices and phosphates as antioxidants in cooked ground beef. J. Food Sci. 71 (5), C292–C297.

Veberg, A., Olsen, E., Nilsen, A.N., Wold, J.P., 2007. Front-face fluorescence measurement of photosensitizers and lipid oxidation products during the photooxidation of butter. J. Dairy Sci. 90 (5), 2189–2199.

Veberg, A., Olsen, E., Vogt, G., Mielnik, M., Nilsen, A.N., Wold, J.P., 2006a. Front face fluorescence spectroscopy—a rapid method to detect early lipid oxidation in freeze stored minced Turkey meat. J. Food Sci. 71 (4), S364–S370.

Veberg, A., Vogt, G., Wold, J.P., 2006b. Fluorescence in aldehyde model systems related to lipid oxidation. LWT Food Sci. Technol. 39 (5), 562–570.

Villarino, B.J., Dy, L.M., Lizada, M.C.C., 2007. Descriptive sensory evaluation of virgin coconut oil and refined, bleached and deodorized coconut oil. LWT Food Sci. Technol. 40 (2), 193–199.

Warner, K., 2009. Oxidative and flavor stability of tortilla chips fried in expeller pressed low linolenic acid soybean oil. J. Food Lipids 16 (2), 133–147.

Wheatley, R.A., 2000. Some recent trends in the analytical chemistry of lipid peroxidation. TrAC Trends Anal. Chem. 19 (10), 617–628.

Woo, A.H., Lindsay, R.C., 1983. Statistical correlation of quantitative flavor intensity assessments and individual free fatty acid measurements for routine detection and prediction of hydrolytic rancidity off-flavors in butter. J. Food Sci. 48 (6), 1761–1766.

Xu, X.-Q., 2000. A new spectrophotometric method for the rapid assessment of deep frying oil quality. J. Am. Oil Chem. Soc. 77 (10), 1083–1086.

Xu, X.-Q., Tran, V., Palmer, M., White, K., Salisbury, P., 1999. Chemical and physical analyses and sensory evaluation of six deep-frying oils. J. Am. Oil Chem. Soc. 76 (9), 1091–1099.

Zajdenwerg, C., Branco, G., Alamed, J., Decker, E., Castro, I., 2011. Correlation between sensory and chemical markers in the evaluation of Brazil nut oxidative shelf life. Eur. Food Res. Technol. 233 (1), 109–116.

Chapter 4

Oxidative Stability and Shelf Life of Vegetable Oils

Asim Syed
Dow AgroSciences LLC, Indianapolis, IN, USA

4.1 INTRODUCTION

Oils and fats are a major component of the human diet, comprising roughly one-third of our calorie intake. A significant number of hours are invested by food companies, academic institutions, and government research organizations on research concerning the nutritional, functional, and sensory properties of oils and fats. Oils function as carriers for fat-soluble vitamins (A, D, E, and K) and provide energy and essential linoleic and linolenic acids that are responsible for growth (Lukaszewicz et al., 2004).

Table 4.1 gives a summary of fatty acid profiles of commercially available natural oils and fats most commonly used in the food industry.

Vegetable oils or plant-derived oils account for about 79% of edible oil consumption worldwide (Fasina et al., 2008). Blended oils in which several vegetable oils are mixed at different proportion are very popular in oriental countries such as China, Thailand, and Japan because of the variety of fatty acids (FAs) applied for different nutritional and processing purposes (Li et al., 2013).

One critical area of research has been the shelf stability of oils as well as of the finished products containing oil as a major component. The reason why this has acquired attention is that all oils and fats degrade over time. This degradation is mostly the result of oxidative and hydrolytic rancidity. Fatty acids break down and generate primary and secondary oxidation products. Most of these products are volatile organic compounds (VOCs,) which are known to cause off-odors and off-flavors. In processed foods where oil is a major component, this off-flavor in oil can impact consumer sensory acceptability of the finished product, thus limiting the product shelf life. Chemical analyses of oils packaged and stored over a period of time are critical for the bottled oil application. When fat is an ingredient in a complex food system (margarines, cookies, French fries, etc.)—instead of just pure fat as in a bottle of oil, the rates and dynamics of fat degradation change significantly and so does the degree of impact of fat degradation on the finished product quality and shelf life. This is primarily due to

Oxidative Stability and Shelf Life of Foods Containing Oils and Fats
http://dx.doi.org/10.1016/B978-1-63067-056-6.00004-5

TABLE 4.1 General Fatty Acid Profiles of Most Commonly Used Food Oils

	Saturated					Monounsaturated	Polyunsaturated	
	Capric	Lauric	Myristic	Palmitic	Stearic	Oleic	Linoleic	α-Linolenic
	Acid	Acid	Acid	Acid	Acid	Acid (n-9)	Acid (n-6)	Acid (n-3)
Oils and Fats	C10:0	C12:0	C14:0	C16:0	C18:0	C18:1	C18:2	C18:3
Natural Solid Fats								
Beef tallow (National Research Council, 1976)	0	0	3	26	14	47	3	1
Butterfat (cow) (National Research Council, 1976)	3	3	11	27	12	29	2	1
Coconut oil	7	48	18	10	3	6	2	0
Lard (pork fat) (National Research Council, 1976)	0	0	2	26	14	44	10	0
Palm oil	0	0	1	45	4	40	10	0
Palm olein	0	0	1	37	4	46	11	0

Palm kernel oil	4	48	16	8	3	15	2	0
Natural Liquid Oils								
Canola	0	0		4	2	62	22	10
Corn	0	0	0	11	2	28	58	1
Cottonseed	0	0	1	22	3	19	54	1
Flaxseed			0	3	7	21	16	53
Palm oil	0	0	1	45	4	40	10	0
Rice bran (Kamel et al., 1985)	0	0	0	22	3	38	35	3
Olive	0	0	0	13	3	71	10	1
Peanut	0	0	0	10	4	49	31	1
Safflower	0	0	0	7	2	13	78	0
Sunflower	0	0	0	7	5	19	68	1

Numbers have been rounded for simplicity; oils without reference were analyzed at Dow AgroSciences Laboratories, Indianapolis, IN.

ingredient–ingredient interactions (fat, water, proteins, sugars, etc.), as well as the sensory properties of the finished product as a whole.

This chapter provides primarily a case study that has been reformatted from a technical report, "Shelf Life Analysis of Cooking Oils," that the author presented at the 105th AOCS Annual Meeting and Expo, May 4–7, 2014, San Antonio, Texas. The author is associated with Dow AgroSciences, where his team conducts food applications research on omega-9 canola and sunflower oils.

4.2 ROLE OF OILS AND FATS IN FOODS

Some of the basic uses of edible oils in foods include home cooking, margarine and shortening, and salad oil. It is critical to choose the right kind of fat when formulating food products. That is because fats impact all aspects of the food's life cycle, from production through consumption. The basic functions of fats in food systems include nutrition, heat exchange, sensory performance, physical structure, and operational efficiency. In heat applications like sautéing and deep frying, heat exchange is a dominant function. In other applications like baked goods, spreads, and icings, the kind of fat used critically impacts the product's physical structure. In all applications, sensory attributes are a key component that is heavily impacted by fats. No matter what fat is used in which foods, one factor that impacts the quality and shelf life of the food product is the oxidative stability of the fat. All fats go through oxidation over time, which can cause quality and shelf-life issues if this factor is not addressed appropriately during product design and testing.

Lipid oxidation is one of the most common causes of deterioration of food quality. Unsaturated fats are oxidized by free radical auto-oxidation, a chain-reaction process catalyzed by the products of the reaction. The susceptibility and the rate of oxidation increase as the number of double bonds in the fatty acid increases. Foods containing fat, such as milk, can undergo a number of subtle chemical and physical changes caused by lipolysis. Lipolysis can be defined as the enzymatic hydrolysis of fats by lipases. The accumulation of the reaction products, especially free fatty acids is responsible for the common off-flavor, frequently referred to as rancidity (Valero et al., 2012).

The main effect that becomes obvious as a result of the oxidation of edible oils is a decrease in sensory quality. In general, this negative characteristic of the product is described by the term "rancid," derived from the Latin word for "stinky," "nasty," or "unfavorable." Depending on the fatty acid composition of the edible oil and the resulting degradation products, characteristics and perceptions of the rancid sensation can be very different (Matthaus, 2010).

4.3 SHELF LIFE AND SHELF STABILITY

Product "shelf life" is defined, according to the American Heritage Dictionary of the English Language, as "the term or period during which a stored commodity remains effective, useful, or suitable for consumption" (Valero et al., 2012).

Oils and fats do not have a definite shelf life per se. The reason is multifold. Shelf life of a food product can be broadly defined as the number of days the quality of the finished product is as good as when it was fresh, when stored at specified storage conditions. This statement seems simple and clear but can generate vast differences of opinions when it comes to the details. Most food companies define shelf life in their own ways and follow their own criteria of defining quality and for considering a product acceptable or unacceptable. The above statement applies to all processed foods, including bottled oils found in local grocery stores. Bottled canola oil sold by two different companies can have different expiration codes. This is because there is no defined measure of acceptability or unacceptability for cooking oils. When it comes to cooking oils for home use, consumers have different levels of tolerance and too much subjectivity is involved. Soybean and canola oils sold as bottled oils are coded as "best by" for 9–12 months. One criterion for bottled oil is that the consumer should not be able to smell any oxidized or rancid aroma during handling and cooking. Again, what that means in terms of the actual acceptable level of oxidation can differ from consumer to consumer. According to the Hygienic Standard of Edible Oils of the People's Republic of China, oils cannot exceed a PV (peroxide value) of 12 mEq/kg oil (Li et al., 2013). These kinds of definitive guidelines would make it easier to measure and establish shelf lives of oil. However, there is no consensus among the global quality and regulatory bodies on this. Also, there are no scientific data that show that PV of 12 mEq/kg oil should be considered the end of shelf life for all edible oils for all applications. One problem with assigning a PV value as the end of shelf life is that PV provides information about peroxides, which are the primary oxidation compound. It does not take into account the secondary oxidation compounds like aldehydes, ketones, etc., which are actually responsible for the off-odors.

The scenario is equally, if not more, ambiguous for bulk oils supplied to the food industries. General minimum criterion in the industry for acceptable bulk RBD (refined bleached and deodorized) oil is PV not more than 1.0 mEq/kg oil, assuming the rest of the quality parameters are within specifications. A lot of oil delivered to a manufacturing plant is usually consumed within 1 week or so. Since oil becomes a component of a complex food system in a formulated food product, like a fried chip, mayonnaise, or a donut, the impact of its freshness and quality on the shelf life of the finished product can be insignificant or critical, depending on the product matrix and the mode of failure. Hence, a 9-month bottle of canola oil with a PV of 15 mEq/kg may be acceptable to some consumers and not to others. An oil blend with a PV of 5 mEq/kg may be acceptable as a component of donut because it will be consumed within a couple of weeks. That may not be true in the case of mayonnaise, which requires a longer shelf life. For all of these reasons, and some that may not have been mentioned here, it is advisable to consider the term of shelf stability instead of shelf life when comparing oils.

Shelf life is an absolute term and applies to a finished product with defined conditions of packaging and storage. In operations, it is expressed as x number of days from the date of manufacture. On product label, it is expressed as a definite "best by" or "use by" date.

Shelf stability, on the other hand, is strictly a comparative term and is used only when comparing two or more oils. It is independent of absolute packaging or storage conditions. As long as the treatment conditions are fixed, no matter what the conditions are, the comparison should be valid. A lot of soybean oil with an OSI (oxidative stability index) of 8 h will be about 20% more stable than another lot of soybean oil (with similar fatty acid profile) with an OSI of 6 h. The information is complete. If the 6-h soybean oil provides a shelf life of 12 weeks to a certain oil-based product, then the 8-h oil will likely provide roughly 16 weeks ("roughly" because the relation between stability of pure oil and its stability in a food product is almost never linear). Similarly, when testing oils for fry life, the critical factor is to have exactly the same testing conditions (fry time, temperature, food-to-oil ratio, etc.) no matter what the conditions are.

4.4 DEGRADATION OF OILS DURING STORAGE

Edible oils consist of about 96% triacylglycerides, composed of different fatty acids (refer to Table 4.1). Some other compounds or groups of compounds, such as free fatty acids, phospholipids, phytosterols, tocopherols, and other antioxidants or waxes, can also be found (Matthaus, 2010).

All oils and fats oxidize during storage. The rate of oxidation depends primarily on inherent factors like fatty acid profile, minor components including moisture and other impurities and quality of processing, as well as external factors like storage temperature, air (oxygen) concentration, and light. Lipid oxidation not only can produce rancid odors, unpleasant flavors, and discoloration but also can decrease the nutritional quality and safety due to degradation products, resulting in harmful effects on human health (Das, 1991). Variations of the natural fatty acid pattern of oil and the positions of fatty acids can affect its susceptibility and resistance to the oxidation process as well as its quality and nutritional value. Edible oils with higher contents of unsaturated fatty acids, especially polyunsaturated fatty acids (PUFAs), are more susceptible to oxidation (Kapich et al., 2010).

Oxidation of edible oils begins with the double bonds of fatty acids in the triacylglycerol molecule. The first products of lipid oxidation are hydroperoxides, which are tasteless and odorless and therefore do not significantly impair the sensory quality of the oil (Reindl and Stan, 1982). The problem is that these compounds are generally unstable, resulting in the formation of secondary oxidation products like ketones, aldehydes, alcohols, lactones, hydrocarbons, esters, or others. Since some of these compounds have a very low threshold value for smell or taste, the degradation of the product becomes obvious very quickly.

The oxidative stabilities of the oils are also influenced by nontriglyceride materials like natural antioxidants, phospholipids, free fatty acids, mono- and diglycerides, polymers, and the position of double bonds of fatty acids in the oil. Position of fatty acids on the triacylglycerol molecule also plays an important role in lipid oxidation. PUFAs in the sn-2 position have been shown to be more susceptible to oxidation than those in the sn-1 or sn-3 positions (Neff and El-Agaimy, 1996).

The pathway of the oxidative degradation of edible fats and oils results in many physical changes in the product, such as increased viscosity, changes in composition, and the formation of degradation products. To measure and assess these changes, a wide range of methods is available for describing the oxidative state of edible oils. Additionally, some methods can be applied to assess the oxidative stability of the product. Oxidative state and oxidative stability should be clearly distinguished: the first describes the product at the present time, and the latter tries to give information about the possible behavior of the product during further storage. In this context, it must be taken into consideration that the measurement of a parameter is one thing, while the interpretation and usefulness of that parameter to describe the real situation are other tasks. The assessment of the actual storage behavior from the obtained results is especially difficult, and misinterpretation can occur (Matthaus, 2010).

4.5 A CASE STUDY ON OIL OXIDATIVE STABILITY

4.5.1 Objective

The objective of the study was to determine the shelf stabilities of multiple cooking oils when stored at ambient and elevated temperatures. The study was designed to understand the extent of degradation that different oils go through during ambient storage, in terms of the standard oil degradation detection methods that are commonly used in the processed food industry. This should also provide a useful comparative data set on vegetable oils to food product developers and food research and development professionals.

4.5.2 Study Design

Six different cooking oils (with PV < 2 mEq/kg) were filled in seven 1-L HDPE (high-density polyethylene) dark amber bottles each, and stored at 21.1 °C (ambient) for 24 months. Headspace (air) was fixed at 5–8% by volume. After every 3 months, one bottle of each oil was removed from the ambient storage and was stored in a freezer at −20 °C. At the end of 24 months, all samples were removed from the freezer and analyzed. All sample preparation, storage, and analyses were conducted at Dow AgroSciences' analytical laboratories in Indianapolis. Samples were analyzed for changes in oxidative stability index (OSI at 110 °C), PV, *para*-anisidine value (*p*AV), total oxidation (TOTOX) (calculated as 2 PV + *p*AV), free fatty acids (FFA), fatty acid profile (FAME), and aroma

TABLE 4.2 Fatty Acid Profiles of the Oils Used in the Study

Fatty Acid Profile	C16:0	C18:0	C18:1	C18:2	C18:3
Soybean	10.6	4.8	22.4	53.7	7.7
Canola	4.1	1.8	62.8	19.9	8.8
Corn	11.4	1.6	27.6	57.1	0.9
High-oleic sunflower	3.0	2.8	85.9	6.4	0.2
Omega-9 canola	3.7	1.8	74.5	14.7	2.1
O-9 RS sunflower	2.2	0.5	92.1	3.8	0.1

intensity by electronic nose (e-nose). Trained panel descriptive sensory analysis was also conducted on four selected oils. Sensory test was conducted at the University of Illinois, Urbana, Illinois.

4.5.3 Materials and Methods

Table 4.2 summarizes the fatty acid profiles of the test oils. Soybean, canola, and corn oils were bought from local grocery stores; high-oleic sunflower oil was supplied by Stratas Foods; omega-9 canola oil was supplied by Bunge Oils, and omega-9 reduced saturation sunflower oil was processed by a DAS contracted comanufacturer.

Table 4.3 shows the list of analyses that were conducted on the oil samples. All analyses other than the sensory test were conducted at DAS laboratories in Indianapolis. The descriptive sensory testing was conducted at the University of Illinois in Urbana. The test details and methodology are summarized in Table 4.5 and Figures 4.7–4.11.

4.5.4 Results

The following tables and figures summarize the results from all the analyses listed above.

4.5.5 Change in Fatty Acid Profile

Table 4.4 shows that after 24 months of ambient storage, the common trend was a slight increase in the relative content of monounsaturated fatty acid and a decrease in that of polyunsaturated fatty acids. In soybean, canola, and

TABLE 4.3 Summary of Analyses Conducted on the Oils in the Shelf Stability Study

Fatty acid composition (FAME)	Modified AOCS official Method Ce 1–62[a]	GC-FID
Oxidative stability index (OSI at 110°C)	AOCS official method Cd 12b-92[b]	Rancimat
Free fatty acid (FFA)	AOCS official method Ca 5a-40	Titration
Peroxide value (PV)	AOCS official method Cd 8b-90	Titration
para-Anisidine value (*p*AV)	AOCS official method Cd 18-90	Spectrophotometer
Total oxidation (TOTOX)	2×PV+*p*AV	Calculation
E-nose aroma intensity	Internal SOP for oil analysis	Alpha-MOS FOX
Sensory description analysis	University of Illinois	Trained panelists

[a]*Modified for high throughput analysis (HTP)*
[b]*Slight changes for clarity in use*

TABLE 4.4 Changes in Fatty Acid Profiles of the Oils Stored for 24 months at 21.1°C

SOY	C16:0	C18:0	C18:1	C18:2	C18:3
0 months	10.53	4.34	21.63	53.66	7.69
6 months	10.50	4.36	22.00	53.36	7.55
12 months	10.61	4.40	22.34	53.20	7.43
18 months	10.63	4.39	22.69	52.92	7.30
24 months	10.65	4.51	23.47	52.35	7.01
Canola	**C16:0**	**C18:0**	**C18:1**	**C18:2**	**C18:3**
0 months	4.07	1.78	61.62	19.90	8.81
6 months	4.18	1.84	62.55	19.33	8.27
12 months	4.07	1.89	63.45	18.85	7.75
18 months	4.09	1.93	64.24	18.43	7.34

Continued

TABLE 4.4 Changes in Fatty Acid Profiles of the Oils Stored for 24 months at 21.1 °C—cont'd

SOY	C16:0	C18:0	C18:1	C18:2	C18:3
24 months	4.03	2.01	65.83	17.52	6.46
Corn	**C16:0**	**C18:0**	**C18:1**	**C18:2**	**C18:3**
0 months	11.42	1.63	27.57	57.06	0.94
6 months	11.39	1.63	27.59	56.99	0.95
12 months	11.49	1.66	27.82	56.70	0.94
18 months	11.49	1.67	27.92	56.55	0.94
24 months	11.67	1.71	28.29	56.01	0.95
HOSO	**C16:0**	**C18:0**	**C18:1**	**C18:2**	**C18:3**
0 months	3.00	2.77	85.94	6.40	0.17
6 months	3.07	2.93	85.93	6.07	0.17
12 months	3.06	2.94	85.95	5.98	0.18
18 months	3.14	2.98	85.91	5.98	0.18
24 months	3.01	2.66	87.14	5.19	0.17
O-9 Canola	**C16:0**	**C18:0**	**C18:1**	**C18:2**	**C18:3**
0 months	3.67	1.82	74.46	14.73	2.07
6 months	3.60	1.84	74.61	14.50	1.99
12 months	3.72	1.83	74.82	14.37	1.94
18 months	3.63	1.83	74.95	14.32	1.90
24 months	3.71	1.83	74.99	14.20	1.88
O-9 RS Sun	**C16:0**	**C18:0**	**C18:1**	**C18:2**	**C18:3**
0 months	2.13	0.51	92.10	3.75	0.11
6 months	2.13	0.50	92.23	3.72	0.10
12 months	2.13	0.50	92.26	3.69	0.10
18 months	2.15	0.50	92.22	3.74	0.10
24 months	2.11	0.48	92.27	3.68	0.10

HOSO, high-oleic sunflower oil; O-9 Canola, omega-9 canola; O-9 RS Sun, omega-9 reduced saturates sunflower oil.

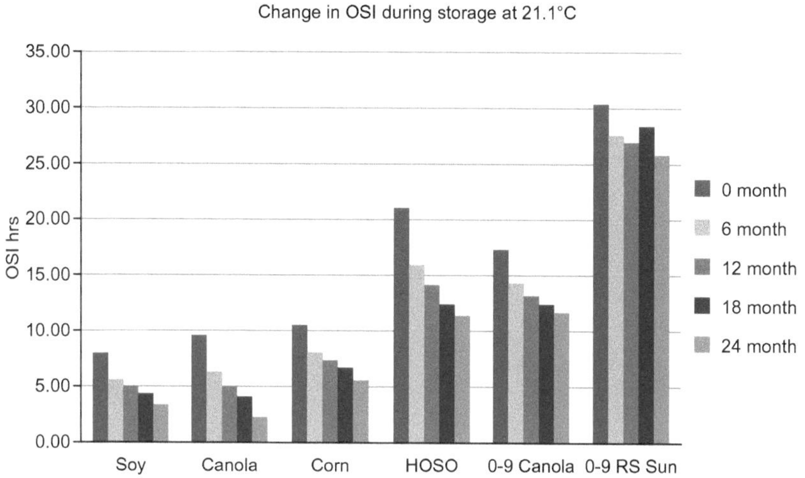

FIGURE 4.1 Change in OSI for oils stored at 21.1 °C for 24 months (HOSO, high oleic sunflower oil; O-9 Canola, Omega-9 Canola; O-9 RS Sun, Omega-9 reduced saturates sunflower oil; OSI, oxidative stability index).

high-oleic sunflower oils, there was an ~1.5–2% increase in oleic acid percent and corresponding decrease in linoleic and linolenic acids. This trend was less obvious in corn, omega-9 canola, and omega-9 RS sunflower oils, where the change in fatty acid profile was miniscule.

4.5.6 Changes in OSI

Figure 4.1 shows the change in OSI (measured at 110 °C) during the storage period. Soybean, canola, and corn oils start at 8, 10, and 10.5 h, respectively, and progressively decline throughout all the following four time points. Most of the decline occurred during the first 6 months. Both omega-9 canola and high oleic sunflower oil (HOSO) show similar decline trend, but the absolute numbers are about twice as high as soy, canola, and corn oils. Omega-9 RS sunflower oil was twice as stable as omega-9 canola oil and showed very little decline in OSI.

4.5.7 Changes in FFA

Figure 4.2 shows the change in FFA during the storage period. As expected, there was no significant change. This is because of low storage temperature, and no addition of moisture or acidity that could cause hydrolysis. After 24 months, FFA in all the oils was <0.05%, which is below the objectionable level of 0.1% in fresh RBD oils. Omega-9 canola showed FFA of 0.06% at 6 months, which could be sampling or analytical error. So, it can be concluded that there was no consequentially significant increase in FFA after 24 months.

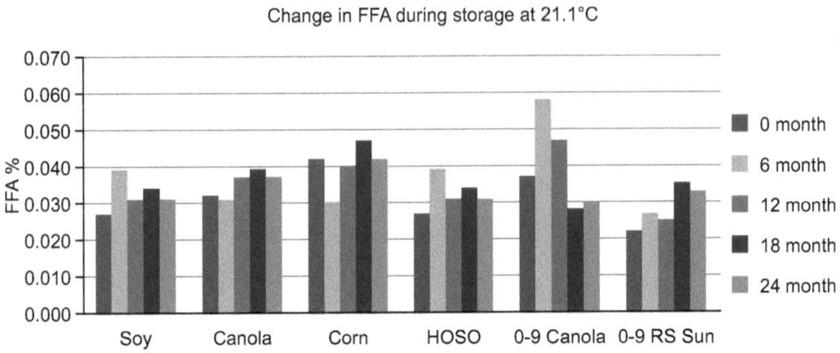

FIGURE 4.2 Change in FFA for oils stored at 21.1 °C for 24 months (HOSO, high oleic sunflower oil; O-9 Canola, Omega-9 Canola; O-9 RS Sun, Omega-9 reduced saturates sunflower oil; FFA, free fatty acids).

4.5.8 Changes in PV, *p*AV, and TOTOX Value

It was predicted that all the oils would progressively oxidize, to different extents, during the study. This would mean consistent increase in both primary and secondary oxidation products, expressed as PV and *p*AV. Figure 4.5 shows the change in TOTOX values [calculated as 2 PV (Figure 4.3) + *p*AV (Figure 4.4)]. TOTOX values consistently increased during the storage for all the oils. Soybean and canola oils show highest levels of oxidation, followed by corn, HOSO, and omega-9 canola oils. Omega-9 RS sunflower oil remained extremely stable and showed very little formation of peroxides and the secondary oxidation compounds. Interestingly, for corn and HOSO oils, *p*AV numbers started high and then went down during the storage, as opposed to the rest of the oils, which show progressive increase in *p*AV. However, the combined values of PV and *p*AV, TOTOX, only increased with time. This suggests that different oils behave differently during storage in terms of the rates of development of primary and secondary oxidation products, and so a more valid measure for comparing the extent of oxidation in oils is TOTOX, as opposed to PV and *p*AV separately.

4.5.9 Changes in Aroma Intensity

Aroma intensity was measured by e-nose (Alpha MOS 4000E-Nose). The instrument is equipped with 18 metal oxide detectors, each of which detects a specific class of volatile organic compounds (VOCs). The sensors are activated in unique patterns when exposed to different VOCs, producing a complex sensor response pattern. The instrument treats a complex mixture of volatiles as an analyte and generates a combined response. A sample is weighed into a vial and heated with shaking, and the VOCs in the headspace are sampled and injected. The amplitude for each sensor identifies the amount present. Odors results from complex mixtures of hundreds, if not thousands, of VOCs emitted by the test

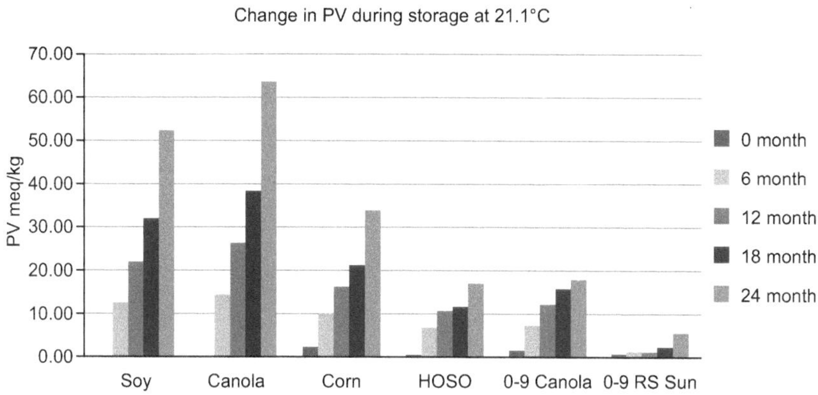

FIGURE 4.3 Change in PV for oils stored at 21.1 °C for 24 months (HOSO, high oleic sunflower oil; O-9 Canola, Omega-9 Canola; O-9 RS Sun, Omega-9 reduced saturates sunflower oil; PV, peroxide value).

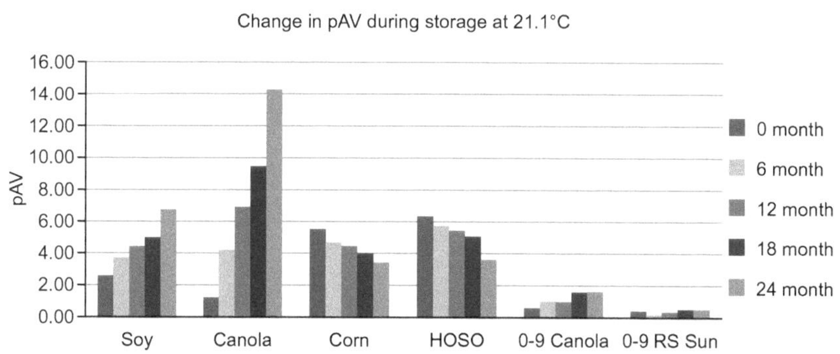

FIGURE 4.4 Change in *p*AV for oils stored at 21.1 °C for 24 months (HOSO, high oleic sunflower oil; O-9 Canola, Omega-9 Canola; O-9 RS Sun, Omega-9 reduced saturates sunflower oil; PAV, para-anisidine value).

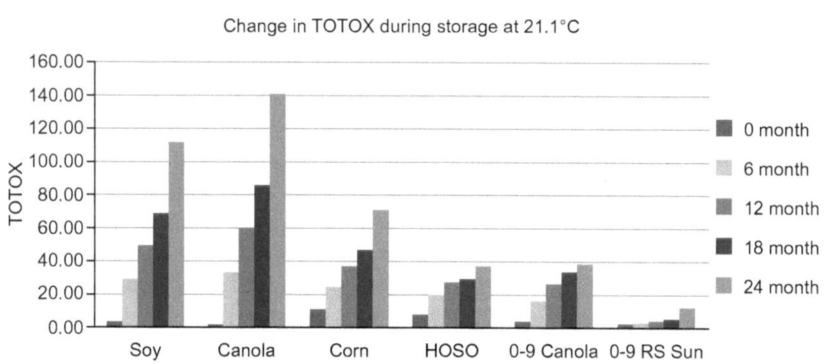

FIGURE 4.5 Change in TOTOX (2PV+*p*AV) values for oils stored at 21.1 °C for 24 months (HOSO, high oleic sunflower oil; O-9 Canola, Omega-9 Canola; O-9 RS Sun, Omega-9 reduced saturates sunflower oil; TOTOX, total oxidation).

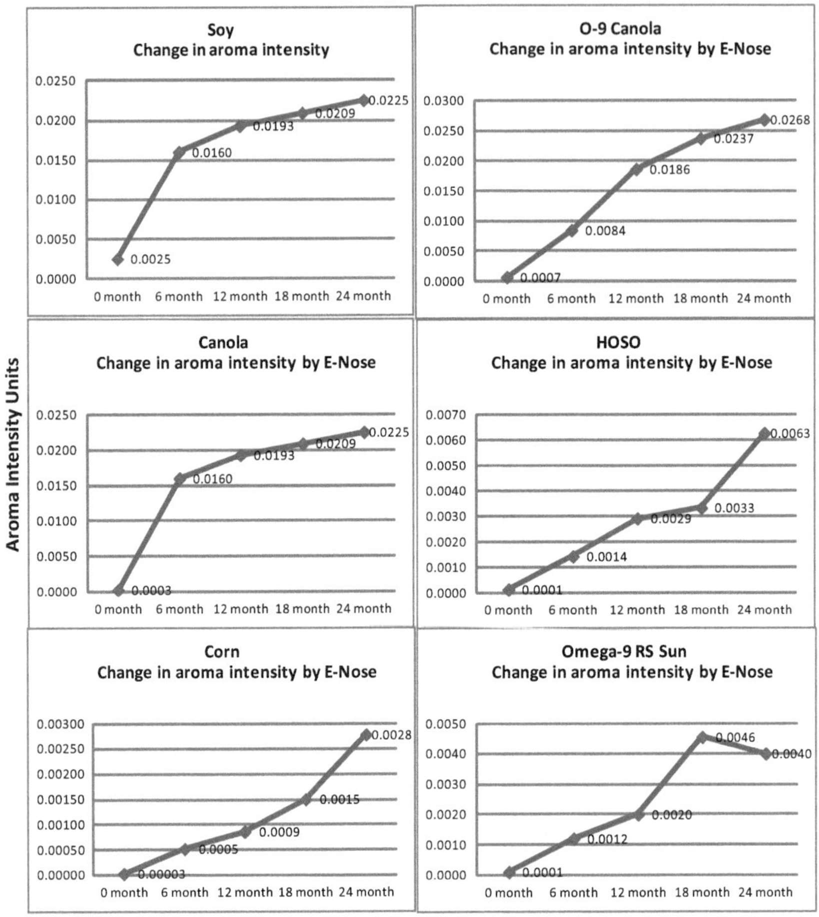

FIGURE 4.6 Change in aroma intensity units for the six oils stored at 21.1 °C for 24 months, by E-nose (Alpha MOS 4000) (HOSO, high oleic sunflower oil; O-9 Canola, Omega-9 Canola; O-9 RS Sun, Omega-9 reduced saturates sunflower oil).

sample. The data are evaluated with multivariate statistical software to generate pattern recognition responses, similar to how the human brain uses pattern recognition to identify aromas.

Figure 4.6 shows the change in aroma intensity (expressed as aroma intensity units on the *y*-axis) for the six oils during the 24-month storage period. Some interesting observations can be made here. The aroma intensity generally increased throughout the storage period. The scales on the *y*-axis are not all the same, except for the two canola oils, or the two sunflower oils, so the charts from different crops should not be compared with each other. It may be hypothesized that the increase in aroma intensity is due to the breakdown and oxidation of the fatty acids, as expressed by the TOTOX values. Since this is a quantitative and not a qualitative measure of VOCs, and since each VOC may have a drastically different human smell threshold, it cannot be simply deduced that higher aroma intensity by e-nose corresponds with a stronger odor as detected by the human nose. Also, the instrument was not calibrated (although it can be) to distinguish between "good" or "bad" aroma as defined by human sensory descriptive test methods.

4.5.10 Changes in Human Sensory Attributes

Four of the six oils were selected for a descriptive sensory analysis to determine the sensory changes in the oils during the 24-month storage period. The sensory test was conducted at the University of Illinois. Below are the test protocol, results, and discussion.

4.5.11 Protocol for Sensory Analyses

Ten subjects were recruited via e-mail distributed through the Department of Food Science and Human Nutrition distribution lists and consisted of University of Illinois students and staff. Panelists' ages ranged from 22 to 42 years. Ten trained panelists took part in the study, in which 100 mL of oil was heated for 20 s in a Sharp Carousel Model R-308JW 1100-Watt microwave, netting a temperature of 43.3 °C. The oil samples were served in 2-oz. plastic cups with lids for evaluation. Panelists generated sensory descriptive attributes, along with reference anchors, separately, for each of the four oils (Table 4.4). Then the panelists rated the four oils from each time point (month 0, 6, 12, 18, and 24) on a 1-to-10 scale for the intensity of each attribute.

Figures 4.7–4.10 show the sensory attributes for soybean, canola, corn, and omega-9 sunflower oil that changed significantly over the 24-month period. Figure 4.11 summarizes the results for all four oils.

Table 4.5 shows the sensory attributes that the trained panel generated for each of the four oils. The panel looked at all oils separately for generating the sensory attribute descriptors.

Figure 4.11 summarizes the attributes that changed significantly in all the four oils during the study. The only attribute change common in corn and soybean oils was color saturation, while rancid aroma was common between canola and soybean oils. Omega-9 RS sunflower oil had only one attribute, umami aftertaste, in common with canola oil. No attribute was common to all the four oils. Soybean oil had the most number of attributes that significantly changed during the study. Corn and omega-9 RS sunflower oils had the least number of attributes that changed. What this might mean for product developers is that, for an oil-based product, when one kind of oil is replaced with another it is important to consider the sensory changes that might occur due to the change in oil. The mechanism of product degradation through the shelf life may also change if the oil in the product formula is changed. This may also mean that a new set of analytical sensory attributives will be required to qualify and monitor product shelf life.

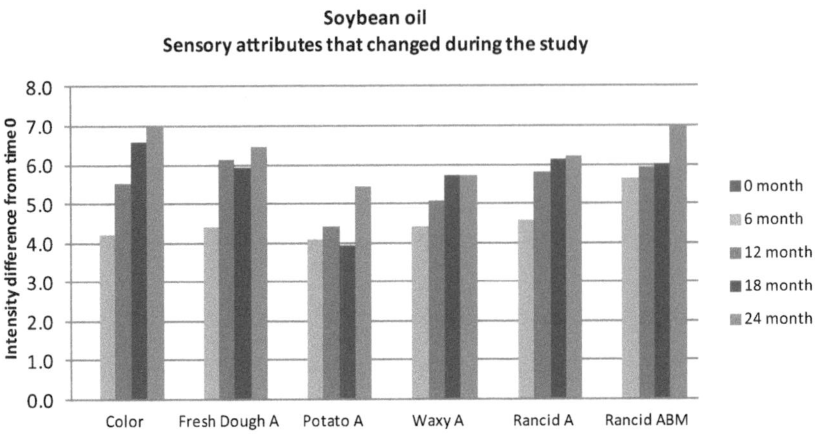

FIGURE 4.7 Sensory attributes for soybean oil that changed significantly during 24-month storage (A, aroma; ABM, aroma by mouth, intensity for month 0 was taken as 0).

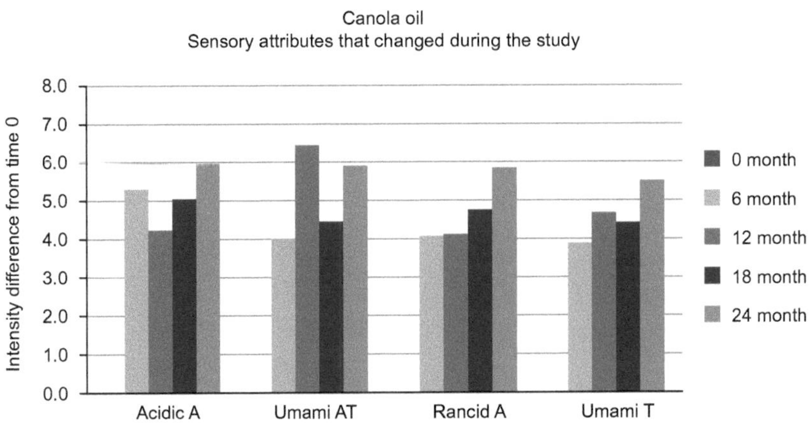

FIGURE 4.8 Sensory attributes for canola oil that changed significantly during 24-month storage (A, aroma; T, taste; AT, aftertaste, intensity for month 0 was taken as 0).

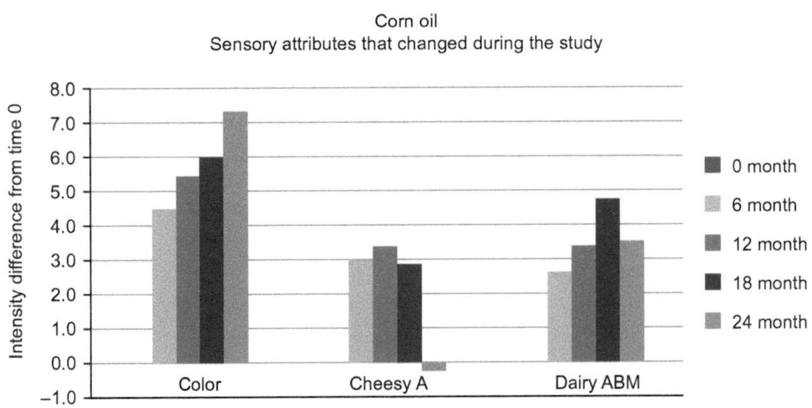

FIGURE 4.9 Sensory attributes for corn oil that changed significantly during 24-month storage (A, aroma; ABM, aroma by mouth, intensity for month 0 was taken as 0).

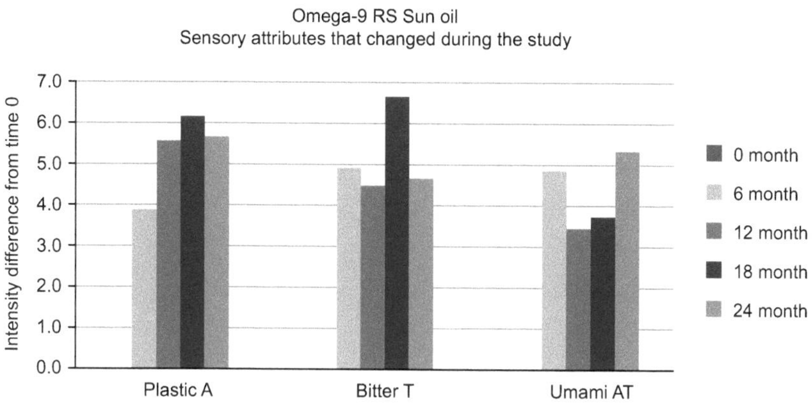

FIGURE 4.10 Sensory attributes for Omega-9 RS Sunflower oil that changed significantly during 24-month storage (A, aroma; T, taste; AT, aftertaste, intensity for month 0 was taken as 0).

Oil	Sensory Attributes Which Changed Significantly During Storage					
Soy Oil	Color Saturation	Fresh Dough Aroma	Potato Aroma	Waxy Aroma	Rancid Aroma	Rancid Aroma By Mouth
Canola Oil	Acidic Aroma	Rancid Aroma	Umami Taste	Umami Aftertaste		
Corn Oil	Color Saturation	Cheesy Aroma	Dairy Aroma By Mouth			
O-9 RS Sunflower Oil	Plastic Aroma	Bitter Taste	Umami Aftertaste			

FIGURE 4.11 Summary of sensory attributes for all the four oils that changed significantly during 24-month storage.

4.5.12 Use of Elevated Temperature Storage for Shelf-Life Prediction

In an effort to generate a simple shelf-life prediction model for the bottled oil application, all of the six oils were also stored at an elevated temperature of 43.3 °C for 18 weeks. Samples were drawn at weeks 0, 3, 7, 13, and 18 and were kept frozen at −20 °C until they were analyzed for OSI. The goal was to understand if the slope of change in OSI during 21.1 °C storage aligns with the slope of change in OSI during 43.3 °C storage.

TABLE 4.5 Lists of Sensory Attributes Generated by the Trained Panel for Each of the Four Oil Types

Sensory Attributes Generated for the Descriptive Analysis				
Attributes	**Soybean**	**Canola**	**Corn**	**Omega-9 RS Sunflower**
Appearance				
	Color saturation	Viscous	Color saturation	None
Aroma				
	Fresh dough	Fresh dough	Cheesy	Plastic
	Waxy	Green	Rancid	Potato
	Potato	Acidic		
	Rancid	Rancid		
Taste				
	Umami	Bitter	Bitter	Bitter
		Umami	Umami	Umami
Aroma-By-Mouth				
	Rancid	Rancid	Rancid	Soylike
	Soylike	Green	Corny	Plastic
	Potato		Dairy	
Texture				
	Fatty	Fatty	Fatty	Viscous
		Viscous	Viscous	
After Taste				
	Umami	Umami	Umami	Umami
		Astringent	Astringent	Astringent
				Dairy

Figure 4.12 shows the changes in OSI for the six oils stored at 43.3 °C. The OSI of soybean, canola, and corn oils start with 8, 10, and 12 h, respectively, and go down progressively at each consecutive time point. The rate of decline in OSI at 43.3 °C is similar to the rate of change in OSI at 21.1 °C. The same is evident for HOSO, omega-9 canola, and omega-9 sunflower oils, but the rates of change were different for these three oils. OSI of HOSO started with 25 h at

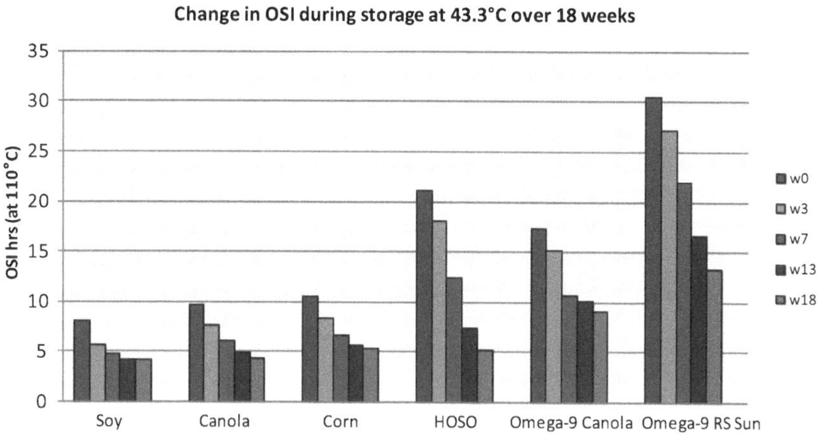

FIGURE 4.12 Change in OSI for all six oils during 18-week storage at 43.3 °C (OSI, oxidative stability index).

time 0 versus 23 h for omega-9 canola, but at the end of 18 weeks, the OSI of HOSO dropped down to 3 h versus 6 h for omega-9 canola. For omega-9 RS sunflower oil, there was no decline in OSI at 3 weeks, and it maintained the highest stability throughout the 18 weeks.

Figure 4.13 shows the rate of decline in OSI in each oil during storage at 21.1 °C for 24 months that is plotted against the rates of decline in OSI during storage at 43.3 °C for 18 weeks. For soybean oil, the trend line of decline in OSI at 21.1 °C follows almost exactly the same path as at 43.3 °C. The same is true for corn and canola oils but to a lesser degree. The three high-oleic oils show very little alignment between the trend lines of decrease in OSI at 21.1 and 43.3 °C.

4.6 DISCUSSION

The study confirmed that during ambient storage under controlled conditions, the defining factor impacting the oxidative stability of oils is the degree of unsaturation. In other words, iodine value, which is an expression of the degree of unsaturation, is a good predictor of relative stability. The higher the level of monounsaturated fatty acid and the lower the level of polyunsaturated fatty acids, the higher is the oxidative stability. OSI and *p*AV are very useful comparative measures of the oxidative states of different oils when stored at room temperature. The slopes of the rate of change in oxidative stability index of oils stored at ambient versus that at elevated temperature were different for different oils. There was no direct correlation between the oxidative breakdown of oils and changes in their sensory attributes, as determined by the trained sensory panel. Oils from a crop species behaved differently from other crop species, in terms of oxidation and sensory characteristics. This could be due to the inherent differences in nontriglyceride minor components that oil from each crop contains.

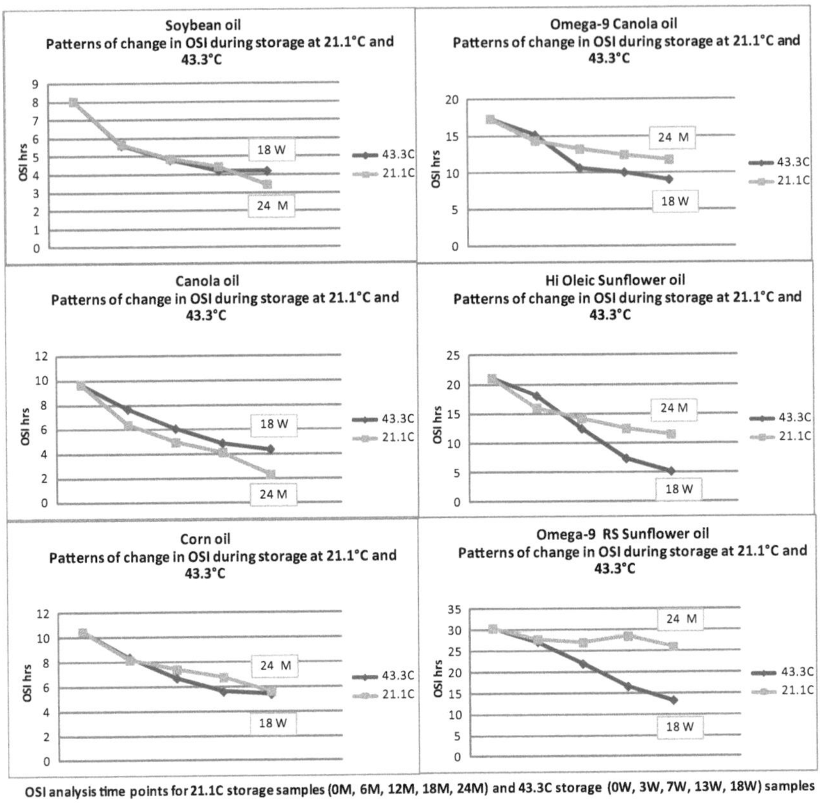

FIGURE 4.13 Comparison of change in OSI for all six oils during storage at 21.1 °C and 43.3 °C. 21.1 °C samples were stored for 24 months and analyzed at 0, 6, 12, 18 and 24 months, 43.3 °C samples were stored for 18 weeks and analyzed at 0, 3, 7, 13, and 18 weeks (OSI, oxidative stability index).

4.7 CONCLUSION

Oils and fats play a critical role in food product development. They influence the physical, chemical, and biological aspects of the foods. Food manufacturers and food scientists are always interested in understanding the shelf life of oils, because the shelf life of oils could influence the shelf life of the product containing that oil. A more instructive concept for oils is shelf stability. Shelf stability is a relative term, which defines the oxidative stability of an oil as it relates to other oils. However, merely comparing oxidative stability of different oils does not provide the complete picture of how these oils will perform in terms of impacting the quality and shelf lives of the finished food products.

The chemical analyses, e-nose, and trained panel sensory data described in this chapter suggest that each kind of oil is a distinct and separate species and that fatty acid profile alone cannot fully define their chemical and sensory properties. Each kind of oil contains numerous minor compounds in differing ratios that could be contributing toward distinct aroma and flavor development. Understanding the chemical degradation behavior of pure oils in a comparative study provides a valuable starting point when choosing between different oils for a specific application. Product developers should always conduct shelf-life analysis on products made with different oils, as the comparative performance of oils in a multi component food product may or may not be a direct reflection of their performance as pure oils. There is not a standard set of criteria that can be applied to determine the shelf life of all the oils. When it comes to oils and fats, it's all relative.

ACKNOWLEDGMENTS

I would like to thank my Dow AgroSciences colleagues for helping with the study design and execution of the study, as well as for literature search, draft review, and suggestions: Tom Patterson, Robin Gunn, Chris Elliot, Mary Evenson, Matt Probst, Steve Wensing, and Anne.

REFERENCES

Das, U.N., 1991. Lipids, lipid peroxidation and human health. Trends Food Sci. Technol. 2, 44–45. http://www.essentialoils.co.za/sesame-oil-analysis.htm.

Fasina, O.O., Craig-Schmidt, M., Colley, Z., Hallman, H., 2008. Predicting melting characteristics of vegetable oils from fatty acid composition. LWT Food Sci. Technol. 41 (8), 1501–1505. http://www.FAO.ORG/docrep/v4700e/v4700e0b.htm referenced December 2014.

Kapich, A.N., Korneichik, T.V., Hatakka, A., Hammel, K.E., 2010. Oxidizability of unsaturated fatty acids and of a non-phenolic lignin structure in the manganese peroxidase-dependent lipid peroxidation system. Enzyme Microb. Technol. 46 (2), 136–140.

Kamel, B.S., Dawson, H., Kakuda, Y., 1985. Characteristics and composition of melon and grape seed oils and cakes. J. Am. Oil Chem. Soc. 62 (5), 881–883. http://dx.doi.org/10.1007/BF02541750.

Li, H., Fan, Y., Li, J., Tang, L., Hu, J., Deng, Z., 2013. Evaluating and predicting the oxidative stability of vegetable oils with different fatty acid compositions. J. Food Sci. 78, H633–H641.

Lukaszewicz, M., Szopa, J., Krasowska, A., 2004. Susceptibility of lipids from different flax cultivars to peroxidation and its lowering by added antioxidants. Food Chem. 88 (2), 225–231.

Matthaus, B., 2010. Oxidation of edible oils, oxidation in plant-based foods and beverages. In: Decker, E., Elias, R., Julian, D. (Eds.), Oxidation in Foods and Beverages and Antioxidant Applications, pp. 183–184.

National Research Council, 1976. Online Edition Fat Content and Composition of Animal Products. Printing and Publishing Office, National Academy of Science, Washington, DC. ISBN: 0-309-02440-4, p. 203.

Neff, W.E., El-Agaimy, M., 1996. Effect of linoleic acid position in triacylglycerols on their oxidative stability. LWT Food Sci. Technol. 29 (8), 772–775.

Reindl, B., Stan, H.-J., 1982. Determination of volatile aldehydes in meat as 2,3-dinitrophenylhydrazones using reversed-phase high-performance liquid chromatography. J. Agric. Food Chem. 30, 849–854.

Valero, A., Carrasco, E., García-Gimeno, R.M., 2012. Principles and methodologies for the determination of shelf life in foods. In: Amer Eissa, A. (Ed.), Trends in Vital Food and Control Engineering. ISBN: 978-953-51-0449-0, pp. 24–26.

Chapter 5

Oxidative Stability and Shelf Life of Fish Oil

Revilija Mozuraityte, Vera Kristinova, Inger B. Standal, Ana K. Carvajal and Marit Aursand
Department of Processing Technology, SINTEF Fisheries and Aquaculture, Trondheim, Norway

5.1 INTRODUCTION

Although fish oils contribute by only 10% to the edible oils commercially available in the world, they play an important role in the modern diet. The difference between fish oils and other sources of oils is that fish oils are rich in long-chain omega-3 polyunsaturated fatty acids (LC omega-3 PUFA) such as docosahexaenoic acid (DHA, 22:6 n-3, or all-*cis*-4,7,10,13,16,19-docosahexaenoic acid), eicosapentaenoic acid (EPA, 20:5 n-3, or all-*cis*-5,8,11,14,17-eicosapentaenoic acid), and docosapentaenoic acid (DPA, 22:5 n-3, or all-*cis*-7,10,13,16,19-docosapentaenoic acid) (Figure 5.1).

Because of the presence of LC omega-3 PUFA, fish oils are documented to have positive effects against cardiovascular and inflammatory diseases, preventing cognitive decline, and others (VKM, 2011a; Calder, 2013). The amounts of these fatty acids in fish oils depend on the raw material selected for oil production, that is, fish species, harvesting season, age, and part of the fish; the diet of the fish (natural diet or aquaculture/fish feed); and the production process (Usydus et al., 2007).

Fish oils in the consumer market are sold in different forms: purified fish oils or concentrates as triglycerides or ethyl esters. The range of total omega-3 fatty acids in the commercial fish oils is typically between 20% and 60%, but for "ultra-high" concentrated fish oils, the level of omega-3 may reach up to 80%. The presence of LC omega-3 PUFA makes fish oils highly prone to autooxidation. This is due to the presence of large number of double bonds (up to six —C=C— double bonds) and bisallylic carbons with low activation energy for hydrogen loss and free radical formation (Shahidi and Zhong, 2010). Thus, usually the oxidative stability of fish oil decreases with an increasing amount of the healthy unsaturated fatty acids (Tengku-Rozaina and Birch, 2013). Fish oils undergo oxidation much faster than oils of vegetable or animal origin and with

Oxidative Stability and Shelf Life of Foods Containing Oils and Fats
http://dx.doi.org/10.1016/B978-1-63067-056-6.00005-7

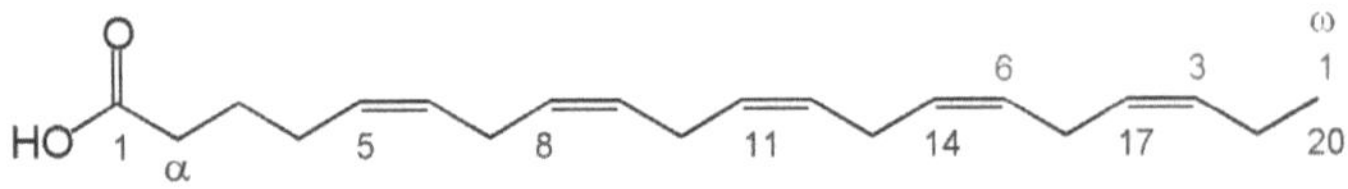

all-cis-5,8,11,14,17-eicosapentaenoic acid (**EPA**, 20:5 *n*-3)

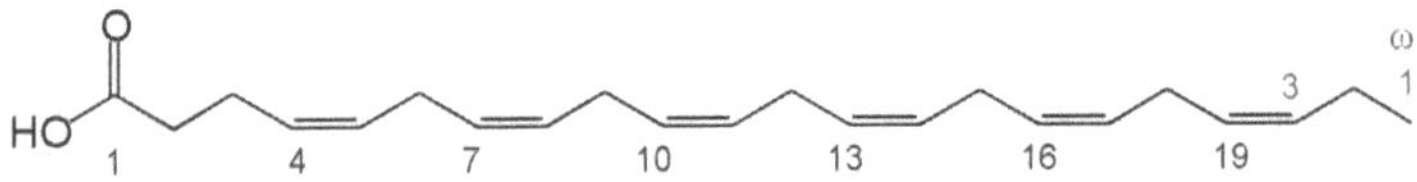

all-cis-4,7,10,13,16,19-docosahexaenoic acid (**DHA**, 22:6 *n*-3)

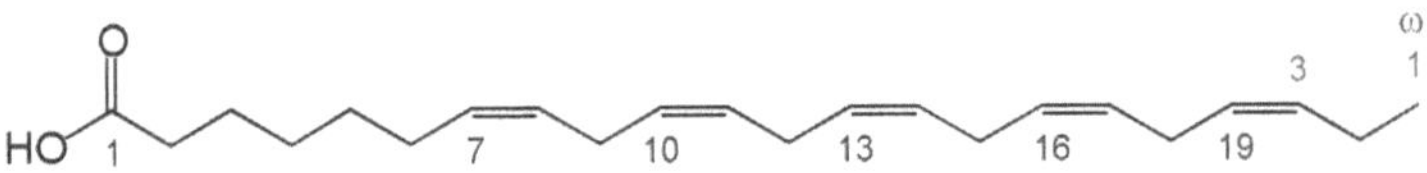

all-cis-7,10,13,16,19-docosapentaenoic acid (**DPA**, 22:5 *n*-3)

FIGURE 5.1 Chemical structure of long-chain omega-3 polyunsaturated fatty acids such as docosahexaenoic acid (DHA, 22:6 n-3, or all-*cis*-4,7,10,13,16,19-docosahexaenoic acid), eicosapentaenoic acid (EPA, 20:5 n-3, or all-*cis*-5,8,11,14,17-eicosapentaenoic acid), and docosapentaenoic acid (DPA, 22:5 n-3, or all-*cis*-7,10,13,16,19-docosapentaenoic acid).

a less-sharp break in the induction period. As oxidative stability is the major factor for the shelf life of the fish oil products, there have been concerns about the safety of oxidized fish oils. Several authorities, like European Food Safety Authority (EFSA, 2010) and the Norwegian Food Safety Authorities (VKM, 2011b), have ordered risk assessment of the health effect of consuming oxidized fish oil. Because of the lack of published studies, no firm conclusion on the health effect of oxidized fish oil could be drawn. However, there is an increasing focus on providing oils with good oxidative quality and stability.

To control and predict the stability of oil, it is important to understand the mechanism of oxidation and main factors that influence oxidation kinetics. Fish oils oxidize via the same mechanism as any other oils—auto-oxidation and photo-oxidation. **Auto-oxidation** is enhanced by heat, light, trace metals, or enzymes, which start the formation of lipid free radicals that react with oxygen to produce lipid hydroperoxides as primary oxidation products. The primary lipid oxidation products of the auto-oxidation cascade are tasteless and odorless fatty acid hydroperoxides (LOOH), and if a polyunsaturated fatty acid is oxidized, each peroxide group is accompanied by a conjugated diene. As the number of double bonds in PUFA increases, more complex mixtures of LOOH may be produced. For example, auto-oxidation of EPA may produce eight types of peroxides; 5-, 8-, 9-, 11-, 12-, 14-, 15-, and 18-LOOH, while in DHA, 10 possible peroxides can be formed; namely 4-, 7-, 8-, 10-, 11-, 13-, 14-, 16-, 17-, and 20-LOOH (Lyberg and Adlercrautz, 2006). These hydroperoxides have either *cis*-, *trans*-, or *trans–trans* configuration, and a complex system of 16 and 20 monohydroperoxide isomers can therefore be expected from the auto-oxidation of EPA and DHA, respectively. However, in the presence of metals and at high temperatures, fatty acid hydroperoxides break down into an array of nonvolatile and volatile secondary oxidation products. Due to the variety of lipid hydroperoxides that can be formed during auto-oxidation of EPA and DHA, a highly complex mixture of secondary oxidation products can be produced after breakdown of these hydroperoxides. Analysis of fish oils by dynamic headspace gas chromatography resulted in more than 150 volatile compounds (Frankel, 2005). Sensory threshold values of volatile decomposition products from LC omega-3 PUFA in fish and fish oil are extremely low (for some, as low as 0.01 ppb) with readily detectable odors at very low oxidation levels (Johnson and Decker, 2015). Volatile compounds in fish oil have been characterized in several studies (Aidos, 2002; Jacobsen and Nielsen, 2007; Olsen et al., 2005), but the data vary between studies due to sample differences (e.g., type of added antioxidants, type of fish oil), oxidation levels, and analytical methods. Important flavor compounds formed during the auto-oxidation of omega-3 fatty acids are 1-penten-3-one, 4-heptenal, 2,4-heptadienal 2,4,7-decatrienal, and others. Formation of these compounds leads to the fishy rancid smell and flavor of oxidized fish oil (Aidos, 2002; Jacobsen and Nielsen, 2007; Kulås et al., 2003).

During homolytic β-cleavage of fatty acid peroxides, aldehydes that are left on triacylglycerol molecule are called core aldehydes. Core aldehydes are nonvolatile compounds that are included in the measured anisidine value of marine oils but have no impact on the sensory quality of the oils. This might be the reason why no studies investigating core aldehydes in detail in fish oil have been found.

Because lipid peroxides and aldehydes are the major oxidation products, peroxide value and anisidine value (PV and AV, respectively) have been chosen as standard analytical methods to characterize the oxidation status of the fish oils. The Global Organization for EPA and DHA (GOED) specifies in their voluntary monograph that the PV of fish oils intended for human consumption should be below 5 mEq/kg and AV should be below 20. These values are given based on palatability and not based on the effect of oxidized oils on human health (Albert et al., 2013). The methods mentioned here are not sensitive enough to measure oxidation products at ppb levels. Thus, it is possible to measure very low PV levels in oils that taste and smell rancid. Even though AV determines the amount of aldehydes, the taste and smell of different aldehydes vary. Therefore, there is still little evidence for correlation between PV and/or AV and sensory assessment (Frankel, 2005; Ritter and Budge, 2012). The fishy flavor often found in fish oil can dissuade people from consuming the oil. In order to protect fish oil from oxidation, it should be optimally produced, stored, and packed. The stability of fish oils is influenced by several factors that will be discussed next. Moreover, to alleviate the oxidation problem, both synthetic and natural antioxidants may be added to fish oils and foods containing them to maintain the quality and extend their shelf life.

5.2 FACTORS INFLUENCING OXIDATIVE STABILITY AND SHELF LIFE OF FISH OILS

Generally, the oxidative stability of fish oils depends on a number of factors, such as fatty acid composition of the oil, presence of impurities that in many cases may act as pro-oxidants, antioxidant protection, access to light and oxygen, and processing and storage temperatures. These factors should be considered when processing, packaging, storing, and transporting the oil and will be discussed next.

5.2.1 Pro-oxidants

Pro-oxidants are compounds that initiate, facilitate, or accelerate lipid oxidation. **Transition metals**, such as iron and copper ions, are common catalysts of lipid oxidation. Crude fish oil usually contains metal ions (a copper concentration in the micrograms-per-kilogram level and iron concentration in the milligrams-per-kilogram level), and trace levels will remain even after refining and deodorization (Jacobsen et al., 2009). Iron and copper are naturally occurring elements (e.g., included as part of protein structures) of the raw material for marine oils, but iron can also contaminate the oil from production equipment. Importantly, even traces of iron (nanomolar levels) are sufficient to cause oxidative instability.

The mechanism of iron-catalyzed lipid peroxidation has been proposed to depend on the presence or absence of preformed lipid hydroperoxides (LOOH) and is divided into LOOH-independent or LOOH-dependent mechanisms. The mechanism of LOOH-independent initiation is still not clear. Transition metals are capable of directly breaking down unsaturated lipids (LH) to alkyl radicals ($L^{\cdot}$), but this reaction occurs slowly and is therefore considered less important in promoting lipid oxidation (Reische et al., 1998). Several studies (Fukuzawa et al., 1988; Mozuraityte et al., 2008; Tadolini et al., 1997) have shown that when peroxides were removed from the lipids, lipid peroxidation was not induced by iron. However, LOOH-dependent oxidation reactions may become dominant even with low amounts of peroxides. Iron catalyzes oxidation via decomposition of lipid peroxides to free radicals by a redox cycling pathway (Figure 5.2). Free lipid radicals will therefore lead to the formation of new peroxides in the presence of oxygen. In the absence of oxygen, metal ions may facilitate the breakdown of preexisting peroxides (reduction of PV value) and formation of secondary oxidation products (β-scission) such as aldehydes (increase in AV). Because metal ions play a crucial role in oxidation reactions, it is important to minimize the contact between these ions and the oil. This can be done by using chelating antioxidants.

The pro-oxidant effect of **free fatty acids (FFAs)** on oxidation of marine oils has also been observed (Aubourg, 2001). A higher degree of pro-oxidative effect was observed for shorter-chain fatty acids (e.g., lauric and myristic fatty acids) compared with longer-chain fatty acids (e.g., arachidic and stearic fatty acid) (Aubourg, 2001). Several mechanisms were proposed for the explanation of

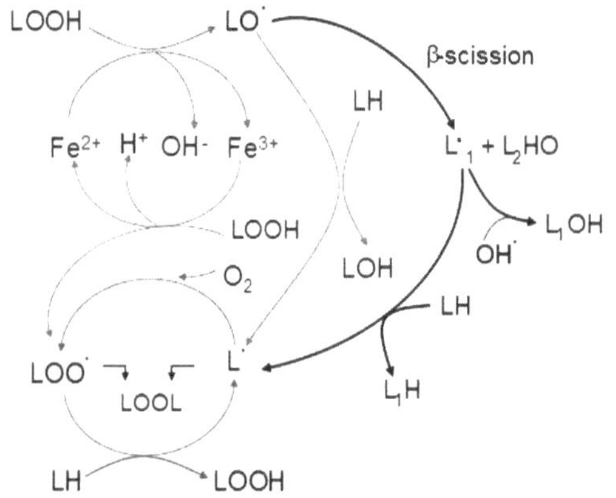

FIGURE 5.2 Iron-catalyzed lipid oxidation (modified from Mozuraityte et al., 2008). Iron catalyzes oxidation via decomposition of lipid peroxides (LOOH) to alkoxyl radicals (LO·) or peroxy radicals (LOO·) through redox cycling pathways. Alkoxyl- and peroxy radicals further propagate lipid peroxidation by chain branching with formation of several peroxides accompanied by oxygen consumption.

pro-oxidative activity of FFAs. The catalytic effect of carboxylic group on the formation of free radicals by the decomposition of hydroperoxides formed in the initial stage of oxidation is one explanation (Aubourg, 2001). Another theory lies in the surface activity of FFAs. The presence of FFAs may reduce the surface tension (Kittipongpittaya et al., 2014), causing an increase of oxygen transfer into the oil. In the presence of trace amounts of water in bulk oils, FFAs can act as surfactants and concentrate at the oil–water interface and accelerate lipid oxidation because surface active lipid hydroperoxides and water-soluble metals are able to migrate to the same location at the interface and lead to lipid peroxide decomposition (Kittipongpittaya et al., 2014).

In the presence of **photo-sensitizers** (e.g., porphyrins, riboflavins, chlorophyll), light can convert the atmospheric triplet oxygen ($^3O_2$) to the singlet state ($^1O_2$) that can directly bind to double bond of fatty acid and form lipid peroxide. Due to this, photo-oxidation leads to the formation of both conjugated and nonconjugated peroxides. Chlorophyll can contaminate fish oil during the preliminary extraction of oil from planktivorous species, such as anchovies, sardines, and mackerels (Indrasena and Barrow, 2010). However, even though during refining some of the photosensitizers may be removed, the pro-oxidative effect of light was higher for refined mackerel liver oil compared with crude oil (Sang and Jin, 2004).

5.3 OXIDATIVE STABILITY OF FISH OILS AND PROCESSING

The chemical composition, quality, and stability of fish oils depend on both the production process and the quality of the raw material (EFSA, 2010). Fish and fish rest raw material is especially vulnerable toward oxidation and degradation because they contain fractions as blood (pro-oxidant) and viscera with high amounts of endogenous enzymes. The presence of endogenous enzymes will lead to the formation of FFAs and reduction in lipid quality and stability. Prolonged storage time results in increased concentration of FFAs, lowering the quality of oil produced from herring (Aidos et al., 2002) and pollock rest raw material (Wu and Bechtel, 2009). To ensure high-quality oil, it is important to process the raw material within a short time. When producing crude oil, the raw material is exposed to high temperatures (90–95 °C) to cause protein coagulation and rupture of the fat depot, allowing the liberation of oil and water. How long the raw material is exposed for high temperatures depends on the processing equipment that is used for cooking. The most common is the use of a continuous steam cooker where the material is moved by a heated rotary screw conveyer, a processing step that takes around 20 min (Bimbo, 2007). Due to exposure to high temperature, oxidation can occur during processing, resulting in a low-quality oil. One way to reduce oxidation during crude oil production is by the addition of antioxidants to the raw material before production (Carvajal et al., 2014), by reducing the presence of oxygen in the production line (flushing with nitrogen) (Carvajal, 2013), or by using equipment that requires less retention period for the raw material and oil at elevated temperatures (e.g., a scraped surface heat exchanger compared with continuous steam cooker with a rotary conveyer).

Before crude fish oils can be used for human consumption, they need to go through several refining steps to remove unwanted components and increase the quality and stability of the oil. The removal of undesirable components while not harming the oil and minimizing oxidation is critical. One of the first refining steps is neutralization, washing of the oil with caustic soda to remove FFAs. Studies have shown that FFAs accelerate the oxidation process reducing the quality and stability of the oil (Chaiyasit et al., 2007). Hence, the content of FFAs is one of the most important factors that influence the overall oil quality and price (Vaisali et al., 2015). Bleaching of the oil by bleaching earth and activated carbon is carried out to increase the oil quality and stability by removal of oxidation products, trace metals, and environmental pollutants. However, bleaching will also result in the removal of natural pigments and antioxidants. A study on processing of sunflower oil showed a reduction of almost 40% in tocopherol concentration during the neutralization, bleaching, and deodorization process (Naz et al., 2011). During deodorization, the oil is steamed under vacuum at high temperatures (150–200 °C) to remove volatiles to enhance the quality of the oil. Although deodorization aims at increasing the quality of the oil by removal of undesirable taste and smell compounds, a simultaneously loss of valuable components can occur. Studies have shown that deodorization of fish oil should not exceed 180 °C, because higher temperatures will increase the development of polymers, cyclic fatty acid monomers (CFAM) and geometrical isomers of EPA and DHA (Fournier et al., 2006).

5.4 STABILIZATION BY THE USE OF ANTIOXIDANTS

Antioxidants can be defined as "any substance that delay, prevent or remove oxidative damage to a target molecule" (Halliwell, 2007). In bulk fish oils, antioxidants have been used to maintain acceptable levels of lipid oxidation, thus maintaining acceptable quality and stability during the shelf life. The most relevant antioxidants for bulk fish oils are radical scavengers, but metal chelators, quenchers of singlet oxygen and reactive oxygen species, may also contribute to maintaining the oxidative stability.

Crude fish oils contain antioxidants naturally, such as tocopherols, carotenoids, amino acids and peptides, ascorbic acid, and ubiquinone. The inherent quantities however usually do not provide sufficient protection during a prolonged storage. Refining processes remove a substantial part of the endogenous compounds, and a common practice is to replenish the lost antioxidants.

To improve stabilization of marine oils, antioxidants, such as organic acids (oxalic acid, citric acid, ethylenediaminetetraacetic acid [EDTA], etc.) and radical scavengers, can be added to the raw material before production of the oil (Breivik and Aanesen, 2002; Carvajal et al., 2014). The addition of acids helps to deactivate or remove free metals and phospholipids, while radical scavengers hinder the propagation of radicals during the separation of the crude oil. Whether the added antioxidants appear in the crude oil depends on their solubility in lipids and the degree of moisture in the oil.

The addition of antioxidants and antioxidant synergists into fish oils is regulated by the Codex Alimentarius General Standard for Food Additives in food category "Lard, tallow, fish oil, and other animal fats" (Codex Alimentarius, 2014). A standard specializing in fish oils ("Proposed Draft Standard for Fish Oils") has been in preparation by the Codex Alimentarius Commission at the time of publishing this book. Table 5.1 summarizes antioxidants approved for fish oils by these standards.

The published data on the effect of antioxidants in bulk fish oils are difficult to compare due to multiple variables, such as the characteristics of both the fish oil and the antioxidant; diverse oxidation conditions, where, for instance, the temperature and the antioxidant concentration play crucial roles; and methods used to assess oxidative stability. Most studies have applied accelerated storage tests to assess the effect of antioxidants, which may or may not reflect the effects at normal storage conditions. In most studies, especially when high temperatures have been used, thermal stability of antioxidants has not been evaluated. The antioxidant purity, concentrations of active components, physical state, and the way of addition are also factors determining the efficacy of an antioxidant.

5.5 SYNTHETIC ANTIOXIDANTS

Several synthetic antioxidants, namely butylated hydroxytoluene (BHT), butylated hydroxyanisole (TBA), *tert*-butylhydroquinone (TBHQ), and propyl gallate (PG), have been approved for preservation of fish oils (Codex Alimentarius, 2014). These compounds exhibit high efficacy at relatively low concentrations (<200 ppm). However, doubts regarding the toxicity and the increasing demand for all natural products from customers have pressed fish oil manufacturers to reduce the use of synthetic additives in recent years and substitute with natural alternatives.

TABLE 5.1 Antioxidants and Antioxidant Synergists Listed by Codex Alimentarius Commission as Additives for Fish Oils

INS	Additive	Maximum Level	Comment
300	Ascorbic acid, L-	GMP	
305	Ascorbyl stearate	2500 mg/kg	
307a, b, c	Tocopherols	6000 mg/kg	Singly or in combination
310	Propyl gallate (PG)	200 mg/kg	Singly or in combination
319	*tert*-Butylhydroquinone (TBHQ)	200 mg/kg	
320	Butylated hydroxyanisol (BHA)	200 mg/kg	
321	Butylated hydroxytoluene (BHT)	200 mg/kg	
160a(i,iii)	β-Carotenes	25 mg/kg	Singly or in combination
160e	Carotenal, beta-apo-8′-	25 mg/kg	
160f	Carotenoic acid, ethyl ester, beta-apo-8′-	25 mg/kg	
160a(ii)	β-Carotene, vegetable	1000 mg/kg	
330	Citric acid	GMP	
384	Isopropyl citrates	200 mg/kg	
388	Thiodipropionic acid	200 mg/kg	
389	Dilauryl thiodipropionate	200 mg/kg	
484	Stearyl citrate	GMP	

GMP, good manufacturing practice.

It has been suggested that for fish oils, TBHQ is much more effective than the other synthetic antioxidants (Ke et al., 1977; Kaitaranta, 1992). In the study of Ke et al. (1977), the order of effectiveness for inhibiting the oxidation in mackerel skin lipids has been found to be TBHQ>α-tocopherol>tempeh oil>BHA>BHT at the concentrations of 0.02% for all synthetic compounds, and 0.1% and 5% for α-tocopherol and tempeh oil, respectively.

5.6 NATURAL ANTIOXIDANTS

In an attempt to replace synthetic antioxidants with the natural ones, various natural substances and antioxidant-rich plant oils and extracts have been tested for fish oil stabilization, such as tocopherols, ascorbic acid/ascorbyl palmitate, flavonoids, extracts from rosemary (carnosol, carnosic acid [CA]) and tea (quercetins), oregano oils (carvacrol), tempeh oil, mustard flour, red chilies, or ground dried leaves of rosemary and bay. Findings of selected studies are presented next.

5.6.1 Tocopherols

Tocopherols are among the most important lipid-soluble natural antioxidants and have been found useful for stabilization of fish oils. However, the antioxidant effects are strongly concentration dependent, and vary with the type of tocopherol (i.e., α-, γ-, and δ-tocopherol) (Kulås and Ackman, 2001a,b).

Tocopherols alone may also exert pro-oxidative effects. In the study of O'Sullivan et al. (2005), both natural and synthetic α-tocopherol significantly increased the generation of peroxide value in crude cod liver oil and white pollock liver oil during 16 days of storage at 30 °C when added at 0.5%.

Rather than tocopherols alone, it has been shown that binary mixtures of tocopherols and ascorbic acid (or ascorbyl palmitate) bring inhibitory effects due to the ability of ascorbic acid to regenerate tocopherol. In the study of Yi et al. (1991), when δ-tocopherol was used together with ascorbic acid in refined sardine oil, adding δ-tocopherol at a fixed concentration of ascorbic acid did not increase the antioxidant effect synergistically. However, when the δ -tocopherol levels were fixed and the ascorbic acid levels were varied, the antioxidant effect rose sigmoidally with increasing levels of ascorbic acid. The authors concluded that at least 0.01–0.02% ascorbic acid is required to obtain a considerable synergistic effect with δ-tocopherol in stabilizing fish oil.

A successful stabilization of fish oil has been achieved with ternary antioxidant systems containing tocopherols, ascorbic acid (or ascorbyl palmitate), and phospholipids (lecithin) (Kamal-Eldin and Yanishlieva, 2002). Due to its amphiphilic character, lecithin forms micro-droplets together with trace water and thus aids solubilization of hydrophilic ascorbic acid within the cores of the micro-droplets. Ascorbyl palmitate is an ester synthesized from ascorbic acid and palmitic acid. The compound is difficult to dissolve in oil directly and is often combined with lecithin to aid solubilization.

Hamilton et al. (1998) studied the effects of tocopherol concentrates (0.2–2.0%), alone and in combination with ascorbyl palmitate (0.1%) and lecithin (0.05%), on the oxidative stability of fish oil stored at 20 °C. The binary systems of ascorbyl palmitate/lecithin and γ/δ- or δ-tocopherol-lecithin were strongly synergistic in delaying oxidation. The ternary blends, however, provided the greatest protection. Fish oil with 2% δ-tocopherol, 0.1% ascorbyl palmitate, and 0.5% lecithin did not oxidize over a period of 6 months at 20 °C. Table 5.2 summarizes findings on the efficacy of ternary mixtures of selected studies.

TABLE 5.2 Selected Results on the Stability of Different Fish Oils Protected by Ternary Antioxidant Systems During Accelerated Stability Tests

Ascorbic Acid (AA) or Ascorbic Palmitate (AP) (ppm)	Tocopherol (ppm)	Lecithin (ppm)	Method and Test Conditions	IP_0 (h)[a]	IP_A (h)[b]	Reference
200 (AA)	4000 (δ-toc, 0.1% water)	3000	1.8 mm film, 30 °C	2.1	46.4	Han et al. (1991)
200 (AA)	1000 (12% α-, 56% γ-, 30% δ-toc)	2000	Rancimat, 90 °C	0.9	24.1	Chang and Wu (1991)
400 (AA)	2000 (δ-toc, 0.1% water)	1000	Rancimat, 80 °C	4.0	40.0	Yi et al.(1991)
1000 (AA)	500 (mixed toc – not defined)	10 000	Fira-Astell, 80 °C	0.5	23.2	Löliger and Saucy (1985)
400 (AP)	200 (δ-toc)	500	Weight gain, 35 °C	9	42	Breivik and Aanesen (2002)

[a] *IP_0, induction period of control (no antioxidant added).*
[b] *IP_A, induction period in the presence of antioxidants.*
Table adapted from Kamal-Eldin and Yanishlieva (2002).

5.6.2 Antioxidant-Rich Extracts—Focus on Rosemary

Rosemary (*Rosmarinus officinalis*) has been shown to have an antioxidant effect in various food matrices (Etter, 2005). In the study of O'Sullivan et al. (2005), rosemary extract (0.5%) followed by tea catechine powder (0.5%) was superior in reducing the production of primary oxidation products (measured as PV) in crude cod liver oil (CLO) during 16 days of storage at 30 °C compared with the same dose of α-tocopherol (natural and synthetic), clove oil (black and white), carvacrol (the main antioxidant in oregano), and mustard flour. On the other hand, in crude white pollock liver oil (WPLO), the tea catechine powder showed the best antioxidant effect, while the rosemary extract was not effective at all. The tea catechine powder also gave the lowest level of the secondary oxidation products (measured as AV) for both the CLO and WPLO. In both oils, α-tocopherol significantly increased the generation of PV, while the black and white clove oils, carvacrol, and mustard flour showed little difference in PV and AV compared with control. The variable antioxidant effects of the different antioxidants in the two crude oils were attributed to the presence of minor components in the oils, such as metals and endogenous tocopherol, and different fatty acid profiles of the two oils.

Two different rosemary extracts effectively inhibited the formation of conjugated dienes, propanal and pentenal, in bulk fish oil during 20 days of storage at 40 °C and concentrations of 300, 500, 600, and 1000 ppm. The main rosemary compounds alone, carnosol and CA, however, showed varying effects. Carnosol showed pro-oxidant activity after 8 and 10 days of oxidation (based on propanal formation), while CA showed antioxidant activity initially (2–6 days) and pro-oxidant activity after 8 and 10 days of oxidation at 30 ppm. CA at 50 ppm was a better inhibitor of pentenal formation than carnosol initially (2–8 days), and after 10 days, carnosol showed no antioxidant activity (Frankel et al., 1996).

Dry ground oregano (0.5%) and dry ground rosemary (0.5%) in mackerel oil stored at 40 °C in the dark had an antioxidant effect (measured by determination of PV) that was comparable to the effects of 200 ppm BHT, while dry bay leaf and red chilies had no significant effect. When the oregano powder was tested at 1%, the activity was comparable to 200 ppm TBHQ (Tsimidiou et al., 1995).

A strong synergistic effect was observed for a mixture of α-tocopherol (0.02%) and rosemary extract (0.05%) in sardine oil stored at 30 °C, possibly due to the regeneration of α-tocopherol. The mixture delayed the onset of oxidation by 5 days compared with either α-tocopherol or rosemary extract alone, and its antioxidant activity was comparable to that of BHA (Wada and Fang, 1992).

In another study of Wada and Fang (1994), a mixture of α-tocopherol and rosemary extract (0.035% + 0.035%) showed significantly stronger antioxidant effect than α-tocopherol alone and rosemary extract alone in inhibition of oxidation of sardine oil. In addition, the effective lifetime of α-tocopherol in samples containing the mixture was 10 days longer than in samples with only α-tocopherol even though the amount of α-tocopherol in the latter case was 0.07%.

The effects of CA extracted from leaves of rosemary of different concentrations (0.1, 0.2, and 0.3 mg/g), tocopherol and a synthetic antioxidant (TBHQ) on oxidative stability of refined fish oil stored for 66 days at different temperatures (30 and 4 °C) were compared in a study by Wang et al. (2011) by measurement of PV, conjugated dienes, FFAs, thiobarbituric acid–reactive substances, and changes in the content of *trans*–fatty acids, aldehyde compounds, and polyunsaturated fatty acids. The results show that the three CA concentrations were more effective in restraining fish oil oxidation, in which a dose–response relationship was observed. The antioxidant activity of CA was stronger than that of tocopherol but still weaker than that of TBHQ. The authors concluded that fish oil supplemented with 0.2 mg/g CA exhibited favorable antioxidant effects and is preferable for effectively avoiding oxidation.

5.6.3 Green Tea Extract and Tea Catechins

Green tea extracts showed a pro-oxidant effect in refined menhaden and seal blubber oil, possibly due to the catalytic effect of chlorophyll. After removal of chlorophyll through a chromatographic technique, the extract showed an excellent antioxidant activity in both oils when applied at levels of 100, 200, 500, and 1000 ppm. The efficacy of the dechlorophyllated extracts (>200 ppm) was much higher than that of BHA and BHT at 200 ppm and of α-tocopherol at 500 ppm but lower than 200 ppm TBHQ (Wanasundara and Shahidi, 1998). Consistently, a mixture of green tea catechins showed a stronger antioxidative activity than BHA and tocopherol in fish oil (Koketsu and Satoh, 1997).

5.6.4 Boldine

Boldine, an alkaloid extracted from leaves of *Peumus boldus*, showed greater antioxidant activity than α-tocopherol, quercetin, BHA, or BHT (all substances at 0.8 g/kg oil) in sardine oil after 36 days of oxidation at 25 °C. Additive effects were observed when mixtures of boldine/quercetin and α-tocopherol/boldine were used. The alkaloid was also effective and exceeded the efficacy of α-tocopherol, quercetin, and BHT against iron-induced oxidation of sardine oil (Valenzuela et al., 1991).

5.6.5 Flavonoids

Nieto et al. (1993) investigated the antioxidant activities of six commercially available flavonoids (catechin, morin, naringin, hesperidin, rutin, and quercetin) and five flavonoids purified from two Chilean native plants against oxidation of partially refined sardine oil and compared them with the activities of α-tocopherol, BHT, and BHA. Catechin, morin, and quercetin and some of the extracted flavonoids showed high inhibitory effects (on either spontaneous or Fe^{2+}-induced oxidation). Among the native flavonoids, the 5,3′,4′-trihydroxy-7-methoxy flavanone (designated as Pt-2) showed the highest antioxidant activity. Mixtures of quercetin or of Pt-2 with α-tocopherol produced better inhibitory effects than each compound alone. Also, when Pt-2 and quercetin were combined (0.3 and 0.7 g/kg oil, respectively), a synergistic antioxidant effect was observed. The authors concluded that flavonoids could be used as natural antioxidants and replace synthetic antioxidants.

5.7 QUALITY RECOMMENDATIONS

For evaluation of oil stability and monitoring of deterioration during storage, commonly used methods include PV, AV, and TBA. More advanced techniques (described next) can be used for more detailed analyses of oxidation products like volatile compound analysis by the use of gas chromatography–mass spectroscopy (GC–MS) analysis and nuclear magnetic resonance (NMR) for detailed analysis of the oil composition and oxidative status. Table 5.3 summarizes recommended criteria for quality and acceptability of oils for human consumption.

Primary oxidation products are usually assessed by PV and/or conjugated dienes. PV analysis can be performed using iodometric titration methods or spectrophotometric ferro methods. The amount of lipid hydroperoxides usually increases in the beginning of an oxidation cascade but decreases later as hydroperoxide breakdown to secondary oxidation products becomes higher than the rate of formation (Gardner, 1983). Thus, PV is an empirical measure of oxidation and is useful for samples that are oxidized to relatively low levels (Frankel, 1998). The measured PVs are relative because they may differ for different methods (Nielsen et al., 2003). AV is used for determination of the amount of aldehydes formed as a result of breakdown of lipid peroxides in the oils. The method has some weaknesses in relation for measurement of fish oil oxidative quality: (1) It is not suitable for salmon oil because the presence of astaxanthin influences the result (RUBIN Report No. 173, 2009); (2) the use of AV for determination of oxidative quality of ethyl esters has also been observed to give overestimated values, and (3) food additives (e.g., color) and/or flavorings (e.g., lemon, orange) in omega-3 fatty acid supplements may have influence on the measurements (RUBIN Report No. 196, 2010; Thorkildsen, 2010).

Volatile oxidation products that are responsible for off-flavors in oils are usually determined by GC–MS. Due to the large number of secondary oxidation products in fish oils, a complicated spectrum is usually observed—typically, more than 100 peaks representing different low molecular weight oxidation compounds. The results vary between studies due to sample differences (e.g., type of added antioxidants and fish oil), degree of oxidation, and chromatographic techniques used (Kamal-Eldin and Yanishlieva, 2002). Because GC methods are widely used for assessing quality of marine oils in scientific lipid oxidation studies, standardization of the method should be performed.

High-resolution NMR has been widely used to determine the composition of fish oils and lipids. NMR is a nondestructive, multicomponent

TABLE 5.3 Recommended Criteria for Oxidative Status of Marine Oil Products for Human Consumption

	Ph. Eur. 7.0 (2011) Maximum Values[b]		GOED Monograph (2006) Maximum Values[c]	
Type of Marine Oil[a]	Peroxide Value[d]	Anisidine Value[e]	Peroxide Value[f]	Anisidine Value[g]
Cod liver oil (type A)	10	30	Not specified	Not specified
Cod liver oil (type B)[h]	10	Not specified		
Cod liver oil, farmed	5	10		
Salmon oil, farmed	5	10		
Fish oil, rich in omega-3 acids (type 1)	10	30	5	20
Fish oil, rich in omega-3 acids (type 2)[i]	5	15		
Omega-3 acid triglycerides	10	30		
Omega-3 acid ethyl esters 60	10	20		
Omega-3 acid ethyl esters 90	10	20	Not specified	Not specified

[a]*See detailed description of the oils in VKM (2011b).*
[b]*Ph. Eur, European Pharmacopoeia.*
[c]*GOED, Global Organization for EPA and DHA.*
[d]*PV (milliequivalents of active oxygen/kg oil) determined according to Ph. Eur. Monograph 2.5.5. Method A.*
[e]*AV determined according to Ph. Eur. Monograph 2.5.36.*
[f]*PV (milliequivalents of active oxygen/kg oil) determined according to AOCS (American Oil Chemists Society) Official Method Cd 8–53.*
[g]*AV determined according to AOCS Official Method Cd 18–90.*
[h]*Cod liver type B is identical to type A except that AV is not given.*
[i]*Type 2 of "Fish oil, rich in omega-3 acids" is intended as medical nutrition for infants and thus lower peroxide- and anisidine values than for type 1 are required.*
Table adapted from VKM (2011b).

technique, which, in one single analysis, gives information on fatty acid composition, lipid classes, and the regiospecific distribution (sn1,3 vs sn2) of fatty acids (Aursand et al., 1993, 1995). NMR has also proved to be a useful technique for detecting minor compounds in oils such as cholesterol, additives or degradation products such as aldehydes, peroxides, and FFAs (Claxson et al., 1994; Satyarthi et al., 2009).

The main advantage of using NMR for evaluating oxidation quality of oils is that it provides quantitative information about hydroperoxide and aldehyde species, without the need for extensive sample preparations. Among the NMR techniques, ^{1}H NMR has been most widely used to study lipid oxidation. The ^{1}H NMR determined ratio between the olefinic (5.1–5.6 ppm) and aliphatic protons (0.5–2.5 ppm) and the ratio between the aliphatic and diallylmethylene (2.6–3.0 ppm) protons in fatty acids have been used as an alternative to the classic PV value. These ratios decrease as the oxidative deterioration proceeds, and a good correlation was found between these ratios and the classic PV value (Saito, 1987). However, for observing small changes of fats and oils, a more specific approach is needed.

Specific oxidation products are previously reported in a wide range of study, such as for vegetable oils (Claxon et al., 1994; Guillen and Goicoechea, 2009) and in fish oils and lipids (Falch et al., 2005a,b, 2007; Tyl et al., 2008; Sacchi et al., 2006; Guillen et al., 2008). Also, substances that interfere with the classic methods (PV/AV) can be identified by ^{1}H NMR (Skiera et al., 2012a). Because chemical shifts and NMR visibility of oxidation products may vary according to solvents applied, protocols for determining and quantifying individual oxidation species such as peroxides (Skiera et al., 2012a), aldehydes (Skiera et al., 2012b), and FFAs have been developed (Skiera et al., 2015).

However, due to the high level of unsaturated fatty acids in marine oils, with the following complex composition of peroxides and aldehydes, there is a need for further work to establish and validate the suggested approaches for quantification of aldehydes and peroxides in marine oils.

Because NMR instruments are not widely available in laboratories conducting lipid analyses (due to the high costs of equipment and the need for skilled operators), the method is not established as routine analyses, but ^{1}H NMR can be carried out with a high degree of automation and gives a rapid fingerprint of the lipid profile. Previously, the detection limit of oxidation products for complex oils was a challenge, but the increasing sensitivity of the NMR instruments (e.g., development of cryo probes) and more defined protocols for different oxidation products have increased the potential of the technique.

5.8 STABILIZATION OF FINAL PRODUCTS

Fish oils are marketed in different forms. Bulk oils can be sold in large containers and barrels (refined oils) and bottles (mainly commercial liver and other fish oils), and the most prevalent are encapsulated oils in gelatin capsules. Oxidation of oils is a dynamic process that develops during storage over time. The rate of oxidation depends on many factors such as composition, physical environment, packaging, and composition of the oils. Therefore, it is a difficult task to predict the oxidative stability of the oils.

5.9 COMPOSITION

The stability of fish oil is dependent on the composition of fatty acids – the more unsaturated fatty acids, the less stable is the oil. Sullivan et al. (2011) obtained lower activation energy for concentrated oils (higher amount shows that less energy is required to initiate the oxidation). Fish oils can be sold in the form of triglycerides and ethyl esters. Several studies show that omega-3 fatty acids are more unstable in the form of ethyl esters compared with triglyceride form (Martin et al., 2012; Ritter et al., 2015). The presence of added antioxidants improves stability of the fish oils in both forms; ethyl esters are still less stable compared with triglycerides (Martin et al., 2012). Because penetration of oxygen in viscous oils is more difficult, this can be an explanation for higher resistance for oxidation of high viscous triglycerides compared with less viscous ethyl-esters (Indrasena and Barrow, 2010).

5.10 STORAGE CONDITIONS

In general, lipid oxidation is a complex pathway of reactions between lipids and oxygen. If one of the reaction substrate is not present, no reaction will occur. Therefore, the concentration of formed peroxides in oil depends on the amount of available oxygen increasing oxidation with increasing amount of dissolved oxygen. Thus, to avoid oxidation the oil should be stored and handled under an inert atmosphere. Oxygen can be present in the headspace of various containers and then dissolved in the oil. Dissolved oxygen is in direct contact with the fatty acids and therefore causes the oxidation. Although the headspace oxygen has less contact with lipids, it can be transported from the headspace and dissolved in the oil, because of the concentration driven force and thus cause the oxidation (Johnson and Decker, 2015). Therefore, maintaining oxygen concentration as low as possible is extremely important for the stability of fish oils. As light promotes oxidation, fish oil should also be protected from UV light.

Encapsulation of marine oils in soft gelatin helps to control oxidative stability of the oil by protecting it against the physical environment of the oil, including oxygen, light, and humidity. However, the physical properties of the encapsulation material as well as the operating conditions of the encapsulation process have a major influence on the oxidative quality and stability of the oil. A gelatin film that coats the oil acts as a barrier against oxygen and light (Bao et al., 2009). Some studies (Shukla and Perkins, 1991; Fantoni et al., 1996; Albert et al., 2015) found oxidized oil in the soft gel fish oil capsules purchased locally. Oxidation compounds in the fish oil could have taken place before encapsulation or after encapsulation. Significant increases in peroxide values in commercially available fish oil capsules during accelerated storage tests (e.g., storage at elevated temperature) suggest that the oxidative stability of the products might be limited despite the thick gelatin capsule cover (Kolanowski, 2010). Glycerol is a central additive for the production of soft gelatin capsules. The oxygen permeability of gelatin films is shown to vary depending on the glycerol amounts. For example, the oxygen permeability decreases by the addition of up to 20% glycerol but increases if more glycerol is added (Kim et al., 2007). The metal ion content of the encapsulation material is an additional factor that may influence the oxidative stability of the capsule content (Shahidi and Kim, 2002).

Packaging used for storage of marine oils is also important for oxidative stability of the oils. The oil should be stored in airtight containers that protect not only from oxygen but also from light. Refined and concentrated oils ready for packaging (e.g., encapsulation) are stored in bulk using airtight and well-filled containers, protected from light, under an inert gas and transported in flexi bags (an alternative package for shipping oils) or in stainless steel or mild steel construction tanks flushed with an inert gas such as nitrogen. The international code of practice for transport of marine oils recommended by Codex Alimentarius applies (CAC/RCP, 2005). The capsules or fish oils are usually packed in a heavily tinted glass or plastic container. In some cases, the bottles containing capsules are packed in cardboard boxes for further protection against light. In a study on oxidative stability of encapsulated fish oil, Pacheco and Regitano-D'Arce (2009) observed that the highest amount of peroxides was formed in fish oil capsules stored in polyvinylchloride (PVC 250 µm) film blister. The best oxidative stability results were observed when encapsulated oil was stored in high-density polyethylene flasks with silica bags. Akhtar et al. (2010) showed that white, red, and yellow edible color pigments added to hydroxypropyl methylcellulose (HPMC) packaging of salmon oil gave good control of photo-oxidation. However, the addition of blue or green color pigments did not provide better protection against photo-oxidation than transparent HPMC films.

Lipid oxidation goes through different steps or reaction pathways depending on temperature conditions. One reason for this is related to the temperature effect on the solubility of oxygen, which decrease with the temperature (Robertson, 2000). Kinetic studies showed similar reaction pathway in the range of <40 °C. Therefore, the prediction of kinetic rate constants at low temperatures based on accelerated tests performed at high temperatures shows limitations (Sullivan et al., 2011; Drusch et al., 2008). Oxidation rate of oils and decomposition of lipid peroxides increase with increasing temperature. As a rule of thumb, the oxidation rate doubles for each 10 °C increase (Jacobsen et al., 2009). Therefore, a 10 °C increase in the storage temperature of an oil product will lead to a 50% decrease in its shelf life. The use of increased temperature (180–200 °C), usually under vacuum conditions, is generally necessary during processing and purification of marine oils. Even if the thermal exposure is under vacuum conditions, the susceptibility of the lipids to oxidation will increase depending on the temperature, exposure time, or level of pro-oxidants. For cod liver oil, a shelf life of 2 months after opening the bottle is usually given by producers. PV values were found to reach 8 mEq/kg after 36 days of typical domestic application (continuous storage at 10 °C in refrigerator with 30-min interval storage at room temperature each day) of cod liver oil in the study by Chol (2005), while

Boran et al. (2006) obtained 8 mEg/kg peroxide after 60–90 days of storage at 4 °C. This showed that storage temperature and introduction of "fresh" oxygen limit the shelf life of fish oil and is of high importance.

5.11 CONCLUSION AND RECOMMENDATIONS

Fish oils are, due to the high content of LC omega-3 PUFA, a demanding material in respect to stabilization against oxidative deterioration and therefore have a unique position within edible oils. Therefore, understanding the chemistry of PUFA oxidation is a crucial prerequisite to develop effective strategies for long-term stabilization of fish oils.

It is clear that several measures have to be taken simultaneously to hinder or prevent and to predict oxidation processes, where factors such as presence of oxygen, storage conditions, minor components in the oil (especially pro-oxidants), added antioxidants, and fish oil source and composition, play crucial roles and must be under control.

The shelf life of the oil is determined by the recommended PV and AV today. However, some of the secondary oxidation products have very low sensory threshold values and may limit palatability of the fish oil. Moreover, some additives can even interfere with the PV and AV measurements. Therefore, better understanding and control of the oxidation and shelf life of fish oil products may be obtained by applying more advance techniques.

In the field of exogenous antioxidants, more research is needed to identify effective antioxidants from natural sources, clarify effective concentration and potential pro-oxidant effects, and identify synergistic interactions in the mixture of different antioxidants. Antioxidant components in rosemary (e.g., carnosol) have gained much attention as effective natural antioxidants in fish oils; however, further research should investigate how rosemary extract affects sensorial attributes in fish oil products.

REFERENCES

Aidos, I., Lourenclo, S., Padt, A., Luten, J.B., Boom, R.M., 2002. Stability of crude herring oil produced from fresh byproducts: influence of temperature during storage. J. Food Sci. 67, 3314–3320.

Aidos, I., 2002. Production of High-quality Fish Oils from Herring Byproducts (Ph.D. thesis). Wageningen University, The Netherlands.

Akhtar, M.J., Jacquot, M., Arab-Tehrany, E., Caiani, C., Linder, M., Desorby, S., 2010. Control of salmon oil photo-oxidation during storage in HPMC packaging film: influence of film colour. Food Chem. 120, 395–401.

Albert, B.B., Cameron-Smith, D., Hofman, P.L., Cutfield, W.S., 2013. Oxidation of marine omega-3 supplements and human health. Biomed. Res. Int. 2013, 1–8. ID 464921.

Albert, B.B., Derraik, J.G.B., Cameron-Smith, D., Hofman, P.L., Tumanov, S., Villas-Boas, S.G., Garg, M.L., Cutfield, W.S., 2015. Fish oil supplements in New Zealand are highly oxidised and do not meet label content of n-3 PUFA. Sci. Rep. 5.

Aubourg, S.P., 2001. Fluorescence study of the pro-oxidant effect of free fatty acids on marine lipids. J. Sci. Food Agric. 81, 385–390.

Aursand, M., Rainuzzo, J.R., Grasdalen, H., 1993. Quantitative high resolution ^{13}C and ^{1}H nuclear magnetic resonance of ω-3 fatty acids from white muscle of Atlantic salmon (*Salmo salar*). J. Am. Oil Chem. Soc. 70, 971–981.

Aursand, M., Jørgensen, L., Grasdalen, H., 1995. Positional distribution of ω-3 fatty acids in marine lipid triacylglycerols by high-resolution ^{13}C nuclear magnetic resonance spectroscopy. J. Am. Oil Chem. Soc. 72, 293–297.

Bao, S., Xu, S., Wang, Z., 2009. Antioxidant activity and properties of gelatin films incorporated with tea polyphenol-loaded chitosan nanoparticles. J. Sci. Agric. 89, 2692–2700.

Bimbo, A.P., 2007. Processing of marine oils. In: Breivik, H. (Ed.), Long-chain Omega-3 Speciality Oils. The Oily Press, Bridgewater, England, pp. 77–109.

Breivik, H., Aanesen, B.A., 2002. Fremgangsmåte Ved Stabilisering Av Mettede Oljer. Norwegian Patent No 312371.

Boran, G., Karacam, H., Boran, M., 2006. Changes in the quality of fish oils due to storage temperature and time. Food Chem. 98, 693–698.

CAC/RCP (Codex Alimentarius Commission/Recommended Code of Practice), 2005. Recommended International Code of Practice for the Storage and Transport of Edible Fats and Oils in Bulk. CAC/RCP 36–1987, Rev.1-1999, Rev.2-2001, Rev.3-2005.

Calder, P.C., 2013. n-3 fatty acids, inflammation and immunity: new mechanisms to explain old actions. Proc. Nutr. Soc. 72, 326–336.

Carvajal, A., Mozuraityte, R., Standal, I., Storrø, I., Aursand, M., 2014. Antioxidants in fish oil production for improved quality. J. Am. Oil Chem. Soc. 91, 1611–1621.

Carvajal, A.K., 2013. Utilization of By-products from Norwegian Spring Spawning Herring for Human Consumption.Trondheim (Ph.D. thesis). The Norwegian University of Science and Technology.

Chaiyasit, W., Elias, R.J., McClements, D.J., Decker, E.A., 2007. Role of physical structures in bulk oils on lipid oxidation. Crit. Rev. Food Sci. Nutr. 47, 299–317.

Chang, S.S., Wu, K., 1991. Composition of Natural Antioxidants for the Stabilization of Polyunsaturated Oils. In US Patent 5,077,069.

Chol, S.P., 2005. Stability and Quality of Fish Oil during Typical Domestic Application. The United Nations University, Fisheries Training Programme. Available at: www.unuftp.is/static/fellows/document/pak05prf.pdf.

Claxson, A.W.D., Hawkes, G.E., Richardson, D.P., Naughton, D.P., Haywood, R.M., Chander, C.L., Atherton, M., Lynch, E.J., Grootveld, M.C., 1994. Generation of lipid-peroxidation products in culinary oils and fats during episodes of thermal stressing - a high-field h-1-NMR study. FEBS Lett. 355, 81–90.

Codex Alimentarius, Revision 2014. CODEX general standard for food additives (CODEX STAN 192-1995). In: Codex Alimentarius Commission. http://www.codexalimentarius.org.

Drusch, S., Grob, N., Schwarz, K., 2008. Efficient stabilization of bulk fish oil rich in long-chain polyunsaturated fatty acids. Eur. J. Lipid Sci. Tech. 110, 351–359.

European Food Safety Authority, 2010. Scientific opinion on fish oil for human consumption. Food Hygiene, including rancidity. EFSA J. 8, 1–48.

Etter, S.C., 2005. *Rosmarinus officinalis* as an antioxidant. J. Herbs Spices Med. Plants 11, 121–159.

Falch, E., Storseth, T.R., Aursand, M., 2005a. HR NMR to study quality changes in marine by-products. In: Engelsen, S.B., Belton, P.S., Jakobsen, H.J. (Eds.), Magnetic Resonance in Food Science: The Multivariate Challenge. The Royal Society of Chemistry, Cambridge, pp. 11–27.

Falch, E., Anthonsen, H.W., Axelson, D.E., Aursand, M., 2005b. Correlation between ^{1}H NMR and traditional methods for determining lipid oxidation of ethyl docosahexaenoate. J. Am. Oil Chem. Soc. 81, 1105–1110.

Falch, E., Storseth, T.R., Aursand, M., 2007. High resolution NMR for studying lipid hydrolysis and esterification in cod (*Gadus morhua*) gonads. Chem. Phys. Lipids 147, 46–57.

Fantoni, C.M., Cuccio, A.P., Barrera-Arellano, D., 1996. Brazilian encapsulated fish oils: oxidative stability and fatty acid composition. J. Am. Oil Chem. Soc. 73, 251–253.

Fournier, V., Destaillats, F., Juanéda, P., Dionisi, F., Lambelet, P., Sébédio, J.-L., Berdeaux, O., 2006. Thermal degradation of long-chain polyunsaturated fatty acids during deodorization of fish oil. Eur. J. Lipid Sci. Technol. 108, 33–42.

Frankel, E.N., Huang, S.-W., Prior, E., Aeschbach, R., 1996. Evaluation of antioxidant activity of rosemary extracts, carnosol and carnosic acid in bulk vegetable oils and fish oil and their emulsions. J. Sci. Food Agric. 72, 201–208.

Frankel, E.N., 1998. Lipid Oxidation. Editorial for the Oily Press, Dundee, Scotland.

Frankel, E.N., 2005. Lipid Oxidation. Editorial for the Oily Press, Dundee, Scotland.

Fukuzawa, K., Tadokoro, T., Kishikawa, K., Mukaib, K., Gebicki, J.M., 1988. Site-specific induction of lipid peroxidation by iron in charged micelles. Arch. Biochem. Biophys. 260, 146–152.

Gardner, H.W., 1983. Effect of lipid hydroperoxides on food components. In: Finley, J.W., Scwass, D.E. (Eds.), Xenobiotics in Foods and Feeds. American Chemical Society Symposium Series, Washington DC, USA, pp. 63–84.

Guillen, M.D., Goicoechea, E., 2009. Oxidation of corn oil at room temperature: primary and secondary oxidation products and determination of their concentration in the oil liquid matrix from H-1 nuclear magnetic resonance data. Food Chem. 116, 183–192.

Guillen, M.D., Carton, I., Goicoechea, E., Uriarte, P.S., 2008. Characterization of cod liver oil by spectroscopic techniques. New approaches for the determination of compositional parameters, acyl groups, and cholesterol from H-1 nuclear magnetic resonance and Fourier transform infrared spectral data. J. Agric. Food Chem. 56, 9072–9079.

Halliwell, B., 2007. Biochemistry of oxidative stress. Biochem. Soc. Trans. 35, 1147–1150.

Hamilton, R.J., Kalu, C., McNeill, G.P., Padley, F.B., Pierce, J.H., 1998. Effects of tocopherols, ascorbyl palmitate, and lecithin on autoxidation of fish oil. J. Am. Oil Chem. Soc. 75, 813–822.

Han, D., Yi, O.-S., Shin, H.-K., 1991. Solubilization of vitamin C in fish oil and synergistic effect with vitamin E in retarding oxidation. J. Am. Oil Chem. Soc. 68, 740–743.

Indrasena, W.M., Barrow, C.J., 2010. Oxidation and stability of food-grade fish oil: role of antioxidants. In: Alasalvar, C., Shahidi, F., Miyashita, K., Wanasundara, U. (Eds.), Handbook of Seafood Quality, Safety and Health Applications. Wiley-Blackwell, Oxford, UK, pp. 488–504.

Jacobsen, C., Nielsen, N.S., 2007. Optimization of oxidative stability of omega-3 enriched foods. In: Breivik, H. (Ed.), Long-Chain Omega-3 Speciality Oils. The Oily Press Bridgewater, UK, pp. 197–217.

Jacobsen, C., Rustad, T., Nielsen, N.S., Falch, E., Janssom, S., Storrø, I., 2009. Processing of marine lipids and factors affecting their quality when used for functional foods. In: Luten, J.B. (Ed.), Marine Functional Food. Academic Publishers, The Netherlands, Wageningen, pp. 89–114.

Johnson, D.R., Decker, E.A., 2015. The role of oxygen in lipid oxidation reactions: a review. Annu. Rev. Food Sci. Technol. 6, 171–190.

Kaitaranta, J., 1992. Control of lipid oxidation in fish oil with various antioxidative compounds. J. Am. Oil Chem. Soc. 69, 810–813.

Kamal-Eldin, A., Yanishlieva, N.V., 2002. N-3 fatty acids for human nutrition: stability considerations. Eur. J. Lipid Sci. Tech. 104, 825–836.

Ke, P.J., Nash, D.M., Ackman, R.G., 1977. Mackerel skin lipids as an unsaturated fat model system for the determination of antioxidative potency of TBHQ and other antioxidant compounds. J. Am. Oil Chem. Soc. 54, 417–420.

Kim, Y.T., Hong, Y.S., Kimmel, R.M., Rho, J.H., Lee, C.H., 2007. New approach for characterization of gelatin biopolymer films using proton behaviour determined by low field ^{1}H NMR spectrometry. J. Agric. Food Chem. 55, 10678–10684.

Kittipongpittaya, K., Panya, A., Cui, L., McClements, D.J., Decker, E.A., 2014. Association colloids formed by multiple surface active minor components and their effect on lipid oxidation in bulk oil. J. Am. Oil Chem. Soc. 91, 1955–1965.

Koketsu, M., Satoh, Y.-I., 1997. Antioxidant activity of green tea polyphenols in edible oils. J. Food Lipids 4, 1–9.

Kolanowski, W., 2010. Omega-3 LC PUFA contents and oxidative stability of encapsulated fish oil dietary supplements. Int. J. Food Prop. 13 (3), 498–511.

Kulås, E., Ackman, R., 2001a. Protection of α-tocopherol in nonpurified and purified fish oil. J. Am. Oil Chem. Soc. 78, 197–203.

Kulås, E., Ackman, R., 2001b. Properties of α-, γ-, and δ-tocopherol in purified fish oil triacylglycerols. J. Am. Oil Chem. Soc. 78, 361–367.

Kulås, E., Olsen, E., Ackman, R., 2003. Oxidation of fish lipids and its inhibition with tocopherols. In: Kamal-Eldin, A. (Ed.), Lipid Oxidation Pathways. AOCS Press, Champaign, Ill, pp. 37–69.

Löliger, J., Saucy, F., 1985. Process for Protecting Fat against Oxidation. US Patent 5,427,814.

Lyberg, A.M., Adlerceutz, P., 2006. Monitoring monohydroperoxides in docosahexaenoic acid using high-performance liquid chromatography. Lipids 41 (1), 67–76.

Martín, D., Terrón, A., Fornari, T., Reglero, G., Torres, C.F., 2012. Oxidative stabilization of ultra-high omega-3 concentrates as ethyl esters or triacylglycerols. Food Res. Int. 45 (1), 336–341.

Mozuraityte, R., Rustad, T., Storrø, I., 2008. The role of iron in peroxidation of polyunsaturated fatty acids in liposomes. J. Agric. Food Chem. 56 (2), 537–543.

Naz, S., Sherazi, S.T.H., Talpur, F., 2011. Changes of total tocopherol and tocopherol species during sunflower oil processing. J. Am. Oil Chem. Soc. 88, 127–132.

Nielsen, N.S., Timm-Heinrich, M., Jacobsen, C., 2003. Comparison of wet-chemical methods for determination of lipid hydroperoxides. J. Food Lipids 10, 35–50.

Nieto, S., Garrido, A., Sanhueza, J., Loyola, L., Morales, G., Leighton, F., Valenzuela, A., 1993. Flavonoids as stabilizers of fish oil: an alternative to synthetic antioxidants. J. Am. Oil Chem. Soc. 70, 773–778.

Olsen, E., Vogt, G., Saarem, K., Greibrok, T., Nilsson, A., 2005. Autoxidation of cod liver oil with tocopherol and ascorbyl palmitate. J. Am. Oil Chem. Soc. 82 (2), 97–103.

O'Sullivan, A., Mayr, A., Shaw, N.B., Murphy, S.C., Kerry, J.P., 2005. Use of natural antioxidants to stabilize fish oil systems. J. Aquat. Food Prod. Technol 14, 75–94.

Pacheco, S.G.A., Regitano-D'Arce, M.A.B., 2009. Encapsulated fish oil oxidative stability stored in different types of packing under ambient conditions. Ciênc. Tecnol. Aliment. [online] 29 (4), 927–932.

Reische, D.W., Lilliard, D.A., Eitenmiller, R.R., 1998. Antioxidants. In: Akoh, C.C., Min, D.B. (Eds.), Food Lipids: Chemistry, Nutrition and Biotechnology. Mercel Dekker, New York, pp. 423–448.

Ritter, J.C.S., Budge, S.M., 2012. Key lipid oxidation products can be used to predict sensory quality of fish oils with different levels of EPA and DHA. Lipids 47 (12), 1169–1179.

Ritter, J.C.S., Budge, S.M., Jovica, F., Reid, A.J.M., 2015. Oxidation rates of triacylglycerol and ethyl ester fish oils. J. Am. Oil Chem. Soc. 92 (4), 561–569.

Robertson, G.l., 2000. Shelf life of packaged foods, its measurements and prediction. In: Brody, A.L., Lords, J.B. (Eds.), Developing New Food Products for a Changing Marketplace. Technomic Publishing, Lancaster, pp. 329–353.

RUBIN Report No 196, 2010. Lite Oksiderte Omega-3 Oljer Og Potensielle Helsefordeler. En Screening Av Omega-3 Oljer Med Hensyn Til Variasjon I Oksidasjonsgrad, Innhold Av Oksidasjonsprodukter Og Effekt På Markørsystemer. Resirkulering og Utnyttelse av organiske Bioprodukter i Norge, Trondheim, pp. 1–62.

RUBIN Report No 173, 2009. Omega-3 oljer fra ferskt marint råstoff. En mulig konkurransestrategi for den norske omega-3 industrien. Resirkulering og Utnyttelse av organiske Bioprodukter i Norge, Trondheim, pp. 1–40.

Sacchi, R., Falcigno, L., Paduano, A., Ambrosino, M.L., Savarese, M., De Giulio, B., Addeo, F., Paolillo, L., 2006. Quantitative evaluation of the aldehydes formed in heated vegetable oils using high resolution proton-NMR spectroscopy. Rivista Italiana Delle Sostanze Grasse 83, 257–263.

Saito, H., 1987. Estimation of the oxidative deterioration of fish oils by measurements of nuclear magnetic-resonance. Agric. Biol. Chem. 51, 3433–3435.

Sang, W., Jin, Z.T., 2004. Lipid oxidation of fish liver oil as affected by light, antioxidants and temperature. J. Food Process. Pres. 28, 1–10.

Satyarthi, J.K., Srinivas, D., Ratnasamy, P., 2009. Estimation of free fatty acid content in oils, fats, and biodiesel by ^{1}H NMR spectroscopy. Energy Fuels 23, 2273–2277.

Shahidi, F., Kim, S.K., 2002. Marine Lipids as Affected by Processing and Their Quality Preservation by Natural Antioxidants, vol. 8216. ACS Symposium Series, pp. 1–13.

Shahidi, F., Zhong, Y., 2010. Lipid oxidation and improving the oxidative stability. Chem. Soc. Rev. 39 (11), 4067–4079.

Shukla, V., Perkins, E., 1991. The presence of oxidative polymeric materials in encapsulated fish oils. Lipids 26 (1), 23–26.

Skiera, C., Steliopoulos, P., Kuballa, T., Holzgrabe, U., Diehl, B., 2012a. H-1-NMR spectroscopy as a new tool in the assessment of the oxidative state in edible oils. J. Am. Oil. Chem. Soc. 89, 1383–1391.

Skiera, C., Steliopoulos, P., Kuballa, T., Holzgrabe, U., Diehl, B., 2012b. H-1 NMR approach as an alternative to the classical p-anisidine value method. Eur. Food Res. Technol. 235, 1101–1105.

Skiera, C., Steliopoulos, P., Kuballa, T., Holzgrabe, U., Diehl, B., 2015. Determination of free fatty acids in edible oils by ^{1}H NMR spectroscopy. Lipid Technol. 24, 279–281.

Sullivan, J.C., Budge, S.M., St-Onge, M., 2011. Modeling the primary oxidation in commercial fish oil preparations. Lipids 46, 87–93.

Tadolini, B., Gabrini, L., Menna, C., Pinna, G.G., Hakim, G., 1997. Iron (III) stimulation of lipid hydroperoxide dependent lipid peroxidation. Free Radic. Res. 27, 563–576.

Tengku-Rozaina, T.M., Birch, E.J., 2013. Physicochemical characterisation and oxidative stability of refined hoki oil, unrefined hoki oil and unrefined tuna oil. Int. J. Food Sci. Tech. 48 (11), 2331–2339.

Thorkildsen, T., 2010. Oksidasjonsnivå I Marine Omega-3 Produkter Tilgjengelig for Norske Forbrukere (Master thesis), Høyskolen i Akershus.

Tsimidou, M., Papavergou, E., Boskou, D., 1995. Evaluation of oregano antioxidant activity in mackerel oil. Food Res. Int. 28, 431–433.

Tyl, C.E., Brecker, L., Wagner, K.H., 2008. H-1 NMR spectroscopy as tool to follow changes in the fatty acids of fish oils. Eur. J. Lipid Sci. Technol. 110, 141–148.

Usydus, Z., Polak-Juszczak, L., Dobrzanski, Z., Malesa-Ciecwierz, M., 2007. Study on the nutritive value of raw fish oils. Pol. J. Food Nutr. Sci. 57 (4(C)), 593–596.

Vaisali, C., Charanyaa, S., Belur, P.D., Regupathi, I., 2015. Refining of edible oils: a critical appraisal of current and potential technologies. Int. J. Food Sci. Technol. 50, 13–23.

Valenzuela, A., Nieto, S., Cassels, B., Speisky, H., 1991. Inhibitory effect of boldine on fish oil oxidation. J. Am. Oil Chem. Soc. 68, 935–937.

VKM, 2011a. Evaluation of negative and positive health effects of n-3 fatty acids as constituents of food supplement and fortified foods. Opinion of Steering Committee of the Norwegian Scientific Commitee for Food Safety, Oslo, Oslo. Available online: http://www.vkm.no/dav/c7a41adb79.pdf.

VKM, 2011b. Description of the processes in the value chain and risk assessment of decomposition substances and oxidation products in fish oils. Opinion of Steering Committee of the Norwegian Scientific Committee for Food Safety, Oslo. Available online: http://www.vkm.no/dav/4be9bee090.pdf.

Wada, S., Fang, X., 1992. The synergistic antioxidant effect of rosemary extract and α-tocopherol in sardine oil model system and frozen-crushed fish meat. J. Food Process. Pres. 16, 263–274.

Wada, S., Fang, X., 1994. Synergistic antioxidant effects of rosemary and α-tocopherol at different storage temperatures and its application for inhibiting dried sardine meat oxidation. Yukagaku 43, 109–115.

Wanasundara, U.N., Shahidi, F., 1998. Antioxidant and pro-oxidant activity of green tea extracts in marine oils. Food Chem. 63, 335–342.

Wang, H., Liu, F., Yang, L., Zu, Y., Wang, H., Qu, S., Zhang, Y., 2011. Oxidative stability of fish oil supplemented with carnosic acid compared with synthetic antioxidants during long-term storage. Food Chem. 128, 93–99.

Wu, T.H., Bechtel, P.J., 2009. Quality of crude oil extracted from aging walleye pollock (*Theragra chalcogramma*) byproducts. J. Am. Oil Chem. Soc. 86, 903–908.

Yi, O.-S., Han, D., Shin, H.-K., 1991. Synergistic antioxidative effects of tocopherol and ascorbic acid in fish oil/lecithin/water system. J. Am. Oil Chem. Soc. 68, 881–883.

Chapter 6

Oxidative Stability and Shelf Life of Bulk Animal Fats and Poultry Fats

Poulson Joseph
Kalsec®, Inc., Kalamazoo, MI, USA

6.1 INTRODUCTION (SOURCE AND USE OF ANIMAL FATS)

Rendering is the process that produces animal fat and poultry fats. It was estimated that the 6.8 million tons of animal and poultry fats were produced globally and more than 50% was contributed from the United States (National Renderers Association, 2014). Recent statistics indicate annual production of approximately 1.4 billion pounds of poultry fat, 1.74 billion pounds of edible tallow, and 6.5 billion pounds of inedible tallow and grease in the United States.

Rendering is the process of using byproducts of the meat/animal/poultry industry through separation of fat from protein components usually termed as "meals." The renderers operate either as part of slaughtering/packaging facilities or independently by using raw materials from other meat-processing facilities, retailers, and restaurants. Rendering process is accomplished by wet rendering (moist cooking process) or dry rendering where the raw material is subjected to heat treatment, either as a batch process or continuous operation, to separate the fat without contact of moisture/steam (National Renderers Association, 2014). The fat derived from rendering process could be either edible or inedible and therefore determines its end use.

For example, various grades of animals fats are described in American Fats and Oils Association (AFOA) Specifications for Tallow and Greases (National Renderers Association, 2014), which range from edible tallow and edible lard to choice white grease, yellow grease, and technical tallow (inedible, Grade 1A). Edible lard and edible tallow are intended for human consumption purpose, and for the same reason, they are subjected to processing under USDA-FSIS (United States Department of Agriculture–Food Safety and Inspection Service) guidelines.

Oxidative Stability and Shelf Life of Foods Containing Oils and Fats
http://dx.doi.org/10.1016/B978-1-63067-056-6.00006-9

The Institutional Meat Purchase Specifications (IMPS) of the USDA states that edible tallow (IMPS No.731) and edible lard (IMPS No. 732) have to comply with the following standards (Table 6.1) set by AFOA (IMPS, 1993).

On a global scenario, the Codex Alimentarius International Food Standards (CODEX STAN, 2014; Section 02.1.3) for edible animal fats (lard, tallow, fish oil, and other animal fats) is outlined as follows: "All animal fats and oils should be derived from animals in good health at the time of slaughter and intended for human consumption. Lard is fat rendered from the fatty tissue of swine. Edible beef fat is obtained from fresh bovine fatty tissue covering the abdominal cavity and surrounding the kidney and heart, and from other compact, undamaged fat tissues. Such fresh fat obtained at the time of slaughter is the 'killing fat'. Prime beef fat (premiere jus or oleo stock) is obtained by low-heat rendering (50–55 °C) of killing fat and selected fat trimmings (cutting fat). Secunda beef fat is a product with typical beef fat odor and taste obtained by rendering (60–65 °C) and purifying beef fat. Rendered pork fat is fat obtained from the tissue and bones of swine. Edible tallow (dripping) is produced by the rendering of fatty tissue (excluding trimmings and cutting fat), attached to muscles and bones of bovine animals or sheep. Fish oils are derived from suitable sources such as herring, sardines, sprat, and anchovies. Other examples include tallow and partially defatted beef or pork fatty tissue."

Apart from the edible purposes, rendered animals fats serve a variety of applications. In general, inedible animal fats are used for the following purposes: (1) as animal feed additives and pet food formulations mainly as caloric source and as palatants, (2) manufacturing personal care products such as soaps and cosmetics, (3) industrial use such as lubricants and oleochemicals, and (4) recently, a major focus has been given on biofuel and energy production

TABLE 6.1 USDA (IMPS) Specifications of Edible Tallow and Lard

Characteristics	Edible Tallow	Lard (Edible)
Titer minimum	41.0	38.0
Free fatty acids (FFA) maximum	0.75%	0.50%
Color (FAC) maximum—Lovibond color 5¼-inch cell	Max 3	Max 1.5 red
Peroxide value	1.0 mEq/kg max (initial)	4.0 mEq/kg max
Moisture, impurities, unsaponifiable matter (MIU)[a]	Moisture: 0.20% max	Moisture: 0.20% max
Insoluble impurities	0.05% max	0.05% max

[a]*MIU: Typically less than 1% for edible fats.*
Source: IMPS (1993), Institutional Meat Purchase Specifications, USDA.

(Mata et al., 2011). The National Renderers Association estimated that the monetary values of U.S. rendered products were approximately U.S. $2.7 billion, which indicates the economic impact of the rendering industry as well as its contribution to the U.S. agricultural economy. Moreover, the recycling aspect of the rendering process is also relevant to animal agricultural industry.

This chapter will mainly focus on the factors influencing the quality of animal-derived fats, oxidative stability of animal and poultry fats, measurement of lipid oxidation, and management of lipid oxidation in animal fats.

6.2 CHARACTERISTICS AND FATTY ACID PROFILE OF ANIMAL FATS AND POULTRY FATS

The physical and chemical characteristics of animal fat primarily depend on the fatty acid profile. Although various factors such as species, feed/diet, animal management practices, and different anatomical locations influence the fatty acid profile, a general range is described in Table 6.2.

Bovine fats are relatively more saturated compared with porcine and poultry fats, which tend to have a greater proportion of unsaturated fatty acids. In all animal species, the diet or feeding regimen influences the fatty acid profile; the effect is noticeable to a greater extent in monogastric animal species such as pigs and poultry (Wood et al., 1999). Nevertheless, ruminants also exhibit variation in fatty acid profile based on their diet, a fact well documented in the scientific literature (Wood et al., 1999; Scollan et al., 2001).

For example, a market sampling study was conducted (unpublished data) to characterize the fatty acid profiles of grass-fed beef as well as typical retail beef, which is grass-fed but grain-finished during stocker phase. The data represented in Table 6.3 indicate that grass-feeding significantly increases the proportion of polyunsaturated fatty acids (PUFA) in beef fat (fat exacted from meat).

The role of fatty acid profile on fat quality and meat product quality has been investigated by several researchers (Wood et al., 1999), including in ducks (Jiang et al., 2012), poultry (Husvéth et al., 2000), pork (Benz et al., 2010, 2011a), and beef (Scollan et al., 2001). In general, the research findings indicate that the feeds composed of more unsaturated fatty acids (UFA) tend to reflect the same trend in animal fats and, subsequently, the meat as well as fat would be more susceptible to lipid oxidation (Lauridsen et al., 1999; Husvéth et al., 2000; Kouba and Mourot, 2011). In recent years, numerous studies have been conducted to evaluate the efficacy of distillers' dried grains with solubles (DDGS) on meat species, including pork, poultry, and beef (Stein and Shurson, 2009; Benz et al., 2010). The authors have concluded that DDGS would lead to greater unsaturation in animal fat, when used at greater than 30% in animal diet.

In addition to the composition of the feed/fodder, quality of the feed could also potentially influence lipid oxidation in animal tissues. Sheehy

TABLE 6.2 Fatty Acid Profiles of Typical Animal Fats and Poultry Fats

	Beef Tallow	Pork Lard	Poultry Fat
Melting point (°C)	40–50	34–44	23–40
Iodine value	25–45	45–75	65–75
Fatty Acid Composition			
Myristic acid (C14:0)	2–3	1–1.5	1–1.5
Palmitic acid (C16:0)	24–28	24–28	20–24
Stearic acid (C18:0)	20–24	13–14	4–6
Saturated fatty acids (%)	46–55	38–43.5	25–31.5
Palmitoleic acid (C16:1)	2–3	2–3	5–9
Oleic acid (C18:1)	40–43	43–47	33–44
Linoleic acid (C18:2)	2–4	8–11	18–20
Linolenic acid (C18:3)	<1	<1	1–2
Unsaturated fatty acids (%)	45–51	54–62	57–75

Adapted with permission from Alm, M; Edible oil processing: Animal fats, AOCS Lipid Library, Copyright (2013) American Oil Chemists' Society.

TABLE 6.3 Fatty Acid Profile (FAME Method[a]) Summary of Grass-Fed Beef versus Grain-Fed Beef

	Grass-Fed Beef (%)	Grain-Fed Beef (%)
Myristic (C14:0)	5.9	10.4
Palmitic (C16:0)	39.2	42.4
Stearic (C18:0)	23.5	9.9
Oleic (C18:1)	27.4	34.7
Linoleic (C18:2)	2.5	2.3
Linolenic (C18:3)	1.4	0.3

[a]*FAME indicates fatty acid methyl esters; gas chromatographic method, AOAC Official Method 963.22.*

et al. (1994) reported that in poultry (chicks), an experimental consumption of thermally oxidized sunflower oil resulted in a reduced alpha-tocopherol status and, therefore, increased susceptibility of tissues to lipid oxidation. Dietary supplementation of certain nutrients (L-arginine) would also critically influence (reduce) fat deposition in meat-type ducks (Wu et al., 2011). The dietary iodine value of the porcine diet is also used as one of the tools to monitor feed quality, since greater dietary iodine value indicates a propensity for softer pig fat (Benz et al., 2011b), with resultant high iodine value indicating more unsaturation in pork fat. To summarize, dietary energy supply and fat source influence the fatty acid pattern of adipose and lean tissues as well as lipogenesis in pigs (Bee et al., 2002).

6.3 LIPID OXIDATION IN ANIMAL-DERIVED FATS

Lipid oxidation in fats is an extensively researched topic and would represent the sequel of chemical changes and reactions when lipids interact with oxygen (McClements and Decker, 2008). Lipid oxidation proceeds in a classic pathway of initiation, propagation, and termination and has been described in detail in the book of food chemistry (Grosch, 1999). The primary lipid oxidation products, hydroperoxides, have minimal impact on the food quality (odor and flavor), while the volatile secondary oxidation products critically influence the flavor and aroma of the food matrices, especially when the flavor thresholds are lower. Therefore, lipid oxidation or fat auto-oxidation leads to rancid flavor or undesirable odors in fats, oils, and fat/oil-containing foods matrices. In meat and meat products, especially precooked products, this phenomenon is described as "warmed-over flavor" or "modified flavor reduction."

Typically, oxidative rancidity is the term used to describe such changes on food flavor/aroma quality that results from chemical deterioration of lipids due to oxidative reactions. On the other hand, hydrolytic reactions also detrimentally influence fat/lipid quality, and these involve either enzymes (lipases) or processing such as frying (high temperature and presence of moisture). The free fatty acids thus derived could also reduce the oxidative stability of the fats/oils and result in off-flavors, termed as hydrolytic rancidity.

Various factors influence lipid oxidation in animal and poultry fats, the primary endogenous factors being fatty acid profile as well as presence of natural antioxidants in fat acquired through feed/fodder. The greater the proportion of unsaturated fatty acids, the higher is the susceptibility to oxidative rancidity. Moreover, pro-oxidants such as oxygen, transition metals such as iron and copper, enzymes, heat, light, etc. also initiate/accelerate lipid oxidation. Trace amounts of iron and water might be present as contaminants in edible oils, for example, Koga and Terao (1994) reported that lard contained iron (0.03 ppm) and water (0.03%).

6.4 MEASURING LIPID OXIDATION

Lipid oxidation or oxidative stability in animal fats/muscle foods could be evaluated using different approaches, namely chemical, physical, and sensory evaluation. Chemical methods include analysis of (1) primary oxidation products (peroxide value [PV] and conjugated dienes) and (2) secondary oxidation products (*para*-anisidine values [*p*AV], thiobarbituric acid reactive substances [TBARS], and gas chromatography–mass spectrometry [GC–MS]). Gas-sensor arrays (electron nose) have been used in association with GC–MS for rapid screening of volatiles (Olsen et al., 2005). Physical methods measure the outcome of oxidative reactions of lipids/fats through changes in conductivity (Rancimat, OSI etc.), pressure/oxygen consumption (oxygen bomb test; Oxipres), and weight gain (lard; Koga and Terao, 1994).

For animal-derived fats, OSI or Rancimat, as well as PV, are the most commonly used analytical tools, regarding lipid oxidation evaluation. Oil Stability Index (OSI) is an AOCS-approved method (Method Cd 12b-92), which is an automated substitution for the Active Oxygen Method (AOM; AOCS Official Method Cd 12-57) for fat stability. TOTOX value (2PV + *p*AV) is also used often in industry to ascertain the oxidative status of animal fats/fat-containing foods. Other tests include rapid testing methods for fats and oils such as PeroxySAFE, which use colorimetric reactions (kit) for measuring peroxides (AOAC certified according to manufacturer). Analytical advances have led to the attempts to use methodologies such as chemiluminescence (Vichi et al., 2001), fluorescence emission (Santas et al., 2014), and Raman spectroscopy, infrared spectroscopy, and magnetic resonance to measure lipid oxidation in fats/food matrices as presented in a recent review by Barriuso et al. (2013). Lipid oxidation evaluation for muscle foods and meat products has been summarized by Ladikos and Lougovois (1990) and Gray and Monahan (1992). Sensory analysis could be

TABLE 6.4 Correlation of Sensory and Analytical Parameters (Partial Least Squares Regression Model) for Early Stages of Oxidative Stability in Pork Back Fat against Storage Time

Parameter	Correlation Coefficient (*r*)	Prediction Errors (RMSEP[a])
Sensory attribute	0.87	4.23
GC–MS (all volatiles)	0.97	2.09
Electronic nose (all sensors)	0.95	2.79
Fluorescence (whole spectra)	0.81	5.00
PV	0.86	4.62
TBARS	No correlation	
Chemiluminescence	No correlation	

[a]*RMSEP, root mean squares of prediction errors.*
Adapted with permission from Olsen et al. (2005), Copyright (2005) American Chemical Society.

used to complement the analytical parameters (Table 6.4), and typically descriptive analysis is undertaken to profile the flavor and aroma changes in fats/fat-containing foods (Olsen et al., 2005).

Olsen et al. (2005) also determined the correlations between various sensory attributes in pork back fat (during storage at −20 °C for 26 weeks) and GC–MS, e-nose, and fluorescence outputs. Interestingly, these analytical approaches revealed high correlation with several sensory attributes, noticeably acidic flavor and paint flavor. The descriptive sensory analysis indicated an increase in the intensity of paint flavor during the storage, whereas acidic flavor intensity reduced during storage. The authors conclude that while dynamic headspace/GC–MS and fluorescence could detect changes in pork back fat during early stages of lipid oxidation compared to sensory panel, the instrumental methods could be used to complement sensory data.

GC–MS approach also revealed that among the volatile secondary oxidation products determined, hexanal is the most abundant compound, which exhibited an increasing trend, as shown in Figure 6.1 (Olsen et al., 2005). The authors determined 35 volatile compounds in pork back fat during storage, including aldehydes, alcohols, ketones, alkanes, alkenes, and others, while the noticeable ones were hexanal, 1-penten-3-ol, 1-octen-3-ol, and 2- and 3-methylbutanal. Interestingly, the analytical parameter for measuring the primary oxidation products (PV) during early stages of lipid oxidation in pork back fat also revealed a similar trend with PV values ranging from 0.27 mEq/kg lipid (week 2) to 2.3 mEq/kg lipid (week 26).

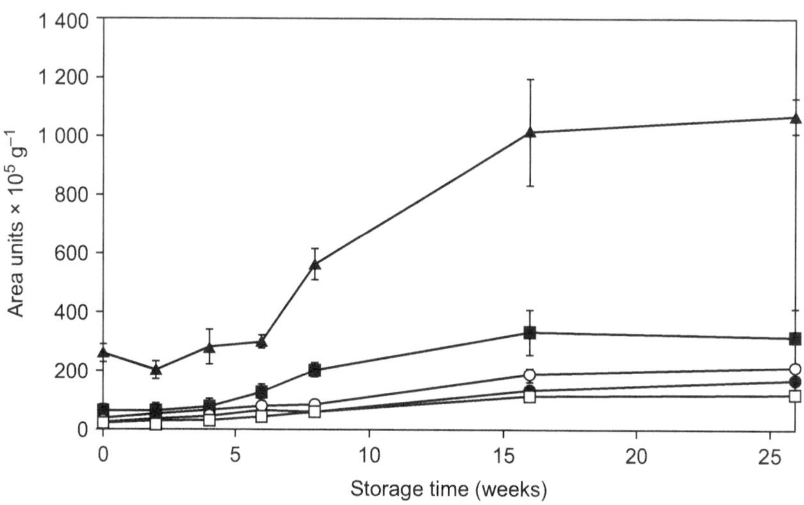

FIGURE 6.1 Major volatile secondary oxidation compounds in pork back fat during frozen storage. ▲, Hexanal; ■, 1-penten-3-ol; □, 1-octen-3-ol; ●, 2-methylbutanal; ○, 3-methylbutanal. *Adapted with permission from Olsen et al. (2005), Copyright (2005) American Chemical Society.*

6.5 MANAGING LIPID OXIDATION IN ANIMAL-DERIVED FATS

Lipid oxidation in animal derived fats is an inevitable process, that cannot be reverted back once initiated, but could be delayed or inhibited by several approaches including incorporation of antioxidants, and adopting packaging strategies, as well as effectively managing storage temperature, raw material quality, and processing conditions.

Antioxidants could be used as a practically efficient strategy to minimize lipid oxidation. The various screening methods for determining the antioxidant potential of compounds is detailed in Table 6.5.

Based on the screening tests (Huang et al., 2005), various antioxidants could be used for stabilizing animal-, including poultry-, derived fats. Synthetic antioxidants have been used in industry for a long time and have been very effective. Recently, there is an increased need for natural antioxidants based on consumer/end product demand. Nevertheless, the selection of an antioxidant depends on various considerations including antioxidant efficacy, solubility, carry through properties/thermal stability, effect of fat on food color and flavor, regulatory guidelines, and market needs and cost.

For example, Figure 6.2 represents data (unpublished) on oxidative stability of fresh rendered lard subjected to antioxidant treatments. The OSI data indicate that combining synergistic antioxidants (2000 ppm) with BHA/BHT (200 ppm) significantly increases the oxidative stability, compared to BHA/BHT by itself. On the other hand, TBHQ was reported to be very effective in extending chicken fat stability, compared to green tea and grape seed extracts (Rababah et al., 2004), as outlined in Figure 6.3.

Rosemary extract has been used in food industry as a natural antioxidant and has been shown to function very effectively in meat products as well as animal-derived fats (Sebranek et al., 2004; Brewer, 2008; Liang and Schwarzer, 1998). The result of oxidative stability of chicken fat from the author's study (unpublished) is represented in Figure 6.4. Interestingly, rosemary extracts increased the oxidative stability in a dose-dependent manner and could be used as an effective natural antioxidant for edible animal fats as well as fats for pet food (coating dry kibble) purposes. Researchers have employed various antioxidants such as phenolic antioxidants (Page, 1979), phosphatidyl derivatives of hydroxytyrosol (Martin et al., 2014), tocotrienols (Kim and Paik, 2014), and phenolic acids and related compounds (Dziedzic and Hudson, 1984) in animals fats or edible fats model systems.

Yeo et al. (2010) used thermally oxidized lard to determine the effectiveness of free-radical scavenging antioxidants (sesamol, BHA, and α-tocopherol) using analytical approaches including conjugated dienoic acid (CDA), *p*AV, and DPPH absorbance. Sesamol demonstrated the greatest antioxidant activity followed by BHA and α-tocopherol. The authors suggest that such model systems could be used as useful indicators for comparing the efficacy of free radical–scavenging antioxidants. In a similar approach, the composition of volatile oxidation products in thermally oxidized lard, with reference to breed,

TABLE 6.5 Screening Methods for Antioxidants

Assay	Type	Principle	References
Antioxidant Screening Methods			
DPPH (2,2-dipheyl-1-picryhydrazyl radical)	Spectrophotometric	Radical scavenging activity decreases the intensity of the solution color at 515 nm	Kedare and Singh (2011)
ORAC (oxygen radical absorbance capacity)	Fluorometric	Oxidation of lipid results in reduced fluorescence	Huang et al. (2002)
TRAP (total radical trapping antioxidant parameter)	Fluorometric	Increase in lag time of fluorescence decay indicates antioxidant activity	Wayner et al. (1985)
TEAC (Trolox equivalence antioxidant capacity)	Spectrophotometric	Slope of linear absorbance over time indicates changes in rate of oxidation	Re et al. (1999)
Ferrozine	Spectrophotometric	Chelation activity interferes with ferrozine-Fe(II) complex color formation, reducing color at 562 nm	Carter (1971)
Hydroperoxide	Spectrophotometric	Primary lipid oxidation products are quantified based on complexing with an iron thiocyanate chromophore, increasing absorption at 510 nm	Decker et al. (2008)
Folin-Ciocalteu	Spectrophotometric	Measures total phenolic content of material based on absorbance at 765 nm compared to Trolox or gallic acid as standard	Singleton et al. (1999)
Antioxidant Efficacy Screening Methods for Animal Derived Fats			
Oil Stability Index (OSI)	Induction time determination	Increase of the induction time of lipid indicates resistance to oxidation	AOCS Official method Cd 12b-92

Continued

TABLE 6.5 Screening Methods for Antioxidants—cont'd

Assay	Type	Principle	References
Peroxide Value (PV)	Titration	Peroxides in lipids react with potassium iodide, increasing the amount of titrant required to quench the reaction as indicated by color indicator	AOCS Official method Cd 8-53
para-Anisidine value (*p*AV)	Spectrophotometric	Measures the amount of aldehydes in lipids by the increase in the absorbance at 350 nm	AOCS Official method Cd 18-90

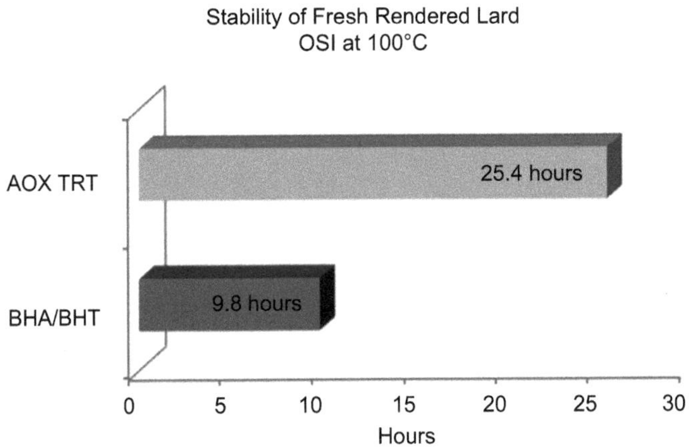

FIGURE 6.2 Oxidative stability of lard as determined by OSI induction times comparing BHA/BHT versus AOX TRT (BHA/BHT+blend of tocopeherols, rosemary extract, and citric acid).

was investigated (Xu et al., 2012) using headspace solid-phase microextraction and GC–MS. Lard of Enshi No. 1 native breed had greater volatile oxidation products than lard from three-way crossbred breed, while the major oxidation products were identified as aldehydes, such as heptanal, octanal, nonanal, 2-heptenal, 2-octenal, 2,4-heptadienal, 2,4-decadienal, and 2-undecenal.

Thermal oxidation of tallow was studied by Song et al. (2010), although the objective was to determine beef flavor precursor compounds. The authors reported that PV and *p*AV values increased drastically during incubation at 140 °C for 2 h, with a corresponding increase in volatile compounds such as hexanal, 1-octen-3-ol, 2,4-decadienal, and 2,4-heptadienal.

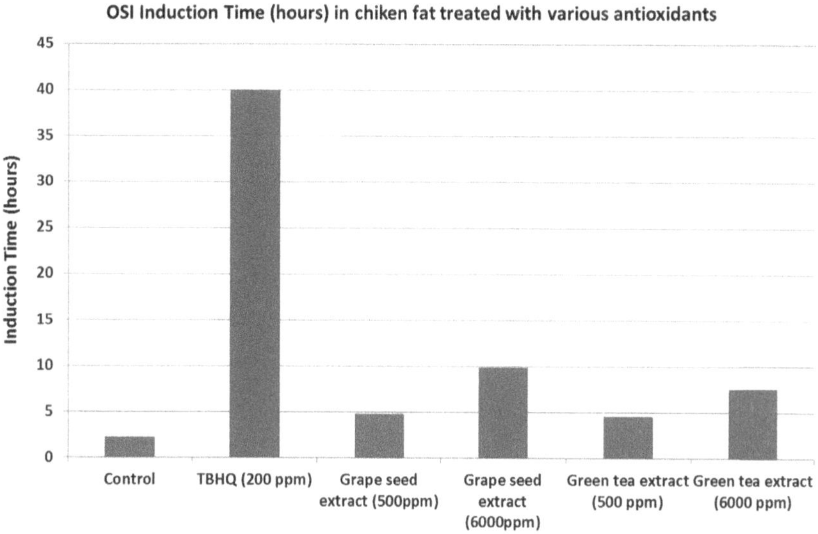

FIGURE 6.3 Antioxidant activities of plant extracts and TBHQ in chicken fat as determined by OSI induction times at 110 °C. *Adapted with permission from Rababah et al. (2004), Copyright (2004) American Chemical Society.*

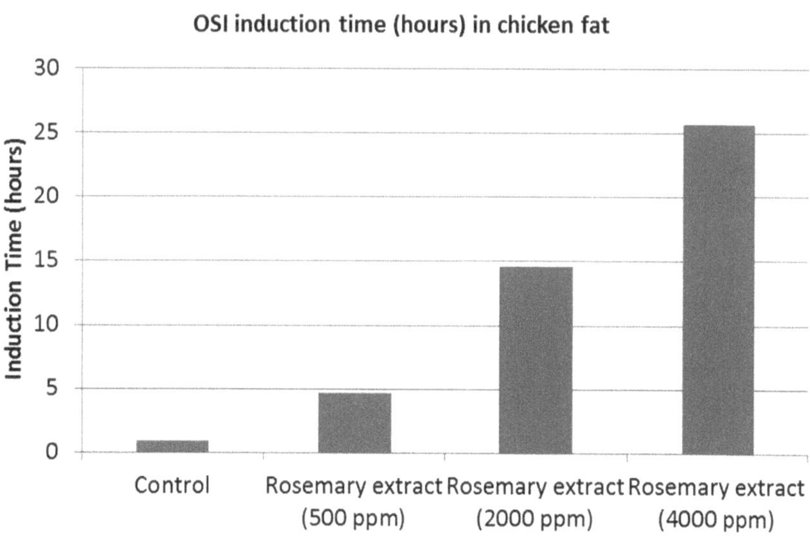

FIGURE 6.4 Oxidative stability of chicken fat with natural antioxidants determined by OSI induction times at 110 °C.

Apart from antioxidants, packaging also influences lipid oxidation in animal fats (Figure 6.5). In an attempt to investigate the effect of packaging and antioxidants on fresh rendered lard, samples from treatments (BHA/BHT versus BHA/BHT + synergistic antioxidants) were stored in containers and held at

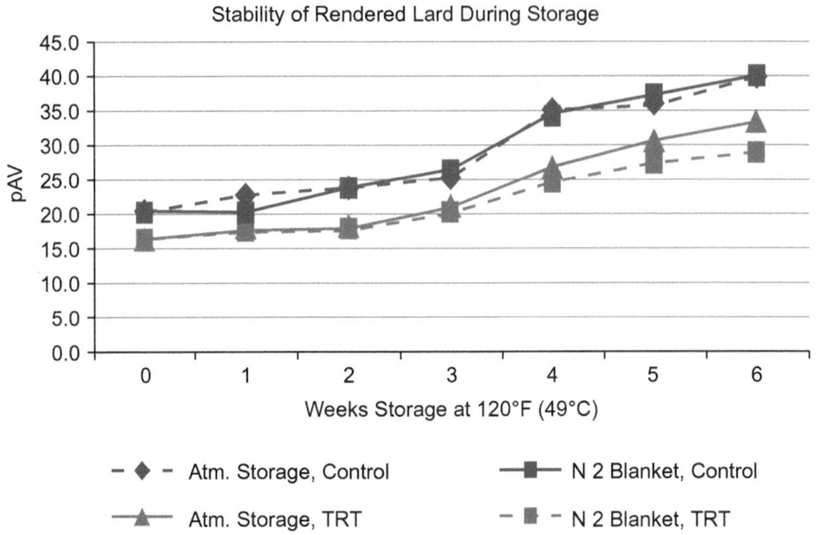

FIGURE 6.5 Oxidative stability of lard during storage as influenced by antioxidants (Control; BHA/BHT versus TRT; BHA/BHT + blend of tocopeherols, rosemary extract, and citric acid) and storage conditions (atmosphere storage versus nitrogen blanket).

120 °F under two different atmospheres (atmospheric storage and storage with nitrogen blanket). The results expressed as *p*AV indicate addition of synergistic antioxidants inhibits oxidation of rendered lard under both storage conditions. Although nitrogen blanketing helped initially in delaying lipid oxidation in lard with BHA/BHT, the synergistic antioxidant treatment was more effective as documented in Figure 6.5.

Others factors that could influence the quality of rendered animal fats are the raw material as well as the processing conditions. Piette et al. (2001) explored this in their investigations on the quality of rendered chicken fat, with respect to raw material handling (homogenized versus ground chicken skin) as well as processing temperatures during extraction. Fat recovery increased with temperature (50 °C vs. 80 °C) and extent of tissue grinding, while the oxidative stability was dependent more on the tissue grind level; homogenized skin treatments displaying better stability (measured as PV values) after 2 months of storage at room temperature. Similar results were obtained during an accelerated testing approach (Schaal oven test) to determine resistance to lipid oxidation and the authors attributed this stability to an increase in tocopeherols (about 30%) in homogenized skin compared to ground skin. The rationale for selecting 80 °C was to mimic the temperature ranges (60–95 °C) of continuous low-temperature rendering process. Although low temperature ranges (50 °C) for extraction results in acceptable quality of edible chicken fat, authors recommend temperatures above 80 °C to ensure microbial safety, as well as homogenizing the chicken skins to offer an additional benefit of improved oxidative stability.

Storage temperature also influences the shelf life of animal derived fats. Recently, Flavia et al. (2014) investigated the shelf life of animal fats (tallow, lard, and chicken fat) during refrigerated and frozen storage conditions. Animal fats were rendered from freshly collected raw materials at 65–70 °C and were subjected to storage at 2 °C and −18 °C with PV and TBARS as the analytical parameters to monitor the shelf life. The authors state that the Romanian regulations have set the limits for acceptability as PV of 10 mEq/kg for edible fats and based on this, the estimated shelf life was 60 days of poultry fat, 90 days for lard, and 150 days for beef tallow during refrigerated storage. On the other hand, frozen storage shelf life was 120, 150, and more than 210 days for poultry fat, lard, and beef tallow, respectively.

6.6 EVALUATION OF SHELF LIFE OF BULK ANIMAL FATS

6.6.1 Real-Time Shelf Life Testing

The real-time shelf life testing of bulk animal fats would be typically undertaken with assessment of PV. GC analysis could also be used to determine the marker compounds of lipid oxidation such as hexanal. Sensory determination would also be helpful to compliment the analytical data.

6.6.2 Accelerated Shelf Life Testing (ASLT)

ASLT is an approach to determine the shelf life of food/matrices under accelerated conditions to predict the shelf life of those products under real-time storage. Temperature is the commonly used and important acceleration factor for determining oxidative rancidity of foods, while oxygen pressure, catalyst addition, and reactant contact have also been researched (Ragnarsson and Labuza, 1977). The temperature dependence of lipid oxidation is attributed to the overall energy of activation. Ragnarsson and Labuza (1977) advise that activation energy may be higher in the presence of antioxidants and have to offer considerations for ASLT testing data obtained at 60–65 °C, since there would be a predictable underestimation of shelf life extension by antioxidants. On the other hand, the authors conclude that ASLT at high temperatures (98–100 °C) is less predictable, resulting in both underestimation and overestimation of shelf life.

Kochar (1993) has concluded that ASLT predictions based on Q_{10} values should be considered carefully since there are different range of values reported of fats and oils (e.g., 2–2.9 and 1.88–2.97), as well as Q_{10} factor decreases with the increasing temperature. Therefore, the closer the temperature of ASLT to the normal storage conditions, the greater will be the accuracy in the prediction of shelf life of fats and fat-containing foods. In general, the equation is summarized as follows (Kochar, 1993):

Q_{10} = shelf life at t/shelf life at $(t+10)$, where t is the temperature in °C
(Q_{10} is not a constant but depends on temperature and activation energy)

Another version is represented as,
$(Q_{10})^{\delta/10}$ = shelf life at t_1/shelf life at t_2, where t is the temperature in °C and $\delta = (t_2 - t_1)$

For example, Kochar (1993) cites the following situation to demonstrate the calculation of shelf life of fat from Q_{10} value. The average Q_{10} factor for fat sample is taken as 2.8 and if, the shelf life at accelerated temperature (37 °C) is 2 months, then calculated shelf life at ambient temperatures (20 °C) would be 11.5 months [$(2.8)^{17/10}$ = shelf life at t_1/2 months]. Alternately at 27 °C, the estimated shelf life would be 5.6 months (2.8 = shelf life at t_1/2 months). Laubli and Bruttel (1986) also determined the Q_{10} values as 1.8 to 2.1 for fats and oils samples based on the induction period coefficients (determined from Rancimat and AOM methods) at 10 °C temperature intervals, ranging from 100 to 120 °C.

6.7 REGULATORY STATUS OF ANTIOXIDANTS FOR ANIMAL FATS

The regulatory status for the usage of antioxidants in animal fats and meats products is specified in Code of Federal Regulations in the United States (9CFR 318.7 for rendered animal fats and 9CFR 381.147 for rendered poultry fats) and is subjected to USDA guidance (Table 6.6), while the U.S. Food and Drug Administration is the regulatory body associated with antioxidants for other food products. The major difference in regulatory approval is pertaining to TBHQ

TABLE 6.6 Regulatory Status of Antioxidants for Animal/Poultry Fats

Antioxidant	USDA/FDA Allowed Levels (ppm)
BHA/BHT[a]	200
BHA	100
BHT	100
TBHQ[a]	200
Propyl gallate	200
Tocopherols	300[b]
Ascorbyl palmitate/stearate	500[b]
Isopropyl citrates	200[b]
Citric acid	GMP[c]
Thiodipropionates	200[b]

[a]*BHA/BHT/TBHQ can be used in combination with single antioxidant not exceeding 100 ppm.*
[b]*Codex Alimentarius guidelines for lard, tallow, fish oil, and other animal fats.*
[c]*GMP represents allowances to "Good Manufacturing Practices."*

that it is not allowed in Canada and Japan. Each member country of European Union establishes maximum guidelines for antioxidants for food applications (TBHQ, BHA, and BHT have E numbers of E319, E320, and E321, respectively, based on Directive 95/2/EC). Therefore, the major synthetic antioxidants used for stabilizing animal and poultry fats include BHA, BHT, and TBHQ. Mixed tocopeherols, citric acid, and natural antioxidants (plant extracts) have also been used effectively to extend the oxidative shelf life of animal fats.

Identification of animal fats: The routine testing, to be noteworthy, is Boehmer number, which is the test used to determine if lard is adulterated with tallow. Numbers greater than 73 indicate that the lard is pure, while those less than 73 indicate contamination (NRA, 2009). Recently, high throughput techniques such as direct analysis in real time ionization–MS and Fourier transform infrared spectroscopy coupled with chemometric tools have been used to determine the authenticity of animal fats (Vaclavik et al., 2011; Kurniawati et al., 2014).

REFERENCES

AOCS. Official Methods and Recommended Practices of the American Oils Chemists Society, 5th Ed., AOCS Press, Champaign.

Barriuso, B., Astiasarán, I., Ansorena, D., 2013. A review of analytical methods measuring lipid oxidation status in foods: a challenging task. Eur. Food Res. Technol. 236 (1), 1–15.

Bee, G., Gebert, S., Messikommer, R., 2002. Effect of dietary energy supply and fat source on the fatty acid pattern of adipose and lean tissues and lipogenesis in the pig. J. Anim. Sci. 80 (6), 1564–1574.

Benz, J.M., Linneen, S.K., Tokach, M.D., Dritz, S.S., Nelssen, J.L., DeRouchey, J.M., Goodband, R.D., Sulabo, R.C., Prusa, K.J., 2010. Effects of dried distillers grains with solubles on carcass fat quality of finishing pigs. J. Anim. Sci. 88 (11), 3666–3682.

Benz, J.M., Tokach, M.D., Dritz, S.S., Nelssen, J.L., DeRouchey, J.M., Sulabo, R.C., Goodband, R.D., 2011a. Effects of choice white grease and soybean oil on growth performance, carcass characteristics, and carcass fat quality of growing-finishing pigs. J. Anim. Sci. 89 (2), 404–413.

Benz, J.M., Tokach, M.D., Dritz, S.S., Nelssen, J.L., Derouchey, J.M., Sulabo, R.C., Goodband, R.D., 2011b. Effects of dietary iodine value product on growth performance and carcass fat quality of finishing pigs. J. Anim. Sci. 89 (5), 1419–1428.

Brewer, S., 2008. Preserving Beef Quality with Natural Antioxidants. National Cattlemen's Beef Association (NCBA) White Paper, NCBA, 9110 E. Nichols Ave., Ste. 300 Centennial, CO, USA.

CODEX STAN, 2014. Codex Alimentarius International Food Standards, General Standard for Food Additives Codex Stan 192–1995. FAO & WHO; Section 02.1.3.

Carter, P., 1971. Spectophotometric determination of serum iron at the submicrogram level with a new reagent. Anal. Biochem. 40, 450–458.

Dziedzic, S.Z., Hudson, B.J.F., 1984. Phenolic acids and related compounds as antioxidants for edible oils. Food Chem. 14 (1), 45–51.

Decker, E.A., et al., 2008. Factors affecting lycopene oxidation in oil in water emulsions. J. Agric. Food Chem. 56, 1408–1414.

Flavia, P., Zorica, V., Delia, B., 2014. Effects of temperature and storage time on the quality of alimentary animal fats. Int. Food Res. J. 21 (4), 1507–1514.

Gray, J.I., Monahan, F.J., 1992. Measurement of lipid oxidation in meat and meat products. Trends Food Sci. Technol. 3, 315–319.

Grosch, B., 1999. Food Chemistry, second ed. in English. Springer-Verlag, Berlin Heidelberg, p. 185.
Huang, D., Ou, B., Prior, R.L., 2005. The chemistry behind antioxidant capacity assays. J. Agric. Food Chem. 53 (6), 1841–1856.
Husvéth, F., Manilla, H.A., Gaál, T., Vajdovich, P., Balogh, N., Wágner, L., Lóth, I., Németh, K., 2000. Effects of saturated and unsaturated fats with vitamin E supplementation on the antioxidant status of broiler chicken tissues. Acta Vet. Hung. 48 (1), 69–79.
Huang, D., et al., 2002. High-throughput assay of oxygen radical absorbance capacity (ORAC) using a multichannel liquid handling system coupled with a microplate fluorescence reader in 96-well format. J. Agric. Food Chem. 50, 4437–4444.
IMPS, 1993. Institutional Meat Purchase Specifications for Variety Meats and Edible By-products-series 700. United States Department of Agriculture, Agricultural Marketing Service, Livestock and Seed Program, Washington, DC.
Jiang, J.F., Song, X.M., Huang, X., Wu, J.L., Zhou, W.D., Zheng, H.C., Jiang, Y.Q., 2012. Effects of alfalfa meal on carcass quality and fat metabolism of Muscovy ducks. Br. Poult. Sci. 53 (5), 681–688.
Kim, H.J., Paik, H.D., 2014. Effect of α-, β-, γ-, and δ-tocotrienol on the singlet oxygen oxidation of lard. J. Am. Oil Chem. Soc. 91 (3), 445–452.
Kochar, S.P., 1993. Deterioration of edible oils, fats and foodstuffs. In: Scott, G. (Ed.), Atmospheric Oxidation and Antioxidants, Vol. II. Elsevier Science Publishers B.V, Amsterdam, The Netherlands.
Koga, T., Terao, J., 1994. Antioxidant activity of a novel phosphatidyl derivative of vitamin E in lard and its model system. J. Agric. Food Chem. 42 (6), 1291–1294.
Kouba, M., Mourot, J., 2011. A review of nutritional effects on fat composition of animal products with special emphasis on n-3 polyunsaturated fatty acids. Biochimie 93 (1), 13–17.
Kurniawati, E., Rohman, A., Triyana, K., 2014. Analysis of lard in meatball broth using Fourier transform infrared spectroscopy and chemometrics. Meat Sci. 96 (1), 94–98.
Kedare, S., Singh, R.P., 2011. Genesis and development of DPPH method of antioxidant assay. J. Food Sci. Technol. 48 (4), 412–422.
Ladikos, D., Lougovois, V., 1990. Lipid oxidation in muscle foods: a review. Food Chem. 35 (4), 295–314.
Laubli, M.W., Bruttel, P.A., 1986. Determination of the oxidative stability of fats and oils: comparison between the active oxygen method (AOCS Cd 12-57) and the Rancimat method. J. Am. Oil Chem. Soc. 63 (6), 792.
Lauridsen, C., Nielsen, J.H., Henckel, P., Sørensen, M.T., 1999. Antioxidative and oxidative status in muscles of pigs fed rapeseed oil, vitamin E, and copper. J. Anim. Sci. 77 (1), 105–115.
Liang, C., Schwarzer, K., 1998. Comparison of four accelerated stability methods for lard and tallow with and without antioxidants. J. Am. Oil Chem. Soc. 75, 1441–1443.
Martin, D., Garcia-Serrano, A., Casado, V., Vázquez, L., Reglero, G., Torres, C.F., 2014. Antioxidant activity of phosphatidyl derivatives of hydroxytyrosol in edible oils. Eur. J. Lipid Sci. Technol. 116 (8), 1035–1043.
Mata, T.M., Cardoso, N., Ornelas, M., Neves, S., Caetano, N.S., 2011. Evaluation of two purification methods of biodiesel from beef tallow, pork lard, and chicken fat. Energy Fuels 25 (10), 4756–4762.
McClements, D.J., Decker, E.A., 2008. Lipids. In: Damodaran, S., Parkin, K.L., Fennema, O.R. (Eds.), Fennema's Food Chemistry, fourth ed. CRC Press, Taylor & Francis Group, FL, USA.
National Renderers Association, 2014. http://www.nationalrenderers.org.
National Rendering Association (NRA), February 2009. A buyer's guide to rendered fats. Render. 14–19. www.rendermagazine.com.
Olsen, E., Vogt, G., Ekeberg, D., Sandbakk, M., Pettersen, J., Nilsson, A., 2005. Analysis of the early stages of lipid oxidation in freeze-stored pork back fat and mechanically recovered poultry meat. J. Agric. Food Chem. 53 (2), 338–348.

Page, B.D., 1979. High performance liquid chromatographic determination of nine phenolic antioxidants in oils, lards, and shortenings. J. Assoc. Off. Anal. Chem. 62 (6), 1239–1246.

Piette, G., Hundt, M., Jacques, L., Lapointe, M., 2001. Influence of extraction temperature on amounts and quality of rendered chicken fat recovered from ground or homogenized skin. Poult. Sci. 80, 496–500.

Rababah, T.M., Hettiarachchy, N.S., Horax, R., 2004. Total phenolics and antioxidant activities of fenugreek, green tea, black tea, grape seed, ginger, rosemary, gotu kola, and ginkgo extracts, vitamin E, and tert-butylhydroquinone. J. Agric. Food Chem. 52, 5183–5186.

Ragnarsson, J.O., Labuza, T.P., 1977. Accelerated shelf life testing for oxidative rancidity in foods—a review. Food Chem. 2 (4), 291–308.

Re, R., et al., 1999. Antioxidant activity applying an improved ABTS radical cation decolorization assay. Free Radic. Biol. Med. 26 (9/10), 1231–1237.

Santas, J., Guzmán, Y.J., Guardiola, F., Rafecas, M., Boub, R., 2014. High-throughput analysis of lipid hydroperoxides in edible oils and fats using the fluorescent reagent diphenyl-1-pyrenylphosphine. Food Chem. 162 (1), 235–241.

Scollan, N.D., Choi, N.J., Kurt, E., Fisher, A.V., Enser, M., Wood, J.D., 2001. Manipulating the fatty acid composition of muscle and adipose tissue in beef cattle. Br. J. Nutr. 85 (1), 115–124.

Sebranek, J.G., Sewalt, V.J.H., Robbins, K.L., Houser, T.A., 2004. Antioxidant Effectiveness of Natural Rosemary Extract in Pork Sausage Animal Industry Report. AS 650, ASL R1852.

Sheehy, P.J., Morrissey, P.A., Flynn, A., 1994. Consumption of thermally-oxidized sunflower oil by chicks reduces alpha-tocopherol status and increases susceptibility of tissues to lipid oxidation. Br. J. Nutr. 71 (1), 53–65.

Song, S., Zhang, X., Hayat, K., Liu, P., Chengsheng, J., Shuqin, X., Xiao, Z., Tian, H., Niu, Y., 2010. Formation of the beef flavour precursors and their correlation with chemical parameters during the controlled thermal oxidation of tallow. Food Chem. 115 (3), 435–443.

Stein, H.H., Shurson, G.C., 2009. Board-invited review: the use and application of distillers dried grains with solubles in swine diets. J. Anim. Sci. 87 (4), 1292–1303.

Singleton, V.L., et al., 1999. Analysis of total phenols and other oxidation substrates and antioxidants by means of Folin-Ciocalteu reagent. Methods Enzymol. 299, 152–178.

Vaclavik, L., Hrbek, V., Cajka, T., Rohlik, B., Pipek, P., Hajslova, J., 2011. Authentication of animal fats using direct analysis in real time (DART) ionization–mass spectrometry and chemometric tools. J. Agric. Food Chem. 59 (11), 5919–5926.

Vichi, S., Zitterl-Eglseer, K., Jugl, M., Franz, C., 2001. Determination of the presence of antioxidants deriving from sage and oregano extracts added to animal fat by means of assessment of the radical scavenging capacity by photochemiluminescence analysis. Nahrung 45 (2), 101–104.

Wood, J.D., Enser, M., Fisher, A.V., Nute, G.R., Richardson, R.I., Sheard, P.R., 1999. Manipulating meat quality and composition. Proc. Nutr. Soc. 58 (2), 363–370.

Wu, L.Y., Fang, Y.J., Guo, X.Y., 2011. Dietary L-arginine supplementation beneficially regulates body fat deposition of meat-type ducks. Br. Poult. Sci. 52 (2), 221–226.

Wayner, D.D., et al., 1985. Quantitative measurement of the total per oxyl radical-trapping antioxidant capacity of human blood plasma by controlled peroxidation. FESB Lett. 187 (1), 33–37.

Xu, Y., Zhao, H., Meng, L., Fan, G., Chen, Q., Pan, S., 2012. Comparison of volatile oxidation products of lard from native breed and three-way crossbred breed under accelerated oxidative conditions. Asian J. Chem. 24 (6), 2401–2406.

Yeo, J.D., Jeong, M.K., Park, C.U., Lee, J., 2010. Comparing antioxidant effectiveness of natural and synthetic free radical scavengers in thermally-oxidized lard using DPPH method. J. Food Sci. 75 (3), C258–C262.

Chapter 7

Oxidative Stability and Shelf Life of Frying Oils and Fried Foods

Hong-Sik Hwang and Jill K. Winkler-Moser
United States Department of Agriculture[a], Agricultural Research Service, National Center for Agricultural Utilization Research, Functional Foods Research, Peoria, IL, USA

7.1 INTRODUCTION

Deep-fat frying is one of the most widespread food preparation techniques used worldwide. It is used to prepare foods quickly in restaurants and cafeterias, to prepare shelf-stable snack foods such as potato and tortilla chips, nuts, convenience foods, and numerous refrigerated and frozen fried foods for retail sales, such as French fries and other fried vegetables, meats, fish, and prepared foods. Potato chips and corn tortilla chips are the two largest categories of fried snack foods. Despite the fact that consumers have become increasingly health conscious, potato chip consumption has increased dramatically over the years, increasing by over 70% from 11.6 lb per person in 1960, to an average of 19.3 lb per person in 2007 (Anonymous, 2008). Tortilla chip consumption in the United States began in the early 1900s in restaurants in southern California, and now the popularity of tortilla chips has sky-rocketed. Sales of the top 10 tortilla chip brands in the United States were greater than three US billion dollars (Statista, 2013).

Deep-fat frying involves the immersion of foods in hot oil, typically at temperatures ranging from 165 °C to 190 °C, and at such high temperatures, frying oil rapidly undergoes chemical reactions including oxidation, polymerization, hydrolysis, and decomposition, resulting in off-flavors, nutritional losses, and other deteriorative changes (Naz et al., 2004). Vegetable oils are rich in monounsaturated fatty acids (MUFAs) and polyunsaturated fatty acids (PUFAs), and their consumption is highly recommended due to benefits to human health such as reducing coronary heart disease risk (Mozaffarian and Clarke, 2009; Micha and Mozaffarian, 2010). However, vegetable oils with higher contents of PUFAs

a. Mention of trade names or commercial products in this publication is solely for the purpose of providing specific information and does not imply recommendation or endorsement by the U.S. Department of Agriculture. USDA is an equal opportunity provider and employer.

Oxidative Stability and Shelf Life of Foods Containing Oils and Fats
http://dx.doi.org/10.1016/B978-1-63067-056-6.00007-0

such as soybean oil have much higher reactivity toward oxidation than animal fats, hydrogenated vegetable oils, and tropical oils containing high contents of saturated fatty acids. Therefore, these saturated fats/oils are much more widely used in restaurants and in food industries for economic and commercial reasons despite their association with cardiovascular disease. The *trans*-fats present in partially hydrogenated vegetable oils are especially associated with many health problems such as cardiovascular disease, elevated blood lipids, inflammation, oxidative stress, endothelial dysfunction, increased body weight, increased insulin resistance, and cancer (Gebauer et al., 2007). The evidence for negative health consequences of consumption of *trans*-fats has become so overwhelming that labeling is mandatory in the United States and in many other countries, while some U.S. municipalities and several countries have even banned *trans*-fats in food products (Downs et al., 2013). As alternatives to hydrogenated oils or oils high in saturated fats, vegetable oils rich in oleic acid, such as regular and high oleic canola oils or mid-oleic and high oleic sunflower oils, have drawn interest due to the higher oxidative stability of MUFAs than PUFAs (Warner and Knowlton, 1997; Bastida and Sanchez-Muniz, 2001; Matthaus, 2005). However, prices of these oils (except for canola oil) are much higher than soybean oil, the most widely used vegetable oil in the United States, due to their limited production. High oleic soybean oils have recently been commercially introduced in the United States, but it may take some time for their production to reach levels significant enough to impact oils used for frying in the United States and worldwide. Therefore, the development of strong antioxidants that suppress oxidation and protect the nutritional quality of highly reactive PUFAs is one of the major solutions to this problem and it will facilitate use of unaltered, healthy, unsaturated oils for frying in the food industry and in restaurants.

Synthetic antioxidants such as butylated hydroxytoluene (BHT), butylated hydroxyanisole (BHA), and *tert*-butylhydroquinone (TBHQ) are known to be strong antioxidants. Various regulatory authorities such as the U.S. Food and Drug Administration (FDA), the European Food Safety Authority (EFSA), and the World Food and Agricultural Association (FAO) have placed limits on the amount of synthetic antioxidants allowed for use in foods and in bulk fats and oils, typically to levels at or below 200 ppm, due to their potential toxicity (Ito et al., 1986). The relatively low concentrations allowed do not provide sufficient oxidative stability to PUFAs under frying conditions. Nontoxic natural antioxidants could potentially be used at higher concentrations than 200 ppm for better protection of frying oil from oxidation. There are numerous natural antioxidants in nature, and there are a huge number of reports, articles, and publications about antioxidant activities of natural substances. In most cases, the evaluation of the antioxidant activity of a new material is conducted by in vitro antioxidant activity tests based on radical scavenging ability at room temperature such as DPPH (2,2-diphenyl-1-picrylhydrazyl) assay, ABTS [2,2′-azino-bis(3-ethylbenzothiazoline-6-sulfonic acid)] assay, FRAP (ferric reducing ability of plasma), and ORAC (oxygen radical absorbance capacity). However, there is no strong correlation between in vitro radical scavenging activity at room temperature and antioxidant activity at frying temperatures (Hwang et al., 2012). For an antioxidant used at frying temperatures, its thermal stability, volatility, reactivity to oil molecules, and radical scavenging ability at frying temperatures should be considered in addition to the radical scavenging ability at room temperature. Other hurdles to the utilization of natural antioxidants in the frying industry are that most natural antioxidants have lower antioxidant activity (Pokorny, 2007) and are more expensive than synthetic antioxidants.

Frying oil is taken up by the food product at levels that differ depending on many different factors. Thus, even after frying, oil oxidation continues during transportation and storage of fried foods, which affects their shelf life, flavor, texture, and nutritional value. Therefore, the stability of antioxidants during the frying process may be equally or even more important than their effect on the stability of the frying oil, as antioxidants taken up by foods with the oil may help prevent further oxidation of these vegetable oils in fried foods (Warner et al., 2003; Melton et al., 1994).

7.2 LIPID OXIDATION IN FRYING OILS AND FRIED FOODS

7.2.1 Frying

Frying is essentially both a heat transfer and mass transfer food operation (Blumenthal, 1991; Dana and Saguy, 2006). Oil at high temperatures transfers heat to the food product, causing the water on the surface and within the food product to turn to steam. Water moves as steam from within the food product to the outside, where it rapidly evaporates. The transfer of heat also facilitates reactions such as starch gelatinization and protein denaturation while cooking the interior of the food. Oil replaces some of the water in fried foods; thus, there is often a correlation between initial water content and oil uptake. There are several mechanisms for oil uptake in fried foods. As water evaporates, a crust forms at the surface of the food product, and pores and cracks may develop due to pressure from steam within the food. The oil may be taken up into these voids during frying if pores or cracks are large, but for very small voids or pores, this mechanism is prevented during frying as long as there is internal pressure in the food from steam. Thus, this mechanism of oil uptake is likely to be more important with foods that have a breading or in yeast- or chemically leavened products that form air bubbles with the evaporation of steam. It has been shown that most oil uptake from frying in uncoated potatoes, potato chips, and tortilla chips occurs after frying, due to a cooling-phase effect (Tseng et al., 1996; Ufheil and Escher, 1996; Dana and Saguy, 2006). As food is removed from the fryer, water vapor condenses and internal pressure from steam decreases, resulting in oil from the surface being sucked into its pores due to a vacuum effect. Therefore, the more oil that adheres to the surface of a food when it is removed from the fryer, the more oil will be taken up by the food unless measures are taken to remove the surface oil before the food cools. Another theory for oil uptake is that surfactant compounds formed in frying oil will enhance the contact between oil and food, resulting in increased oil uptake. This is one aspect of the surfactant theory of frying described later.

There are two basic types of frying processes: "continuous" and "intermittent" (also known as batch or discontinuous). Continuous fryers are used for large-scale production of fried foods. In continuous fryers, oil and raw food product are fed into a long fryer and the food is carried by a conveyer belt through the fryer and removed at the other end (Gupta et al., 2004). The oil removed by the product is often automatically replaced to keep the oil at a constant level. Most potato chips and tortilla chips are manufactured in this manner, although there are some manufacturers that use batch fryers to fry snack chips. Intermittent frying is the style of frying that is often used by restaurants, cafeterias, and home frying. A kettle heats the oil, and the product is added in batches and removed when the food is fully cooked. The oil is often replenished only after a significant amount of oil is lost, or at regular time intervals, for example, daily in a restaurant. There are significant differences between these styles of frying. In continuous fryers, the oil is replenished as it is removed by the constant stream of product. Oil turnover rate, which is the amount of time that it takes for all of the oil in a fryer to theoretically be removed by product, and replaced with fresh oil, is typically anywhere from 3 to 8 h (Gupta et al., 2004). Thus, oil hydrolysis and oxidation products reach a steady state rather than accumulating, because they are constantly removed with the oil by the food product. On the other hand, because oil is not continuously being replaced in batch frying, and because there are often long periods of time, especially in a restaurant or cafeteria setting, where the oil is not used, oil degradation compounds tend to accumulate over time, and the oil often degrades to a point where it eventually has to be completely discarded and replaced.

Despite the fact that humans have been frying for thousands of years (Stier, 2004), the science of frying is not completely understood. Blumenthal (1991) introduced the surfactant theory of frying, which correlates observations of changes in the quality of fried foods with changes in the chemical composition of the frying oil (Figure 7.1). According to the surfactant theory of frying, when fresh oil is first used for frying, it is composed mainly of hydrophobic triacylglycerols, which do not easily interact with the mostly aqueous food product. During this stage, called "break in," the fried food quality is lower, meaning it has a bland flavor, does not develop good color, and is undercooked, because the hydrophobic oil does not come into full contact with the food long enough to efficiently transfer heat to the center. As frying time proceeds, the triacylglycerols in the oil are transformed to increasingly polar compounds through mechanisms including hydrolysis and oxidation, which increases the surfactancy of the oil and results in better heat transfer, allowing proper cooking and oil uptake. This is referred to as the "fresh" and "optimum" stages of frying. Eventually, as oil continues to break down, especially during intermittent frying, the oil enters the "degrading" and

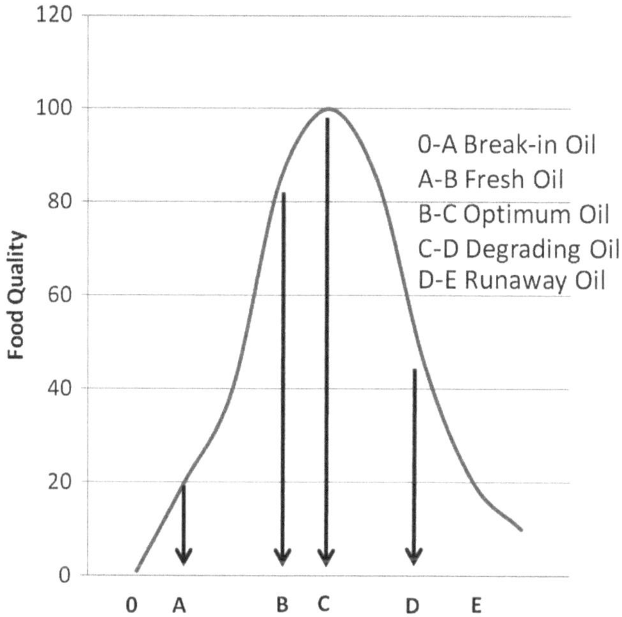

FIGURE 7.1 The relation of frying oil quality to food quality according to Blumenthal's surfactant theory of frying (1991). *Adapted with permission from Blumenthal (1991).*

"runaway" stages, where the amount of surfactant compounds is so high that the food is overcooked and takes up too much oil, becoming greasy. Blumenthal's work was one of the first attempts to understand the science explaining frying. However, Dana and Saguy (2006) argue that the evidence does not fully support this theory. Instead, they propose that increased oil uptake during the end stages of frying is actually produced by increased oil viscosity due to polymerization, which increases adherence of oil when removed from the fryer, and prevents oil from being sucked into pores during the cooling phase, leading to a greasy appearance. Whatever the mechanism or mechanisms are for oil uptake, it is clear that the oxidative deterioration of oil affects the quality of fried foods, and efforts to maintain foods at their optimum quality must focus on preventing the oil from deteriorating. Thus, we will take a closer look at the deteriorative reactions that occur in oils during frying and their impact on fried food stability and shelf life.

7.2.2 Oxidative Characteristics of the Oil during Frying

During frying, the major reactions occurring in the oil include hydrolysis, thermal reactions, and oxidation. These reactions are occurring simultaneously, resulting in the formation of hundreds of different compounds. Moisture release from food is one of the main drivers of hydrolysis, resulting in the release of free fatty acids, diglycerides, and monoglycerides. The high amount of thermal energy in frying can also drive many reactions in oil. Thermal scission, where the high thermal energy cleaves bonds in fatty acids to yield free radicals, was introduced in the review of thermal degradation of oil by Nawar (1969), but these reactions have largely been ignored. Recent studies have shown that thermal scission reactions are predominant over traditional oxidation in early frying, resulting in a homologous series of volatile alkanes and alkenes and their corresponding oxidation products, hydroperoxides, aldehydes, carboxylic acids, ketones, and epoxides (Qin, 2011; Tian, 2013). These studies have provided evidence that terminal peroxides formed on the alkyl radicals produced by thermal scission may be the driving force and the main source of radicals initiating further oxidation during frying. Thermal reactions can also produce triacylglycerol dimers and polymers with C—C bonds (Márquez-Ruiz et al., 1995). During lipid oxidation, radicals are formed by hydrogen abstraction from lipid molecules, which react with oxygen atoms to produce peroxides. Peroxides further undergo reactions including isomerization, cyclization, and polymerization to produce various secondary oxidation products including volatile and nonvolatile compounds (Schaich, 2012). Volatile compounds include alkanals, 2-alkenals, 2,4-alkadienals, acids, alcohols, ketones, alkylfurans, and lactones (Goicoechea and Guillén, 2014). Volatile aldehydes with different unsaturation degrees are often used to determine oxidation levels of oil and fried foods, typically using gas chromatography (Petersen et al., 2012; Poyato et al., 2014), but many are volatilized or stripped from oils by steam during frying. While volatile compounds are not accumulated to a great extent in fried food, amounts of nonvolatile compounds in fried food increase as frying continues, resulting in increased toxicity and deteriorated sensory value of food. Nonvolatile products include polymeric triglycerides and monomeric triglycerides that contain oxidized or cyclic fatty acyls and/or various breakdown products (Urbancic et al., 2014). The oxidation reactions lead to significant changes in the physical characteristics of frying oils during the frying process. Color becomes darker, smoking point decreases, viscosity increases due to polymerization, and dielectricity ascends due to the formation of polar degradation products (Weisshaar, 2014).

Frying also causes degradation of the minor components in vegetable oils, such as tocopherols and phytosterols. Tocopherols, which are the primary antioxidant in most oils, are quickly degraded through reaction with the free radicals that form during frying and by thermal degradation (Aladedunye, 2014). Tocopherols play a major role in protecting vegetable oils from degradation during frying (Warner et al., 2003; Warner and Moser, 2009). Phytosterols, which are the main unsaponifiable component in most vegetable oils, are less susceptible to degradation than tocopherols, but they are also degraded by oxidation and dehydration reactions during frying (Dutta and Savage, 2002; Soupas et al., 2004; Winkler et al., 2007).

7.2.3 Oxidative Characteristics of Fried Foods

7.2.3.1 Effect of Frying Oil Oxidation on Texture, Flavor, and Shelf Life of Fried Foods

The reactions that frying oils undergo contribute significantly to the flavor and texture of fried foods. As previously mentioned, when oil used for frying is excessively degraded, the resulting food product quality deteriorates as it may become overcooked, burned, and greasy (Blumenthal, 1991). This will impact the texture and mouth feel of different fried food products in different ways. The characteristic crisp and crunchy texture of fried potato and tortilla chips is due to dehydration crust formation at the surface of the chips. Potato chip crispness, as evaluated by a trained sensory panel, and hardness, measured instrumentally, were not different when fried in canola, soybean, cottonseed, or partially hydrogenated canola oil (Hawrysh et al., 1996). Tortilla chips increased in hardness when fried in soybean oil used for frying for up to 60h, but the increase was not significant (Tseng et al., 1996). Oil content also increased slightly, but not significantly in these tortilla chips, but it was found that as oil degradation increased, more oil was found on the surface of the chips and less oil was absorbed into the chips, likely due to the increased viscosity of the oil.

The flavor of fried foods can be produced by the frying oil, by constituents in the food, and by the interaction between oils and foods (Chang et al., 1977). Fresh refined frying oils are typically bland because they are composed mainly of triacylglycerols with little to no volatile compounds. During frying, volatile and semivolatile compounds are formed by the thermal, oxidative, and hydrolytic mechanisms discussed earlier. Most of the volatile compounds evaporate during frying, but some stay in the oil and are absorbed with the oil by the food, contributing to the flavor. It takes some time for oil to develop optimal deep-fried flavor during frying (Blumenthal, 1991), and the development of optimal flavor corresponds to measures of total polar compounds (Pokorny, 1989). As oil degradation progresses during frying, the flavor quality of fried foods deteriorates, due to the accumulation of a greater number of volatile and semivolatile compounds. Numerous studies have identified the volatile compounds produced during heating and frying in a variety of oils (Chang et al., 1977; Nawar et al., 1977; Neff et al., 2000; Warner et al., 2001; Qin, 2011), and because every oil is composed of a unique combination of similar fatty acids, many of the same compounds are identified in different oils, but at different ratios due to the different ratios of the originating fatty acid. Hundreds of compounds have been identified in frying oils (Chang et al., 1977; Nawar et al., 1977). Correlations of flavor quality of fried foods with oil composition have indicated that frying oils with high linoleic acid content contribute to better flavor quality compared to hydrogenated oils and high oleic oils (Warner and Mounts, 1993; Warner and Knowlton, 1997). Oils with high linoleic acid composition produce alkenals, dienals, and lactones that contribute to typical fried food flavor (Chang et al., 1977; Nawar et al., 1977; May et al., 1978). Warner et al. (2001), quantified seven compounds in heated trilinolein that were found to have a deep-fried odor as determined by GC-O, where trained sensory judges evaluated the odor of peaks as they eluted from a GC column: (E,E)-2, 4-decadienal, (E)-2-heptenal, (E)-2-octenal, (E,Z/Z,E)-2,4-decadienal, (E,E)-2,4-nonadienal, and (E,Z)-2,4-nonadienal, and (E,E)-2,4-octadienal. Levels of these compounds were much lower in triolein, which may explain why flavor quality tends to be lower, and foods fried in high-oleic oils tend to have lower flavor ratings. Neff et al. (2000) described the odor characteristics of many of the other volatiles produced in heated triolein and trilinolein. The odor characteristics varied widely, from grassy and plastic, to acrid or fruity. Thus, when volatiles contributing to deep-fried flavor break down, other flavors may be predominant and be responsible for the low flavor quality of foods fried in oils that are in the degrading or runaway stages.

7.2.3.2 Effect of Frying on the Shelf Life of Fried Foods

Fried foods absorb the frying oil and oxidation products in the oil, during frying. Hydroperoxides and other oxidation products in the oil can catalyze further lipid oxidation; thus, fried foods are highly susceptible to oxidation during transport and storage. It is mainly the volatile compounds formed during autooxidation in storage that cause off-flavors and odors that shorten the shelf life of fried foods. This is a major concern for snack food producers, who need to produce products with long shelf lives, because months may pass between production, distribution, and sales to when the consumer actually opens and consumes the product. Ideally, the snack products will have the same characteristic deep-fried flavors and crunch after storage that they had right after frying. As mentioned, the linoleic acid that is required for production of volatiles with deep fried aroma is also more susceptible to oxidation during storage. The shelf life of fried foods is affected by factors such as oil composition, level of degradation, and content of antioxidants remaining; thus, it cannot easily be predicted without testing (Márquiz-Ruiz et al., 1999).

Lipid oxidation products may also react with protein, starches, and other food components (Schaich, 2014). There are limited data on the co-oxidation lipids with other food components in fried foods and the effects of these reactions on flavor, texture, and shelf life. Dong (2011) demonstrated by polyacrylamide gel electrophoresis of extracted proteins from fried and baked corn chips, that storage at 40 °C and 60 °C resulted in increasing levels of protein cross-linking. The extent of cross-linking was increased with increased temperatures; in addition, cross-linking at both temperatures was much more pronounced in fried tortilla chips compared to tortilla chips that were baked and that had lower oil content. In earlier storage times, disulfide bonds were predominant in both types of chips, but in later storage times, nondisulfide linkages between proteins formed. Interestingly, in late stages of oxidation, some of the peptide bonds broke down. Investigation of protein carbonyl compounds also indicated that thermal processing, both baking and frying, induced oxidation of proteins, but they did not appear to increase significantly during storage. In addition, almost complete loss of tryptophan was noted in baked and fried chips. Analysis of Schiff base products indicated that they formed with lysine during storage, but that levels were not great, indicating that this is not the major form of protein cross-linking in oxidized tortilla chips. Lipid extractability was also reduced during storage, indicating that oxidized lipids may have been irreversibly bound to proteins or starch.

7.2.3.3 Analyzing and Assessing the Oxidative Stability and Shelf Life of Frying Oils and Fried Foods

Methodologies used for testing the stability of oils and antioxidant effectiveness in frying oils and fried foods vary widely due to the lack of any standardized protocols (Márquez-Ruiz et al., 2014). In addition, few studies include analysis of the impact on fried food stability. Accurate assessment of lipid oxidation in fried foods will enhance research efforts to develop effective new and natural antioxidants as replacements for undesirable synthetic antioxidants that are currently added to some frying oils. Proper assessment of lipid oxidation in fried foods will also enable food processors to determine whether formulation changes, oil changes, or the addition of antioxidants has a significant impact on shelf life and whether changes in formulation are cost-effective.

The quality of freshly fried foods prepared in restaurants, cafeterias, donut shops, and other retail establishments is mostly affected by the ingredients and the oil quality, and shelf-stability is not a concern because the foods are consumed quickly after preparation. However, packaged fried foods, such as fried chips, nuts, and Asian-style fried and dried noodles, require a longer shelf life at ambient temperatures. Because these foods typically have a low water activity and relatively high lipid content, the major shelf life concerns are nonmicrobial, with lipid oxidation being one of the most important reactions affecting their quality and shelf life. Lipid oxidation may also affect the shelf life of frozen fried foods, such as par-fried French fries, breaded and fried meats, poultry, seafoods, and vegetables. However, because these foods are stored at freezer temperatures, other concerns such as the activity of degradative enzymes and freezer burn due to moisture loss and freeze–thaw cycles may be more important issues affecting their shelf life. Therefore, this section will focus on strategies and methods used for the evaluation of frying oil stability, efficacy of antioxidants used for frying, and the evaluation of the oxidative stability of shelf-stable fried foods.

7.2.4 Frying Oils

It is well-known that lipid oxidation intermediates and products can initiate oxidation and affect oil degradation rates. Thus, when comparing the stability of oils during frying, or for evaluating the effectiveness of antioxidants in frying oils and in fried foods, it is important to start with fresh, unoxidized oils.

If possible, the oils should be obtained directly from processors or from wholesale distributors close to the date of manufacture. The oils should be assessed for their initial peroxide value and *p*-anisidine value or similar measure of secondary oxidation. Peroxide value of the oil should be at or below the manufacturer's specification (usually 0.5 mEq/kg). However, oils that are oxidized prior to processing may have artificially low peroxide value due to destruction or removal of peroxides during refining and deodorization; thus, measurement of the *p*-anisidine or other secondary oxidation markers will rule out any abuse. While oxidative stability index (OSI) is not highly correlated with stability during frying, OSI values below the manufacturers specification are another good indicator of initial oxidative status or prior abuse.

7.2.5 Fried Food Products

In a food manufacturing environment (i.e., research and development), the food product used for frying tests should be the same as those produced or prepared. In other research environments, for example, antioxidant discovery, potatoes or French fries are often used for measuring frying oil stability and when evaluating antioxidant effectiveness in frying oils. Fresh potatoes sliced into chips or French fries are ideal because they are relatively uniform in composition, they have low fat content to interfere with the treatment oil composition, they have a bland (potato) flavor and thus the oil flavor comes through easily, and they have a relatively high moisture content that mimics many fried food products and results in a more rapid abuse of the oil. Commercially prepared par-fried frozen French fries are also often used, especially for large-scale frying studies; however, when these are used, care must be taken to consider the oil content and composition in the par-fried potatoes, which can affect the oil and product flavor and stability, due to oil migration, especially after multiple fry cycles. Potato chips or tortilla chips are often used for storage studies. Uniform tortilla chips can be prepared by purchasing commercial corn tortillas and cutting into uniform pieces (Warner and Moser, 2009).

7.2.6 Frying Conditions

The heating or frying conditions used in studies may be the factor with the widest amount of variation. It is generally agreed that the conditions used should either consist of frying or mimic the frying process as closely as possible (Márquez-Ruiz et al., 2014). Temperature range for frying is typically between 165 and 190 °C. There are two major types of frying: industrial continuous frying and discontinuous (or intermittent) frying. In continuous frying, food travels continuously on a conveyer belt through a fryer; the oil temperature is constantly monitored and maintained, and the oil is continuously filtered, while fresh oil is replenished in the fryer at approximately the rate of oil removal by fried food (Gupta et al., 2004). This continuous removal of used oil by food and replenishment by fresh oil, along with the continuous protection of the oil by a blanket of steam from the food, results in low rates of accumulation of oil degradation products. The industrial continuous frying process is difficult to mimic unless a pilot-scale industrial fryer is available.

Restaurants, industrial batch fryers, and consumer frying are examples of discontinuous (or intermittent) frying. In discontinuous frying, the oil temperature cycles with the addition of food product, and unlike continuous frying, oil often sits unprotected by frying food at high temperatures and is only replenished periodically, allowing degradation products to accumulate. Most frying studies attempt to mimic the harsher conditions of discontinuous frying. Frying of control and treatment oils is conducted at the same time or using a statistically valid experimental design. Frying is typically conducted at a defined interval (e.g., every 30 min) for 8 or 12 h per day, often over several days. Oil is periodically sampled and tested until one or more of the oils achieves a level of degradation nearing the end of the fry life, measured as described later. Such fry tests are tedious and labor intensive, and limitations in the supply of oils or antioxidants for a study, especially when new oils or natural antioxidants are being tested, often hamper the ability of researchers to conduct studies using traditional fryers. Several model systems have been developed to mimic frying at a smaller scale. Winkler-Moser et al. (2012) developed a miniature frying protocol using vials held in a temperature-controlled aluminum heating block and periodic frying with small potato cubes. This protocol allowed for reliable and repeatable testing using small quantities of natural antioxidants but still requires 1 to 2 days of frying, depending on the oil type. Warner and Moser (2009) used a glass dish on a temperature-controlled hot plate to fry small quantities of tortilla chips for sensory and headspace analysis during storage. Several other protocols that mimic the discontinuous frying process are summarized by Márquez-Ruiz et al. (2014).

7.2.7 Assessment and Analysis of Frying Oil Degradation

The chemical changes that take place in frying oils cause physical as well as sensory changes to the oil and fried foods. For example, triacylglycerol polymerization reactions cause an increase in oil viscosity, which results in increased oil foaming and increased oil adsorbed to the surface of foods, which affects appearance and texture. Hydrolysis can decrease oil smoke point, while the free fatty acids can form soaps giving a bitter taste to foods, and volatile compounds absorbed by the food products affect the flavor in both positive and negative ways. Thus, there are a number of methods—chemical, physical, and sensory—that could potentially be used to assess frying oil degradation. One of the challenges in comparing results across the literature of frying studies with different oils or of the impact of antioxidants during frying is the lack of consistency in analytical methods used for the analysis of frying oil degradation. Researchers often attempt to use the same measures of lipid oxidation that are used in oxidation studies at lower temperatures, but for various reasons, most of these methods are inappropriate for assessing frying oil degradation. For example, hydroperoxides are extremely labile at frying temperatures, so if they are measured during a frying study, results will be inconsistent, and will not correlate with the physical and sensory changes that occur. Free fatty acids have historically been measured during frying, but because in most cases the levels of fatty acids that form are low and they do not rise consistently during frying, hydrolysis has less of an impact on frying oil quality compared to oxidative and thermal degradation. Volatile products quickly evaporate during frying, so these compounds are not very good indicators of frying oil degradation unless the volatiles above a fryer can be trapped and analyzed, which is impractical except for in strict research settings (Tian, 2013). The nonvolatile compounds tend to accumulate in the oil, especially under discontinuous frying conditions, and thus these compounds are measured to assess frying oil deterioration. The major nonvolatile compounds include oxygenated and nonoxygenated dimers and polymers of triacylglycerols and oxidized monomeric triacylglycerols. The oxidized triacylglycerols consist of one or more oxidized fatty acid with, for example, a ketone, hydroxyl, or epoxy group (Márquez-Ruiz et al., 1995; Zhang et al., 2012). While a number of analytical methods exist for analyzing individual components or degradation compounds, it is well established that the best two chemical methods for analysis of frying oil degradation are the determination of total polar compounds (TPC, also called total polar materials, TPM) and the determination of dimerized and polymerized triacylglycerols (DPTG) (Gertz and Kochhar, 2000). In fact, many European countries have established regulatory limits of 24–27% (w/w) TPC and/or 10–16% DPTG (Firestone, 2007).

TPC is the percentage by weight of oxidized monomers, dimerized and polymerized triacylglycerols, along with any other nonvolatile or semivolatile polar compounds. The standardized methods (AOCS, 2011a; IUPAC, 1987a) entail the separation of oil into polar and nonpolar fractions by silica gel column chromatography. A weighed sample is applied to a silica gel column, and the nonpolar fraction is eluted with a mixture of petroleum ether and diethyl ether into a tared round-bottom flask. After drying completely, the weight of the nonpolar fraction is determined and the percentage of polar compounds is determined by difference. The standard methods for TPC take a considerable amount of time and use a relatively large quantity of organic solvents. Several micromethods using smaller columns or solid-phase extraction columns (Márquiz-Ruiz et al., 1996; Schulte, 2004), along with several high-performance liquid chromatography (HPLC) methods (Kaufmann et al., 2001; Caldwell et al., 2011), have been developed for more rapid analysis while reducing the amount of sample, solvent, and time required. The German Society for Fat Science has recently adopted a new standard method that uses Fourier-transform near infrared spectroscopy (FT-NIR) for determination of TPC, DPTG, acid value (free fatty acids), and anisidine value (Gertz et al., 2013). In addition, rapid testing kits and several sensors have been developed to allow TPC determination in restaurants or factories (Weisshaar, 2014).

The standard methods for DPTG determination employ HPLC using a gel permeation (size exclusion) column (AOCS, 2011b; IUPAC, 1987b). The oil, dissolved in a solvent, is separated into monomer, dimer, trimer, and polymer peaks on the size exclusion column (Figure 7.2). DPTG are calculated as the area% of dimer, trimer, and polymer peaks divided by the total peak area. TPC and DPTG have been shown to be highly correlated; for example, Marmesat et al. (2007) demonstrated with 105 samples of used fats and oils from domestic, industrial, restaurant, and catering service fryers, that the correlation coefficient between TPC and DPTG was 0.9536. Therefore, it is recommended that at least one or both values are determined in frying oils. Rapid testing kits and sensors that have been developed to replace traditional laboratory methods for analysis of TPC and other chemical analyses are discussed by Weisshaar (2014).

The most common physical methods used to assess frying oil quality include measuring oil viscosity, color, and foaming. Oil viscosity is correlated with DPTG, but different oils have different intrinsic viscosity, so the direct correlation between these two measurements would need to be established for each

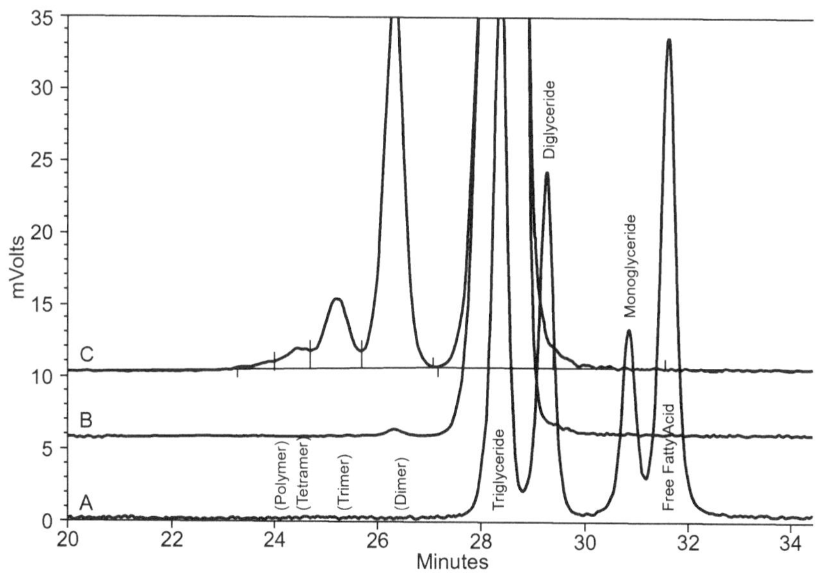

FIGURE 7.2 Example of HPLC-size exclusion chromatography of frying oils used to determine DPTG. Chromatogram A: standards of free fatty acid, monoglyceride, diglyceride, and triglyceride. Chromatogram B: Chromatogram of a fresh oil showing mostly triacylglycerols and a very small level of dimerized triacylglycerols. Chromatogram C: Chromatogram of an oil used for frying showing the intact triacylglycerols as well as dimerized, trimerized, and polymerized triacylglycerols. *Reprinted with permission from: Winkler, J.K., Warner, K.A., 2008. Eur. J. Lipid Sci. Technol. 110, 455–464.*

oil type. Oil becomes darker during frying, but oil color is not directly related to oil degradation, and the color is highly affected by the type of food used for frying. However, if oil color becomes too dark, it will affect the appearance of the food product. In addition, when testing antioxidants, it is important to establish whether oil color is impacted by the additive, so measuring oil color during frying is a good practice in comparative studies. Oil foaming is caused by polymers and polar components that form during frying, such as diglycerides or other surface active substances. Oil foaming is not directly related to frying oil quality (Weisshaar, 2014); however, if foaming becomes too prominent, there is a risk that the oil will boil over, so the oil would need to be discarded. When establishing or determining the end point of frying oil, it is suggested that several measurements should be taken, an end point determined for each, then the average of the time it takes for oils of different treatments to reach those established end points can be used to determine the fry-life of an individual oil or treatment (i.e., antioxidant) (O'Brien, 1998).

Sensory evaluation is considered the best method for assessing the quality of fried foods, as well as assessing lipid oxidation during storage of fried foods. Chemical methods such as headspace analysis may be less time and resource intensive, but product shelf life is based on consumer ability to detect when a product is rancid; thus, chemical markers must be correlated to odor or flavor quality in order to establish a product's shelf life. Sensory evaluation techniques can be broadly categorized as either affective or analytical (Warner, 1985). Affective testing includes consumer testing such as preference or acceptance testing (Meilgaard et al., 2007) and may be important in determining the quality of freshly fried foods or, for example, evaluating the effect of a change in oil on the acceptance of foods fried in the new oil. In this case, panels may include a large number of randomly selected, untrained consumers who are asked to evaluate their preference between several products or their overall liking of a product. Analytical sensory testing includes difference tests and descriptive tests. The objective of difference testing is to determine if trained or untrained panelists can detect overall differences or differences in certain attributes between products. Descriptive analysis of food products usually entails a smaller, carefully selected and trained panel that will rate the intensity of color, texture, mouth feel, flavors, and/or odor attributes in a product (Meilgaard et al., 2007).

Preference testing is typically carried out by presenting paired products and asking panel members, after tasting both, to indicate which product they prefer and why. Acceptance, or overall liking of a product, is typically carried out using either hedonic scales that are anchored either verbally (e.g., "dislike extremely" to "like extremely") or by using pictures (i.e., smiley or frown faces), indicating the relative level of unacceptability to acceptability. Panelists are asked to rate their like or dislike of a product using the scales. The hedonic scales can be converted to numerical values, and products compared by *t*-tests or other statistical analyses. Detailed testing procedures and methods for analysis of results are provided by the American Society for Testing Materials (ASTM, 1979) and by Meilgaard et al. (2007).

Affective testing is most commonly used for determination of acceptability of freshly fried foods (Melton, 2006) and can be used to compare different frying oils. However, affective testing can be limited in its usefulness because it may be highly affected by wide variations in consumer background and preferences. For example, Melton et al. (1994) found in consumer testing of the likability of codfish fried in either fresh or spent (used) frying oil that one group of consumers liked the fish fried in fresh oil, while another group liked the fish fried in used oil better, while another group showed no preference in the likability.

Analytical testing such as difference testing or descriptive analysis is more commonly used for research evaluations of oils used for frying or evaluation of antioxidant efficacy of frying oils or fried foods. Difference testing is less commonly used in research compared to descriptive analysis. Difference tests such as the triangle test may be useful for selecting members of a trained panel, by determining if potential panel members are able to detect differences between fresh and oxidized samples. The triangle test was used to determine whether trained panelists could detect differences in flavor of potato chips fried in high oleic safflower, cottonseed, and hydrogenated vegetable oils with or without added propyl gallate (Fuller et al., 1971). Triangle tests and paired comparison difference testing were used by Ravelli et al. (2010) to select panel members, as well as to determine the discard point for soybean oil used to fry potato chips. This demonstrates that difference testing may be useful in settings where there are limited resources (time and/or financial) to train and maintain a dedicated panel for more thorough descriptive analysis.

Descriptive analysis is the most commonly used type of testing method for analysis of frying oils, fried foods, and storage stability of fried foods. The AOCS has developed official methods with quality and intensity rating scales for oils, along with an oil lexicon and reference standards (AOCS, 2011c). Similar rating scales have been adapted for use in fried foods. For example, Warner (1995) has used a trained panel to evaluate French-fried potatoes as well as tortilla chips for intensities of individual flavors including deep-fried, potato, stale, cardboard, waxy, hydrogenated, and fishy, using a 10-point intensity scale with 0 = no intensity and 10 = strong intensity.

Warner et al. (1997) illustrate how chemical and sensory analysis methods can be used to study the correlation of chemical measurements with sensory analysis to assess the stability of fried food during storage (Figure 7.3(A–D)). In this study, cottonseed oil (CSO), high oleic sunflower oil (HOSUN), and two blends (67:33 CSO/HOSUN and 33:67 CSO/HOSUN) were used to fry potato chips to determine the effect of different ratios of oleic acid with linoleic acid on fresh and aged potato chip flavor quality. The ratio of oleic acid to linoleic acid in the four oils was 0.30, 1.20, 3.61, and 6.44 for CSO, HOSUN, 67:33 CSO/HOSUN, and 33:67 CSO/HOSUN, respectively. Potato chips were fried

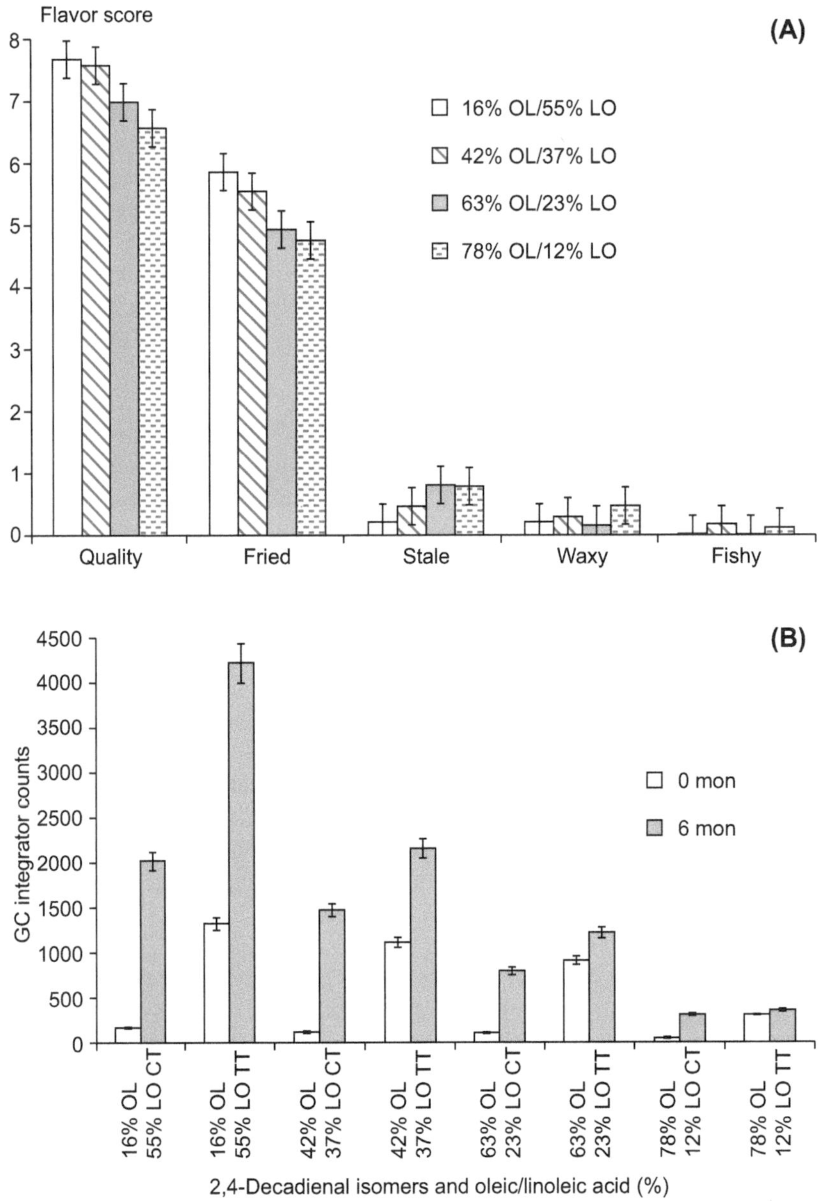

FIGURE 7.3 Sensory and headspace analysis of fresh and stored (25 °C) potato chips fried in cottonseed oil, high oleic sunflower oil, or 67:33 or 33:67 blends of the cottonseed/high oleic sunflower to achieve the shown ratios of oleic acid (OL) to linoleic acid (LO). Results are from chips sampled after 12 h of frying. (A) Flavor scores for fresh potato chips; (B) Headspace levels of *cis, trans* (CT) and *trans, trans* (TT)-2,4-decadienal isomers in fresh chips or chips stored for 6 months at 25 °C; (C) Flavor scores for potato chips stored for 6 months at 25 °C; (D) Headspace levels of selected saturated aldehydes in potato chips stored for 6 months at 25 °C. *Reprinted with permission from Warner et al. (1997).*

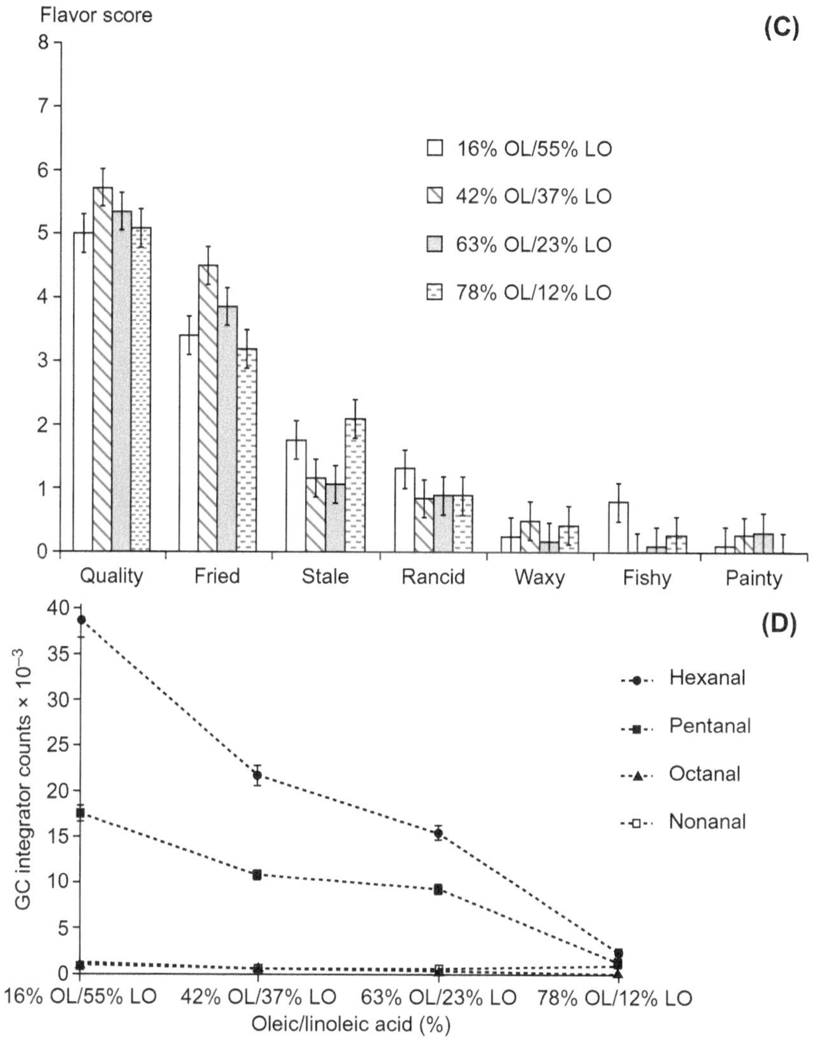

FIGURE 7.3 Cont'd

in a pilot-scale continuous fryer for 18 h over 2 days, and oil and potato chips were collected at 0, 3, 6, 12, 15, and 18 h. Headspace and sensory analysis were conducted on fresh chips at each time point as well as for chips stored in sealed bags for up to 6 months. Graphs shown (Figure 7.3(A–D)) are for fresh or aged chips sampled after 12 h of frying. The overall quality and fried food flavor of chips decreased with decreasing ratios of linoleic to oleic acid as shown in Figure 7.3(A). This was explained to be likely due to higher levels of volatile compounds with fried food aroma produced in oils with lower oleic/linoleic acid ratios. Headspace measurements of *t,t*-2, 4-decadienal and *c,t*-2,4-decadienal, two compounds known to have fried food aroma, were consistent with the flavor trend (Figure 7.3(B), white bars). After chips were stored for 6 months, overall quality and fried food aroma decreased, while stale, rancid, and waxy aromas increased (Figure 7.3(C)). At this stage, chips fried in CSO or HOSUN had lower overall quality and fried food flavor, as well as stale flavor. Chips fried in CSO had highest levels of rancid and fishy flavors. While headspace 2,4-decadienal isomers increased in all of the chips after 6 months (Figure 7.3(B), gray bars), integrator counts of headspace aldehydes such as hexanal and pentanal, which contribute to rancid odors, were higher (Figure 7.3(D)). These two aldehydes were highest in CSO and lowest in HOSUN, overall indicating that rancid and fishy flavors from headspace volatiles may have overtaken the deep-fried aroma of the chips fried in CSO and aged for 6 months, while lack of any aroma compounds may have contributed to the lower flavor scores of chips fried in HOSUN and stored for 6 months. While discard point (exact shelf life) of the chips was not determined in this example, in a practical setting the shelf life could be assigned to a point where panelist scores for overall quality drop below a given average score.

7.2.7.1 Managing Lipid Oxidation during Frying and Storage

The major factor dictating the oxidation rate is degree of oil unsaturation, in other words, the fatty acid composition of oil. Because oxidation is initiated by radicals formed by hydrogen abstract from oil molecules, the hydrogen with the weakest bond has the highest vulnerability. The energy required to break carbon-hydrogen bond of —CH_2— having two double bonds in both sides (e.g., carbon 11 of linoleic acid) is 50 kcal/mol, while energies for hydrogen abstraction with one double bond and no double bond are 75 and 100 kcal/mol, respectively (Min and Boff, 2002). Due to these different bond energies, the oxidation rate correlates well with the degree of oil unsaturation (Marmesat et al., 2012).

In addition to degree of oil unsaturation, there are many other factors affecting quality of oil during frying such as original oil quality (e.g., free fatty acids increase thermal oxidation of oil), endogenous and added antioxidants, frying conditions (e.g., temperature, frying time, and light), replenishment with fresh oil, composition of fried foods, type of fryer, and dissolved oxygen content in oil (Choe and Min, 2007). While moisture in food causes hydrolysis of oil, it could form a steam blanket over the fryer, which reduces contact with air and, subsequently, decreases oil oxidation (Danna et al., 2003; Kochhar, Gertz, 2004). Negishi et al. (2003) reported that a fryer with a greater ratio of oil depth to oil surface area showed a slower oxidation. Flushing oil with nitrogen or carbon dioxide also reduced oxidation of oil (Przybylski and Eskin, 1988). Vacuum frying not only could lower the frying temperature due to the consequent reduced boiling point of water but also reduce the concentration of oxygen over frying oil to reduce oxidation of oil (Shyu et al., 1998). While there are some other ways to control oxidation of frying oil, protecting frying oil from oxidation using antioxidants, especially, natural antioxidants will be discussed in this section.

7.2.7.2 Antioxidants for Frying Oil

As previously mentioned, there has been a renewed focus on the need for antioxidants with improved antioxidant activity in frying oils due to the desire to replace synthetic antioxidants with safer alternatives and to reduce the amount of saturated and *trans*-fats in frying oils. Two comprehensive reviews have recently been published on antioxidants, both synthetic and natural, that have been tested for effectiveness in frying or heated oils (Aladedunye, 2014; Márquez-Ruiz et al., 2014). Many studies have reported the properties of natural antioxidants at frying temperatures, which include tocopherols, tocotrienols, polyphenols, flavonoids, sesame lignans, phytosterols, phenolic acids, carotenoids, phospholipids, ascorbic acid, oryzanols, propolis, extracts from olive tree leaves, tea leaves and other plants, and catechin (Boskou, 2011; Aladedunye, 2014; Márquez-Ruiz et al., 2014). Vegetable oils contain a variety of endogenous minor components and some of these components such as tocopherols, polyphenols, phospholipids, carotenoids, squalene, and certain sterols have antioxidant activity and are beneficial to oil stability during frying (Rossi et al., 2007). Among them, tocopherols are found at relatively high concentrations in oil and take the major role in preventing oxidation of oil. Tocopherols are the most well recognized effective natural antioxidants at frying temperatures and are occasionally added as additives for further protection of frying oil from oxidation (Warner and Moser, 2009). Even though a large number of natural antioxidants have been reported, only a few natural antioxidants have been commercialized in the frying industry. Along with tocopherols, antioxidant systems containing rosemary extract, sage extract, and catechins as the major active ingredients with additional antioxidants such as ascorbic acid and emulsifiers are widely used in the frying industry. Rosemary and sage extracts are relatively thermally stable at frying temperatures and have strong antioxidant activity, which is attributed to the major antioxidant components in them, carnosic acid and carnosol (Lalas and Dourtoglou, 2003; Rau et al., 2006).

Other promising natural antioxidants are corn steryl ferulates and oryzanol that are obtained from agricultural co-products, corn fiber and rice bran, respectively. Corn steryl ferulates showed stronger antioxidant activity and better thermal stability than oryzanol during the frying process at 180 °C with soybean oil (Winkler-Moser et al., 2012), which is consistent with the previous study by Wang et al. (2002) conducted with FAME (fatty acid methyl esters) at 90 °C and with soybean oil at 100 °C. The stronger antioxidant activity of corn steryl ferulates was recently explained by the higher saturation of the sterol parts in corn steryl ferulates than in oryzanol (Winkler-Moser et al., 2015). In this study conducted with several synthesized steryl ferulates, steryl ferulates with a saturated sterol group had the best antioxidant activity, followed by sterols with one double bond in the C5 position. A dimethyl group at C4 as well as a cyclopropane group at C9 and C19, as found in oryzanol, negatively affected antioxidant activity in frying oils.

7.2.8 Lignans as New Natural Antioxidants – Structural Effect on Radical Scavenging Ability

Lignans occur in a wide range of plant families showing a variety of biological activities including antitumor, anti-inflammatory, hypolipidemic, cyclic adenosine monophosphate phosphodiesterase inhibitory, stress-reducing, anti-complementary, Ca^{2+} antagonist, liver protection, antiallergy, and relaxation activities (Deyama and Nishibe, 2010; Fuss, 2003). Lignans are 8,8-coupled dimers of coniferyl alcohol according to IUPAC nomenclature and generally have antioxidant activity due to the phenol moiety in their molecular structure. Sesamol, a monophenolic compound, is not considered as a lignan by this definition, but it is commonly regarded as a sesame lignan along with sesamin and sesamolin. Niemeyer and Metzler (2003) investigated structural effects of several lignans on radical scavenging ability using ferric reducing/antioxidant power (FRAP) assay. This study showed that the lignans containing a methoxy or hydroxyl group at the *ortho* position to the phenolic OH group were far more active than other lignans without this functional group. Another study (Hwang et al., 2012) was also conducted with lignans and sesamol (Figure 7.4) for their radical scavenging abilities toward ABTS [2,2′-azinobis-(3-ethylbenzothiazoline-6-sulfonic acid)]. TEAC (Trolox equivalent antioxidant capacity) values per phenolic moiety of the lignans, sesamol, and TBHQ are compared in Table 7.1. Sesamin without a phenolic OH did not show inhibition, and sesamol showed the highest TEAC value among other lignans. As reported by Niemeyer and Metzler (2003), NDGA (nordihydroguaiaretic acid), PINO [(+)-pinoresinol], and SECO [(−)-secoisolariciresinol] with a hydroxyl group or methoxyl group at *ortho* position of the phenolic OH showed much higher radical scavenging activities than END (enterodiol) due to stabilization of the

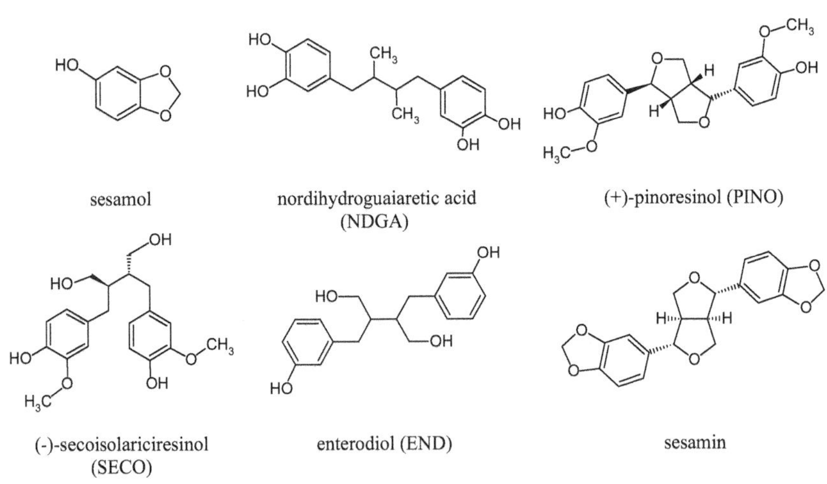

FIGURE 7.4 Molecular structure of sesamol and lignans.

TABLE 7.1 Trolox Equivalent Antioxidant Capacity (TEAC) of Lignans and Sesamol Measured by Scavenging of ABTS Radical and Oxidation Results after Heating Soybean Oil with Antioxidant (0.0012 mEq Phenolic Moiety/g oil) at 180 °C for 8 h

		Oxidation of Soybean Oil		
Antioxidant	TEAC per Phenolic Moiety	PTAG (%)	Loss of Bisallylic Protons (%)	Loss of Olefinic Protons (%)
Sesamol	1.80	4.36	12.09	8.01
NDGA	0.95	4.06	11.65	8.45
PINO	0.30	4.55	12.29	8.71
SECO	0.61	4.06	11.52	8.24
END	0.09	4.06	13.69	9.85
Sesamin	No inhibition	4.48	11.33	7.77
TBHQ	1.13	3.75	9.94	6.95
Control	–	5.48	13.84	10.01

NDGA, nordihydroguaiaretic acid; PINO, (+)-pinoresinol; SECO, (–)-secoisolariciresinol; END, enterodiol.

phenoxy radical by an oxygen atom at *ortho* or *para* position through resonance structures. Not only the molecular structure of the phenolic part but also the molecular structure between two phenolic parts significantly affected the antioxidation activity, which was observed in comparison between PINO and SECO bearing the same phenolic moiety. In a separate experiment, two model compounds, 4-ethylcatechol with an *ortho*-hydroxyl group and 2-methoxy-4-propylphenol with an *ortho*-methoxy group, showed the same TEAC value (1.11), indicating the same radical scavenging ability of these two different phenolic moieties. Therefore, the stronger activity of NDGA than PINO and SECO is also attributed to the molecular structure between the two phenolic parts. Niemeyer and Metzler (2003) also found stronger radical scavenging activity of NDGA over SECO.

7.2.9 Antioxidation Activity of Lignans in Heated Oil

Several research groups tested lignans, mainly sesame lignans, including sesamol, sesamin, and sesamolin, and observed their significant antioxidant activities during heating of both vegetable oils and methyl linoleate at frying temperatures (Lee et al., 2008, 2010; Hemalatha and Ghafoorunissa, 2007; Lee and Choe, 2006). If the radical scavenging ability tested at room temperature shows a strong correlation with the antioxidant activity of a lignan in oil at frying temperatures, it can be used to predict antioxidant activities of lignans for frying oils because its test method is relatively fast, easy, and convenient compared to heating and frying tests. However, the results in Table 7.1 show no strong correlation between the radical scavenging ability toward ABTS radical and three indications of oil oxidation, PTAG (polymerized triacylglycerols), loss of bisallylic protons, and loss of olefinic protons after heating soybean oil containing a lignan or sesamol (0.0012 mEq phenolic moiety/g oil) at 180 °C for 8 h (Hwang et al., 2012). Radical scavenging ability of a lignan gives only a rough idea of antioxidant activity in the heating test. As found in radical scavenging ability tests, the heating study also showed that oxygen atom located at *ortho* or *para* position to the phenolic OH group of the lignan significantly increased antioxidant activity. However, all lignans and sesamol showed weaker activities than TBHQ. This result is somewhat contradictory to the previous reports by Yeo et al. (2010, 2011) that sesamol showed the similar conjugated dienoic acid and *p*-anisidine values to TBHQ in lard heated at 180 °C for 80 min. Unlike its slight radical scavenging activity observed at room temperature, END (enterodiol) did not show any antioxidant activity in the heating test, indicating that an oxygen atom at *ortho* or *para* position to the phenolic OH is essential for antioxidant activity at frying conditions. Instead, sesamin showing no radical scavenging ability toward ABTS radicals due to the absence of the phenolic OH showed a significant antioxidant activity in the heating experiments. The methylenedioxybenzene group in sesamin may act like a phenol group at high temperature, or seamine may thermally decompose to generate phenolic species. Lee et al. (2008) also reported that sesamin showed positive effects on reducing conjugated dienoic acid, conserving α-tocopherol and reducing the *p*-anisidine value of methyl linoleate at 180 °C. However, sesamin showed pro-oxidant activity at increased concentrations (0.0024–0.012 mEq/g oil) (Hwang et al., 2012). At these higher concentrations, sesamol performed better than other lignans, and 0.0060 mEq of sesamol showed better antioxidant activity than 0.0012 mEq TBHQ.

In addition to the antioxidant activity, potential toxicity of lignans must be considered before the practical application of a lignan because, for example, nordihydroguaiaretic acid (NDGA) has been reported to show renal toxicity and hepatotoxicity by chronic use (Arteaga et al., 2005). Even sesamol that has been consumed by people for thousands of years has not been listed in the U.S. Food and Drug Administration (FDA)'s inventory of generally recognized as safe (GRAS) substances or in the GRAS Notice Inventory. Hirose et al. (1997) has raised a concern on sesamol that slightly increased incidences of fore-stomach papillomas were observed when mice were fed with 0.4% sesamol in their food. However, several other natural and common antioxidants such as caffeic acid and catechol were also found to be positive for tumors in this study, and use at such a high level of the diet for a long time is very unlikely. By all means, all of the available data must be evaluated and a notice must be filed to the FDA of GRAS before using any lignan or sesamol as a food ingredient.

7.2.10 Sesamol as Antioxidant for Frying Oil

Sesame is considered to be the oldest oilseed crop known to man and is now grown in many parts of the world including the United States (Ram et al., 1990). The seed contains 50–60% oil and oil contains up to 0.02% (wt.) sesamol and 0.12% (wt.) sesamolin, the precursor of sesamol (Budowski et al., 1950). In addition to its promising antioxidant activity, sesamol also has many important biological activities and health-promoting benefits such as inducing growth arrest and apoptosis in cancer and cardiovascular cells and enhancing vascular fibrinolytic capacity (Jacklin et al., 2003). Its price is relatively low ranging: about $60/kg for the synthetic one to about $100/kg for the natural one.

Because sesamol showed promising results in the heating study, a frying study was conducted with sesamol in soybean oil at 180 °C using a miniaturized frying device where pieces of potato were fried with small-scale frying equipment (Hwang et al., 2013). The antioxidant activity for sesamol observed in this frying experiment was not as good as demonstrated in previous heating studies. The concentration of sesamol checked with HPLC sharply decreased during the frying process and became close to 0% after frying at 180 °C for 8 h. Thermogravimetric analysis (TGA) of a neat sesamol using a linear temperature ramp (5 °C/min) from 25 °C to 180 °C under nitrogen showed fast evaporation resulting in 99% weight loss at 128 °C. Warner et al. (1986) reported that 50% total volatiles, which include decomposed and unaltered molecules, were observed after 1 h at 195–200 °C in their isotope-labeled experiment with TBHQ in lard. Hamama and Nawar (1991) also reported 20% unaltered volatiles of TBHQ in 1 h at 185 °C in the experiment with TBHQ placed in a test tube. The concentration decrease of an antioxidant limits the efficacy of the antioxidant in high temperature applications and reactions with radicals, decomposition, and evaporation are known to be three major reasons for this concentration decrease (Frankel, 1993; Marmesat et al., 2010; Lin et al., 1981). Antioxidants deactivate radicals by two major mechanisms, hydrogen atom transfer and single-electron transfer (Prior et al., 2005). In either mechanism, the antioxidant becomes a radical that eventually decomposes. However, when an antioxidant is lost by decomposition or evaporation before it reacts with radicals, it will result in decreased antioxidant activity.

Considering the lower molecular weight of sesamol (138.12 g/mol) than TBHQ (166.22 g/mol), faster evaporation is expected for sesamol and, in fact, the TGA experiment showed that sesamol had higher volatility. Therefore, volatility of sesamol is problematic for the practical application of sesamol as antioxidant for frying, and finding a method to solve this problem would be critical to utilizing this natural antioxidant. One possible solution for this problem is a multiple addition approach, in which the same amount of sesamol is added portion by portion to keep the antioxidant level more consistent. When 0.083% sesamol was added every hour during 8-h frying at 180 °C, its concentration throughout the frying process was maintained at 438–569 μg/g (Hwang et al., 2013). In contrast, when 0.66% sesamol was added at the beginning of frying, the concentration of sesamol went down to 321 μg/g at 4 h and 6 μg/g at 8 h. By keeping the sesamol concentration at a certain level throughout the frying process using the multiple addition methods, formation of PTAG, a typical indication of oil oxidation, was reduced by 18% from the single addition method after 8-h frying at 180 °C.

7.2.11 Reduced Volatility of Sesamol by an Additive

A recent study showed that evaporation of sesamol could be reduced by an additive that binds to the antioxidant (Hwang et al., 2014). The additive can bind to sesamol by the interaction between the phenolic OH group of sesamol and functional groups of the additive such as a hydroxyl, amino, amide, ether, sulfate (SO_3^-), carboxylate (CO_2^-), phosphate (PO_4^-), ammonium, or thioether. A 1:1 M mixture of sesamol and the additive was tested for sesamol volatility by TGA under nitrogen atmosphere. A number of additives with a long alkyl chain and at least one polar group are widely used as food additives, generally as emulsifiers. Six such kinds of additives were tested to show much higher sesamol levels retained at the temperature where 20% sesamol remained in the TGA experiment with sesamol alone (Table 7.2). While the experimental results were excellent (63.7–93.6% sesamol retained) when sesamol:additive mixtures were tested as neat samples, the effectiveness of these additives on reducing volatility of sesamol was not as impressive when they were dissolved in soybean oil. Although the mixtures seem to form very stable adducts, they may exist in an equilibrium of association and dissociation when added to oil and, therefore, their effects are diminished. The noticeable effects by cetyl alcohol, containing one hydroxyl group, evidence that the hydroxyl group can effectively bind sesamol to reduce its volatility. In general, additives with more than one hydroxyl group such as glyceryl stearate, sorbitan tristearate, and mono-/diglyceride reduced sesamol volatility more effectively than cetyl alcohol containing one hydroxyl group.

Polysorbates 20, 60, and 80 are fatty acids (lauric acid, stearic acid, and oleic acid, respectively) containing about 20 oxyethylene (CH_2CH_2O) groups and are used as food additives. Polysorbates showed strong binding of sesamol indicating that ether groups can effectively bind sesamol. Because binding occurs at the polyether groups, the fatty acid chain part did not affect the binding force as shown in the very similar results by three polysorbates. Two food additives, sodium stearoyl lactylate and lecithin, and one nonfood additive, dioctyl sodium sulfosuccinate, that contain fully charged functional groups were also found to reduce sesamol volatility. Naturally occurring amino acids, which exist as zwitterions, were also effective in reducing sesamol volatility. Amino acids with more than one amino group such as L-lysine and L-arginine showed better

TABLE 7.2 TGA Experiment Results: Sesamol Retained (%) at 154.7 °C for the 1:1 M Ratio Mixture of Sesamol and Additive (1:1 Weight Ratio for Polysorbates) and at 189.8 °C for the Mixture in Soybean Oil

	Sesamol Retained (%)	
Additive	Sesamol + Additive	Sesamol + Additive in Soybean Oil
None (control, sesamol only)	20.0	20.0
Long Alkyl Chain with Polar Groups		
Cetyl alcohol	76.8	33.6
Glyceryl stearate	89.7	39.5
Glycol stearate	87.4	32.9
Sorbitan monostearate	63.7	59.3
Sorbitan tristearate	96.8	57.7
Mono-/diglyceride	93.6	42.1
Polymers with Ether Groups		
Polysorbate 20	79.8	27.8
Polysorbate 60	79.6	27.9
Polysorbate 80	76.9	24.3
Charged Groups		
Sodium stearoyl lactylate	86.4	27.3
Lecithin	96.0	61.0
Dioctyl sodium sulfosuccinate	93.1	43.2
Amino Acids and Dipeptide		
DL-Phenylalanine	64.6	28.7
L-Tryptophan	72.7	35.7
DL-Methionine	67.1	40.6
L-Glutamine	68.9	29.4
L-Arginine	89.0	56.1
L-Histidine	60.7	28.9
L-Lysine	87.4	57.3
L-Carnosine	50.0	37.3

Continued

TABLE 7.2 TGA Experiment Results: Sesamol Retained (%) at 154.7 °C for the 1:1 M Ratio Mixture of Sesamol and Additive (1:1 Weight Ratio for Polysorbates) and at 189.8 °C for the Mixture in Soybean Oil—cont'd

	Sesamol Retained (%)	
Additive	**Sesamol + Additive**	**Sesamol + Additive in Soybean Oil**
Other Additives		
Chitosan	35.7	37.4
Octadecylamine	73.4	47.7

effects than DL-phenylalanine containing only one amine group indicating that the extra amino group in the molecule provides a stronger binding force. When there are more than one amine group in amino acid, one amine group will exist as ammonium ion (N^+) and the other as a neutral amine group that could bind strongly with the phenolic OH of sesamol. Chitosan, an insoluble additive, showed a significant effect evidencing that interactions by hydroxyl groups and amino groups on the surface of the insoluble additive can reduce the volatility of sesamol. Nonedible compound, octadecylamine, was tested to examine the effect of a single amino group, and the results showed that the binding force of the amino group to sesamol was very strong.

It is believed that the low molecular weight sesamol binds an additive to form an adduct of which the molecular weight is the sum of the two compounds' molecular weights. The increased molecular weight of the adduct reduces its volatility. It is also believed that when the additive has extra polar groups (e.g., amino acid) in addition to the polar group binding sesamol, these additional polar groups may provide additional interactions between adducts, which further reduce volatility of the adducts.

The interaction of two compounds such as hydrogen bonding can be detected by proton NMR. Litwinienko et al. (2002) observed the hydrogen bonding between a fatty acid ester and a phenolic compound by monitoring downfield shift of the phenolic OH peak in the ^1H NMR spectrum. Similarly, as shown in Figure 7.5, when octylamine was added to sesamol at ratios of 0.25, 0.5, 1.0, and 2.0, all the peaks including the benzene ring proton peaks (Ha, Hb, and Hc) and the methylene group peak (Hd) moved upfield, indicating the increased electron density in the molecule (Hwang et al., 2014). The benzene ring of sesamol becomes richer in electron density when the phenolic OH proton of sesamol is involved in the hydrogen bonding, with the amino group becoming similar to a phenoxide ion and, consequently, the peaks move upfield. The methylene protons (Hd) shifted the most, indicating that the highest electron density was built

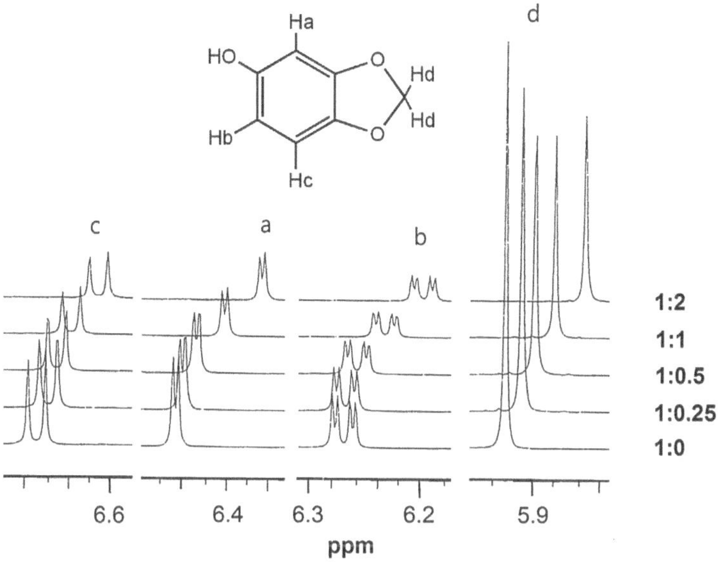

FIGURE 7.5 ^{1}H NMR spectra of mixtures of sesamol and octylamine at ratios of 1:0, 1:0.25, 1:0.5, 1:1, and 1:2.

on the oxygen atoms next to the methylene group when hydrogen bonding was formed. The same pattern of proton NMR peak shifts was observed with cetyl alcohol indicating the hydrogen bonding with the alcohol group in cetyl alcohol, although it was not as strong as that with the amino group in octylamine. Polysorbate 20 showed slightly greater proton peak shifts than cetyl alcohol in the NMR study. Because polysorbate 20 has many ether groups and much fewer hydroxyl groups than cetyl alcohol, this result suggests that ether groups can form relatively strong hydrogen bonding.

The interaction between sesamol and an additive and the consequent volatility reduction improved the antioxidant activity of sesamol during the heating experiment conducted with soybean oil containing 830 ppm sesamol and the molar equivalent amount of an additive at 180 °C for 8 h (Table 7.3). The 1:1 M (or weight) mixture of 830 ppm sesamol and an additive such as mono-/diglycerides, polysorbate 20, and L-carnosine exhibited stronger antioxidant activity than 200 ppm TBHQ. Cetyl alcohol, glycol stearate, and L-histidine did not improve the antioxidant activity of sesamol. One factor that should be considered in choosing the additive is the antioxidant activity of the additive, which will be an additional effect. The other factor is that the interaction between sesamol and the additive could reduce the antioxidant activity of sesamol. Valgimigli et al. (1996) reported a much slower hydrogen atom abstraction from α-tocopherol in water than in benzene due to the stronger hydrogen bonding of α-tocopherol with water than with benzene. While there are other factors that can affect the antioxidant activity of sesamol, it was confirmed that the reduced

TABLE 7.3 Heating Test Results with the Mixtures of 830 ppm Sesamol and the Same Molar Equivalent Additive (the Same Weight for Polysorbate 20)

Additive	Percent Loss of Olefinic H	Percent Loss of Bisallylic H	Percent PTAG	Percent Retained Sesamol (4 h)
None (sesamol only)	13.17	18.76	7.86	8.6
Mono-/diglycerides	10.98	15.86	6.32	15.1
Cetyl alcohol	13.66	19.34	8.58	9.2
Glycol stearate	13.45	19.05	8.09	13.6
Polysorbate 20	11.08	16.00	6.34	17.2
L-Histidine	13.61	19.35	8.65	9.4
L-Carnosine	11.03	15.90	6.55	16.0
Control (soybean oil only)	15.56	22.08	10.94	–
200 ppm TBHQ[a]	11.45	17.14	8.88	–

[a]*Contains only 200 ppm TBHQ (no sesamol, no additive) in soybean oil.*

volatility of sesamol is one major factor to improve its antioxidant activity by the correlation test between the oxidation results at 8 h and the sesamol concentration retained at 4 h (no sesamol was detected after 8 h) (Hwang et al., 2014). R^2 values (coefficient of determination) of the correlation between the sesamol concentration at 4 h and the oxidation of soybean oil at 8 h were 0.74, 0.75, and 0.76 for percent loss olefinic proton, percent loss of bisallylic proton, and percent PTAG, respectively. Selecting an additive that can effectively reduce sesamol volatility and has its own antioxidant activity is a good strategy to develop new antioxidant systems.

7.2.12 Reduced Volatility of Synthetic Antioxidants

Although natural antioxidants are safer and preferred over synthetic antioxidants, it may take a long time to completely replace synthetic antioxidants with natural antioxidants due to their relatively lower antioxidant activities, possible negative effects on sensory properties (Pokorny, 2007) and higher costs. Until then, synthetic antioxidants will still be widely used. The most widely used synthetic antioxidants, TBHQ, BHA, and BHT, are also substantially volatile (Warner et al., 1986).

Additives reducing volatility of sesamol were also evaluated for these synthetic antioxidants in a separate study (Hwang and Winkler-Moser, 2014). The TGA experiments conducted with a neat mixture of antioxidant and additive and with the mixture in soybean oil showed that these additives significantly reduced volatilities of TBHQ, BHA, and BHT. The concentration of the antioxidant was well retained when an additive was added during heating of soybean oil at 180 °C. The proton NMR study also showed a strong hydrogen bonding between BHT and cetyl alcohol, octylamine, polysorbate 60, and dioctyl sodium sulfosuccinate. The heating study with 200 ppm TBHQ in soybean oil at 180 °C for 8 h showed a significantly improved antioxidant activity by addition of L-lysine, L-arginine, sorbitan tristearate, mono-/diglycerides, polysorbate 60, lecithin, and L-carnosine. The correlation analysis showed that the THBQ concentration retained during heating showed moderate to strong correlations with loss of olefinic protons, loss of bisallylic protons, and percent PTAG of heated oil (R^2 values were 0.85, 0.84, and 0.63, respectively). The heating experiments with BHT and BHA also showed that the additives effectively enhanced activities of these antioxidants while the effects were not as marked as TBHQ due to their inherently low antioxidant activities at high temperatures (Allam and Mohamed, 2002). Additives having antioxidant activities by themselves always showed much higher antioxidation activities when added to an antioxidant. Therefore, additives with their own antioxidant activities and synergistic effects with the antioxidant are desirable for the practical development of a new antioxidant system in addition to the ability to reduce volatility of the antioxidant.

REFERENCES

Aladedunye, F., 2014. Natural antioxidants as stabilizers for frying oils. Eur. J. Lipid Sci. Technol. 116, 688–706.

Allam, S.S.M., Mohamed, H.M.A., 2002. Thermal stability of some commercial natural and synthetic antioxidants and their mixtures. J. Food Lipids 9, 277–293.

American Society for Testing and Materials (ASTM) STP 682, 1979. In: Schaefer, E.E. (Ed.), Manual on Consumer Sensory Evaluation. ASTM, Philadelphia, PA.

Anonymous, April 2008. Potato chip consumption up despite focus on healthy foods. Food Bus. News 29, 23.

AOCS, 2011a. Determination of polar compounds in frying fats. In: Firestone, D. (Ed.), Official Methods and Recommended Practices of the AOCS, sixth ed. AOCS Official Method, Cd 20–91, Champaign, IL. Second printing.

AOCS, 2011b. Determination of polymerized triglycerides by gel-permeation HPLC. In: Firestone, D. (Ed.), Official Methods and Recommended Practices of the AOCS, sixth ed. AOCS Official Method, Cd 22–91, Champaign, IL. Second printing.

AOCS, 2011c. Flavor panel evaluation of vegetable oils. In: Firestone, D. (Ed.), Official Methods and Recommended Practices of the AOCS, sixth ed. AOCS Official Method, Cd 22–91, Champaign, IL. Second printing.

Arteaga, S., Andrade-Cetto, A., Cardenas, R., 2005. *Larrea tridentata* (Creosote bush), an abundant plant of Mexican and US-American deserts and its metabolite nordihydroguaiaretic acid. J. Ethnopharmacol. 98, 231–239.

Bastida, S., Sanchez-Muniz, F.J., 2001. Thermal oxidation of olive oil, sunflower oil and a mix of both oils during forty discontinuous domestic fryings of different foods. Food Sci. Technol. 7, 15–21.

Blumenthal, M., 1991. A new look at the chemistry and physics of deep-fat frying. Food Technol. 45, 68–71.

Boskou, D., 2011. Nonnutrient antioxidants and stability of frying oils. In: Boskou, D., Elmadfa, I. (Eds.), Frying of Food. CRC Press, Boca Raton, FL, pp. 199–224.

Budowski, P., O'connor, R.T., Field, E.T., 1950. Sesame oil. IV. Determination of free and bound sesamol. J. Am. Oil Chem. Soc. 27, 307–310.

Caldwell, J., Cooke, B., Greer, M., 2011. High performance liquid chromatography-size exclusion chromatography for rapid analysis of total polar compounds in used frying oils. J. Am. Oil Chem. Soc. 88, 1669–1674.

Chang, S.S., Peterson, R.J., Ho, C.-T., 1977. Chemistry of deep fat fried flavor. In: Supran, M.K. (Ed.), Lipids as a Source of Flavor. American Chemical Society, Washington, DC, pp. 18–41.

Choe, E., Min, D.B., 2007. Chemistry of deep-fat frying oils. J. Food Sci. 72, R77–R86.

Dana, D., Blumenthal, M.M., Saguy, I.S., 2003. The protective role of water injection on oil quality in deep fat frying conditions. Eur. Food Res. Technol. 217, 104–109.

Dana, D., Saguy, I.S., 2006. Review: mechanism of oil uptake during deep-fat frying and the surfactant effect-theory and myth. Adv. Colloid Interface Sci. 128-130, 267–272.

Deyama, T., Nishibe, S., 2010. Pharmacological properties of lignans. In: Heitner, C., Dimmel, D.R., Schmidt, J.A. (Eds.), Lignin and Lignans: Advances in Chemistry. CRC Press, Boca Raton, pp. 585–629.

Dong, Y., 2011. Protein Modifications in Baked versus Fried Tortilla Chips (M.S. thesis). Rutgers, The State University of New Jersey, New Brunswick.

Downs, S.M., Thow, A.M., Leeder, S.R., 2013. The effectiveness of policies for reducing dietary trans fat: a systematic review of the evidence. Bull. World Health Organ. 91, 262–269.

Dutta, P.C., Savage, G.P., 2002. Formation and content of phytosterol oxidation products in foods. In: Guardiola, F., Dutta, P.C., Codony, R., Savage, G.P. (Eds.), Cholesterol and Phytosterol Oxidation Products in Foods. AOCS Press, Champaign, pp. 319–334.

Firestone, D., 2007. Regulation of frying fat and oil. In: Erickson, M.D. (Ed.), Deep Frying: Chemistry, Nutrition, and Practical Applications. AOCS Press, Urbana, IL, pp. 373–386.

Frankel, E.N., 1993. In search of better methods to evaluate natural antioxidants and oxidative stability in food lipids. Trends Food Sci. Technol. 4, 220–225.

Fuller, G., Guadagni, G.G., Weaver, M.L., Notter, G., Horvat, R.J., 1971. Evaluation of oleic safflower oil in frying of potato chips. J. Food Sci. 36, 43–44.

Fuss, E., 2003. Lignans in plant cell and organ cultures: an overview. Phytochem. Rev. 2, 307–320.

Gebauer, S.K., Psota, T.L., Kris-Etherton, P.M., 2007. The diversity of health effects of individual trans fatty acid isomers. Lipids 42, 787–799.

Gertz, C., Kochhar, S.P., 2000. Proceedings of the 3rd international symposium of deep-fat frying – final recommendations. Eur. J. Lipid Sci. Technol. 102, 306–308.

Gertz, C., Fiebig, H.-J., Hancock, J.N.S., 2013. FT-near infrared (NIR) spectroscopy-screening analysis of used frying fats and oils for rapid determination of polar compounds, polymerized triacylglycerols, acid value and anisidine value [DGF C-VI 21a (13)]. Eur. J. Lipid Sci. Technol. 115, 1193–1197.

Goicoechea, E., Guillén, M.D., 2014. Volatile compounds generated in corn oil stored at room temperature. Presence of toxic compounds. Eur. J. Lipid Sci. Technol. 116, 395–406.

Gupta, M.K., Grant, R., Stier, R.F., 2004. Critical factors in the selection of an industrial fryer. In: Gupta, M.K., Warner, K., White, P.J. (Eds.), Frying Technology and Practices. AOCS Press, Urbana, IL, pp. 91–190.

Hawrysh, Z.J., Erin, M.K., Kim, S., Hardin, R., 1996. Quality and stability of potato chips fried in canola, partially hydrogenated canola, soybean and cottonseed oils. J. Food Qual. 19, 107–120.

Hamama, A., Nawar, W., 1991. Thermal decomposition of some phenolic antioxidants. J. Agric. Food Chem. 39, 1063–1069.

Hemalatha, S., Ghafoorunissa, 2007. Sesame lignans enhance the thermal stability of edible vegetable oils. Food Chem. 105, 1076–1085.

Hirose, M., Takesada, Y., Tanaka, H., Tamano, S., Kato, T., Shirai, T., 1997. Carcinogenicity of antioxidants BHA, caffeic acid, sesamol, 4-methoxyphenol and catechol at low doses, either alone or in combination and modulation of their effects in a rat medium term multi organ carcinogenesis model. Carcinogenesis 19, 207–212.

Hwang, H.S., Winkler-Moser, J.K., Liu, X.L., 2012. Structural effect of lignans and sesamol on polymerization of soybean oil at frying temperature. J. Am. Oil Chem. Soc. 89, 1067–1076.

Hwang, H.S., Winkler-Moser, J.K., Bakota, E.L., Berhow, M.A., Liu, S.X., 2013. Antioxidant activity of sesamol in soybean oil under frying conditions. J. Am. Oil Chem. Soc. 90, 659–666.

Hwang, H.S., Winkler-Moser, J.K., Vermilion, K., Liu, S.X., 2014. Enhancing antioxidant activity of sesamol at frying temperature by addition of additives through reducing volatility. J. Food Sci. 79, C2164–C2173.

Hwang, H.S., Winkler-Moser, J.K., 2014. Food additives reducing volatility of antioxidants at frying temperature. J. Am. Oil Chem. Soc. 91, 1745–1761.

Ito, N., Hirose, M., Fukushima, G., Tauda, H., Shira, T., Tatematsu, M., 1986. Studies on antioxidant. Their carcinogenic and modifying effects on chemical carcinogenesis. Food Chem. Toxical. 24, 1071–1081.

IUPAC. Standard Method 2.507, 1987a. Determination of polar compounds in frying fats. In: Standard Methods for the Analysis of Oils, Fats and Derivatives, seventh ed. International Union of Pure and Applied Chemistry, Blackwell, Oxford.

IUPAC. Standard Method 2.508, 1987b. Determination of polymerized triglycerides in oils and fats by high performance liquid chromatography. In: Standard Methods for the Analysis of Oils, Fats and Derivatives, seventh ed. International Union of Pure and Applied Chemistry, Blackwell, Oxford.

Jacklin, A., Ratledge, C., Welham, K., Bilko, D., Newton, C.J., 2003. The sesame seed oil constituent, sesamol, induces growth arrest and apoptosis of cancer and cardiovascular cells. Ann. N. Y. Acad. Sci. 1010, 374–380.

Kaufmann, A., Ryser, B., Suter, B., 2001. HPLC with evaporative light scattering detection for the determination of polar compounds in used frying oils. Eur. Food Res. Technol. 213, 372–376.

Kochhar, S.P., Gertz, C., 2004. New theoretical and practical aspects of the frying process. Eur. J. Lipid Sci. Technol. 106, 722–727.

Lalas, S., Dourtoglou, V., 2003. Use of rosemary extract in preventing oxidation during deep-fat frying of potato chips. J. Am. Oil Chem. Soc. 80, 579–583.

Lee, J., Choe, E., 2006. Extraction of lignan compounds from roasted sesame oil and their effects on the autoxidation of methyl linoleate. J. Food Sci. 71, C431–C436.

Lee, J., Lee, Y., Choe, E., 2008. Effects of sesamol, sesamin, and sesamolin extracted from roasted sesame oil on the thermal oxidation of methyl linoleate. LWT - Food Sci. Technol. 41, 1871–1875.

Lee, S.W., Keung, M.K., Park, M.H., Lee, S.Y., Lee, J., 2010. Effects of roasting conditions of sesame seeds on the oxidative stability of pressed oil during thermal oxidation. Food Chem. 118, 681–685.

Lin, F.S., Warner, C.R., Fazio, T., 1981. Alteration of phenolic antioxidants in heated vegetable oils. J. Am. Oil Chem. Soc. 58, 789–791.

Litwinienko, G., Megiel, E., Wojnicz, M., 2002. Hydrogen bonding between phenols and fatty acid esters: ^{1}H NMR study and ab initio calculations. Org. Lett. 4, 2425–2428.

Marmesat, S., Rodrigues, E., Velasco, J., Dobarganes, C., 2007. Quality of used frying fats and oils: comparison of rapid tests based on chemical and physical properties. Int. J. Food Sci. Technol. 42, 601–608.

Marmesat, S., Morales, A., Velasco, J., Dobarganes, M.C., 2010. Action and fate of natural and synthetic antioxidants during frying. Grasas Aceites 61, 333–340.

Marmesat, S., Morales, A., Velasco, J., Dobarganes, M.C., 2012. Influence of fatty acid composition on chemical changes in blends of sunflower oils during thermoxidation and frying. Food Chem. 135, 2333–2339.

Márquez-Ruiz, G., Tasioula-Margari, M., Dobarganes, M.C., 1995. Quantitation and distribution of altered fatty acids in frying fats. J. Am. Oil Chem. Soc. 72, 1171–1176.

Márquez-Ruiz, G., Jorge, N., Martín-Polvillo, M., Dobarganes, M.C., 1996. Rapid, quantitative determination of polar compounds in fats and oils by solid-phase extraction and size exclusion chromatography using monostearin as an internal standard. J. Chromat. A 749, 55–60.

Márquez-Ruiz, G., Polvillo, M., Jorge, N., Ruiz Méndez, M.V., Dobarganes, M.C., 1999. Influence of frying oil quality and natural tocopherol content on oxidative stability of fried potatoes. J. Am. Oil Chem. Soc. 76, 421–425.

Márquez-Ruiz, G., Ruiz-Méndez, M.V., Velasco, J., 2014. Antioxidants in frying: analysis and evaluation of efficacy. Eur. J. Lipid Sci. Technol. 116, 1441–1450.

Matthaus, B., 2005. Utilization of high-oleic rapeseed oil for deep-fat frying of French fries compared to other commonly used edible oils. Eur. J. Lipid Sci. Technol. 108, 200–211.

May, W.A., Peterson, R.J., Chang, S.S., 1978. Synthesis of some unsaturated lactones and their relationship to deep-fat fried flavor. J. Food Sci. 43, 1248–1252.

Meilgaard, M.C., Vance, G.V., Carr, B.T., 2007. Sensory Evaluation Techniques, fourth ed. CRC Press, Taylor & Francis Group, Boca Raton, FL, pp. 1–170.

Melton, S.L., Jafar, S., Sykes, D., Trigiano, M.K., 1994. Review of stability measurements for frying oils and fried food flavor. J. Am. Oil Chem. Soc. 71, 1301–1308.

Melton, S.L., 2006. Sensory evaluation of frying fat and deep-fried products. In: Erickson, M.D. (Ed.), Deep Frying: Chemistry, Nutrition, and Practical Applications (Chapter 20). AOCS Press, Urbana, IL, pp. 359–370.

Micha, R., Mozaffarian, D., 2010. Saturated fat and cardiometabolic risk factors, coronary heart disease, stroke, and diabetes: a fresh look at the evidence. Lipids 45, 893–905.

Min, D.B., Boff, J.M., 2002. Chemistry and reaction of singlet oxygen in foods. Comp. Rev. Food Sci. Technol. 1, 58–72.

Mozaffarian, D., Clarke, R., 2009. Quantitative effects on cardiovascular risk factors and coronary heart disease risk of replacing partially hydrogenated vegetable oils with other fats. Eur. J. Clin. Nutr. 63, S22–S33.

Nawar, W.W., 1969. Thermal degradation of lipids: a review. J. Agric. Food Chem. 17, 18–21.

Nawar, W.W., Bradley, S.J., Lomanno, S.S., Richardson, G.G., Whiteman, R.C., 1977. Volatiles from frying fats: a comparative study. In: Supran, M.K. (Ed.), Lipids as a Source of Flavor. American Chemical Society, Washington, DC, pp. 42–55.

Naz, S., Sheikh, H., Siddiqi, R., Sayeed, S.A., 2004. Oxidative stability of olive, corn and soybean oil under different conditions. Food Chem. 88, 253–259.

Neff, W.E., Warner, K., Byrdwell, W.C., 2000. Odor significance of undesirable degradation compounds in heated triolein and trilinolein. J. Am. Oil Chem. Soc. 77, 1303–1313.

Negishi, H., Nishida, M., Endo, Y., Fujimoto, K., 2003. Effect of a modified deep-fat fryer on chemical and physical characteristics of frying oil. J. Am. Oil Chem. Soc. 80, 163–166.

Nicmeyer, H.B., Metzler, M., 2003. Differences in the antioxidant activity of plant and mammalian lignans. J. Food Eng. 56, 255–256.

O'Brien, R.D., 1998. Fats and Oils: Formulating and Processing for Applications. Technomic Publishing Co., Lancaster, PA.

Poyato, C., Ansorena, D., Navarro-Blasco, I., Astiasarán, I., 2014. A novel approach to monitor the oxidation process of different types of heated oils by using chemometric tools. Food Res. Int. 57, 152–161.

Petersen, K., Kleeberg, K., Jahreis, G., Busch-Stockfisch, M., Fritsche, J., 2012. Comparison of analytical and sensory lipid oxidation parameters in conventional and high-oleic rapeseed oil. Eur. J. Lipid Sci. Technol. 114, 1193–1203.

Pokorny, J., 1989. Flavor chemistry of deep fat frying in oil. In: Min, D.B., Smouse, T.H. (Eds.), Flavor Chemistry of Lipid Foods. AOCS Press, Champaign, pp. 113–155.

Pokorny, J., 2007. Are natural antioxidants better – and safer –than synthetic antioxidants? Eur. J. Lipid Sci. Technol. 109, 629–642.

Prior, R.L., Wu, X., Schaich, K., 2005. Standardized methods for the determination of antioxidant capacity and phenolics in foods and dietary supplements. J. Agric. Food Chem. 53, 4290–4302.

Przybylski, R., Eskin, N.A.M., 1988. A comparative study on the effectiveness of nitrogen or carbon dioxide flushing in preventing oxidation during the heating of oil. J. Am. Oil Chem. Soc. 65, 629–633.

Qin, X., 2011. Thermal Desorption Studies of Corn Oil at Frying Temperatures: Thermal Scission versus Autoxidation (thesis). The State University of New Jersey, New Brunswick.

Ram, R., Catlin, D., Romero, J., Cowley, C., 1990. Sesame: new approaches for crop improvement. In: Janick, J., Simon, J.E. (Eds.), Advances in New Crops. Timber Press, Portland, OR, pp. 225–228.

Ravelli, D., Matsuoka, C.R., Modesta, R.C.D., de Souza Vieira, T.M.F., Regitano-d'Arce, M.A.B., 2010. Determination of deep frying soybean oil disposal point by a sensory method. J. Am. Oil Chem. Soc. 87, 515–520.

Rau, O., Wurglics, M., Paulke, A., Zitzkowski, J., Meindl, N., Bock, A., Dingermann, T., Abdel-Tawab, M., Schubert-Zsilavecz, M., 2006. Carnosic acid and carnosol, phenolic diterpene compounds of the labiate herbs rosemary and sage, are activators of the human peroxisome proliferator-activated receptor gamma. Planta Med. 72, 881–887.

Rossi, M., Alamprese, C., Ratti, S., 2007. Tocopherols and tocotrienols as free radical-scavengers in refined vegetable oils and their stability during deep-fat frying. Food Chem. 102, 812–817.

Schaich, K.M., 2012. Thinking outside the classical chain reaction box of lipid oxidation. Lipid Technol. 24, 55–58.

Schaich, K.M., 2014. Lipid co-oxidation of proteins: one size does not fit all. Inform 62, 4251–4260.

Shulte, E., 2004. Economical micromethod for determination of polar components in frying fats. Eur. J. Lipid Sci. Technol. 106, 772–776.

Shyu, S.-L., Hau, L.-B., Hwang, L.S., 1998. Effect of vacuum frying on the oxidative stability of oils. J. Am. Oil Chem. Soc. 75, 1393–1398.

Soupas, L., Juntunen, L., Lampi, A.M., Piironen, V., 2004. Effects of sterol structure, temperature, and lipid medium on phytosterol oxidation. J. Agric. Food Chem. 52, 6485–6491.

Statista, 2013. Sales of the Leading 10 Tortilla and Tostada Chip Brands of the United States in 2013. (in million U.S. dollars) http://www.statista.com/statistics/188233/top-tortilla-tostada-chip-brands/ (accessed 03/2015).

Stier, R.F., 2004. Frying as a science-an introduction. Eur. J. Lipid Sci. Technol. 106, 715–721.

Tian, X., 2013. Chemical Factors Affecting Degradation Processes of Vegetable Oils during Frying (Ph.D. thesis). The State University of New Jersey, New Brunswick.

Tseng, Y.-C., Moreira, R., Sun, X., 1996. Total frying-use time effects on soybean oil deterioration and on tortilla chip quality. Int. J. Food Sci. Technol. 31, 287–294.

Ufheil, G., Escher, F., 1996. Dynamics of oil uptake during deep-fat frying of potato slices. LWT - Food Sci. Technol. 29, 640–644.

Urbancic, S., Kolar, M.H., Dimitrijevic, D., Demsar, L., Vidrih, R., 2014. Stabilisation of sunflower oil and reduction of acrylamide formation of potato with rosemary extract during deep-fat frying. LWT - Food Sci. Technol. 57, 671–678.

Valgimigli, L., Ingold, K.U., Lusztyk, J., 1996. Antioxidant activities of vitamin E analogues in water and a Kamlet-Taft β-value for water. J. Am. Chem. Soc. 118, 3545–3549.

Wang, T., Kevin, B., Hicks, K.B., Robert Moreau, R., 2002. Antioxidant activity of phytosterols, oryzanol, and other phytosterol conjugates. J. Am. Oil Chem. Soc. 79, 1201–1206.

Warner, K., 1985. Sensory evaluation of flavor quality of oils. In: Min, D.B., Smouse, T.H. (Eds.), Flavor Chemistry of Fats and Oils. American Oil Chemists' Society, Champaign, IL, pp. 207–221.

Warner, C.R., Daniels, D.H., Lin, F.S., Joe, F.L., Fazio, T., 1986. Fate of antioxidants and antioxidant-derived products in deep-fat frying and cookie baking. J. Agric. Food Chem. 34, 1–5.

Warner, K., Mounts, T.L., 1993. Frying stability of soybean and canola oils with modified fatty acid compositions. J. Am. Oil Chem. Soc. 70, 983–988.

Warner, K., 1995. Sensory evaluation of oils and fat-containing foods. In: Warner, K., Eskin, N.A.M. (Eds.), Methods to Assess Quality and Stability of Oils and Fat-Containing Foods. AOCS Press, Champaign, IL, pp. 49–75.

Warner, K., Knowlton, S., 1997. Frying quality and oxidative stability of high-oleic corn oils. J. Am. Oil Chem. Soc. 74, 1317–1322.

Warner, K., Orr, P., Glynn, M., 1997. Effect of fatty acid composition of oils on flavor and stability of fried foods. J. Am. Oil Chem. Soc. 74, 347–356.

Warner, K., Neff, W.E., Byrdwell, W.C., Gardner, H.W., 2001. Effect of oleic and linoleic acids on the production of deep-fried odor in heated triolein and trilinolein. J. Agric. Food Chem. 49, 899–905.

Warner, K., Neff, W.E., Eller, F., 2003. Enhancing quality and oxidative stability of aged fried food with γ-tocopherol. J. Agric. Food Chem. 51, 623–627.

Warner, K., Moser, J., 2009. Frying stability of purified mid-oleic sunflower oil triacylglycerols with added pure tocopherols and tocopherol mixtures. J. Am. Oil Chem. Soc. 86, 1199–1207.

Weisshaar, R., 2014. Quality control of used deep–frying oils. Eur. J. Lipid Sci. Technol. 116, 716–722.

Winkler, J.K., Warner, K., Glynn, M.T., 2007. Effect of deep-fat frying on phytosterol content in oils with differing fatty acid composition. J. Am. Oil Chem. Soc. 84, 1023–1030.

Winkler-Moser, J.K., Rennick, K.A., Palmquist, D.A., Berhow, M.A., Vaughn, S.F., 2012. Comparison of the impact of γ -oryzanol and corn steryl ferulates on the polymerization of soybean oil during frying. J. Am. Oil Chem. Soc. 89, 243–252.

Winkler-Moser, J.K., Hwang, H.-S., Bakota, E.L., Palmquist, D., 2015. Synthesis of steryl ferulates with various sterol structures and comparison of their antioxidant activity. Food Chem. 169, 92–101.

Yeo, J.D., Jeong, M.K., Park, C.U., Lee, J., 2010. Comparing antioxidant effectiveness of natural and synthetic free radical scavengers in thermally-oxidized lard using DPPH method. J. Food Sci. 75, C258–C262.

Yeo, J.D., Park, J., Lee, J., 2011. Evaluation of antioxidant capacity of sesamol and free radical scavengers at different heating temperatures. Eur. J. Lipid Technol. 113, 910–915.

Zhang, Q., Saleh, A.S.M., Chen, J., Shen, Q., 2012. Chemical alterations taken place during deep-fat frying based on certain reaction products: a review. Chem. Phys. Lipids 165, 662–681.

Chapter 8

Oxidative Stability and Shelf Life of Food Emulsions

Charlotte Jacobsen
National Food Institute, Technical University of Denmark, Kgs. Lyngby, Denmark

8.1 INTRODUCTION

Oil-in-water (o/w) and water-in-oil (w/o) emulsions are important colloid structures in food products, as many foods exist as emulsions or have been in an emulsified state at some time during their existence (McClements, 2008). The o/w emulsions such as milk, mayonnaise, and salad dressing constitute a particularly large—and widely consumed—group of such complex, heterophasic food systems, but also w/o emulsions, including margarine and butter, are consumed in large amounts. In both types of emulsions, the oil and the aqueous phases are separated by an interface composed of amphiphilic compounds (emulsifiers). Emulsions face two different types of stability problems during storage: physical instability due to creaming, flocculation, and so on and instability due to lipid oxidation. This chapter will discuss the latter type of instability. Lipid oxidation is an interfacial phenomenon, and because emulsions have a large interfacial area, they are particularly susceptible to oxidation. Lipid oxidation is induced by transition metals, light, temperature, and oxygen and is also affected by pH, droplet size, and numerous other factors related to the composition of the emulsion. Importantly, the oxidative stability of food emulsions is greatly dependent on the chemical composition of the lipid phase. Apart from a few important exceptions (Miyashita et al., 1993, 1994), the more unsaturated the lipid, the more susceptible it is to lipid oxidation. During the past decade, polyunsaturated fatty acids (PUFA) such as long-chain n-3 PUFA have increasingly been incorporated into emulsified foods in order to address recent nutritional recommendations (EFSA Panel on Dietetic Products, 2010). The increased use of n-3 PUFA has resulted in new challenges for the food industry with respect to avoiding lipid oxidation. Besides altering the sensory (rancidity) and nutritional quality (loss of vitamins and essential fatty acids) of food, lipid oxidation will result in highly reactive and toxic compounds (e.g., malondialdehyde, 4-hydroxynonenal), which may pose a danger to consumers. Therefore, both academia and industry are increasingly addressing the challenge of preventing

Oxidative Stability and Shelf Life of Foods Containing Oils and Fats
http://dx.doi.org/10.1016/B978-1-63067-056-6.00008-2

lipid oxidation in food emulsions. This chapter will summarize the most important factors influencing lipid oxidation in emulsions as well as the methods used to assess oxidative stability and will then discuss different strategies to prevent lipid oxidation in a range of food emulsions.

8.2 EMULSIONS AND PHYSICAL STABILIZATION

Food emulsions are composed of two immiscible liquids and an interface. Food emulsions can exist as either o/w emulsions for which the oil is dispersed in the aqueous phase or as w/o emulsions for which water is dispersed in the oil phase. Emulsions are thermodynamically unstable systems, but they can be physically stabilized to avoid the immediate separation into oil and aqueous phases. The physical stabilization is provided by texture modifiers (e.g., hydrocolloids), emulsifying agents, or solid particles (pickering agents) (Figure 8.1). Emulsifiers are surface active molecules with amphiphilic properties, which can interact with the oil–water interface and reduce surface tension. Emulsifiers for food use are thus either macromolecules, such as proteins unfolding at the interface, or smaller surfactant molecules. In food emulsions, different emulsifiers and texture-modifying agents are often used together to provide physical stabilization. For example, in mayonnaise,

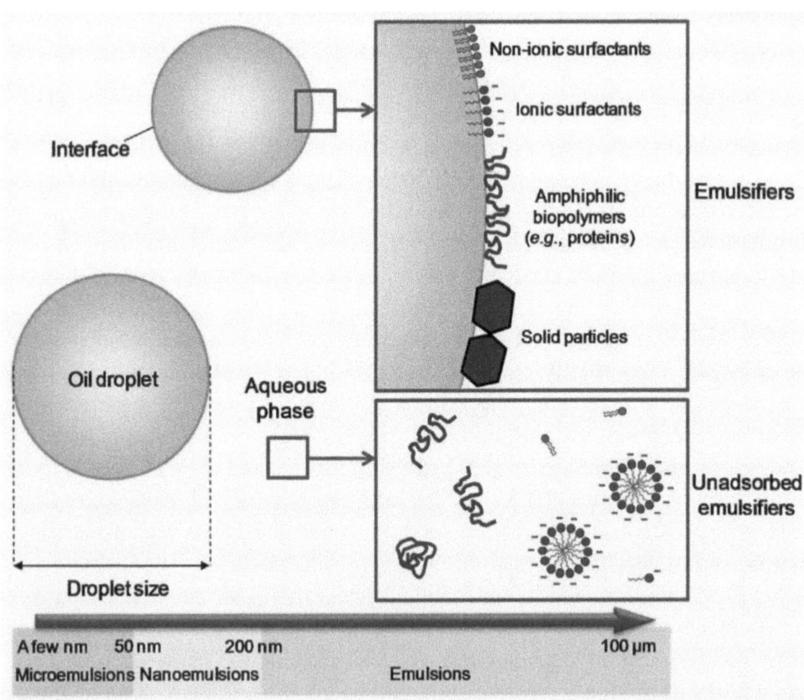

FIGURE 8.1 Oil-in-water emulsion and examples on the location of different emulsifiers at the interface or in the aqueous phase. *From Berton-Carabin et al. (2014).*

egg yolks consisting of both proteins and phospholipids are used together with hydrocolloids such as guar gum and xanthan gum to stabilize the emulsion.

Surfactants (low molecular weight emulsifiers) includes a wide range of small amphiphilic molecules such as phospholipids, free fatty acids, mono- and diacylglycerols, polyoxyethylene sorbitan esters (Tweens), polysorbates, sorbitan esters (Spans), sucrose esters, citric, lactic, and acetic acid esters of mono- and diacylglycerols. Surfactants can be anionic, cationic, zwitterionic, or nonionic.

High molecular weight emulsifiers include proteins, which provide interfacial activity due to the presence of hydrophilic and hydrophobic domains distributed along their polymeric chain. When adsorbed at the interface, proteins will form a strong viscoelastic film around the droplet and they will adopt a conformation, which allows the hydrophobic domain to be close to the oil phase and the hydrophilic domain to be close to the aqueous phase. This will result in "trains," "loops," and "tails" bending the molecule. Proteins can adopt an extended (e.g., casein) or globular conformation (e.g., whey proteins) at the interface.

The structural conformations and the emulsifying properties of, for example, whey proteins are very sensitive toward different treatments, such as homogenization, a change in pH, or heating (Fang and Dalgleish, 1997, 1998; Hunt and Dalgleish, 1994; Lee et al., 2007; Stapelfeldt and Skibsted, 1999).

8.2.1 The Interface

The interfacial layer between oil and water droplets is only a few nanometers thick. Nevertheless, it may comprise a significant volume of the total droplet (McClements and Decker, 2000). Low molecular weight surfactants form a monolayer at the interface and will cover almost 100% of the interface. In contrast, high molecular weight emulsifiers can form either a monolayer or multilayers if present in high concentrations. Moreover, only a part of the emulsifier will be in direct contact with the oil phase. The remaining segments will be located in the aqueous phase. Thereby, only approximately 30–40% of the interface is covered by emulsifier (Dickinson, 1992). For these reasons, low molecular weight surfactants are more packed at the surface than, for example, proteins, but the thickness of the interfacial layer provided by surfactants will be smaller (0.5–1 nm) than that provided by proteins (1–15 nm).

Most of the proteins that are used as emulsifiers are water soluble, and for that reason unadsorbed protein will mainly be located in the aqueous phase. Likewise, unadsorbed surfactant will form micelles in the aqueous phase if the concentration of unadsorbed surfactant is above the critical micelle concentration. Berton-Carabin et al. (2014), in a recent review, estimated the concentrations and proportions of unadsorbed emulsifiers in the aqueous phase of o/w emulsions in a large number of studies in the literature, which had used either surfactants or proteins as emulsifiers. Their estimations showed that the great majority (76%) of the formulations gave emulsions in which more than 50% of the emulsifier was unadsorbed. In 28 out of 80 formulations, more than 90% of the emulsifier was unadsorbed. As will be discussed later, the unadsorbed emulsifier can play a crucial role in lipid oxidation.

8.3 LIPID OXIDATION IN EMULSIONS

8.3.1 Lipid Oxidation in Emulsions versus Bulk Oil

Lipid oxidation in emulsions is very complex as it may include oxidation or electron transfer events in all the different phases of the systems. This makes the mechanisms of lipid oxidation in emulsions very different and significantly more complex than lipid oxidation in bulk oil systems. Hence, although the basic oxidation reactions of lipids present in emulsions are exactly the same as those of lipids in bulk oils, the factors affecting lipid oxidation in emulsions differ significantly between emulsions and bulk oil systems. In Figure 8.2, the most important factors influencing lipid oxidation in o/w emulsions are summarized and they will be discussed in the following. Some of these factors will have the same effects in w/o emulsions, but too few studies have been conducted in w/o emulsions to make firm conclusions on this matter.

Oxidation generally occurs faster in o/w emulsions than in bulk oils (Frankel et al., 2002). This may be because of the relatively harsh processing conditions required for emulsion fabrication, including exposure to air (oxygen), direct production of free radicals by acoustic cavitation in the case of sonication (Jana et al., 1990; Riesz and Kondo, 1992), and, in some cases, high temperature. Furthermore, the creation of interfacial area between the oil and the aqueous phase may promote the contacts between PUFA in the oil phase and pro-oxidants

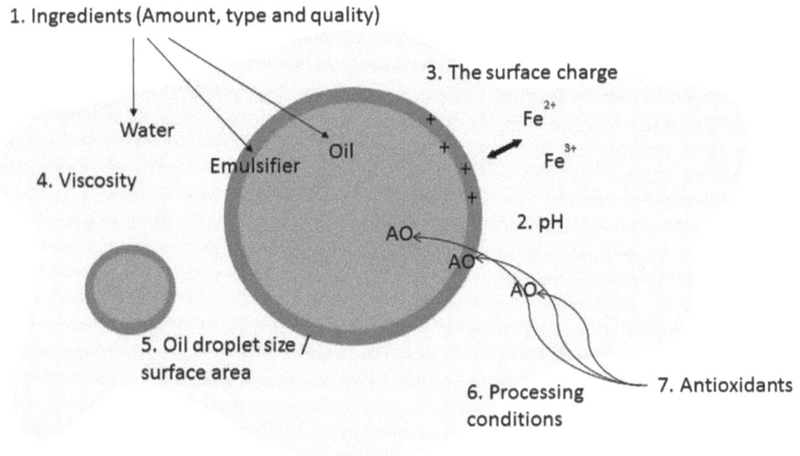

FIGURE 8.2 Overview of factors influencing lipid oxidation in food emulsions. *Reprinted from Horn (2012).*

such as trace metal ions in the aqueous phase. Thus, in o/w emulsions, pro-oxidant concentrations are often highest in the aqueous phase, and therefore oxidation may be initiated in the aqueous phase or at the interface rather than in the oil phase "inside" the oil droplets. Oxygen, pro-oxidant metals, and water-soluble antioxidants have to diffuse through the aqueous phase before reaching the oil droplets or the o/w interface. Therefore, differences in solubility properties, mobility, and mass transfer rates of transition metals and antioxidants through different phases may affect oxidation significantly in emulsions. An important property of oxygen is that it is more soluble in oil than in water ([O_2 oil]=ca. 1.2 mmol/L [O_2 aqua]=ca. 0.25 mmol/L) (Battino, 1987; Weast et al., 1987), and this may also influence oxidation rates.

8.3.2 Droplet Size Distribution

Since lipid oxidation often involves the interaction at the oil–water interface between water soluble pro-oxidants and the lipids located in the oil phase, lipid oxidation is regarded as an interfacial phenomenon. For that reason, lipid oxidation could be expected to be depending on size of the total interfacial area, which is determined by the size distribution of the oil droplet (the smaller the droplets, the larger is the total interfacial area). However, contradicting reports are available on the effect of the droplet size on oxidation in different food systems (Jacobsen et al., 2000; Lethuaut et al., 2002; Nakaya et al., 2005; Let et al., 2007a). The available data thus suggest that other factors than the mere interfacial area are of significance for oxidation. Such other factors may include the thickness and charge of the interfacial boundary (Donnelly et al., 1998; Hu et al., 2003a,b) and the metal binding capacity of the emulsifier (Jacobsen et al., 1999). Berton-Carabin et al. (2014) discussed whether differences in the amount of unadsorbed emulsifier could have played a role in some of the studies that showed difference in oxidative stability according to droplet size but concluded that this did not seem to be the case. They also stated that the contradicting results on the effect of the droplet size on lipid oxidation in different dispersed systems underline the fact that droplet size and interfacial layer probably influence lipid oxidation in o/w emulsions through several mechanisms, which are not fully understood.

8.3.3 Emulsifier Type and pH

Emulsifiers are able to affect lipid oxidation in different ways. Apart from influencing the structure and thickness of the interface, emulsifiers will also affect the charge of the interface. In protein-stabilized emulsions, pH will usually be either below or above the pI of the protein in order to avoid coalescence of droplets. Hence, the emulsified droplets may have either a positive or a negative surface charge. The surface charge of emulsion droplets is important for lipid oxidation catalyzed by trace metal ions, such as Fe^{2+}. When having a negative surface charge, emulsion droplets will attract the highly pro-oxidative positively charged trace metals and bring them closer to the oil, thereby enhancing lipid oxidation (Mei et al., 1998). In contrast, positively charged droplets may repel trace metals—notably Fe^{2+} and Fe^{3+}—and oxidation is likely to be reduced (Mei et al., 1998). Furthermore, the solubility of trace metals generally increases at decreasing pH. The increased trace metal solubility then potentially promotes oxidation. Since different emulsifiers are used in different food emulsions, and pH values also vary in different types of food emulsions, the significance of the emulsifier and the pH varies in different systems. In turn, this means that different oxidation mechanisms prevail in different emulsion systems. Berton-Carabin et al. (2014) summarized a large number of studies in which the effect of pH had been evaluated in different simple o/w emulsions. They found that in some studies, the oxidative stability of emulsions prepared at neutral pH was better than in emulsions at acidic pH, whereas the opposite was the case in other studies even if the same emulsifier and thereby the same surface charge were obtained. They also referred to a study of Faraji et al. (2004), who had studied the effect of pH in whey protein-stabilized emulsions that had been washed to remove unadsorbed emulsifier and in the same emulsions without washing. In that study, it was found that the effect of pH was much larger in the washed emulsions than in the emulsions that contained unadsorbed emulsifier. From these findings, Berton-Carabin et al. (2014) concluded that there are important interactions between the intrinsic effects of pH on oxidation and the effects of unadsorbed proteins.

Several studies have compared the effect of the type of emulsifier on lipid oxidation. In many studies, proteins have been shown to provide better protection against oxidation than surfactants (Haahr and Jacobsen, 2008; Donnelly et al., 1998). However, the relation between the oxidation rate of different types of emulsifiers is also highly dependent on other factors such as oil content, pH, emulsifier concentration, and presence of iron. In a recent study, Nielsen et al. (2013) compared the effect of milk protein (casein or whey), soy phospholipid (Lec), and two mixtures of phospholipids and milk proteins with low (MPL20) or high (MPL75) contents of milk phospholipid on lipid oxidation in 5% fish o/w emulsions. The emulsifier-to-oil ratio was either 0.04 or 0.15. Two different pH values were studied (3.0 and 7.0), and emulsions were prepared with or without iron. Lipid oxidation in 5% fish o/w emulsions was significantly affected by emulsifier type, emulsifier concentration, pH, and iron addition. Generally, casein emulsions were the most oxidatively stable, followed by emulsions stabilized by whey protein, MPL20, Lec, and MPL75 in increasing order of oxidative instability. Moreover, a high emulsifier concentration in general resulted in more physically and oxidatively stable emulsions at pH 7.0, particularly in emulsions with iron. At pH 3.0, the effect of increasing the emulsifier concentration depended on both the presence of iron and the emulsifier type. In the presence of iron, a high emulsifier concentration generally decreased the peroxide value (PV), whereas the effect on volatile oxidation products was less pronounced. In the absence of iron, oxidation was more pronounced at high concentrations of emulsifier for whey proteins and casein when evaluated from both PV and volatiles, whereas PV data suggested a protective effect of increasing emulsifier concentration for Lec, MPL20, and MPL75, but the volatiles data showed the opposite. In addition, when data were compared with data from a study on high fat emulsions with 70% fish oil from the same laboratory (Horn et al., 2011, 2012b), it was observed that the above-mentioned differences affected emulsions differently depending on oil content.

8.3.4 Effect of Unadsorbed Emulsifier

In most emulsion studies, there is an excess of emulsifier and the unadsorbed emulsifier will usually be present in the aqueous phase. Unadsorbed protein emulsifiers may strongly interact with water-soluble free radicals and metal ions (Donnelly et al., 1998; Ponginebby et al., 1999; Faraji et al., 2004). Micelles in the aqueous phase are also able to trap antioxidants, hydroperoxides, fatty acids, and iron. Therefore, excess proteins will affect lipid oxidation. In many cases, the unadsorbed protein will act as an antioxidant due to its ability to chelate metals or scavenge free radicals, as reviewed by Berton-Carabin et al. (2014). Likewise, unadsorbed surfactants present in the aqueous phase in concentrations above the critical micelle concentration can form colloid structures such as micelles, mixed micelles, or very small droplets. Ponginebbi et al. (1999) showed that excess Tween 20 had a protective effect on lipid oxidation in linoleic acid emulsions. A reduced oxidation rate in the presence of excess Tween 20 was also observed in rapeseed o/w emulsions in a study by Berton et al. (2011).

Most of the studies in which proteins have been shown to provide better oxidative stability in o/w emulsions than surfactants have been conducted with high emulsifier concentrations and thereby a high proportion of unadsorbed surfactant. Berton et al. (2011) showed that when emulsions were formulated to contain lower levels of unadsorbed emulsifiers, proteins were less effective in protecting emulsions against lipid oxidation than Tween 20, Tween 80, and Citrem. They suggested that this could be due to the more structurally homogeneous and compact interfacial films formed by surfactants than protein-based films. However, the physical stability of emulsions with such low concentrations of emulsifier is poor and, therefore, under realistic conditions, proteins may still be the best choice if one wants to obtain high oxidative stability.

8.3.5 Effect of Processing Conditions

Emulsification can be obtained by different processes such as high-pressure homogenization, microfluidization, mixing (e.g., rotor-stator principle), and membrane homogenization. Only few systematic studies have been carried out in which lipid oxidation has been compared in emulsions produced under similar conditions in different homogenization devices. The obtainable droplet sizes in different homogenization equipments vary, and for that reason the type of homogenization can affect lipid oxidation. Moreover, different high-pressure homogenization equipment has been shown to differ with respect to the generation of heat (Mao et al., 2010), which is another factor that can potentially influence lipid oxidation.

Horn et al. (2012a) compared the effect of using a two-stage valve homogenizer or a microfluidizer on lipid oxidation in 10% fish o/w emulsions. Emulsions were optimized to have similar oil droplet size distributions. Results showed that the type of high-pressure homogenizer used, influenced lipid oxidation when emulsions were prepared with whey proteins as emulsifiers but not when prepared with caseinate. In the whey protein-stabilized emulsions, the oxidative stability was increased when prepared on a microfluidizer and less protein was present in the aqueous phase, and thereby more protein was present at the interface when the microfluidizer was used. It was also suggested that a more favorable composition of the interface was obtained. For casein-stabilized emulsions, it was suggested that the casein proteins in the aqueous phase, and its metal chelating effect, rather than the actual composition of protein components at the interface, may be the most important factor determining its oxidative stability.

Apart from the emulsification principle, homogenization conditions could potentially influence lipid oxidation as demonstrated by Horn et al. (2013). In their study on high-pressure homogenization of fish o/w emulsions, homogenization was conducted at two different pressures (5 MPa or 22.5 MPa) and at two different temperatures (22 °C and 72 °C). Emulsions were prepared with either a combination of α-lactalbumin and β-lactoglobulin or with a combination of sodium caseinate and β-lactoglobulin. Results showed that an increase in pressure increased the oxidative stability of emulsions with caseinate and β-lactoglobulin, whereas it decreased the oxidative stability of emulsions with α-lactalbumin and β-lactoglobulin. For both types of emulsions, the partitioning of proteins between the interface and the aqueous phase appeared to be important for the oxidative stability.

8.4 ANTIOXIDANTS

The addition of antioxidants is a possible means of inhibiting lipid oxidation, but due to the complexity of food emulsions, a large number of factors can affect their efficacy. In emulsions, the antioxidant will partition into the oil phase, into the water phase, or at the interface, and this will influence its efficacy. Several studies have observed a relationship between antioxidant partitioning and antioxidant efficacy, which is also referred to as "the polar paradox," as proposed by Frankel et al. (1994) and Porter (1993). According to the polar paradox, polar antioxidants such as ascorbic acid are more active in nonpolar media like neat oils than their more nonpolar counterparts (e.g., ascorbyl palmitate), whereas nonpolar antioxidants are more active in more polar systems like emulsions. For several years, attempts have been made to use the polar paradox theory to predict antioxidant effects in emulsified systems. However, in several cases the experimental data did not agree with the predictions made from this theory. Recent research on lipophilization of polar antioxidants has led to the so-called cut-off theory, which encompasses and expands the polar paradox theory (Laguerre et al., 2009, 2013). In brief, this theory states that the efficacy of polar antioxidants such as ascorbic acid in emulsions can be increased by lipophilizing the antioxidants up to a certain length of the alkyl group esterified to the polar antioxidant. Beyond the optimal chain length of the alkyl group, the efficacy of the lipophilized antioxidant will collapse. The cut-off theory has been confirmed in simple o/w emulsions with antioxidants such as rosmarinic acid (Laguerre et al., 2010), chlorogenic acid (Laguerre et al., 2009), and dihydrocaffeic acid (Sørensen et al., 2012b).

8.5 MEASUREMENT OF OXIDATION AND SHELF LIFE STABILITY IN FOOD EMULSIONS

It is possible to use PV or AV to measure lipid oxidation in food emulsions. While PV can provide useful information about the initial stages of lipid oxidation, the AV method will in many cases be too insensitive to provide useful results. Moreover, often AV will not be able to measure changes in composition of volatile oxidation products responsible for changes in off-flavor. For this reason, AV often correlates poorly to sensory data in food emulsions (Jacobsen, 1999). Therefore, AV cannot be recommended for measuring lipid oxidation in complex food emulsions. A better alternative is to use gas chromatography (GC) based techniques such as GC–mass spectrometry (MS). Different techniques are available to collect volatile oxidation products for GC-MS analysis, including static headspace, dynamic headspace (DHS), and solid phase microextraction (SPME). Static headspace is the simplest but also the least sensitive technique among the three. Often, it cannot measure aldehydes and ketones with a longer chain length than six or seven. SPME is increasingly being used for emulsion studies. However, this technique must be used with caution when analyzing complex food emulsions. This is due to the fact that volatile oxidation products will compete for space on the SPME fiber and if the emulsion contains a complex mixture of volatiles erroneous results can be obtained. This was illustrated by Lu et al. (2013), who compared DHS GC-MS (with Tenax GR as the adsorbing material) with SPME GC-MS for analysis of marine phospholipid emulsions using a CAR/PDMS fiber for the SPME. As shown in Figure 8.3, the MPW emulsion had higher concentration of volatiles after both 16 and 32 days when collected by DHS. In contrast, MPW had the highest content of volatiles after 16 days, but not after 32 days when volatiles were collected by SPME. Further scrutinizing of the data obtained showed that a so-called Strecker aldehyde, 3-methylbutanal, had formed in high concentration in the MPW emulsion but

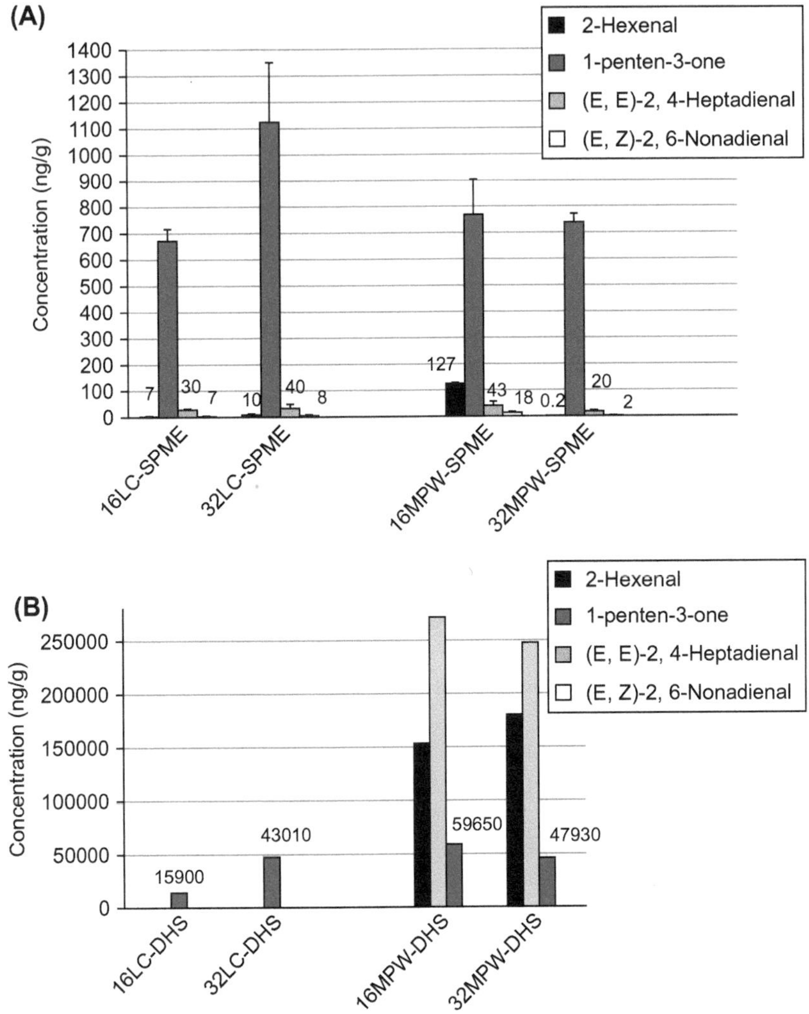

FIGURE 8.3 Comparison of n-3–derived volatiles in emulsions (LC and MPW) extracted by two different methods (A) SPME and (B) DHS after 16 days and 32 days storage (SD < 10%). *Reprinted from Lu et al. (2013).*

not in the LC emulsion. When SPME was used, 3-methylbutanal outcompeted other volatiles for space on the fiber, whereas a similar problem did not occur when using DHS since the total surface area in the Tenax tube was much larger.

A range of different aldehydes, ketones, and alcohols can be analyzed in food emulsions by GC-MS. In the literature, hexanal is often used as a marker for lipid oxidation in food emulsions rich in omega-6 fatty acids, whereas propanal is used in emulsions rich in omega-3 fatty acids. However, it is recommended to use more than one aldehyde for measuring secondary oxidation products in complex food emulsions. This is particularly the case if the aim is to be able to correlate the GC results to sensory data and/or if the aim is to evaluate the efficacy of antioxidants. In the case of the former, propanal is not a good marker for lipid oxidation as it has little impact on odor and flavor. Moreover, formation of off-odors is often due to a combination of different volatiles and cannot be explained by only one volatile. In the case of antioxidant studies, it has been shown that the same antioxidant can prevent the formation of some volatile oxidation products but promote the formation of other volatiles (Jacobsen et al., 2008). If only one volatile oxidation product is measured, there is a risk that wrong conclusions are drawn. For lipid oxidation in food emulsions rich in omega-3 fatty acids, good markers for lipid oxidation are 1-penten-3-one, 1-penten-3-ol, 4-heptenal, 2,4-heptadienal, and 2,6-nonadienal (Venkateshwarlu et al., 2004).

Oxidative stability in food emulsions can be determined by accelerated studies carried out at increased temperatures. However, temperatures have to be kept sufficiently low to prevent phase separation of the emulsion. For mayonnaise, this usually means that the temperature has to be kept well below 50–60 °C. Results from studies carried out at elevated temperatures should always be confirmed by studies carried out at realistic storage temperatures and sensory analysis should be included if the final product is aimed for human consumption.

8.6 STRATEGIES TO PREVENT OXIDATION IN FOOD EMULSIONS

A number of different approaches can be used to prevent or at least reduce the rate of lipid oxidation in real food emulsions. These approaches include: using the best possible quality of oil and emulsifier, optimization of the recipe, selecting the right type of emulsification equipment and optimizing the emulsification conditions, optimization of antioxidant addition, as well as optimization of packaging and storage conditions. In the following examples, the application of such approaches in different food emulsions will be given.

8.6.1 Mayonnaise

Mayonnaise is an o/w emulsion with a high oil content (70–80%) and a low pH (~4.0). Jacobsen et al. (1999, 2001) observed that in fish oil–enriched mayonnaise lipid oxidation increased when pH was decreased from 6.0 to 3.8. In order to obtain an acceptable microbial stability and sensory properties, pH has to be kept at approximately 4.0 and the approach to increase pH to 6.0 is therefore not viable. The high oxidative stability obtained at pH 6.0 in mayonnaise is related to the fact that egg yolk is used as an emulsifier. The iron content of egg yolk is high and at low pH iron becomes dissociated from phosvitin to which it is bound. Thereby, iron becomes more active as a catalyst of oxidation (Jacobsen et al., 1999, 2001). Hence, the low pH in combination with the iron-rich egg yolk is the driving force for the rapid oxidation processes in fish oil–enriched mayonnaise. Another strategy to improve oxidative stability of mayonnaise could therefore be to replace egg yolk by a protein with lower content of iron. This was attempted by Sørensen et al. (2010a), who replaced egg yolk with milk protein. However, the quality of the milk protein was poor, i.e., it was already oxidized before use, and this was probably the reason why milk protein was not able to increase the oxidative stability of mayonnaise.

A number of commercially available antioxidants have been evaluated in fish oil–enriched mayonnaise. So far, EDTA (6–75 mg/kg) has been found to be the most efficient antioxidant in mayonnaise (Jacobsen et al., 2008). This is due to its excellent metal chelating properties and the fact that oxidation is catalyzed by iron from egg yolk in mayonnaise. TBHQ has also been demonstrated to be efficient, particularly in combination with EDTA (Jafar et al., 1994). Other antioxidants such as propyl gallate, gallic acid, tocopherol, ascorbic acid, or a mixture of ascorbic acid (8.6% w/w), lecithin (86.2% w/w), and tocopherol (5.2% w/w) (the so-called A/L/T system) have also been evaluated in fish oil–enriched mayonnaise as reviewed by Jacobsen et al. (2008). However, these antioxidants had either little effect or were pro-oxidants. This was particularly the case for antioxidant systems containing ascorbic acid. This phenomenon was suggested to be due to the ability of ascorbic acid to promote the release of iron from egg yolk located at the oil–water interface. The released iron would then be able to decompose preexisting lipid hydroperoxides located near the oil–water interface or in the aqueous phase.

Due to its synthetic nature, EDTA is not well accepted by industry or consumers. Therefore, attempts to use natural antioxidants to prevent lipid oxidation in mayonnaise have also been made. Other metal chelators such as lactoferrin (700–2800 mg/kg) and phytic acid (15–116 mg/kg) are unfortunately far less effective in preventing lipid oxidation in mayonnaise (Nielsen et al., 2004). However, a recent study on the effect of ethyl acetate and water extracts from the brown algae *Fucus vesiculosus* in fish oil–enriched mayonnaise found that particularly the water extracts efficiently reduced lipid oxidation when added in a concentration of 2 g/kg (Hermund et al., 2015). The high efficacy of the extracts was related to their high total phenolic content, high radical scavenging activity, moderate or high metal chelating ability, and high carotenoid content.

A study has also been carried out to investigate the "cut-off" theory in mayonnaise (Alemán et al., 2015). In this study, the effect of lipophilization of caffeic acid was investigated in fish oil–enriched mayonnaise (Figure 8.4). As expected, lipophilized caffeates showed a higher antioxidant capacity than the native phenolic compound, caffeic acid. Caffeates with short to medium chain (C4, C8, and C12) were effective in mayonnaise enriched with fish oil. Further increase of the alkyl chain length led to a collapse in the antioxidant capacity

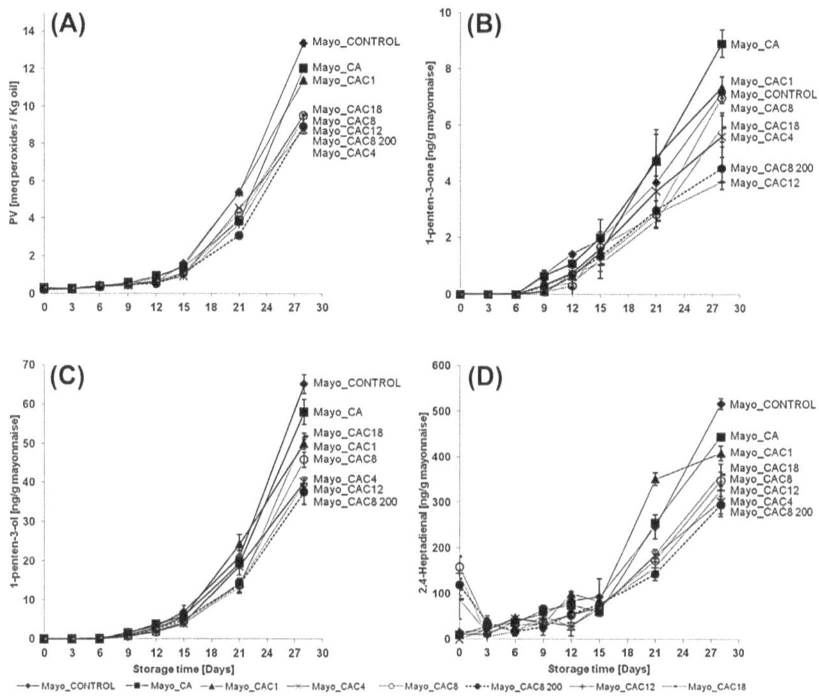

FIGURE 8.4 Concentration of peroxides measured as PV [mEq peroxides/kg oil] (A), concentration of 1-penten-3-one (B), 1-penten-3-ol (C), and 2,4-heptadienal (D) [ng/g mayonnaise] in the different fish oil–enriched mayonnaises during storage at 20 °C. Error bars indicate SD of the measurements ($n=2$ for PV and $n=3$ for volatiles compounds). *From Alemán et al. (2015).*

of the esterified phenolic compound, thus supporting the cut-off effect theory. This finding may be used to design new efficient antioxidants based on natural compounds.

8.6.2 Mayonnaise-Based Salads

Sørensen et al. (2010b) investigated whether the ingredient composition could affect lipid oxidation in fish oil–enriched shrimp and tuna salad. In shrimp salad, it was found that shrimps promoted oxidation, whereas asparagus reduced oxidation, possibly due to the content of phenolic compounds in asparagus or to the fact that they contained the metal chelator, citric acid, which had been added by the producer of canned asparagus. The effect of the ingredients in tuna salad was more unclear. Sørensen et al. (2010b) reported that addition of 1% of oregano, thyme, or rosemary to fish oil–enriched tuna salads slowed down the formation of volatiles during storage, indicating that addition of spices could increase the oxidative stability compared with the standard tuna salad with fish oil. On the basis of the volatiles data, it could be concluded that oregano had the strongest antioxidative effect, followed by rosemary and thyme. However, the dry spices led to undesirable flavors from the spices in the traditional tuna salads and further studies are therefore needed to evaluate whether extracts from such spices could be used as antioxidants.

8.6.3 Dressing

A recent study evaluated the effect of ascorbic acid and EDTA in salad dressings based on cold-pressed high-oleic acid (80%) or medium-oleic acid (40%) sunflower oil (both containing α- and β-tocopherol) or refined high linoleic acid (67%) containing no β-tocopherol but only α-tocopherol (Pavlovic et al., 2012). The high- and medium-oleic sunflower oil had a ratio between α- and β-tocopherol of ~1:1. The dressing was prepared with whey protein (1.2% in the final product) as emulsifier. The dressings prepared with high-oleic sunflower oil without addition of exogenous antioxidants were oxidatively stable after 3 months of storage at 25 °C, whereas salad dressings prepared with cold-pressed medium-oleic α- and β-tocopherol oil were more unstable. In contrast to the above-mentioned findings in mayonnaise, oxidation in the latter salad dressing was significantly inhibited by both ascorbic acid (0.50 g/kg) and EDTA (0.075 g/kg). In salad dressings containing linoleic acid α-tocopherol sunflower oil, oxidation was significantly reduced by ascorbic acid (0.25–4.00 g/kg). EDTA (0.075 g/kg) was slightly less efficient than ascorbic acid in this dressing after 12 months of storage at 4 °C. Pavlovic et al. (2012) concluded that cold-pressed high- or medium-oleic sunflower oils with increased β-tocopherol content were of superior oxidative quality compared with the refined linoleic/α-tocopherol oil in whey-based salad dressings. The different effects of ascorbic acid in mayonnaise and dressing may be explained by the fact that different emulsifiers were used. Whey proteins contain much lower levels of iron than egg yolk. Therefore, metal catalyzed oxidation and the ability of ascorbic acid to reduce Fe^{3+} to Fe^{2+} may be less important in dressing than in mayonnaise. In contrast, ascorbic acid's primary role in dressing seems to be to scavenge free radicals. However, this needs further investigation.

Wine grape pomace (WGP) has been evaluated as an antioxidant in two different commercial salad dressings, namely Italian and Thousand Island, representing the liquid and creaming type, respectively (Tseng and Zhao, 2013). Three different forms of WGP were evaluated, including dried whole grape pomace (WP), pomace liquid extract (LE), and freeze-dried liquid extract (FDE). Irrespective of the type of WGP, PV in the Italian dressing was reduced to less than 33% of the value found in the control at the end of 4 weeks of storage at 4 °C. In Thousand Island samples, WGP reduced the PV to approximately 60% of the value observed in the control, and again no differences among the different WGP fortified samples ($P > 0.05$) was observed.

Rosemary extract (RE) at different concentrations (0, 2, 4, and 6 g/L) was evaluated as an antioxidant in dressing consisting of sunflower oil, garlic, and parsley (Frutos and Hernandez-Herrero, 2005). The dressing was then treated at 140 °C for 1 h to simulate the process of baking and was later stored in an oven for 10 days at 50 °C, to promote oxidation and to evaluate the antioxidant efficacy of the RE. An antioxidant effect of RE on PV formation was observed from day 8 and on. The most efficient concentration was 0.4 g/L. Sensory tests performed on bread with the oil and garlic/parsley dressing revealed no rancid taste during the initial 6 days of storage at 50 °C. At the two lowest concentrations, the trained sensory panel could not detect any taste of RE. Hence, RE can be used as an efficient antioxidant in this type of dressing intended for spreads on bread.

Abdalla and Roozen (2001) studied the antioxidative activity of sage and oregano in different carrier systems in salad dressing. The spices were either dissolved in ethanol or homogenized with olive oil, or they were encapsulated in liposomes by ultrasonification or microfluidization. The samples were stored in the dark at ambient temperature and at 40 °C or with light exposure at ambient temperature for the ethanol- and olive oil–solubilized antioxidants or in the dark at ambient temperature, at 40 °C and at 60 °C for the encapsulated antioxidants. When olive oil was used as a carrier for oregano and sage extracts, the antioxidative effects were better than for the extracts dissolved in ethanol during storage in the dark at ambient temperature and at 40 °C. This may be due to synergistic effects of phenolic compounds present in the olive oil. However, light exposure changed the antioxidative effect of plant extracts into a pro-oxidative effect, particularly when the plant extracts were added with olive oil as carrier. The authors suggested that this may be due to photosensitizers such as chlorophyll present in the olive oil and plant extracts. Sage liposomes prepared by micofluidization showed high antioxidative effects similar to butylated hydroxytoluene (BHT) liposomes in salad dressings during storage in the dark at ambient temperature and at 40 °C. In contrast, at 60 °C none of the liposomes with the plant extracts were as efficient in preventing lipid oxidation as liposomes prepared with BHT. However, the sage liposomes prepared by microfluidization tended to be more efficient than the other liposomes (Abdalla and Roozen, 2001).

Riboflavin is a natural part of vitamin B_2 and is naturally present in several food products. Moreover, riboflavin tetrabutylate, a fat-soluble derivative of riboflavin, is used as a food additive to enrich foods with vitamin B_2 in many foods. However, riboflavin is considered to have pro-oxidative effect on certain foods in the presence of light. The effect of different levels of riboflavin (0–100 mg/kg) was investigated in 30% oil dressings, which were stored under light (2500 lux) at 25 °C for 5 days (Lee et al., 2014). Surprisingly, an increase in the concentration of added riboflavin resulted in a decrease in lipid oxidation (both PV and volatile oxidation products) of salad dressing, indicating that riboflavin acted as an antioxidant in the salad dressing under light storage conditions. The authors suggested that the presence of spices in the dressing could have protected riboflavin against oxidation, and that this in turn leads to protection of the lipids.

Let et al. (2007b) studied the effect of antioxidants in salad dressing prepared with 25% fat of which 40% was fish oil. Whey protein was used as emulsifier and a mixture of guar gum, xanthan gum, and acetylated distarch adipate was added as a stabilizer. γ-Tocopherol, ascorbyl palmitate, and EDTA were tested as antioxidants either alone or in combination. Similar to the findings for mayonnaise, EDTA was a very efficient antioxidant, which inhibited oxidation by approximately 80% at both concentrations evaluated (10 and 50 mg/kg). Moreover, γ-tocopherol was found to exert an intermediate antioxidative effect (ca. 40% inhibition) on peroxide formation irrespective of the addition level (Let et al., 2007b). However, for the volatiles, a concentration-dependent effect was observed. The lowest concentration (22 mg γ-tocopherol/kg product) resulted in the highest protection (66% inhibition). Likewise, a concentration-dependent effect was observed for ascorbyl palmitate. Ascorbyl palmitate slightly reduced oxidation at low concentrations (5 mg/kg), whereas it acted as a pro-oxidant in high concentrations (30 mg/kg). When used in combination with ascorbyl palmitate and tocopherol, the efficacy of EDTA could be further improved (Let et al., 2007b). The strong effects of the metal chelator EDTA and the poor effects of tocopherol and ascorbyl palmitate when added alone suggest that metal-catalyzed oxidation is also very important in this type of salad dressing.

8.6.4 Dairy Products

So far, this chapter has only discussed oxidative rancidity. However, rancidity may also be caused by lipid hydrolysis and both reactions are important in nonpasteurized milk. In this food emulsion, lipolytic enzymes will give rise to hydrolytic rancidity and oxidation of milk lipids will lead to oxidative rancidity. Flavor defects caused by lipid oxidation are usually referred to as "oxidized flavor" in milk and cream and as "metallic" or "tallowy" in butter (Frankel, 2005). Oxidized flavors are very easily detected in fluid milk even at very low oxidation levels (PV < 1 mEq/kg) (Frankel, 2005). This is partly due to the fact that milk has a mild flavor itself, which cannot easily mask the off-flavor of the volatile oxidation products. Exposure of milk or butterfat to sunlight produces burnt, light-activated, or sunlight off-flavors (Frankel, 2005). This flavor is caused by photo-oxidative degradation of sulfur containing amino acids including methionine, which leads to formation of carbonyls and methional. Riboflavin, which is present in substantial amounts in milk, plays an important role as a photosensitizer in this process.

8.6.4.1 Protection against Photo-oxidation

It is possible to protect dairy products against sunlight-induced lipid oxidation by packaging in materials with a light barrier. This was illustrated in a study by Larsen et al. (2009), who investigated photo-oxidation of sour cream packaged in cups with different light barrier properties. The sour cream was light exposed for 36 h with a standard fluorescent light tube, simulating storage conditions in grocery stores. Three different cups were evaluated: a white cup, a cup with medium light barrier, and a cup with high light barrier. As expected, the cup with the high light barrier resulted in the best protection against photo-oxidation. A sensory panel could thus not distinguish between sour cream stored in the high light barrier cup and sour cream stored in the dark. In contrast, they found significantly higher intensities of off-odors/flavors (sunlight, rancid, acidic) in sour cream stored in white cups or in cups with medium light barrier. Fluorescence measurements of the sour cream showed that photosensitizers were intact in the sour cream stored in the high light barrier cup, whereas they were somewhat degraded in the sour creams stored in the white cup or in the medium light barrier cup.

Similarly, development of a "stale" flavor in noncultured sweet cream butter during storage at 4 °C and −20 °C could be prevented if the butter was packaged in foil but not when packaged in wax paper (Lozano et al., 2007).

Milk can also be protected against photo-oxidation by the addition of antioxidants. This was illustrated in a study by Jung (2011) on the effects of green tea catechin (GTC, catechin content 61.4%) on the formation of hydroperoxide, malonaldehyde, and volatile compounds, as well as losses of retinol and tocopherol in whole milk during light illumination (5500 lux at sample level). GTC was highly effective in protecting milk against photo-oxidation in a dose-dependent manner. GTC added in a concentration of 25 mg/kg efficiently reduced lipid oxidation during 30-h light illumination. GTC (25 mg/kg) also significantly suppressed the losses of retinol and α-tocopherol. Comparison of the protective activity of GTC (25 mg/kg) with that of different concentrations of ascorbic acid showed that GTC was at least 4 times more effective in protecting milk against photo-oxidation than ascorbic acid, irrespective of the concentration of ascorbic acid. Jung (2011) therefore suggested that GTC could be added to milk for prevention of photo-oxidation.

8.6.4.2 Homogenization Conditions

Let et al. (2007a) and Sørensen et al. (2007) demonstrated that lipid oxidation in fish oil–enriched milk could be significantly reduced if the homogenization conditions were optimized. In both studies, it was observed that high temperature and high pressure (72 °C–22.5 MPa) resulted in less lipid oxidation than when a lower pressure and lower temperature (50 °C–5 MPa) were used. These unexpected results could be explained as follows: High temperature resulted in an increase in β-lactoglobulin adsorbed at the oil–water interface, and this was even more pronounced with higher pressure (Sørensen et al., 2007). In contrast, less casein seemed to be present at the oil–water interface with increasing pressure. It was therefore hypothesized that a combination of more β-lactoglobulin and less casein at the oil–water interface (i.e., more casein in the aqueous phase) was responsible for the increased oxidative stability. Hence, these results demonstrated that the composition of the interface is very important and that thermal oxidation may not necessarily trigger lipid oxidation. Rather, in the case of milk emulsions, this high temperature can result in unfolding of the proteins at the interface, which in turn gives the highest protection against oxidation. It was interesting that homogenization at high temperature and pressure also led to the smallest droplets. Despite this large total interfacial area, lipid oxidation was low, again demonstrating that other factors than the mere interfacial area may be more important for the oxidation rate.

8.6.4.3 Prevention of Lipid Auto-Oxidation in Dairy Products by Antioxidant Addition

Some antioxidants have been shown to be able to efficiently prevent oxidation in fish oil–enriched milk emulsions, whereas other antioxidants were less efficient, as will be discussed next.

Let et al. (2003, 2005) observed that EDTA could reduce lipid oxidation in milk enriched with 1.5% fish oil, whereas EDTA was less efficient when a lower concentration of high-quality fish oil (0.5% fish oil) was used. Timm-Heinrich et al. (2004) showed that in strawberry-flavored milk drink containing 5% fat of which 0.5% (absolute value) was enzyme-modified fish oil, EDTA effectively reduced oxidative flavor deterioration.

The antioxidant effect of another metal chelator, lactoferrin (a milk protein), added in different concentrations (500–2000 mg/kg) in the same type of milk drink prepared with enzyme-modified fish oil or sunflower oil was evaluated by Nielsen et al. (2004). They found that even at its optimum addition levels (1000 mg/kg), lactoferrin only slightly reduced oxidation, whereas it had slightly pro-oxidative effects at other concentration levels. The reason for the poor ability of lactoferrin to inhibit lipid oxidation may be that it does not efficiently chelate iron, that iron-catalyzed oxidation is not so important in this system, or that lactoferrin was not heat stable (Nielsen et al., 2004).

Pro-oxidative effects have also been observed for a combination of α- and γ-tocopherol in milk enriched with 0.5% fish oil (Let et al., 2005). The pro-oxidative effects of the combination of α- and γ-tocopherol was suggested to be due to a too high total concentration of α-tocopherol in this emulsion, resulting from the high level of endogenous tocopherol in the fish oil used for the milk system. In contrast, pure α-tocopherol in a concentration of 1.1 mg/kg or γ-tocopherol in concentrations of 0.8 or 3.3 mg/kg had weak antioxidative effects. When γ-tocopherol in an intermediate concentration of 1.7 mg/kg was used alone a much stronger antioxidative effect was observed.

A strong antioxidative effect was also observed when ascorbyl palmitate in a concentration of only 1.5 mg/kg in milk with 1.5% fat was used (Let et al., 2005). Interestingly, when ascorbyl palmitate was added to the strawberry-flavored milk drink system, ambiguous results were obtained. Ascorbyl palmitate was thus found to be able to reduce the formation of lipid hydroperoxides and certain volatiles such as heptadienal when added in high concentrations (15 and 30 mg/kg). In contrast, ascorbyl palmitate promoted the formation of hexenal and 2,6-nonadienal as well as fishy off-flavors when added in concentrations of 7.5 mg/kg and above, but reduced the formation of these two volatiles when added at a lower concentration of 3.75 mg/kg (Jacobsen et al., 2008). The different efficacies observed for ascorbyl palmitate in dependence of the composition of the milk emulsions could be due to the different ascorbyl palmitate concentrations applied or to the different compositions of the two milk systems (Jacobsen et al., 2008). The milk drink contained mono- and diglycerides, carageenan, and guar gum, which may interact at the o/w interface and thereby reduce the ability of ascorbyl palmitate to regenerate tocopherol, whereas no additives were added to the milk with 1.5% fat.

The antioxidative properties of GTE and RE were recently evaluated in milk enriched with 1.5% fish oil (Sørensen, 2010). Both types of extracts efficiently reduced lipid oxidation as evaluated by PV and formation of volatile oxidation products. Unfortunately, addition of the plant extracts resulted in undesirable off-flavors of rosemary and green tea and they can therefore not be used as antioxidants in this food system without further improvement of their flavor properties.

Another omega-3 PUFA (2 g linseed oil/100 g)-enriched dairy product, namely sterilized dairy beverage, was studied by Giroux et al. (2010), who evaluated the antioxidative effect of Maillard products. Maillard products were obtained by heating aqueous dispersions of milk protein and sugars at 110 °C for 10 min. Nonheated blends of protein and sugar were evaluated as controls. Various proportions of lactose, sucrose, and their monosaccharide mixtures glucose/galactose and glucose/fructose were evaluated for the production of Maillard products. These preparations were added at low concentration (5 mL/100 mL) to the omega-3 PUFA–enriched dairy beverages. The milk beverage contained 3.6 g protein/100 g and was supplemented with 0.001 g/100 g Fe as $FeSO_4$. Nonheated protein–sugar blends did not reduce lipid oxidation during sterilization, whereas preheated blends containing monosaccharides were highly effective in reducing lipid oxidation. The authors therefore concluded that adding a small fraction of milk protein as preheated protein–sugar blend in functional dairy beverage formulations is a promising approach to prevent the oxidation of omega-3 PUFA during sterilization treatment (Giroux et al., 2010).

The antioxidative effect of other natural antioxidants, namely oregano extract and oregano essential oil, was evaluated in a dairy beverage formulation consisting of skim milk powder and water. This dairy product was also supplemented with $FeSO_4$ and linseed oil (2 g/100 g beverage) (Boroski et al., 2012). They found that both oregano extract and oregano essential oil could reduce heat- or light-induced lipid oxidation. The oregano extract was more efficient than oregano essential oil.

The effects of lipophilization of polar antioxidants on their antioxidative effects in fish oil–enriched milk have been evaluated by Sørensen et al. (2012a, 2015) and Aléman et al. (2015). In the first study, the antioxidative effects of dihydrocaffeic acid, rutin, two dihydrocaffeates (octyl and oleyl caffeate), and two rutin esters (rutin laurate and rutin palmitate) were evaluated. For both types of compounds, the medium-chain ester was a better antioxidant than the long-chain ester and the phenol (Sørensen et al., 2012a). In the second study, caffeic acid and its esters were evaluated. It was observed that the short-chain caffeates (with chain lengths C1 or C4) more efficiently inhibited lipid oxidation than caffeic acid or caffeates with longer chain length (C8–C20) (Alemán et al., 2015). It was somewhat surprising that the optimum chain length was so short. In a third study, Sørensen et al. (2015) evaluated the antioxidative effect of ferulic acid and its esters. Again, the strongest antioxidative effect was observed for the short chain esters (C1 and C4). In contrast to the previous study at the chosen antioxidant concentration, a strong pro-oxidative effect was observed with medium alkyl ferulates (i.e., octyl ferulate and dodecyl ferulate), whereas the longer alkyl ferulates (C16 and C18) did not perform as strong pro-oxidants. The authors proposed that both the exact conformation of the ferulates and their partitioning could influence their efficacy.

Yogurt has previously been described as a food system with high oxidative stability due to its content of antioxidative peptides formed during its fermentation (Farvin et al., 2010a,b). Nevertheless, Tseng and Zhao (2013) demonstrated that it was possible to use WGP to reduce lipid oxidation in yogurt produced from reduced fat milk (2% milk fat) with 4% sugar and fermented by a combination of *Streptococcus thermophilus*, *Lactobacillus delbrueckii* subsp. *Bulgaricus*, *Lactobacillus acidophilus*, and *Bifidobacterium lactis*. At the end of 3 weeks of storage, 3% of dried whole powder WP had lower PV than the control and yogurts prepared with lower concentrations of WP. However, sensory data suggested that addition of 1% WP resulted in better sensory perception than higher levels of WP (Tseng and Zhao, 2013).

Huvaere et al. (2011) investigated the effects of GTE and EDTA on light induced lipid oxidation in reduced fat soft cheeses (0.2% or 6% fat). The antioxidants were either used individually or in combination. Addition of 50 mg/kg EDTA to samples with higher fat content (6%) was ineffective as antioxidant, but 750 mg/kg GTE (alone or in combination with EDTA) strongly reduced levels of hexanal and heptanal. As the accumulation of primary lipid hydroperoxides was not affected by GTE, the antioxidative activity was ascribed to the scavenging of hexanal and heptanal precursors by GTE. The authors concluded that the presence of GTE effectively reduced formation of lipid-derived aldehydes in light-exposed reduced fat soft cheese, possibly resulting in improved flavor stability.

8.6.5 Margarine and Spreads

Margarine and spreads are w/o emulsions in which lipid oxidation takes place in the continuous liquid oil phase. Fat crystals are also dispersed in the liquid oil phase. Therefore, margarine and spreads have a very different physical structure compared with the other types of emulsions discussed in this chapter. Due to the differences in the physical structure of margarine and spreads compared with o/w emulsions, lipid oxidation mechanisms may be different in margarine and spreads. Unfortunately, there are only few studies available in the scientific literature on the effect of other factors than antioxidant addition. A few examples on the effect of different antioxidants in margarine and spreads are summarized here.

Filip et al. (2009) evaluated the antioxidant efficacy of L-(+)-ascorbic acid, ascorbyl palmitate, and DL-α-tocopherol in margarine. They observed that increasing polarity and decreasing molecular size of the antioxidant had a positive influence on lipid oxidation. Thus, DL-α-tocopherol was the least effective antioxidant, whereas ascorbic acid was the most effective antioxidant. Ascorbyl palmitate had similar effects as ascorbic acid, although it was less efficient. The authors concluded that 0.02% ascorbic acid or ascorbyl palmitate reduced the extent of lipid oxidation in margarine when the margarine contained endogenous tocopherols in the fat blend used (Filip et al., 2009). According to the polar paradox one would expect ascorbyl palmitate to be more efficient in emulsions than ascorbic acid, but w/o emulsions behave more like bulk oils with respect to the effects of antioxidants and the findings from this study seem to confirm this hypothesis.

Kamkar et al. (2011) evaluated the addition of individual tocopherol homologs (α, γ, and δ) to the oil phase of margarine in concentrations of 0, 100, 250, and 500 mg/kg. The antioxidative activity was evaluated through oven test (at $60 \pm 1\,^{\circ}C$ in 5 days and measuring PV and AV) and Rancimat test ($110 \pm 1\,^{\circ}C$). Antioxidative activity increased progressively as concentration of δ-tocopherol increased. As the concentration of γ-tocopherol increased from 0 to 100, and 250 mg/kg, oxidation of oil phase decreased. At 500 mg/kg, α-tocopherol acted as a pro-oxidant and γ-tocopherol did not have any noticeable effect, but δ-tocopherol acted as an antioxidant. The order of antioxidative activity of tocopherols was: α- < γ- < δ- (Kamkar et al., 2011). The results of this study thus confirmed the findings by Filip et al. (2009) who found that α-tocopherol was not an efficient antioxidant in margarine.

The effectiveness of fat- and water-soluble antioxidants on the oxidative stability of omega-3 PUFA-rich table spreads, produced using novel multiple emulsion technology was recently investigated (O'Dwyer et al., 2012). Table spreads were produced by dispersing an o/w emulsion (500 g/kg 85 camelina/15 fish oil blend) in a hardstock/rapeseed oil blend, using sodium caseinate and polyglycerol polyricinoleate as emulsifiers. The water-soluble antioxidant was GTE added in a concentration of 0.75 g/kg, and the oil-soluble antioxidant was α-tocopherol (250 mg/kg). Combinations of both antioxidants were also evaluated. Similar to the findings by Filip et al. (2009) and Kamkar et al. (2011), α-tocopherol could not efficiently inhibit oxidation as it resulted in higher levels of lipid hydroperoxide values than the control. In contrast, GTE significantly reduced peroxide values during storage at 5 °C. The authors suggested that GTE improved the oxidative stability of omega-3 PUFA–rich o/w and o/w/o emulsions as a result of catechins binding to the protein during storage. Moreover, they proposed that α-tocopherol levels incorporated into emulsions may have been too high, as a pro-oxidant effect was observed (O'Dwyer et al., 2012).

8.7 CONCLUSIONS

Lipid oxidation in food emulsions is influenced by a large number of factors such as droplet size, composition of the interfacial layer including its electrical charge, colloid structures in the aqueous phase, pH, composition of other ingredients than oil and water, etc. These factors probably influence lipid oxidation in o/w emulsions through several mechanisms, which are not fully understood. Likewise, the effect of antioxidants in food emulsions is also influenced by several factors such as the concentration, polarity, and conformation of the antioxidant, as well as its partitioning between the different phases and interaction with the interface and colloid structures in the aqueous phase. Therefore, the effect of the same antioxidant can vary substantially from food system to food system. Hence, it is impossible to predict the antioxidant efficacy in one food system from the effects observed in another system. Nevertheless, this chapter has demonstrated that GTE/green tea catechins exerted strong antioxidative effects in food emulsions with very different compositions (whole milk, soft cheese, spreads) and with different oxidation mechanisms (photo-oxidation vs auto-oxidation). This indicates that GTE contain antioxidative compounds, which exert their action by different mechanisms. More studies are required to fully understand how these and other natural extracts from land and marine plants exert their antioxidative effect in real food emulsions. More studies are also needed on the effect of lipophilized antioxidants in food emulsions in order to understand which factors determine the optimal chain length of these so-called phenolipids.

REFERENCES

Abdalla, A.E., Roozen, J.P., 2001. The effects of stabilised extracts of sage and oregano on the oxidation of salad dressings. Eur. Food Res. Technol. 212, 551–560.

Alemán, M., Bou, R., Guardiola, F., Durand, E., Villeneuve, P., Jacobsen, C., Sørensen, A.-D.M., 2015. Antioxidative effect of lipophilized caffeic acid in fish oil enriched mayonnaise and milk. Food Chem. 167, 236–244.

Battino, R. (Ed.), 1987. Oxygen and Ozone. IUPAC Solubility Data Series, vol. 7. Pergamon, Oxford.

Berton-Carabin, C., Ropers, M.-H., Genot, C., 2014. Lipid oxidation in oil-in-water emulsions: involvement of the interfacial layer. Compr. Rev. Food Sci. Food Saf. 13, 945–977.

Berton, C., Ropers, M.H., Viau, M., Genot, C., 2011. Contribution of the interfacial layer to the protection of emulsified lipids against oxidation. J. Agric. Food Chem. 59, 5052–5061.

Boroski, M., Giroux, H.J., Sabik, H., Petit, H.V., Visentainer, J.V., Matumoto-Pintro, P.T., Britten, M., 2012. Use of oregano extract and essential oil as antioxidants in functional dairy beverage formulations. LWT – Food Sci. Technol. 47, 167–174.

Dickinson, E., 1992. An Introduction to Food Colloids. Oxford University Press, Oxford.

Donnelly, J.L., Decker, E.A., McClements, D.J., 1998. Iron-catalyzed oxidation of menhaden oil as affected by emulsifiers. J. Food Sci. 63, 997–1000.

EFSA Panel on Dietetic Products, Nutrition and Allergies (NDA), 2010. Scientific opinion on dietary reference values for fats, including saturated fatty acids, polyunsaturated fatty acids, monounsaturated fatty acids, trans-fatty acids, and cholesterol. EFSA J. 8, 1461–1568.

Fang, Y., Dalgleish, D.G., 1998. The conformation of alpha-lactalbumin as a function of pH, heat treatment and adsorption at hydrophobic surfaces studied by FTIR. Food Hydrocolloids 12, 121–126.

Fang, Y., Dalgleish, D.G., 1997. Conformation of beta-lactoglobulin studied by FTIR: effect of pH, temperature, and adsorption to the oil-water interface. J. Colloid Interface Sci. 196, 292–298.

Faraji, H., McClements, D.J., Decker, E.A., 2004. Role of continuous phase protein on the oxidative stability of fish oil-in-water emulsions. J. Agric. Food Chem. 52, 4558–4564.

Farvin, K.H.S., Baron, C.P., Nielsen, N.S., Jacobsen, C., 2010a. Antioxidant activity of yoghurt peptides: Part 1 – in vitro assays and evaluation in ω-3 enriched milk. Food Chem. 123, 1081–1089.

Farvin, K.H.S., Baron, C.P., Nielsen, N.S., Otte, J., Jacobsen, C., 2010b. Antioxidant activity of yoghurt peptides: Part 2 – characterisation of peptide fractions. Food Chem. 123, 1090–1097.

Filip, V., Hradkova, I., Smidrkal, J., 2009. Antioxidants in margarine emulsions. Czech J. Food Sci. 27, S9–S11.

Frankel, E.N., 2005. Lipid Oxidation. The Oily Press Dundee, Scotland.

Frankel, E.N., Satue-Gracia, T., Meyer, A.S., German, J.B., 2002. Oxidative stability of fish and algae oils containing long-chain polyunsaturated fatty acids in bulk and in oil-in-water emulsions. J. Agric. Food Chem. 50, 2094–2099.

Frankel, E.N., Huang, S.-W., Kanner, J., German, J.B., 1994. Interfacial phenomena in the evaluation of antioxidants: bulk oils vs emulsions. J. Agric. Food Chem. 42, 1054–1059.

Frutos, M.J., Hernandez-Herrero, J.A., 2005. Effects of rosemary extract (*Rosmarinus officinalis*) on the stability of bread with an oil, garlic and parsley dressing. LWT – Food Sci. Technol. 38, 651–655.

Giroux, H.J., Houde, J., Britten, M., 2010. Use of heated milk protein-sugar blends as antioxidant in dairy beverages enriched with linseed oil. LWT – Food Sci. Technol. 43, 1373–1378.

Haahr, A.-M., Jacobsen, C., 2008. Emulsifier type, metal chelation and pH affect oxidative stability of omega-3 enriched emulsions. Eur. J. Lipid Sci. Techol. 110, 949–961.

Hermund, D.B., Yesiltas, B., Honold, P., Jónsdóttir, R., Kristinsson, H.G., Jacobsen, C., 2015. Characterisation and antioxidant evaluation of Icelandic *F. vesiculosus* extracts in vitro and in fish oil-enriched milk and mayonnaise. J. Funct. Foods (accepted).

Horn, A.F., 2012. Factors Influencing the Effect of Milk Based Emulsifiers on Lipid Oxidation in Omega-3 Emulsions (Ph.D. thesis). Technical University of Denmark.

Horn, A.F., Barouh, N., Nielsen, N.S., Baron, C.P., Jacobsen, C., 2013. Homogenization pressure and temperature affect composition of proteins in the aqueous phase and oxidative stability of fish oil-in-water emulsions. J. Am. Oil Chem. Soc. 90, 1541–1550.

Horn, A.F., Nielsen, N.S., Jensen, L.S., Horsewell, A., Jacobsen, C., 2012a. The choice of homogenization equipment affects lipid oxidation in emulsions. Food Chem. 134, 803–810.

Horn, A.F., Nielsen, N.S., Jacobsen, C., 2012b. Iron-mediated lipid oxidation in 70% fish oil-in-water emulsions: effect of emulsifier type and pH. Internat. J. Food Sci. Technol. 47, 1097–1108.

Horn, A.F., Nielsen, N.S., Andersen, U., Søgaard, L.H., Horsewell, A., Jacobsen, C., 2011. Oxidative stability of 70% fish oil-in-water emulsions: impact of emulsifiers and pH. Eur. J. Lipid Sci. Technol. 113, 1243–1257.

Hu, M., McClements, D.J., Decker, E.A., 2003a. Lipid oxidation in corn oil-in-water emulsions stabilized by casein, whey protein isolate, and soy protein isolate. J. Agric. Food Chem. 51, 1696–1700.

Hu, M., McClements, D.J., Decker, E.A., 2003b. Impact of whey protein emulsifiers on the oxidative stability of salmon oil-in-water emulsions. J. Agric. Food Chem. 51, 1435–1439.

Hunt, J.A., Dalgleish, D.G., 1994. Effect of pH on the stability and surface-composition of emulsions made with whey-protein isolate. J. Agric. Food Chem. 42, 2131–2135.

Huvaere, K., Nielsen, J.H., Bakman, M., Hammershøj, M., Skibsted, L.H., Sørensen, J., Vognsen, L., Dalsgaard, T.K., 2011. Antioxidant properties of green tea extract protect reduced fat soft cheese against oxidation induced by light exposure. J. Agric. Food Chem. 59, 8718–8723.

Jacobsen, C., Let, M.B., Nielsen, N.S., Meyer, A.S., 2008. Antioxidant strategies for preventing oxidative flavour deterioration of foods enriched with n-3 polyunsaturated lipids: a comparative evaluation. Trends Food Sci. Technol. 19 (2), 76–93.

Jacobsen, C., Timm, M., Meyer, A.S., 2001. Oxidation in fish oil enriched mayonnaise: ascorbic acid and low pH increase oxidative deterioration. J. Agric. Food Chem. 49, 3947–3956.

Jacobsen, C., Hartvigsen, K., Lund, P., Thomsen, M.K., Skibsted, L., Adler-Nissen, J., Hølmer, G., Meyer, A.S., 2000. Oxidation in fish oil enriched mayonnaise: 3. Assessment of the influence of the emulsion structure on oxidation by discriminant partial least squares regression analysis. Eur. Food Res. Technol. 211, 86–98.

Jacobsen, C., Adler-Nissen, J., Meyer, A.S., 1999. The effect of ascorbic acid on iron release from the emulsifier interface and on the oxidative flavor deterioration in fish oil enriched mayonnaise. J. Agric. Food Chem. 47, 4917–4926.

Jacobsen, C., 1999. Sensory impact of lipid oxidation in complex food systems. Fett/Lipid 12, 484–492.

Jafar, S.S., Hultin, H.O., Bimbo, A.P., Crowther, J.B., Barlow, S.M., 1994. Stabilization by antioxidants of mayonnaise made from fish oil. J. Food Lipids 1, 295–311.

Jana, A.K., Agarwal, S., Chatterjee, S.N., 1990. Membrane lipid-peroxidation by ultrasound – mechanisms and implications. J. Biosci. (Bangalore) 15, 211–215.

Jung, M.Y., 2011. Effects of green tea catechin on the lipid oxidation, volatile compound formation, and losses of retinol and α-tocopherol in whole milk during light illumination as compared with ascorbic acid. Food Sci. Biotechnol. 20, 1425–1434.

Kamkar, A., Azizkhani, M., Nejad, A., 2011. Effects of tocopherols on oxidative stability of margarine. J. Chem. Soc. Pak. 33, 134–137.

Laguerre, M., Sørensen, A.-D.M., Bayrasy, C., Lecomte, J., Jacobsen, C., Decker, E.A., Villeneuve, P., 2013. Role of hydrophobicity on antioxidant activity in lipid dispersions from the polar paradox to the cut-off theory. In: Logan, A., Nienaber, U., Pan, X. (Eds.), Lipid Oxidation: Challenges in Food Systems. AOCS Press, Champaign, IL, pp. 261–296.

Laguerre, M., Giraldo, L.J.L., Lecomte, J., Figueroa-Espinoza, M.C., Baréa, B., Weiss, J., Decker, E.A., Villeneuve, P., 2010. Relationship between hydrophobocity and antioxidant ability of "phenolipids" in emulsion: a parabolic effect of the chain length of rosmarinate esters. J. Agric. Food Chem. 58, 2869–2876.

Laguerre, M., Giraldo, L.J.L., Lecomte, J., Figueroa-Espinoza, M.C., Barea, B., Weiss, J., Decker, E.A., Villeneuve, P., 2009. Chain length affects antioxidant properties of chlorogenate esters in emulsion: the cutoff theory behind the polar paradox. J. Agric. Food Chem. 57, 11335–11342.

Larsen, H., Tellefsen, S.B.G., Dahl, A.V., 2009. Quality of sour cream packaged in cups with different light barrier properties measured by fluorescence spectroscopy and sensory analysis. J. Food Sci. 74 (8), S345–S350.

Lee, S.H., Lefevre, T., Subirade, M., Paquin, P., 2007. Changes and roles of secondary structures of whey protein for the formation of protein membrane at soy oil/water interface under high-pressure homogenization. J. Agric. Food Chem. 55, 10924–10931.

Lee, Y.-H., Lee, J., Min, D.B., Pascall, M.A., 2014. Effect of riboflavin on the photo-oxidative stability of vegetable oil in salad dressing. Food Chem. 152, 349–354.

Let, M.B., Jacobsen, C., Sørensen, A.D.M., Meyer, A.S., 2007a. Homogenization conditions affects the oxidative stability of fish oil enriched milk emulsions: lipid oxidation. J. Agric. Food Chem. 55, 1773–1780.

Let, M.B., Jacobsen, C., Meyer, A.S., 2007b. Ascorbyl palmitate, gamma-tocopherol, and EDTA affect lipid oxidation in fish oil enriched salad dressing differently. J. Agric. Food Chem. 55, 2369–2375.

Let, M.B., Jacobsen, C., Pham, K.A., Meyer, A.S., 2005. Protection against oxidation of fish oil-enriched milk emulsions through addition of rapeseed oil or antioxidants. J. Agric. Food Chem. 53, 5429–5437.

Let, M.B., Jacobsen, C., Frankel, E.N., Meyer, A.S., 2003. Oxidative flavour deterioration of fish oil enriched milk: effects of oil type and Ethylenediaminetetraacetate (EDTA). Eur. J. Lipid Sci. Technol. 105, 518–528.

Lethuaut, L., Metro, F., Genot, C., 2002. Effect of droplet size on lipid oxidation rates of oil-in-water emulsions stabilized by protein. J. Am. Oil Chem. Soc. 79, 425–430.

Lozano, P.R., Miracle, E.R., Krause, A.J., Drake, M., Cadwallader, K.R., 2007. Effect of cold storage and packaging material on the major aroma components of sweet cream butter. J. Agric. Food Chem. 55, 7840–7846.

Lu, F.S.H., Nielsen, N.S., Jacobsen, C., 2013. Comparison of two methods for extraction of volatiles from marine PL emulsions. Eur. J. Lipid Sci. Technol. 115, 246–251.

Mao, L.K., Yang, J., Xu, D.X., Yuan, F., Gao, Y.X., 2010. Effects of homogenization models and emulsifiers on the physicochemical properties of β-carotene nanoemulsions. J. Dispersion Sci. Technol. 31, 986–993.

McClements, D.J., Decker, E.A., 2000. Lipid oxidation in oil-in-water emulsions: impact of molecular environment on chemical reactions in heterogeneous food systems. J. Food Sci. 65, 1270–1282.

McClements, D.J., 2008. Lipid-based emulsions and emulsifiers. In: Akoh, C.K., Min, D.B. (Eds.), Food Lipids. Chemistry, Nutrition, and Biotechnology, third ed. CRC Press, Boca Raton, pp. 63–97.

Mei, L.Y., Decker, E.A., McClements, D.J., 1998. Evidence of iron association with emulsion droplets and its impact on lipid oxidation. J. Agric. Food Chem. 46, 5072–5077.

Miyashita, K., Tateda, T., Ota, T., 1994. Oxidative stability of polyunsaturated fatty acids in aqueous solution. Biosci. Biotechnol. Biochem. 57, 1638–1640.

Miyashita, K., Nara, E., Ota, T., 1993. Oxidative stability of free fatty acids mixtures from soybean, linseed, and sardine oils in aqueous solution. Biosci. Biotechnol. Biochem. 60, 315–318.

Nakaya, K., Ushio, H., Matsukawa, S., Shimizu, M., Ohshima, T., 2005. Effects of droplet size on the oxidative stability of oil-in-water emulsions. Lipids 40, 501–507.

Nielsen, N.S., Horn, A.F., Jacobsen, C., 2013. Effect of emulsifier type, pH and iron on oxidative stability of 5% fish oil-in-water emulsions. Eur. J. Lipid Sci. Technol. 115, 874–889.

Nielsen, N.S., Petersen, A., Meyer, A.S., Timm-Heinrich, M., Jacobsen, C., 2004. Effects of lactoferrin, phytic acid and EDTA on oxidation in two food emulsions enriched with long chain polyunsaturated fatty acids. J. Agric. Food Chem. 52, 7690–7699.

O'Dwyer, S.P., Beirne, D.O., Eidhin, D.N., Kennedy, B.T.O., 2012. Effects of green tea extract and alpha-tocopherol on the lipid oxidation rate of omega-3 oils, incorporated into table spreads, prepared using multiple emulsion technology. J. Food Sci. 77, N58–N65.

Pavlovic, M.D., Pucarevic, M., Micovic, V., Miroslav Zivic, M., Zlatanovic, S., Gorjanovic, S., Gvozdenovic, J., 2012. Influence of sunflower oil qualities and antioxidants on oxidative stability on whey-based salad dressings. Acta Chim. Slov. 59, 42–49.

Ponginebbi, L., Nawar, W.W., Chinachoti, P., 1999. Oxidation of linoleic acid in emulsions: effect of substrate, emulsifier, and sugar concentration. J. Am. Oil Chem. Soc. 76, 131–138.

Porter, W.L., 1993. Paradoxical behavior of antioxidants in food and biological systems. Toxicol. Ind. Health 9, 93–122.

Riesz, P., Kondo, T., 1992. Free-radical formation induced by ultrasound and its biological implications. Free Radic. Biol. Med. 13, 247–270.

Stapelfeldt, H., Skibsted, L.H., 1999. Pressure denaturation and aggregation of beta-lactoglobulin studied by intrinsic fluorescence depolarization, Rayleigh scattering, radiationless energy transfer and hydrophobic fluoroprobing. J. Dairy Res. 66, 545–558.

Sørensen, A.-D.M., Lyneborg, K.S., Villeneuve, P., Jacobsen, C., 2015. Alkyl chain length impacts the antioxidative effect of lipophilized ferulic acid in fish oil enriched milk. J. Funct. Foods 18, 959–967. http://dx.doi.org/10.1016/j.jff.2014.04.008.

Sørensen, A.-D.M., Petersen, L.K., de Diego, S., Nielsen, N.S., Lue, B.-M., Yang, Z., Xu, X., Jacobsen, C., 2012a. The antioxidative effect of lipophilized rutin and dihydrocaffeic acid in fish-oil-enriched milk. Eur. J. Lipid Sci. Technol. 114, 434–445.

Sørensen, A.-D.M., Nielsen, N.S., Yang, Z., Xu, X., Jacobsen, C., 2012b. The effect of lipohilization of dihydrocaffeic acid on its antioxidative properties in fish-oil-enriched emulsion. Eur. J. Lipid Sci. Technol. 114, 134–145.

Sørensen, A.-D.M., Nielsen, N.S., Hyldig, G., Jacobsen, C., 2010a. The influence of emulsifier type in fish oil enriched light mayonnaise. Eur. J. Lipid Sci. Technol. 112, 1012–1023.

Sørensen, A.D.M., Nielsen, N.S., Jacobsen, C., 2010b. Oxidative stability of fish oil enriched mayonnaise based salads. Eur. J. Lipid Sci. Technol. 112, 476–487.

Sørensen, A.-D.M., 2010. The Influence of Ingredients or Lipophilized Antioxidants on the Oxidative Stability of Fish Oil Enriched Food Systems (Ph.D. thesis). National Food Institute, Technical University of Denmark.

Sørensen, A.D.M., Baron, C.P., Let, M.B., Brüggemann, D., Pedersen, L.R.L., Jacobsen, C., 2007. Homogenization conditions affects the oxidative stability of fish oil enriched milk emulsions: oxidation linked to changes in the protein composition at the oil-water interface. J. Agric. Food Chem. 55, 1781–1789.

Timm-Heinrich, M., Xu, X., Nielsen, N.S., Jacobsen, C., 2004. Oxidative stability of mayonnaise and milk drink produced with structured lipids based on fish oil and caprylic acid. Eur. Food Res. Technol. 219, 32–41.

Tseng, A., Zhao, Y.Y., 2013. Wine grape pomace as antioxidant dietary fibre for enhancing nutritional value and improving storability of yogurt and salad dressing. Food Chem. 138, 356–365.

Venkateshwarlu, G., Let, M.B., Meyer, A.S., Jacobsen, C., 2004. Chemical and olfactometric characterization of volatile flavor compounds in a fish oil enriched milk emulsion. J. Agric. Food Chem. 52, 311–317.

Weast, R.C., Astle, M.J., Beyer, W.H., 1987. CRC Handbook of Chemistry and Physics, 67th ed. CRC Press, Boca Raton, FL.

Chapter 9

Oxidative Stability and Shelf Life of Low-Moisture Foods

Min Hu
DuPont Nutrition and Health, New Century, KS, USA

9.1 INTRODUCTION

A variety of low-moisture or dry foods and ingredients have been widely applied in the food and pet food industry. This is due to the fact that the low-moisture or dry foods and ingredients are more oxidatively stable and have a longer shelf life than the liquid counterparts due to the removal of water, minimizing the microbial growth and chemical deterioration including lipid oxidation. Low-moisture foods are also more convenient to use and cheaper to transport because of reduced weight and volume.

Foods are generally categorized into low-, intermediate-, and high-moisture foods based on their water activity (a_w) values and overall moisture contents by weight (Knechtges, 2012). According to the author, low-moisture foods may have a_w values of 0.00–0.60 and less than 25% moisture content; intermediate-moisture foods have a_w values of 0.60–0.85 and 15–50% moisture content; and high-moisture foods a_w values of greater than 0.85 and more than 50% moisture content. Examples of low-moisture foods are whole milk powders (WMPs) with a_w being around 0.20, egg powders around 0.30, noodle around 0.50, crackers about 0.10, and dry kibble about 0.35–0.55. Examples of intermediate-moisture foods are flour with a_w being 0.70, raw peanuts 0.60–0.70, and dried fruit 0.70. Most fresh foods are high-moisture ones, with the moisture contents between 60% and 95% and a_w close to 1 (Nelson and Labuza, 1992). According to Nelson and Labuza (1992), the deterioration of high-moisture foods is generally caused by microbes or enzymes, which could deteriorate the foods first, since auto-oxidation (except enzyme-mediated oxidation) is not the principal source of deterioration of the high-moisture food. In this chapter, we will focus on low-moisture (or dry) foods.

In the past few decades, lipid oxidation in bulk oils/fats and food emulsions containing oils and fats has been widely studied (Choe and Min, 2006; McClments and Decker, 2000; Chaiyasit et al., 2007; Laguerre et al., 2015), and

Oxidative Stability and Shelf Life of Foods Containing Oils and Fats
http://dx.doi.org/10.1016/B978-1-63067-056-6.00009-4

the oxidation of neat lipids is relatively well understood (Skibsted, 2010). But the progress of research on the mechanism of lipid oxidation in low-moisture (dry) foods has been relatively slow, and the analytical and research methods involving low-moisture foods need to be further studied and developed (Eldin, 2010). The main reason could be that the complex, multiphase low-moisture foods are different from monophase bulk oils/fats and multiphase food emulsions in chemical composition, molecular structure, physical state, and physicochemical property. Thus, lipid oxidation in solid phase such as low-moisture foods may not be the same as that in liquid phase such as bulk oil/fat, oil-in-water, and water-in-oil emulsion (Hu, 2014). Obviously, a deep understanding of the oxidative stability and shelf life of low-moisture foods is needed in order to develop various oxidatively stable low-moisture food products with expected shelf life.

There are a variety of oil/fat-containing low-moisture foods in the food and pet food market. Table 9.1 summarizes the principal low-moisture (or dry) foods and ingredients. In general, low-moisture foods containing oils/fats may be roughly divided into six groups. The first group is the dried powder products produced by drying food emulsions or mixing different powder ingredients.

TABLE 9.1 Main Low-Moisture (Dry) Foods and Ingredients Containing Oils and Fats

Name	Examples
Dried powders	Milk powders, egg powders, microencapsulated oil powders, infant formula powders, soup powders
Extruded foods	Breakfast cereals (e.g., corn and potato chips, tortilla chips, multigrain chips), snack foods, baby foods, dry pet foods
Baked foods	Crackers, cookies, biscuits
Fried foods	Fried breakfast cereals (e.g., fried potato and corn chips, tortilla chips), fried snacks, fried chicken, fish, meatballs
Protein meals	Fish meals, poultry meals, chicken meals, chicken byproduct meals, meat and bone meals, lamb meals, pork meals, fish powders
Nuts and nut products	Pecans, walnuts, peanuts
Dehydrated foods	Dehydrated soup powders, meat, ham, mutton, beef, fish, shrimp, instant noodles
Dried microalgae	Dried laver, *Phaeodactylum tricornutum*, and *Spirulina platensis*

Microencapsulated oils and fish powders (Artharn et al., 2009; Rahman et al., 2009; Shaviklo et al., 2010), infant formula powders (Kazemzadeh, 2001), dried laver (Choe and Oh, 2013), and dried *Phaeodactylum tricornutum* (Ryckebosch et al., 2011).

Milk and dried microencapsulated oil powders are made by drying the food emulsions using freeze- or spray-drying technology and dried eggs or yolk powders by dehydrating liquid eggs or yolks. The microstructures of freeze-dried powders may be different from those of spray-dried ones. Dehydrated soup powders are another type of powder products produced by mixing different powder ingredients. The microstructure of the soup powder products is distinct from that of the emulsion-based powder products. Thus, they may have different oxidative stability and shelf life.

The second group is the dried foods with various shapes and sizes made by extrusion and drying process technologies such as frying and baking processes. Examples are breakfast cereals, snack foods, baked foods, and dry pet foods. A potato chip can be produced using three different processing methods, thereby producing three varying matrices. Extruded, fried, and baked potato chips have different microstructures (matrices), physicochemical properties, oxidative stability, and shelf life. In the food industry, these processing technologies may be applied together to produce a dried product. Crackers, cookies, and biscuits are baked foods and will be discussed in Chapter 12. Protein meals produced by cooking and drying belong to the third group, while dried meat products, such as sausages, cured hams, dried fish, beef, mutton, and shrimp, belong to the fourth. The fifth group consists of processed nuts and plant seeds such as almonds, chestnuts, and peanuts. The last group is the dried microalgae. We will, in this chapter, mainly discuss the oxidative stability and shelf life of milk, egg, and encapsulated oil powders, extruded breakfast cereals and snack foods, protein meals, and dry pet food.

9.2 IMPORTANT FACTORS IMPACTING OXIDATIVE STABILITY AND SHELF LIFE OF LOW-MOISTURE FOODS

Unlike bulk oils/fats and food emulsions, low-moisture foods have complicated microstructures and varying lipid distributions. In addition, a number of oil/fat-containing low-moisture foods contain proteins and carbohydrates as well as transition metals. Lipid co-oxidation with proteins makes the lipid oxidation mechanism of low-moisture foods more complicated and makes the oxidative stability and shelf life of the low-moisture foods more difficult to evaluate and predict. Further, chemical reactions in solid foods may be different from those in liquid foods such as bulk oils/fats, oil-in-water, and water-in-oil emulsions. Diffusion rate in solid phase is lower than that in liquid phase. Therefore, lipid oxidation in low-moisture foods may not always be the same as that in liquid foods, and the dry foods may not always follow the rules of liquid foods.

Internal factors impacting lipid oxidation in low-moisture foods may include oil/fat type, degree of unsaturation, matrix type (microstructures and physical states), molecular mobility, lipid distribution, ratio of surface area to volume of lipid, a_w, moisture content, glass transition temperature, pH, quality of raw ingredients, transition metal ions, enzymes, interaction between lipid and other components such as protein and amino acids, as well as pro-oxidants such as transition metals and enzymes, and antioxidants such as natural antioxidants, such as mixed tocopherols, rosemary extract, green tea extract (GTE), grape seed extract (GSE), and synthetic antioxidants, such as BHA, BHT, TBHQ, and PG. We will, in this chapter, principally discuss matrix types, lipid distribution, a_w, moisture content, glass transition temperature, and transition metal, as well as lipid co-oxidation with protein and amino acids.

As discussed earlier, the matrices of dry powders produced with food emulsions (e.g., milk and encapsulated oil powders) are different from those produced by mixing different powder ingredients (e.g., dried cauliflower soup powder). The matrix of egg powder is different from that of milk powder. For formulated microencapsulated oil powders, completely different matrices would be formed if varying carbohydrates such as maltodextrins, lactose, cyclodextrins, gums, cellulose and starch, and proteins such as sodium casinate, whey protein isolate, gelatin, and others are selected to encapsulate oils (Velasco et al., 2003). Different matrices in dry foods may constitute varying lipid distributions and porous microstructures, thereby leading to different oxidative stability in dry foods. For instance, extruded, fried, and baked corn chips are distinct in matrix, microstructure, and lipid distribution, thus resulting in different oxidative stability and shelf life. The matrix of the extrudate produced with starch-based structure-forming materials is different from that produced with protein-based structure-forming materials. The microstructure of processed peanuts also differs from those of processed pecans and walnuts. Therefore, low-moisture food matrices significantly impact the oxidative stability and shelf life of the low-moisture foods.

According to Fritsch (1994), there are three cases for lipid distribution (location) in oil/fat-containing foods. The first case consists of the foods with continuous lipid phase in which carbohydrate and protein globules are dispersed in continuous phase lipid matrix and the oxygen in the air is in direct contact with the lipids (Fritsch, 1994). The oxidation rate in this type of food is similar to that of the same lipids without the dispersed components such as bulk oil or fat. The author highlighted that the critical parameters impacting the lipid oxidation rates in this type of food are the type and surface of the lipid. Peanut butter is an example of such a system. The second case consists of the foods without continuous lipid phase in which lipid globules are dispersed in carbohydrate and protein matrix (Fritsch, 1994). The author also highlighted that the oxidation rate in this type of food depends on the total lipid surface area and the ability of the carbohydrate and protein matrix to transport and activate oxygen. The important variables influencing the oxidation rates are a_w, the ratio of free to bound lipids, and pH (Fritsch, 1994). Peanut, a high-fat food with a noncontinuous lipid phase, is an example of the system. According to Fritsch (1994), it is critical to distinguish the free and bound (encapsulated) fat in the low-moisture foods. Hexane-extractable lipids are considered free ones on the surface of the lipid droplets. The bound lipids are the ones that cannot be extracted by hexane or petroleum ether but can be extracted after acid hydrolysis. The last case is the foods with both continuous and noncontinuous lipid phase (Fritsch, 1994). Examples are encapsulated fish oil powders, fried potato and corn chips, and dry pet food (Velasco et al., 2006; Hu, 2014).

Water activity (a_w) appreciably influences the oxidative stability of oil/fat-containing low-moisture foods, particularly powder foods. According to food stability map (Labuza, 1971), lipid oxidation rate are lowest at a_w between 0.2 and 0.3 for most dry foods, where water exists in the form of monolayer at which water is bound tightly to the food surface and thus cannot act as an aqueous phase and is not available for reactants and catalysts. But lipid oxidation rate increases appreciably when a_w is greater than 0.4 or lower than 0.2. Actually, the oxidative stability of low-moisture foods does not always follow the food stability map, or rather, it is more product-related (Jensen and Risbo, 2007; Raitio et al., 2011).The majority of the low- and intermediate-moisture foods (with an a_w between 0.4 and 0.7) did not show accelerated lipid oxidation at a_w values above the monolayer (Fritsch, 1994). For instance, high-protein nutrition bar is an intermediate-moisture food with a_w between 0.6 and 0.85 at room temperature. The shelf life of high protein nutrition bar was evaluated to be 7–26 months at 23 °C depending on bar systems (Rao et al., 2013). Extremely high lipid oxidation rate in fresh or tissue foods occurs at a_w near 1.0 due to lipoxygenase-catalyzed oxidation reactions.

Water activity may not fully explain lipid oxidation rate of low- and intermediate-moisture foods. Hence, the polymer glass transition theory was put forward to explain lipid oxidation and predict the oxidative stability of low- and intermediate-moisture foods. The glass transition theory considers physical state of water and solid systems and the influence of polymers on lipid oxidation. Many dry foods contain biopolymers such as proteins, starch, and gums. According to Nelson and Labuza (1992), amorphous polymers exist in their glass state at low temperature and low moisture, while polymers are in the rubbery state at high temperature and high moisture. The glass transition temperature (T_g) is the one at which amorphous regions of a polymer change from a glass state to a rubbery state (Nelson and Labuza, 1992). According to the authors, lipids are dispersed in the free volume in the carbohydrates and protein polymer matrix. In the glass state, there is less free volume and diffusion may be restricted. Thus, lipid oxidation rates would decrease. However, there is more free volume in the rubbery state. Thus, lipid oxidation would increase. Further, if the polymer system undergoes collapse in the rubbery state resulting in the shrinkage of the polymer matrix, lipid oxidation rate would decrease due to the reduction in free volume and the decrease of the pore size at the surface of the matrix (Nelson and Labuza, 1992). The authors indicated that lipid oxidation rates should be higher when the system is above glass transition temperature than below the temperature due to an increase of free volume. Nelson and Labuza (1992) recommended that food formulation was modified to bring the food into the glass state or increase the glass transition temperature of the food matrix. When predicting the shelf life of low-moisture foods containing starches, proteins and monosaccharides, the amount of water, the a_w and the state of water need to be considered. Actually, food stability map and glass transition theory cannot fully explain oxidative stability of various low-moisture foods. Therefore, new research and theories are needed.

The quality of raw ingredients affects initial product quality and oxidative stability of the product. For instance, initial levels of hexanal of WMP from four production facilities ranged from 0.5 to 50 μg/kg, with a mean of 13 μg/kg. The hexanal content of WMP at 12 month storage at 21 °C and 45% RH varied greatly, ranging from 490 to 4400 μg/kg (Lloyd et al., 2009a). In addition, the effect of transition metal ions such as ferrous iron and ferric iron on lipid oxidation in low-moisture foods is complex and may be different from that in bulk oils/fats and food emulsions. In milk powders containing 1–1.5 ppm of iron and 0.24–0.47 ppm of Cu, the transition metal ion shows pro-oxidant activity (Lloyd et al., 2009a). However, in the extruded food, ferrous iron did increase the oxidative stability of uncoated extrudate based on the evaluation of PV, when higher level (50 ppm) of ferrous acetate was added (Rao and Artz, 1989). The research on dry pet food in our laboratory also showed that adding high level of ferrous iron to kibble matrix did increase the oxidative stability and shelf life of the dry pet foods based on both PV and hexanal data (data not shown). The reason why ferrous iron increased the oxidative stability of extruded foods will be discussed in the following section.

Lipid co-oxidation with protein in low-moisture foods has drawn a great deal of attention over the past decade, since a number of low-moisture foods contain both lipid and proteins or amino acids, and the oxidative stability and shelf life study of the low-moisture foods have challenged food scientists in food and pet food industry as well as professors at universities. Clearly, lipid primary oxidation product hydroperoxide and secondary oxidation products aldehydes and ketones as well as free radicals, such as alkoxyl radicals, are very reactive. These reactive compounds and free radicals can react with proteins or amino acids to cause protein oxidation (Dong, 2011; Raitio et al., 2011; Schaich, 2008). The oxidatively modified proteins are formed by protein cross-linking, amino acid side chain modification, and protein fragmentation (Pokorny et al., 2005; Wanibadullah, 2013; Skibsted, 2010). Thus, after lipid co-oxidation with proteins or amino acids in the protein-containing dry foods, the concentrations of hydroperoxide and volatile oxidation products such as hexanals and 2,4-decadienal may decrease or may not change significantly, compared to the oxidation of bulk oil/fat. Moreover, the aldehydes such as hexanal and pentenal may further oxidize to form hexanoic and pentonoic acid, leading to the concentration decrease of the aldehydes at the late stage of oxidation. Further, a dry food may contain more than one oil or fat-containing ingredient with different oxidative stability and freshness when used to produce the dry food. In other words, dry foods may consist of mixed oils/fats with varying shelf life and freshness. In addition, a dry food may contain both continuous and noncontinuous lipid phase, such as fried corn chips and dry pet foods. Therefore, the PV and hexanal level curves of these dry foods during storage may be different from those of bulk oils or fats.

External factors influencing lipid oxidation of dry foods may include temperature, light, relative humidity (RH), processing and storage conditions, and packaging (including gas composition). The influence of external factors on lipid oxidation in bulk oils/fats and food emulsions was well documented (Choe and Min, 2006; McClments and Decker, 2000; Frankel, 2005). Although the impact of most external factors on lipid oxidation in low-moisture foods is similar to that in bulk oils/fats and food emulsions, different processing methods such as extruding, frying, and baking would produce low-moisture foods with varying matrices and lipid distributions, thus affecting the oxidative stability and shelf life of the low-moisture foods significantly. The more severe heat treatment, when producing WMP, resulted in more Maillard reaction products, which underwent Strecker degradation to form Strecker aldehydes (Lloyd et al., 2009a,b). High temperature also accelerated lipid oxidation, thus producing more aldehydes from lipid oxidation.

9.3 METHODS TO INCREASE OXIDATION STABILITY AND EXTEND SHELF LIFE OF LOW-MOISTURE FOODS

Because a number of internal and external factors impact lipid oxidation of low-moisture foods, our main focus should be on effectively controlling these factors to minimize lipid oxidation and extend the shelf life. First, a number of low-moisture foods contain oils or fats to increase flavor and essential fatty acids of the low-moisture foods. Fish, algal, and flaxseed oils contain a large amount of polyunsaturated fatty acids such as DHA, EPA, and ALA (alpha-linolenic acid) that are more readily oxidized than animal fats such as pork and chicken fat and plant oils such as soybean, canola, and corn oils. Thus, blending oils containing n-3 fatty acids with animal fats and/or plant oils may be an effective method to increase the oxidative stability and shelf life of low-moisture foods containing n-3 fatty acids. Blends of fish oil, plant oil, and chicken or pork fat have been used in food and pet food industries. Also mid- and high-oleic vegetable oils can be used to mix with flaxseed oil or algal oil to increase the oxidative stability of the blends containing flaxseed or algal oil. Second, for the dry foods containing both continuous and noncontinuous fat phase such as frying cereal chips and dry kibbles, antioxidant should be added to both continuous and noncontinuous fat phase, or inside of the chips and kibble and the surface of the chips and kibble, in order to manage the lipid oxidation and extend the shelf life.

Second, it is critical to control a_w, RH, and glass transition temperature of dried food powders. Milk powders were oxidatively stable at a_w between 0.1 and 0.24 (Lloyd et al., 2009a). Spray-dried milk base for baby food was the most oxidatively stable at a_w near 0.24 (Roozen and Linssen, 1992). But the highest cholesterol oxidation appeared in egg powders (a_w=0.17–0.87) with the lowest a_w during storage (Obara et al., 2006). Increasing RH (0%, 11%, 33%, 54%, and 75%)

improved the oxidative stability of the encapsulated sunflower oil powders. The greatest improvement of the oxidative stability of the encapsulated oil powders was seen in the glass state powders at RH 54% (Damerau et al., 2014a). Dry kibble was oxidative stable at a_w 0.35 to 0.55 (unpublished data). Glass transition temperature should be considered in stability prediction for polymeric systems. It is feasible to formulate food powders to bring the powders into glass state and increase the glass transition temperature, so that the oxidative stability of the powders can be increased. Further, it is critical to decrease the temperature and avoid light and inactivate lipid enzymes during processing, shipping, and storage. In addition, it is also essential to order high-quality raw ingredients such as bulk oils/fats and oil/fat-containing ingredients. Food manufacturers need to check the quality of bulk oils/fats and oil/fat-based ingredients and apply high-quality ingredients for production of dry foods. If nonfresh oil/fat-containing ingredients or bulk oils are used to produce dry foods, not only will the shelf life be decreasing but it will also be difficult to understand and predict the oxidative stability and shelf life of the products. Moreover, selecting appropriate antioxidants or antioxidant blends to be applied to dry foods is very important for food companies to produce high-quality oxidatively stable dried food products. An antioxidant may have different efficacy in varying matrices of dry foods. As for dry foods containing both continuous and noncontinuous lipid phase, antioxidant(s) should be added to both phases in order to extend the shelf life of the dry foods. The antioxidants can be either synthetic antioxidants such as BHA, BHT, TBHQ, and PG (propyl gallate) or natural antioxidants such as tocopherols, caffeic acid, catechin, quercetin, green tea, rosemary, and pomegranate extract.

Finally, lipid oxidation needs oxygen in the air. Thus, oxygen absorbents, modified atmosphere, and active packaging are used in food industry. Small bags of oxygen absorbent are usually added to the packaged bags containing dried foods. In the past few years, various modified atmosphere and active packages have drawn a great deal of attention. The modified atmosphere packaging is a technology of modifying the composition of the internal atmosphere of a food package to extend the shelf life of the products. The modification may lower the amount of oxygen from around 21% to 0%, in order to decrease the rate of lipid oxidation. The removed oxygen can be replaced with nitrogen or carbon dioxide; 80% nitrogen, 20% carbon dioxide, or vacuum can be selected for packaging. Active packaging involves the incorporation of antioxidants to food packaging materials and releases the antioxidants by a controlled mechanism of diffusion (Sanches-Siva et al., 2014).

9.4 HOW TO EVALUATE THE OXIDATIVE STABILITY AND SHELF LIFE OF LOW-MOISTURE FOODS

Lipid oxidation reaction of low-moisture foods is very complicated. Thus, we need different analytical methods to evaluate the oxidative stability and shelf life of various low-moisture foods. A number of physical and chemical analytical methods can be used to assess the oxidative stability and shelf life of the low-moisture foods. The available analytical methods are Oxipres (or oxygen bomb); OSI (Oxidative Stability Index); modified Schaal oven test; hydroperoxide value (PV); volatiles such as hexanal or 2.4-decadienal, hexanoic acid, propanal, 2,4-heptadienal, and propanoic acid; TBARS; headspace oxygen; *para*-anisidine value (*p*AV), TOTOX ([2×PV]+AV); and antioxidant and fatty acid concentration change. Moreover, techniques such as GCO (gas chromatography olfactometry), electronic nose and tongue, as well as sensory evaluation can also be used. Each method has its advantages and limitations (see Table 1 in the Introduction). Therefore, it is essential to select appropriate analytical methods to assess a specific low-moisture food and to correlate the oxidative stability data with sensory data. If an inappropriate method is used, an incorrect conclusion may be drawn.

To quickly assess the oxidative stability and shelf life of a low-moisture food, an accelerated stability test and a sensory test are generally needed for the assessment. First, Oxipres or oxygen bomb is applied to measuring induction period (IP) of the low-moisture food at different temperatures, such as 60, 70, 80, and 90 °C. Then, the Arrhenius equation or Q_{10} is used to extrapolate the actual (or ambient) IP value at lower temperature from the IPs at high temperatures, in order to predict the shelf life of the low-moisture foods under the actual (or ambient) storage condition (more information can be found in the Introduction). It should be pointed out that care must be taken when using IPs at high temperatures to extrapolate actual (or ambient) IP value, since the mechanism and activation energy of lipid oxidation in low-moisture foods at high temperatures may be different from those at actual (or ambient) temperatures, the linear correlation of IPs between the high tested temperatures and the actual storage temperature may be lost (Verleyen et al., 2005). Further, high temperature may change a_w and moisture content and modify original structures of dry foods, particularly the dry foods containing lipids, proteins, and polysaccharides, so that the glass transition temperature may be changed. In addition, Maillard reaction in some dehydrated foods may occur when temperature increases. As a result, the products of the Maillard reaction may inhibit lipid oxidation. All these factors will have big impacts on the analytical results. Finally, some antioxidants are heat-sensitive or labile. The oxidative stability of low-moisture foods containing heat-sensitive antioxidants would be underevaluated if the test is conducted at high temperatures. Also, the accelerated test at high temperature may underestimate the efficacy of heat sensitive antioxidant (*s*) in the low-moisture foods. It is therefore recommended that sensory test be conducted at the same time and sensory data be used to correlate with IP data to make a final conclusion. Preferably, the correlation between accelerated stability data and ambient stability ones should be set up. Some laboratories use organic solvents to extract lipid from dry foods and then run OSI at high temperatures such as 120, 110, and 100 °C to measure IPs of the extracted lipids. This method is not recommended because the oxidation of extracted bulk lipid is different from that of the lipid in the dry food matrices. It was reported that the relative performance of rosemary extract evaluated by OSI was 57, 93, and 122 at 100, 80, and 60 °C, respectively. But the relative performance of mixed tocopherols was 100 at 100, 80, and 60 °C (Van Dyck et al., 2005). This suggests that mixed tocopherols were much more efficient than rosemary extract at 100 °C and still efficient at 80 °C. However, the rosemary extract was more efficient than mixed tocopherols at 60 °C based on OSI evaluation.

Second, accelerated storage tests at 98.6–106 °F (37–41 °C) and 50–70% RH and actual storage tests such as at 72 °F (22.2 °C) and 50% RH may be selected for shelf life evaluation of dry or low-moisture foods. PV and volatiles such as hexanal or 2.4-dicadienal, propanal or 2,4-heptadienal or 1-penten-3-one, hexanoic, and propanoic acid may be measured over storage. Meanwhile, the sensory test should be also conducted. Hexanal or 2,4-decadienal, hexanoic acid are used to evaluate the dry foods containing n-6 fatty acids such as the foods containing pork and chicken fat, soybean, canola, and corn oil. Propanal or 2,4-heptadienal, 1-penten-3-one, and propanoic acid to assess the dry foods containing n-3 fatty acids such as the foods containing fish, algal, flaxseed, and krill oil. For breakfast cereals and snack foods, hexanoic or propanoic acid may be measured to monitor the long-term shelf life of these dry foods, since the hexanal or propanal level may be decreasing or no obvious change over long-term storage (Paradiso et al., 2008, 2014). The frequency of sample analysis and length of accelerated and actual storage tests are dependent on types of samples and expected shelf life. The final conclusion of shelf life study will be drawn based on the sensory data and oxidative stability data such as hexanal or hexanoic acid level changes over storage.

For milk powders and extruded chips, the evolution of PV and hexanal concentration during storage could be used for evaluating oxidative stability and shelf life (Jena and Das, 2012; Gray et al., 2008). A good correlation between PV and hexanal was established (Lloyd et al., 2009a). In this case, TOTOX may be used to comprehensively assess the oxidative stability and shelf life of the dry foods. As one can see from the Introduction, there are several TOTOX values being defined. For low-moisture foods, TOTOX value = a PV + b hexanal might be used to predict the shelf life. The ratio (a/b) of PV to hexanal needs to be determined by the correlation between PV and hexanal as well as sensory data, because varying food matrices and different measuring methods may have appreciable impacts on PV and hexanal values. Once the ratio is determined by regression analysis, the cut-off level of TOTOX value, which roughly evaluates if the dry food is rancid or not, will be determined by the correlation with sensory data.

For some dry foods without continuous lipid phase or containing very low level of lipid, PV may not be suitable for assessing the oxidative stability. In this case, hexanal and sensory data can be selected to evaluate the shelf life of the dry foods.

For dehydrated cauliflower powder (Raitio et al., 2011) and some of protein meals and dry pet foods, PV and hexanal levels may not be increasing, probably decreasing or may be no obvious changes during storage is observed (see Figure 2(B) in the Introduction). In this case, we may need to measure hexanoic acid, protein oxidation, and conduct sensory analysis for the shelf life evaluation. In addition, the prior experience in shelf life study would be critical for the evaluation.

Concentration change of fatty acids and antioxidants in bulk oils/fats during storage can be an indicator assessing the oxidative stability and shelf life of bulk oils and fats. But the method may not work well when measuring a low-fat dry food. For instance, it is feasible to analyze concentration change of an antioxidant such as α-tocopherol in encapsulated oil powders over storage to predict the oxidative stability and shelf life of the powders (Marquez-Ruiz et al., 2003). However, the method may not be suitable for some dry kibbles, since the tocopherols level in the tested dry kibble did not change or change slightly over storage, but the kibbles were gradually oxidized during the storage as evidenced by the increase of hexanal level (unpublished data).

Measuring headspace oxygen level in a sealed container containing dry foods is another method for evaluating the oxidative stability of the dry foods. This method may work for some dry foods but may not work for others, because the oxygen may react with other components except lipids, or other components may absorb the oxygen in the sealed container containing the dry foods.

Although the modified Schaal oven test may work well for the shelf life evaluation of bulk oils or fats, it needs further modifications for assessing dry foods, such as measuring hexanal over storage, and decreasing storage temperature from 60 °C to a lower temperature such as 40 °C or 37 °C.

It should be pointed out that TBARS and *p*-anisidine methods may not be suitable for some of the dry foods containing both lipid and proteins, such as protein meals and dry kibbles, due to the interference of other colors from the dry foods and protein meals, even though TBARS works for meat and meat products and *p*-anisidine for bulk oils or fats.

In the food and pet food industries, a business decision may be made based on the comparison of the oxidative stability data. When evaluating the oxidative stability of dry foods using PV, hexanal, and Oxipres, the data obtained only from same analytical method can be comparable. For instance, there are several methods to measure PV, such as AOCS titration, Fox II, IDF method, and SafTest. The PV would be different if varying methods are used for the measurement. Also, the PVs with same unit such as mEq/kg sample or mEq/kg fat or mmol/kg sample can be used for comparison. Further, there are several methods for hexanal or propanal analysis, such as static and dynamic headspace, SPEM, and direct injection. Different hexanal data could be obtained if varying GC methods are used. In addition, even though the same GC analytical method is used, different analytical results can be obtained if varying pretreatment methods (e.g., different heating temperature and extracting solvent) are used for treating samples before GC analysis. Therefore, care should be taken when using hexanal or propanal data for comparison of the oxidative stability and shelf life of dry foods.

9.5 OXIDATIVE STABILITY AND SHELF LIFE OF DIFFERENT DRY FOOD PRODUCTS

9.5.1 Milk Powders

Milk is considered a whole food designed by nature to provide macronutrients and micronutrients as well as bioactive compounds (Zhu and Damodaran, 2011). Due to short shelf life and high shipping costs of bovine milk, dry milk powders have been a main focus for the dairy industry. The short shelf life of bovine milk is because it contains high level of water (around 88%) and available macronutrients for microbial growth (Park and Drake, 2014). Various dehydration or drying technologies are used to produce dried milk powders with extended shelf life. As a functional ingredient, milk powders are used to produce a variety of food products because of their nutritional and functional properties, such as confectionary such as chocolate, dairy products, dry sauce, soups and soup mixes, baked foods, meats, and animal feed. Thus, the oxidative stability of milk powders impacts the oxidative stability and shelf life of these milk powder–derived products.

Dried dairy ingredients are composed of milk powders and protein powders, such as whey protein concentrate (WPC) and whey protein isolate (WPI). We will, in this section, focus on the oxidative stability and shelf life of milk powders. According to U.S. Dairy Export Council (USDEC, 2005), the moisture level of milk powders is less than 5%. Milk powders consist of nonfat dry milk (NFDM), skim milk powder (SMP), whole milk powder (WMP), and buttermilk powder (BMP). SMP and NFDM are produced from the pasteurized skim milk containing less than 1.5% fat by weight and at least 34% protein by weight. WMP is from the pasteurized whole milk and has 26–40% fat by weight. BMP is from the buttermilk and contains greater than 4.5% milk fat.

There are, generally, three ways to dehydrate milk to produce dried milk powders: spray drying, freeze drying, and roller drying. But spray drying is a widely used method and determines surface and bulk compositions of milk powders, and it has an important impact on lipid distribution in dried milk particles. Lipid oxidation in milk powders has been intensively studied (Park and Drake, 2014; Lloyd et al., 2009a,b). According to the authors, lipid distribution has a pronounced influence on the oxidative stability, flavor release, and flavor stability of milk powders. After a milk fat emulsion was made by a homogenizer, the microencapsulated fat powders were produced by spray-drying the milk fat emulsion. Keogh (2005) commented the microstructure of spray-dried microencapsulated fat powder particles. Clearly, the vacuoles would have an impact on the oxidative stability of the encapsulated fat.

Free fat in dried milk powders is not entirely coated by amphiphilic molecules or protected by a matrix of carbohydrates and proteins during spray drying (Vignolles et al., 2007). During spray drying, the fat, protein, and lactose reorient themselves. The fat and protein migrate to the surface due to their hydrophobicity, and the lactose migrates to the center due to its hydrophilic nature (Kim et al., 2009; Park and Drake, 2014). Hence, the surface composition of dried milk powders is different from the interior or bulk composition. Much more fat is located on the surface of the dried particles than that located in the interior of the particles. According to Kim et al. (2002), for SMP, the bulk composition of lactose, protein, and fat was 58%, 41%, and 1%, respectively, but the surface composition of lactose, protein, and fat was 36%, 46%, and 18%, respectively. For WMP, fat represented 98% of surface composition. Surface free fat on dried milk powder particles may be readily oxidized compared to emulsified fat in the interior of the particles (Park and Drake, 2014). This is because the free fat on the surface contains more unsaturated fatty acids, such as oleic, linoleic, and linolenic acid, and is exposed to oxygen. The emulsified fat is encapsulated with proteins and phospholipids that may have antioxidant activity. Granelli et al. (1996) also reported that higher levels of surface fat correlated to increased oxidation in dairy powders.

9.5.2 How to Evaluate the Oxidative Stability and Shelf Life of Milk Powders

While there are a number of analytical methods monitoring the oxidative stability, commonly used methods are: headspace volatile compounds analysis by GC–MS and SPME (solid phase microextraction) (Lloyd et al., 2009a,b), such as hexanal, 2-methylbutanal, 3-methylbutanal, 2-heptanone, heptanal, 1-octen-3-ol, octanal, 3-octen-2-one, and nonanal, PV, headspace oxygen concentration (Jena and Das, 2012; Lloyd et al., 2009b), TBA (Fenaille et al., 2001), and *p*AV (Roozen and Linssen, 1992). Oxipres may not be suitable for quickly assessing the oxidative stability of milk powders due to high temperature being required for the analysis. But the combination of sensory data and volatile compounds measured by GC headspace analysis provides a valuable method that can predict lipid oxidation and grassy and painty flavors (Lloyd et al., 2009a,b). Thus, a better evaluation method is to measure hexanal level change and conduct sensory test over ambient storage at 22 °C or 37 °C. Higher temperatures such as 50 °C are not recommended for storage test. In addition, lipid co-oxidation with protein and protein oxidation, such as milk protein–lipid adducts, protein–protein cross-linking, protein scission, and amino acid damage, should be considered when evaluating the oxidative stability of milk powders (Roozen and Linssen, 1992).

Lloyd et al. (2009a) reported that WMP were collected from four U.S. production facilities five times over a 1-year period. Each sample was analyzed for sensory profile, volatiles, color, a_w, and moisture throughout a 12-month storage at 21 °C and 35% RH. These WMPs had an a_w of 0.11–0.24, moisture content 2–3.3%, fat content 26.2–32%, free fat content 1.1–7.7%, iron content 0.2–3.1 ppm, and copper content 0.1–0.9 ppm. According to the author, because lipid oxidation and Maillard reaction are two main deterioration reactions in WMP, the volatile compounds of lipid oxidation and Maillard reaction were analyzed with SPME–GC–MS, with the samples being heated in salty water at 40 °C for 30 min before analysis. Dimethyl sulfide, 2-methylbutanal, 3-methylbutanal, hexanal, 2-heptanone, heptanal, 1-octen-3-ol, octanal, 3-octen-2-one, and nonanal were quantified. They found that painty and grassy flavors were correlated to increased levels of hexanal, heptanal, and octanal. The best predictors of grass flavor were hexanal, nonanal, 2-heptanone, and 1-octen-3-ol, while the best predictors of painty flavor were hexanal, 2-methylbutanal, 3-methylbutanal, octanal, and 3-octen-2-one. According to the authors, the prediction equations for grassy and painty are:

Grassy = 0.92 + (0.00058 × hexanal) + (0.024 × 2-heptanone) + (0.073 × 1-octen-3-ol) − (0.083 × nonanal), $R^2 = 0.38$, $P < 0.0001$.

Painty = 0.47 − (0.086 × 3-methylbutanal) + (0.070 × 2-methylbutanal) + (0.00073 × hexanal) + (0.018 × octanal) − (0.022 × 3-octen-2-one), $R^2 = 0.61$, $P < 0.0001$.

The authors also found that hexanal was a key compound in the predictive equation for both grassy and painty flavors. For the optimum quality, hexanal content of WMP should be less than 50 μg/kg. If hexanal or nonanal concentrations were greater than 600 or 4 μg/kg, respectively, painty flavors would be evident.

Figure 9.1 shows the mean hexanal level change (ppb) in WMP from four facilities in the United States during a 12-month storage at 21 °C and 45% RH (Lloyd et al., 2009a). The data in Figure 9.1 indicate that the hexanal level

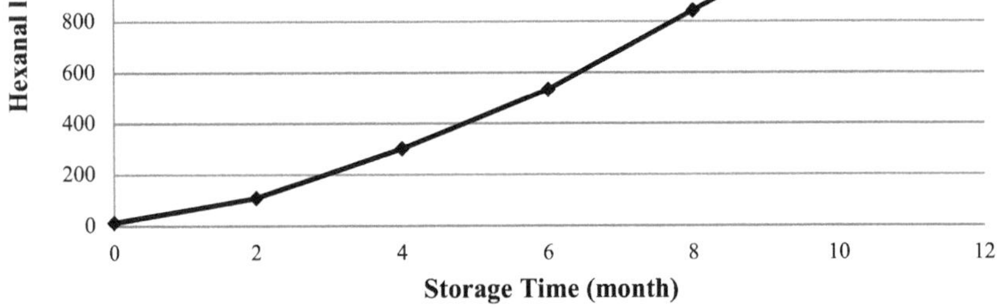

FIGURE 9.1 Mean hexanal level change in WMP from four facilities in the United States over a 12-month storage at 21 °C and 45% RH. *Source: Lloyd et al. (2009a). Curve drawn based on data on Table 9.2 of the reference.*

increased as the storage time increased, reaching to the maximum at 12 months. Indeed, varying hexanal data can be obtained if different hexanal analytical and pretreatment methods are used. Therefore, a standard hexanal analytical method is indeed required for food industry to conduct comparative tests. Table 9.2 shows the concentration changes of 10 volatile compounds in WMP during 1-year storage (Lloyd et al., 2009b). Hexanal level increased from 0 to 10 months. After that, its level decreased until 12 months. Obviously, hexanal is a very useful indicator not only for oxidative stability evaluation but also as a key compound to predict grassy and painty flavors of WMP, although PV was reported to be correlated ($P<0.0001$) with hexanal ($r=0.91$) and painty flavors ($r=0.86$) (Lloyd et al., 2009b).

Shelf life of WMP was reported to range from 3 months to 3 years (Carunchia Whetstine and Drake, 2007; VanMil and Jans, 1991; U.S. Dairy Export Council, 2005). The big difference could be attributed to several reasons, such as varying samples produced from different companies, varying raw materials, processing parameters, storage conditions, and evaluation methods. According to the U.S. Dairy Export Council (2005), the U.S. industry standard for shelf life of WMP is from 6 to 9 months under storage at less than 27 °C and less than 65% RH. Lloyd et al. (2009b) reported that the optimum shelf life of WMP from a flavor standpoint was about 3 months. They found that after 6 months of storage, painty off-flavors in WMP were detected by consumers in the chocolate-containing WMP. Nitrogen flushing not only remarkably enhanced the storage stability of WMP by preventing the development of painty flavor but was also more effective than decreasing storage temperature (from 23 °C to 2 °C) in extending the shelf life. The authors concluded that WMP stored in nitrogen-flushed packaging for up to 1 year can be used to manufacture milk chocolate with a high level of consumer acceptance.

As discussed earlier, higher levels of surface free fat has been correlated to increased oxidation in dairy powders. However, Lloyd et al. (2009a) noted that their research did not show any differences in the stability of the WMP samples containing various levels of free fat, which was in agreement with Buma's research result (1971). Therefore, more research is needed to investigate if the surface free fat is increasing or decreasing the oxidative stability of WMP. Indeed, for the shelf life study, the milk powders containing total fat, other than surface free fat, are usually used for analyzing volatiles.

9.5.3 How to Increase Oxidative Stability and Extend Shelf Life of Milk Powders

Initial milk quality; processing conditions including heat treatment; particle size of milk powders; storage conditions including temperature and RH, a_w; moisture content; packaging including gas composition and concentration, oxygen and light exposure, as well as antioxidants added, significantly impact lipid oxidation of milk powders (Lloyd et al., 2009b; Park & Drake, 2014). Therefore,

TABLE 9.2 Concentrations (μg/kg) of Selected Volatile Compounds* in Whole Milk Powder During 1 Year of Storage

Storage Time (mo)	0	2	3	4	6	8	10	12
Dimethyl sulfide	25.8 ± 6.8^{ab}	12.0 ± 1.0^{c}	6.4 ± 0.5^{c}	11.2 ± 1.0^{c}	13.2 ± 0.7^{c}	24.3 ± 2.7^{b}	36.8 ± 6.5^{a}	32.4 ± 6.4^{ab}
3-Methyl butanal	5.1 ± 1.4^{bc}	3.7 ± 0.4^{c}	1.8 ± 0.2^{d}	3.9 ± 0.3^{c}	4.5 ± 0.4^{c}	7.2 ± 0.5^{a}	6.5 ± 0.7^{ab}	6.5 ± 0.9^{ab}
2-Methyl butanal	3.5 ± 0.9^{bc}	2.4 ± 0.3^{c}	1.9 ± 0.2^{c}	3.5 ± 0.3^{bc}	4.4 ± 0.5^{b}	7.4 ± 0.4^{a}	8.8 ± 1.0^{a}	8.3 ± 1.0^{a}
Hexanal	20.7 ± 2.1^{c}	57.9 ± 9.1^{c}	118.4 ± 37.6^{c}	167.2 ± 52.2^{c}	421.7 ± 120.3^{bc}	605.6 ± 139.7^{b}	1174.9 ± 259.5^{a}	664.7 ± 167.9^{b}
2-Heptanone	10.5 ± 2.3^{cd}	7.2 ± 1.0^{d}	6.7 ± 0.4^{d}	7.6 ± 0.8^{d}	9.1 ± 1.1^{d}	16.3 ± 1.9^{bc}	18.7 ± 3.4^{b}	25.2 ± 3.9^{a}
Heptanal	2.7 ± 0.3^{d}	7.0 ± 1.0^{d}	12.1 ± 3.3^{d}	15.9 ± 4.3^{cd}	40.0 ± 10.5^{cd}	52.0 ± 11.0^{bc}	113.0 ± 23.8^{a}	84.7 ± 4.0^{ab}
1-Octen-3-ol	0.01 ± 0.0^{c}	0.4 ± 0.0^{c}	0.6 ± 0.1^{c}	0.5 ± 0.1^{c}	1.4 ± 0.2^{b}	2.2 ± 0.2^{b}	4.10 ± 8^{a}	4.0 ± 0.9^{a}
Octanal	3.7 ± 1.0^{e}	8.8 ± 0.7^{de}	8.7 ± 0.4^{de}	10.2 ± 0.6^{cde}	13.3 ± 1.5^{cd}	16.2 ± 1.9^{c}	35.9 ± 4.4^{a}	28.2 ± 5.0^{b}
3-Octen-2-one	0.2 ± 0.0^{b}	0.7 ± 0.1^{b}	1.1 ± 0.3^{b}	1.3 ± 0.4^{b}	3.4 ± 1.2^{b}	5.0 ± 1.4^{b}	12.7 ± 3.2^{a}	12.2 ± 4.1^{a}
Nonanal	0.8 ± 0.2^{d}	0.9 ± 0.1^{d}	1.3 ± 0.2^{cd}	1.3 ± 0.2^{cd}	2.7 ± 0.6^{cd}	3.30 ± 0.6^{bc}	6.5 ± 1.3^{a}	5.2 ± 1.4^{ab}

[a–e]Different superscript letters within a row indicate significant differences ($P < 0.05$).
**Data are given as means ± SEM.*
Source: Lloyd et al. (2009b).

focus should be on optimizing the processing and storage conditions to manage the oxidative stability and extend the shelf life of milk powders. Van Mil and Jans (1991) reported that the WMP had 3-year shelf life. The optimal processing and storage conditions were moderate heat treatment of 95 °C for 2 min before spray drying, moisture content ≥2.8%, storage temperature ≤20 °C, and air- and moisture-proof oxygen-reduced packaging. Adding natural antioxidants such as mixed tocopherols may increase the shelf life of dry milk powders.

Processing conditions, particularly spray-dry parameters, determine particle surface and wall composition, surface free fat, and protein binding, thus affecting lipid oxidation. As the oxidative stability decreases, off-flavors of milk powders will increase, since lipid oxidation and Maillard reaction are the main reasons for the flavor stability of milk powders. Flavor is the principal factor influencing consumer acceptance for dried milk powders (Park and Drake, 2014; Lloyd et al., 2009b). Therefore, controlling lipid oxidation is critical for ensuring the quality of dried milk powders. In the dairy industry, carefully controlling spray-dry parameters is an effective method to produce high quality of milk powders, which have strong oxidative stability and long shelf life. For instance, processing parameters such as inlet and outlet air temperature, feed solids concentration, and atomizing conditions can be manipulated to decrease the surface free fat. Increasing inlet air temperatures increased the particle size, thus decreasing surface free fat in WMP, whereas increasing outlet air temperatures increased free fat (De Vilder et al., 1976; Kelly et al., 2002). Surface free fat in WMP may be susceptible to lipid oxidation leading to off-flavor development. Preheating treatment is an important unit operation in the manufacture of dried milk powders, because preheating milk proteins would cause milk proteins denature, and the denatured milk proteins would have a greater ability to encapsulate the fat. As a result, surface free fat level would decrease. It is also important to control processing conditions to obtain a milk powder with a_w value being close to 0.2, which was recommended as the optimal a_w to preserve WMP quality during 25 °C storage (Stapelfeldt et al., 1997; Schuck et al., 2007). The optimal a_w was also reported to be 0.2 for increasing shelf life of spray-dried milk base for baby food (Roozen and Linssen, 1992). Van Mil and Jans (1991) noticed that the WMP with 3% moisture had a higher flavor score and lower PV than the one with 2.7% or 2.4% moisture during 36 months of storage.

In addition, packaging atmosphere and storage temperature have impacts on the oxidative stability and shelf life of milk powders. WMP storage tests were conducted at low, ambient, and high temperatures (Lloyd et al., 2009b; Van Mil and Jans, 1991; Stapelfeldt et al., 1997). Lloyd et al. (2009b) investigated the effect of packaging atmosphere, storage temperature, and time on WMP shelf life using sensory, GC–MS–SPME, PV, and headspace oxygen analytical methods. Two commercial batches of WMP were repackaged in plastic laminate pouches with air or nitrogen and stored at 2 °C or 23 °C for 1 year. Volatile compounds such as hexanal, heptanal, octanal, and nonanal in WMP were measured at 0, 2, 3, 4, 6, 8, 10, and 12 months using SPME GC–MS. Their experimental results indicate that the air-stored WMP had higher level of PVs, volatile compounds, and grassy and painty flavors than the nitrogen-flushed WMP. Storage temperatures (2 °C and 23 °C) did not affect levels of volatiles derived from lipid oxidation. Therefore, nitrogen flushing had a much greater effect than storage temperature in preventing the formation of hexanal, a key indicator associated with n-6 fatty acids oxidation and cardboard, grassy, and painty off-flavors (Carunchia Whetstine and Drake, 2007).

Lee et al. (2012) also reported a comprehensive stability evaluation of iron-fortified milk powder. The result suggests that Fe(III)-fortified formula had the lowest stability in aerobic condition but its stability improved significantly with vacuum/anaerobic packaging. In addition, iron-fortified milk powder–based infant formulas were stabilized by forming an iron–lactoferrin complex (Ueno et al., 2012) or adjusting a_w to 0.2 (Roozen and Linssen, 1992) due to free iron, a potential oxidation catalyst, speeding up auto-oxidation in dried milk powder and its derived products.

9.6 DRIED EGG POWDERS

Eggs are suitably used as food ingredients for different applications in food industry due to their unique compositions of proteins and lipids. The great consumption of industrialized egg-containing foods such as bakery products and pasta has imposed the need to transform liquid eggs into egg powders (Mazalli and Bragagnolo, 2007; Pignoli et al., 2009). Dried egg powders are fully dehydrated eggs. They are produced with liquid eggs by spray drying or freeze drying in the same way as milk powders are made. Dried egg powders have many advantages over liquid eggs, such as lower price, much longer shelf life, convenient storage and shipment, and easy handling. Egg powders can be divided into three different categories: whole egg, egg yolk, and egg white powders. The compositions of protein, moisture, fat, and cholesterol of whole egg, egg yolk, and egg white powders are showed on Table 9.3.Whole egg and egg yolk powders contain high levels of lipid and cholesterol, with whole egg powders being 40.9 g lipid/100 g and 1.5 g cholesterol/100 g and egg yolk powders 55.8 g lipid/100 g and 2.1 g cholesterol/100 g. According to USDA nutrient database for standard reference (2010), whole egg and egg yolk powders contain 13.50% and 18.13% of total saturated fatty acids, 17.10% and 23.70% of total monounsaturated fatty acids, and 6.86% and 8.76% of total polyunsaturated fatty acids, respectively. The predominant polyunsaturated fatty acids in whole egg and egg yolk powders are 18:2 linoleic fatty acids, with whole egg powders being 5.96% and egg yolk powders 7.56%. We will principally discuss the oxidation of fat and cholesterol of whole egg and egg yolk powders, since the egg white powders contain a very low level of fat.

Because whole egg and egg yolk powders contain phospholipids, triacylglycerides, and cholesterols, the lipid oxidation of whole egg and egg yolk powders is complicated and forms various undesirable compounds, such as primary

TABLE 9.3 Assayed Egg Nutrition Value (USDA Nutrient Database for Standard Reference, Released November 13, 1999)

	Liquid/Frozen						Dried		
Nutrients (per 100g)	Whole Eggs	Yolk	Sugared Yolk	Salted Yolk	Salted Whole Eggs	White	Whole Eggs	Yolk	Stabilized White
Protein (g)	12.56	15.86	13.80	14.00	10.97	10.90	47.35	34.25	82.40
Moisture (g)	76.15	52.31	51.25	50.80	67.83	87.57	3.10	2.95	8.54
Fat (g; total lipid)	9.51	26.54	22.75	23.00	10.07	0.17	40.95	55.80	0.04
Ash (g)	1.06	1.71	1.40	10.60	10.30	0.63	3.65	3.40	4.55
Carbohydrate (g)	0.72	3.59	10.80	1.60	0.83	0.73	4.95	3.60	4.47
Free glucose (g)	<0.1	0.2	N/A	N/A	N/A	0.3	0.3	<0.1	0
Calories (kcal)	143	322	307	274	138	52	594	666	376
Cholesterol (mg)	372	1085	959	955	387	0	1507	2052	0

and secondary oxidation products, and cholesterol oxidation products (COPs) or cholesterol oxides or oxysterols. COPs may be involved in lipid metabolism, various chronic and degenerative diseases such as cancer, aging, and human atherosclerosis, and disturbance of cell functionality (Schroepfer, 2000; Pignoli et al., 2009). Therefore, it is critical to prevent all three lipids in whole egg and egg yolk powders from oxidation during process and storage.

As unique food ingredients, whole egg and egg yolk powders have been widely used to produce a variety of food products, such as pasta (spaghetti), bakery foods, mayonnaise, salad dressing, high-protein nutrition bars, and high-protein flour dough. Thus, the lipid oxidation and shelf life of these food products are closely related to those of whole egg or egg yolk powders. Therefore, it is essential to study the oxidative stability and shelf life of whole egg, egg yolk powders, and food products containing these egg powders.

Figure 9.2 shows the chemical structures of main oxidation products of cholesterol. Cholesterol contains one double bond between C5 and C6. The cholesterol oxidation process probably undergoes the same free radical mechanism as the lipid oxidation (Smith, 1987; Smith and Johnson, 1989). Cholesterol oxidation is initiated by the abstraction of the allylic C-7 hydrogen, then the peroxide radical, 7-hydroperoxide, and alkoxyl radical are formed. Alkoxy radical can be further decomposing to produce 7α- and 7β-hydroxycholesterols and 7-ketocholesterol, as well as cholesterol α- and β-epoxides.

Hydroperoxide value (PV), conjugated dienes (CD), TBARS, and the concentration change of fatty acids and tocopherols over storage have been used to evaluate the oxidative stability of phospholipids and triacylglycerides in whole egg and egg yolk powders (Verardo et al., 2010). GC–MS or FID is used to assess the COPs in whole egg and egg yolk powders (Boselli et al., 2002). 7-Hydroxy-cholesterol was a classic marker of cholesterol oxidation. The commonly used oxidative stability markers are COPs, TBARS, and PV (Obara et al., 2006).

A number of factors impact the oxidative stability and shelf life of whole egg and egg yolk powders as well as the food products containing these egg powders. Examples of the impact factors are processing and storage conditions, such as temperature, thermal processing, drying process and types, oxygen concentration, a_w, light, temperature, package, and rearing and feeding systems (Verardo et al., 2010; Obara et al., 2006; Mazalli and Bragagnolo, 2007; Rao et al., 2013; Meynier et al., 2014; Pignoli et al., 2009). Various strategies have been taken to prevent the lipid oxidation and extend the shelf life of whole milk and egg yolk powders as well as the powder-related products.

Obra et al. (2006) reported the effect of a_w in spray- and freeze-dried whole egg and egg yolk powders on cholesterol oxidation during 3 months of ambient storage. The first group of powders was not moisturized, with 1.37% and 4.05% water content for freeze-dried and spray-dried whole egg powders, respectively, and 1.67% and 2.78% water content for freeze-dried and spray-dried egg yolk powders, respectively. The second and the third groups were moisturized to 8% and 12% of water content. According to the authors, egg

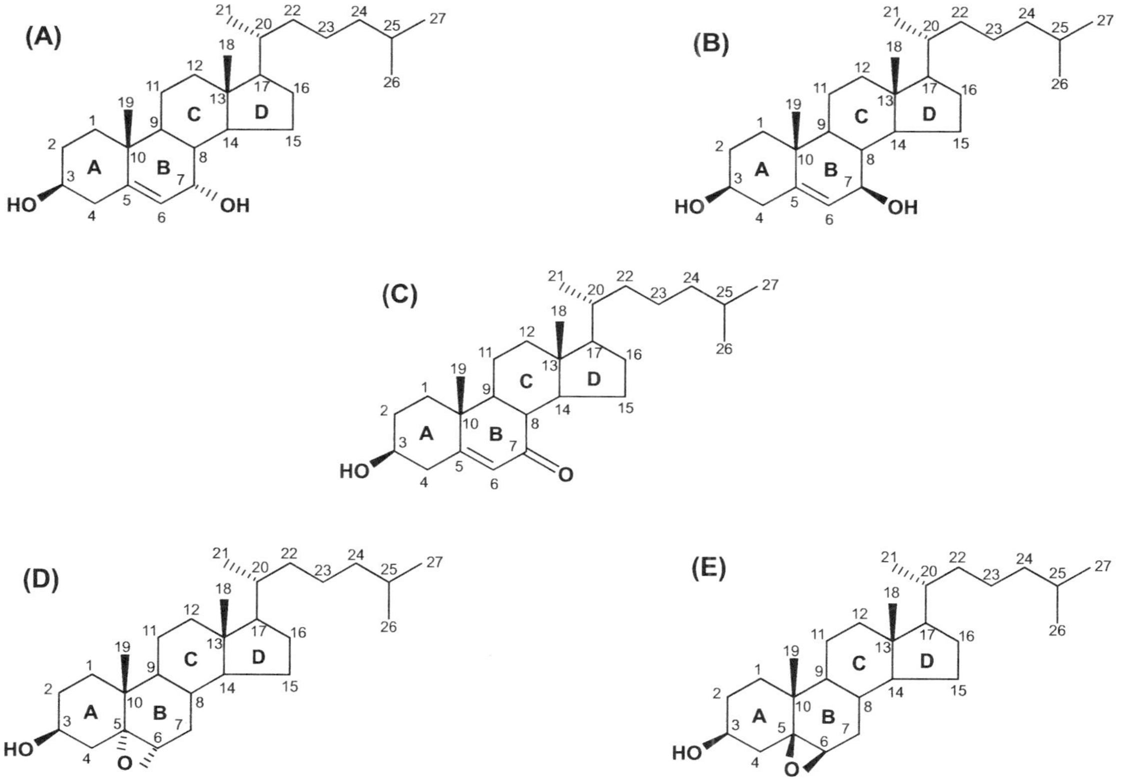

FIGURE 9.2 Structures of oxidized cholesterols: (A) 7α-hydroxycholesterol (7α-HC), (B) 7β-hydroxycholesterol (7-β-HC), (C) 7-ketocholesterol (7-KC), (D) 5α,6α-epoxycholesterol (5α,6α-EP or α-CE), and (E) 5β,6β-epoxycholesterol (5β,6β-EP or β-CE).

yolk powders had higher a_w than whole egg powders. GC-FID was used to analyze concentration changes of COPs, 5α,6α-epoxycholesterol (5α,6α-EP), 7α-hydroxycholesterol (7α-HC), 5β,6β-epoxycholesterol (5β,6β-EP), 7-keto-cholesterol (7-KC), and 7β-hydroxycholesterol (7β-HC), over storage. Water activity in powders had significant influence on COPs accumulation. The authors concluded that during 3-month ambient storage, COP accumulation appeared the highest in the samples with the lowest a_w, such as 0.17 and 0.32 for freeze-dried and spray-dried whole egg powders (not moisturized), respectively. Whole egg powders had higher COPs than egg yolk powder. Spray-dried egg powders had higher concentrations of COPs than freeze-dried ones. The highest amount of COPs in the tested samples were 5,6-EP isomers. Therefore, controlling a_w of whole egg and egg yolk powders is critical to extend their shelf life.

Rao et al. (2013) investigated the storage stability of a commercial hen egg yolk powder in dry and intermediate-moisture food (IMF) matrices such as high-protein nutrition bars. Dry egg yolk was produced with 21.4 MPa of pressure of spray-drying, inlet temperature was 185 °C. Pasteurization temperature and time were 62 °C and 4 min. Dry egg yolk contained 36% protein and 60% lipid. Its a_w and moisture content were 0.27% and 2.1%. According to the authors, in order to maximize the shelf life, the optimal moisture content of the dry egg yolk should be at its GAB (Guggeuheim-Anderson-de Boer) monolayer moisture value (m_0, ~2.3 g H_2O/100 g solids), at which its a_w is about 0.25 at 25 °C. Thus, for the optimum stability, dry egg yolk should be dried to no less than 2 or more than 3 g water/100 g solids to minimize lipid oxidation (Rao et al., 2013). The moisture sorption isotherms of dry egg powders can be used to select the optimum packaging material needed to maintain the powders below an a_w for a desired shelf life at a selected external temperature and humidity of storage (Rao et al., 2013).

Rao et al. (2013) used the dry egg powder to produce intermediate-moisture foods (IMF), a high-nutrition bar model system. The storage stability of dried egg yolk powders and 10 IMF systems (a_w ~ 0.6) containing 5% glycerol, 10% shortening, 35% protein, and 50% sweetener were studied. According to the authors, overall, the storage stability of dry egg yolk in dry and IMF matrices, during the storage at three different temperatures (23, 35, and 45 °C), was mainly controlled by disulfide bond interaction, Maillard reaction, and lipid oxidation. Their test results show that by replacing 25% of dry egg white in an IMF model system with dry egg powder, the rate of bar hardening was significantly lower than that of the models with only dry egg white at all temperatures due to the softening effect of the fat in dry egg yolk.

In addition, Rao et al. (2013) used log Q_{10} and the change in rate Q_A to predict the shelf life of high-protein bar model system at 23 °C, based on color and hardness. The two equations can be also used to predict the shelf life based on oxidative stability data, although the authors did not do it due to lack of data on oxidative stability.

Verardo et al. (2010) studied the oxidative stability of three raw egg ingredients and the influence of storage conditions on cholesterol oxidation in dried egg pasta (spaghetti) produced with the three raw egg ingredients: pasteurized eggs from hens breed with organic methods (POE), pasteurized eggs from conventional breeding (PCE), and spray-dried pasteurized eggs (SPCE) obtained from conventional breeding. SPCE and POE showed the highest and lowest PV and content of cholesterol oxidation products, respectively. After POE, PCE, and SPCE were used to produce dried egg spaghetti, the three spaghetti products were stored at room temperature for 12 months using typical pasta packaging (polypropylene foil) under light and in the dark. The changes of PVs and five cholesterol oxidation products (Figure 9.2) concentrations were measured during 1 year of dark and light storage at room temperature.

According to the authors, while PV and the concentration of cholesterol oxidation products of POE, PCE, and SPCE significantly varied, the PVs of pasta produced with POE, PCE, and SPCE were in the range of 3.0–3.5 mEq of O_2/kg of fat, showing no differences. Samples stored in the dark did not show any significant difference in PVs. However, the PVs of PCE-, SPCE-, and POE-containing spaghetti stored under the light increased by 742.8%, 846.7%, and 625.7%, respectively. Further, COP value was ~50 μg of COPs/g fat on day 0. During the storage, the COPs increased significantly (Verardo et al., 2010). Authors' test result shows that the total COPs levels of spaghetti produced with PCE, SPCE, and POE stored in the dark were 2.0, 2.0, and 1.5 times lower than that of the samples stored under the light, respectively. Accordingly, the packaging with light barrier is critical to prevent the light-induced cholesterol oxidation in pasta (Verardo et al., 2010). In summary, selecting a suitable breeding method and processing and packaging material with a light barrier is essential to decrease cholesterols and fatty acid oxidation and extend the shell life of the products containing egg powders.

Meynier et al. (2014) reported the impact of the feed composition and processing parameters on lipid oxidation of the shell eggs and the three different egg powders. Two batches of shell eggs were processed: one issued from hens fed a standard diet and another receiving a diet enriched in extruded linseed containing linolenic acid. Whole egg, standard egg yolk (control) and n-3 fatty acid–enriched egg yolk powders were stored at 15 and 30 °C for 9 months. CD, malondialdehyde (MDA), tocopherols, lutein and zeaxanthin, COPs, and hydroxyalkenals were quantified during storage and used to assess fatty acid and cholesterol oxidation. According to the authors, egg yolk lipids were composed of approximately three-quarters triacylglycerols (TAG) and one-quarter phospholipids (PL), regardless of the hen diet. But the three tested egg powders contain varying levels of triacylglycerides and phospholipids, which contain different levels of 18:3 n-3 and 22:6 n-3 fatty acids and varying levels of lutein, zeaxanthin, and tocopherols.

The order of oxidative stability during 9-month storage was n-3 fatty acid–enriched egg yolk powders > standard egg yolk powders > whole egg powders, based on CD. But the order of oxidative stability was: standard egg yolk powders > n-3 fatty acid–enriched egg yolk powder > whole egg powders, based on MDA (Meynier et al., 2014). Clearly, the oxidative stability of whole egg powders was lower than that of n-3 fatty acid–enriched egg yolk and standard egg yolk powders. According to the authors, the reason may be attributed to the egg yolk powders having high level of phospholipids and phosvitin (an iron-binding chelate). In addition, the long-chain PUFA such as DHA in egg yolks was esterified with phospholipids, which were mainly located in the high-density lipoproteins. The esterified PUFA may be more oxidatively stable than the free PUFA. Further, lutein, zeaxanthin, and tocopherols may also contribute to antioxidant activity of the egg yolk powders. The oxidative stability of standard egg yolk powders was greater than that of n-3 fatty acid–enriched egg yolk powders (based on MDA), because the enriched egg yolk powders contained a much higher level of n-3 fatty acids and a slightly lower level of mixed tocopherols, although the enriched egg yolk powders had a much higher level of lutein and zeaxanthin.

The authors finally concluded that no cholesterol oxidation products were detected in the egg yolks during different steps of processing and storage, even in the enriched eggs. The lipid oxidation remained at a low level throughout the process, even if the quantities of primary and secondary oxidation products slightly increased after pasteurization, drying, and on storage. In addition, while sensory test showed an increase in the overall aroma in all the stored powders, off-flavors and rancid notes were not detected during 9-month storage. Therefore, n-3 fatty acid–enriched egg yolk powders, like standard egg yolk powder, had a remarkable oxidative stability (Meynier et al., 2014).

Influences of process, storage, and package on the lipid oxidation of egg powders were also investigated by other researchers. COP levels of egg yolk powders were found to vary from 8 to 311 mg/kg (Lercker and Rodriguez-Estrada, 2000), which depended on the type of drying process and the drying and storage conditions such as temperature and oxygen partial pressure. Spray-dried egg yolks had more COPs than freeze-dried ones (Pignoli et al., 2009). The cholesterol oxidation rate in egg yolk powders was found to vary from 0.52% to 1.46%, being higher in spray-dried egg yolks than in freeze-dried ones (Pie et al., 1990). Vacuum-packaged egg powder in aluminum foil without exposure to light prevented the loss of a large amount of PUFAs, especially C20:4 n6 and C22:6 n3.

Future research may be needed to investigate physical location of PL, TAG, and cholesterols in egg yolk and whole egg powders. Headspace volatile compounds such as hexanal and propanal (for egg powders containing n-3 fatty acids), together with COPs, may be used as indicators to monitor the oxidative stability of PL, TAG, and cholesterols in egg powders.

9.7 MICROENCAPSULATED OIL POWDERS

Unlike the milk powders that are produced by dehydrating natural emulsified milks, the dried microencapsulated oils are made by drying formulated food emulsions. Physically and oxidatively stable oil-in-water emulsions are first prepared, and then the formulated emulsions are spray-dried or freeze-dried to produce microencapsulated oil powders in which oil droplets are surrounded by a dry matrix of proteins and/or carbohydrates (Marquez-Ruiz et al., 2003). There are a variety of microencapsulated oil powders, such as encapsulated fish, flaxseed, algal and sunflower oils, and carotenoids powders, as well as infant formula powders containing the encapsulated oils. These microencapsulated oil powders have been widely applied to food industry as foods and food ingredients. Microencapsulated processing would protect the oxidation-sensitive oils from oxidation, mask or preserve flavors and aromas, and transfer bioactive liquid lipids into easily handle powdery solids (Matsuno and Adachi, 1993). The commonly used oils/fats for encapsulation are marine fish, algal, flaxseed, linseed, sunflower and extra virgin olive oils, milk fat, seed oils, and lipophilic bioactive compounds such as lycopene, astaxanthin, lutein, β-carotene and plant sterols, as well as the blend of oils. The matrix materials applied to encapsulating oils are proteins, such as sodium caseinate, whey protein isolate, pea protein and gelatin, and carbohydrates such as gum arabic, pectin, cellulose, alginate, chitosan, maltodextrins, corn syrup solids, lactose, and cyclodextrins. Electrostatic layer-by-layer deposition technologies have been used to produce a variety of thick multilayer interfacial membranes to physically stabilize emulsions and inhibit lipid oxidation in oil-in-water emulsions and encapsulate flavor compounds in the recent years. The pea protein–pectin complexes (Gharsallaoui et al., 2012) and lecithin–chitosan complexes (Shaw et al., 2007; Klinkesorn et al., 2005) were reported as the multilayer interfacial membrances to stabilize encapsulated oil powders. In addition, a novel method for stabilizing n-3 fatty acid (DHA) by ultrathin electrosprayed zein prolamine was reported (Torres-Giner et al., 2001).

There are, generally, two methods reported to produce microencapsulated oil powders. The first one is to produce a physically stable formulated oil-in-water emulsion made with high-pressure homogenizer or microfluidizer, then to dehydrate the emulsion using spray-drying or freeze-drying. The second one is to make a pre-emulsion using a high speed blender or mixer and then to dry the pre-emulsion. A homogeneous and physically stable encapsulated oil powder would be obtained when drying a physically stable oil-in-water emulsion. A pre-emulsion made with a high-speed blender, other than a high-pressure homogenizer, may not be physically stable. Thus, a homogeneous and physically stable encapsulated oil powder may not be easily obtained when dehydrating a pre-emulsion. As far as evaluating and comparing the oxidative stability and shelf life of dried encapsulated oils, we may need to notice what kind of emulsion is used to produce the encapsulated oil powders.

It is critical to understand the location of the oil in the matrix of the encapsulated oil powders and the impact of the oil distribution on lipid oxidation in the encapsulated oil powders. As discussed earlier regarding the milk powders, the encapsulated oil globules and vacuoles could be embedded in the wall or inert filler. The distribution of the oil in the powder particles may not be continuous. Therefore, the oil in an encapsulated oil powder constitutes a discontinuous or disperse phase and the oil distribution in the encapsulated oil powders is heterogeneous.

The oxidation of oil in encapsulated oil powder is complicated and the oxidation of free oil may be different from that of encapsulated oil depending on a number of factors such as encapsulated matrix, antioxidants, and RH. The oxidation of the milk fat encapsulated in a matrix of maltodextrin and sodium caseinate was slower than that of the free oil fraction either under the light or in the dark (Hardas et al., 2002). But for the relatively stable oils such as the oils containing naturally occurring and added antioxidants, the oxidation was remarkably faster in the encapsulated oil fraction than in free oil fraction (Velasco et al., 2006). Distinguishing free oil and encapsulated oil and monitoring the oxidative stability of both oils are helpful in understanding the oil oxidation in encapsulated oil powders and extending the shelf life of the powders.

GC headspace propanal or 2,4-heptadienal, hexanal or 2,4-decadienal analysis, and HPSEC might be more reliable methods for evaluating the oxidative stability of encapsulated oil powders, although a number of analytical methods are available, such as determination of PV, CD, TBARS, headspace oxygen consumption, fluorescence ratio (Barrett et al., 2011), propanal and hexanal, oxidized triglyceride monomers, dimers and oligomers by high-performance size-exclusion chromatography (HPSEC), loss of tocopherols and unoxidized substrates such as EPA and DHA. HPSEC can analyze nonvolatile lipid oxidation products such as oxidized triglyceride monomers, dimmers, and oligomers (Marquez-Ruiz et al., 2003), indicating the degree of oxidation. The GC- headspace method was successfully applied to measuring the hexanal level changes of WMP over 1-year storage (Lloyd et al., 2009a,b). Thus, GC headspace propanal or 2,4-heptadienal, monadienal, hexanal, or 2,4-decadienal analytical methods are recommended for the food industry to monitor the oxidative stability of encapsulated oil powders. The effect of HS-SPME (headspace solid-phase microextraction) extract and humidity on the release of volatile lipid oxidation products from spray-dried emulsion powders has recently been reported (Damerau et al., 2014a,b). The highest overall release was obtained at water contents of 3.1% and 5.2% (RH 11% and 33%). Eighteen volatile lipid oxidation products in the encapsulated oil powders were quantified using HS-SPME (40 °C and 250 rpm) and GC–MS. But no water or salty water was added to the powders for pretreatment. In addition, the fluorescence method may be a good one for evaluating solid samples, such as encapsulated oil powders containing amino acids or proteins, because the powder sample can be directly measured without pretreatment (Barrett et al., 2011). The principle of the measurement is based on the fluorescent molecule Schiff base formation due to volatile unsaturated aldehydes such as MDA reacting with free amine groups derived from amino acids or proteins.

One of the most important factors greatly influencing analytical results is the pretreatment of samples before using GC–FID or MS headspace method to analyze the samples. If hexanal and propanal are encapsulated in the inner wall of the dried carbohydrate and/or protein martrix, and pretreatment method cannot effectively force them to liberate, the concentrations of hexanal and propanal analyzed might be lower. This may be the case when heating encapsulated oil powder without adding water or salty water for pretreatment, since protein such as caseinate, and oligosaccharides and monosaccharides such as maltodextrins and corn syrup solids, can bind volatile compounds through hydrophobic interactions and hydrogen bonds. It is therefore recommended that water or salty water be used for pretreatment when heating. In this case, the encapsulated hexanal or propanal can be liberated much more completely. Another factor is heating temperature: 40, 50, 60, 70, 80, and 90 °C could be selected. The higher pretreatment temperature may facilitate the liberation of the encapsulated volatile compounds from the matrices. On the other hand, the higher pretreatment temperature may also facilitate further oxidation of the encapsulated oil powders. Therefore, it is important to test an appropriate heating temperature for a special dry powder sample.

Lipid extraction is also important when the lipid oxidation in encapsulated oil powders is evaluated by AOCS hydroperoxide method or OSI. If the encapsulated powders contain both neutral oil triglycerides and polar fat phospholipids, it is critical to extract oil and fat as completely as possible when the extracted oil and fat are used for PV or OSI measurement. In some of the literature and in the practice of the food industry, hexane or petroleum ether is used for the extraction of oil/fat. If that is the case, the encapsulated oil may not be completely extracted, which would have a negative impact on the evaluation of oxidative stability of the encapsulated oil powders. In order to extract encapsulated oil, the Rose-Gottlieb method (Richardson, 1985) or Folch extraction method (using chloroform-methanol [2:1 by volume]) is usually used. The disadvantage of Folch methods is using very toxic chloroform and complicated procedures. It is critical to select appropriate solvent(s) to extract oils from encapsulated oil powders in order to evaluate the oxidative stability accurately. Currently, the standard GC headspace hexanal and propanal analytical methods are not established. Hence, it is very difficult to compare and evaluate the oxidative stability and shelf life of encapsulated oil powders and other foods containing oils and/or fats in different laboratories.

Water activity, RH, and glass transition temperature (Tg) have great impacts on the oxidative stability of encapsulated oil powders. As discussed previously, lipid oxidation rate is the lowest at a_w being close to the water monolayer, which corresponds to 0.2 and 0.3 for most foods (Labuza, 1980). However, the monolayer theory meets challenges when predicting some of dried foods such as encapsulated oil powders. At lower moisture content, the presence of pores and cracks in the encapsulated oil powders may facilitate the transportation of oxygen, thereby increasing the oxidation rates (Damerau et al., 2014a,b). Actually, the relationship between oxidation rates and a_w is complex. The state of water and Tg should be also considered when predicting the oxidative stability and shelf life. Although at the temperature above the glass transition, increase of molecular mobility would lead to an increase of lipid oxidation (Labrousee et al., 1992), the opposite results were also reported (Grattard et al., 2002; Selim et al., 2000). Therefore, the monolayer theory and glass transition theory fail to predict the lipid stability of various dry foods.

It is a great challenge to assess the oxidative stability and shelf life of encapsulated oil powders and to draw general conclusions, because a number of factors influence the oxidative stability and shelf life of the powders. The first one is the matrix. There are a variety of matrices available because a number of proteins and carbohydrates can be used to encapsulate oils. Different matrices constitute varying microstructures that impact oil distribution and lipid oxidation. Varying concentrations of carbohydrates and proteins have impacts on physicochemical properties of encapsulated powders. The second one is the emulsion used for encapsulation. Two types of oil-in-water emulsions, pre-emulsion and physically stable emulsion, are used to produce encapsulated oil powders. Oil distribution and microstructure in encapsulated oil powders may be different in pre-emulsion–based and stable emulsion–based encapsulated powders. In addition, processing conditions such as drying types, spray-drying and freeze drying; storage conditions, such as storage temperature and RH; and package, such as vacuum and modified atmosphere package, affect the oxidative stability of the encapsulated oil powders. Further, as discussed earlier, a_w, Tg, and size of porosity can greatly affect the oxidation of the powders. Finally, different extraction, pretreatment, and analytical methods significantly influence the analytical result and the assessment of oxidative stability.

The prediction of shelf life of microencapsulated oil powders or the microencapsulated oil–based food products can be conducted using the Rancimat test. Rubilar et al. (2012) reported that OSI measured the IP of the extracted oil from a soup powder enriched with microencapsulated linseed oil at temperatures 80, 90, 100, 110, and 120 °C. The equation, $\log(\mathrm{OSI}_{20}) = a\,(T) + b$, was used to predict the actual (20 °C) shelf life of the soup powder.

9.7.1 Microencapsulated Oil Powders Produced with Pre-Emulsions

Chemical reaction in solid powders may be different from that in liquid. Physical state, molecular mobility, and the amount of water in the matrices of encapsulated oil powders may have impacts on the lipid oxidation rate of encapsulated oils (Grattard et al., 2002). According to the authors, encapsulated flaxseed oil powders were prepared using a Polytron PCU-2 high-speed blender (stirring for 2 min) to make a pre-emulsion and then freeze-drying the pre-emulsion. We call the pre-emulsions since a high-speed lab mixer, other than a high-pressure homogenizer, was used for preparing the emulsions. The encapsulated flaxseed oil powders contained maltodexteins with varying DEs (dextrose equivalents), lecithin, and xanthan gum. The authors used electron spin resonance (ESR) to measure the oxidation rate of the encapsulated flaxseed oil, indicating that the encapsulated oil oxidation reaction was first order. Oxidation was accelerated at lower a_w (0.15) and high a_w (0.6 and 0.8), but the oxidation rate was slow at intermediate a_w (0.3 and 0.4) for the sample containing maltodextrins DE 40 (Grattard et al., 2002). The maltodextrins with high DEs may have different influence on the structure and the oxidative stability of encapsulated oils.

According to Grattard et al. (2002), relatively low molecular mobility would be one of the characteristics of glassy matrices in encapsulated oil powders. A diffusion-controlled chemical reaction rate could be relatively slow in a glass matrix. As a_w or temperature increased, the molecular mobility and diffusion rate would increase. Although the T_g was used to predict oxidation rate, the T_g of the maltodextrin matrix did not seem to control the rate of lipid oxidation in freeze-dried emulsions (Grattard et al., 2002). The authors concluded that no apparent relation between molecular mobility and lipid oxidation rate were observed in the various freeze-dried maltodextrin matrices. Amount of water, much more than its mobility, could be a key parameter controlling lipid oxidation rate.

Velasco et al. (2000, 2003, 2004, 2006, 2009) conducted a lot of research on lipid oxidation in the encapsulated oil powders. The main focus of their studies was on the oxidation of free and encapsulated oil in the encapsulated oil powders produced with pre-emulsions. The oxidation in dried encapsulated fish oils produced with pre-emulsions was much faster under the light than in the dark. In the dark, the oxidation was triggered first in the free oil fraction of the samples without adding antioxidant blend. Difference in lipid oxidation between free oil and encapsulated oil in encapsulated oil powders produced with pre-emulsions was reported by Velasco and coworkers (Velasco et al., 2000, 2003, 2006, 2009). According to the authors, the oxidation pattern of the free oil was very similar to that in bulk oils, showing a clear end of the induction period or lag phase as marked by the initiation of polymerization and the exhaustion of tocopherols. However, the encapsulated oil oxidation is unusual. There was no clear induction period or lag phase. High content of advanced oxidized polymers was found in the samples but the samples still contained a high level of tocopherols (Velasco et al., 2006). The authors indicated that the encapsulated oil droplets had different oxidation status. Some of the oil droplets were at lower oxidation stages due to the protection of the tocopehrol and the impacts of other factors, such as oxygen availability, molecular mobility, size, and distribution of oil globules embedded in the wall. But others were at advanced oxidation stages due to the depletion of the tocopherol and the influence of other factors. These cases could be partly attributed to the encapsulated oil powders being produced with pre-emulsions.

Velasco et al. (2006) studied the difference of oxidative stability between free oil and encapsulated oil by analyzing the concentration changes of nonvolatile lipid oxidation products and α-tocopherol. HPSEC with refraction index detector was used to quantify the nonvolatile lipid oxidation products, oxidized triglyceride monomers, dimmers, and oligomers. According to the authors, dried microencapsulated oils (DMOs) were prepared with sunflower oil with tocopherols (DMSO), sunflower oil without tocopherols (DMSO-ST), fish oil without antioxidants (DMFO), and fish oil containing antioxidant blend (DMFO-ALT), ascorbic acid (0.03% w/w), lecithin (0.5% w/w), and α-tocopherol (0.03% w/w). The pre-emulsions containing sodium caseinate and D-lactose were prepared using a lab mixer at 10,000 rpm for 5 min. Then the pre-emulsions were freeze-dried to produce the encapsulated oils powders. The powder samples were stored at 30 °C in the dark at 32% RH for up to 380 days. The authors concluded that the oxidative stability of DMSO was much better than that of DMSO-ST based on the evolution of nonvolatile lipid oxidation products and α-tocopherol. The oxidative stability of DMFO-ALT was much better than that of DMFO. The free oil in DMSO-ST was more readily oxidized than that of the encapsulated oil in DMSO-ST. The same trend was observed for the free and encapsulated oils in DMFO. However, the encapsulated oil in DMSO (with mixed tocophorols) was much more easily oxidized than the free oil in DMSO (with mixed tocopherols). The same conclusion can be drawn for the DMFO-ALT (Velasco et al., 2006).

According to Velasco et al. (2006, 2009), when the oil in dried encapsulated oil powders contained antioxidant or antioxidant blend, which was the case in the food industry, lipid oxidation rate was slow due to the addition of antioxidant. In this case, the encapsulated oil oxidized more rapidly than the free oil. The reason may be attributed to the oxidation rate of the encapsulated oil being slower than the rate of oxygen supply, since the oxygen diffusion through the solid matrix or the wall was not a limiting factor for the oxidation of the encapsulated oil fraction when the oxidation rate was slow (Velasco et al., 2006, 2009). Further, the ratio of surface area to volume of the free oil might be smaller than that of the encapsulated oil in the dried powder particles. When the oil in encapsulated oil powders contained no antioxidant, lipid oxidation rate was very fast. Oxygen diffusion through the solid matrix became the determining factor for the oxidation of the encapsulated oil fraction (Velasco et al., 2006, 2009). Fast lipid oxidation needs more oxygen, but the solid matrix limited the oxygen diffusion, although the encapsulated oil had a high ratio of surface area to volume. In this case, the consumption of oxygen was faster than was the oxygen supply. As a result, lipid oxidation rate of the encapsulated oil decreased.

Velasco et al. (2006) reported that for the bulk sunflower oil, an excellent linear correlation ($r^2 = 0.996$) was established between the amount of oxidized triglyceride monomers (ox-TGM) and the PV before the oxidation accelerated. With the oxidation speeding up, more and more ox-TGM accumulated as the secondary oxidation products (dimers and oligomers) increased. This oxidation pattern was also found in other bulk oils and foods with a continuous oil phase, such as fried foods (Marquez-Ruiz et al., 1999).

Velasco et al. (2009) also evaluated the effect of RH on oxidative stability of the free and encapsulated sunflower and fish oil fractions of the dried microencapsulated oils during the long-term storage at 30 °C in the dark. The results show that the lipid oxidation rate of encapsulated oil under 0% RH atmosphere was faster than that of the encapsulated oil under 32% RH atmosphere by measuring the concentration changes of the total nonvolatile lipid oxidation products and α-tocopherol over storage at 30 °C.

In the food industry, the optimum water content or a_w of a product of encapsulated oil powders needs to be tested. The encapsulated oil powders may be packaged with vacuum or nitrogen or modified atmosphere.

9.7.2 Microencapsulated Oil Powders Produced with Physically Stable Emulsions

Compared with the oil in oil-in-water emulsions, the microencapsulated oils may be oxidatively stable. But the encapsulated oil powders are still readily oxidized probably due to the inability to obtain 100% efficiency of encapsulation, large ratio of surface area to volume of encapsulated oil, and varying size of vacuoles in the powder particles. Therefore, the encapsulated oil powders produced with the emulsions stabilized by monolayer or multilayer interfacial membrane have been developed to further increase physical and oxidative stability of encapsulated oil powders.

9.7.3 Microencapsulated Oil Powders Produced with the Emulsions Stabilized with Multilayer Interfacial Membrane

Varying concentrations of corn syrup solids (CSS) were used to study both physical and oxidative stability of the menhaden oil encapsulated with two-layer interfacial membranes (Shaw et al., 2007). The emulsion used to spray-dry was a lecithin-chitosan multilayer emulsion made with a two-stage high-pressure homogenizer. Scanning electron microscopy pictures showed that the spray-dried powders containing 1% or 2% CSS were extremely clumping and coalescent, indicating that the 1 and 2 wt% CSS emulsions were not fully microencapsulated. But the spray-dried powders containing 5, 10, and 20 wt% CSS had discrete and individual particles, suggesting a full microencapsulation (Shaw et al., 2007).

Shaw et al. (2007) evaluated the oxidative stability of the microencapsulated fish oil powders by measuring PV and TBARS. The order of oxidative stability of the encapsulated fish oil powders with different levels of CSS was: 5% CSS > 20% CSS > 10% CSS > 2% CSS = 1% CSS. According to the authors, the order may be mainly attributed to the concentration of CSS and reduction ability of the CSS. Therefore, it is important to test an optimal level of CSS in order to produce physically and oxidatively stable encapsulated fish oil powders using the lecithin-chitosan multilayer emulsion.

Effects of RH, temperature, and antioxidants on the dispersibility, oxidative stability, and color of spray-dried microencapsulated tuna oil were investigated by Klinkesorn et al. (2005). Encapsulated dry powders were also produced using a lecithin-chitosan multilayer emulsion with corn syrup solid (CSS, DE=36) as wall material. Total lipid extracted from the encapsulated powders was used to evaluate the oxidative stability of the encapsulated powders stored at 37 °C and 11%, 33%, and 52% RH, respectively. The authors highlighted that the encapsulated oil powders containing multilayer interfacial membrane significantly improved the oxidative stability of tuna oil compared to bulk tuna oil, as evaluated by the formations of both PV and TBARS over a 32-day storage at 37 °C. According to the authors, the encapsulation of spray-dried tuna oil with the multilayer interfacial membrane was more effective at retarding oxidation than was 500 ppm mixed tocopherols in the bulk oil. In addition, the oxidative stability of the encapsulated powders under varying RH atmosphere was on the order of 52% RH > 32% RH > 11% RH. The author indicated that the result did not follow the classic oxidation patterns of minimum oxidation at intermediate activity (a_w 0.2–0.4) for many foods but instead at high a_w (52% RH) for encapsulated tuna oil powder. The authors believed that the low oxidation rate at high a_w or high RH may be attributed to structure change of the powder particles resulting in reduced dispersibility and less porosity, sugar crystallization and Maillard reaction. Further, the blend of 500 ppm mixed tocopheols and 12 μmol/L EDTA was effective at inhibiting lipid oxidation of the encapsulated tuna oil powders stored at 37 °C for 47 days at above-mentioned three RH levels (Klinkesorn et al., 2005).

Therefore, in order to increase the oxidative stability and shelf life of encapsulated oil powders, it is essential to select a physically and oxidatively stable emulsion containing lecithin-chitosan multilayer interfacial membrane to produce encapsulated fish oil powders with an appropriate level of filler, such as CSS and lactose. Adding antioxidant (e.g., mixed tocopherols) or antioxidant blend (e.g., blend of ascorbic acid/lecithin/tocopherols) to the encapsulated oil powders is also critical to extend their shelf life. In addition, the physical stable multilayer interfacial membrane can be also prepared using pectin, alginate, gum acacia, modified starches, pea protein, beta-lactoglobulin, caseinate, carrageenan, and gelatin.

Similarly, the physically stable emulsions stabilized by pea protein–pectin layers were spray-dried to produce the encapsulated flavor with maltodextrins as wall material (Gharsallaoui et al., 2012). The test showed that the volatile compounds were more retained in dry emulsions stabilized by pea protein isolate (PPI)–pectin complex than those stabilized by PPI alone. The pea protein–pectin multilayer was able to greatly improve physical stability and flavor retention of dried emulsion droplets. This emulsion system can also be used to encapsulate oils containing n-3 fatty acids to increase the oxidative stability and shelf life of the encapsulated oil powders. The narrow acidic pH application range such as pH 3, could be the disadvantage when applying the emulsions stabilized with the multilayer interfacial membranes and the encapsulated oil powders produced with such emulsions to the food industry.

9.7.4 Microencapsulated Oil Powders Made with the Emulsions Stabilized with Monolayer Interfacial Membrane

Sugar beet pectin–stabilized emulsions (pH 3) containing fish oil and a blend of fish oil and extra virgin olive oil (1:1 w/w) were applied to produce microencapsulated oil powders using spray-drying technology (Polavarapu et al., 2011). Although the microencapsulation efficiencies were high, deterioration in microcapsule wall integrity and an increase of oil droplet size were observed over 3-month storage at 25 °C. Lipid oxidation was evaluated by measuring propanal, hexanal, and free fatty acid concentrations over storage. The authors concluded that lipid oxidation increased with both increased oil load and storage duration but was independent of oil composition. This conclusion was based on EPA and DHA concentration changes during storage, because the authors suggested that the lower propanal level in fish oil microcapsules with 50% w/w oil at the 3-month storage did not necessarily indicate a lower extent of oxidation, as propanal was an intermediate product of oxidation. The author indicated that propanal was oxidized to propanoic acid and its concentration leveled off when the rate of conversion to the acid exceeded its rate of formation from hydroperoxides. The authors also suggested that the extent of oxidation of EPA and DHA was determined to provide an independent assessment of oxidation of these fatty acids. It is true that the concentrations of primary and secondary oxidation products could change over storage since lipid oxidation is a dynamic free radical chain reaction. Propanal can be oxidized to form propanoic acid and hexanal to form hexanoic acid. Meanwhile, propanal and hexanal can be also formed as the oxidation continues. Thus we may find an answer for the issue by measuring propanoic acid concentration. Clearly, a lot of hexanal and propanal data in the published papers and in our laboratory showed that hexanal and propanal worked well as oxidation markers in the encapsulated oil powders (e.g., milk powders) and could be better than other evaluation methods. Therefore, we may need standard hexanal and propanal analytical methods so that it is possible to compare and evaluate the data of hexanal and propanal in dried foods.

Second, Maillard reaction and lipid co-oxidation with protein may have impacts on primary and secondary oxidation products, since dried glucose syrup was a wall material and the pectin contained 4.9% protein. Furthermore, what would happen if the ambient storage test was conducted for more than 3 months? Finally, in a storage test of dried cauliflower soup powder, 11 free fatty acids concentrations did not change between 0 and 12-week storage, but oxidation did occur (Raitio et al., 2011).

According to Polavarapu et al. (2011), Oxipres (at 80 °C) test indicates that microencapsulated powder containing the oil blend (fish oil and extra virgin olive oil) showed better oxidative stability than microencapsulated powder containing fish oil. The oxidative stability order based on Oxipres at 80 °C was: bulk oil > microencapsulated oil > emulsion. But the oxidative stability of the microencapsulated oil assessed at 80 °C using Oxipres did not correlate with the one of the microencapsulated oil evaluated by analyzing EPA and DHA concentration change over 3-month storage at 25 °C. The different results could be due to high measurement temperature being used in Oxipres test. Therefore, care should be taken when the data obtained under ambient and accelerated storage and measurement conditions are used to evaluate the oxidative stability of encapsulated oil powders.

Impact of RH and cross-linking of Na-caseinate on the oxidative stability of spray-dried emulsions with sunflower oil in a Na-caseinate–maltodextrin matrix was studied by Damerau et al. (2014a,b). Two model spray-dried emulsions containing sunflower oil, maltodextrin, and either non–cross-linked (NCL) or cross-linked (CL) Na-caseinate were stored at varying relative humidities (0%, 11%, 33%, 54%, and 75%) for 29 weeks at 22 °C, in the dark in acrylic desiccator cabinets. The authors evaluated the oxidative stability of two model powders using PV, α-tocopherol, and hexanal. The data indicate that increasing RH improved the oxidative stability of the encapsulated oil powders, which was principally attributed to the structure change of the powder particles such as collapse and caking at high RH (Damerau et al., 2014a,b).

According to Damerau et al. (2014a,b), PV data measured by spectrophotometric method suggested that generally the PVs of free oil were up to two times more than those of total oils at all five RHs, because no antioxidant was added to the sunflower oil. Although the PVs were higher at RH around 0% than at RH 11% during 29-week storage, the hexanal levels were just opposite. The authors indicated that higher level of hexanal at RH 11% may be attributed to hexanal being more easily released from the matrix at higher RH due to the low solubility of hexanal in water. Further, the PVs were much higher at RH 33% than at RH 75%, but hexanal levels at both RH 33% and RH 75% were nearly the same and remained very low. As both dried powders showed comparable PVs and α-tocopherol losses in total lipids during storage at RH 11% and 33%, the comparable hexanal amounts were expected. Instead, almost no hexanal formation was found in the powders during storage at RH 33% (Damerau et al., 2014a,b). According to the authors, the possible explanation is that hydrogen bond binding at the aqueous interface between hydroperoxide and water at RH 33% may prevent hydroperoxide from decomposing to hexanal. In addition, metal catalysts may be inactivated by hydration at or near the monolayer water content. Furthermore, lipid co-oxidation with protein may affect balanced concentrations of hydroperoxide and hexanal. Finally, hexanal measurement method used may not completely release hexanal in the matrix. Or the hexanal was further oxidized to form hexanoic acid. At RH 54%, hexanal levels in the NCL were higher than those in the CL, and PVs showed the same trend. This research indicates that the cross-linking of protein only slightly improved the oxidative stability (Damerau et al., 2014a,b).

9.7.4.1 How to Manage the Oxidative Stability and Extend the Shelf Life of Encapsulated Oil Powders

The strategies of increasing the oxidative stability and extending the shelf life of encapsulated oil powders might be (1) controlling water content such as 4 to 5 wt%,

(2) adding antioxidants such as mixed tocopherols, blend of natural antioxidants, and EDTA, (3) using physical stable emulsions containing multilayer and monolayer interfacial membrane to produce encapsulated oil powders, (4) using vacuum, nitrogen, or active package, and (5) storing samples at lower temperature and intermediate RH atmosphere.

9.8 EXTRUDED BREAKFAST CEREALS AND SNACK FOODS

Breakfast cereals and snack foods are usually produced using extrusion, baking, and frying technologies. Although two or three technologies may be applied together to make these food products, we principally focus, in this section, on the extruded breakfast cereals and snack foods and their oxidative stability and shelf life. Fried foods, including fried breakfast cereals, have been discussed in Chapter 7.

Extrusion cooking technology can be used to process a variety of foods, such as breakfast cereals, snack foods, pasta, fruit- and meat-based cereals, baby foods, dry and semi-moist pet foods, and animal feeds. But breakfast cereals are among the most well-known products (Paradiso et al., 2008). According to *Food Business News* (Schroeder, 2014), the global breakfast cereal market, valued at \$32.5 US billion in 2012, is expected to grow at a compound annual growth rate of 4% for the next several years, reaching \$43.2 US billion in 2019.

Extruders are categorized based on the number of screw. Single- and twin-screw extruders are most commonly used in food and pet food industry (Riaz, 2001). Food extrusion is a high-temperature-short-time (HTST) cooking process (Moscicki and Zuilichem, 2011). The main ingredients used to produce breakfast cereals and snack foods are corn, potato, oats, wheat, rice, tapioca, soy flour, meats (for some of the products), sugar, salt, fruit powders, plant extracts, antioxidants, emulsifiers, vegetable oils, and water (Guy, 2001a). The ingredient mix is transferred to a preconditioner (or directly transferred to an extruder) and then moves through an extruder at a defined temperature, pressure, and moisture level, being forced through a spiral-shaped screw and the die of the extruder. Extruded breakfast cereals are processed in high temperature and relatively low-moisture conditions compared to conventional baking or dough process. There are several processing technologies used to produce breakfast cereals. The basic processing procedure is as follows (Paradiso et al., 2014):

> Premixing→Steam Precooking in Pre-Conditioner→Extrusion→Shredding→Flaking→Toasting (Drying)→(Coating)→Cooling→Packaging.

The basic procedure to produce the extruded snack foods is as follows (Guy, 2001b; Moscicki and Zuilichem, 2011):

> Premixing Dry Ingredients→Extruding or Sheeting and Cutting (formation of a dough and product pieces)→Frying or Microwaving or Drying by a Roller Dryer→Tumbled, Dusted, or Sprayed with Flavoring→Packaging.

9.8.1 Characteristics of the Matrices of Extruded Breakfast Cereals

The microstructure characteristics of the extruded food matrices are porous and expanded. Different microstructures may be formed if varying cereals are used as ingredients for extrusion, since cereals are composed of different levels and types of carbohydrates, fibers, proteins, and lipids. In general, the microstructure of extruded food products is porous and with air pocket (cell) due to expansion (Hu et al., 1996), and this is the result of various interactions among starch, protein, lipid, and water under the high temperature, pressure, and shear. Starch–protein interactions can be critical in determining the texture and microstructure of extruded products (Izzo and Ho, 1989). The natural biopolymers are transformed into a fluid melt in the defined temperature and moisture ranges during the extrusion. Because cereals contain high level of starches, a continuous phase of starch biopolymers could be formed and several dispersed phases such as protein and lipid dispersed phases would lie within the continuous phase (Guy, 2001). In addition, Gray et al. (2008) studied model rubbery and glassy starch-based matrices. The rubbery starch-based matrix (around 30% moisture (wet weight basis [wwb]), $a_w=0.95$) was produced by extruding waxy maize starch and free fatty acid, and the glass starch-based matrix (around 10% moisture [wwb], $a_w=0.3$) was made by freeze-drying the rubbery matrix. The authors indicated that lipid oxidation was negligible in the glassy samples without surface micro-cracks based on hexanal level change over 300 h storage at 50 °C but was fast when micro-cracks were present. The authors highlighted that linoleic acid was effectively protected against oxidation in the starch extrudates held in the glassy state and devoid of micro-cracks. Therefore, the microstructure of extruded matrix has big impact on the lipid distribution and the oxidation of the extruded foods.

Several chemical and physical changes occur during the extrusion due to high temperature and pressure, which could have effects on the matrix microstructure and the oxidative stability and shelf life of extruded foods. These changes may include starch gelatinization (Camire et al., 1990), protein denaturation, enzymes inactivation such as lipo-oxygenase and lipase (Rao and Artz, 1989), lipid binding within the starch (Artz and Rao, 1993; Strange and Schaich, 2000), iron level increase because of the wearing of the screw and barrel during extrusion (Artz and Rao, 1993), endogenous antioxidant tocopherol and carotenoid levels decrease, increased surface area due to expansion (Camire et al., 2005), and Maillard reaction. Some of the changes such as lipid binding with starch, increase of iron level (to be discussed later) and Maillard reaction could decrease oxidation of the lipid. But others, such as enzyme inactivation, increased surface of lipid, and decrease of concentration of endogenous and exogenous (added) antioxidants, would increase lipid oxidation. In addition, a_w, water content, and extrusion temperature also affect the oxidative stability of extruded food products (Artz and Rao, 1993; Paradiso et al., 2014).

9.8.2 Characteristics of Lipid Oxidation in Extruded Breakfast Cereals

The oil in extruded cereals derives generally from cereals, but vegetable oil can be also added to produce ballets before extrusion (Rababah et al., 2011). Cornmeal contained about 70% starch, 9% proteins, 8% moisture, and 5% lipids (Price, 1975). A large percentage of the lipids were nonpolar lipid triglycerides, and a small percentage of the lipids were polar lipid, with around two-thirds being phospholipids and one-third glycolipids (Weber, 1978). While the lipid content of extruded breakfast cereals usually does not exceed 6–7%, they are susceptible to oxidation due to the lower moisture content, high surface area, and potential metal contamination during extrusion, in addition to porous microstructure (Paradiso et al., 2014). However, like the coated dry kibble, the surface of the fried breakfast cereals and snack foods is covered with vegetable oil. Thus, the oxidation of the fried breakfast cereals and snack foods would be different from that of nonfried extruded breakfast cereals and snack foods.

Is iron in the extruded food matrix pro-oxidant or antioxidant? It is well documented that transition metals such as iron and copper speed up the lipid oxidation in bulk oils and oil-in-water and water-in-oil emulsions (McClemnets and Decker, 2000). However, Rao and Artz (1989) reported that 50 ppm ferrous acetate (dry weight) inhibited lipid oxidation of extruded corn starch containing added soybean oil compared to both control and 50 ppm BHA (oil weight) based on the assessment of PV, as the ground extrudate samples were stored at 37 °C in air for 40 days. Camire and Dougherty (1998) tested the effect of added phenolic compounds on the stability of extruded corn, indicating that iron sequestration did not appear to contribute to antioxidant effects of phenolics. Further, an experiment in our laboratory was conducted to test the impact of 200 ppm ferrous iron in the extruded dog food compared to the control without adding ferrous iron. PV and hexanal concentration change of the samples were monitored during the 9-month ambient (22 °C, 50% RH) and accelerated (at 37 °C, 60% RH) storage. The data indicate that 200 ppm ferrous iron did increase the oxidative stability of the coated dog food at both storage conditions (data not shown).

The antioxidant activity of ferrous iron in the extruded matrices could be attributed to several reasons. First, the antioxidant property of the ferrous iron may be limited to the extruded dry foods containing high levels of ferrous iron (relative to the hydroperoxides) such as ≥50 ppm, since the concentration of most transition metals as catalyst was 0.1–5 ppm (Rao and Artz, 1989), and the catalyst-to-inhibitor conversion was solvent dependent (Black, 1978). According to Black, when the studied system was polar media or solvent, the relatively low concentration (relative to hydroperoxides) of iron would preferentially react with hydroperoxides to speed up lipid oxidation, which was the case in the metal-catalyzed oxidation in oil-in-water-emulsions. However, if the studied system was nonpolar media or solvent such as an extruded matrix (Rao and Artz, 1989), an excess of iron relative to the hydroperoxides would result in inhibition of lipid oxidation, since there is catalytic concentration range for iron-catalyzed auto-oxidation. As the metal concentration is increased above the hydroperoxide concentration, inhibition will occur for nonpolar media (Black, 1978). Second, the lipid oxidation reactions in solid matrices may be not the same as those in liquid (e.g., oil-in-water emulsions containing polar media). It is well known that the redox potential change or free energy change (ΔG) of a lipid oxidation reaction is used to predict whether the redox reaction of the lipid can be occurring or not thermodynamically. However, the standard redox potential is generally measured at the standard condition and in the water solution. Therefore, the redox potential of a molecule in water solution would be different from that in solid matrices.

9.8.3 Effective Analytical Methods for Assessing the Shelf Life of Extruded and Baked Breakfast Cereals

Several analytical methods were reported to evaluate the oxidative stability and shelf life of the extruded breakfast cereals (Paradiso et al., 2014, 2008, 2009; Jamora and Rhee, 2002; Nissen et al., 2002; Camire and Dougherty, 1998; Viscidi et al., 2004), such as TBARS, ESR, PV, CD, oxidized aldehydes, and organic acids such as hexanal, pentanal, hexanoic, and pentanoic acid (Paradiso et al., 2008), protein oxidation products such as protein carbonyl, thiols, and tryptophan (Dong, 2011; Viljanen et al., 2004), as well as sensory tests. It was confirmed that TBARS was not an effective analytical method to assess the oxidative stability of the extruded products containing corn starch (Jamora and Rhee, 2002). But monitoring both hexanal and hexanoic acid level change during storage seemed to be a good method to evaluate the long-term shelf life of extruded breakfast cereals. Paradiso et al. (2008, 2009, 2014) used SPME–GC–MS to measure a series of volatile oxidation products, in which hexanoic acid was considered to play a key role in off-flavors related to long-term storage of extruded breakfast cereals. A good correlation between the peak area ratio of hexanoic acid and the mean sensory scores of the samples was confirmed. Figure 9.3 shows the evolution of hexanal and hexanoic acid in the extruded corn flakes without adding tocopherols over a 1-year storage. Obviously, the trend is that the hexanal ratio of peak area/total peak area percent decreased during 1-year storage, however, the hexanoic acid ratio of peak area/total peak area percent increased gradually. Therefore, hexanoic acid peak area or concentration would be a good marker to assess the shelf life of the extruded corn flakes. While the authors used semi-quantitative data (electronic units of the integrated peak area) for the evaluation, it is feasible that the quantitative data can be obtained by modifying the method.

Data reported and the research in our laboratory indicate that hexanal, 2,4-decadienal (for n-6 fatty acid oxidation), and propanal, 2,4-heptadienal, 1-penten-3-one, 1-penten-3-ol (for n-3 fatty acids oxidation) were better indicators for evaluating the shelf life of extruded foods and dry or low-moisture foods.

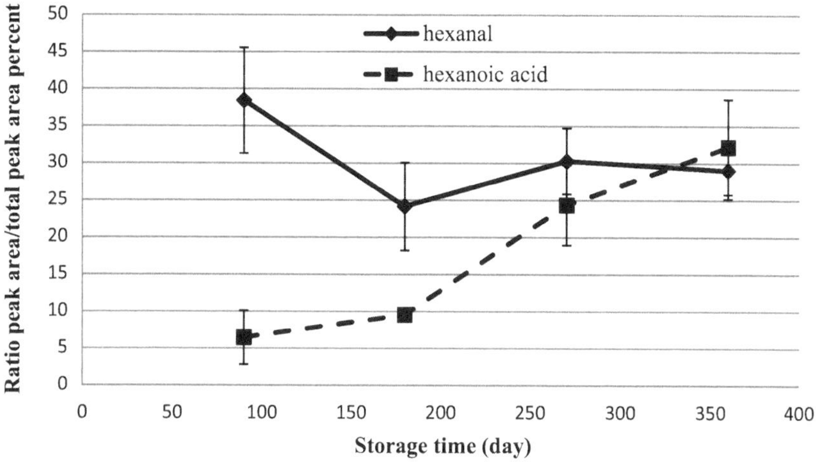

FIGURE 9.3 Evoluation of hexanal and hexanoic acid of breakfast cereals without tocopherols over 1-year storage. *Source: Paradiso et al. (2008). Curves drawn based on data in Table 9.2 of the reference.*

For some of the dried foods, hexanal level could be decreasing or only slight change over long-term storage. In this case, hexanoic acid may need to be monitored. Therefore, the headspace hexanal and hexanoic acid analysis, together with sensory test, are recommended to evaluate the shelf life of extruded and baked products.

9.8.4 Case Studies

Camire and Dougherty (1998), Camire et al. (2005, 2007), and Viscidi et al. (2004) investigated the impact of individual natural antioxidants and the antioxidants in fruit and vegetable powders and plant extracts on the oxidative stability of extruded corns, oat cereals, and other breakfast cereals. Extruded corn meals containing 200 and 1000 ppm of BHT (control), cinnamic acid, and vanillin, respectively, were fried in corn oil for 2 min at 204 °C. The fried samples were stored at 35 °C for 12 weeks for oxidative stability study (Camire and Dougherty, 1998). The data indicate that 200 ppm cinnamic acid and vanillin worked better than 200 ppm BHT and were comparable to 1000 ppm BHT over 12-week storage based on PV and CD. Viscidi et al. (2004) studied the effect of 1000 ppm of benzoin, catechin, chlorogenic acid, ferulic acid, and quercetin on the oxidative stability of the extruded oat cereals (2% lipid, dry mass) containing sucrose, respectively. The extrusion decreased total phenolics by 24–46%. The authors concluded that benzoin, chlorogenic acid, quercetin retarded hexanal formation in the extruded oat cereals at 12-week of storage at 35 °C, but only quercetin was effective up to 24 weeks. Benzoin was the most effective in controlling PV and CD at week 24. In addition, the chlorogenic acid sample had less grassy odor than did the control. Samples containing ferulic acid, and 1.0 g/kg quercetin were rated lower in painty flavor than was the control.

Extruded yellow cornmeal containing 0.02% ascorbic acid and quercetin, 2% spray-dried ginkgo extract, onion powder, potato peels, and wheat bran were studied respectively (Camire et al., 2005). The tested samples contained 5% moisture and were stored in trilaminate bags at 25 °C for 11 weeks. No obvious difference in hexanal levels between the control (no antioxidant added) and tested samples was observed. Finally, extruded corn breakfast cereals containing 84.3% degermed white corn meal, 14.3% sucrose, 0.4% citric acid, and 1% dehydrated fruit powders (blueberry, craneberry, concord grape, and raspberry) were produced using a laboratory-scale twin-screw extruder (Camire et al., 2007). These cereal samples with 5% moisture were stored in opaque bags for a 11-week shelf life study at room temperature. Six volatile compounds including hexanal were monitored using GC-FID headspace analysis. The authors concluded that overall, hexanal levels in the extruded corn meals containing Concord grape and raspberry were significantly lower than those containing no antioxidant (control).

Potato flakes (1% lipid, $a_w = 0.4$) containing 200 ppm rosemary, coffee, grape seed skin extracts, and 500 ppm tea extract were packed in aluminized sachets with nitrogen flush or aluminum cans and stored at 22 °C for 52 or 59 weeks (Nissen et al., 2002). ESR, TBA_{532}, TBA_{450}, propanal, hexanal, ethane, and pentane concentration changes were analyzed during storage. Sensory evaluation was also conducted. According to the authors, free radical levels measured by ESR were confirmed to be a marker both for early detection of lipid oxidation during the storage and for the evaluation of efficacy of different antioxidants, due to the significant increase of the level of free radicals during the first 2-week storage, followed by an obvious decrease over a 2- to 22-week storage to reach a steady state level. The authors also observed that the can headspace concentrations of ethane and pentane gradually increased over 59-week storage. However, the other markers of the secondary oxidation products such as TBA, hexanal and propanal did not change significantly during the storage. The data of ESR suggested that the efficacy order of antioxidants in potato flakes was: rosemary extract > grape skin extract = tea extract > coffee extract > control (Nissen et al., 2002).

Paradisco et al. (2008, 2009) investigated the effect of mixed tocopherols, rosemary extract and the blend of mixed tocopherols and ascorbic acid on the shelf life of corn flakes based on volatile lipid oxidation products and sensory evaluation. The corn flakes were produced on an industrious scale and stored in commercial packaging under shelf conditions for 1 year. The authors highlighted that headspace hexanoic acid, together with sensory test, was used to assess the shelf life of the corn flake products, since no significant change in hexanal level over storage was observed. The result indicated that mixed tocopherols increased the oxidative stability and extended the shelf life of corn flakes by controlling further oxidation of hexanal to form hexanoic acid. The authors concluded that the percentage of hexanoic acid positively correlated with sensory data. The efficacy of mixed tocopherols (125 and 187.5 ppm) was better than that of rosemary extract (810 ppm), and the efficacy of the blend of mixed tocopherols (125 ppm) and ascorbic acid (400 ppm) was improved compared to that of mixed tocopherols (125 ppm). Paradisco et al. (2014) further studied the impact of some packaging materials and natural mixed tocopherols on the sensory properties and shelf life of corn flakes. The corn flakes contained around 2.5% fat and 187.5 ppm of natural mixed tocopherols. There were three different packages: coextruded polypropylene/polypropylene cast coupled film, coextruded HDPE (high-density polyethylene) with an ionomer layer, and coextruded polypropylene/coextruded metallized polypropylene coupled film. The authors indicated that packaging type had a significant impact on the sensory profile of aged corn flakes. Polypropylene, particularly the high-density polyethylene, improved the sensory profile of the corn flakes. But the metallized polypropylene favored the highest level of oxidation products and sensory defects. The authors concluded that the most obvious improvement was observed in the flakes containing mixed tocophereols stored in polypropylene packaging.

GSE was also applied to corn chips to increase their shelf life (Rababah et al., 2011). The extruded corn chips contained 100 g corn flour, 0.5 g shortening, and 0.93 g salt and water. Then, 200, 400, and 800 ppm of GSE and 200 ppm of BHT were mixed well with shortening and other ingredients before extrusion. The authors used PV, OSI, color, and sensory tests conducted by a trained panel and the consumer to assess the shelf life. The products were stored at room temperature for 90 days. The chemical analysis indicated that corn chips samples contained less than 2% moisture, around 5.4% fat, 6.2% protein, and 6.6% fiber. The authors observed that there was almost no change in fat and protein contents over 90 days of storage at room temperature for all the samples. PV increased markedly over 90-day storage for the control and the sample containing 200 ppm of GSE, but there was no obvious change in PV for the samples containing 400 and 800 ppm of GSE and 200 ppm of BHT. Sensory scores on rancidity were in agreement with PV. The authors indicated that adding GSE (400 and 800 ppm) and BHT (200 ppm) could extend the shelf life of the corn chips, and 400 and 800 ppm of GSE could replace 200 ppm of BHT. In addition, the authors measured the IP of the shortening treated with three different levels of GSE and 200 ppm of BHT using OSI at 110 °C. IP of the shortening containing BHT (18.8 h) was much larger than those of the shortenings containing varying levels of GSE (1.6–3.9 h). This result was quite distinct from the ones of PV and sensory evaluation of the corn chips containing the shortenings in which GSEs and BHT were added. Nissen et al. (2002) also reported that the tested antioxidants worked well in rapeseed bulk oil, but they did not work well in cauliflower soup powder containing the rapeseed oil. This is because the efficacy of antioxidants is dependent on the matrix. Accordingly, we need to be very careful when OSI and Oxipres data were used to assess the oxidative stability and shelf life of bulk oils and extruded products.

Another type of extruded product is composed of the blend of corn starch and ground meats such as goat, lamb, mutton, spent hen meat, and beef (Rhee et al., 1999). The well-expanded extruded products contained 0.41–1.36% lipid, 0.32–1.87% moisture, 6.48–7.30% protein, and less than 0.12 a_w. The samples were stored aerobically at 37 °C for up to 120 days. According to the authors, PV data show that lipid oxidation rate was slower in the products containing goat, lamb, and mutton meat but high in the products containing beef and chicken over storage. The blend of corn flour, defatted soy flour, and minced catfish was also reported to produce extruded snake foods (Rhee et al., 2004). The final extrudates contained ≤6% fat, ~2.5% moisture, and 11.5% protein.

In summary, the matrices of extruded breakfast cereal and snack foods are expanded and porous. The lipid level in the extruded breakfast cereals and snack foods is relatively low, and the ratio of surface area to the volume of lipid is high. Therefore, the extruded breakfast cereal and snack foods are readily oxidized. We may need to add antioxidants to extruded breakfast cereals and snack foods to extend their shelf life. Compared to BHT, mixed tocopherols, rosemary extract, and GSE can increase the shelf life of the extruded breakfast cereals and snack foods. High levels of ferrous iron may increase the oxidative stability of extruded breakfast cereals and snack foods, but more research may be needed to further confirm it. The evolution curve of some of the extruded foods over storage may not be the same as that of bulk oil, with the decrease or slight change of PV and hexanal and without lag phase. This could be owing to oil/fat-containing ingredients in extruded products having been oxidized or slightly oxidized before extrusion, varying oils/fats in extruded products having different "freshness" and oxidative stability, lipid co-oxidation with protein or amino acids, and hexanal or propanal having been further oxidized to form hexanonic or propanoic acid. It is critical to select appropriate lipid oxidation markers to assess the shelf life of extruded products. An incorrect conclusion could be made if an inappropriate marker is used to evaluate the oxidative stability. Hexanal and hexanonic acid, together with sensory score, would be very useful markers for monitoring long-term oxidative stability and shelf life of extruded breakfast cereals, snack foods, and other extruded foods.

9.9 DRY PET FOOD

9.9.1 Dry Pet Food and Ingredients

Pet foods can be generally segmented into dry, wet (canned or moist), semi-moist, treats (or snacks), and raw/frozen/refrigerated dog and cat foods. Dry pet food may contain around 5–12% moisture, semi-moist pet food 22–35% moisture, and wet pet food more than 65% moisture. Dry pet food is the largest segment in the pet food market. Different pet foods have varying oxidative stability and shelf life. This section will focus on the oxidative stability and shelf life of dry dog and cat food. Why is the oxidative stability and shelf life study critical for dry pet food?

This is because lipid oxidation may change the color, taste, and odor of dry pet food, leading to decrease in nutrients, palatability, and the quality of the dry pet food. Therefore, a pet food company must make sure that the developed dry pet food products have their claimed shelf life.

Dry pet food is composed of a number of ingredients that contain carbohydrates such as polysaccharides, oligosaccharides, monosaccharides, and fibers; proteins such as animal and plant proteins; amino acids; lipids such as plant oils and animal fats; vitamins; minerals; and preservatives such as antioxidants and antimicrobials (Hussein, 2003). Generally, dry dog and cat diets may contain up to 50% starch, 25–70% (dry matter) protein, and less than 22% lipid (Tran et al., 2008). The selection of ingredients for producing dry products is based on their nutrient content, digestibility, palatability, functionality, availability, cost, and label (Thompson, 2008). These selected ingredients provide balance or special nutrients and palatability for the companion animals.

While dry pet food consists of a number of ingredients, lipid oxidation in dry pet food is closely related to some of the ingredients such as oils/fats and oil/fat-containing ingredients. Commonly used ingredients containing oils or fats in pet food industry are plant and animal oils and animal fats, such as soybean, canola, sunflower, corn, flaxseed, palm, algal and fish oils, and chicken, poultry, pork fat or CWG (choice white grease), and beef tallow. Dry pet food contains relatively high level of oils and fats to increase the palatability of the products and provide essential fatty acids, nutrients, and energy for pets. In addition, lipid oxidation in dry pet food is closely related to the oxidation of ingredients containing both lipid and proteins or amino acids. This is because lipid co-oxidation with protein or amino acids may also contribute to rancidity of the dry kibbles. Widely used ingredients containing both fats/oils and proteins or amino acids are protein meals such as chicken, poultry byproduct, pork, beef, lamb, dried egg, fish, meat and bone meals, and dried distillers grains with solubles (DDGS, Thompson et al., 2008; Folador et al., 2006). Chicken or poultry fats and poultry meals are produced via the rendering process. Protein meals are subject to oxidation due to higher levels of fats and proteins and amino acids, as well as high ratio of surface area to volume of fat. Lipid co-oxidation with protein in protein meals such as chicken byproduct meal and dry kibble further decrease the oxidative stability of the protein meals and dry kibble (Sara, 2013). Except starches, corn gluten, pea, flour, wheat, rice, and soybean meals also contain proteins and oils. Finally, liquid and dry palatants are readily oxidized. In particular, liquid palatants are oil-in-water emulsions that are prone to oxidation.

9.10 CHARACTERISTICS OF EXTRUDED KIBBLE

As discussed previously, extrusion is a high-temperature short-time cooking process. The kibble matrix is produced by mixing a number of formulated dry and liquid ingredients together. The mix is transferred to a pre-conditioner for pre-cooking, where liquid ingredients and meat can also be added, and then moved through an extruder at a defined temperature, pressure, and moisture level, being forced through a spiral-shaped screw and the die of the extruder. Following the extrusion, the extruded kibble is dried and cooled. The cooled, uncoated kibble is enrobed with oils/fats and palatants and then packaged.

When pet food ingredients are exposed to heat, shear, moisture, and specific mechanical energy during extrusion (Plattner, 2005), starch polymers are dispersed and degraded to form a continuum in the melt fluid. Starch polymer cell walls recoil and stiffen as they cool to stabilize the extrudate structure (Guy, 2001). Starch gelatinization and melting help the kibble to bind and cause the expansion of the kibble during extrusion. Meanwhile, the protein denatures, and the lipid in the ingredients such as chicken meal interacts with the starch and protein. The gelatinization of starch, denaturation of protein, and the interactions (i.e., covalent cross-linking and noncovalent association) among the starch, protein, and lipid generate the porous, expanded, and multiphase microstructure of the kibble matrix. Also, "air bubbles" could be distributed in the kibble matrix (Hu, 2014; Schaich and Rebello, 1999; Wang, 2000). Therefore, the structure characteristics of extruded kibble matrix are porous, expanded, with air bubbles.

According to Guy (2001), when the structure-forming materials are based on starch, a continuous phase of starch polymers is formed for starch-rich recipes, while several dispersed phases lie within the continuous phase. For instance, proteins and lipids form the separate (dispersed) phases within the continuous starch phase (Guy, 2001). The lipids in the matrix may be derived from protein meals, added oil or fat, and other ingredients. There are usually two kinds of lipids, neutral lipid triglycerides and polar lipids phospholipids and glycolipids, in the kibble matrix. Monoglycerides, free fatty acid, and lysophosphatidylcholine could be bound to amylose during extrusion (Guzman et al., 1992; Camire et al., 1990). When the structure-forming materials are based on protein, a continuous phase of protein polymers is formed for protein-rich recipes with more than 40% w/w protein and 30–40% w/w moisture (Guy, 2001). Currently, one of the pet food trends is grain-free. Thus, a number of ingredients containing high protein have been applied to producing dry pet food.

Dry kibble is an extruded product characterized by a porous and expanded structure and is roughly divided into two parts: the kibble matrix (core) and the kibble surface coated with oils/fats and palatant(s). It is critical to understand the lipid distribution in dry kibble. The lipid in kibble matrix may be named the noncontinuous lipid phase, while the lipid on the coating surface is considered a continuous lipid phase. This would be the case when higher levels of oils/fats are coated to the surface of the kibble. Moreover, the lipid in the matrix of the extrudate can be also divided into the free and bound fat. Free lipid is the ones extracted with petroleum ether at room temperature. Bound lipid is the one extracted by a mixture of chloroform and methanol after the free lipid is extracted (Guzman et al., 1992). It was reported that the bound lipid in the matrix was less susceptible to oxidation than the free lipid (Artz and Rao, 1993).

What is the difference between the extruded kibbles and the extruded breakfast cereals or snack foods? In general, dry pet food may contain much more ingredients than the extruded breakfast cereals. Thus, there are more interactions and more complicated microstructures in dry kibbles than in extruded breakfast cereals. The microstructures of dry kibble are likely less porous and expanded than the breakfast cereals. Also, the fat content of dry kibble is more than that of the breakfast cereals. In addition, dry pet food is coated with fats or oils and palatants, but the extruded cereals and snack foods, except fried corn or potato chips, may not be enrobed with oils and fats. Different compositions and microstructures of the two extrudates lead to varying oxidative stability and shelf life.

9.11 CHARACTERISTICS OF LIPID OXIDATION IN DRY PET FOOD

Although no literature reports the free radicals are formed in dry kibble, the free radicals generated in the extrudates during extrusion and storage were reported (Schaich and Rebello, 1999). Also, PV, hexanal, and hexanoic acid concentration changes in dry kibble suggest the existence of free radicals in the dry kibble during extrusion and storage. ESR spectra showed dominant broad singlets from nitrogen-centered radicals originating from heat-induced peptide scission and reactions of lipid radicals with side-chain amino group in the extrudates containing wheat flour proteins (Schaich and Rebello, 1999; Partridge et al., 2003). The formation and transfer of the free radicals in extrudates are involved in molecular structure changes that occur during extrusion and lead to protein scission and cross-linking. The free radicals can be formed by extrusion heat and shear, as well as oxidized lipids (Schaich and Rebello, 1999). The lipid content of wheat flour is very low, but the fatty acids in wheat flour are predominantly unsaturated and are readily oxidized. Lipid alkoxyl and peroxyl radicals transfer radicals to amino acid side chains with nitrogen and sulfur functional groups (Schaich, 1980).

Lipid co-oxidation with protein in dry pet food was monitored by PV and protein carbonyl over storage. The result shows that protein and lipid oxidation rates increased during storage as indicated by gradual increase of protein carbonyl value and PV (Sara, 2013).

It was reported that the lipid oxidation rate on the surface of the kibble was different from that in the matrix (core) of the kibble (Hu, 2014). Because the kibble matrix is porous and the lipids in the matrix are noncontinuous, lipid oxidation may occur at a higher rate in the matrix than on the coated kibble surface. Coated oils/fats, by contrast, are basically continuous, particularly when a high level of oils/fats is used for enrobing, and a higher level of antioxidants may be added to the coated oils/fats to provide further protection against lipid oxidation.

As one can see from Figure 9.4, hexanal levels in three commercially available dry dog foods were evaluated after being stored for 56 and 68 days. Data were collected from the kibble surface, the kibble core (matrix), and the whole kibble. For dry pet food 1, the hexanal level in the kibble core was slightly higher than that of the surface. The core of dry pet foods 2 and 3 showed higher hexanal levels in comparison to the surface and whole kibble.

Furthermore, a commercial product with fish oil was selected to evaluate the oxidative stability of kibble core (matrix) and kibble surface. Propanal was used as a marker for the evaluation of the oxidative stability of the product, since the dry product contained fish oil. Figure 9.5(A) and (B) indicate that the PV of the kibble core was much larger than that of the kibble surface, the hexanal level of the kibble core was two times larger than that of kibble surface, and propanal level of kibble core was also larger than that of kibble surface. The data confirm that the kibble core (matrix) is more readily oxidized than the kibble surface. The higher lipid oxidation rate in kibble matrix may be attributed to the kibble's porous structure, noncontinuous lipid distribution, high ratio of surface area to volume of the lipid, and the inclusion of "air bubbles."

It was found that in the certain range of the coated oils and/or fats concentration, the oxidation rate of the kibble decreased as the increase of the coated oils and/or fats concentration. This could be due to high level of oils and/or fats forming more homogeneous continuous lipid phase on the kibble surface to prevent air from diffusing into the core of the kibble. It was also found that the coated liquid palatants had a great impact on the oxidation rate of the dry kibble. Clearly, liquid palatants are oil-in-water emulsions that are readily oxidized compared to bulk oils or fats. The oxidation rate increased when the lower level of oils and/or fats and high level of liquid palatant were applied to enrobing.

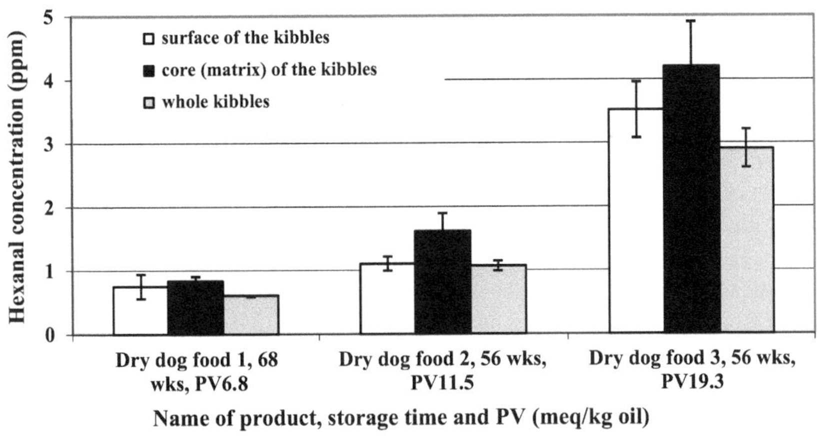

FIGURE 9.4 Hexanal concentration of kibble surface, core, and whole kibbles of three commercial dry dog foods without fish oil. *Source: Hu (2014).*

Therefore, the oxidation rate of dry kibble depends on matrix microstructure, diffusion of O_2, lipid distribution in matrix, and the level of coated lipid and palatant on the kibble surface.

Lipid oxidation in dry pet food may be the one of mixed oils/fats, since a dry pet food may contain different oils/fats for coating such as chicken fat, canola, and fish oil. In addition, the dry pet food may contain varying protein meals such as chicken, pork, fish, and lamb meal in kibble matrix. These oils and fats and oil/fat-containing ingredients may have different oxidative stability and varying degree of freshness while used to produce dry pet food, which may have negative impact on the oxidative stability of the products and increases the difficulty in explaining the oxidative stability data.

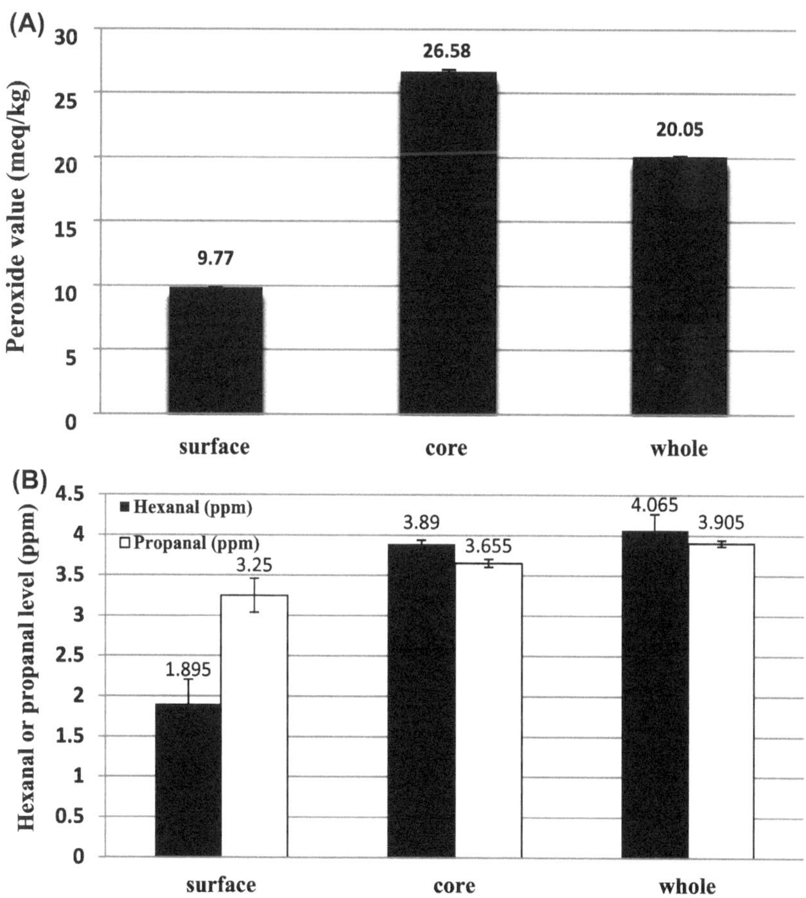

FIGURE 9.5 (A) PV of kibble surface, core (matrix) and whole kibble of a commercial dry pet food containing fish oil stored for 125 days. (B) Hexanal and propanal level of kibble surface, core (matrix) and whole kibble of a commercial dry pet food containing fish oil stored for 125 days. *Source: Hu (2014).*

As discussed in the extruded breakfast cereals, high level of ferrous iron may not be a pro-oxidant in the uncoated extruded corn meal and starch products. The data in our laboratory indicated that higher level of ferrous iron might increase oxidative stability of the coated kibble as evaluated by hexanal and PV during the storage (data not shown), while ferrous iron speeds up the lipid oxidation in bulk oils and food emulsions. In fact, dry pet food contains relatively high level of ferrous iron to meet nutritional requirements for company animals. For instance, the iron level in some of the dry pet foods may be around 140–200 ppm.

9.11.1 Effect of Processing Conditions and Lipid Type and Concentration on Lipid Oxidation in Dry Pet Food

Processing conditions and lipid type have impacts on the lipid oxidation in extruded pet food. The pet food is generally extruded under the temperatures between 80 and 200 °C for 20 to 270 s, with the moisture levels of 10–25% (Rooijen et al., 2013; Hand et al., 2010). Rao and Artz (1989) studied the extrudates containing cornmeal–soybean oil and corn starch–soybean oil with different extrusion temperatures, 125, 150, and 175 °C. While the uncoated extrudates used for testing were not real pet food matrix, the data showed that the PV, CD, and oxodiene values increased as the increase of the extrusion temperature. Tran et al. (2011) investigated the impact of drying temperature and time for a canine diet extruded with a 4- or 84-mm die on physical and nutritional quality indicators. The results show that drying kibble below 160 °C minimized the negative effects of the drying process on nutrients. Drying temperature of 200 °C decreased the concentration of linolenic and linoleic acid and increased that of oleic acid, probably being indicative for lipid oxidation during the extrusion (Tran et al., 2011).

Lin et al. (1998) reported the effects of processing conditions (feed moisture and screw speed) and lipid type (beef tallow and poultry fat) and content (0–0.75 g/kg) on the lipid oxidation in uncoated dry kibble. The beef tallow or poultry fat containing the added synthetic antioxidants was mixed with corn meal, soybean meal, and sodium chloride before extrusion. The uncoated kibbles with 8% moisture were packed in bags made of low density polyethylene film and were stored in cardboard boxes at 23 °C for 14 months. The authors noticed that the lipid oxidation rates were the highest in the uncoated kibble without added fat, followed by the one containing poultry fat and then containing beef tallow. Increasing the fat content decreased the rate of lipid oxidation of uncoated kibble during storage. The authors also noticed that feed moisture level had the same impact as the fat level. The uncoated kibble with a high moisture content resulted in a lower lipid oxidation rate. The uncoated kibble extruded at 300 rpm had an appreciably higher lipid oxidation rate than the ones extruded at 200 and 400 rpm (Lin et al., 1998). The degree of expansion or porosity of the uncoated extruded kibble is the principle factor influencing lipid oxidation of the uncoated kibble. It should be pointed out that all the oxidative stability evaluation in this study was only based on TRARS.

Finally, there are several changes that may occur during the extrusion. Some of the changes are desirable but others are undesirable. Extrusion processing may inactivate lipoxygenase and lipase in some of the ingredients, which would have positive impact on lipid oxidation and shelf life (Lin et al., 1998); extrusion may form amylose–lipid complex that might slightly prevent the lipid oxidation in the kibble matrix (Artz and Razo, 1993; Lin et al., 1997); Maillard reaction products may be formed, which would increase the oxidative stability of lipid since the Maillard reaction products might be antioxidative (Rooijen et al., 2013). On the other hand, extrusion increases the porosity of the kibble thus decreasing the oxidative stability; high temperature during extrusion may reduce the oxidative stability of extrudate (kibble). Further, extrusion processing can produce kibbles with different size and shape, with small size and larger surface area probably increasing the lipid oxidation.

9.11.2 Challenges in Assessing the Oxidative Stability and Shelf Life of Dry Pet Food

Different analytical methods may be applied to assessing the oxidative stability of dry pet food and ingredients, such as bulk oils, fats, protein meals, palatants, and dry dog and cat foods. Unlike evaluating human food, the assessment of dry pet food is more complicated. Not only oxidative stability data such as hexanal and PV, but also sensory data by human panel and palatability data by animal panel are needed in order to comprehensively evaluate the oxidative stability and shelf life and make a business decision. Because some ingredients may be oxidized before producing dry pet food and lipid co-oxidation with protein may occur in dry pet food and ingredients, hexanonic acid (for n-6 fatty acid–containing dry pet food) or propanoic acid (for n-3 fatty acid–containing dry pet food), as well as protein carbonyl measurements may also be needed to assess the long-term shelf life of dry pet food. Another challenge is that it is very difficult to find literature reports on evaluating the oxidative stability of dry pet food.

While a number of methods are available (Aldrich, 2004), the effective analytical methods are limited. OSI and Oxipres can be used to roughly evaluate the oxidative stability and shelf life of bulk oils, fats, protein meals, and dry kibbles. As discussed previously, care should be taken when the OSI and Oxipres IP (induction period) data tested at high temperatures are used to predict the shelf life of dry pet food and ingredients in actual (ambient) storage conditions. TBARS may not be suitable for assessing the shelf life of dry pet food (Gross et al., 1994; Lin et al., 1998). TBARS tests on dry pet food in our laboratory also indicated that it was not an accurate method for evaluating the oxidative stability of dry pet food (data not shown). Even though PV was used to evaluate the oxidative stability and the correlation between PV and palatability data was found (Gross et al., 1994), the PV was not considered a good marker to assess the shelf life of dry pet food. For the dry pet food, measuring hexanal and hexanonic acid and/or propanal and propanoic acid over storage, together with sensory and palatability test, would be a better method to evaluate the long-term shelf life of dry pet food.

Ambient storage test at around 22 °C and accelerated storage test at 37 or 40 °C are usually used for shelf life study. The correlation between ambient and accelerated test may need to be established based on the PV and hexanal data at 22 °C (ambient) and 37 or 40 °C (accelerated). Also, the correlation between oxidative stability markers such as hexanal and intake ratio data from two-bowl palatability test may also need to be established.

Palatability test is usually used as a critical standard test to comprehensively assess the taste, aroma, texture, size, and shape of the kibble in pet food industry (Trivedi and Benning, 2003). An oxidatively stable kibble is the prerequisite of good palatability. Two-bowl (pan) consumption test is generally used to evaluate the palatability (PAL) of dry pet food. After the two-bowl test data were treated statistically, a high intake ratio may indicate a better palatability for a dry pet food. A control diet and tested one are provided for pets in the two-bowl preference test. PAL win means that the pets prefer the tested diet to the control one. PAL loss is just opposite. In order to establish the correlation between the oxidative stability markers such as propanal, hexanal, and TOTOX (a combination of hexanal and PV) and intake ratio (or palatability), palatability and oxidative stability test for a dog food containing fish oil were conducted over 3-month storage at 40 °C. The results are showed in Figure 9.6. The propanal concentration change or TOTOX change was used for assessment since the two-bowl test needs two diets. Propanal concentration change was defined as the

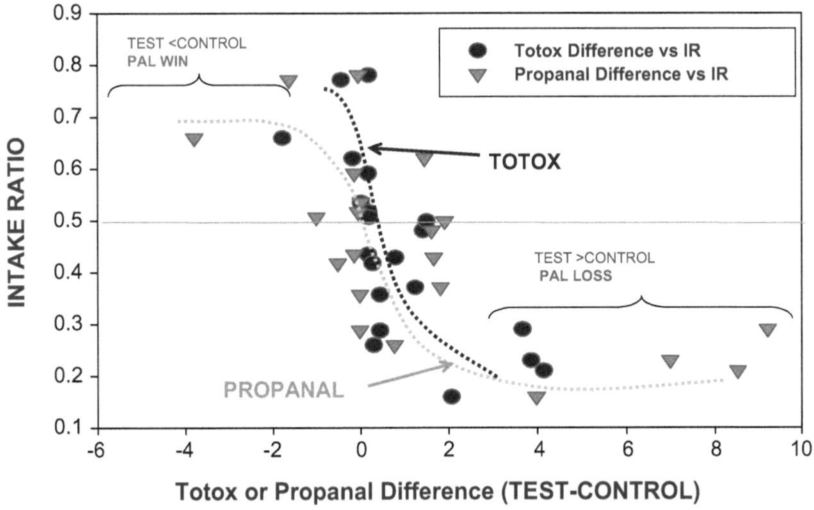

FIGURE 9.6 Correlation between intake ratio and TOTOX or propanal difference.

propanal concentration of the tested diet minus the one of the control diet. The definition of TOTOX change was the same. Obviously, the propanal concentration change was better than TOTOX change in assessing the oxidative stability since the diets contained high level of fish oil. When the propanal concentration change of the diets was larger than or equal to 4, the intake ratio larger than 0.3, it was called the PAL loss. That means that the pets would prefer the control diet, and the oxidative stability of control diet was better than that of tested diet. However, when the propanal concentration change was less than –2, the intake ratio greater than 0.6, it is called the PAL win. That indicates that the pets would prefer the tested diet, whose oxidative stability is better than that of control diet. Therefore, the oxidative stability marker propanal concentration change could predict the oxidative stability of the diets containing fish oil and the palatability of the diets.

9.11.3 Strategies of Increasing the Shelf Life of Dry Pet Food

High-quality oxidatively stable ingredients containing oils and fats are the prerequisite of producing high-quality dry pet food. However, it is a challenge for the pet food industry to guarantee the high quality of the ingredients containing oils and fats due to the challenge to manage the variables such as raw material quality, processing condition, season change, and temperature change in transportation. Thus, it is critical to select and keep high-quality ingredients until they are used to produce dry pet food.

As discussed previously, dry kibble has a porous microstructure. The kibble matrix contains noncontinuous lipid phase and liquid palatants are readily oxidized. In addition, the requirement for the shelf life of dry pet food has been increased from 9 months or 1 year to 1.5 years, even 2 years. Further, the current trend of pet food is natural, functional, and grain-free. Thus, the natural antioxidants and antioxidant blends are needed. Finally, even though an antioxidant works well in a bulk oil, it may not work well in a palatant, protein meal, and dry kibble. Therefore, it is essential to select appropriate antioxidants or antioxidant blends that are added to a specific dry pet food to increase the oxidative stability and extend shelf life of the dry pet food. Since kibble matrix is susceptible to oxidation, appropriate antioxidants are needed to add not only to coating oils and/or fats but also to kibble matrix or core. Further, the appropriate antioxidants are also needed in palatants, particularly in liquid palatants.

The following is an example of the application of antioxidants to dry dog food. Gross et al. (1994) tested the impact of three different antioxidant blends on the oxidative stability of extruded dog food. The three antioxidant blends were as follows: (1) 0.0133% ethoxyquin in kibble matrix and 0.076% Tenox 2 (20% BHA, 6% propyl gallate, 4% citric acid, and 70% propylene glycol) in animal fat (EX/BHA), (2) 0.04% Tenox GT-1 (50% mixed tocopherols and 50% vegetable oil) in kibble matrix and 0.02% Tenox GT-1 in animal fat (TC/TC), and (3) 0.16% Ronoxan FS-20 (5% ascorbyl palmitate, 1% α-tocopherol,

32% lecithin, and 62% vegetable oil) in kibble matrix and 0.02% Tenox GT-1(ATL/TC) in animal fat.

The three coated kibble samples containing different antioxidant blends were placed in multi-wall bags with a plastic inner liner and stored at 48.8 °C for 16 weeks and at 22.2 °C for 12 months. TBARS and PV were measured during the storage. The palatability test of the dog foods was conducted when stored for 16 weeks at 48.8 °C and stored for 5 and 12 months at 22.2 °C. The authors concluded that TBARS did not work in assessing oxidative stability, but PV worked. Further, PV data were closely correlated with palatability data. According to the authors, in both ambient and accelerated storage test, dogs always consumed more of the foods with the lowest PV. For products stored at 48.8 °C, PV increased from week 4 to week 16 for both TC/TC and ATL/TC blend treatments. But PVs of the dog foods treated with TC/TC and ATL/TC were significantly higher than those treated with EX/BHA. In the ambient storage test (22.2 °C), the PVs were also higher for the TC/TC and the ATL/TC treatments compared to the EX/BHA treatment after 4 and 12 weeks of storage, respectively (Gross et al., 1994). It is not difficult to understand that EX/BHA blend worked well, since it is well known that ethoxyquin is very efficient in pet food and animal feed, and BHA, propyl gallate, and citric acid had a synergistic effect. But they are synthetic antioxidants, except citric acid if it is produced by fermentation technology. Ethoxyquin may be toxic and its usage level was suggested to be decreased in pet food and feed from 150 ppm to 75 ppm, according to AFFCO (Deffenbaugh, 2003). TC (tocopherols) is a natural antioxidant. If the mixed-tocopherols level increased, the efficacy in dry dog food would increase. ATL/TC worked better than TC/TC in the coated dog food because of the synergistic effect of ascorbic palmitate, α-tocopherol, and lecithin and higher level of ascorbic palmitate (compared to mixed tocopehrols) being used.

9.12 CONCLUSION

We have addressed critical factors impacting the oxidative stability and shelf life of five principal low-moisture or dry foods, milk powders, egg powders, microencapsulated oil powders, extruded breakfast cereals and snack foods, as well as dry pet foods. How to evaluate and manage the oxidative stability and shelf life of lower-moisture foods have been also discussed. The microstructure and lipid distribution of low-moisture foods represent one of the critical factors influencing oxidative stability and shelf life. Milk powders are produced by drying natural homogenized milk emulsions, while encapsulated oil powders by drying formulated oil-in-water emulsions. Egg powders are manufactured by dehydrating natural eggs. They are all powders. However, the extruded dry pet food, breakfast cereals, and snack foods are produced with extruders. They are not powders but solids with varying shapes and size. As discussed previously, compared to the breakfast cereals and snack foods, extruded dry pet food is usually composed of a number of varying ingredients and is coated with oils and/or fats and palatants.

Thus, the lipid distribution in dry kibble is different, with the coated lipid on the surface of the kibble and the other lipid in the kibble matrix or core. All these low-moisture or dry foods have varying compositions, structures, lipid distributions, and physicochemical properties, thereby leading to different oxidative stability and shelf life. However, they are solid. The lipid oxidation in solid may be different from that in liquid such as bulk oil and food emulsions. More research should be conducted to understand the difference in the microstructures, oxidation mechanisms, and physicochemical properties of the low-moisture foods in order to increase the oxidative stability and extend the shelf life of the low-moisture foods.

REFERENCES

Aldrich, G., June 2004. Shelf-life: lab prediction vs reality. PetFood Ind. 2, 12–14.

Artharn, A., Prodpran, T., Benjakul, S., 2009. Round scad protein-based film: storage stability and its effectiveness for shelf-life extension of dried fish powder. LWT - Food Sci. Technol. 42, 1238–1244.

Artz, W.E., Rao, S.K., 1993. Lipid oxidation in extruded products. In: Parliment, T.H., Morello, M.J., McGorrin, R.J. (Eds.), Thermally Generated Flavors. ACS, Washington DC, pp. 296–314.

Barrett, A., Porter, W.L., Marando, G., Chinachoti, P., 2011. Effect of various antioxidants, antioxidant levels, and encapsulation on the stability of fish and flaxseed oils: assessment by fluorometric analysis. J. Food Process. Preservation 35, 349–358.

Black, J.F., 1978. Metal-catalyzed autoxidation. The unrecognized consequences of metal-hydroperoxide complex formation. J. Am. Chem. Soc. 100 (2), 527–535.

Boselli, E., Bonoli, M., Caboni, M.F., Lercker, G., 2002. Determination of cholesterol oxidation products in the supercritical carbon dioxide extract of egg yolk powder: comparison with conventional liquid solvent extraction methods. Eur. Food Res. Technol. 215, 72–75.

Buma, T.J., 1971. Free fat in spray-dried whole milk. 4. Significance of free fat for other properties of practical importance. Neth. Milk Dairy J. 25, 88–106.

Camire, M.E., Camire, A., Mrumhar, K., 1990. Chemical and nutritional change in foods during extrusion. Crit. Rev. Food Sci. Nutr. 29 (1), 35–57.

Camire, M.E., Dougherty, M.P., 1998. Added phenolic compounds enhance lipid stability in extruded corn. J. Food Sci. 63 (3), 516–518.

Camire, M.E., Dougherty, M.P., Briggs, J.L., 2005. Antioxidant-rich foods retard lipid oxidation in extruded corn. Cereal Chem. 82 (6), 666–670.

Camire, M.E., Dougherty, M.P., Briggs, J.L., 2007. Functionality of fruit powders in extruded corn breakfast cereals. Food Chem. 101, 765–770.

Carunchia Whetstine, M.E., Drake, M.A., 2007. The flavor and flavor stability of skim and whole milk powders. In: Cadwallader, K.R., Drake, M.A., McGorrin, R. (Eds.), Flavor of Dairy Products. ACS Publishing, Washington, DC, pp. 217–251.

Chaiyasit, W., Elias, R.J., McClments, D.J., Decker, E.A., 2007. Role of physical structure in bulk oils on lipid oxidation. Crit. Rev. Food Sci. Nutr. 47, 299–317.

Choe, E., Min, D.B., 2006. Mechanisms and factors for edible oil oxidation. Compr. Rev. Food Sci. Food Saf. 5, 169–186.

Choe, E., Oh, S.J., 2013. Effects of water activity on the lipid oxidation and antioxidants of dried laver (*Porphyra*) during storage in the dark. J. Food Sci. 78 (8), 1144–1151.

Damerau, A., Moisio, T., Partanen, R., Forssell, P., Lampi, A.-M., Piironen, V., 2014a. Interfacial protein engineering for spray-dried emulsions–part II: oxidative stability. Food Chem. 144, 57–64.

Damerau, A., Pimwalee Kamlang-ek, P., Moisio, T., Lampi, A.-M., Piironen, V., 2014b. Effect of SPME extraction conditions and humidity on the release of volatile lipid oxidation products from spray-dried emulsions. Food Chem. 157, 1–9.

Deffenbaugh, L., 2003. Palatability keys. In: Kvamme, J.L., Phillips, T.D. (Eds.), Petfood Technology. Watt Publishing Co., Mt. Morris, Illinois, pp. 194–202.

Dong, Y., October 2011. Protein Modifications in Baked versus Fried Tortilla Chips (M.S. thesis). The State University of New Jersey, Rutgers.

De Vilder, J., Martens, R., Naudts, M., 1976. Influence of process variables on some whole milk powder characteristics. Milchwissenschaft 31, 396–401.

Eldin, A.K., 2010. Methods to determine the extent of lipid oxidation in foods. In: Decker, E.A., Elias, R.J., McClements, D.J. (Eds.), Oxidation in Foods and Beverage and Antioxidant Applications. Understanding mechanisms of Oxidation and Antioxidant Activity, vol. 1. Woodhead Publishing Limited, Cambridge, Cambridgeshire, UK, pp. 181–195.

Fenaille, F., Mottier, P., Turesky, R.J., Ali, S., Guy, P.A., 2001. Comparison of analytical techniques to quantify malondialdehyde in milk powders. J. Chromatogr. A. 921, 237–245.

Folador, J.F., Karr-Lilienthal, L.K., Parsons, C.M., Bauer, L.L., Utterback, P.L., Schasteen, C.S., Bechtel, P.J., Fahey Jr., G.C., 2006. Fish meals, fish components, and fish protein hydrolysates as potential ingredients in pet foods. J. Anim. Sci. 84, 2752–2765.

Fritsch, C.W., 1994. Lipid oxidation- the other dimensions. INFORM 5 (4), 423–436.

Frankel, E.N., 2005. Lipid Oxidation, second ed. PJ Barnes and Associates, Bridgwater, England.

Gharsallaoui, A., Roudaut, G., Beney, L., Chambin, O., Voilley, A., Saurel, R., 2012. Properties of spray-dried food flavours microencapsulated with two-layered membranes: roles of interfacial interactions and water. Food Chem. 132, 1713–1720.

Granelli, K., Faldt, P., Appelqvist, L.A., Bergenstahl, B., 1996. Influence of surface structure on cholesterol oxidation in model food powders. J. Sci. Food Agric. 71 (1), 75–82.

Grattard, N., Salaun, F., Champion, D., Roudaut, G., Le Meste, M., 2002. Influence of physical state and molecular mobility of freeze-dried maltodextrin matrices on the oxidation rate of encapsulated lipids. J. Food Sci. 67 (8), 3002–3010.

Gray, Q.A., Bowen, S.E., Farhat, I., Hill, A.E., 2008. Lipid oxidation in glassy and rubbery-state starch extrudates. Food Chem. 106, 227–234.

Gross, K.L., Brolunger, R., Thawnghmung, P., Collings, G.F., 1994. Effect of three different preservative systems on the stability of extruded dog food subjected to ambient and high temperature storage. J. Nutr. 124, 2638S–2642S.

Guy, R., 2001a. Raw materials for extrusion cooking. In: Guy, R. (Ed.), Extrusion Cooking : Technology and Applications. Woodhead Publishing Limited, Cambridge, England, pp. 5–29.

Guy, R., 2001b. Snack foods. In: Guy, R. (Ed.), Extrusion Cooking: Technology and Applications. Woodhead Publishing Limited, Cambridge, England, pp. 161–181.

Guzman, L.B., Lee, T.-C., Chichester, C.O., 1992. Lipid binding during extrusion cooking. In: Kokini, J.L., Ho, C.-T., Karwe, M.V. (Eds.), Food Extrusion Science and Technology. Marcel Dekker, New York, pp. 427–436.

Hand, M.S., Thatcher, C.D., Remillard, R.L., Roudebush, P., Novotry, B.J. (Eds.), 2010. Commercial Pet Foods, In: Small Animal Clinical Nutrition, fifth ed., Topeka, Mark Morris Institute, pp. 156–190.

Hardas, N., Danviriyakul, S., Foley, J.L., Nawar, W.W., Chinachoti, P., 2002. Effect of relative humidity on the oxidative and physical stability of encapsulated milk fat. J. Am. Oil Chem. Soc. 79, 151–158.

Hu, L., Huff, H.E., Heymann, H., Hsieh, F., 1996. Effects of emulsifier and soy fiber addition on sensory properties of corn meal extrudate. J. Food Qual. 19, 57–77.

Hu, M., 2014. Lipid oxidation in dry pet foods. INFORM 25 (1), 7–10.

Hussein, H.S., 2003. Basic nutrient requirements for healthy adult dogs and nutrient requirements for healthy adult cats. In: Kvamme, J.L., Phillips, T.D. (Eds.), Petfood Technology. Watt Publishing Co., Mt. Morris, Illinois, pp. 2–28.

Izzo, M.T., Ho, C.T., 1989. Protein-lipid interaction during single-screw extrusion of zein and corn oil. Cereal Chem. 66 (1), 47–51.

Jamora, J.J., Rhee, K.S., 2002. Storage stability of extruded products from blends of meat and nonmeat ingredients: evaluation methods and antioxidative effects of onion, carrot, and oat ingredients. J. Food Sci. 67 (5), 1654–1659.

Jena, S., Das, H., 2012. Shelf life prediction of aluminum foil laminated polyethylene packed vacuum dried coconut milk powder. J. Food Eng. 108, 135–142.

Jensen, P., Risbo, J., 2007. Oxidative stability of snack and cereal products in relation to moisture sorption. Food Chem. 103, 717–724.

Kazemzadeh, M., 2001. Baby foods. In: Guy, R. (Ed.), Extrusion Cooking, Technology and Applications. Woodhead Publishing Limited, Cambridge, England, pp. 182–196.

Keogh, M.K., 2005. Spray–dried microencapsulated fat powders. In: Onwulata, C. (Ed.), Encapsulated and Powdered Foods. CRC Press, Taylor & Francis Group, Boca Raton, FL, p. 478.

Kim, E.H.J., Chen, X.D., Pearce, D., 2002. Surface characterization of four industrial spray-dried dairy powders in relation to chemical composition, structure and wetting property. Colloids Surf. B: Biointerfaces 26 (3), 197–212.

Kim, E.H.J., Chen, X.D., Pearce, D., 2009. Surface composition of industrial spray-dried milk powders. 1. Development of surface composition during manufacture. J. Food Eng. 94 (2), 163–168.

Klinkesorn, U., Sopyanodra, P., Chinachoti, P., McClements, D.J., Decker, E.A., 2005. Stability of spray-dried tuna oil emulsions encapsulated with two-layered interfacial membranes. J. Agric. Food Chem. 53, 8361–8371.

Knechtges, P.L. (Ed.), 2012. Food Safety: Theory and Practice. Jones & Bartlett Learning, Burlington, MA, p. 246.

Kelly, J., Kelly, P.M., Harrington, D., 2002. Influence of processing variables on the physicochemical properties of spray dried fat-based milk powders. Lait 82, 401–412.

Labrousse, S., Roos, Y.H., Karel, M., 1992. Collapse and crystallization in amorphous matrices with encapsulated compounds. Sci. Aliments 12, 757–769.

Labuza, T.P., 1971. Kinetics of lipid oxidation in foods. CRC Crit. Rev. Food Technol. 2, 355–405.

Labuza, T.P., 1980. Effect of water activity on reaction kinetics of food deterioration. Food Technol. 34 (4), 36–59.

Laguerre, M., Bayrasy, C., Panya, A., Weiss, J., McClments, D.J., Lecomte, J., Decker, E.A., Villeneuve, P., 2015. What makes good antioxidants in lipid-based systems? The next theories beyond the polar paradox. Crit. Rev. Food Sci. Nutr. 55, 183–201.

Lee, T.-A., Ho, J.-H., Khoo, S.K., Chow, C.-F., 2012. Comprehensive stability evaluation of iron-fortified milk powder. Food Sci. Technol. Res. 18 (3), 419–428.

Lercker, G., Rodriguez-Estrada, M.T., 2000. Cholesterol oxidation: presence of 7-ketocholesterol in different food products. J. Food Compos. Anal. 13, 625–631.

Lin, S., Hsieh, F., Huff, H.E., 1997. Effects of lipids and processing conditions on degree of starch gelatinization of extruded dry pet food. LWT - Food Sci. Technol. 30, 754–761.

Lin, S., Hsieh, F., Huff, H.E., 1998. Effects of lipids and processing conditions on lipid oxidation of extruded dry pet food during storage. Anim. Feed Sci. Technol. 71, 283–294.

Lloyd, M.A., Drake, M.A., Gerard, P.D., 2009a. Flavor variability and flavor stability of U.S.-produced whole milk powder. J. Food Sci. 74 (7), 334–343.

Lloyd, M.A., Hess, S.J., Drake, M.A., 2009b. Effect of nitrogen flushing and storage temperature on flavor and shelf-life of whole milk powder. J. Dairy Sci. 92, 2409–2422.

Marquez-Ruiz, G., Martin-Polvillo, M., Jorge, N., Ruiz-Mendez, M.V., Dobarganes, M.C., 1999. Influence of used frying oil quality and natural tocopherol content on oxidative stability of fried potatoes. J. Am. Oil Chem. Soc. 76, 421–425.

Marquez-Ruiz, G., Velasco, J., Dobarganes, C., 2003. Oxidation in dried encapsulated oils. In: Kamal-Eldin, A. (Ed.), Lipid Oxidation Pathways. AOCS Press, Champaign, Illinois, pp. 245–264.

Matsuno, R., Adachi, S., 1993. Lipid encapsulation techniques and applications to food. Trends Food Sci. Technol. 41, 256–261.

Mazalli, M.R., Bragagnolo, N., 2007. Effect of storage on cholesterol oxide formation and fatty acid alterations in egg powder. J. Agric. Food Chem. 55, 2743–2748.

Meynier, A., Leborgne, C., Viau, M., Schuck, P., Guichardant, M., Rannou, C., Anton, M., 2014. N-3 fatty acid enriched eggs and production of egg yolk powders: an increased risk of lipid oxidation? Food Chem. 153, 94–100.

McClements, D.J., Decker, E.A., 2000. Lipid oxidation in oil-in-water emulsions: impact of molecular environment on chemical reactions in heterogeneous food systems. J. Food Sci. 65 (8), 1270–1282.

Moscicki, L., Zuilichem, D.J.V., 2001. Extrusion-cooking and related technique. In: Moscicki, L. (Ed.), Extrusion-Cooking Techniques: Applications, Theory and Sustainability. WILEY-VCH Verlag GmbH & Co. KGaA, Weinheim, pp. 1–2.

Nelson, K.A., Labuza, T.P., 1992. Relationship between water and lipid oxidation rates: water activity and glass transition theory. In: St Angelo, A.J. (Ed.), Lipid Oxidation in Food. ACS, Washington, DC, pp. 93–103.

Nissena, L.R., Huynh-Bab, T., Petersena, M.A., Bertelsena, G., Skibsted, L.H., 2002. Potential use of electron spin resonance spectroscopy for evaluating the oxidative status of potato flakes. Food Chem. 79, 387–394.

Obara, A., Obiedzinski, M., Kołczak, T., 2006. The effect of water activity on cholesterol oxidation in spray- and freeze-dried egg powders. Food Chem. 95, 173–179.

Paradiso, V.M., Caponio, F., Summo, C., Gomes, T., 2014. Influence of some packaging materials and of natural tocopherols on the sensory properties of breakfast cereals. Food Sci. Technol. Int. 20 (3), 161–170.

Paradiso, V.M., Summo, C., Antonio Trani, A., Caponio, F., 2008. An effort to improve the shelf life of breakfast cereals using natural mixed tocopherols. J. Cereal Sci. 47, 322–330.

Paradiso, V.M., Summo, C., Pasqualone, A., Caponio, F., 2009. Evaluation of different natural antioxidants as affecting volatile lipid oxidation products related to off-flavours in corn flakes. Food Chem. 113, 543–549.

Park, C.W., Drake, M.A., 2014. The distribution of fat in dried dairy particles determines flavor release and flavor stability. J. Food Sci. 79 (4), 452–459.

Partridge, M.A.K., Jiang, Y., Skerritt, J.H., Schaich, K.M., 2003. Immunochemical and electrophoretic analysis of the modification of wheat proteins in extruded flour products. Cereal Chem. 80 (6), 791–798.

Pie, J.E., Spahis, K., Seillan, C., 1990. Evaluation of oxidative degradation of cholesterol in food and food ingredients: identification and quantification of cholesterol oxides. J. Agric. Food Chem. 38, 973–979.

Pignoli, G., Rodriguez-Estrada, M.T., Mandrioli, M., Barbanti, L., Rizzi, L., Lercker, G., 2009. Effects of different rearing and feeding systems on lipid oxidation and antioxidant capacity of freeze-dried egg yolks. J. Agric. Food Chem. 57, 11517–11527.

Plattner, B., 2005. Extrusion for palatability. PetFood Ind. 9, 12–16.

Pokorny, J., Kolakowska, A., Bienkiewicz, G., 2005. Analysis of interaction products of oxidized lipids with amino acids, proteins and carbohydrates. In: Kamal-Eldin, A., Pokorny, J. (Eds.), Analysis of Lipid Oxidation. AOCS Press, Urbana, Illinois, pp. 263–280.

Polavarapu, S., Oliver, C.M., Ajlouni, S., Augustin, M.A., 2011. Physicochemical characterisation and oxidative stability of fish oil and fish oil-extra virgin olive oil microencapsulated by sugar beet pectin. Food Chem. 127, 1694–1705.

Price, P.B., 1975. Lipids of several cereal grains. J. Am. Oil Chem. Soc. 52, 490.

Rababah, T.M., Yucel, S., Ereifej, K.I., Alhamad, M.N., Al-Mahasneh, M.A., Yang, W., Muhammad, A.H., Ismaeal, K., 2011. Effect of grape seed extracts on the physicochemical and sensory properties of corn chips during storage. J. Am. Oil Chem. Soc. 88, 631–637.

Rahman, M.S., Al-Belushi, R.M., Guizani, N., Al-Saidi, G.S., Soussi, B., 2009. Fat oxidation in freeze-dried grouper during storage at different temperatures and moisture contents. Food Chem. 114, 1257–1264.

Raitio, R., Orlien, V., Skibsted, L.H., 2011. Storage stability of cauliflower soup powder: the effect of lipid oxidation and protein degradation reactions. Food Chem. 128, 371–379.

Rao, Q.C., Fisher, M.C., Guo, M.F., Labuza, T.P., 2013. Storage stability of a commercial hen egg yolk powder in dry and intermediate-moisture food matrices. J. Agric. Food Chem. 61, 8676–8686.

Rao, S.K., Artz, W.E., 1989. Effect of extrusion on lipid oxidation. J. Food Sci. 54 (6), 1580–1583.

Rhee, K.S., Chol, S.H., Pradahn, A.M., 1999. Composition, storage stability and sensory properties of expanded extrudates from blends of corn starch and goat meat, lamb, mutton, spent fowl meat, or beef. Meat Sci. 52, 135–141.

Rhee, K.S., Kim, E.S., Kim, B.K., Jung, B.M., Rhee, K.C., 2004. Extrusion of minced catfish with corn and defatted soy flours for snack foods. J. Food Process. Preservation 28, 288–301.

Riaz, M.N., 2001. Selecting the right extruder. In: Guy, R. (Ed.), Extrusion Cooking: Technology and Applications. Woodhead Publishing Limited, Cambridge, England, pp. 29–49.

Richardson, G.H., 1985. Standard Methods for the Examination of Dairy Products, fifteenth ed. American Public Health Association, Washington, DC, p. 358.

Rooijen, C.V., Bosch, G., Van Der Poel, A.F.B., Wierenga, P.A., Alexander, L., Hendriks, W.H., 2013. The Maillard reaction and pet food processing: effects on nutritive value and pet health. Nutr. Res. Rev. 26, 130–148.

Roozen, J.P., Linssen, J.P.H., 1992. Factors affecting lipid autoxidation of a spray-dried milk base for baby food. In: Lipid Oxidation in Food. ACS, Washington, DC, pp. 302–309.

Rubilar, M., Morales, E., Contreras, K., Ceballos, C., Acevedo, F., Villarroel, M., Shene, C., 2012. Development of a soup powder enriched with microencapsulated linseed oil as a source of omega-3 fatty acids. Eur. J. Lipid Sci. Technol. 114, 423–433.

Ryckebosch, E., Muylaert, K., Eeckhout, M., Ruyssen, T., Fouber, I., 2011. Influence of drying and storage on lipid and carotenoid stability of the microalga *Phaeodactylum tricornutum*. J. Agric. Food Chem. 59, 11063–11069.

Sanches-Silvaa, A., Costaa, D., Albuquerquea, T.G., Buonocored, G.G., Ramose, F., Castilhoe, M.C., Machadog, A.V., Costaa, H.S., 2014. Trends in the use of natural antioxidants in active food packaging: a review. Food Additives Contaminants: Part A 31 (3), 374–395.

Sara, C., May 2013. Protein oxidation and pet food. In: Oral Presentation at 104 AOCS Annual Meeting, Montreal, Canada.

Schaich, K.M., 2008. Co-oxidations of oxidizing lipids: reactions with proteins. In: Kamal-Eldin, A., Min, D. (Eds.), Lipid Oxidation Pathways, Chapter 8, vol. 2. AOCS Press, pp. 183–274.

Schaich, K.M., 1980. Free radical initiation in proteins and amino acids by ionizing and ultraviolet radiations and lipid oxidation–part III: free radical transfer from oxidizing lipids. Crit. Rev. Food Sci. Nutr. 13, 189–244.

Schaich, K.N., Rebello, C.A., 1999. Extrusion chemistry of wheat flour proteins: I. Free radical formation. Cereal Chem. 76 (5), 748–755.

Schroeder, E., February 14, 2014. Report: Global Breakfast Cereal Market to Reach $43.3 Billion by 2019. Food Business News.

Schroepfer, G.J., 2000. Oxysterols: modulators of cholesterol metabolism and other processes. Physiol. Rev. 80, 361–554.

Schuck, P., Mejean, S., Dolivet, A., Jeantet, R., Bhandari, B., 2007. Keeping quality of dairy ingredients. Lait 87, 481–488.

Selim, K., Tsimidou, M., Biliaderis, C.G., 2000. Kinetics studies of degradation of saffron carotenoids encapsulated in amorphous polymer matrices. Food Chem. 71, 199–206.

Shaviklo, G.R., Thorkelsson, G., Arason, S., Kristinssonb, H.G., Sveinsdottir, K., 2010. The influence of additives and drying methods on quality attributes of fish protein powder made from saithe (*Pollachius virens*). J. Sci. Food Agric. 90, 2133–2143.

Shaw, L.A., McClements, D.J., Decker, E.A., 2007. Spray-Dried multilayered emulsions as a delivery method for n-3 fatty acids into food systems. J. Agric. Food Chem. 55, 3112–3119.

Skibsted, L.H., 2010. Understanding oxidation processes in foods. In: Decker, E.A., Elias, R.J., McClements, D.J. (Eds.), Oxidation in Foods and Beverage and Antioxidant Applications. Understanding Mechanisms of Oxidation and Antioxidant Activity, vol. 1. Woodhead Publishing Limited, Cambridge, Cambridgeshire, UK, pp. 3–35.

Smith, L.L., 1987. In: Smith, L.L. (Ed.), Cholesterol Autoxidation. vol. 44. Plenum Press, New York, pp. 87–125.

Smith, L.L., Johnson, B.H., 1989. Biological activities of oxysterols. Free Rad. Biol. Med. 7, 285–332.

Stapelfeldt, H., Nielsen, B.R., Skibsted, L.H., 1997. Effect of heat treatment, water activity and storage temperature on the oxidative stability of whole milk powder. Intl. Dairy J. 7, 331–339.

Strange, E.D., Schaich, K.M., 2000. Extraction of lipids from extruded corn-soy blends. J. Food Lipids 7, 217–224.

Thompson, A., 2008. Ingredients: where pet food starts. Top. Company Anim. Med. 23 (3), 127–132.

Thompson, K., Rawles, S., Metts, L.S., Smith, R.G., Wimsatt, A., Gannam, A.L., Twibell, R.G., Hohnson, R.B., Brady, Y.J., Webster, A.D., 2008. Digestibility of dry matter, protein, lipid, and organic matter of two fish meals, two poultry by-product meals, soybean meal, and distiller's dried grains with solubles in practical diets for sunshine bass, *Morone chrysops* ×*M. saxatilis*. J. World Aquacult. Soc. 39 (3), 352–363.

Torres-Giner, S., Martinez-Abad, A., Ocio, M.J., Lagsaron, J.M., 2001. Stabilization of a nutraceutical omega-3 fatty acid by encapsulation in ultrathin electrosprayed zein prolamine. J. Food Sci. 75, 69–79.

Tran, Q.D., Hendriks, W.H., Van der Poel, A.F.B., 2011. Effects of drying temperature and time of a canine diet extruded with a 4 or 8 mm die on physical and nutritional quality indicators. Anim. Feed Sci. Technol. 165, 258–264.

Tran, Q.D., Hendriks, W.H., Van der Poel, A.F.B., 2008. Mini-review effects of extrusion processing on nutrients in dry pet food. J. Sci. Food Agric. 88, 1487–1493.

Trivedi, N., Benning, J., 2003. Preserving palatabilitys. In: Kvamme, J.L., Phillips, T.D. (Eds.), PetFood Technology. Watt Publishing Co., Mt. Morris, Illinois, pp. 178–193.

Ueno, H.M., Shiota, M., Ueda, N., Isogai, T., Kobayashi, T., 2012. Iron–Lactoferrin complex reduces iron- catalyzed off-flavor formation in powdered milk with added fish oil. J. Food Sci. 77 (8), 853–858.

U.S. Dairy Export Council (USDEC), 2005. In: Lagrange, V. (Ed.), Reference Manual for U.S. Milk Products, revised ed. Arlington, VA, pp. 21–24.

USDA Nutrient Data Base for Standard Reference, Release 23, Assayed Egg Nutrition Value, November 2010. http://ndb.nal.usda.gov(accessed 15.09.14.).

Van Dyck, S.M.O., Verleyen, T., Dooghe, W., Teunckens, A., Adams, C.A., 2005. Free radical generation assays: new methodology for accelerated oxidation studies at low temperature in complex food matrices. J. Agric. Food Chem. 53 (4), 887–892.

Van Mil, P., Jans, J.A., 1991. Storage stability of whole milk powder: effects of process and storage conditions on product properties. Neth. Milk Dairy J. 45, 145–167.

Velasco, J., Dobarganes, M.C., Marquez-Ruiz, G., 2004. Antioxidant activity of phenolic compounds in sunflower oil-in-water emulsions containing sodium caseinate and lactose. Eur. J. Lipid Sci. Technol. 106, 325–333.

Velasco, J., Dobarganes, M.C., Marquez-Ruiz, G., 2000. Oxidation of free and encapsulated oil fractions in dried microencapsulated oils. Grasas y Aceites 51, 439–446.

Velasco, J., Dobarganes, M.C., Marquez-Ruiz, G., 2003. Variables affecting lipid oxidation in dried microencapsulated oils. Grasas y Aceites 54, 304–314.

Velasco, J., Holgado, F., Dobarganes, C., Marquez-Ruiz, G., 2009. Influence of relative humidity on oxidation of the free and encapsulated oil fractions in freeze-dried microencapsulated oils. Food Res. Int. 42, 1492–1500.

Velasco, J., Marmesat, S., Dobarganes, C., Marquez-Ruiz, G., 2006. Heterogeneous aspects of lipid oxidation in dried microencapsulated oils. J. Agric. Food Chem. 54, 1722–1729.

Verardo, V., Pasini, F., Iafelice, G., Cristinamessia, M., Marconi, E., Caroni, M.F., 2010. Influence of storage conditions on cholesterol oxidation in dried egg pasta. J. Agric. Food Chem. 58, 3586–3590.

Verleyen, T., Van Dyck, S., Adams, C.A., 2005. Accelerated stability tests. In: Kamal-Eldin, A., Pokorny, J. (Eds.), Analysis of Lipid Oxidation. AOCS Press, Urbana, Illinois, pp. 210–233.

Vignolles, M., Jeantet, R., Lopez, C., Schuck, P., 2007. Free fat, surface fat and dairy powders: interactions between process and product. A review. Lait 87, 187–236.

Viljanen, K., Kivikari, R., Heinonen, M., 2004. Protein-lipid interactions during liposome oxidation with added anthocyanin and other phenolic compounds. J. Agri. Food Chem. 52, 1104–1111.

Viscidia, K.A., Doughertyb, M.P., Briggsc, J., Camire, M.E., 2004. Complex phenolic compounds reduce lipid oxidation in extruded oat cereals. LWT - Food Sci. Technol 37, 789–796.

Wang, S., 2000. Protein-Starch Interactions during Wheat Flour Extrusion: Formation and Effects on Textural Properties of Extrudates (Ph.D. thesis). Rutgers University, New Brunswick, NJ.

Wanibadullah, W., October 2013. Lipid Protein Interactions in Peanut Butter (Ph.D. thesis). The State University of New Jersey, Rutgers.

Weber, E.J., 1978. Corn lipids. Cereal Chem. 55, 572.

Zhu, D., Damodaran, S., 2011. Composition, thermotropic properties, and oxidative stability of freeze-dried and spray-dried milk fat globule membrane isolated from cheese whey. J. Agric. Food Chem. 59, 8931–8938.

Chapter 10

Oxidative Stability and Shelf Life of Meat and Meat Products

Fereidoon Shahidi
Department of Biochemistry, Memorial University of Newfoundland, St. John's, NL, Canada

10.1 INTRODUCTION

Meat and meat products are susceptible to oxidative changes, in both fresh and cooked states. In the fresh form, they are affected by storage and packaging as well as other factors related to the species of the animal, the cut of meat, and its fat and hemoprotein contents. Upon cooking, the oxidative stability and shelf life are affected again by the species, cut of meat, type of meat (solid, ground, mechanically deboned), as well as the heat processing method employed (boiling, frying, barbecuing, and curing, among others).

Oxidative changes in meat and meat products may lead to the development of off-flavors and off-odors as well as discoloration and other quality deteriorations. The mechanisms of action involved in oxidative alteration of muscle foods are therefore of critical importance, and in rendering adverse effects on proteins, lipids, pigments, and nucleic acids, some of which may have little, if any, sensory effects. However, the deleterious effects of oxidative changes may affect health and could be involved in the modulation of gene expression and the inflammatory response (Laguerre et al., 2007).

10.2 OXIDATION OF MEAT AND MEAT PRODUCTS

Lipid oxidation is a complex phenomenon induced by oxygen in the presence of initiators such as heat, free radicals, light, photosensitizing pigments, and metal ions. It occurs via different reaction pathways: (1) auto-oxidation mediated by free radicals, (2) photo-oxidation, (3) enzymatic oxidation, and (4) thermal oxidation. The first two types of oxidation consist of a combination of reactions involving triplet oxygen ($^3O_2$), which could be considered as a ground-state, and the singlet oxygen ($^1O_2$), which corresponds to an excited state, of the oxygen molecule. There are many sources of $^1O_2$ but its presence is often coupled with UV photonic impact in the presence of photosensitizers such as riboflavin, chlorophylls, and erythrosine. Briefly, the photo-oxidation mechanism is dependent

Oxidative Stability and Shelf Life of Foods Containing Oils and Fats
http://dx.doi.org/10.1016/B978-1-63067-056-6.00010-0

on the type of photosensitizer. The singlet oxygen is electrophilic and can thus bind directly to carbon–carbon double bonds, leading to hydroperoxide formation. However, this so-called nonradical photo-oxidation seems to be a minor reaction in comparison with $^3O_2$-induced radical chain auto-oxidation. Photo-oxidation is important when dealing with edible oils, such as extra virgin olive oil, which is rich in chlorophylls, under display light of the supermarket and when stored in clear bottles (Sikorska et al., 2004; Usuki et al., 1984; Shahidi and Wanasundara, 1992). It has thus been suggested that photo-oxidation mainly generates hydroperoxides that break down into free radicals that could initiate auto-oxidation reactions. In meat and meat products, auto-oxidation is therefore a key mechanism in lipid oxidation. It mainly generates hydroperoxides and volatile compounds, generally through a three-phase process (initiation, propagation, and termination) (Shahidi and Wanasundara, 1992).

The initiation reaction may begin via external physical agents such as heat, ionizing radiation, or a photonic impact in the ultraviolet spectrum and by chemical agents such as transition metal ions, free radicals, and metalloproteins. Thus, L1H produces a lipid free radical $L1^{\bullet}$, which has an unpaired electron. The lipid radicals are highly unstable, short-lived intermediates that are stabilized by abstracting a hydrogen atom from another chemical species. The oxidation process remains slow during this phase. At the end of the initiation period, oxidation suddenly accelerates, oxygen consumption becomes high, and the peroxide content increases substantially. Lipid oxidation is primarily initiated by hydroxyl ($^{\bullet}OH$) and hydroperoxyl ($HOO^{\bullet}$) radicals, as well as lipid alkoxyl ($LO^{\bullet}$) and peroxyl ($LOO^{\bullet}$) radicals (Shahidi and Wanasundara, 1992).

In aerobic environments, the ($L1^{\bullet}$) radical centered on the carbon molecule and formed during the initiation phase reacts very quickly with triplet oxygen to generate different radical species, including the peroxyl radical $L1OO^{\bullet}$. This reaction has very low activation energy and a high rate constant, so the concentration of $L1OO^{\bullet}$ becomes much higher than the $L1^{\bullet}$ content in all oxygen-bearing systems. The peroxyl radical then abstracts a hydrogen atom from another unsaturated lipid molecule L2H to form hydroperoxide (primary oxidation product) and another radical, the $L2^{\bullet}$ radical. This is a so-called self-sustained radical chain reaction and thus propagates at a high rate and rapidly peaks (Barclay and Ingold, 1981). Many hydroperoxide isomers are thus formed during this phase. Hydroperoxide formation from unsaturated fatty acids is generally accompanied by stabilization of the radical state via double-bond rearrangement (electron delocalization), which gives rise to conjugated dienes and trienes.

The oxidation process then continues with the transformation of hydroperoxides into secondary nonradical oxidation compounds. The main hydroperoxide decomposition mechanism involves scission of the double bond adjacent to the hydroperoxyl group, leading to the formation of hydrocarbons, aldehydes, alcohols, and volatile ketones, among others. Other nonvolatile secondary products are also formed, including nonvolatile aldehydes, oxidized triacylglycerols, and their polymers (Shahidi and Naczk, 2004). Decomposition of primary oxidation compounds involves a complex mechanism in which a single hydroperoxide can generate several types of volatile or nonvolatile molecules. The type of byproducts obtained after fatty acid oxidation is determined by the hydroperoxide composition and by the type of scission of double bonds in the fatty acid chain. The reaction can also terminate after polymer formation.

Lipid oxidation makes both cooked and uncooked meat rancid (Gennadios et al., 1997) and reduces consumer acceptability. Undesirable changes in color, flavor, and nutritive value occur as meat lipids oxidize and interact with other meat constituents, such as pigments, other proteins, carbohydrates, and vitamins (Love and Pearson, 1971). Lipid oxidation is positively correlated to pigment oxidation. The correlation of lipid oxidation with a decrease in redness of uncooked meat is related to oxidation of myoglobin after generation of oxymyoglobin and formation of metmyoglobin. The mechanism of lipid oxidation in meat is dependent on extrinsic and intrinsic factors (Liu et al., 1995). The control of lipid oxidation in meat and meat products has become increasingly important with greater production and consumption of precooked meat items for institutional and home use (Love and Pearson, 1971). Unsaturated lipid substrates are the prime target of oxidation in meat and meat products. It is known that the reaction of radicals with proteins and peptides in the presence of oxygen gives rise to alterations in both their backbone structure and of the amino acid side chains. These oxidative changes include cleavage of peptide bonds, modification of amino acid side chains, and formation of covalent intermolecular cross-linked protein derivatives. Some of the most general amino acid modifications are the formation of protein carbonyl groups and protein hydroperoxides, while cross-linking has mostly been described as formation of disulfide and dityrosine through the loss of cysteine and tyrosine residues, respectively (Lund et al., 2011).

Protein oxidation caused by reactive oxygen species in muscle proteins could be implicated in the loss of their functionality and, therefore, in some loss of the quality of meat, poultry, and fish. Oxidation of proteins in processed meat products leads to reduced water-holding capacity and texture-forming ability. This has been regarded as a major cause for altered digestibility and hence lesser nutritional value of oxidized proteins (Xiong, 2000; Rowe et al., 2004; Prior et al., 2005). However, no flavor change is associated with protein oxidation and hence its organoleptic properties remain mostly unaffected. For this reason, this chapter does not discuss protein oxidation in muscle food products to any great extent.

Different studies have shown that the amount of metal ions, such as iron from heme compounds, copper, and heavy metals, that are present in enzymes and metalloproteins or those migrated from the processing equipment, either by abrasion or due to acidic dissolution of metals from their surfaces could promote lipid oxidation in meat (Jacobsen et al., 2008). Exposure of meat to oxygen, light, and temperature, as well as preservative and processing operations, such as chilling, freezing, inclusion of additives (salt, nitrite/nitrate, ascorbates, and spices), cooking, irradiation, high-pressure treatment, and packaging, could influence the extent of oxidation. Thus, oxidation is one of the biggest economic problems in the meat industry. It compromises the nutritional quality, shortens shelf life, increases toxicity, and decreases the market value of meat and meat products (Sampels, 2013).

The type of packaging material employed for raw fresh meat should be selected in such a way that it is air permeable and moisture impermeable, such as polyvinyl chloride. This would allow the oxygen to interact with myoglobin to produce oxymyoglobin which has a desirable cherry red color while preventing moisture loss. Furthermore, modified atmosphere packaging (MAP) may be used to replace part of the air with carbon dioxide. For cooked meat, however, MAP with or without antioxidants may be employed. The MAP that may include various proportion of carbon dioxide might be combined with the use of spice, spice oil, or spice oleoresin such as that of oregano (Chouliara et al., 2007). Detailed discussion of MAP and its use in meat packaging is beyond the scope of this mini-review but can be found elsewhere (McMillin, 2008; Skandamis and Nychas, 2002).

10.3 MEASUREMENT OF OXIDATIVE STABILITY

Several methods are available to measure the same or different indicators of the progress of oxidation in muscle foods. Lipid oxidation is a multistep reaction with a wide range of products and product groups. Conventional measurements include assessment of total hydroperoxide levels and a selection of indicators of secondary oxidation (hydroperoxide decomposition) products usually by simple titermetric or spectroscopic techniques. Chromatographic techniques (gas chromatography and/or high-performance liquid chromatography together with different detectors including mass spectroscopy) are useful for more detailed description of the reaction products.

Hydroperoxides are considered as primary oxidation products that generate the most relevant information on lipoperoxidation intensity. Hydroperoxide measurement has been the focus of the most developmental work for measuring primary oxidation products. A conventional standardized method for quantifying total hydroperoxides and hydrogen peroxide involves iodometric assay (Lea, 1931). This is one of the oldest and most commonly used methods for assessing lipid substrate oxidation. In acidic medium, hydroperoxides and ROOR compounds react with the iodide ion to generate iodine, which is then titrated using a sodium thiosulfate solution in the presence of a starch solution as indicator.

The peroxide value (PV) is considered to represent the quantity of active oxygen (in mg) contained in 1 g of lipid and which could oxidize potassium iodide, followed by iodine release. However, the products of lipid oxidation are often measured by spectroscopic means. The advantage of using spectroscopic methods is mainly related to the fact that in most assays can be made with no or minimal sample treatment (Kamal-Eldin, 2003; Shahidi and Zhong, 2007). Over 90% of hydroperoxides formed by lipoperoxidation have a conjugated diene system resulting from stabilization of the radical by double bond rearrangement. Oxidation of polyunsaturated fatty acids is accompanied by increased ultraviolet absorption. Fatty acids with conjugated unsaturation absorb strongly in the region 230–375 nm, diene unsaturation at 234 nm, and triene unsaturation at 268 nm (Shahidi and Zhong, 2007). The magnitude of change is not readily related to the degree of oxidation because the effects on the various unsaturated fatty acids vary in quality and magnitude. However, the changes in the ultraviolet spectrum of a given substance can be used as a relative measurement of oxidation. Lipids containing linoleic acid or more highly unsaturated fatty acids are oxidized to conjugated diene systems that can be measured by ultraviolet absorption at 234 nm (Gray et al., 1996). Thus, measurement of conjugated dienes is a facile way to measure primary products of meat oxidation.

The thiobarbituric acid (TBA) test is a well-known method for both screening and monitoring of secondary lipid oxidation status and quality of meats and meat products. It involves reacting the TBA reagent with malondialdehyde produced by lipid hydroperoxide decomposition to form a pink-colored chromophore with absorption maximum at 532 nm (Janero, 1990). The TBA number is defined as the quantity of malondialdehyde (in mg) present in 1 kg of sample. Many oxidizable substrates have been used in determination of TBA-reactive substances (TBARS), including free fatty acids, low density lipoprotein, and fluids (urine, serum) from cells or tissues as well as in muscle foods, in general (Dahle et al., 1962; Korchazhkina et al., 2003).

10.3.1 Gas Chromatography (GC) and GC–Mass Spectroscopy (MS)

GC and GC–MS can be used for the analysis of volatile lipid oxidation products. As an example, hexanal and 2,4-decadienal have been used to monitor oxidation of meat, mainly due to the presence of linoleic acid (Shahidi and Pegg, 1994a,b). He and Shahidi (1997) found a correlation between hexanal and TBARS values in meat. A similar correlation existed between TBARS and sensory quality of meat (Ahn et al., 2002). The advantage of using hexanal is that it is a specific compound arising from breakdown of specific fatty acids, namely linoleic and arachidonic acids, but TBARS values are nonspecific and are produced from different fatty acids (Shahidi et al., 1987). In addition, other compounds present in the meat formulations may give positive results with the TBA reagent and hence the values so obtained may not be realistic in the true sense. Nonetheless, TBARS values, represented as MDA equivalents are useful for assessing the oxidative state of meat and meat products, especially when comparison is made among different samples, and trends so obtained are dependable. In case of seafood, where omega-3 fatty acids are a main contributor, propanal would serve as a useful indicator for assessing their oxidative stability. Other products that possess different omega-3 fatty acids, namely alpha-linolenic acid (ALA), eicosapentaenoic acid (EPA), and docosahexaenoic acid (DHA) also afford propanal as a specific secondary oxidation product. It should also be mentioned that MDA is formed as a main product of oxidation of lipids and lipid-containing foods when the associated fatty acids possess three or more double bonds. The direct analysis of MDA is possible only when an HPLC method is employed (Zhong et al., 2007).

10.4 SHELF LIFE OF MEAT, MEAT PRODUCTS, AND OTHER MUSCLE FOODS

The shelf life of meat is dependent on the animal species and includes differences in breed and muscle fiber type (Insausti et al., 2000; Berg, 2001). Before the animal is slaughtered, external factors such as diet and stress also affect the shelf life of meat (Pfalzgraf et al., 1995); storage conditions such as time, temperature, and packaging atmosphere are among variables that affect the length of time meat products will remain acceptable to consumers (Insausti et al., 2000).

There are specific characteristics that indicate to the consumer whether a meat product is old. These include purge accumulation, color deterioration, rancidity, and microbial spoilage. Purge results when protein in meat loses its ability to bind water (Berg, 2001). Water binding is affected by the pH, the muscle type, the species, and the packaging system employed (Berg, 2001; Hernandez et al., 1998; Aberle et al., 2001; Mead, 2004). Color deterioration results from oxidation of the meat pigment (myoglobin) and is affected by many factors including the breed, the muscle type, the species, the pH, aging, the packaging environment, the storage temperature, bacterial growth, exposure to light, and processing conditions, among others. Rancidity results from lipid oxidation, as already mentioned (Gennadios et al., 1997).

In poultry, the hygienic quality and health safety must particularly be ensured until the expiry date. It is well known that poultry meat is liable to rapid deterioration when kept under inadequate storage conditions. Consequently, poultry meat and meat products are often the cause of alimentary infections (Kozailnski et al., 2012). The investigations of the etiology of food poisoning show that contamination with pathogenic bacteria primarily refers to *Salmonella* spp., *Campylobacter* spp., *Staphylococcus* spp., *Listeria* spp., *Eromonas* spp., as well as *Yersinia euterocolitica, Escherichia coli,* and *Clostridium perfringens* (Bailey et al., 2001; Capita et al., 2002).

The short shelf life of poultry meat at refrigerator temperature can be associated with its composition but also with spoilage microorganisms present during poultry rearing and primary production. These microorganisms can multiply at a relatively low temperature, and the result of their metabolic activity is manifested as product spoilage (Singh, 1993); hence, they are the most important factors that influence the shelf life of chicken meat. The shelf life of poultry meat depends on the initial number of microorganisms, which emphasizes the importance of hygienic conditions and control during various stages of the production process (Yashoda et al., 2001).

Decomposition processes are manifested by a change in specific sensory properties of meat. In a majority of cases, the sensory changes and the degree of contamination with microorganisms, and their biochemical activity, correlates with the meat ammonia content (Baeza, 2004). This is also important in gadoid fish during the storage of fresh products that may lead to production of trimethylamine and dimethylamine from the osmoregulator trimethylamine oxide. It is also important to mention warmed-over-flavor development in reheated cooked poultry as it might be experienced by the consumers (Shahidi et al., 1987; Sitz et al., 2005; Shi and Ho, 1994). In the case of seafood, the high content of highly unsaturated omega-3 fatty acids, particularly in fatty fish, is of much importance. This issue was mentioned in a cursory manner and its detailed coverage is beyond the scope of this chapter.

10.5 PREVENTION OF OXIDATION IN MEAT AND MEAT PRODUCTS USING NATURAL ANTIOXIDANTS

Natural antioxidants can be applied either through dietary means by feeding the animals or through technological strategies to reduce or prevent oxidative process in muscle food. In dietary manipulations, antioxidants are generally introduced into the muscle via animal feeding or possibly direct use. It has been shown that inclusion of natural antioxidants in animal diet not only slows down oxidation but also greatly improves meat quality when compared with those fed diets devoid of antioxidants. In this, dietary inclusion of tocopherols or direct addition into meat may render stability to the products (Shahidi and Pegg, 1992, 1993; Shahidi and Alexander, 1998). However, the content of antioxidant compounds in plant materials varies considerably and hence their dosage application in diets and meat products varies (Moyo et al., 2012; Nkukwana et al., 2014). Antioxidant compounds are usually added at a moderate dosage level, since a high level of inclusion may mechanistically cause adverse effects through prooxidative action (Martin and Appel, 2010). Technological strategies involve the application of antioxidants directly into the meat and meat products or through coating packaging materials with antioxidants or plant extracts to improve the oxidative stability of the products.

Most natural antioxidants are obtained from plant sources, such as culinary herbs, spices, vegetables, as well as fruits and more generally oilseed and protein extender products, which may also contain antioxidants (Shahidi and Zhong, 2010). In spices used in meats, rosemary and sage are most popular. In addition, green tea extracts may be used successfully in meats to arrest oxidation (Table 10.1). Senanayake (2013) has reported the use of rosemary, rosemary oleoresin, or rosemary extracts in raw structured beef steaks, cooked restructured beef, frankfurters, restructured chicken nuggets, cooked beef patties, and cooked meatballs as well as boned turkey meat. In these, the TBARS values of

TABLE 10.1 Effect of Various Additives on the Inhibition of TBARS (mg of MDA Equivalents per kg of Meat)

Antioxidants	Concentration (ppm)	Sample	Storage (days)				References
			0	5	7	14	
Green tea catechins	10	Roasted turkey burgers	0.48	–	1.75	1.9	Senanayake (2013)
Green tea catechins	20	Roasted turkey burgers	0.3	–	1	0.7	Senanayake (2013)
Ascorbic acid	500	Cooked comminuted pork	0.5	–	4.8	5.7	Shahidi and Pegg (1992, 1993)
Ascorbic acid	2000	Cooked comminuted pork	0.2	–	0.8	0.9	Shahidi and Pegg (1992, 1993)
Citric acid	500	Cooked comminuted pork	1.3	–	7.1	8.5	Shahidi and Pegg (1992, 1993)
Citric acid	3000	Cooked comminuted pork	0.3	–	2.3	5.2	Shahidi and Pegg (1992, 1993)
Tocopherol	400	Cooked comminuted pork	2.7	–	7.9	10	Shahidi and Pegg (1992, 1993)
Ascorbic acid+tocopherol	2000+400	Cooked comminuted pork	0.1	–	0.6	0.8	Shahidi and Pegg (1992, 1993)
Ascorbic acid+citric acid	500+1000	Cooked comminuted pork	0.1	–	0.5	1	Shahidi and Pegg (1992, 1993)
Ascorbic acid+citric acid	2000+3000	Cooked comminuted pork	0.1	–	0.3	0.4	Shahidi and Pegg (1992, 1993)

BHA	124	White muscle of mackerel	3.0	3.8	3.8	–	He and Shahidi (1997)
BHT	152	White muscle of mackerel	4.3	6.1	6.1	–	He and Shahidi (1997)
TBHQ	115	White muscle of mackerel	3.0	3.7	4.0	–	He and Shahidi (1997)
α-Tocopherol	296	White muscle of mackerel	7.3	8.1	8.9	–	He and Shahidi (1997)
EC	200	White muscle of mackerel	5.8	6.5	6.6	–	He and Shahidi (1997)
ECG	304	White muscle of mackerel	3.4	3.9	3.7	–	He and Shahidi (1997)
Rosemary	200	Cooked comminuted pork	3.62	6.91	10.11	8.77	Shahidi et al. (1995)
Rosemary	2000	Cooked comminuted pork	0.6	0.97	1.12	1.71	Shahidi et al. (1995)
Clove	200	Cooked comminuted pork	0.89	2.95	4.96	4.87	Shahidi et al. (1995)
Clove	2000	Cooked comminuted pork	0.08	0.19	0.25	0.38	Shahidi et al. (1995)
Sage	200	Cooked comminuted pork	3.0	5.64	8.9	8.3	Shahidi et al. (1995)

Continued

TABLE 10.1 Effect of Various Additives on the Inhibition of TBARS (mg of MDA Equivalents per kg of Meat)—cont'd

Antioxidants	Concentration (ppm)	Sample	Storage (days)				References
			0	5	7	14	
Sage	2000	Cooked comminuted pork	0.18	0.29	0.48	0.58	Shahidi et al. (1995)
Ginger	200	Cooked comminuted pork	4.17	7.52	8.9	10.04	Shahidi et al. (1995)
Ginger	2000	Cooked comminuted pork	1.14	3.66	5.83	6.14	Shahidi et al. (1995)
Thyme	200	Cooked comminuted pork	3.28	6.15	9.47	9.58	Shahidi et al. (1995)
Thyme	2000	Cooked comminuted pork	0.36	2.21	4.12	3.76	Shahidi et al. (1995)
Oregano	200	Cooked comminuted pork	2.51	6.4	8.28	8.47	Shahidi et al. (1995)
Oregano	2000	Cooked comminuted pork	0.32	0.61	1.42	1.32	Shahidi et al. (1995)
BHT	30	Cooked comminuted pork	1.6	3.12	5.0	4.73	Shahidi et al. (1995)

BHT	200	Cooked comminuted pork	0.5	0.76	0.91	0.98	Shahidi et al. (1995)
TBHQ	30	Cooked comminuted pork	0.13	0.3	0.34	0.38	Shahidi et al. (1995)
TBHQ	200	Cooked comminuted pork	0.13	0.18	0.26	0.5	Shahidi et al. (1995)
Propyl gallate	50	Beef steak	–	1.8	–	–	St.Angelo et al. (1992)
EDTA	100	Beef steak	–	7.2	–	–	St.Angelo et al. (1992)
Propyl gallate + EDTA	50 + 100	Beef steak	–	1.7	–	–	St.Angelo et al. (1992)
[a]$NaNO_2$	100	Fish sausage	0.4	0.6	0.65	0.75	Flick et al. (1992)
[a]$NaNO_2$ + sodium ascorbate	100 + 1000	Fish Sausage	0.5	0.7	0.7	0.65	Flick et al. (1992)

[a]Packed with PVC, polyvinyl chloride; BHA, butylated hydroxyanisole; BHT, butylated hydroxytoluene; TBHQ, tert-butylhydroquinone; EC, epicatechin; ECG, epicatechin gallate; and EDTA, ethylenediaminetetraacetic acid.

TABLE 10.2 Effect of Green Tea Extract and Rosemary Extract on the Inhibition of Hexanal (mg of Hexanal/kg of Meat) Formation in Chilled (4 °C) Roasted Beef Burgers

	Storage (days)		
	0	5	14
Control	0.3	2.4	3.0
Green tea extract (100 ppm)	0.2	0.3	0.2
Rosemary extract (375 ppm)	0.3	0.5	0.7

Adapted from Senanayake (2013).

products were reduced by up to 82% during refrigerated or frozen storage for different time periods (Table 10.2).

In cured meats, nitrite is the main compound responsible for protection and shelf life extension of products, which usually is about 5–6 weeks under refrigerated conditions. Nitrite provides the desirable pink color to products and is responsible for the particular flavor and oxidative as well as microbial stability of cured products. The permitted level of nitrite in processed meats is 156 ppm, but due to potential problems associated with the presence of unreacted nitrite in meats, its current use in meats as practiced by the user industry is 120 ppm. In addition, some processors use celery juice, which contains nitrate by careful converting it to nitrite and call it nitrite-free, although this is a myth as some products have even higher amount of residual nitrite.

In cured meat products, the adjuncts used have their own effect on oxidation. Thus, the type of salt constituents may affect the stability of meat products considerably, depending on the nature of both their cation and anion, as demonstrated by Wettasinghe and Shahidi (1996). In addition, cooked cured meat pigment was found to have its own antioxidant activity in meats. Furthermore, ascorbic acid, phosphate and polyphosphates, and other adjunct, alone or in combination exert antioxidant effects (Emanuel et al., 1961; Shahidi, 2009). Combination of such adjuncts as an alternate to nitrite curing of meat with pre-formed cooked cured meat pigment was successful (Shahidi and Pegg, 1992, 1993).

Synthetic antioxidants, such as butylated hydroxyanisole (BHA), butylated hydroxytoluene (BHT), and *tert*-butylhydroquinone (TBHQ), have been used in inhibiting meat oxidation. The addition level is recommended to be no more than 200 ppm on fat level and 30 ppm on the meat weight basis. Chelators such as ethylenediaminetetraacetic acid (EDTA) and polyphosphates may also provide protection to meat and meat products against oxidation, presumably by chelating prooxidant metal ions such as those of iron (Table 10.3) (Shahidi et al., 1986).

Another important protection provided to meats against oxidation is smoke. Although direct use of wood smoke has been practiced in the past, nowadays liquid smoke is used in different applications. Phenolic compounds present in

TABLE 10.3 Effect of Metal Chelators on Lipid Oxidation (TBA Numbers: mg MDA Eq/kg Sample) in Cooked Ground Pork Catalyzed by Metals and Heme Compounds Stored at 4 °C

			With Chelator	
Treatments	Concentration (ppm)	Without Chelator	Na_2EDTA (500 ppm)	Sodium Tripolyphosphate (3000 ppm)
Control		2.90–3.15	0.07	0.22
Fe^{2+}	30	4.80	0.09	0.27
Fe^{3+}	30	4.60	0.08	0.37
Cu^{+}	30	3.55	0.35	0.32
Cu^{2+}	30	3.20	0.31	0.38
Myoglobin	30	4.20	0.07	0.28
Hemoglobin	30	4.33	0.10	0.30
Hemin	30	4.35	0.07	0.33
Cooked cured meat pigment (CCMP)	12	0.42	0.30	0.30
Sodium ascorbate (SA)	550	–	–	0.27
SA + TBHQ	550 + 30	–	–	0.20
SA + TBHQ + CCMP	550 + 30 + 12	–	–	0.34

Adapted from Shahidi and Hong (1991).

smoke not only provide a desirable taste and flavor to the product but also help in rendering antioxidant and antimicrobial effect to such processed meat and fish products (Hollenbeck, 1994). He and Shahidi (1997) reported that the potency of catechins in the prevention of oxidation in fish meat was in the decreasing order EGCG ≈ ECG > EGC > EC. In addition, studies have shown that in comparison with tocopherol, catechins were generally more effective in controlling oxidation of meat lipids (Shahidi and Alexander, 1998; Shahidi et al., 1997).

10.5.1 Antimicrobial Activities of Antioxidants in Meat Preservation

Meat wastage is caused by microbial and chemical spoilage with the consequence of foodborne illnesses, economic loss, and food insecurity. Meat spoilage bacteria can be reduced by applying natural antioxidants directly into the meat products. Different studies have shown the use of natural compounds such as organic acids and essential oils for decontamination of beef, pork, and poultry products against *Salmonella* (Mani-López et al., 2012; Sant'Ana et al., 2014).

The presence and level of different phytochemicals such as phenolic acids, flavonoids, alkaloids, saponins, tannins, terpenes, and, more specifically, carvacrol and thymol, among others, which have antimicrobial activities, would be useful (Sharma et al., 2012). Further studies may be needed on combinations and application of different natural antioxidants to reduce meat spoilage and to extend their shelf life, as these will greatly help to reduce financial loss and labor cost, ensure safety, and ultimately improve the functional properties of meat and meat products.

REFERENCES

Aberle, E.D., Forrest, J.C., Gerrard, D.E., Mills, E.W., 2001. Principles of Meat Science, fourth ed. Kendal/Hunt, Dubuque, Iowa.

Ahn, J., Grün, I.U., Fernando, L.N., 2002. Antioxidant properties of natural plant extracts containing polyphenolic compounds in cooked ground beef. J. Food Sci. 67, 1364–1369.

Baeza, E., 2004. Measuring quality parameters. In: Mead, G.S. (Ed.), Poultry Meat Processing and Quality. Woodhead Publishing, Oxford, UK, pp. 304–332.

Bailey, J.S., Stern, N.J., Fedorka-Cray, P.S., Craven, N.A., Cox, D.E., Cosby, S., Ladely, M.T., 2001. Sources and movement of Salmonella through integrated poultry operations: a multistage epidemiological investigation. J. Food Prot. 63, 1690–1697.

Barclay, L.R.C., Ingold, K.U., 1981. Autoxidation of biological molecules. The autoxidation of a model membrane. A comparison of the autoxidation of egg lecithin phosphatidylcholine in water and chlorobenzene. J. Am. Oil Chem. Soc. 103, 6478–6485.

Berg, E.P., 2001. Influence of Stress on Composition and Quality of Meat, Poultry, and Meat Products. Accessed online http://www.fass.org/fass01/pdfs/Berg.pdf.

Capita, R.C., Alonso-Calleja, M., Prieto, M., Garcia-Fernandez, D.C., Moreno, B., 2002. Incidence and pathogenicity of *Yersinia* spp. isolates from poultry in Spain. Food Microbiol. 19, 295–301.

Chouliara, E., Karatapanis, A., Savvaidis, I.N., Kontominas, M.G., 2007. Combined effect of oregano essential oil and modified atmosphere packaging on shelf life extension of fresh chicken breast meat, stored at 4 °C. Food Microbiol. 24, 607–617.

Dahle, L.K., Hill, E.G., Holman, R.T., 1962. The thiobarbituric acid reaction and the autoxidations of polyunsaturated fatty acid methyl esters. Arch. Biochem. Biophys. 98, 253–261.

Emanuel, N., Lyaskovskaya, Y., Yu, N., 1961. The Inhibition of Fat Oxidation Processes. Pergamon Press Ltd, Oxford, UK, p. 340.

Flick, G.J., Hong, G.P., Knobl, G.M., 1992. Lipid Oxidation of Seafood during Storage. Lipid Oxidation in Food, ACS Symposium Series 500. American Chemical Society, Washington, DC, pp. 183–207.

Gennadios, A., Hanna, M.A., Kurth, L.B., 1997. Application of edible coatings on meats, poultry, and seafoods: a review. LWT Food Sci. Technol. 30, 337–350.

Gray, J.I., Gomaa, E.A., Buckley, D.J., 1996. Oxidative quality and shelf life of meats. Meat Sci. 43, S111–S123.

He, Y., Shahidi, F., 1997. Antioxidant activity of green tea and its catechins in a fish meat model system. J. Agric. Food Chem. 45, 4262–4266.

Hernandez, P., Navarro, J.L., Toldra, F., 1998. Lipid composition and lipolytic enzyme activities in porcine skeletal muscles with different oxidative pattern. Meat Sci. 49, 1–10.

Hollenbeck, C.M., 1994. Contribution of smoke flavourings to processed meats. In: Shahidi, F. (Ed.), Flavor of Meat and Meat Products. Springer, pp. 199–209.

Insausti, K., Beriain, M.J., Purroy, A., Alberti, P., Gorraiz, C., Alzueta, M.J., 2000. Shelf life of beef from local Spanish cattle breeds stored under modified atmosphere. Meat Sci. 57, 273–281.

Jacobsen, C., Undeland, I., Storrø, I., Rustad, T., Hedges, N., Medina, I., 2008. Preventing lipid oxidation in seafood, in improving seafood products for the consumer. In: Borresen, T. (Ed.), Improving Seafood Products for the Consumer. Woodhead Publishing, Cambridge, UK, pp. 426–460.

Janero, D., 1990. Malondialdehyde and thiobarbituric acid-reactivity as diagnostic indices of lipid peroxidation and peroxidative tissue injury. Free Rad. Biol. Med. 9, 515–540.

Kamal-Eldin, A., 2003. Lipid Oxidation Pathways. AOCS Press, p. 323.

Korchazhkina, O., Exley, C., Andrew, S.S., 2003. Measurement by reversed-phasehigh-performance liquid chromatography of malondialdehyde in normal humanurine following derivatisation with 2, 4-dinitrophenylhydrazine. J. Chromatogr. B 794, 353–362.

Kozailnski, L., Fleck, Z.C., Kozatinski, Z., Filipoviel, I., Mitak, M., Bratulies, M., Mikug, T., 2012. Evaluation of shelf life of pre-packed cut poultry meat. Veterinarski Arhiv. 82, 47–58.

Laguerre, M., Lecomte, J., Villeneuve, P., 2007. Evaluation of the ability of antioxidants to counteract lipid oxidation: existing methods, new trends and challenges. Prog. Lipid Res. 46, 244–282.

Lea, C.H., 1931. The Effect of Light on the Oxidation of Fats, 108. Proc. Roy. Soc., London, pp. 175–189.

Liu, Q., Lanari, M.C., Schaefer, D.M., 1995. A review of dietary vitamin E supplementation for improvement of beef quality. J. Anim. Sci. 73, 3131–3140.

Love, J.D., Pearson, A.M., 1971. Lipid oxidation in meat and meat products—a review. J. Am. Oil Chem. Soc. 10, 547–549.

Lund, M.N., Heinonen, M., Baron, P.C., Estevez, M., 2011. Protein oxidation in muscle foods: a review. Mol. Nutr. Food Res. 55, 83–95.

Mani-López, E., Garcia, H.S., López-Malo, A., 2012. Organic acids as antimicrobials to control Salmonella in meat and poultry products. Food Res. Int. 45, 713–721.

Martin, K.A., Appel, C.L., 2010. Polyphenols as dietary supplements: a double-edged sword. Nutr. Diet. Suppl. 2, 1–12.

McMillin, K.W., 2008. Where is MAP going? A review and future potential of modified atmosphere packaging for meat. Meat Sci. 80, 43–65.

Mead, G.M., 2004. Keeping poultry meat fresh. Food Sci. Technol. 19, 20–21.

Moyo, B., Oyedemi, S., Masika, P.J., Muchenje, V., 2012. Polyphenolic content and antioxidant properties of *Moringa oleifera* leaf extracts and enzymatic activity of liver from goats supplemented with *Moringa oleifera* leaves/sunflower seed cake. Meat Sci. 91, 441–447.

Nkukwana, T.T., Muchenje, V., Masika, P.J., Hoffman, L.C., Dzama, K., Descalzo, A.M., 2014. Fatty acid composition and oxidative stability of breast meat from broiler chickens supplemented with *Moringa oleifera* leaf meal over a period of refrigeration. Food Chem. 142, 255–261.

Pfalzgraf, A., Frigg, M., Steinhart, H., 1995. α-tocopherol contents and lipid oxidation in pork muscle and adipose tissue during storage. J. Agric. Food Chem. 43, 1339–1342.

Prior, R.L., Wu, X.L., Schaich, K., 2005. Standardized methods for the determination of antioxidant capacity and phenolics in food and dietary supplements. J. Agric. Food Chem. 53, 4290–4302.

Rowe, L.J., Maddock, K.R., Lonergan, S.M., Huff-Lonergan, E., 2004. Influence of early postmortem protein oxidation on beef quality. J. Anim. Sci. 82, 785–793.

Sampels, S., 2013. Oxidation and antioxidants in fish and meat from farm to fork. In: Muzzalupo, I. (Ed.), Food Industry, pp. 114–144 (Intechopen.com).

Sant'Ana, A.S., Franco, B.D.G.M., Schaffner, D.W., 2014. Risk of infection with Salmonella and *Listeria monocytogenes* due to consumption of ready-to-eat leafy vegetables in Brazil. Food Control 42, 1–8.

Senanayake, N.S.P.J., 2013. Rosemary and green tea extracts as natural antioxidants: chemistry, technology, and applications. J. Funct. Foods 5, 1529–1541.

Shahidi, F., Alexander, D.M., 1998. Green tea catechins as inhibitors of oxidation of meat lipids. J. Food Lipids 5, 125–133.

Shahidi, F., Hong, C., 1991. Role of metal ions and heme pigments in autoxidation of heat-processed meat products. Food Chem. 42, 339–346.

Shahidi, F., Naczk, M., 2004. Phenolics in Food and Nutraceuticals. CRC Press, Boca Raton, FL, pp. 1–558.

Shahidi, F., Pegg, R.B., 1992. Nitrite-free meat curing systems: update and review. Food Chem. 43, 185–191.

Shahidi, F., and Pegg, R.B., 1993. Powdered cooked cured-meat pigment which is a non-nitrite meat preservative. US Patent No. 5230915. Washington, DC: US Patent and Trademark Office.

Shahidi, F., Pegg, R.B., 1994a. Hexanal as an indicator of meat flavor deterioration. J. Food Lipids 1, 177–186.

Shahidi, F., Pegg, R.B., 1994b. Hexanal as an indicator of flavor deterioration of meat and meat products. In: Ho, C.-T., Hartman, T.G. (Eds.), Lipids in Food Flavours. ACS Symposium Series 558, American Chemical Society, Washington, DC, pp. 256–279.

Shahidi, F., Wanasundara, P.K.J.P.D., 1992. Phenolic antioxidants. Crit. Rev. Food Sci. Nutr. 32, 67–103.

Shahidi, F., Zhong, Y., 2007. Measurement of antioxidant activity in food and biological systems. In: Shahidi, F., Ho, C.T. (Eds.), Antioxidant Measurement and Applications. ACS Symposium Series 956, American Chemical Society, Washington, DC, pp. 36–66.

Shahidi, F., Zhong, Y., 2010. Novel antioxidants in food quality preservation and health promotion. Eur. J. Lipid Sci. Technol. 112, 930–940.

Shahidi, F., Rubin, L.J., Diosady, L.L., Kassum, N., Fong, J.C., Wood, D.F., 1986. Effect of sequestering agents on lipid oxidation in cooked meats. Food Chem. 21, 145–152.

Shahidi, F., Yun, J., Rubin, L.J., Wood, D.F., 1987. The hexanal content as an indicator of oxidative stability and flavor acceptability in cooked ground pork. Can. Int. Food Sci. Technol. J. 20, 104–106.

Shahidi, F., Pegg, R.B., Saleemi, Z.O., 1995. Stabilization of meat lipids with ground spices. J. Food Lipids 2, 145–153.

Shahidi, F., Wanasundara, U.N., He, Y., Shukla, V.K.S., 1997. Marine lipids and their stabilization with green tea catechins. In: Shahidi, F., Cadwallader, K.R. (Eds.), Flavor and Lipid Chemistry of Seafoods. ACS Symposium Series 674; American Chemical Society, Washington, DC, pp. 186–197.

Shahidi, F., 2009. Prevention of lipid oxidation in muscle foods by nitrite and nitrite-free compositions. In: St Angelo, A.J. (Ed.), Lipid Oxidation in Food. ACS Symposium Series 500, Amercian Chemical Society, Washington, DC, pp. 161–182.

Sharma, A.K., Gangwar, M., Tilak, R., Nath, G., Sinha, A.S.K., Tripathi, Y.B., 2012. Comparative in vitro antimicrobial and phytochemical evaluation of methanolic extract of root, stem and leaf of *Jatropha curcas* Linn. Pharmacogn. J. 4, 34–40.

Shi, H., Ho, C.T., 1994. The flavour of poultry meat. In: Shahidi, F. (Ed.), Flavour of Meat and Meat Products. Blackie Academic and Professional, Glasgow, pp. 52–69.

Sikorska, E., Romaniuk, A., Khmelinskii, I.V., Herance, R., Bourdelande, J.L., Sikorski, M., Kozioł, J., 2004. Characterization of edible oils using total luminescence spectroscopy. J. Fluoresc. 14, 25–35.

Singh, H., 1993. Extension of shelf life of meats and fish by irradiation. In: Charalambous, G. (Ed.), Shelf Life Studies of Foods and Beverages. Chemical. Biological Physical and Nutritional Aspects. Elsevier Science Publisher B.V., The Netherlands.

Sitz, B.M., Calkins, C.R., Feuz, D.M., Umberger, W.J., Eskridge, K.M., 2005. Consumer sensory acceptance and value of domestic, Canadian, and Australian grass-fed beef steaks. J. Anim. Sci. 83, 2863–2868.

Skandamis, P.N., Nychas, G.-J.E., 2002. Preservation of fresh meat with active and modified atmosphere packaging conditions. Int. J. Food Microbiol. 79, 35–45.

St Angelo, A.J., Spanier, A.M., Bett, K.L., 1992. Chemical and sensory evaluation of flavor in untreated and antioxidant-treated meat. In: St Angelo, A.J. (Ed.), Lipid Oxidation in Food. ACS Symposium Series-American Chemical Society, New York, NY, pp. 140–160.

Usuki, R., Endo, Y., Kaneda, T., 1984. Prooxidant activities of chlorophylls and pheophytins on the photooxidation of edible oils. Agric. Biol. Chem. 48, 991–994.

Wettasinghe, M., Shahidi, F., 1996. Oxidative stability of cooked comminuted lean pork as affected by alkali and alkali earth halides. J. Food Sci. 61, 1160–1164.

Xiong, Y.L., 2000. Protein oxidation and implications for muscle food quality. In: Decker, E., Faustman, C., Lopze-Bote, C.J. (Eds.), Antioxidants in Muscle Foods: Nutritional Strategies to Improve Quality. John Wiley and Son Publishing, New York, NY, pp. 85–111.

Yashoda, K.P., Sachiondra, N.M., Sakhare, P.Z., Rao, D.N., 2001. Microbiological quality of broiler chicken carcasses processed hygienically in small scale poultry processing unit. J. Food Qual. 24, 249–259.

Zhong, Y., Madhujith, T., Mahfouz, N., Shahidi, F., 2007. Compositional characteristics of muscle and visceral oil from steelhead trout and their oxidative stability. Food Chem. 104, 602–608.

Chapter 11

Oxidative Stability of Seafood

Ingrid Undeland
Biology and Biological Engineering – Food and Nutrition Science, Chalmers University of Technology, Gothenburg, Sweden

11.1 GENERAL INTRODUCTION

The oxidative stability of seafood lipids is a large research area with a long history. Reflecting this, many excellent reviews have already been published, with examples being those of Flick et al. (1992), Hultin (1994), Erickson (1997), Harris and Tall (1994), Ashton (2002), Bao and Oshima (2014), Raghavan and Kristinsson (2014), and Karlsdottir et al. (2014a), all with somewhat different focuses. Research trends have indeed changed over the years, with links between oxidation and fatty acid (FA) patterns, lipid classes, and endogenous pro- and antioxidants being in focus in the 1960s–1990s, and then the big trend has been the evaluation of various natural antioxidants as a means to inhibit seafood lipid oxidation. When it comes to basic mechanisms, the greatest steps ahead in the last 10–15 years have been taken within the area of hemoglobin (Hb)-mediated lipid oxidation in seafood. Along with an increased interest in alternative aquatic biomasses (i.e., beyond fish fillets) as sources of food and food ingredients, a significant number of reports have also emerged on oxidation development during value-adding processing such as protein isolation and hydrolysis production. A handful of reports addressing oxidation in algae and krill biomasses are also available. In this chapter, a more general introduction on the consequences of oxidation for seafood quality is first given, followed by a listing of the main methods used to detect oxidation in seafood. Endogenous factors affecting oxidation development in fish are then addressed: lipids, pro- and antioxidants, with a somewhat deeper focus on the mechanisms for Hb-mediated oxidation and on lipid/antioxidant alterations resulting from modulations of aquafeed. External factors affecting lipid oxidation are then put in focus through a series of common handling and processing steps to which seafood is subjected, and then the focus is on prevention of seafood oxidation using plant- or muscle-derived antioxidants. The bulk of the studies reviewed here are from research on fish muscle, and work done in model systems simpler than washed fish mince are excluded so as to not interfere with the chapter on marine oils.

Oxidative Stability and Shelf Life of Foods Containing Oils and Fats
http://dx.doi.org/10.1016/B978-1-63067-056-6.00011-2

11.2 THE ROLE OF LIPID OXIDATION FOR THE QUALITY OF SEAFOOD

The highly unsaturated character of lipids in aquatic raw materials, the presence of strong pro-oxidative systems, and, in many cases, an adaption in vivo to cold habitat temperatures and/or high pressure (HP) make lipid oxidation proceed fast during postmortem handling, storage, and processing of seafood products. The lipid oxidation reaction can affect sensory and nutritional characteristics of seafood in many different ways and can be regarded the prime chemical reaction jeopardizing a sustainable use of a food resource known to be subjected to a lot of environmental stress. It will be evident from this chapter that the aquatic raw materials that are the most underutilized for food production, such as certain small pelagic fish species (herring, sardines, anchovy) and rest-raw materials from the filleting industry, are those with the highest inherent susceptibility to oxidation due, for example, to their highly vascularized muscle, the abundance of enzymes and lipids, and sometimes fast decrease in muscle pH.

The most obvious consequence of uncontrolled lipid oxidation in fish and shellfish is a loss of the "fresh fish/shellfish aroma," often described as green, planty, melon-like notes, and the development of rancid odors, often described as painty, rancid, fishy, cod liver oil–like, or metallic. Paradoxically, the fresh fish aroma in part comes from low levels of volatile 6-, 8-, and 9-carbon carbonyls and alcohols arising in lipoxygenase (LOX)-catalyzed oxidation of long-chain polyunsaturated FAs (LC PUFA) (Lindsay, 1994). The rancid odors are linked to aldehydes like *E*-2-pentenal, *E*-2-hexenal, *Z*-4-heptenal, (*E,E*)-2,4-heptadienal, 2,4,7-decatrienal, 1-penten-3-ol, 1-octen-3-ol, 1,5-octadien-3-one, 1-penten-3-one, and 2,6-nonadienal, with some of the latter having very low odor thresholds (Josephson et al., 1984; Milo and Grosch, 1993; Iglesias and Medina, 2008). Specific patterns of volatiles give various types of seafood-specific rancid aromas.

However, rancidity in seafood is much more than an aroma problem. The color also suffers according to several different chemical reactions, one being the conversion of reduced heme proteins into oxidized ones (met-Hb and met-myoglobin (Mb)), triggered by the formation of lipid oxidation products (Chaijan and Undeland, 2015). The result is a conversion of the fresh reddish/pink color to grayish-brown shades (Wetterskog and Undeland, 2004). Also, complete destruction of the heme ring (protoporphyrin IX) during lipid oxidation bleaches the reddish color. Another reaction that can affect the redness of seafood is the bleaching of carotenoids like astaxanthin and cantaxanthin in salmonoid fish during their co-oxidation with PUFA, something that can be driven by LOX (Hard, 2002). Due to interactions between reactive lipid-derived aldehydes like alkanals, alkenals, 4-hydroxy-2-alkenals, keto-alkenals, and alkanedials (dialdehydes) with amino groups like lysine or arginine residues in proteins/peptides, nucleic acids, and phospholipids (PL), Schiff base adducts can also form (Chaouqy et al., 2008; Domingues et al., 2013) (see Figure 11.1). Malondialdehyde (MDA) and 4-hydroxy-trans-2-hexenal (HHE), both common products of *n*-3 PUFA, are

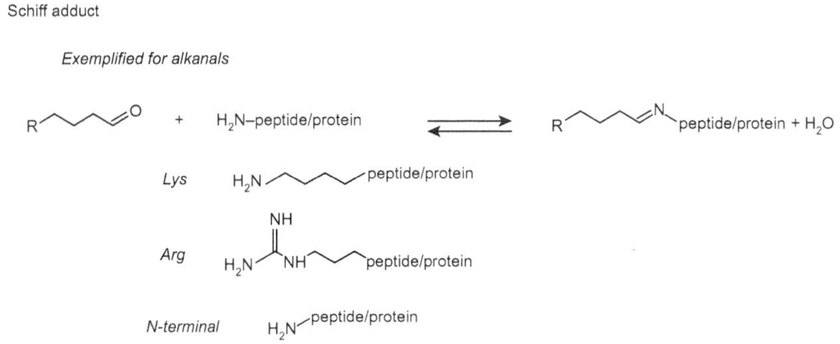

FIGURE 11.1 Formation of Schiff adducts between alkanals derived from lipid peroxidation products and Lys, Arg, and N-terminal amines of peptides. *Reprinted with permission from Domingues et al. (2013).*

pointed out as particularly reactive according to this pathway (Domingues et al., 2013). Schiff bases, which have the general formula $R^1R^2C = NR^3$, that is, an N,N^1-disubstituted 1-amino-3-iminopropene, can be stabilized to pyrroles, which further polymerize into yellow-brownish melanoidin-like pigments (Zamora and Hidalgo, 2005). Therefore, yellowness/brownness development, sometimes seen on a fish fillet surface as "rust," can also be a consequence of lipid oxidation.

Texture changes can also appear as a result of cross-reactions between aldehydes with proteins but also from the reaction of lipid free radicals with proteins. Underlying reasons are, for example, the formation of intra-/intermolecular S—S-bridges or increased protein hydrophobicity yielding increased toughness and decreased water-holding capacity, respectively (Bertram et al., 2007). Domingues et al. (2013) listed that aldehydes that can induce protein cross-linking include alkenals, 4-hydroxy-2-alkenals, keto-alkenals, and dialdehydes (see Table 11.1). Table 11.1 also summarizes which classes of aldehydes that can form Schiff bases and Michael adducts, the latter being the result of the reaction of electrophilic carbon located in the β position with respect to an α,β-unsaturated carbonyl group, with the nitrogen lone pair of the histidine, or lysine, or the sulfhydryl group of cysteine (Domingues et al., 2013). Odor compounds in the form of protein carbonyls can also form in the cross-reactions between lipid oxidation products and proteins (Lund et al., 2011). As summarized by Bizzozero et al. (2007), three kinds of PUFA oxidation products can yield protein carbonyls: (1) dialdehydes such as MDA, which can form carbonyl derivatives with lysine residues, (2) α,β-unsaturated aldehydes (e.g., HNE, HHE), which undergo a Michael addition reaction with the ε-amino group of lysine, the thiol group of cysteine residues, and the imidazole ring of histidine residues, and (3) lipid hydroperoxides, which via metal-induced cleavage form alkoxyl and peroxyl radicals that can react with amino acid side chains to form carbonyls. A small portion can also arise from the decomposition of side chain hydroperoxides of valine and leucine (Fu and Dean, 1997).

TABLE 11.1 Type of Adducts, Schiff and Michael Adducts, Formed Dependently on the Structural Characteristics of Each Distinct Class of Aldehydes

	Alkanals	Alkenal	4-Hydroxy-2-Alkenal	Keto-Alkenals	Alkanedial (Dialdehydes)
Schiff adducts	X	X	X	X	X
Michael adducts		X	X	X	
Proteins cross-linking		X	X	X	X

Reprinted with permission from Dominguez et al. (2013).

The significance of lipid oxidation for the nutritional quality of seafood is not clear cut. In short, the nutritional quality can be affected by losses of nutrients, by development of potentially toxic oxidation products (Gabrielsson et al., 2014), and by decreased digestibility as a result of the mentioned protein cross-links (Sante-Lhoutellier et al., 2007). Nutrient losses include decreases in endogenous antioxidants/vitamins/pro-vitamins such as α-tocopherol, ascorbic acid, and carotenoids (Choubert and Baccaunaud, 2006; Passi et al., 2005). As examples, α-tocopherol levels had completely disappeared after 1 day during ice storage of Hb-enriched washed cod mince (Larsson and Undeland, 2010) and after 12 days in ice stored herring fillets (Undeland et al., 1999).The latter study also revealed a nearly linear loss of ascorbic acid, which has also been found by others (Tuckey et al., 2009). In the study by Passi et al. (2005), ubiquinol and ascorbic acid disappeared almost completely in light muscle of different species of Mediterranean fish within 48 h on ice. Decreases in the substrate LC n-3 PUFA as a result of oxidation indeed also occur, but whether such changes are measurable differs between studies. No decreases were found during refrigerated and frozen storage of skin-on mackerel and cod fillets (Xing et al., 1993), while 6 days' ice storage of sardines induced 13.2% and 5.5% decrease of EPA and DHA, respectively (Chaijan et al., 2006). A crucial question is however whether lipid oxidation reduces LC n-3 PUFA to a point where the nutritional value of the seafood product is seriously affected *before* the point where it is regarded unacceptable from a sensory point of view. The same question applies to the rising levels of reactive oxidation products in seafood (lipid free radicals, peroxides, aldehydes, etc.)—do they reach levels of potential toxicity while the seafood product is still interesting from a sensory point of view? As mentioned, certain unsaturated aldehydes formed in seafood have extremely low odor thresholds and thus can ruin the sensory impression already at 0.001 ppb levels

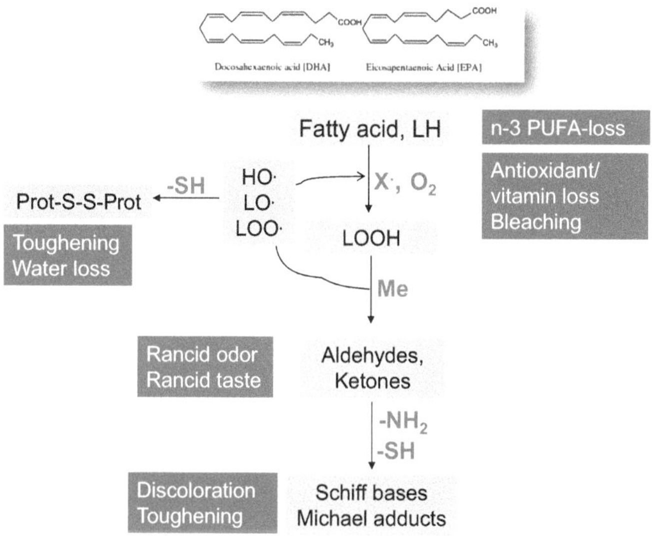

FIGURE 11.2 Simplified version of the lipid oxidation reaction in seafood with different quality decreasing features appearing at each reaction step shown in the dark gray boxes. LH, fatty acid; X˙, any radical with higher redox potential than the substrate fatty acids; Me, metal, which can be, for example, hemoglobin, myoglobin, or low molecular iron/copper.

(odor threshold reported for 1,5-octadien-3-one in aqueous solution (Swoboda and Peers, 1977; Josephson et al., 1983). This point significantly distinguishes seafood from fish oil capsules as a source of lipid oxidation products, the latter being more or less insensitive to sensory characteristics, but still having recommendations on maximum levels of oxidation products the encapsulated oil should not exceed. No such recommendations exist for seafood, and best-before dates are instead mainly set out from a bacteriological perspective (chilled products) or sensory perspective (frozen products). In frozen seafood, where lipid oxidation plays a major role for shelf life, best-before-dates normally ranges from 12 to 24 months at conventional freezing temperature (−18 °C to −25 °C) depending, for example, on the type of packaging applied (e.g., glazing vs no glazing, vacuum vs. no vacuum). Figure 11.2 summarizes the main steps of the lipid oxidation reaction indicating how these steps affect seafood quality.

11.3 COMMON MARKERS OF LIPID OXIDATION IN SEAFOODS

The most common analytical strategies by which lipid oxidation is followed in seafood to a large extent reflect the quality changes described earlier. Analysis of volatiles by gas chromatography–mass spectrometry (GC–MS) is a core technology in oxidation research giving valuable specific information on the types of secondary oxidation products formed under various settings. Some studies have suggested propanal as a particularly valuable oxidation marker for seafood since it develops in systems enriched in the n-3 PUFAs α-linolenic acid (ALA), eicosapentaenoic acid (EPA), and docosahexaenoic acid (DHA) (Shahidi, 2001). However, high levels of alcohols, ketones, unsaturated aldehydes, and furans have also commonly been reported during oxidation of seafood raw materials, for example, 1-penten-3-ol in and 2,4-heptadienal in mackerel mince (Eymard et al., 2009), 1-penten-3-ol and 2-penten-1-ol in herring fillets (Sampels et al., 2010), as well as 1-penten-3-ol and 2-pentyl-furan in Mb-enriched washed sea bass mince (Thiansilakul et al., 2011). Results from the latter study are exemplified in Table 11.2, showing how 13 of 15 analyzed volatiles reached higher levels with addition of extra Mb than without extra Mb. Sensory analysis of rancid odor using linear scales is also quite widespread (see, for example,

TABLE 11.2 Volatile Compounds in Washed Seabass Mince without and with the Addition of Myoglobin (2 mg/g) from Eastern Little Tuna at pH 6.0 after 8 Days of Refrigerated Storage

	Concentration (%)	
Compounds	**Washed Mince**	**Washed Mince + Mb**
Hexanal	2.52	11.00
Heptanal	ND	2.02
2-Pentyl furan	1.29	7.08
Octanal	ND	2.61
Nonanal	ND	3.93
E–2-Octenal	ND	1.91
1-Octen-3-ol	ND	11.54
1-Heptanol	ND	1.31
Benzaldehyde	ND	1.43
Octyl formate	ND	1.60
2-Undecanone	ND	3.09
Ethyl benzaldehyde	ND	0.73
Tetradecanal	0.35	ND
Pentadecanal	ND	1.12
Octadecanal	0.81	ND

ND, nondetectable.
Reprinted with permission from Thiansilakul et al. (2011).

Medina et al., 2009; Halldorsdottir et al., 2014a; Sannaveerappa et al., 2014), but the methodology is indeed dependent on the accessibility to a trained panel.

Measuring color changes as a way of following lipid oxidation in seafood has increased in the last 15 years, especially the loss of redness, that is, a decrease in a^*-value, which is linked to the large focus on heme proteins as lipid oxidation catalysts in this time period. Figure 11.3 illustrates typical kinetics for redness loss in ice-stored Hb-fortified washed cod mince (Wetterskog and Undeland, 2004). In all "oxidizing" samples of this study, a^* values correlated well with thiobarbituric acid reactive substances test (TBARS) and painty odor development, $r=-0.95$ and -0.77, respectively. Yellowness increase (increase in b^*-value) is also quite commonly used, as well as brown color formation, determined, for example, at 400 and 450 nm in chloroform extracts (Hassan et al., 1999). A browning ratio can be expressed between the wavelength assessments (A_{450}/A_{400}).

In addition to the above, the classic peroxide value (PV) and TBARS are the true core methods for primary and secondary products, respectively. One or the other of these methods appears in almost any report on seafood lipid oxidation. Varieties in the methodology of course exist, with the ferric thiocyanate method (Santha and Decker, 1994), being a common alternative to the

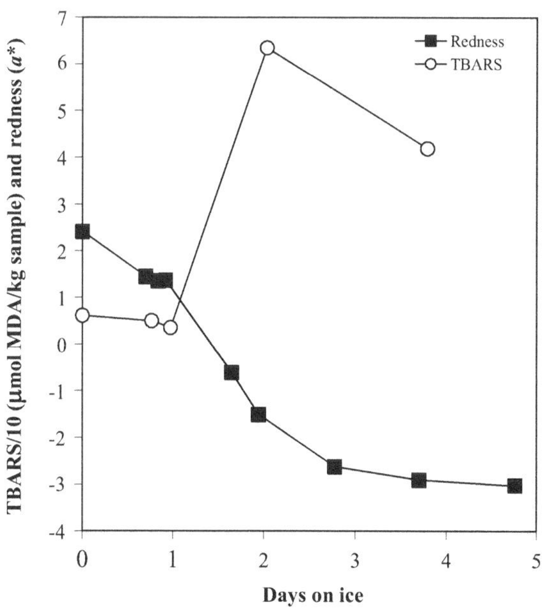

FIGURE 11.3 Hb-mediated changes in TBARS and redness (a^* value) during ice storage of washed cod mince prepared at physiological moisture (81%). Oxidation was initiated with 15 μmol/L trout Hb, and 200 ppm of streptomycin was added to prevent bacterial growth. The final pH of the model system was 6.5. *Reprinted with permission from Wetterskog and Undeland (2004). Copyright (2004) American Chemical Society.*

classic PV determination by iodometric titration. TBARS methods vary, for example, with respect to the type of extracts that are analyzed: methanol-water or trichloroacetic acid (TCA), and with respect to the exclusion or inclusion of a hydrolysis step leading to analysis of free or total TBARS, respectively. Some studies have applied specific analyses of MDA with HPLC or GC as an alternative to the cruder TBARS test (Mendes et al., 2009). Other common methods are spectrophotometric analyses of conjugated dienes (CD) (absorbance at 234 nm, A_{234}), or conjugated trienes (CT) (A_{268}) in lipid extracts. Analyses of fluorescent Schiff bases in aqueous or organic solvent extracts have also been relatively widespread (Aubourg, 2001). Examples of applied excitation/emission wavelengths are 327/415 and 393/463 nm (Aubourg and Medina, 1999; Aubourg, 2001).

Many studies have also incorporated losses of endogenous antioxidants like the tocopherols (Larsson and Undeland, 2010; Farvin et al., 2012) to follow oxidation. To follow cross-reactions between lipid oxidation products and proteins, the number of studies where protein oxidation has been followed in parallel with lipid oxidation has also gradually increased in the last years. Markers used have been, for example, protein carbonyls using the reaction with dinitrophenylhydrazine (DNPH) (Levine et al., 1990, 1994), tryptophan and tyrosine residues (Farvin et al., 2012), as well as SH-group content and protein cross-linking (Chanarat et al., 2014).

11.4 INTRINSIC FACTORS AFFECTING LIPID OXIDATION

11.4.1 The Role of Lipid Level and Lipid Composition

The two major groups of muscle lipids are the phospholipids (PL) and triacylglycerols (TG). The PL give structure and fluidity to membranes and are found at relatively constant levels, in white muscle at 0.5–1% (w/w) (Ackman and Ratnayake, 1992) and in dark muscle at slightly higher levels due to the presence of more mitochondria (Ackman, 1992). The TG are found both in adipose tissues and integrated into muscle tissue. In the muscle tissue, the TG are located in the endomysial area outside the muscle cell and in the interior of muscle cells; the latter is particularly true for fatty species (Driedzic and Hochachka, 1978). The intracellular TG are surrounded by a PL monolayer (Zweytick et al., 2000). In contrast to membrane lipids, TG levels vary widely with species and environmental conditions. Lean fish usually contain <2% lipids (e.g., cod, haddock, pollack), medium fat fish 2–8% lipids, and high-fat fish, like mackerel, farmed salmon, and sometimes herring, >8% (Ashton, 2002). For some fish species, like the small pelagic species, it can, however, be difficult to make clear categorizations. For example, the total fat content of herring can vary from 0.5% to 30% depending on season and geographic location (Ackman, 1979).

Within the meat research field, it has been stated already in the 1950s that membrane lipids are the main substrates in muscle lipid oxidation (Younathan and Watts, 1959). Although indications are given in earlier literature (Tichivangana and Morrissey, 1982), no such clear statements have been made regarding fish, possibly since a substantial part of the fish TG are integrated within muscle cells. Instead, rapid oxidation in the muscle of fatty fish species has traditionally been attributed to its high total lipid content, that is, TG level (Aubourg et al., 1999). However, research in the past 15 years clearly indicates that the type and level of pro-oxidants in such fish appear to be of greater importance than the actual lipid content. As an example, strong rancid odor developed in a reduced lipid (~0.1%) washed cod muscle system fortified with whole trout blood (Richards and Hultin, 2002). The rate and extent of rancidity were not increased by the presence of >6 times more membrane lipids. Also, Undeland et al. (2002) found the same rapid development of Hb-mediated TBARS and painty odor in a washed cod mince model system (0.7% fat) with and without the presence of up to 15% menhaden oil. Without Hb, development of TBARS and painty odor was slow in the washed cod mince both with and without added oil. Nearly doubling the Hb concentration approximately doubled the extent of lipid oxidation with and without added oil. This indicated that Hb was crucial for the oxidation reaction. The absence of effect of added oil suggests that membrane lipids and/or preformed membrane lipid hydroperoxides provided sufficient substrate in Hb-catalyzed oxidation of washed minced cod muscle.

In this context, it is important to highlight that PL are more unsaturated than TG, and also that the surface area they expose to the aqueous phase is ~50–100 times greater than the oil droplet surface areas on a weight basis (Hultin, 1994). Most pro-oxidants are located in the aqueous phase, and therefore the location of PL in the membrane is crucial for lipid–pro-oxidant interactions. That oxidation of PL can be further transferred to the TG fraction of muscle was indicated by Slabyj and Hultin (1984) who found that NADH-driven, iron-catalyzed enzymatic lipid oxidation of herring membrane lipids (microsomes) accelerated oxidation of herring TG emulsified in the microsomal suspension.

The role of lipolysis, in either the TG or PL fraction, for development of lipid oxidation in fish is well covered in the review by Ashton (2002). However, it deserves to be mentioned that the buildup of free FAs (FFAs) can be substantial in postmortem fish muscle, something that plays a role for lipid oxidation development. As an example, Karlsdottir et al. (2014c) documented a raise in FFA from ~15 g/100 g to ~35 g/100 g in saithe during 18 months' frozen storage, with more FFAs accumulating in the dark than in the light muscle. Further, Chaijan et al. (2006) reported an increase from ~1 to 7 g FFA/100 g in sardine fillets during 15 days of ice storage. It is generally accepted that FFAs oxidize more rapidly than esterified FAs, especially when the oxidation is catalyzed by enzymes such as LOX (Ashton, 2002). However, some earlier studies indicate that this might not be true for both TAG and PL-lipolysis. Incubating isolated flounder sarcoplasmic reticulum with phospholipase A2 (PLA2) reduced both enzymatic and nonenzymatic oxidation, which the authors ascribed structural rearrangement of the membrane FAs following hydrolysis, limiting the free radical chain propagation

(Schewfelt and Hultin, 1983). Recently, Tatiyaborworntham and Richards (2015) also showed how adding PLA2 from pig pancreas to Hb-fortified cod mince suppressed TBARS development during ice storage; at day 5 TBARS levels were 85% lower with PLA2.

Along with the growth of the aquaculture industry, intense research is currently on-going to substitute fish-based aquafeed ingredients, for example, with vegetable-based ingredients, something that reduces the levels of the LC *n*-3 PUFAs EPA, DPA, and DHA. The obvious question has thus been raised how this reduced unsaturation affects the oxidative stability of cultivated fish muscle lipids. A number of studies have been addressing this issue. For example, Wan Ahmad et al. (2013) found that completely replacing dietary fish oil with palm or poultry oil in the diet of the warm water fish species barramundi (*Lates calcarifer*), decreased the *n*-3 PUFA content of the fish fillet lipids from 23.2% of total FA to 2.3–3.6%. As a consequence, TBARS in muscle samples from fish fed the palm or poultry oil–based diets developed slower during storage at 4 °C for up to 8 days compared with barramundi fed fish oil. In agreement with this study, Ng and Bahurmiz (2009) found that muscle TBARS concentration of red hybrid tilapia (*Oreochromis* sp.) stored at −20 °C developed slower in fish that had been fed either crude palm oil, palm FA distillate, or refined, bleached, deodorized palm olein compared with fish-fed fish oil. Baron et al. (2009) fed rainbow trout (*Oncorhynchus mykiss*) a diet containing either fish oil or rapeseed oil, whereupon the muscle of fish oil–fed fish contained ~20% of total FAs as EPA and DHA while the vegetable oil–fed fish contained ~9%. Also here, resulting fish fillets was frozen stored (−20 °C up to 22 months), and sensory data, PV, and levels of volatiles revealed that the fish-fed fish oil were slightly more oxidized than fish-fed vegetable oil. From the same laboratory, a later study on rainbow trout revealed that replacing fish oil with vegetable oil reduced formation of primary oxidation products, while the effect on secondary oxidation products differed depending on the types of volatiles being addressed. Also, data indicated that compounds present in the marine ingredients might have had an effect on protein oxidation, although effects were not clear (Timm-Henrich et al., 2013). Thus, it is clear that a more saturated FA profile of farmed fish has a stabilizing effect when it comes to lipid oxidation development, but that the profile of volatile oxidation products may change. Both Ng and Bahurmiz and Wan Ahmad et al. concluded from their studies that the degree of unsaturation of the dietary lipid, rather than the antioxidant content of the diet, determines the oxidative stability of the resultant fish tissue.

11.4.2 The Role of Endogenous Pro-oxidants in Seafood

According to earlier chapters, a key step in the initiation of lipid oxidation is the conversion of ground state oxygen, $^3O_2$, to active oxygen species (the superoxide anion radical, $O_2^{\bullet-}$, hydroperoxyl radical, $HO_2^{\bullet}$ hydrogen peroxide, H_2O_2, and the extremely reactive hydroxyl radicals, $^{\bullet}OH$). In contrast to $^3O_2$, these forms can directly react with FAs generating lipid free radicals. In postmortem seafood muscle tissue, both low molecular weight (LMW) and heme-bound iron are involved in formation of active oxygen species, for example, through the Fenton–Haber–Weiss reactions:

$$Fe^{2+} + H_2O_2 \rightarrow Fe^{3+} + OH^- + {}^{\bullet}OH \qquad \text{(Fenton reaction)} \qquad (1)$$

$$Fe^{3+} + O_2^{\bullet-} \rightarrow Fe^{2+} + O_2$$

$$\underline{Fe^{2+} + H_2O_2 \rightarrow Fe^{3+} + OH^- + {}^{\bullet}OH}$$

$$O_2^{\bullet-} + H_2O_2 \rightarrow O_2 + OH^- + {}^{\bullet}OH \qquad \text{(Haber–Weiss reaction)} \qquad (2)$$

It should be stressed that the most reactive radicals in biological tissue, like $^{\bullet}OH$, seems to be the least selective (Davies, 2005; Buettner, 1993). $^{\bullet}OH$ is produced in the aqueous phase of the cell, and it is believed that this radical reacts, for example, with proteins before it reaches the hydrophobic interior of the membranes (Davies, 2005). Other radicals, such as protein radicals, with lower redox potentials then mediate the oxidation process into the lipid phase. Although detailed descriptions are beyond the scope of the present chapter, this highlights the importance of also considering protein oxidation in parallel with lipid oxidation to fully understand the oxidation mechanisms operating in seafood tissue.

Ground state oxygen $^3O_2$ can also be activated in the presence of light and certain pigments (photosensitizers, S) via the conversion into singlet oxygen, $^1O_2$ (Davies and Truscott, 2001). This species can directly add to a C=C double bond according to a cyclo-addition mechanism, generating peroxy radicals, an event normally entitled photo-oxidation Type II. The underlying mechanism is excitation of the photosensitizer molecule by light, whereupon the sensitizer can transfer its extra energy to $^3O_2$, yielding $^1O_2$ (see reaction 3). Among natural compounds that can act as photosensitizers in muscle tissue are porphyrins and riboflavin. Whang and Peng (1988), however, described how porphyrins containing transition metals (e.g., heme iron) generally are not sensitizers for singlet oxygen production.

$$S + h\nu \rightarrow S^* + {}^3O_2 \rightarrow S + {}^1O_2$$

$$^1O_2 + LH \rightarrow LOOH \text{ (Photooxidation Type II)} \qquad (3)$$

In fish, also a third type of oxidation initiation can occur via LOX, which are iron-containing enzymes being situated in the cell cytosol or microsomal fraction (Harris and Tall, 1989). LOX catalyzes the insertion of one molecule of $^3O_2$ into an unsaturated FA containing a 1,4-*cis*-pentadiene group (Belitz and Grosch, 1987). LOX have been identified in skin (Mohri et al., 1999), gills

TABLE 11.3 Main Endogenous Pro-oxidants/Pro-oxidative Systems Present in Seafood Muscle

Low Molecular Weight Metals	Reducing Systems	Heme Proteins	Enzymes
Iron	Superoxide	Myoglobin	Lipoxygenses
Copper	Ascorbate	Hemoglobin	Cyclooxygenase
	Mitochondrial enzymes or Microsomal enzymes	Cytochromes	Peroxidases (e.g., *myeloperoxidase*)

(Liu and Sun Pan, 2004), and muscle (Stodolnik et al., 2005; Banerjee, 2006) of various fish. Also, another enzyme, myeloperoxidase, which has been isolated from trout leukocytes, can initiate lipid oxidation in the presence of hydrogen peroxide and halides such as chloride, bromide, and iodide (Kanner and Kinsella, 1983). This way of initiation might be critical during processing of fish, when the contact between air and blood is increased or following the addition of salt. In the presence of myeloperoxidase, the chloride ion can be converted to a radical that could add directly to a double bond or abstract a hydrogen atom (Kanner and Kinsella, 1983).

Further to the above, reinitiations of lipid oxidation occur as soon as lipid hydroperoxides are cleaved into free radicals and secondary products in hemolytic or heterolytic scission reactions. Heme and LMW trace elements are here crucial catalysts. Table 11.3 summarizes the main endogenous pro-oxidants of seafood muscle.

11.4.2.1 Heme-Proteins as Pro-oxidants

Both Hb and Mb appear to control the onset of oxidation via several mechanisms (Richards and Hultin, 2002; Maqsood et al., 2012; Baron and Andersen, 2002). Hb is a tetrameric molecule with allosteric O_2-binding properties while Mb is a monomer without allosterism. According to O'Brien et al. (1992), the ratio between Mb and Hb in dark (i.e., slow) fish muscle is roughly equal. Thiansilakul et al. (2012) recently showed that Mb and Hb accounted for 61% and 39%, respectively, of the total heme protein in bighead carp (*Hypophthalmichthys nobilis*) dark muscle. Corresponding data for mackerel dark muscle were 65% and 35% (Richards and Hultin, 2002). Slight differences between studies may appear as the level of blood in fish muscle increases with higher stress level just prior to slaughter (Olsen et al., 2008). In light (i.e., fast) fish muscle, Hb is known to be the dominating heme protein (O'Brien et al., 1992; Richards and Hultin, 2002), but also here, absolute levels are largely dependent on the state of rest/stress prior to slaughter. Hb-levels are always lower in light than in dark fish muscle as capillaries cover 18–25% of the dark muscle fiber surface, whereas only 0.2–9.0% is covered in white muscle (Soldatov, 2006). Compared with dark muscle fibers from mammals, fish dark muscle has a significantly higher capillary density (Johnston, 1983).

Many authors have observed how the heme iron content decreases during ice storage of fish muscle due to heme (protoporphyrin IX) breakdown, with a parallel increase in nonheme iron content (Lee et al., 2015; Maqsood and Benjakul, 2010b; Benjakul and Bauer, 2001; Chaijan and Undeland, 2015). Heme-reductions have in most cases been closely correlated to lipid oxidation development, pointing at lipophilic free radicals being involved in the breakage of the protoporphyrin IX moiety (Nagababu and Rifkind, 2004). Protoporphyrin IX destruction of trout met-Hb added to washed cod was more rapid compared with that of bovine met-Hb (Lee et al., 2015). After 42 h at 2 °C, the protoporphryin moiety of trout Hb was barely detectable while nearly all the bovine protoporphryin IX was still detected at 42 h of storage. We (Sannaveerappa et al., 2014; Chaijan and Undeland, 2015) have also observed that heme/heme proteins bind more firmly to muscle tissue as a result of their oxidation (i.e., met-formation) and lipid oxidation progression. Therefore, the documented reduced Hb/heme levels along with lipid oxidation may also be an analytical artifact due to incomplete extraction.

It has repeatedly been shown that the natural pH-drop that rapidly takes place in fish muscle postmortem, particularly in certain pelagic dark muscle species (e.g., from ~7 to ~6.4 in herring), activates Hb as a pro-oxidant by multiple ionizations (Undeland et al., 2004; Sannaveerappa et al., 2014; Aranda et al., 2009). This feature may, together with the abundance of Hb in these fishes, be a key behind the susceptibility of pelagic dark muscle species to lipid oxidation. The reduced pH will induce Hb deoxygenation (the so-called "Bohr effect"), Hb-auto-oxidation, and heme loss due to protonations of: (1) the distal histidine, which disrupts stabilization of the bound O_2, (2) bound O_2 allowing dissociation of $HOO^•$ and formation of met-Hb (i.e., Hb-Fe^{3+}), (3) the proximal histidine, breaking the coordination bond with the heme-iron, and (4) the heme-propionates, disrupting favorable electrostatic interactions with amino acids at position 10 in the E-helix and position three of the CD turn of the Hb β subunit (Aranda et al., 2009). Deoxy-Hb has in itself been suggested to act as a strong pro-oxidant due to the domed structure (Richards and Hultin, 2000), but particularly, it is a critical species since it is more susceptible than oxy-Hb to formation of met-Hb, and subsequently to heme loss (Aranda et al., 2009). It should here be mentioned that met-Hb formation also can be stimulated by preformed lipid oxidation products such as lipid hydroperoxides and aldehydes, for example, *trans*-2-pentenal (Maestre et al., 2009).

The highly catalytic role of met-Hb and met-Mb is linked to their ability to generate active oxygen species like $O_2^{•-}$, $HO^{2•}$, H_2O_2, and $^•OH$, of which H_2O_2 can react with met-Hb or met-Mb to form a hypervalent perferryl-Hb (Fe^{4+}=O) radical capable of initiating lipid oxidation via hydrogen abstraction (Kanner and Harel, 1985). The reaction of H_2O_2 with reduced Hb (i.e., deoxy-Hb and oxy-Hb) results in the formation of ferryl-Hb without a globin-based radical (Winterbourn, 1990; Giulivi and Davies, 1994). Perferryl Hb/Mb can also undergo a one-electron reduction to form ferryl Hb/Mb. Ferryl Mb has a relatively long half-life (~30 s) and cleaves preformed protein and lipid hydroperoxides to generate peroxyl radicals (LOO•) that can further propagate lipid oxidation (Carlsen et al., 2005). It was recently shown by Lee et al. (2015) that adding reduced trout Hb to washed cod mince muscle resulted in ferryl Hb formation, confirming that as trout Hb auto-oxidized in washed muscle, the resulting superoxide radicals were converted to H_2O_2 (Halliwell and Gutteridge, 1999). Ferryl-Hb formation did not take place in the washed cod mince when trout met-Hb was added. The latter was attributed to the absence of any O_2 ligand in the distal heme pocket, thereby preventing generation of $O_2^{\bullet -}$ and H_2O_2. In addition, H_2O_2 was more reactive with trout deoxy-Hb compared with trout oxy-Hb and met-Hb (Gabbianelli et al., 2004). Collectively, these results suggested that deoxy-Hb facilitates ferryl-Hb formation.

Lipid hydroperoxide cleavage into LOO• and LO• by met-Hb/met-Mb or released hemin is also a very important catalytic mechanisms for lipid oxidation in seafood (Ryter and Tyrrell, 2000). A series of studies in the last 10 years (Richards et al., 2005; Grunwald and Richards, 2006; Cai et al., 2013) have revealed that heme/hemin may actually be the most critical peroxide breaking species, and that the heme release itself may be one of the most critical events in seafood lipid oxidation development. The low polarity of the free hemin is expected to aid dissolving it into the hydrophobic interior of membranes where lipid hydroperoxides are located, thereby reacting with them to produce lipid radicals and secondary lipid oxidation products (Van der Zee et al., 1996).

The actual heme/hemin release from fish Hb occurs as the binding to the proximate histidine gets weakened (Everse and Hsia, 1997), and in relation to this event, the very elegant studies of Aranda et al. (2009) and later Cai and Richards (2012) deserve to be mentioned. Aranda et al. (2009) for the first time determined the crystal structures of trout IV Hb and perch Hb and performed new crystal structures of bovine Hb to explain the higher pro-oxidative activity of the fish Hb compared with mammalian Hb. It was illustrated that a major difference between bovine and fish Hb is the 50- to 100-fold faster auto-oxidation and hemin release rates in the fish Hb. Five specific amino acid replacements at or near the heme pocket, in the CD corner and along the E-helix, according to different mechanisms appeared to cause the increased susceptibility of fish Hb to oxidative degradation (see Figure 11.4 and Table 11.4).

In earlier studies, fish Mbs have been reported to be more sensitive to auto-oxidation compared with mammalian Mbs (Livingston and Brown, 1981), suggesting structural differences between the different Mbs, which may also affect hemin affinity. In the study of Lee et al. (2015), ApoH64Y/V68F was used to measure hemin dissociation from horse met-Mb compared with bighead carp met-Mb. Hemin dissociation from the fish met-Mb

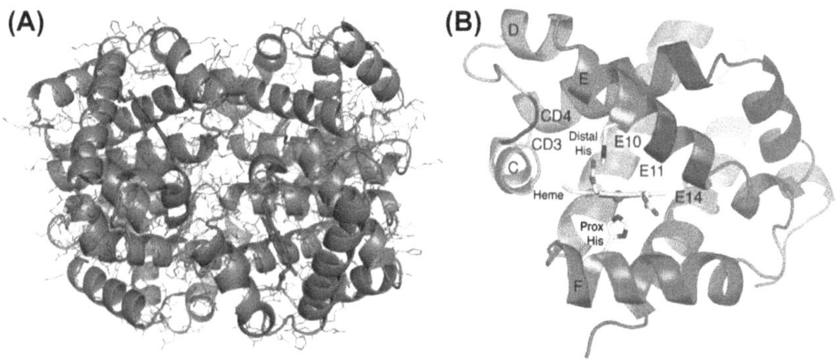

FIGURE 11.4 (A) Tetrameric structure of bovine Hb at pH 5.7 (heme groups are shown in red (dark gray in print versions)). (B) Highlighted amino acid differences in the E helix and CD turn. The structure of the trout IV β subunit is shown. The amino acid residues listed in Table 11.4 are highlighted in red (dark gray in print versions) and labeled. The C, D, E, and F helices and the CD turn are labeled along with the heme and proximal and distal histidine residues. *Reprinted with permission from Aranda et al. (2009).*

TABLE 11.4 Amino Acid Differences in the E Helix and CD Turn of Bovine, Trout IV and Perch Hbs

Residue Position	Bovine α/β Amino Acids	Trout α/β Amino Acids	Perch α/β Amino Acids
E10	Lys/Lys	Thr/Thr	Thr/Thr
E11	Val/Val	Ile/Val	Ile/Ile
CD4	Phe/Phe	Trp/Phe	Trp/Phe
CD3	His/Ser	His/Thr	His/Asn
E14	Ala/Ser	Asn/Gly	Gly/Gly

Amino acid changes are listed in the table for both the α and β subunits of each Hb.
Adapted with permission from Aranda et al. (2009).

was up to 3.3-fold greater compared with the mammalian met-Mb during storage at pH 5.5 and 4 °C. Hemin loss was, however, negligible from both heme proteins at pH 6.0 and 4 °C.

Apart from comparisons between fish and mammalian Hbs (e.g., Park et al., 2013; Aranda et al., 2009) and Mbs (Lee et al., 2015), a series of comparisons have been done during the last 10–15 years on the pro-oxidative role of Hbs from different fish species. Most of these have been carried out in washed fish mince model systems. In general, results have pointed at Hbs from fish adapted to cold/deep water like cod and pollock are more pro-oxidative than Hbs from very active migratory pelagic species like herring and mackerel, in turn being more active than Hbs from nonmigratory Hbs from warmer/shallower waters like trout, flounder, and sea bass (e.g., Richards and Hultin, 2003; Undeland et al., 2004; Aranda et al., 2009; Maqsood and Benjakul, 2011a; Maestre et al., 2009). It has earlier been found that oxy-Hb from deep-sea fish more quickly auto-oxidize than oxy-Hb from species living in more shallow waters, which was linked to the adaptation to low temperature (Wilson and Knowles, 1987). It is also generally known that heme proteins of fish have a more flexible and unstable protein conformation than warm-blooded animals due to adaptation to cold water temperatures rather than to HP (Chantai et al., 1996). Further, Hbs from highly active species are noted for having very strong Bohr effects (Riggs, 1970). Altogether, this may explain the faster met-Hb formation and hemin losses observed by the most pro-oxidative Hbs. Thus, not only the amount of Hb in fish dictates its instability but also the nature of Hb.

Further to this, a handful of studies have also compared the pro-oxidative role of different types of Hbs from the same species (e.g., trout catodic Hb I and anodic Hb IV) in washed fish minces with results uniformly pointing at Hb IV as the most catalytic (Sannaveerappa et al., 2014; Richards et al., 2005). More rapid hemin release from the anionic Hb compared with the catodic one appeared to be a primary facilitator of lipid oxidation.

Several comparisons have been made regarding pro-oxidative characteristics of Hb versus Mb but with contradicting results. The few comparisons made in fish systems (washed fish minces) have shown that trout Hb was more pro-oxidative than trout Mb (Richards et al., 2005) and that bighead carp (*Hypothalimichthys nobilis*) Hb was more pro-oxidative than Mb from the same species. The latter was attributed to the lower hemin affinity of Hb than for Mb (Thiansilakul et al., 2012). Already in 1969, it was also shown that TBARS values were elevated 14-fold after day one on ice in unwashed cod containing added Hb compared with added Mb at an equivalent level of heme (Castell and Bishop, 1969). However, when comparing the ferryl species of Hb and Mb, the latter was formed more quickly (Kanner and Harel, 1985) and has also been ascribed to be the most active pro-oxidant of the two (Everse and Hsia, 1997). Lee et al. (2015) compared the ability of horse met-Mb to promote lipid oxidation in washed cod mince with that of cross-linked horse Mb at equivalent heme concentration. The background was that perferryl Mb earlier has been noted to cross-link at acidic conditions (Reeder et al., 2002). The results revealed that cross-linked Mb was a potent catalyst of lipid oxidation compared with met-Mb, and the authors discussed how the cross-linking appears to be due to a tyrosine residue forming a covalent bond with the porphyrin moiety, inducing a peroxidase activity that can readily oxidize lipids. It was concluded in the report that the major form of Mb that oxidizes lipids may be cross-linked Mb rather than perferryl- or ferryl-Mb species. Finally, it can be mentioned that a "pseudolipoxygenase" activity of Mb and Hb has been discussed in a few reports. It is, however, supposed to be distinct from LOX activity in the sense that it is not decreased by denaturation, and is even enhanced by heating, urea treatment, and acid splitting (Carlsen et al., 2005, Kühn et al., 1981).

11.4.2.2 Low Molecular Weight Iron and Other Transition Metals

As stated earlier, LMW iron and other transition metals like copper can also initiate lipid oxidation in a variety of ways, primarily by interacting with oxygen to generate $O_2^{\bullet -}$, by participation in the Fenton reaction generating $^{\bullet}OH$, and by breaking down preformed hydroperoxides into radicals such as $LO^{\bullet}$ and $LOO^{\bullet}$. Generally, Fe is well suited to catalyze redox-reactions as it has a number of different oxidation states that enable it to transfer electrons. Since the reduced forms of LMW-trace metals have been reported as the most potent oxidation catalysts (Chaiyasit et al., 2007), the presence of reducing compounds is important for the action of transition metals. According to the excellent review by Hultin (1994), where transition metal-induced oxidation of fish is well covered, superoxide and ascorbate are the two major reductants in fish tissue. It was here described that the concentration of ascorbate present in fish muscle varies around 60–100 μmol/L, a range where it effectively reduces ferric iron (Decker and Hultin, 1990b). Ascorbate data reported from more recent studies on Atlantic mackerel (Maestre et al., 2011) and herring (Sannaveerappa et al., 2007) muscle falls close to this range: 14–60 μmol/L and 35–113 μmol/L, respectively. According to further work done by the Hultin research group, which has not been taken up again in recent years, iron can also be reduced by enzyme systems associated with both mitochondria (Luo, 1987) and microsomes (i.e., the sarcoplasmic reticulum, SR) (Mc Donald and Hultin, 1987; Slabyj and Hultin, 1982) using electrons from NADH.

Many studies have shown how the levels of LMW-Fe can increase in postmortem fish and shellfish muscle due to Mb-/Hb-hemin breakdown (e.g., Wongwichian et al., 2015; Chaijan et al., 2005; Decker and Hultin, 1990a). Fe liberation from the storage protein ferritin (Decker and Welch, 1990) and contamination during processing are other sources of LMW-Fe in fish muscle. Fe is usually chelated in the tissue, for example, with ATP, ADP, amino acids (e.g., histidine), peptides, or ascorbate, due to its poor solubility >pH 3 (Hultin, 1994). Chelating agents have twofold roles in the sense that they render Fe soluble but also that they may modify the redox-potential of the Fe.

In simpler lipid systems such as bulk oil, liposomes, emulsions, and isolated SR, Fe has proved to have a strong role as catalyst. As an example, Erickson and Hultin (1988) showed that Fe and equimolar levels of a variety of nucleotides effectively stimulated enzymatic oxidation of fish SR. However, from several studies in fish muscle systems, there is little clear evidence that endogenous levels of LMW-Fe contribute significantly as an oxidation catalyst. Several studies seem to suggest that heme is a stronger pro-oxidant than LMW-Fe (Chiu et al., 1996; Undeland et al., 2002). For example, 15 μmol/L Fe^{2+}–ADP did not induce any oxidation of washed minced cod with and without added menhaden oil, which 3 and 7.5 μmol/L Hb did (Undeland et al., 2002). In a recent study on the comparison of pro-oxidative effects of several LMW metals, LMW-Cu(I/II), LMW-Fe(II/III), and LMW-Cd (0, 5, 25 ppm) in minced cuttlefish muscle subjected to several freeze–thaw cycles, results showed that Fe(II) induced TBARS development far more effectively and its pro-oxidative effect was concentration dependent. However, the obtained results suggested that while LMW-Fe can induce TBARS and yellowness in fish muscle, Cu rather promotes oxidation of protein (Thanonkaew et al., 2006).

11.4.2.3 Lipoxygenase

LOX is a nonheme iron containing dioxygenase that stereospecifically incorporates one molecule of oxygen into the 1,4-pentadiene-system of (free) PUFA, forming a conjugated FA hydroperoxide. The mechanism is supposed to be radical based; the FA is oxidized by ferric iron at the active site, forming a pentadienyl upon abstraction of H. The resulting pentadienyl radical intermediate reacts with oxygen, subsequently forming the hydroperoxide (Maqsood, 2010). The LOX enzyme is activated by the lipid hydroperoxide products, and both the FA substrate and product are competitive inhibitors. According to Qiu et al. (2013), LOX are also able to cleave C–C bonds in the vicinity of hydroperoxy groups, yielding aldehydes and alkanes. An atypical LOX, cyclooxygenase (COX), introduces two oxygen molecules into unsaturated FAs (Hultin, 1994). LOX and COX produce numerous lipid oxidation products in vivo with important physiological roles in animal tissues. Within a given organism, the expression of LOX can, however, vary widely between different tissues and with age and environmental factors. LOX activity toward free (Harris and Tall, 1989) and esterified (Mohri et al., 1992) FAs has been observed in fish. Wang et al. (2012) reported how the LOX distribution in grass carp gradually increased in the order skin and muscle < gill < viscera < blood, indicating that it is very important to remove the viscera and bleed fish completely during fish-product processing to decrease the risk of lipid oxidation caused by LOX. The exact contribution of LOX, and tentatively COX, as stimulators of lipid oxidation in seafood is not clear, and few studies have been carried out in situ. During operations like filleting, de-skinning, and grinding, there are, however, large chances for contamination of skin LOX into the muscle and potentially also of gill LOX. Several authors have described earlier how LOX is responsible for production of specific volatile compounds in fish tissue (German et al., 1985; Josephson et al., 1987; Saeed and Howell, 2001), which are important for both the fresh and spoiled seafood odor (Hsieh and Kinsella, 1989; Lindsay, 1990).

Medina et al. (1999) studied isolated LOX from sardine (*Sardina pilchardus*) and herring (*Clupea harengus*) during chilled storage. The hydroperoxide degradation products produced after LOX-induced oxidation of arachidonic acid (AA) and DHA were identified to be 12- and 16-hydroxy acids. The LOX were concentrated in the skin tissue of fish and were active for up to 48 h of chilled storage, which was shorter than the activity of heme proteins. The authors therefore concluded that the detrimental effects of transitional metals and heme proteins on seafood quality are greater than effects due to LOX. Contrary to this, LOX activities of Nile tilapia skin increased throughout 18 days of ice storage at (by 568%) (Sae-leaw et al., 2013). The authors discussed that the increase in LOX activity was due either to increase in extractable LOX or the activation of LOX in skin during the extended storage in ice. Increase in LOX activities during frozen storage was also observed in mackerel muscle and Baltic herring muscle (Samson et al., 2001; Stodolnik et al., 2005).

Fu et al. (2009), who studied and compared the effect of LOX (primarily the 12-LOX form) and Hb from silver carp on lipid oxidation and off-odor development in washed silver carp (*Hypophthalmichthys molitrix*) mince during ice storage, found that LOX caused faster lipid oxidation than Hb during initial storage, but yielded lower PV and TBARS values during late storage. This was possibly due to the fact that the LOX is unstable and can be inactivated by hydroperoxides (Hsiu and Pan, 1996). Sulfhydryl groups have been shown to be important for LOX to maintain its activities (Hsiu and Pan, 1996; Kuribayashi et al., 2002). Formed hydroperoxides may oxidize thiol groups, inactivating LOX in later stages of the reaction. The LOX was affiliated with strong fishy odor, while Hb resulted in severe "oxidized oil odor." Based on volatile analyses (see Table 11.5), the authors hypothesized that the 2,4-heptadienal (*E,E*) contributed to the fishy odor, while hexanal and nonanal were responsible for oxidized oil odor. The Hb system contained significantly more hexanal than the LOX system (28.45 vs 3.55 mg/kg).

Some of the earlier studies of LOX-induced lipid oxidation in fish and in food in general were reviewed, for example, by German et al. in 1985 and 1992.

11.4.3 Endogenous Antioxidants in Fish Muscle

Muscle is well suited for preventing oxidative attacks in vivo, a capacity that, however, ceases postmortem. Important endogenous oxidation inhibitors in muscle, including those from seafood, have been discussed in depth by Hultin (1994) and Chan et al. (1994) but also to some extent by Ashton (2002) and Raghavan and Kristinsson (2014). Generally, oxidation inhibitors in muscle act by removing active reduced forms of oxygen, reducing peroxides, and converting transition metals to inactive forms or scavenging radicals. Superoxide dismutase (SOD) is a metalloprotein, situated in the cytosol or mitochondrial fraction of the cell, which converts $O_2^{\bullet -}$ into O_2 and H_2O_2. The H_2O_2 produced is then removed by catalase and peroxidase with the production of water (Hultin, 1994). The selenium-containing GSH-px reduces a number of peroxides to the corresponding alcohols while oxidizing 2 mol of glutathione (GSH), thus breaking the lipid oxidation chain reaction (Mourente et al., 2007). Oxidized GSH (GSSG) is then reduced back by GSH reductase

TABLE 11.5 Content (mg/kg) and Odor Quality of Volatile Compounds Induced by Silver Carp (*Hypophthalmichthys molitrix*) Hemoglobin (Hb) and Lipoxygenase (LOX) in Washed Silver Carp Mince

Name	Odor Quality	Hb	LOX
Ethyl acetate	Fruity	14.70	1.50
Ethanol	Fruity	33.11	41.30
2,3-Butanedione	Buttery	9.83	11.24
Hexanal	Grassy	28.45	3.55
1-Penten-3-ol	Earthy	2.04	ND
Undecane	ND	3.58	5.09
Styrene	ND	1.40	0.54
1-Octen-3-one	Mushroom like	0.31	ND
Nonanal	Oxidated oil	9.69	2.02
2,4-Hexadienal	Grassy	1.24	0.16
2-Decen-1-ol (*E*)	Fatty	0.75	0.39
1-Octen-3-ol	Earthy	2.17	ND
2,4-Heptadienal (*E*,*E*)	Fishy	0.26	1.75
3,5-Octadien-2-one (*E*,*E*)	Fatty	1.66	0.12
2,4-Octadienal (*E*,*E*)	Fatty	0.33	ND
2-Undecanone	Grassy	3.52	2.71
1-Dodecanol	Grassy	0.92	ND
Heptadecane	ND	14.76	17.70
2-Methyl hexadecanol	Grassy	0.57	1.13
1-Undecanol	ND	0.19	0.37
Dodecanoic acid	ND	0.23	0.26

ND, not detected.
Adapted with permission from Fu et al. (2009).

at the expense of NAD(P)H. Inhibitors reacting with transition metals generally work according to oxidation or chelation, the former being preferred in extracellular fluids, for example, ferroxidases, such as ceruloplasmin present in blood serum (Kanner et al., 1988), while chelation is favored intracellularly (Halliwell and Gutteridge, 1986). Whether chelation is stabilizing or destabilizing depends, for example, on the valency state of the metal that is stabilized, the site of the metal that is occupied, the type of electron transfer they mediate, and the chelator hydrophobicity; these factors affect the binding of the metal to membranes and macromolecules (Kathirvel and Richards, 2012; Weers and Thomason, 1990). Important chelators in fish muscle are nucleotides, amino acids, peptides, and organic acids. Histidine-containing dipeptides (e.g., carnosine, anserine) are particularly acknowledged for their high efficiency in forming inactive metal complexes (Chan et al., 1994). However, these dipeptides can also scavenge free radicals (Chan et al., 1994) and act as sinks for reactive aldehydes (Xie et al., 2013).

The lipid-soluble tocopherols (α, β, γ, and δ) are the most well-known chain-breaking antioxidants in fish tissue and are situated in the interior of membranes with their oxidizable chromanol entity located near the surface (Polat et al., 2013; Hultin, 1994). Each tocopherol can donate two electrons as a chain-breaking antioxidant before it needs to be recycled (Buettner, 1993). It has been reported that 97–100% of the total tocopherol in the tissues of several fish species consists of the α-isomer (Polat et al., 2013), that this is the most active antioxidative form in vivo (Wang and Quinn, 1999), and that the tocopherol levels in fish reach a peak during the spawning season (Syväoja et al., 1985). The contribution of α-tocopherol to the oxidative stability of skeletal muscle is largely influenced by diet (see later). Another lipid-soluble antioxidant present in fish is ubiquinol (Passi et al., 2005; Hamre et al., 2010); the only endogenously synthesized lipid with redox function in vertebrates. Ubiquinol is a reduced form of coenzyme Q (ubiquinone) that is located in membranes, including the inner mitochondrial membrane (Ernster and Dallner, 1995). It has been suggested that in mitochondria, ubiquinol prevents oxidation by regenerating the tocopheroxyl radical (Hamre et al., 2010). Carotenoid pigments such as astaxanthin and canthaxantin which are added to salmon feed in order to give the characteristic red color, could protect the slaughtered fish muscle against oxidation, for example, by singlet oxygen and radical quenching (Waagbo et al., 2003; Jensen et al., 1998). In the study by Baron et al. (2009), where the diet of rainbow trout (*Oncorhynchus mykiss*) contained either fish oil or rapeseed oil and with or without 200 mg/kg carotenoid (either astaxanthin or canthaxanthin), results showed that canthaxanthin effectively protected both protein and lipid against oxidation during frozen storage while astaxanthin did not seem to have a clear and systematic effect.

Further to the mentioned metal chelators, known aqueous antioxidants in fish tissue include, for example, ascorbate, uric acid, and GSH. Ascorbate is multifunctional and can act as an electron donor, a metal chelator, a scavenger of active oxygen species, and a free-radical scavenger (Frankel, 1996; St Angelo, 1996). Its ability to recycle the tocopheroxyl radical; thus, its action as an electron donor, is well accepted. As described by Hamre et al. (2010), the produced dehydroascorbic acid can be reduced back to ascorbate by GSH, where after GSSG is reduced by NADPH produced in the pentose phosphate shunt. Thus, the recycling of dietary antioxidants may ultimately be coupled to

TABLE 11.6 Main Endogenous Inhibitors of Lipid Oxidation in Seafood Muscle Divided into Preventive Inhibitors (e.g., Reducing Agents, Metal Chelators, Oxygen Scavengers) and Chain-Breaking Antioxidants with, for Example, Radical Scavenging Effects

Preventive Inhibitors	Chain-Breaking Antioxidants
Superoxide dismutase	Tocopherols
Catalase	Ubiquinol
Peroxidases	Carotenoids
Ferroxidases	Ascorbate
Nucleotides	Glutathione peroxidase
Amino acids	Uric acid
Peptides	
Organic acids	
Ascorbate	
Gluthathione	
Uric acid	

energy metabolism, explaining why it ceases postmortem. Uric acid, present, for example, in plasma, can like ascorbic acid prevent oxidation via numerous mechanisms: by scavenging $^{\cdot}OH$ radicals, lipid radicals, singlet oxygen, and oxo-heme oxidants during its own oxidation to allantoin (Davies et al., 1986). Without being oxidized itself, it can also protect ascorbic acid from oxidation. In addition to the mentioned roles of GSH in oxidation prevention, it has been ascribed the capacity of reducing ferryl-Mb (Romero et al., 1992). Table 11.6 summarizes the main inhibitors of lipid oxidation in seafood muscle divided into preventive inhibitors and chain-breaking antioxidants.

Passi et al. (2005) elegantly monitored the decrease in ascorbate, GSH, ubiquinol, total coenzyme-Q, and tocopherol along with development of TBARS, protein carbonyls, and DNA-oxidation products (8OHdG) in ice-stored and frozen-stored (−30°C) gutted Mediterranean fish. Figure 11.5 shows how ascorbate and ubiquinol disappeared first in *Pagellus erithrinus* during ice storage, followed by (in order) total CoQ, tocopherol, and GSH. TBARS, protein carbonyls, and 8OHdG took off between days three and five and coincided with the ascorbate and ubiquinol emptying. After 6- and 12-month storage at −30°C, only ascorbate and ubiquinol degraded: by 61% and 82%, as well as by 61% and 87% after 6 and 12 months, respectively.

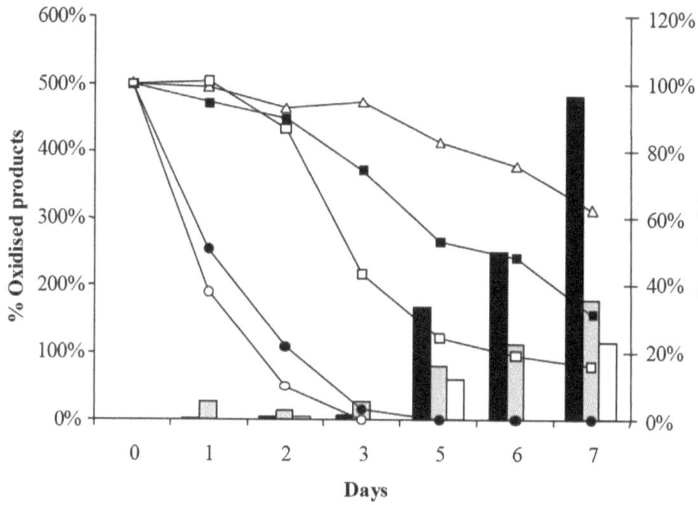

FIGURE 11.5 Biochemical modifications in light muscle of *Pagellus erithrinus* during storage on ice for 7 days. Lines, consumption of antioxidants (ascorbate ●, ubiquinol ○, total CoQ □, vitamin E ■ and GSH Δ; Histograms, oxidation product development; Black, TBARS; gray, Protein carbonyls; white, 8OHdG. *Reprinted with permission from Passi et al. (2005).*

Attempts have been carried out to increase postmortem stability of farmed fish fillets by fortifying the aquafeed with antioxidants, particularly tocopherol. Inconsistent results have, however, emerged. In the study of Kamireddy et al. (2011), rainbow trout were fed a diet with either 200 or 5000 vitamin E mg/kg for the last 9 weeks, that is, in the finishing diet. High–vitamin E diet increased fillet α-tocopherol level I muscle from 33 to 155 mg/kg. The high-tocopherol diet did not affect zero-time TBARS values in the trout muscle at any week of the feeding trial. Also after 14 days of storage of slaughtered fillets at 2 °C, beneficial effects of high vitamin E levels on lipid oxidation development were not evident. In addition, Chaiyapechara et al. (2003) observed no significant differences in TBARS of nonstored muscle of rainbow trout fed 300 and 1500 mg vitamin E/kg diet. However, after storage for 7 days at 4 °C and after storage for up to 24 week at −30 °C, fillet TBARS were lower from fish fed either 15% or 30% lipid and vitamin E at 1500 mg/kg diet compared with the fish fed 30% lipid and 500 mg vitamin E/kg diet. Also, in turbot stored on ice under retail conditions for 9 days, dietary supplementation with 500 and 1000 mg vitamin E/kg decreased lipid oxidation compared with supplementation with 100 mg vitamin E/kg. However, no significant differences were observed between the two high–vitamin E diets. Supplementing α-tocopheryl acetate (300 and 5000 mg/kg diet) in a trout finishing diet was done to minimize lipid oxidation in oven-cooked fillets and hot-smoked products (Jittinandana et al., 2006). Feeding diets containing 5000 mg vitamin E/kg increased muscle α-tocopherol content, which minimized lipid oxidation in oven-cooked fillets produced from fresh and 7-day refrigerated fillets and in hot-smoked products following 8 weeks' refrigerated storage.

In the studies of Chen et al. (2007, 2008), a trout diet was supplemented with 0, 8.5, or 15 g/100 g of flaxseed oil and further enhanced with 0, 400, and 900 mg/kg of α-tocopheryl acetate. The highest supplement level yielded highest α-tocopherol content in fillets, and it was shown that regardless of flaxseed oil level in the trout diet, the high–α-tocopheryl acetate supplementation level rendered the lowest TBARS values in muscle right after slaughter. However, in this study, the dietary tocopherol did not retard lipid oxidation during ice storage of fillets.

11.5 OXIDATION IN DIFFERENT PARTS OF FISH

It is evident from earlier sections that different parts of the fish differ in their lipid levels/lipid compositions and pro- and antioxidants and have different access to oxygen. Thus, the rate of lipid oxidation in different anatomical parts will not be uniform. Knowledge about this is extremely important in sampling and in the comparison of data between different studies. (It can be expected that differences are larger in certain pelagic fatty fish species like mackerel and herring due to their high amounts of dark muscle, being rich in both mitochondria, storage lipids, Hb, and Mb.) In 1998, two related studies were published on this topic, and one later came in 2004. In that of Icekson et al. (1998), mackerel were kept frozen in the round at −18 °C and, at regular intervals, TBARS and fat content were determined in three dorsal (cranial, medial, and caudal) and three ventral (cranial, medial, and caudal) samples of approximately the same length (see Figure 11.6(A)). Results showed that the TBARS expressed on a wet weight basis was similar in the ventral and dorsal sections; however, when it was expressed on a lipid basis, the leaner dorsal sections gave ~50% higher TBARS than the ventral sections in the cranial and medial areas. In the caudal area, dorsal and ventral sections did not show a significant difference in TBARS, both being higher than values in cranial and medial sections. In conclusion, the caudal sections had the highest levels of lipid oxidation. Different results were found by Halamíčková et al. (2003), who followed MDA development in craniocaudal and dorsoventral directions (in total eight different sections, see Figure 11.6(B)), as well as in surface and deep muscle layers of common tench (*Tinca tinca* L.) stored at −18 °C. Highest MDA-levels were found in the cranial segments, which was ascribed, for example, Hb and LOX contamination from the gill tissue. Higher MDA was also found in the ventral parts of the first three segments, which, among other things was ascribed the presence of ovaries and the fact that estrogen biosynthesis is associated with free-radical formation. Most surface samples had nonsignificantly higher MDA-levels than deep muscle samples.

In the study of Undeland et al. (1998b), different horizontal layers ("under skin," "middle part," and "inner part," see Figure 11.6(C)) of herring fillets

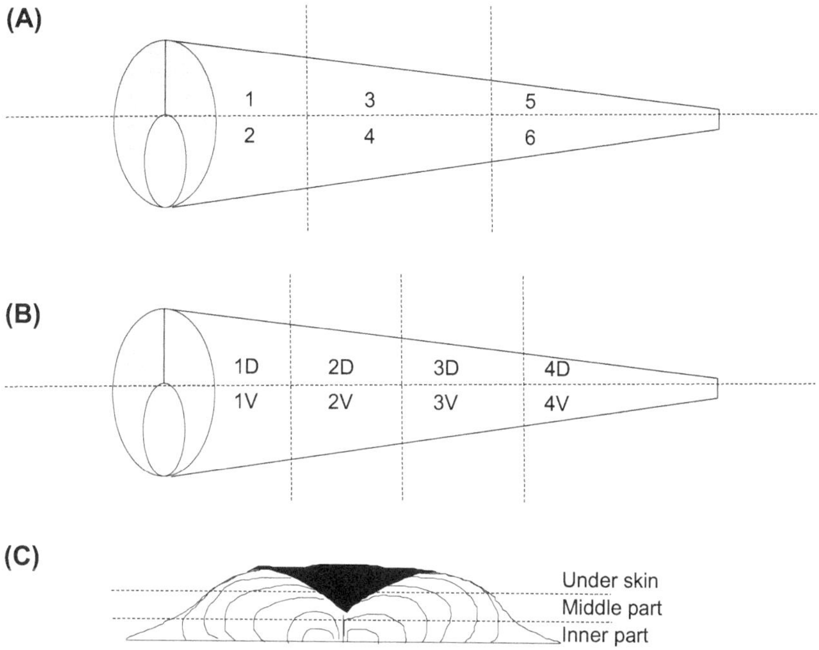

FIGURE 11.6 Fish sections studied with respect to lipid oxidation development by (A) Icekson et al. (1998), (B) Halamíčková et al. (2003), and (C) Undeland et al. (1998b).

stored at −18 °C with or without skin were compared as an attempt to see how deep into a small fish fillet the oxidation reaction proceeds during frozen storage (Undeland et al., 1998b). It was found that the presence of skin during storage significantly suppressed oxidation in the under skin layer of the fillet, and, to some extent, also in the middle part. However, in both skin-on and skinned fillets, the under skin layer oxidized significantly faster than the inner and middle parts. Among reasons given were the abundance of dark muscle situated under the skin, contamination of LOX from skin tissue, and a lower α-tocopherol/lipid ratio compared with in the other parts.

Several studies have focused on differences in oxidation between dark and light muscles (see Figure 11.6(C)), and in some cases also between muscle and skin. In our own study from 1997 (Undeland et al., 1997), lipid oxidation development in light muscle, dark muscle, and skin of herring stored either as an intact fillet or when stored as separate tissues was followed. When dark muscle, light muscle, and skin were stored *separately* from each other, the three tissues oxidized according to dark muscle > skin > light muscle, which was the same ranking order as was found for levels of iron, copper, and total aqueous pro-oxidative activity. However, when the tissues were stored as an *intact fillet*, the oxidation order changed to skin > dark muscle > light muscle, indicating that the skin and light muscle protected the dark muscle, for example, from oxygen. Also in earlier studies on pelagic fish (Ke et al., 1977), it was found that the skin with its subcutaneous fat layer was very sensitive to oxidation during frozen storage. When skin, light muscle and dark muscle from mackerel stored 2 months at −15, −30, and −40 °C was analyzed for TBARS, the skin sample (including subcutaneous fat) was found to be eight times faster than in the white and dark muscles. Higher sensitivity of dark than light muscle tissue to oxidation has also been found in tuna (Sohn and Ohshima, 2010) and yellowtail (*Seriola quinqueradiata*) (Sakai et al., 1998) as well as lean fish species: saithe (*Pollachius virens*) and hoki (*Macruronus novaezelandiae*) (Dulavik et al., 1998; Karlsdottir et al., 2014b,c). TBARS or MDA, PV, and reduction in α-tocopherol were all more pronounced in dark muscle during ice and frozen storage. In the study of Sakai it was interesting to notify that protein carbonyls did not differ between light and dark muscle while MDA did. In the study of Dulavik et al., it was shown that industrial deep-skinning improved cold storage stability of saithe fillets.

11.6 EXTERNAL FACTORS INFLUENCING LIPID OXIDATION IN SEAFOOD

11.6.1 Preslaughter and Slaughter Stress

Under aquaculture settings, it is possible to control the slaughter procedure and the level of stress that fish is subjected to prior to slaughter. As a way of controlling quality, a few studies have tried to link the level of stress antemortem to postmortem development of lipid oxidation in the muscle. The stress per se can: (1) speed up rigor mortis onset and resolution, (2) reduce muscle pH and ATP/IMP ratios, and (3) increase blood cortisol as well as blood metabolites involved in metabolism of glucose, lactate, and osmolality (Ribas et al., 2007). Also, the total levels of blood in muscle increases, the latter since the elevated stress can redistribute blood from the intestines to the muscle to prepare for increased muscle activity and escape (Olsen et al., 2008). Based on earlier sections, both higher Hb-levels and lower muscle pH as a result of stress could theoretically stimulate oxidation, but also other theories have been outlined, such as higher proteolytic activity and faster conversion of xanthine dehydrogenase to xanthine oxidase as a result of faster ATP degradation, the latter of which can lead to $^{\bullet}OH$ formation in the presence of redox iron and reintroduced oxygen (Hultin, 1992).

Matos et al. (2010) assessed the independent effects of harvesting stress and slaughter method on TBARS in gilthead seabream flesh. The fish was subjected to profound anesthesia with the lipophilic phenolic antioxidant isoeugenol (PA, low stress condition) and net crowding (NC, high stress condition). The slaughter methods were either lethal anesthesia using AQUI-S™ (270 μL/L) or immersion in ice–salt water slurry in a proportion of 4:1. Slaughter method did not show any significant effect on TBARS, while harvesting stress at the time of death did. Surprisingly, fish from the profound anesthesia treatment presented higher muscle TBARS. As shown by Atsumi et al. (2005) in vitro, isoeugenol can also act as a pro-oxidant under certain conditions of oxidative stress, enhancing ROS levels, possibly through benzyl radical formation. Also, Tuckey et al. (2009) investigated the role of rested harvesting and exposure to isoeugenol on oxidation and loss of tissue antioxidants in fillets of chinook salmon (*Oncorhyncus tshawytscha*). The study groups were rested isoeugenol harvest, exhausted harvest, and an exhausted isoeugenol harvest. Immediately after slaughter, no differences in TBARS were seen, but after 32-h storage at 15 °C, both the rested and exhausted groups with isoeugenol had significantly lower TBARS levels in the fillets. The actual harvest thus had no effect on lipid oxidation, which was the same for increases in protein carbonyls and losses of vitamin C.

The effects of several slaughter methods on the quality of fresh and smoked trout and fresh gilthead seabream were evaluated by Giuffrida et al. (2007) during storage at 2 °C. Electrically stunned trout had slower ATP depletion of raw muscle and lower lipid oxidation in a smoked product during storage. Gilthead seabream immersed in an ice slurry after the harvest showed a more regular ATP depletion than in fish exposed to CO_2. This group showed slower ATP depletion and lower lipid oxidation of muscle during storage. Thus, for both smoked trout and fresh gilthead seabream, lipid oxidation appeared to be more intense in those groups that had greater ATP depletion. It was discussed that the observed faster ATP consumption could have reduced nucleotides such as NAD^+ and $NADP^+$, which are involved in the regeneration of various pro-oxidant substances. Also, it was hypothesized that the rapid conversion of xanthine dehydrogenase to xanthine oxidase, induced by rapid consumption of ATP, was involved. Tejada and Huidobro (2002) studied the relation between slaughter stress and lipid oxidation in *Sparus aurata.* The authors found a slightly, but not significantly higher, TBARS level in fish killed by a more stressful commercial method (immersion in ice plus water) compared with less stressful percussive stunning. In the study of Secci et al. (2016) results were more clear; killing trout by asphyxia in air instead of percussion resulted in higher endogenous oxidative stress (e.g., hydroperoxides and EPA-derived prostaglandins) and higher TBARS development during subsequent storage of fillets at −10 °C (see Figure 11.7).

In the work by Morzel and Van de Vis (2003), eels (*Anguilla anguilla* L.) were slaughtered using a commercial method (desliming in dry salt followed by evisceration) and an alternative method (stunning and killing in water with a combination of electricity and oxygen removal). Eels slaughtered commercially exhibited higher level of muscular activity before death. In postmortem raw fillets, the alternative slaughter led to less lipid oxidation, and also to slower muscle ATP depletion, a redder, firmer flesh with a higher pH, and less myofibrillar proteolysis. The lower degree of tissue disintegration may have prevented contact, for example, between lipids and pro-oxidants.

Altogether, most of these studies point at a stimulating effect of antemortem stress upon subsequent lipid oxidation during storage, but different mechanisms

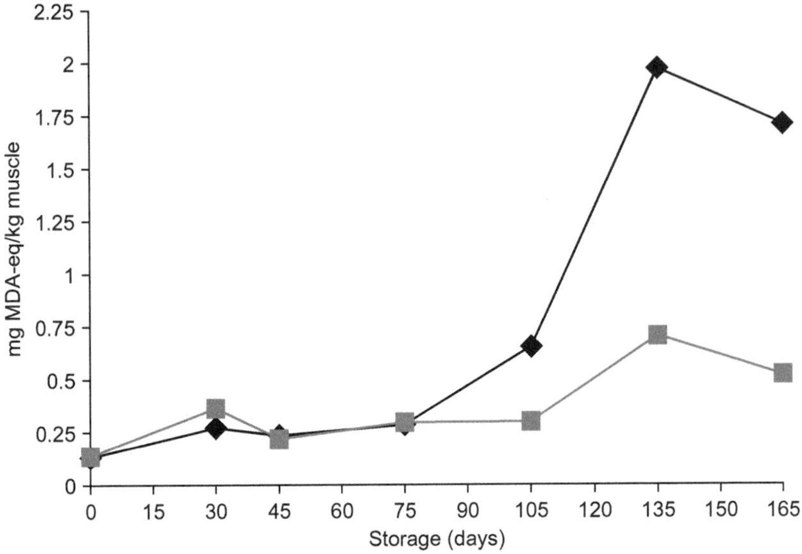

FIGURE 11.7 Secondary lipid oxidation products during 165 days of frozen storage (−10 °C), expressed as mg MDA-equivalents/kg muscle, in fillet of trout slaughtered by asphyxia (black line) or percussion (gray line). *Reprinted with permission from Secci et al. (2016).*

are outlined to explain this. Minimizing stress is thus not only an ethical issue but also can extend shelf life.

11.6.2 Bleeding

In many places, it is currently regulated by law that all commercially caught fish, both wild and farmed, should be bled. However, some species like the small pelagics are excluded from this recommendation, which is indeed a paradox as these fish species, according to the preceding, are expected to need it the most. Bleeding is normally done via tail-/gill-cutting and/or regular gutting followed by leaching in cold sea water or fresh water. A few studies have shown a reducing effect from bleeding on lipid oxidation during subsequent storage (Richards and Hultin, 2002; Tretsven and Patten, 1981; Sakai et al., 2006; Maqsood and Benjakul, 2011c), but in other studies, no such positive effect has been shown (Ruff et al., 2002; Sohn et al., 2007; Porter et al., 1992). According to the literature, regular bleeding normally removes ≤50% blood, with poor blood removal often resulting from too long holding of the fish prior to bleeding (Tretsven and Patten, 1981) or too high temperature. Both these features induce coagulation. Fish blood, especially from marine species, coagulates very quickly, often within 0–30 min (Fänge, 1994). Olsen et al. (2008) found that salmon cooled at 4 °C–0.5 °C had blood coagulating after 33 and 61 min, respectively. The following factors also render bleeding difficult:

(1) fish only has one-third of the blood pressure of humans, (2) fish light muscle (i.e., 75–95% of the total muscle w/w) has poor blood circulation, and (3) the stressful situation during catch causes vascular contraction, capturing the blood in muscle (Olsen et al., 2008). Richards et al. (1998) performed a related study on the wash-away of blood from fillets prepared from very fresh (rigor) fish or somewhat aged mackerel. Washing improved the quality of fillets from the rigor fish but not from aged mackerel, which was attributed to the greater amount of blood removal from the fresher mackerel by washing due to less coagulation.

11.6.3 Washing Fish Mince Such as Surimi Production

Washing fish mince is a key step in the production of surimi and is traditionally applied to remove fat, pro-oxidants and aroma compounds, thus yielding a fish mince more neutral in color and flavor. In a series of studies, it has, however, become evident that not only pro-oxidants are removed in the washing steps, but also endogenous antioxidants, which can negatively affect the stability against oxidation. As an example of this, Tokur et al. (2006) followed chemical and sensory qualities of fish fingers produced from unwashed and water washed (1.5:2, mince:water) mirror carp (*Cyprinus carpio*) mince during storage at −18 °C and found that TBARS became significantly higher in washed than unwashed mince. It should be mentioned that pH was lower in the washed mince, which could trigger oxidation. Also in a study of crude and washed mince from whiting (*Merlangius merlangus euxinus*), higher TBARS developed in the end of a refrigerated storage period in the washed mince (Köse et al., 2006). In addition our own data have pointed in this direction (Undeland et al., 2003). PV and painty odor developed ~4 days faster in washed cod mince (3×3 vol water and buffer) than in unwashed cod mince during ice storage. Both minces were enriched with Hb and adjusted to pH 6.6. When adding back an aqueous fraction (press juice) isolated from the unwashed cod mince to the washed mince at two- to sixfold dilutions, Hb-mediated PV and painty odor development were either delayed or completely prevented. Also in herring muscle mince, oxidation proceeded faster after washing than in the corresponding nonwashed minces (Undeland et al., 1999). In the latter study, losses of tocopherol and GSH-px during washing were recorded along with wash-out of Hb, Fe, Cu, and Zn. When tested in an iron-enriched linoleic acid emulsion, the herring mince wash water showed a net antioxidative capacity, supporting the oxidation data. Deeper investigations of the aqueous antioxidants in fish muscle press juice are discussed in a later section of this chapter.

Eymard et al. (2009) studied lipid oxidation during processing and storage of unwashed/washed horse mackerel (*Trachurus trachurus*) minces. Briefly, the mince (M0) was washed once (M1), twice (M2), or three times (M3) with three volumes of water, mimicking the surimi production. The different products were characterized (i.e., lipid content, protein, water, iron, FA profile, and tocopherol content) and analyzed for PV and volatiles; the latter was done immediately after washing and after up to 96h at 5 °C. Lipid and protein oxidation markers indicated that both lipid and protein oxidation took place during the washing process, and the ranking for oxidation levels at time 0 was as follows: $M0 < M1 < M2 \leq M3$, with M0 being significantly less oxidized than M3. An imbalance in the initial pro-oxidant–antioxidant equilibrium in the washed muscle tissue was given as explanation. However, during storage at 5 °C, the oxidation development order was changed into $M0 > M1 > M2 \geq M3$. To explain this, it was discussed that oxidation in M2 and M3 had already during processing reached a steady state and did not increase further.

A few studies from the 1990s aimed at minimizing lipid oxidation during washing of mackerel mince. In one of them (Kelleher et al., 1992), an antioxidative system was specifically designed to maintain heme iron reduced, nonheme iron as an inactive chelate, and to provide free-radical scavenging in the aqueous and lipid phases. This system, consisting of erythorbate, sodium tripolyphosphate (STPP), ethylenediaminetetraacetic acid (EDTA), and propyl gallate, was most effective when added early in the washing process. Washed mackerel light muscle mince in the presence of the antioxidative system was stable during 24 weeks at −20 °C. It was also found that postmortem age of the mackerel muscle had little effect on initial odor quality of the washed mince and that washed mince prepared from mackerel dark muscle showed greater propensity to oxidize than did the mince from light muscle. In another similar study, avoidance of added sodium chloride, and exclusion of oxygen during washing also proved to be efficient strategies. Antioxidant additions during washing were also tested by Eymard et al. (2010). Horse mackerel mince was here washed with water ($3 \times 1{:}3$) or water enriched with propyl gallate, caffeic acid (100 mg/kg), or spermine (400 mg/kg). Results showed that washing with caffeic acid and propyl gallate completely inhibited PV and volatiles during subsequent refrigerated storage and that spermine lowered oxidation by ~50%.

As an attempt to remove oxidation substrates and pro-oxidants from fish mince prior to hydrolysis making (see later), mince from brown stripe red snapper (*Lutjanus vitta*) was subjected to: (1) washing (2×5 vol of water); (2) cellular membrane separation (1×9 vol of 8 mmol $CaCl_2$ with 5 mmol citric acid), (3) washing followed by cellular membrane separation, and (4) cellular membrane separation followed by washing (Khantaphant et al., 2011). Among the resulting minces, the one subjected to membrane separation with subsequent washing (i.e., 4) became the most stable toward oxidation as measured by TBARS during 9 days at 4 °C. This mince also contained the least Mb, PL, heme iron, and nonheme iron. All treated minces were more stable than the control mince.

The studies listed point at effects of washing being complex as numerous compositional parameters vary simultaneously. Results generally differ depending on whether the washed minces have been stored without further treatment or whether they have been enriched with pro-oxidants after washing to be used, for example, as research model systems. It is, however, clear that aqueous antioxidants are washed away along with pro-oxidants and neutral fat, with remaining minces having a higher relative ratio of membrane lipids and, in some cases, a less favorable pro- to antioxidant balance. Limited dilution during washing appears to favor the action of pro-oxidants while additions of antioxidants in wash waters appear to reintroduce oxidative stability to the mince.

11.6.4 Oxidation during pH-Shift Processing

In early 2000, alternative methods for concentrating fish proteins were developed by Hultin and Kelleher (e.g., Hultin and Kelleher, 1999, 2000, 2001), which were based on acid or alkaline protein solubilization in water followed by centrifugation to remove insolubles, isoelectric protein precipitation, and protein dewatering ("pH-shift processing"). The method has gained increasing popularity since it is applicable directly on complex aquatic raw materials as whole fish or filleting cut-offs (back-bones, heads, etc.) without mechanic preseparation of muscle tissue. Based on the abundance of blood, enzymes, and lipids in this kind of raw materials, there has been an interest in the degree of lipid oxidation developing during the pH-shift process. As during fish mince washing, the relatively large additions of water dilute the aqueous antioxidative defense system but, further, the subjection to low pH in solubilization ($pH \leq 3$) and precipitation (~pH 5.5) can be critical for activation of heme proteins as catalysts. To study lipid oxidation as a function of acid and alkaline pH-shift processing, Kristinsson and Liang (2006) investigated TBARS in unprocessed ground Atlantic croaker (*Micropogonias undulates*) as well as in acid- and alkali-processed isolates made thereof. Conventional surimi was also produced. Samples were stored on ice for up to 14 days. The highest initial TBARS value was found in the acid-processed protein isolate followed by the ground muscle and then finally the alkali-processed protein isolate. The conventional surimi sample obtained lower TBARS during processing compared with the initial ground raw material. Over the storage period studied, the alkali-processed protein isolate and the conventional surimi maintained a low TBARS value while the other samples, especially the acid-produced isolate, had a substantial increase in TBARS. Also in a study on channel catfish, the same research team (Kristinsson et al., 2005) followed TBARS both after acid/alkaline pH-shift processing and after classic surimi production. It was found that the acid-aided process led to higher yellowness than the other two processes. However, with this fish species, all processes led to minimal levels of lipid oxidation as assessed by TBARS.

Similar to the study on Atlantic croaker, significant oxidation was documented after acid pH-shift processing of herring fillets (Undeland et al., 2005). In this study, the herring fillets were subjected to different versions of acid pH-shift process, with and without a prewash of the mince with volumes of water, with different holding times at the solubilization pH (2.7) and with or without additions of erythorbate (0.2%), STPP (0.2%), and EDTA (0.044%). Antioxidants were added in the prewash or during the homogenization step. Significant lipid oxidation, as measured by TBARS and sensory analysis, took place during the process without antioxidants added, which was not significantly affected by a prewash or varied exposure time to pH 2.7. Erythorbate alone, or in combination with STPP/EDTA, significantly reduced lipid oxidation during processing if added in the prewash or homogenization step. During subsequent ice storage, better stability was gained when antioxidants were added in both of these steps and when EDTA was used instead of STPP.

Shaviklo et al. (2012) studied TBARS values in freeze-dried alkali-produced pH-shift isolates made from saithe (*Pollachius virens*) fillets. Compared to freeze-dried saithe mince, TBARS in the freeze-dried isolate was >10 times higher, indicating that also the alkaline process can be critical if no antioxidants are added. The subjection to relatively low pH during precipitation (pH 5.5) was discussed as a reason.

Thus, although both acid and alkaline pH-shift processing can initiate lipid oxidation, dark muscle fish species combined with acid pH-shift processing appears to be a specifically sensitive case, but the problem can be circumvented by combinations of antioxidants. The topic has been reviewed in detail by Nolsøe and Undeland (2009).

11.6.5 Oxidation during Fish Protein Hydrolysis Production

As a way to use seafood co-products in a more sustainable manner, there has also been a large interest in fish protein hydrolysate production using enzymes. Application of the resulting hydrolysates into foods has, however, been held back by the fact the hydrolysates can be relatively oxidized, partly due to the heat and pH-changes applied during hydrolysis making in combination with the presence of lipids and pro-oxidants in complex raw materials (Halldorsdottir et al., 2014a,b). Oxidation affects flavor negatively and bring in lipid radicals/ oxidation product to the fish or shellfish muscle to be stabilized and/or nutritionally enriched by the hydrolysate product.

That lipid oxidation develops to a lower extent during fish protein hydrolysis production when less Hb present was recently shown by Yarnpakdee et al. (2014) and Halldorsdottir et al. (2013). The former group produced protein hydrolysates from Nile tilapia protein isolates with and without added oxy-Hb and met-Hb. When oxy-Hb or met-Hb was incorporated, marked increases in fishy odor, nonheme iron content, PV, and TBARS were observed, especially within the first 60 min of hydrolysis and particularly with met-Hb. Also in the final hydrolysate, Hb clearly stimulated volatiles and color changes. In the study of Halldorsdottir et al. (2013), a washed cod mince model system with and without Hb and natural antioxidants (L-ascorbic acid and *Fucus vesiculosus* extract) was subjected to hydrolysis with Protease P "Amano" 6. Results indicated that high TBARS values developed during the hydrolysis process with Hb but not without it, but that the two antioxidant strategies could prevent Hb-mediated oxidation. A later study by this group confirmed the usefulness of a brown seaweed extract as an antioxidant in hydrolysis making. They produced fish protein hydrolysate from cod bone mince in the absence and presence of a *Fucus vesiculosus* ethyl acetate extract (Halldorsdottir et al., 2013). Results show that the *Fucus* extract contributed to less rancid taste and slightly less TBARS in hydrolysates, particularly after freeze-drying. Other antioxidants were evaluated by Yarnpakdee et al. (2012), who addressed lipid oxidation and fishy odor development in protein hydrolysate from fresh and ice-stored (18 days) Nile tilapia (*Oreochromis niloticus*). Ice storage for 18 days of the final hydrolysates raised PV and TBARS, reduced heme iron content, increased nonheme iron content, reduced PL content, and elevated FFA. Oxidation and fishy odor were more pronounced if hydrolysates were produced from ice-stored tilapia than from fresh tilapia. Adding 2 mmol/L EDTA and 100 μmol/L Trolox during the hydrolysis process markedly lowered lipid oxidation and fishy odor. Other strategies than antioxidant additions applied to produce a more stable fish protein hydrolysate have been to prewash the fish mince, to remove cellular membranes, and to concentrate fish proteins by pH-shift process prior to hydrolysis making. These strategies lower, for example, levels of lipids and heme proteins. When Indian mackerel mince and Nile tilapia mince were subjected to prewashing (2 × 5 vol of water), membrane removal with Ca^{2+} and citric acid as well as alkaline solubilization/precipitation process, alone or in combination, the combined preprocessing (referred to as W–MR–Al) resulted in the lowest remaining Mb, heme iron, total lipid, and phospholipid contents (Yarnpakdee et al., 2012b,c). When this raw material, together with crude mince, was hydrolyzed using Alcalase for up to 120 min, hydrolysate from W–MR–Al had lower PV, TBARS, and nonheme iron content throughout hydrolysis period compared with crude mince.

Thus, as for pH-shift processing, strategies have to be applied to prevent oxidation from occurring during hydrolysis making; several examples have been outlined in the literature.

11.6.6 High Hydrostatic Pressure Treatment

It has been described that HP treatment deteriorates hydrolytic endogenous enzymes, for example, lipases, phospholipases, peroxidases, and LOX, which could stabilize seafood products against oxidation (Murchie et al., 2005). A series of studies have addressed this question.

Vázquez et al. (2013) subjected Atlantic mackerel (*Scomber scombrus*) to HP treatment (150, 300, and 450 MPa with holding times of 0.0, 2.5, and 5.0 min) where after PV, TBARS, browning compound formation, FFA release, and polyene content were analyzed during 3 months of storage at −10 °C. A marked inhibition of browning, that is, tertiary lipid oxidation product formation, during storage was observed along with inhibition of free FA content when increasing the pressure level or the pressure holding time of the HP treatment. However, no effects on PV and TBARS were observed. In another study, HP processing at low temperatures combined with modified atmosphere packing (MAP) (50% O_2 + 50% CO_2) was evaluated with respect to shelf life of salmon at 5 °C (Amanatidou et al., 2000). Initial TBARS were not significantly affected by any of the treatments compared with fresh samples. However, after 14 days of storage at 5 °C, TBARS increased significantly for all samples, but especially for MAP samples and samples treated with compressed gases. Also, Yagiz et al. (2007), who studied the effect of HP processing on quality of rainbow trout and mahi mahi during cold storage, documented some pro-oxidative effects following subjection to HP. Skinless fillets treated with different pressures (150, 300, 450, and 600 MPa for 15 min) were stored for up to 6 days at 4 °C after the dark muscle was analyzed. TBARS of trout dark muscle increased with increased pressure, whereas the same trend was not seen for mahi mahi dark muscle, where maximum oxidation was found at 300 MPa and then declined; thus, species responded differently. The best HP conditions when looking at all study parameters—lipid oxidation, microbial load, and color changes—were found to be 300 MPa for rainbow trout and 450 MPa for mahi mahi. The specific inactivation of endogenous LOX in a crude silver carp (*Hypophthalmichthys molitrix*) extract during thermal (50–80 °C, ambient pressure) and HP + thermal (300–700 MPa at 50–80 °C) process conditions were investigated by Qui et al. (2013). LOX inactivation followed first-order kinetics at all pressure–temperature combinations used. Moreover, LOX inactivation rates became less pressure dependent with increasing pressure at all temperatures studied and showed less temperature dependence of the enzyme inactivation rate at various HP levels tested.

It appears from these studies that HP treatment can have inactivating effects on some pro-oxidants but that the net effect of HP still can be pro-oxidative under some circumstances. Thus, careful optimizations have to be done to obtain good results on chemical, physical, and microbial parameters.

11.6.7 CO Treatment

Carbon monoxide (CO) has been used to stabilize the color of fish muscle during frozen storage based on the ability of CO to replace O_2 in heme proteins forming the CO counterpart. Tuna is a very common species to treat with this method. In one study, yellowfin tuna steaks were subjected to treatments with 4% CO, 18% CO, and 100% CO as well as filtered smoke containing 18% CO for 48 h, followed by 30 days of freezing and subsequent cold storage (4 °C) after thawing (Kristinsson et al., 2006). The filtered smoke, 18% CO, and 100% CO treatments led to reductions in formation of TBARS during freezing and subsequent cold storage, possibly due to increased stability of heme proteins. Pivarnik et al. (2013) compared changes in the quality profiles of CO-treated and untreated tilapia fillets stored at room temperature, refrigerated temperature, and on ice. At time 0, PV and TBARS were lower for CO-treated samples and continued to have lower trends throughout all storage temperatures.

Redness (a^*) color values were greater in CO-treated tilapia at time 0; however, the treated product showed a more rapid decline in a^* throughout all storage temperatures.

11.6.8 Reduced Oxygen and MAP

Removing oxygen is known to be a very powerful way of reducing the degree of lipid oxidation in seafood (e.g., Khalil and Mansour, 1998; Rostamzad et al., 2010). Flick et al. (1992) stated that vacuum packaging generally gives a greater reduction of lipid oxidation than the presence of additives. In later years, MAP has emerged as a very popular alternative way of preserving seafood for the retail market. Although the largest focus has been on preventing microbial growth, a number of studies have also addressed the effect of MAP on lipid oxidation since this reaction can become a problem in MAP containing significant amounts of O_2.

TBARS of rainbow trout (*O. mykiss*) fillets packed in air, vacuum, and MAP (100% CO_2, 2.5% O_2+7.5% N_2+90% CO_2 and 30% O_2+30% N_2+40% CO_2) were followed during storage at 4 °C (Arashisaret al., 2004). After 6 days, TBARS increased the fastest in the samples containing 30% O_2 followed by controls containing air and samples with 2.5% O_2. Fillets containing 100% CO_2 and vacuum packed fillets were the most stable. Using hake slices packaged in MAP, similar results were reported by Pastoriza et al. (1996). Along the same lines, Gimenez et al. (2002) reported that lipid oxidation in rainbow trout fillets was significantly higher in gas packages with 20% and 30% O_2 than in those with 10% O_2. In contrast to this, Masniyom et al. (2002), who studied the effect of CO_2-enriched atmosphere in combination with either oxygen or nitrogen on the keeping quality of seabass slices at 4 °C, found that TBARS increased as the CO_2 concentration increased. Explanations given were that carbonic acid formed could denature heme proteins causing heme release, and also the antioxidant enzymes like GSH-px could be inactivated.

A series of studies have also been conducted on combined effects of adding various natural antioxidants and MAP (e.g., Goulas and Kontominas, 2007; Maqsood and Benjakul, 2010a; Kostaki et al., 2009). The first of them showed that while MAP (40% CO_2/30% O_2/30% N_2) increased TBARS in salted seabream fillets compared with air-packed fillets during 4 °C storage; oregano oil prevented this MAP-induced oxidation (Goulas and Kontominas, 2007). In the study of Maqsood and Benjakul, 2010a), MAP (60% N_2/35% CO_2/5% O_2) and tannic acid showed a synergistic preventive effect on lipid oxidation in refrigerated striped catfish.

11.6.9 Edible Coatings

From a lipid oxidation perspective, several benefits can be expected from treating seafood with edible coatings (Gennadios et al., 1997). The prime advantage is indeed the reduced oxygen access, but also the prevention of moisture loss is important since dried-out surfaces can increase the area available for oxidative attack. Further, edible coatings can be fortified with antioxidants as a way to directly treat product surfaces. This specific application is further discussed under the section on natural antioxidants. Examples of biopolymers used in edible coatings include gelatin, blood protein, feather keratin, fish myofibrillar protein, and the polysaccharide chitosan produced from crustacean shells. Biopolymers for coating seafood could thus be retrieved from seafood processing byproducts, something that supports sustainable use of the whole catch or harvest.

Successful combinations of seafood raw materials and coating materials are discussed in the review of Gennadios et al. (1997) and later by Bonillilla et al. (2012). In the latter review, most studies, however, consisted of coatings with antioxidants, for example, (1) rosemary/oregano essential oils added into gelatin films ± chitosan (e.g., Gómez-Estaca et al., 2009), (2), vitamin E (plus fish oil!) added into chitosan coatings (Duan et al., 2010), and (3) cinnamon oil added into chitosan coatings (Ojagh et al., 2010). Jeon et al. (2002), however, with good results, applied pure chitosan-based coatings to herring and cod fillets, and Lopez-Caballero et al. (2005) added chitosan and gelatin coatings to fish patties. It is discussed in some of the mentioned studies that both the inherent antioxidant activity of films and the actual oxygen barrier they provide are responsible for the reduced lipid oxidation development.

11.6.10 Glazing

Advantages of glazing—that is, the application of a thin layer of ice onto the surface of frozen seafood—are very similar to the advantages listed earlier for other types of edible coatings. Glazing with water has been repeatedly shown to slow down the development of lipid oxidation in frozen seafood (e.g., Mesarčová et al., 2013). Also in this area, an emerging trend has been to incorporate natural antioxidant extracts into the glazing liquids, with examples being tocopherols, propyl gallate, or rosemary extract (Tseng et al., 2005), hawthorn (*Crataegus oxyacantha* L.) and agrimony (*Agrimonia eupatoria* L.) (Mesarčová et al., 2013), flaxseed aqueous extract (Stodolnik et al., 2005), and green tea and grape seed extracts (Yerlikaya and Gokoglu, 2010). Fortifying glazing solutions with antioxidants have generally improved oxidative stability further compared with water glaze (e.g., Tseng et al., 2005).

11.6.11 Cooking

Along with an increasing popularity of ready meals, knowledge on how precooking affects levels of lipid oxidation products in seafood, and the oxidative stability during subsequent storage, is important to maximize product quality. Heat can have numerous effects on pro-/antioxidants in seafood, with the main ones being inactivation of enzymes and denaturation of heme proteins with subsequent loss of heme or LMW iron. During the 1970s and 1980s, many studies were carried out on the role of lipid oxidation in development of warmed over flavor (WOF) of cooked fish as a result of heme protein changes (e.g., Ke et al., 1978; Tichivangana and Morrisey, 1984). In a more recent study, the significance of heat for the state of heme proteins was studied by adding heme protein variants with a 975-fold range of hemin affinities to washed cod mince, with and without subsequent heat treatment at 88 °C until the mince had reached 80 °C (Grunwald and Richards, 2006b). The heme variants promoted TBARS with equivalent efficacy in cooked washed cod mince contrary to results in uncooked washed cod mince. The authors concluded that cooking temperatures apparently denature the globin and release hemin reactant to such an extent that the impact of hemin affinity on lipid oxidation observed in the raw state is negated in the cooked state.

Karlsdottir et al. (2014c) addressed the role of the fish starting quality for the effects of a cooking step. Lipid decomposition of saithe (*Pollachius virens*) light and dark muscles was monitored during frozen storage at −25 °C of noncooked fish (stored up to 18 months) and cooked products. The cooked samples had first been stored 0, 6, and 12 months and then cooking was done, and they were refrozen and stored at −25 °C for 12 months. In noncooked saithe, enzymatic lipolysis was the driving factor influencing the quality. Cooking, however, inhibited FFA formation and instead induced formation of PV and TBARS. This behavior was more evident in samples cooked after long storage periods of the raw materials. The authors hereby concluded that the initial quality of the raw material before cooking is critical with regard to oxidative stability of cooked fish products.

The influence of seven cooking methods (boiling, conventional baking, microwave baking, grilling, deep-frying in soybean oil, canola oil, or partially hydrogenated vegetable oil) on the level of oxidation products in silver catfish (*Rhamdia quelen*) fillets was evaluated by Weber et al. (2008). CD levels and PV decreased for all fried samples during the process, which was ascribed heat-induced primary product decomposition but remained constant in the samples subjected to the other cooking methods. Boiling and baking increased TBARS values, while this was not the case for grilling and frying. Different cooking methods were also studied by Naseri et al. (2010). Steaming, oven baking, and microwave cooking were applied to carp fillets, followed by immediate analyses of chemical composition and lipid quality. It was found that TBARS increased for oven- and microwave-cooked fillets and remained constant in the steam-cooked samples. CD and browning color formation levels significantly increased in the oven-baked fillets. Further, the effect of boiling, steaming, microwaving, grilling, pan-frying, and deep-frying on lipid oxidation and composition of grass carp (*Ctenopharyngodon idellus*) fillets were investigated by Zhang et al. (2013). As in the study by Weber et al. (2008), PV decreased during pan-frying and deep-frying but remained constant in the samples subjected to the other four cooking methods. Microwave baking and grilling increased the *p*-anisidine values and resulted in higher "lipid oxidative extents." It was noted that all treatments reduced the moisture contents. Using herring mince as a model, the combined role of a surimi-like washing, cooking time (38 and 54 min), and cooking temperature (55 and 100 °C) on PV, CD, CT, and fluorescent oxidation products during subsequent frozen storage (−18 °C) was studied (Undeland et al., 1998a). The best lipid stability was obtained at minimum cooking time and at lower cooking temperature which we ascribed heat inactivation of catalytic enzymes, without simultaneous activation, for example, of heme proteins. In a parallel trial also including raw mince, it was found that the order of oxidation development was mince subjected to 100 °C > raw mince > mince subjected to 55 °C. Washing reduced these benefits from precooking by removal of pro-oxidative enzymes and through a reduction in the amount of antioxidants as well as a relative increase in PL and FFA in the fat. In another study (Khalil and Mansour, 1998), carp fillets cooked to an internal temperature of 70 °C and packaged either with or without vacuum, obtained higher TBARS than uncooked fillets during storage at ~5 °C for 16 days. Changes critical for the onset of lipid oxidation thus appear to take place just above 55–60 °C. In general, exact mechanistic conclusions regarding effects from cooking on seafood lipid oxidation can, however, be difficult to draw out of the same reason as was discussed for washing—that is, that numerous compositional changes appear simultaneously. Examples are losses of moisture, fat, endogenous antioxidants, and sarcoplasmic proteins and/or uptake of cooking oils, the former of which can have a concentrating effect and that latter of which can "dilute" levels of oxidation products originating from the fish tissue.

11.7 PREVENTING LIPID OXIDATION IN SEAFOOD BY ADDED NATURAL ANTIOXIDANTS

In the last 10–20 years, the interest has increased dramatically within the food industry in replacing synthetic food additives with natural ones, and particularly, the replacement of synthetic antioxidants has been in focus. Within the scientific community, this has been reflected by an incredible number of scientific studies investigating whether a certain all-natural extract can prevent lipid oxidation in a certain food product. The seafood segment, with its built-in high susceptibility to lipid oxidation, has indeed not been an exception. By far the largest bulk of studies in this direction have addressed plant-derived antioxidants, primarily polyphenols of various kinds. However, some research has also been dedicated the use of (seafood) muscle-derived extracts/fractions/compounds, such as protein hydrolysates. Both of these fields will be discussed later.

11.7.1 Addition of Plant-Derived Antioxidants

A series of reviews on the uses of phenols/polyphenols in seafood stabilization have been published in, for example, Gómez-Guillén and Montero García (2007),

Maqsood et al. (2013), and Medina et al. (2012), the latter specifically addressing the hydroxycinnamic acid caffeic acid. It is generally well recognized that the radical scavenging activity of phenolic compounds is due to their efficient hydrogen-donating ability. The ability of the phenol ring(s) in delocalizing electrons and thereby yielding a stable radical then prevents further transmission of the radical attack. The stability of the intermediate radical and the hydrogen-donating ability of phenols/polyphenols are linked to the degree of derivatization, that is, the number of methyl and/or hydroxyl groups. Among efficient phenolic radical scavengers in biological systems are, for example, the flavonoid molecules like catechin, luteolin, and quercetin. However, that some flavonoids are very prone to polymerization affect their radical-scavenging activity. When the degree of polymerization exceeds a critical value, hydroxyl group availability decreases and consequently also the radical scavenging activity.

Generally, the activity found for phenols/polyphenols has been related with both the molecular structures and polarity (Jacobsen et al., 2008). As a result, the activity of these natural compounds is greatly influenced by the lipid substrate of the seafood matrix to which they are added. Some polyphenols are very effective in fish oils, but not in fish muscle or in fish oil emulsified products, and vice versa. The correct incorporation of the polyphenols into the sensitive oxidizeable sites of fish muscle is also a factor that largely affects their antioxidant activity. This can be controlled by the polarity of the vehicle solvent (Raghavan and Hultin, 2005). Medina et al. (2007, 2009) described how the capacity of phenolic acids for donating electrons showed a high correlation with their ability to retard lipid oxidation in fish muscle. In contrast, their ability for chelating metals or their polarity was not correlated with their inhibiting activities. The full antioxidant capacity of polyphenols following addition into muscle foods can be somewhat reduced by the high reactivity of these compounds toward proteins, yielding soluble or insoluble complexes that render active groups less accessible for scavenging lipid radicals (Arts et al., 2001). These kinds of interactions may also reduce the bioavailability of both proteins and phenolic compounds, which should be taken into consideration when applying phenols/polyphenols to prevent rancidity. Further, the bitter taste and strong color of many polyphenol-rich extracts can also have limiting effects (Gómez-Guillén and Montero García, 2007); unfortunately, these aspects have often been overlooked in research studies. Nevertheless, there are a lot of promising results on very efficient prevention of lipid oxidation in seafood by phenolic compounds that defend the great interest. Results from a selection of studies from the last 10 years will be highlighted later, grouped according to pure compounds vs extracts, and according to the strategy of adding the plant-derived extract into the seafood product.

11.7.1.1 Pure Phenols/Polyphenols

An extensive series of studies have been performed on the role of caffeic acid (3,4-dihydroxycinnamic acid) and related hydroxycinnamic acids (see Figure 11.8) in preventing lipid oxidation in fish muscle. Examples are the studies

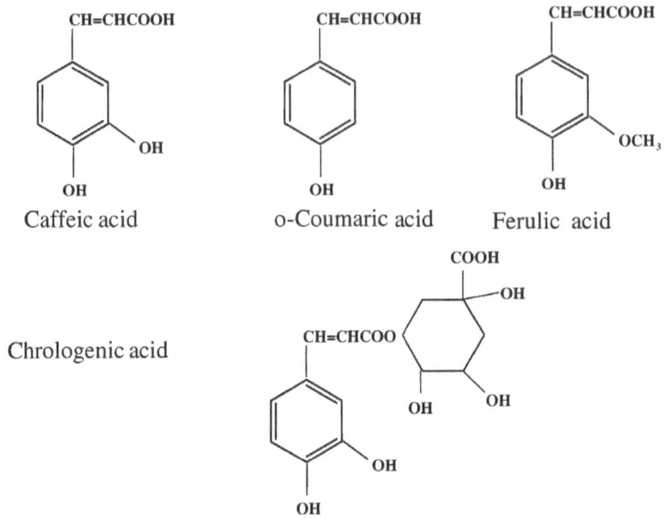

FIGURE 11.8 Molecular structures of hydroxycinnamic acids; caffeic acid, *o*-coumaric acid, ferulic acid, and chlorogenic acid. *Reprinted with permission from Medina et al. (2007).*

of Medina et al. (2007, 2009). In minced horse mackerel, caffeic acid (10–100 ppm) prevented oxidation during ice storage in a concentration-dependent manner (Medina et al., 2007). Its activity was followed by that of ferulic acid and then *o*-coumaric acid. The efficacy of caffeic acid was highly related to its capacity of donating 12.2 mol electrons/mg antioxidant. According to Medina et al. (2009), the antioxidant effectiveness of caffeic acid showed a significant reduction when the lipid level of fish mince was ≥2%, which was ascribed that a phenolic compound-to-lipid ratio of 1:200 is needed for optimal antioxidant effect. The effect of caffeic acid efficiency was also lower under frozen than refrigerated storage, which was linked to diffusion being a limiting factor. Using an Hb-enriched washed cod mince model system, caffeic acid efficiently inhibited formation of peroxides, rancid odor, and redness loss at 50 ppm or more during ice storage (Larsson and Undeland, 2010). In this study, caffeic acid (≥100 ppm) also inhibited the formation of protein carbonyls and loss of protein salt solubility as well as the loss of α-tocopherol. Coumaric acid (50–200 ppm) was not effective for inhibiting lipid oxidation.

Maqsood and Benjakul (2010b) studied the impact of catechin, caffeic acid, ferulic acid, and tannic acid on lipid oxidation of menhaden oil-in-water emulsion (100 mg/L) and mackerel mince (100 mg/kg) during ice storage. Tannic acid showed the highest efficacy in preventing PV, CD, and TBARS for both model systems. Metal chelation, radical scavenging, and LOX prevention were given as explanations. Ferulic acid seemed the least preventive for lipid oxidation development. In a later study by the same team (Maqsood and Benjakul, 2011b), lipid oxidation mediated by tilapia Hb was monitored in ice-stored washed Asian seabass mince with and without added tannic acid (200 and

400 ppm). Addition of 400 ppm tannic acid reduced PV, TBARS, and volatile lipid oxidation products and yielded lower fishy odor score, compared with the control sample and that with added 200 ppm tannic acid.

As a follow-up of the many studies addressing strong antioxidative capacity of caffeic acid in fish mince, Park et al. (2013) specifically studied the role of caffeic acid in Hb-mediated lipid oxidation of washed fish mince to further elucidate mechanisms. It was found that caffeic acid accelerated met-Hb formation at pH 5.8 and 25 °C, which was attributed to electron donation from caffeic acid to liganded O_2 in Hb. However, caffeic acid inhibited the critical subsequent hemin dissociation from met-Hb, which could be a key event for its antioxidative mechanism. Using a washed cod mince model system, perch Hb was rapidly converted to met-Hb and readily promoted PV development and hexanal formation, the latter two events being inhibited by caffeic acid. Once met-Hb formation occurred in the washed cod mince, caffeic acid appeared to maintain the heme protein as met-Hb during the remainder of iced storage, possibly through binding. Caffeic acid also facilitated the transfer of perch Hb from the aqueous phase to the insoluble components of washed cod muscle. Another study specifically addressing the effects of polyphenols on the pro-oxidative role of Hb was that of Gebicka and Banasiak (2009); they studied reactions of met- and oxy-Hb with hydrogen peroxide in the presence of catechin, quercetin, and rutin. These flavonoids accelerated reduction of ferryl-Hb to met-Hb with rate constants of the reactions of ferryl-Hb with catechin, quercetin, and rutin being in the order of 102 $M^{-1} s^{-1}$, that is, similar to the rate constants of ferryl-Hb with intracellular reducing compounds like urate or ascorbate. According to the authors, the beneficial effect of flavonoids against oxidative damage of Hb caused by hydroperoxides, earlier reported in the literature, is probably, at least in part, connected with the ability of flavonoids to scavenge ferryl-Hb.

The influence of polymerization and galloylation (content of esterified gallates) of oligomeric catechins (proanthocyanidins) on their antioxidative capacity in pelagic fish muscle was evaluated by Pazos et al. (2010). Compounds used were extracted from pine or grape. Proanthocyanidins with medium size (2–3 monomeric units) and low galloylation degree (0.15–0.25 gallate group/molecule) most efficiently inhibited lipid oxidation in pelagic fish muscle. In the study by Lee et al. (2006), fractions enriched in phenolic acids (Fraction 1), anthocyanins (Fraction 2), flavonols (Fractions 3 and 4), and proanthocyanidins (Fractions 5 and 6) were prepared from cranberry powder. It was found that Fractions 1 and 6 had lower levels of polyphenols per mg dry weight than Fractions 2, 3, 4, and 5, which were nearly equivalent. Addition of cranberry fractions at a level of 74 μmol quercetin equivalents per kg of Hb-enriched washed cod mince extended the induction time of TBARS formation in the order: Fraction 1, Fraction 3, Fraction 4 > Fraction 2 > Fraction 5 > Fraction 6. As described by the authors, their results suggest that oligomeric polyphenols (e.g., proanthocyanidins) were least effective at inhibiting Hb-mediated lipid oxidation in washed cod muscle compared with the other classes of carnberry polyphenolics.

The ability of the different cranberry fractions to scavenge DPPH radicals did not reflect their relative TBARS inhibiting ability in washed cod mince. Quercetin was tentatively identified as a component in cranberry that was especially effective at inhibiting Hb-mediated lipid oxidation. The ability of flavonol and proanthocyanidin-enriched fractions to inhibit Hb-mediated lipid oxidation, despite efforts to wash away the added polyphenolics prior to Hb addition, confirmed that these classes of polyphenolics bind efficiently to insoluble components of washed cod mince such as proteins.

A few relatively recent studies have also tested the effects of polyphenols specifically on fish LOX activity (Banerjee, 2006) and LOX-induced oxidation of FAs in vitro (Liu and Sun Pan, 2004). In the study by Banerjee (2006), the specific effect of commercially available green tea polyphenols (catehin, epicatechin, epicatechin gallate, epigallocaechin gallate, epigallocatechin gallate) on isolated mackerel muscle LOX was studied together with caffeic acid and some more conventional antioxidants like BHA (butylated hydroxyanisole), BHT (butylated hydroxytoluene), esculetin (a specific LOX inhibitor), ascorbic acid, and EDTA. Also, nordihydroguaiaretic acid (NDGA) was included. The strongest green tea–derived inhibitor was epigallocatechin gallate (IC_{50} value 0.13 nmol/L); all the tea catechins showed a *mixed* noncompetitive-type inhibition. The other antioxidants gave IC_{50} values between 0.02 and >50 μmol/L in the inhibition of mackerel muscle LOX, with NDGA, BHA, and BHT having lowest values. It was discussed that the effective prevention by polyphenols was based on radical scavenging rather than specific LOX inhibition. In the study of Liu and Sun Pan (2004), oxidation of arachidonic acid and linoleic acid in vitro was initiated by LOX from gills of tilapia and gray mullet, and an inhibitory effect was found from a water extract of green tea.

11.7.1.2 Plant-Derived Extracts or Concentrates

Within this category, the complex mixture of phenolic compounds and other molecules being present in extracts/concentrates renders full interpretation of results difficult. The specific solvent used to prepare the extract indeed affects the polarity of molecules being extracted and their partitioning into the fish muscle matrix. Among common solvents used are ethyl acetate, ethanol, and water. The mode of application also influences data as it affects how well extracts/concentrates reach the oxidizable sites. Broadly categorized, there are two ways by which additions into fish muscle have been made: mixing into a mince or treating the surface of fillets or round fish by dipping, icing, glazing, or coating.

Within the first category, Wang et al. (2010) studied how an extract from the brown seaweed *Fucus vesiculosus* and its fractions could prevent Hb-catalyzed lipid oxidation in a washed cod mince system and in pH-shift process produced cod protein isolates during ice storage. The extract and fractions were also characterized in terms of total phlorotannin content (TPC), DPPH radical scavenging activity, ferrous ion–chelating ability, and reducing power. Rancid odor, TBARS, redness loss, and volatiles were followed during ice storage. In both systems, a phlorotannin-enriched ethyl acetate fraction showed higher inhibitory effect than a crude 80% ethanol extract. The addition of oligomeric phlorotannin-rich subfraction completely inhibited the initiation of oxidation in both systems throughout 8 days on ice. Its effectiveness at 300 mg/kg level was comparable to that of 100 mg/kg propyl gallate (PG). Although a polymeric phlorotannin-rich subfraction (called "LH-5") had similar level of TPC and chemical antioxidant activities as oligomeric subfraction "LH-2," it was far less efficient in model systems. According to the authors, the results suggested that other factors rather than the intrinsic reactivity toward radicals could be responsible for the inhibitory effect of phlorotannins on lipid oxidation in fish muscle. Other relatively recent studies (Cerantola et al., 2006; Kim et al., 2004) have indicated that oligomeric and polymeric phlorotannins are more potent antioxidants than the monomer phloroglucinol. Further, oligomerization of phloroglucinol appears to be crucial for the radical scavenging capacity of phlorotannins.

Grape-derived extracts and fibers have been addressed in some studies. One is that of Sánchez-Alonso et al. (2007) in which a white grape dietary fiber concentrate (WGDF) was found to completely inhibit Hb-mediated development of rancid odor, TBARS, and redness loss in a washed cod muscle mince with/without 10% added herring oil during ice storage. An ethanol extract of the WGDF had the same high antioxidant capacity as whole WGDF, while the ethanol-extracted fiber residue had some preventive effect. Another study on grape-derived antioxidants addressed both mixing into mince and the treatment of fillet surfaces (Pazos et al., 2005). A phenolic extract and a purified procyanidin fraction from grape (*Vitis vinifera*) byproducts were added at 0.01% (w/w) to minced mackerel (*Scomber scombrus*) and horse mackerel (*Trauchurus trauchurus*) fillets, before freezing at 10 °C with the focus to study preservation of endogenous antioxidants. Grape polyphenols inhibited depletion of endogenous α-tocopherol, ubiquinone-10, and total glutathione. Grape polyphenols were similar to propyl gallate in the preservation of ubiquinone in both minced and filleted muscle and total glutathione in minced muscle. Total glutathione in the fillets was better maintained by propyl gallate than grape polyphenols. α-Tocopherol was most efficiently preserved by grape polyphenols and propyl gallate; its loss decreased according to control > phenolic extract > purified fraction > propyl gallate. The depletion of α-tocopherol was highly correlated with the evolution of lipid oxidation.

Although not directly plant derived, it should be mentioned here that a series of studies also have tested caramelization products, that is, maillard reaction products (MRPs) as natural antioxidants in fish muscle. In one study (Benjakul et al., 2005), caramelization products from different sugars (glucose, fructose, ribose, and xylose) prepared by heating the sugar solutions at

100 °C (pH 7 and 10) were evaluated in a model system consisting of minced Pacific saury (*Cololabis saira*). Results showed that TBARS development was retarded with the addition of fructose caramelization products during ice storage; fructose also exhibited the highest in vitro radical scavenging effect and reducing power. Fructose caramelization products also showed ferrous iron–chelating potential.

Among studies only including surface treatments, Feng et al. (2012) evaluated a tea polyphenol dip treatment (in a 0.2% solution, w/v) after a wash with ozone (O_3) water. This method effectively reduced lipid oxidation in black sea bream (*Sparus macrocephalus*) during refrigerated storage. The efficiency of tea polyphenol + O_3 treatment was better than that of tea polyphenol treatment or O_3 treatment alone. Following soaking of mackerel fillets for 3 min in pre-chilled quince (*Cydonia oblonga Miller*) extract mainly consisting of procyanidin B dimer (50.8%) and hydroxycinnamic acids (36.62%), PV and TBARS were efficiently prevented during subsequent ice storage compared with water-soaked controls (Fattouch et al., 2008). Pezeshk et al. (2011) studied the effect of aqueous solutions of turmeric extract (T, 1.5%; v/v), shallot extract (Sh, 1.5%; v/v), or T + Sh extracts in combination (1.5% + 1.5% v/v) on the quality of vacuum-packaged gutted rainbow trout (*O. mykiss*) during storage at 4 ± 1 °C for 20 days. Fish samples were dipped for 30 min in the three extracts or in distilled water (control). During storage, PV and TBARS developed significantly slower in fish dipped in extracts compared with the control. Differences were evident after 5 and 10 days respectively for PV vs. TBARS.

The effects of turmeric extract (T), shallot extract (Sh), and their combination (T + Sh) on the quality of vacuum-packaged rainbow trout (*O. mykiss*) were examined during refrigerated storage (4 ± 1 °C) over a period of 20 days. Fish samples were divided into four batches; three batches were treated by dipping for 30 min in aqueous solution of T (1.5%; v/v), Sh (1.5%; v/v), or T + Sh combination (1.5% + 1.5% v/v), while the fourth batch was dipped in distilled water as a control sample. The control and the treated fish samples were analyzed periodically for microbiological (total viable count, psychrotrophic count), chemical (total volatile base nitrogen (TVB-N), PV, and TBA value), and sensory characteristics. The results indicated that the effects of the T, Sh, and T + Sh on the fish samples were to enable the good-quality characteristics to be retained longer and to extend the shelf life during the refrigerated storage.

Antioxidant additions into icing systems have also been evaluated in some studies. The effect of ice-containing thyme (0.04% w/v), oregano (0.03% w/v), and clove (0.02% w/v) ethanol extracts on lipid oxidation in anchovy (*Engraulis encrasicholus*) was assessed by Bensid et al. (2014). Results showed a certain antioxidant effect on fish kept under plant extract icing systems, as measured by PV and TBARS. In a related study, ice prepared from aqueous extracts of oregano (*Origanum vulgare*) and rosemary (*Rosmarinus officinalis*) leaves was applied as chilling system for Chilean jack mackerel (*Trachurus murphyi*) and compared with traditional ice (Quitral et al., 2009). A marked antioxidant effect could be detected with fish kept under both plant-extract icing systems for a 23-day chilling period, according to PV and TBARS formation.

Another mode of surface treatment was evaluated by Arfat et al. (2015). Seabass slices were here wrapped with fish protein isolate (FPI)/fish skin gelatin (FSG) films incorporated with 3% ZnO nanoparticles (ZnONP) (w/w, based on protein content) and 100% basil leaf essential oil (BEO) (w/w, based on protein content) during storage of 12 days at 4 °C. Lowered increases in PV and TBARS values were found in FPI/FSG-ZnO + BEO film–wrapped samples, compared with the other ones.

11.7.2 Addition of Muscle-Based Antioxidants

Peptides, primarily in the form of protein hydrolysates, but also in pure form (e.g., anserine, carnosine), have also been the subject of increasing interest as all-natural antioxidants in seafood. Mechanisms for prevention are here different, with, for example, metal chelation capacity and reducing ability dominating over radical scavenging (Jónsdóttir et al., 2015). However, in cases where crude protein hydrolysates are applied, it is important to keep in mind that many other muscle-derived antioxidants may follow the peptides, such as tocopherol, ubiquinol, ascorbic acid, uric acid, and glutathione. It can be speculated that this whole cocktail of muscle-derived antioxidants play a role for the observed effects.

11.7.2.1 Protein Hydrolysates

The many documentations of in vitro metal chelating, radical scavenging, and reducing capacities of fish protein hydrolysates, well reviewed by Najafian and Babji (2012), have led to several recent studies of fish protein hydrolysates as antioxidative ingredients in fish mince or washed fish mince enriched with fish Hb. As in the case of plant-derived extracts, hydrolysates have also in some cases been used in surface treatments.

The ability of freeze-dried grass carp protein hydrolysate (0.5%, 2%, and 4% w/w) to inhibit lipid oxidation in grass carp mince during frozen storage (−10 °C) was studied by Li et al. (2013). After 5 weeks of storage, PV, CD, and TBARS were only 51.1%, 49.7%, and 34.5%, respectively, at the 2% fortification level compared with control. At the 0.5% and 4% levels, inhibition was less efficient. The metal chelating activity and reducing power of the hydrolysate were stable from 40 °C to 100 °C and from pH 3 to 9, indicating that it should be stable during various types of seafood processing/cooking. Further, the hydrolysate could still scavenge 63.8% 1,1-diphenyl-2-picrylhydrazyl (DPPH) radical and chelate 70.5% ferrous ion in the presence of glucose, sucrose, or NaCl.

Tilapia protein hydrolysates produced from proteins made by the alkaline pH-shift process version were studied by Raghavan and Kristinsson (2008). After addition into a washed fish mince model system, the hydrolysates prepared using Cryotin F (a mixture of proteinases) were most effective in inhibiting TBARS and PV development and the hydrolysates prepared using Flavourzyme (a mixture of endo- and exo-peptidases) and Neutrase (a bacterial protease) were least effective. However, in an aqueous solution, hydrolysates prepared using Flavourzyme were most effective in scavenging DPPH radicals and in chelating ferrous ions. In general, LMW tilapia protein hydrolysates were better antioxidants than high-molecular-weight (HMW) hydrolysates (i.e., smaller peptides were most efficient). The authors stressed that the ability of different enzyme-catalyzed protein hydrolysates to scavenge DPPH radicals in vitro was not reflected in their ability to inhibit oxidation in a washed tilapia mince model system. In another study from the same group (Theodore et al., 2008), hydrolyzed pH-shift produced protein isolate from catfish was separated into hydrolysates and soluble supernatants, both then being tested for in vitro antioxidant capacity and ability to prevent Hb-mediated oxidation of washed tilapia mince. Hydrolysate samples showed higher DPPH radical scavenging ability and Fe^{3+}-reducing ability, while supernatant samples had higher metal chelating ability. In the washed tilapia mince model, the ability of catfish protein hydrolysates and their corresponding supernatants in inhibiting the formation of TBARS was very similar, and in both cases 15% and 30% hydrolysis was more efficient than 5% hydrolysis, confirming the earlier data. We (Jónsdóttir et al., 2015) recently tested the addition of cod protein hydrolysate (0.7% w/w) produced from a cod protein isolate made by the alkaline pH-shift process, into an Hb-enriched washed cod mince model or an iron-fortified cod liver oil emulsion. In vitro tests revealed high metal chelating capacity of the hydrolysate and a good efficacy in reducing peroxides and rancid odor in the cod liver oil emulsion. However, in the cod muscle model, no preventive effect was seen on rancid odor or redness loss, while PV and TBARS were delayed by only 1 day. Despite the attempt to produce a more stable protein hydrolysate by first concentrating the proteins by pH-shift processing, it had elevated TBARS values, $37.3 \pm 0.8\,\mu mol/kg$, which could have negatively affected its antioxidative power.

Among studies addressing surface treatments is the one by Qui et al. (2014). Here, silver carp protein hydrolysate was made from isopropanol-defatted carp meat using alcalase. Sierra (*Scomberomorus niphonius*) fish fillets were then immersed in a 1.5% solution of the hydrolysate, or in control solutions containing 0.5% carnosine solution, 0.02% propyl gallate or distilled water, for 1 h at 4 °C and were subsequently stored at −18 °C for 150 days. Results showed that the hydrolysate could significantly reduce CD formation, TBARS value, and fluorescent compounds during the storage and was comparable to carnosine. The hydrolysate could also prevent EPA/DHA loss and preserve endogenous α-tocopherol contents of the sierra fillets. The best protection against oxidation in this study was, however, provided by the 0.02% PG solution. Dipping in tilapia protein hydrolysates was evaluated by Dekkers et al. (2011) as a way to prevent oxidation of mahi mahi red muscle. Alkali-solubilized tilapia protein isolate was hydrolyzed using Flavourzyme to 13% degree of hydrolysis and then the whole tilapia protein hydrolysate and an ultrafiltered fraction <10 kDa were used as dip solutions for 2 or 4 min. Results showed that dip treatments significantly decreased the formation of PV and TBARS over a 90-h storage at 4 °C; no significant difference was seen between whole hydrolysate and <10-kDa fractions and between 2- and 4-min treatments.

11.7.2.2 Fish Muscle Press Juice

Attempts to include a press juice from fish muscle, for example, cod and herring muscle, as an antioxidative liquid into Hb-enriched washed fish minces have been tested in a series of studies (e.g., Undeland et al., 2003; Sannaveerappa et al., 2007), with some of the results already being outlined in the washing section. However, following the documentation that cod press juice efficiently inhibited Hb-mediated oxidation of washed cod mince, we performed a deeper investigation on the underlying mechanisms (Undeland et al., 2003). Results from painty odor development in washed cod mince are shown in Figure 11.9(A). The inhibitory substances were found to be heat stable, less effective at reduced pH, and sensitive to prestorage of the press juice. The latter could indicate the involvement of unstable nucleotides and/or reducing agents in the inhibition. The <1-kDa fractions of whole and heated press juices were found to be as inhibitory as the unfractionated press juices, pointing at LMW substances being responsible for the inhibition. This was further confirmed by the fact that inhibition by the unheated, heated, and ultrafiltered (30-kDa) press juices was lost after dialysis with a 3.5-kDa membrane. The same antioxidative properties were found in heated haddock, dab, and winter flounder muscle press juices. In a series of later studies (e.g., Sannaveerappa et al., 2007), press juice from herring muscle was also carefully evaluated for its antioxidative properties using a similar Hb-enriched washed cod mince model; see results from rancid odor development in Figure 11.9(B). Also here, whole press juice and the LMW fraction significantly extended the oxidation lag phase, whereas an HWM fraction of press juice was significantly less efficient than the whole and LMW-press juice samples. Heat had a somewhat more destructive effect on the herring aqueous antioxidants than was the case for white fish. A series of LMW compounds such as ascorbic acid, uric acid, phosphates, spermine, trimethyl amine oxide, inorganic phosphate, chloride, and 2,3-diphosphoglycerate were also tested as potential preventive candidates of press juice, but none of them alone gave the same antioxidant capacity as press juice, indicating that the specific cocktail of LMW muscle–derived molecules provided strong effect. An important role of ascorbic acid for the antioxidative activity of herring press juice was, however, proved. It has generally been an interesting notification in the studies on herring press juice that strong net antioxidative power is provided despite a significant level of Hb (up to 12 μmol/L) present in the press juice.

Following the positive results from herring press juice, a recent evaluation was performed on glazing of herring fillets with herring press juice (Cavonius and Undeland, 2011). Results showed better protection of the fillets toward PV and TBARS development during 1 year of frozen storage (−18 °C) as compared

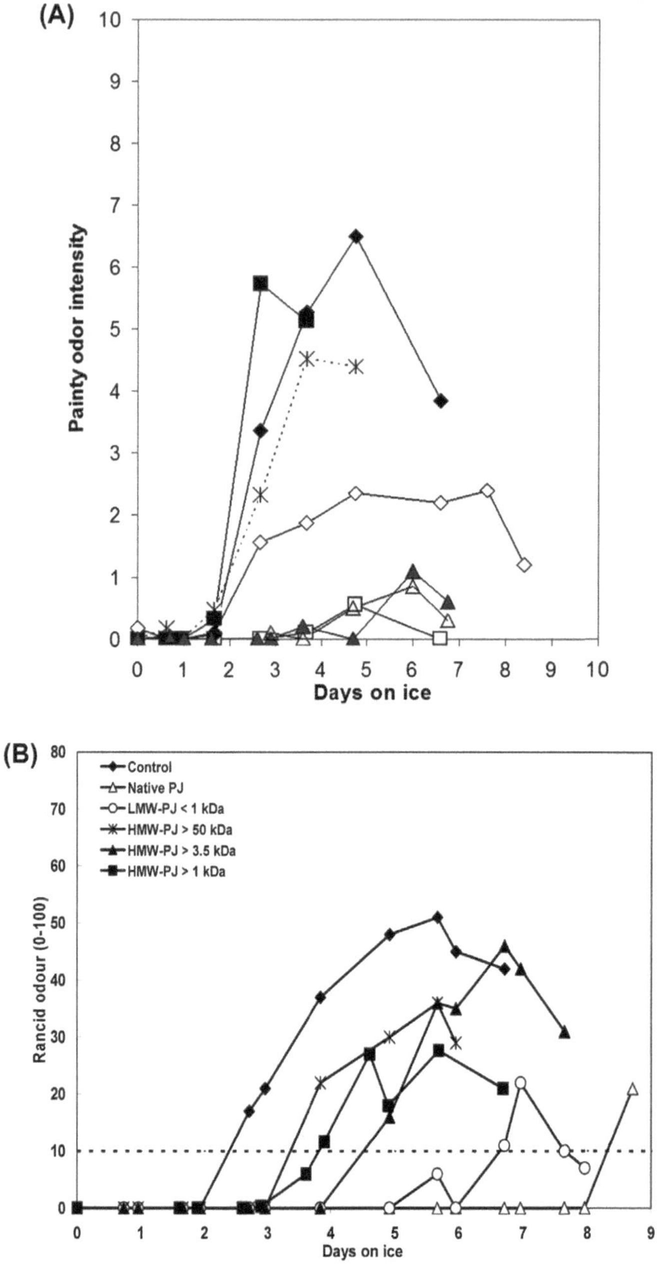

FIGURE 11.9 (A) Painty odor development in washed cod mince fortified with: 50 mM phosphate buffer (control) (*), unheated cod press juice (▲), heated cod press juice (△), unheated cod press juice stored for 48 h at 4 °C (□), >3.5-kDa fraction of unheated cod press juice (■), heated cod press juice stored for 48 h at 4 °C (◇), >3.5-kDa fraction of heated cod press juice (◆). Buffer and press juices were added at levels raising the moisture content of the extra dry washed mince (75%) to the same value as that of the unwashed mince (82%). This diluted the press juice 3.2 times. Oxidation was initiated with 6 μmol trout Hb/kg washed mince. The final pH of the model system was 6.55–6.65. *(Reprinted with permission from Undeland et al. (2003).)* (B) Rancid odor development in washed cod mince fortified with 50 mmol/L phosphate buffer (control) or different fractions of herring PJ: native PJ, LMW-PJ fraction (<1 kDa), HMW-PJ fraction (>1 kDa, >3.5 kDa, and >50 kDa). Oxidation was catalyzed by adding 15 μmol/L trout Hb. The final pH and moisture of the models were 6.3% and 81%, respectively. The dotted line in a indicates a rancid odor intensity of 10 of 100, which we used to define the rancid odor lag phase. *(Reprinted with permission from Sannaveerappa et al. (2007).)*

with glazing in water. Including a third study group glazed with phosphate buffer at pH 6.6 indicated that retaining the muscle surface pH above a level where Hb/Mb becomes oxidized might have played a role in the protection offered by press juice.

11.8 OXIDATION IN ALTERNATIVE SEAFOOD BIOMASSES: MICROALGAE, MACROALGAE, AND KRILL

Due to their high content of PUFA, antioxidants, and proteins, microalgae hold a lot of potential for nutritional and stabilizing applications. However, while a long series of studies have addressed the addition of algae-derived antioxidants in seafood to prevent their oxidation, few studies have addressed oxidative changes in the algae biomass itself, something that seriously can affect the flavor of the final product.

In one study (van Durme et al., 2013) the aroma properties of four species of marine microalgae (*Botryococcus braunii, Rhodomonas*, *Tetraselmis* species, and *Nannochloropsis oculata*) and one fresh water microalga (*Chlorella vulgaris*) were studied by correlating data on the volatile composition with sensory evaluations. Multivariate data processing revealed that microalgal samples having a seafood-like odor character contain high levels of sulfuric compounds (dimethyl disulfide, dimethyl trisulfide, and methional), diketones, α-ionone, and β-ionone. However, fresh green, fruity flavors were linked with typical aldehydes such as 2,4-alkadienals and 2,4,6-alkatrienals. According to the authors, enzymatic lipid oxidation thus explains some of the aroma formation mechanisms operating.

Several seaweeds contain significant levels of lipids: up to ~10% per dry weight (w/w). In contrast to the abundance of phospholipids in the membranes of animals and yeast, the main lipid class of algae membranes is glycoglycerolipids (GL), which are rich in ALA, EPA, and AA. Despite the high levels of PUFA, for example, brown seaweed lipids have been documented as surprisingly stable toward oxidation, something that has been ascribed both the abundance of antioxidants (e.g., fucoxanthin) and also the presence of PUFA in their GL forms. The chloroplast GL are especially important in the photosynthetic membranes. Chloroplast GL are always exposed to oxidative stress due to their exceptionally high level of PUFAs and light energy absorption for photosynthesis. To control the level of ROS and to protect the cells, seaweeds possess a number of antioxidants (ascorbate, glutathione, phenolic compounds, tocopherols, and carotenoids) and enzymes (SOD, catalase, ascorbate peroxidase, and glutathione reductase) to scavenge ROS and regenerate the active forms of antioxidants. In brown seaweed and products thereof, GL along with fucoxanthin proved to have exceptionally high stability, as shown, for example, in dried brown seaweed products (Sugimura et al., 2012). When dried *U. pinnatifida* (wakame), one of the most popular edible brown seaweeds in Japan, was stored at 50 °C, there was a slight decrease in the PUFA content. There was no increase in the PV of the extracted lipids after the incubation of the dried material for 210 days at 50 °C. Also, Prabhasankar et al. (2009) reported the sensory, chemical, and structural properties of cooked pasta containing wakame powder. The heat process involved in pasta preparation (3 h at 75 °C) did not destroy fucoxanthin or oxidize PUFA in the pasta. Also in seaweed-enriched baked scones, no oxidation of *n*-3 and *n*-6 PUFAs of brown seaweed lipids was found (Sugimura et al., 2012; Miyashita et al., 2013).

High oxidative stability of GL was also shown by Yamaguchi et al. (2012), who compared the oxidative stability of GL from brown edible seaweed (*Laminaria japonica*) with that of phosphatidylcholines (PCs) from salmon roe and TG from soybean oil or sardine oil. All the lipids were subjected to autoxidation after removing oxidants and/or endogenous antioxidants such as chlorophylls, tocopherols, and carotenoids. The oxidative stability of the lipids decreased with increasing number of bisallylic positions in the molecule. Due to the higher mean number of bisallylic positions, salmon roe PC and sardine oil TG were oxidized more rapidly than soybean oil TG. Brown edible seaweed GL showed the same oxidative stability as that found in soybean oil TG, although the mean number of bisallylic positions of both GLs was much higher than that of soybean oil TG and approached the number found in sardine oil TG and salmon roe PC. According to the authors, this study indicated the important effect of galactosyl and sulfoquinovosyl moieties for the oxidative stability of GL (Yamaguchi et al., 2012).

As found by Choe and Oh (2013), the PV and CD contents of dried laver lipids increased during storage as a_w increased from 0.11 to 0.30, 0.51, 0.75, and 0.89, whereas the contents of polyphenols, α-tocopherol, and porphyran in dried laver showed the reverse phenomenon. At an a_w of 0.51, lipid oxidation and antioxidant degradation in dried laver sharply increased. The degree of lipid oxidation of dried laver was more dependent on the concentration of α-tocopherol than that of either polyphenols or porphyran during storage in the dark. The authors suggested that the quality of dried laver can be improved by preserving tocopherols as much as possible while decreasing a_w during storage.

When dried seaweed (*Porphyra yezoensis Ueda*) was exposed to light for 24 h in a cold room (5 °C), characteristic isomers, including both the 20-carbon FA isomers 6-OOH and 17-OOH as well as the 18-carbon FA isomers 10-OOH and 12-OOH, were detected in the light-exposed sample but were not found in the control (Pan et al., 2004). These results clearly showed that singlet oxygen oxidation of lipids occurred in the seaweed exposed to light. The authors concluded that this lipid oxidation was catalyzed by chlorophyll as a photosensitizer in seaweed.

Further to algae, krill (*Eupausia superba*) has been an aquatic biomass receiving a lot of attention, not least in relation to extraction of krill oil. However, also the whole biomass is interesting from a nutritional viewpoint. Headspace volatile compounds of cold-stored and freeze-dried krill were analyzed to investigate their flavor qualities using a system combining a dynamic headspace isolator, an automatic thermal desorber, and a GC–MS (Park and Lee, 2014). Levels of PUFA oxidation products such as aldehydes, alcohols, and ketones, which are known to give seafood a rancid smell because of their low flavor threshold values, increased during cold storage of fresh krill. Notably, levels of 2-methylpropanal, 3-methylbutanal, 2-methylbutanal, and 2-butanone increased during its storage. These compounds can be considered indexes of off-odor according to freshness degradation during storage. By contrast, in freeze-dried krill powder, levels of aldehydes, ketones, and aromatic compounds decreased rapidly. Only alcohols, which did not greatly affect the food flavor, were isolated in large amounts. It was confirmed that levels of oxidized compounds of krill increased during cold storage but decreased in freeze-dried krill.

11.9 CONCLUSIONS

Lipid oxidation is one of the main threats to seafood quality since this highly complex reaction negatively affects all of the quality attributes being important for a food product: odor, taste, color, texture, and nutritional properties. The most common markers currently used to follow oxidative changes in seafood are PV for primary oxidation products and TBARS for secondary products. However, many studies have also followed color changes, CD, volatiles, and fluorescent tertiary products. The specific research question indeed dictates the type of products to be monitored and the requirement for sensitivity and precision. Intrinsic factors determining the rate of oxidation development are degree of FA unsaturation, the balance between endogenous pro- and antioxidants, and features such as pH and degree of muscle disintegration. The impact of FA unsaturation for oxidation development has in the last years reached a second peak of interest due to the great focus on replacing fish-based ingredients with plant-based ones in the feed of aquacultured fish. When it comes to main pro-oxidants in seafood, the LOX enzymes, LMW trace elements, and heme proteins have been most in focus. Current knowledge points at Hb/blood contamination of seafood tissue as the main threat to lipid oxidation progression. Met-Hb formation and subsequent heme detachment appears to be especially critical events for the onset of oxidation. On the antioxidant side, the importance of enzymes (SOD, catalase, GSH-px), membrane-bound radical scavengers (tocopherol, ubiquinol), and LMW compounds with numerous functions (e.g., ascorbate, uric acid, GSH, certain peptides, and amino acids) have been well documented over the years. Since activity and recycling of the mentioned compounds depend on energy metabolism, their preventive effect, however, ceases relatively fast postmortem, paving the way for pro-oxidants to exert their full power. Thus, short (preprocessing) storage times are highly recommended to take advantage of the endogenous antioxidant defense system in seafood products. Enhancing, for example, the tocopherol levels through addition into aquafeed has in some studies proved to delay oxidation during subsequent storage.

Since lipids and pro- and antioxidants are all unevenly distributed in seafood tissue and since oxygen access differs, lipid oxidation develops with different rates in different morphological sections. Dark muscle tissue and adjacent regions right under the skin are particularly sensitive, especially in deskinned products. Keeping the skin on for as long as possible thus serves as a valuable natural packaging. Contradicting results have been reported regarding whether cranial or caudal segments are the most sensitive to oxidation, which could be linked to species and exact cutting schemes used.

A long series of external factors can affect lipid oxidation development in fish/shellfish throughout the value-adding chain. Already the harvest/slaughter step can have a certain impact on stability; a number of studies have shown how increased stress during slaughter enhances oxidation development in the muscle and why rested harvest, including some specific anesthesia, can have a stabilizing effect. Bleeding, which at some places and for some species is regulated by law, has in a number of studies been shown to decrease oxidation during subsequent storage. However, other studies have not been able to detect any difference between bled and unbled fish. Generally, conventional bleeding only removes ≤50% blood, for example, due to fast coagulation, low blood pressure, and stress-induced contraction. In the production of surimi, washing of fish mince is the key step to concentrate myofibrillar proteins, and although one of the intentions is to stabilize the proteins from biochemical degradation, washing per se has in some cases proven negative from a lipid oxidation perspective. Several studies have reported how washing stimulates subsequent oxidation development, which appears linked to the washout of endogenous antioxidants and tentatively neutral fat, leaving behind a mince with an elevated pro-oxidant–to–membrane PL ratio. Other studies, in which the washing has been more extensive and/or antioxidants have been added to the washing water, have, however, yielded less oxidation during subsequent storage compared with crude mince. It thus is a trade-off between water consumption and degree of stabilization. Other ways of concentrating fish proteins or peptides that have gained increasing interest in the last 10 years are pH-shift processing and enzymatic hydrolysis. The former process can be carried out by either acid or alkaline protein solubilization in water. Both of these processes are unique in that they can be applied on complex raw materials like frames and whole fish, however such raw materials also bring in large levels of pro-oxidants. Acid pH-shift processing of dark muscle fish has in some studies triggered lipid oxidation, most likely due to acid activation of Hb and dilution of endogenous antioxidants into the added water. The problem has however been suppressed by fortifying the water with a mixture of antioxidants prior to pH adjustments. In hydrolysis making, the heat and pH adjustments applied might explain why oxidation has been found to increase during the hydrolysis process, especially when significant amounts of heme proteins are present. Preremoval of pro-oxidants or oxidation substrates (e.g., membranes) by washing and precipitation, respectively, has, together with antioxidant additions, aided in stabilizing lipids present in the hydrolysates.

Among strategies applied to stabilize round or filleted fish from enzymatic reactions is HP treatment. A few studies have addressed this process with respect to lipid oxidation, but results are not straightforward. HP seems to inactivate, for example, LOX, but under some circumstances the net effect from HP has been pro-oxidative. Species also appear to respond differently to this treatment. To stabilize color, for example, of tuna muscle, CO is sometimes applied. The two studies reviewed here showed a reduction in lipid oxidation following treatment of tuna and tilapia with CO. To preserve refrigerated seafood for the retail marked from microbial growth, MAP has increasingly been used. With respect to lipid oxidation, MAP has however proven problematic in cases where elevated oxygen levels (≥30%) are applied and for that reason the method becomes a tradeoff between microbial and chemical quality. To protect surfaces of seafood, for example, from oxidation and moisture loss, edible coatings and glazing are possible techniques to apply, with the latter being by far the most used one commercially. In both cases, there are numerous studies on the incorporation of antioxidants into the coating/glazing media. When applied without such additions, coatings consisting of, for example, chitosan and gelatin have delayed lipid oxidation in fish fillets and patties, and water glazes have repeatedly been shown to reduce oxidation development in frozen seafood. In studies of fortified coatings/glazes, the added antioxidative compound/extract has generally improved the oxidative stability further compared with without additions. Surface treatments thus seem to be very secure strategies to delay oxidation as long as sensory or visual aspects caused by the coating materials/additives are taken into consideration.

The link between cooking/heat treatment of seafood and lipid oxidation is complex and depends on the cooking procedure (steaming, grilling, frying, etc.), the final temperature reached in the muscle and whether the seafood item is stored again after the cooking. Water, protein, and fat loss, combined with cooking fat uptake, can completely change the system composition. Generally, enzymes are deactivated by heat, but when heme proteins denature, free hemin can trigger oxidation. Some studies have shown stabilizing effects from mild heat, but pro-oxidative effects from stronger heat. Since many oxidation products are highly heat sensitive (e.g., peroxides), the oxidation product profile will look very different immediately after cooking and after a period of storage.

The area within seafood lipid oxidation research that has increased the most in the last 10–20 years is the addition of natural antioxidants to seafood raw materials or final seafood products, with some types of applications already being mentioned. Broadly categorized, there are studies on pure compounds versus complex extracts coming either from plant or muscle sources. The largest bulk of literature within the first category is on pure phenols/polyphenols and plant extracts. Hydroxycinnamic acids like caffeic acid and flavonoids like catechins have obtained particular attention and have also proved to be highly efficient in various fish muscle systems. In addition to radical scavenging, reducing capacity, and metal chelation, deactivation of ferryl Hb and LOX inactivation have been outlined as potential mechanisms. Low galloylation and polymerization has in the case of proanthocyanidins proved to be positive. Among the examples of plant extracts highlighted in this review are, for example, those from brown seaweed, grapes (here also the fiber residue has been tested), tea, and herbs. Modes of addition have been mixing into a disintegrated product or treating a surface by dipping, icing, glazing, or coating. In almost all studies, extracts have delayed oxidation compared with controls, but relatively little attention has been paid to sensory and visual aspects of these often highly flavored ad colored extracts with strong reactivity toward proteins. Within the category of muscle-based extracts, protein hydrolysates have in several studies been mixed into minces or used in dipping solutions with good results. Smaller peptides have in some cases been more promising than larger ones. One study, however, revealed that hydrolysate additions can elevate the oxidation product levels due to inherent oxidation of the hydrolysate itself. Fish muscle press juice has also in a series of studies proved highly effective against Hb-mediated oxidation in fish mince, which has been ascribed its unique cocktail of antioxidants tailor-made for protecting muscle against oxidation. As discussed, little attention has been given sensory aspects of adding muscle-derived antioxidants.

Only a handful of studies have addressed lipid oxidation in alternative aquatic raw materials like micro-/macroalgae or krill during storage, although these biomasses have received increasing attention as novel sources of food or of food/feed/nutraceutical ingredients. Much more focus in research has been on their properties as antioxidants, being especially true for algae. From a lipid perspective, algae are, however, very interesting from the point of view that their PUFA to a large extent are bound in GL, which appear more stable against oxidation than TG and PL. Lipids in algae also coexist with powerful antioxidants as mentioned earlier. As for fish, lipid oxidation, however, appears to be involved in generation of the typical flavor profile of algae, not least the appealing notes. Susceptibility to photo-oxidation in some seaweed has been highlighted due to the abundance of chlorophyll. In the one study found on oxidation in krill biomass, fresh krill developed typical Strecker degradation products (e.g., 3-/2-methylbutanal) during cold storage, while in freeze-dried krill, aldehydes, and ketones decreased over time; only alcohols accumulated.

Research areas that deserve attention in the coming years include, for example, the development of novel strategies for efficient blood removal and heme protein stabilization, which would be particularly needed in the processing of small pelagic species, but also in the value adding of complex blood-rich seafood byproducts. On the antioxidant side, better insight into how the endogenous antioxidant defense system of seafood muscle can be preserved for a longer period of time would be of great value. Although the identification of exogenous all-natural extracts have been successfully explored in the last years, simultaneous investigations of the sensory consequences of such additions, and their effects on, for example, protein functionality and digestibility would need further attention for successful applicability in industry. To the last, the oxidation mechanisms operating in all the novel types of marine biomasses currently being explored for food must be put in focus to provide consumers with a positive first experience. Just like for fish and shellfish, the early post-harvest/postlanding handling is most likely crucial for the oxidative stability, but we need to know according to which reactions. Finally, we need to know how integrated processing—biorefining—of sensitive marine biomasses will affect oxidation and how this approach can provide us with new possibilities for stabilizing the biomass.

ACKNOWLEDGMENTS

Hanna Harrysson and Mehdi Abdollahi at Chalmers University of Technology, Unit of Food and Nutrition Science, are kindly thanked for help with the reference list.

REFERENCES

Ackman, R.G., 1979. Fish lipids. Part 1. In: Connell, U. (Ed.), Advances in Fish Science and Technology. Fishing New Books Ltd., Farham, Surrey, U.K, pp. 86–103.

Ackman, R.G., Ratnayake, W.M.N., 1992. Nonenzymatic oxidation of seafood lipids. In: Flick, G.J., Martin, R.E. (Eds.), Advances in Seafood Biochemistry. Composition and Quality. Technomic, Lancaster, pp. 245–267.

Ackman, R.G., 1992. Bioavailability of omega-3 polyunsaturated fatty acids. In: Flick, G.J., Martin, R.E. (Eds.), Advances in Seafood Biochemistry. Composition and Quality. Technomic, Lancaster, pp. 269–288.

Ahmad, W.A.W., Stone, D.A., Schuller, K.A., 2013. Dietary fish oil replacement with palm or poultry oil increases fillet oxidative stability and decreases liver glutathione peroxidase activity in barramundi (*Lates calcarifer*). Fish Physiol. Biochem. 39, 1631–1640.

Amanatidou, A., Schlüter, O., Lemkau, K., Gorris, L., Smid, E., Knorr, D., 2000. Effect of combined application of high pressure treatment and modified atmospheres on the shelf life of fresh Atlantic salmon. Innovative Food Sci. Emerging. Tech. 1, 87–98.

Aranda, R., Cai, H., Worley, C.E., Levin, E.J., Li, R., Olson, J.S., Ph Jr., G.N., Richards, M.P., 2009. Structural analysis of fish versus mammalian hemoglobins: effect of the heme pocket environment on autooxidation and hemin loss. Proteins Struct. Funct. Bioinform. 75, 217–230.

Arashisar, Ş., Hisar, O., Kaya, M., Yanik, T., 2004. Effects of modified atmosphere and vacuum packaging on microbiological and chemical properties of rainbow trout (*Oncorynchus mykiss*) fillets. Int. J. Food Microbiol. 97, 209–214.

Arfat, Y.A., Benjakul, S., Vongkamjan, K., Sumpavapol, P., Yarnpakdee, S., 2015. Shelf-life extension of refrigerated sea bass slices wrapped with fish protein isolate/fish skin gelatin-ZnO nanocomposite film incorporated with basil leaf essential oil. J. Food Sci. Technol. http://dx.doi.org/10.1007/s13197-014-1706-y.

Arts, M.J., Haenen, G.R., Voss, H.P., Bast, A., 2001. Masking of antioxidant capacity by the interaction of flavonoids with protein. Food Chem. Toxicol. 39, 787–791.

Ashton, I., 2002. Understanding lipid oxidation in fish. In: Bremner, M.A. (Ed.), Safety and Quality Issues in Fish Processing. Woodhead Publishing Ltd, Cambridge, UK, pp. 254–285.

Atsumi, T., Fujisawa, S., Tonosaki, K., 2005. A comparative study of the antioxidant/prooxidant activities of eugenol and isoeugenol with various concentrations and oxidation conditions. Toxicol. Vitro. 19, 1025–1033.

Aubourg, S.P., 2001. Fluorescence study of the pro-oxidant effect of free fatty acids on marine lipids. J. Sci. Food Agric. 81, 385–390.

Aubourg, S.P., Medina, I., 1999. Influence of storage time and temperature on lipid deterioration during cod (*Gadus morhua*) and haddock (*Melanogrammus aeglefinus*) frozen storage. J. Sci. Food Agric. 79, 1943–1948.

Aubourg, S.P., Rey-Mansilla, M., Sotelo, C.G., 1999. Differential lipid damage in various muscle zones of frozen hake (*Merluccius merluccius*). Z. Lebensm.-Unters. Forsch 208, 189–193.

St Angelo, A.J., 1996. Lipid oxidation in foods. Crit. Rev. Food Sci. Nutr. 36, 175–226.

Banerjee, S., 2006. Inhibition of mackerel (*Scomber scombrus*) muscle lipoxygenase by green tea polyphenols. Food Res. Int. 39, 486–491.

Bao, H.N.D., Ohshima, T., 2014. Strategies to minimize lipid oxidation of aquatic food products post harvest. In: Kristinsson, H.G. (Ed.), Antioxidants and Functional Components in Aquatic Foods. John Wiley & Sons Ltd., West Sussex, UK, pp. 95–125.

Baron, C.P., Andersen, H.J., 2002. Myoglobin-induced lipid oxidation. A review. J. Agric. Food Chem. 50, 3887–3897.

Baron, C.P., Hyldig, G., Jacobsen, C., 2009. Does feed composition affect oxidation of rainbow trout (*Oncorhynchus mykiss*) during frozen storage? J. Agric. Food Chem. 57, 4185–4194.

Belitz, H.-D., Grosch, W., 1987. Lipids. In: Food Chemistry. Springer, Berlin, pp. 128–200.

Benjakul, S., Bauer, F., 2001. Biochemical and physicochemical changes in catfish (*Silurus glanis* Linne) muscle as influenced by different freeze–thaw cycles. Food Chem. 72, 207–217.

Benjakul, S., Visessanguan, W., Phongkanpai, V., Tanaka, M., 2005. Antioxidative activity of caramelisation products and their preventive effect on lipid oxidation in fish mince. Food Chem. 90, 231–239.

Bensid, A., Ucar, Y., Bendeddouche, B., Özogul, F., 2014. Effect of the icing with thyme, oregano and clove extracts on quality parameters of gutted and beheaded anchovy (*Engraulis encrasicholus*) during chilled storage. Food Chem. 145, 681–686.

Bertram, H.C., Kristensen, M., Østdal, H., Baron, C.P., Young, J.F., Andersen, H.J., 2007. Does oxidation affect the water functionality of myofibrillar proteins? J. Agric. Food Chem. 55, 2342–2348.

Bizzozero, O.A., Reyes, S., Ziegler, J., Smerjac, S., 2007. Lipid peroxidation scavengers prevent the carbonylation of cytoskeletal brain proteins induced by glutathione depletion. Neurochem. Res. 32, 2114–2122.

Bonilla, J., Atarés, L., Vargas, M., Chiralt, A., 2012. Edible films and coatings to prevent the detrimental effect of oxygen on food quality: possibilities and limitations. J. Food Eng. 110, 208–213.

Buettner, G.R., 1993. The pecking order of free radicals and antioxidants: lipid peroxidation, α-tocopherol, and ascorbate. Arch. Biochem. Biophys. 300, 535–543.

Cai, H., Grunwald, E.W., Park, S.Y., Lei, B., Richards, M.P., 2013. Lipid oxidation in trout muscle is strongly inhibited by a protein that specifically binds hemin released from hemoglobin. J. Agric. Food Chem. 61, 4180–4187.

Cai, H., Richards, M.P., 2012. Site E14 in hemoglobins and myoglobins: a key residue that affects hemin loss and lipid oxidation capacity. J. Agric. Food Chem. 60, 7729–7734.

Carlsen, C.U., Møller, J.K., Skibsted, L.H., 2005. Heme-iron in lipid oxidation. Coord. Chem. Rev. 249, 485–498.

Castell, C.H., Bishop, D.M., 1969. Effect of hematin compounds on the development of rancidity in muscle of cod, flounder, scallops, and lobster. J. Fish. Res. Board Can. 26, 2299–2309.

Cavonius, L., Undeland, I., 2011. Press Juice from herring (*Clupea harengus*) as a glaze for frozen herring fillets. Poster presented at the 41st annual WEFTA (West European Fish Technologists Association) meeting. Gothenburg, Sweden, September 27–30, 2011.

Cerantola, S., Breton, F., Ar Gall, E., Deslandes, E., 2006. Co-occurrence and antioxidant activities of fucol and fucophlorethol classes of polymeric phenols in *Fucus spiralis*. Bot. Mar 49, 47–351.

Chaijan, M., Undeland, I., 2015. Development of a new method for determination of total heme protein in fish muscle. Food Chem. 173, 1133–1141.

Chaijan, M., Benjakul, S., Visessanguan, W., Faustman, C., 2005. Changes of pigments and color in sardine (*Sardinella gibbosa*) and mackerel (*Rastrelliger kanagurta*) muscle during iced storage. Food Chem. 93, 607–617.

Chaijan, M., Benjakul, S., Visessanguan, W., Faustman, C., 2006. Changes of lipids in sardine (*Sardinella gibbosa*) muscle during iced storage. Food Chem. 99, 83–91.

Chaiyapechara, S., Casten, M.T., Hardy, R.W., Dong, F.M., 2003. Fish performance, fillet characteristics, and health assessment index of rainbow trout (*Oncorhynchus mykiss*) fed diets containing adequate and high concentrations of lipid and vitamin E. Aquaculture 219, 715–738.

Chanarat, S., Benjakul, S., Xiong, Y.L., 2014. Physicochemical changes of myosin and gelling properties of washed tilapia mince as influenced by oxidative stress and microbial transglutaminase. J. Food Sci. Technol. 1–13.

Chantai, S., Ogawa, M., Tamiya, T., Tsuchiya, T., 1996. Studies on thermal denaturation of fish myoglobins using differential scanning calorimetry, circular dichroism, and tryptophan fluorescence. Fish. Sci. 62, 927–932.

Chan, K.M., Decker, E.A., Feustman, C., 1994. Endogenous skeletal muscle antioxidants. Crit. Rev. Food Sci. Nutr. 34, 403–426.

Chaouqy, N.-E., Gallardo, J.M., El Marrakchi, A., Aubourg, S.P., 2008. Lipid damage development in anchovy (*Engraulis encrasicholus*) muscle during storage under refrigerated conditions. Grasas y Aceites 59, 309–315.

Chen, Y.C., Nguyen, J., Semmens, K., Beamer, S., Jaczynski, J., 2007. Physicochemical changes in omega-3-enhanced farmed rainbowtrout (*Oncorhynchus mykiss*) muscle during refrigerated storage. Food Chem. 104, 1143–1152.

Chaiyasit, W., Elias, R.J., Mc Clements, D.J., Decker, E.A., 2007. Role of physical structures in bulk oils on lipid oxidation. Crit. Rev. Food Sci. Nutr. 47, 299–317.

Chen, Y.-C., Nguyen, J., Semmens, K., Beamer, S., Jaczynski, J., 2008. Effects of dietary alpha-tocopheryl acetate on lipid oxidation and alpha-tocopherol content of novel omega-3-enhanced farmed rainbow trout (*Oncorhynchus mykiss*) fillets. LWT - Food Sci. Technol. 41, 244–253.

Chiu, D.T.-Y., Van Den Berg, J., Kuypers, F.A., Hung, I.-J., Wei, J.-S., Liu, T.Z., 1996. Correlation of membrane lipid peroxidation with oxidation of hemoglobin variants: possibly related to the rates of hemin release. Free Radic. Biol. Med. 21, 89–95.

Choe, E., Oh, S., 2013. Effects of water activity on the lipid oxidation and antioxidants of dried laver (Porphyra) during storage in the dark. J. Food Sci. 78, 1144–1151.

Choubert, G., Baccaunaud, M., 2006. Colour changes of fillets of rainbow trout (*Oncorhynchus mykiss* W.) fed astaxanthin or canthaxanthin during storage under controlled or modified atmosphere. LWT - Food Sci. Technol. 39, 1203–1213.

Davies, K.J., Sevanian, A., Muakkassah-Kelly, S.F., Hochstein, P., 1986. Uric acid-iron ion complexes. A new aspect of the antioxidant functions of uric acid. Biochem. J. 235, 747–754.

Davies, M.J., 2005. The oxidative environment and protein damage. Biochim. Biophys. Acta Protein Proteonomics 1703, 93–109.

Davies, M.J., Truscott, R.J., 2001. Photo-oxidation of proteins and its role in cataractogenesis. J. Photochem. Photobiol. B Biol. 63, 114–125.

Decker, E.A., Hultin, H.O., 1990a. Nonenzymic catalysts of lipid oxidation in mackerel Ordinary muscle. J. Food Sci. 55, 951–953.

Decker, E.A., Hultin, H.O., 1990b. Factors influencing catalysis of lipid oxidation by the soluble fraction of mackerel muscle. J. Food Sci. 55, 947–950.

Decker, E.A., Welch, B., 1990. Role of ferritin as a lipid oxidation catalyst in muscle food. J. Agric. Food Chem. 38, 674–677.

Dekkers, E., Raghavan, S., Kristinsson, H.G., Marshall, M.R., 2011. Oxidative stability of mahi mahi red muscle dipped in tilapia protein hydrolysates. Food Chem. 124, 640–645.

Domingues, R.M., Domingues, P., Melo, T., Pérez-Sala, D., Reis, A., Spickett, C.M., 2013. Lipoxidation adducts with peptides and proteins: deleterious modifications or signaling mechanisms? J. Proteomics 92, 110–131.

Driedzic, W.R., Hochachka, P.W., 1978. Metabolism in fish during exercise. In: Hoar, W.S., Randall, D.J. (Eds.), Fish Physiology. Academic Press, New York, pp. 503–536.

Duan, J., Cherian, G., Zhao, Y., 2010. Quality enhancement in fresh and frozen lingcod (*Ophiodon elongates*) fillets by employment of fish oil incorporated chitosan coatings. Food Chem. 119, 524–532.

Dulavik, B., Sørensen, N.K., Barstad, H., Horvli, O., Olsen, R.L., 1998. Oxidative stability of frozen light and dark muscles of saithe (*Pollachius virens* l.). J. Food Lipids 5 (3), 233–245.

Erickson, M.C., Hultin, H.O., 1988. A unique role of histidine in Fe-catalyzed lipid oxidation by fish sarcoplasmic reticulum. In: Simic, M.G., Taylor, K.A., Ward, J.R., von Sonntag, E. (Eds.), Oxygen Radicals in Biology and Medicine. Plenum Press, New, York, pp. 307–312.

Erickson, M.C., 1997. Lipid oxidation: flavor and nutritional quality deterioration in frozen foods. In: Erickson, M.C., Hung, Y.C., (Eds.), Quality in Frozen Food, vol. 1. Springer US, Dordrecht, pp. 141–173.

Ernster, L., Dallner, G., 1995. Biochemical, physiologiscal and medical aspects of ubiquinone function. Biochem. Biophys. Acta 1271, 195–204.

Everse, J., Hsia, N., 1997. The toxicities of native and modified hemoglobins. Free Radic. Biol. Med. 22, 1075–1099.

Eymard, S., Baron, C.P., Jacobsen, C., 2009. Oxidation of lipid and protein in horse mackerel (*Trachurus trachurus*) mince and washed minces during processing and storage. Food Chem. 114, 57–65.

Eymard, S., Jacobsen, C., Baron, C.P., 2010. Assessment of washing with antioxidant on the oxidative stability of fatty fish mince during processing and storage. J. Agric. Food Chem. 58, 6182–6189.

Farvin, K.S., Grejsen, H.D., Jacobsen, C., 2012. Potato peel extract as a natural antioxidant in chilled storage of minced horse mackerel (*Trachurus trachurus*): effect on lipid and protein oxidation. Food Chem. 131, 843–851.

Fattouch, S., Sadok, S., Raboudi-Fattouch, F., Slama, M.B., 2008. Damage inhibition during refrigerated storage of mackerel (*Scomber scombrus*) fillets by a presoaking in quince (*Cydonia oblonga*) polyphenolic extract. Int. J. Food Sci. Technol. 43 (11), 2056–2064.

Feng, L., Jiang, T., Wang, Y., Li, J., 2012. Effects of tea polyphenol coating combined with ozone water washing on the storage quality of black sea bream (*Sparus macrocephalus*). Food Chem. 135, 2915–2921.

Flick, G.J., Hong, G.P., Knobl, G.M., 1992. Lipid oxidation of seafood during storage. In: Allen, J., Angelo, St (Eds.), Lipid Oxidation in Food, vol. 500. American Chemical Society, Washington, DC, pp. 183–207.

Frankel, E.N., 1996. Antioxidants in lipid foods and their impact on food quality. Food Chem. 57, 51–55.

Fu, X., Xu, S., Wang, Z., 2009. Kinetics of lipid oxidation and off-odor formation in silver carp mince: the effect of lipoxygenase and hemoglobin. Food Res. Int. 42, 85–90.

Fu, S.-L., Dean, R.T., 1997. Structural characterization of the products of hydroxyl-radical damage to leucine and their detection on proteins. Biochem. J. 324, 41–48.

Fänge, R., 1994. Blood cells, haemopoiesis and lymphomyeloid tissues in fish. Fish Shellfish Immunol. 4, 405–411.

Gabbianelli, R., Zolese, G., Bertoli, E., Falcioni, G., 2004. Correlation between functional and structural changes of reduced and oxidized trout hemoglobins I and IV at different pHs. Eur. J. Biochem. 271, 1971–1979.

Gabrielsson, B., Andersson, N., Undeland, I., 2014. Influence of fish consumption and some of its individual constituents on oxidative stress in cells, animals and humans. In: Kristinsson, H.G. (Ed.), Antioxidants and Functional Components in Aquatic Foods. John Wiley & Sons Ltd., West Sussex, pp. 175–217.

Gebicka, L., Banasiak, E., 2009. Flavonoids as reductants of ferryl hemoglobin. Acta Biochim. Pol. 56, 509–513.

Gennadios, A., Hanna, M.A., Kurth, L.B., 1997. Application of edible coatings on meats, poultry and seafoods: a review. LWT - Food Sci. Technol. 30, 337–350.

German, J.B., Chen, S.E., Kinsella, J.E., 1985. Lipid oxidation in fish tissue. Enzymic initiation via lipoxygenase. J. Agric. Food Chem. 33, 680–683.

German, J.B., Zhang, H., Berger, R., 1992. Role of lipoxygenases in lipid oxidation in foods. In: ACS Symposium Series, vol. 500. American Chemical Society (USA), pp. 74–92.

Gimenez, B., Roncales, P., Beltran, J.A., 2002. Modified atmosphere packaging of filleted rainbow trout. J. Sci. Food Agric. 82, 1154–1159.

Giuffrida, A., Pennisi, L., Ziino, G., Fortino, L., Valvo, G., Marino, S., Panebianco, A., 2007. Influence of slaughtering method on some aspects of quality of gilthead seabream and smoked rainbow trout. Vet. Res. Comm. 31, 437–446.

Giulivi, C., Davies, K.J., 1994. Hydrogen peroxide-mediated ferrylhemoglobin generation in vitro and in red blood cells. Meth. Enzymol. 231, 490–496.

Gómez-Estaca, J., Bravo, L., Gómez-Guillén, M.C., Alemán, A., Montero, P., 2009. Antioxidant properties of tuna-skin and bovine-hide gelatine films induced by the addition of oregano and Rosemary extracts. Food Chem. 112, 18–25.

Goulas, A.E., Kontominas, M.G., 2007. Combined effect of light salting, modified atmosphere packaging and oregano essential oil on the shelf-life of sea bream (*Sparus aurata*): biochemical and sensory attributes. Food Chem. 100, 287–296.

Gómez-Guillén, M.C., Montero García, P., 2007. Polyphenol uses in seafood conservation. Am. J. Food Technol. 2, 593–601.

Grunwald, E.W., Richards, M.P., 2006a. Studies with myoglobin variants indicate that released hemin is the primary promoter of lipid oxidation in washed fish muscle. J. Agric. Food Chem. 54, 4452–4460.

Grunwald, E.W., Richards, M.P., 2006b. Mechanisms of heme protein-mediated lipid oxidation using hemoglobin and myoglobin variants in raw and heated washed muscle. J. Agric. Food Chem. 54, 8271–8280.

Haard, N., 2002. The role of enzymes in determining seafood color, flavor and texture. In: Bremner, A. (Ed.), Safety and Quality Issues in Fish Processing. Woodhead Publishing Limited, Cambridge, pp. 220–253.

Halamíčková, A., Vorlová, L., Smutná, M., Svobodová, Z., Buchtová, H., 2003. Comparison of malondialdehyde concentrations in different muscle areas of tench *Tinca tinca* L. Fish Physiol. Biochem. 29, 305–312.

Halldorsdottir, S.M., Kristinsson, H.G., Sveinsdottir, H., Thorkelsson, G., Hamaguchi, P.Y., 2013. The effect of natural antioxidants on haemoglobin-mediated lipid oxidation during enzymatic hydrolysis of cod protein. Food Chem. 141, 914–919.

Halldorsdottir, S.M., Sveinsdottir, H., Freysdottir, J., Kristinsson, H.G., 2014a. Oxidative processes during enzymatic hydrolysis of cod protein and their influence on antioxidant and immunomodulating ability. Food Chem. 142, 201–209.

Halldorsdottir, S.M., Sveinsdottir, H., Gudmundsdottir, A., Thorkelsson, G., Kristinsson, H.G., 2014b. High quality fish protein hydrolysates prepared from by-product material with *Fucus vesiculosus* extract. J. Funct. Foods 9, 10–17.

Halliwell, B., Gutteridge, J.M., 1986. Oxygen free radicals and iron in relation to biology and medicine: some problems and concepts. Arch. Biochem. Biophys. 246, 501–514.

Halliwell, B., Gutteridge, J.M.C., 1999. Free Radicals in Biology and Medicine, third ed. Oxford University Press, New York.

Hamre, K., Torstensen, B.E., Maage, A., Waagbø, R., Berge, R.K., Albrektsen, S., 2010. Effects of dietary lipid, vitamins and minerals on total amounts and redox status of glutathione and ubiquinone in tissues of Atlantic salmon (*Salmo salar*): a multivariate approach. Brit. J. Nutr. 104, 980–988.

Harris, P., Tall, J., 1994. Substrate specificity of mackerel flesh lipopoxygenase. J. Food Sci. 59, 504–506.

Harris, P., Tall, J., 1989. Rancidity in fish. In: Rancidity in Foods. Blackie A&P, London, pp. 256–272.

Hassan, I.M., Khallaf, M.F., Abd-Al Fattah, L.E., Yasin, N.M., 1999. Quality criteria, expiration period and marketing loss estimations of pre-treated and cold stored mullet fish. Grasas y Aceites 50, 208–217.

Hsieh, R.J., Kinsella, J.E., 1989. Lipoxygenase generation of specific volatile flavor carbonyl compounds in fish tissues. J. Agric. Food Chem. 37, 279–286.

Hsu, H.H., Pan, B.S., 1996. Effects of protector and hydroxyapatite partial purification on stability of lipoxygenase from gray mullet gill. J. Agric. Food Chem. 44, 741–745.

Hultin, H.O., 1992. Lipid oxidation in fish muscle. In: Advances in Seafood Biochemistry. Composition and Quality. Technomic Publishing Company, Lancaster, Pennsylvania, pp. 99–122.

Hultin, H.O., 1994. Oxidation of lipids in seafoods. In: Seafoods Chemistry, Processing; Technology and Quality. Blackie A&P, London, pp. 49–74.

Hultin, H.O., Kelleher, S.D. Protein composition isolated from a muscle source. U.S. Patent 6,005,073, December 21, 1999.

Hultin, H.O., Kelleher, S.D. High efficiency alkaline protein extraction. U.S. Patent 6,136,959, October 24, 2000.

Hultin, H.O., Kelleher, S.D. Process for isolating a protein composition from a muscle source and protein composition. U.S. Patent 6,288,216B1, September 11, 2001.

Icekson, I., Drabkin, V., Aizendorf, S., Gelman, A., 1998. Lipid oxidation levels in different parts of the mackerel, *Scomber scombrus*. J. Aquat. Food Prod. Technol. 7, 17–29.

Iglesias, J., Medina, I., 2008. Solid-phase microextraction method for the determination of volatile compounds associated to oxidation of fish muscle. J. Chrom. 1192, 9–16.

Jacobsen, C., Undeland, I., Storrø, I., Rustad, T., Hedges, N., Medina, I., 2008. Preventing lipid oxidation in seafood. In: Børresen, T. (Ed.), Improving Seafood Products for the Consumer. Woodhead Publishing Limited, Cambridge, UK, pp. 426–453.

Jensen, C., Birk, E., Jokumsen, A., Skibsted, L.H., Bertelsen, G., 1998. Effect of dietary level of fat, R-tocopherol and astaxanthin on colour and lipid oxidation during storage of frozen rainbow trout (*Oncorhynchus mykiss*) and during chill storage of smoked trout. Z. Lebensm. Unters. Forsch. A 207, 189–196.

Jeon, Y.-J., Kamil, J.Y., Shahidi, F., 2002. Chitosan as an edible invisible film for quality preservation of herring and Atlantic cod. J. Agric. Food Chem. 50, 5167–5178.

Jittinandana, S., Kenney, P.B., Slider, S.D., Hankins, J.A., 2006. Effect of high dietary vitamin E on lipid stability of oven-cooked and hot-smoked trout fillets. J. Food Sci. 71, C130–C136.

Johnston, I.A., 1983. On the design of fish myotomal muscles. Mar. Freshwat. Behav. Physiol. 9, 83–98.

Jónsdóttir, R., Geirsdóttir, M., Hamaguchi, P., Jamnik, P., Kristinsson, H., Undeland, I., 2015. The ability of in vitro antioxidant assays to predict the efficiency of a cod protein hydrolysate and brown seaweed extract to prevent oxidation in marine food model systems. J. Sci. Food Agric. http://dx.doi.org/10.1002/jsfa.7328.

Josephson, D.B., Lindsay, R.C., Stuiber, D.A., 1983. Identification of compounds characterizing the aroma of fresh whitefish (*Coregonus clupeaformis*). J. Agric. Food Chem. 31, 326–330.

Josephson, D.B., Lindsay, R.C., Stuiber, D.A., 1984. Variations in the occurrences of enzymically derived volatile aroma compounds in salt and freshwater fish. J. Agric. Food Chem. 32, 1344–1347.

Josephson, D.B., Lindsay, R.C., Stuiber, D.A., 1987. Enzymic hydroperoxide initiated effects in fresh fish. J. Food Sci. 52, 596–600.

Kamireddy, N., Jittinandana, S., Kenney, P., Slider, S., Kiser, R., Mazik, P., Hankins, J., 2011. Effect of dietary vitamin E supplementation and refrigerated storage on quality of rainbow trout fillets. J. Food Sci. 76, S233–S241.

Kanner, J., Harel, S., 1985. Initiation of membranal lipid peroxidation by activated metmyoglobin and methemoglobin. Arch. Biochem. Biophys. 237, 314–321.

Kanner, J., Kinsella, J.E., 1983. Lipid deterioration initiated by phagocytic cells in muscle foods: beta-carotene destruction of a myeloperoxidase-hydrogen peroxide-halide system. J. Agric. Food Chem. 31, 370–376.

Kanner, J., Sofer, F., Harel, S., Doll, L., 1988. Antioxidant activity of ceruloplasmin in muscle membrane and in situ lipid peroxidation. J. Agric. Food Chem. 36, 415–417.

Karlsdottir, M.G., Sveinsdottir, K., Kristinsson, H.G., Villot, D., Craft, B.D., Arason, S., 2014c. Effect of thermal treatment and frozen storage on lipid decomposition of light and dark muscles of saithe (*Pollachius virens*). Food Chem. 164, 476–484.

Karlsdottir, M.G., Sveinsdottir, K., Kristinsson, H.G., Villot, D., Craft, B.D., Arason, S., 2014b. Effects of temperature during frozen storage on lipid deterioration of saithe (*Pollachius virens*) and hoki (*Macruronus novaezelandiae*) muscles. Food Chem 156, 234–242.

Karlsdottir, M.G., Petty, H.T., Kristinsson, H.G., 2014a. Oxidation in aquatic foods and analysis methods. In: Kristinsson, H.G. (Ed.), Antioxidants and Functional Components in Aquatic Foods. John Wiley & Sons Ltd., Chichester, pp. 11–22.

Kathirvel, P., Richards, M.P., 2012. Effect of a membrane permeable metal chelator on iron and hemoglobin-mediated lipid oxidation in washed fish muscle. Food Res. Int. 48, 346–352.

Ke, P., Ackman, R.G., Linke, B., Nash, D., 1977. Differential lipid oxidation in various parts of frozen mackerel. Int. J. Food Sci. Technol. 12, 37–47.

Ke, P.J., Linke, B.A., Ackman, R.G., 1978. Acceleration of lipid oxidation in frozen mackerel fillet by pretreatement with microwave heating. J. Food Sci. 43, 38–40.

Kelleher, S.D., Silva, L.A., Hultin, H.O., Wilhelm, K.A., 1992. Inhibition of lipid oxidation during processing of washed, minced Atlantic mackerel. J. Food Sci. 57, 1103–1119.

Khalil, A.H., Mansour, E.H., 1998. Control of lipid oxidation in cooked and uncooked refrigerated carp fillets by antioxidant and packaging combinations. J. Agric. Food Chem. 46, 1158–1162.

Khantaphant, S., Benjakul, S., Ghomi, M.R., 2011. The effects of pretreatments on antioxidative activities of protein hydrolysate from the muscle of brownstripe red snapper (*Lutjanus vitta*). LWT - Food Sci. Technol. 44, 1139–1148.

Kim, J.A., Lee, J.M., Shin, D.B., Lee, N.H., 2004. The antioxidant activity and tyrosinase inhibitory activity of phlorotannins in *Ecklonia cava*. Food Sci. Biotechnol. 13, 476–480.

Köse, S., Boran, M., Boran, G., 2006. Storage properties of refrigerated whiting mince after mincing by three different methods. Food Chem. 99, 129–135.

Kostaki, M., Giatrakou, V., Savvaidis, I.N., Kontominas, M.G., 2009. Combined effect of MAP and thyme essential oil on the microbiological, chemical and sensory attributes of organically aquacultured sea bass (*Dicentrarchus labrax*) fillets. Food Microbiol. 26, 475–482.

Kristinsson, H., Liang, Y., 2006. Effect of pH-shift processing and surimi processing on Atlantic croaker (*Micropogonias undulates*) muscle proteins. J. Food Sci. 71, C304–C312.

Kristinsson, H., Ludlow, N., Balaban, M., Otwell, W., Welt, B., 2006. Muscle quality of yellowfin tuna (*Thunnus albacares*) steaks after treatment with carbon monoxide gases and filtered wood smoke. J. Aquat. Food Prod. Technol. 15, 49–67.

Kristinsson, H.G., Theodore, A.E., Demir, N., Ingadottir, B., 2005. A comparative study between acid- and alkali-aided processing and surimi processing for the recovery of proteins from channel catfish muscle. J. Food Sci. 70, 298–306.

Kühn, H., Götze, R., Schewe, T., Rapoport, S.M., 1981. Quasi-lipoxygenase activity of haemoglobin. Eur. J. Biochem. 120 (1), 161–168.

Kuribayashi, T., Kaise, H., Uno, C., Hara, T., Hayakawa, T., Joh, T., 2002. Purification and characterization of lipoxygenase from *Pleurotus ostreatus*. J. Agric. Food Chem. 50, 1247–1253.

Larsson, K.J., Undeland, I.K., 2010. Effect of caffeic acid on haemoglobin-mediated lipid and protein oxidation in washed cod mince during ice and frozen storage. J. Sci. Food Agric. 90, 2531–2540.

Lee, C.H., Krueger, C.G., Reed, J.D., Richards, M.P., 2006. Inhibition of hemoglobin-mediated lipid oxidation in washed fish muscle by cranberry components. Food Chem. 99 (3), 591–599.

Lee, S.K., Tatiyaborworntham, N., Grunwald, E.W., Richards, M.P., 2015. Myoglobin and haemoglobin-mediated lipid oxidation in washed muscle: observations on crosslinking, ferryl formation, porphyrin degradation, and haemin loss rate. Food Chem. 167, 258–263.

Levine, R.L., Garland, D., Oliver, C.N., Amici, A., Climent, I., Lenz, A.-G., Ahn, B.-W., Shaltiel, S., Stadtman, E.R., 1990. Determination of carbonyl content in oxidatively modified proteins. Meth. Enzymol. 186, 464–478.

Levine, R.L., Williams, J.A., Stadtman, E.R., Shacter, E., 1994. Carbonyl assays for determination of oxidatively modified proteins. Meth. Enzymol. 223, 346–357.

Li, X., Luo, Y., You, J., Shen, H., 2013. Stability of papain-treated grass carp (*Ctenopharyngodon idellus*) protein hydrolysate during food processing and its ability to inhibit lipid oxidation in frozen fish mince. J. Food Sci. Tech. 1–7.

Lindsay, R., 1990. Fish flavors. Food Rev. Int. 6, 437–455.

Lindsay, R.C., 1994. Flavour of fish in Seafood. In: Shahidi, F., Botta, J.R. (Eds.), Chemistry, Processing Technology and Quality, vol. 1. Springer, Dordrecht, pp. 75–84.

Liu, Y.J., Pan, B.S., 2004. Inhibition of fish gill lipoxygenase and blood thinning effects of green tea extract. J. Agric. Food Chem. 52, 4860–4864.

Livingston, D.J., Brown, W.D., 1981. The chemistry of myoglobin and its reactions [Meat pigments, food quality indices]. Food Technol. (USA).

López-Caballero, M., Gómez-Guillén, M., Pérez-Mateos, M., Montero, P., 2005. A chitosan–gelatin blend as a coating for fish patties. Food Hydrocoll. 19, 303–311.

Lund, M.N., Heinonen, M., Baron, C.P., Estevez, M., 2011. Protein oxidation in muscle foods: a review. Mol. Nutr. Food Res. 55, 83–95.

Luo, S.-W.J., 1987. NAD(P)H Dependent Lipid Peroxidation in Trout Muscle Mitochondria (Ph.D. thesis). University of Massachusetts.

Maestre, R., Pazos, M., Medina, I., 2009. Involvement of methemoglobin (MetHb) formation and hemin loss in the pro-oxidant activity of fish hemoglobins. J. Agric. Food Chem. 57, 7013–7021.

Maestre, R., Pazos, M., Medina, I., 2011. Role of the raw composition of pelagic fish muscle on the development of lipid oxidation and rancidity during storage. J. Agric. Food Chem. 59 (11), 6284–6291.

Maqsood, S., 2010. Maximized Uses of Phenolic Compounds in Retardation of Lipid Oxidation and Shelf Life Extension of Fish and Fish Products (Ph.D. thesis). Prince of Songkla University.

Maqsood, S., Benjakul, S., 2010a. Synergistic effect of tannic acid and modified atmospheric packaging on the prevention of lipid oxidation and quality losses of refrigerated striped catfish slices. Food Chem. 121, 29–38.

Maqsood, S., Benjakul, S., 2010b. Comparative studies of four different phenolic compounds on in vitro antioxidative activity and the preventive effect on lipid oxidation of fish oil emulsion and fish mince. Food Chem. 119, 123–132.

Maqsood, S., Benjakul, S., 2011a. Comparative studies on molecular changes and pro-oxidative activity of haemoglobin from different fish species as influenced by pH. Food Chem. 124, 875–883.

Maqsood, S., Benjakul, S., 2011b. Retardation of haemoglobin-mediated lipid oxidation of Asian sea bass muscle by tannic acid during iced storage. Food Chem. 124, 1056–1062.

Maqsood, S., Benjakul, S., 2011c. Effect of bleeding on lipid oxidation and quality changes of Asian sea bass (*Lates calcarifer*) muscle during iced storage. Food Chem. 124, 459–467.

Maqsood, S., Benjakul, S., Shahidi, F., 2013. Emerging role of phenolic compounds as natural food additives in fish and fish products. Crit. Rev. Food Sci. Nutr. 53, 162–179.

Maqsood, S., Benjakul, S., Kamal-Eldin, A., 2012. Haemoglobin-mediated lipid oxidation in the fish muscle: a review. Trends Food Sci. Technol. 28, 33–43.

Masniyom, P., Benjakul, S., Visessanguan, W., 2002. Shelf-life extension of refrigerated sea bass slices under modified atmosphere packaging. J. Sci. Food Agric. 82, 873–880.

Matos, E., Gonçalves, A., Nunes, M.L., Dinis, M.T., Dias, J., 2010. Effect of harvesting stress and slaughter conditions on selected flesh quality criteria of gilthead seabream (*Sparus aurata*). Aquaculture 305, 66–72.

McDonald, R., Hultin, H.O., 1987. Some characteristics of the enzymic lipid peroxidation system in the microsomal fraction of flounder skeletal muscle. J. Food Sci. 52, 15–21.

Medina, I., Saeed, S., Howell, N., 1999. Enzymatic oxidative activity in sardine (*Sardina pilchardus*) and herring (*Clupea harengus*) during chilling and correlation with quality. Eur. Food Res. Tech. 210, 34–38.

Medina, I., Gallardo, J.M., González, M.J., Lois, S., Hedges, N., 2007. Effect of molecular strycture of phenolic families as hydroxycinnamic acids and catechins on their antioxidant effectiveness in minced fish muscle. J. Agric. Food Chem. 55, 3889–3895.

Medina, I., González, M.J., Iglesias, J., Hedges, N.D., 2009. Effect of hydroxycinnamic acids on lipid oxidation and protein changes as well as water holding capacity in frozen minced horse mackerel white muscle. Food Chem. 114, 881–888.

Medina, I., Undeland, I., Larsson, K., Storrø, I., Rustad, T., Jacobsen, C., Kristinová, V., Gallardo, J.M., 2012. Activity of caffeic acid in different fish lipid matrices: a review. Food Chem. 131, 730–740.

Mendes, R., Cardoso, C., Pestana, C., 2009. Measurement of malondialdehyde in fish: a comparison study between HPLC methods and the traditional spectrophotometric test. Food Chem. 112 (4), 1038–1045.

Mesarčová, L., Marcinčák, S., Nagy, J., Popelka, P., Bača, M., Jevinová, P., Marcinčáková, D., 2013. Effect of glaze and selected herbal extracts on lipid oxidation and sensory properties of frozen Atlantic herrings (*Clupea harengus* L.). Acta Alimentaria 42, 236–244.

Milo, C., Grosch, W., 1993. Changes in the odorants of boiled trout (*Salmo fario*) as affected by the storage of the raw material. J. Agric. Food Chem. 41, 2076–2081.

Miyashita, K., Mikami, N., Hosokawa, M., 2013. Chemical and nutritional characteristics of brown seaweed lipids: a review. J. Funct. Foods 5, 1507–1517.

Mohri, S., Cho, S.Y., Endo, Y., Fujimoto, K., 1992. Linoleate 13 (S)-lipoxygenase in sardine skin. J. Agric. Food Chem. 40, 573–576.

Mohri, S., Tokuori, K., Endo, Y., Fujimoto, K., 1999. Prooxidant activities in fish skin extracts and effects of some antioxidants and inhibitors on their activities. Fish. Sci. 65, 269–273.

Morzel, M., Van De Vis, H., 2003. Effect of the slaughter method on the quality of raw and smoked eels (*Anguilla anguilla* L.). Aquacult. Res. 34, 1–11.

Murchie, L.W., Cruz-Romero, M., Kerry, J.P., Linton, M., Patterson, M.F., Smiddy, M., Kelly, A.L., 2005. High pressure processing of shellfish: a review of microbiological and other quality aspects. Innovative Food Sci. Emerging Technol. 6, 257–270.

Mourente, G., Bell, J.G., Tocher, D.R., 2007. Does dietary tocopherol level affect fatty acid metabolism in fish? Fish Physiol. Biochem. 33 (3), 269–280.

Nagababu, E., Rifkind, J.M., 2004. Heme degradation by reactive oxygen species. Antioxid. Redox Signal 6, 967–978.

Najafian, L., Babji, A., 2012. A review of fish-derived antioxidant and antimicrobial peptides: their production, assessment, and applications. Peptides 33, 178–185.

Naseri, M., Rezaei, M., Moieni, S., Hosseni, H., Eskandari, S., 2010. Effect of different precooking methods on chemical composition and lipid damage of silver carp (*Hypophthalmichthys molitrix*) muscle. Int. J. Food Sci. Technol. 45, 1973–1979.

Ng, W.-K., Bahurmiz, O.M., 2009. The impact of dietary oil source and frozen storage on the physical, chemical and sensorial quality of fillets from market-size red hybrid tilapia, *Oreochromis* sp. Food Chem. 113, 1041–1048.

Nolsøe, H., Undeland, I., 2009. The acid and alkaline solubilization process for the isolation of muscle proteins: state of the art. Food Bioprocess Technol. 2, 1–27.

O'Brien, P.J., Shen, H., McCutcheon, L.J., O'Grady, M., Byrne, P.J., Ferguson, H.W., Mirsalimi, M.S., Julian, R.J., Sargeant, J.M., Tremblay, R.R., 1992. Rapid, simple and sensitive microassay for skeletal and cardiac muscle myoglobin and hemoglobin: use in various animals indicates functional role of myohemoproteins. Mol. Cell. Biochem. 112, 45–52.

Olsen, S.H., Sørensen, N.K., Larsen, R., Elvevoll, E.O., Nilsen, H., 2008. Impact of pre-slaughter stress on residual blood in fillet portions of farmed Atlantic cod (*Gadus morhua*)—measured chemically and by visible and near-infrared spectroscopy. Aquaculture 284, 90–97.

Ojagh, S.M., Rezaei, M., Razavi, S.H., Hosseini, S.M.H., 2010. Development and evaluation of a novel biodegradable film made from chitosan and cinnamon essential oil with low affinity toward water. Food Chem. 122, 161–166.

Pan, X., Ushio, H., Ohshima, T., 2004. Photo-oxidation of lipids impregnated on the surface of dried seaweed (*Porphyra yezoensis* Ueda). Hydroperoxide distribution. J. Am. Oil Chem. Soc. 81, 765–771.

Park, J.Y., Lee, Y.B., 2014. Analysis of headspace volatile compounds in cold-stored and freeze-dried krill eupausia superba. Fish. Aqua. Sci. 17, 189–195.

Park, S.Y., Undeland, I., Sannaveerappa, T., Richards, M.P., 2013. Novel interactions of caffeic acid with different hemoglobins: effects on discoloration and lipid oxidation in different washed muscles. Meat Sci. 95, 110–117.

Passi, S., Cataudella, S., Tiano, L., Littarru, G.P., 2005. Dynamics of lipid oxidation and antioxidant depletion in Mediterranean fish stored at different temperatures. Biofactors 25, 241–254.

Pastoriza, L., Sampedro, G., Herrera, J.J., Cabo, M.L., 1996. Effect of modified atmosphere on shelf-life of iced fresh hake slices. J. Sci. Food Agric. 71, 541–547.

Pazos, M., González, M.J., Gallardo, J.M., Torres, J.L., Medina, I., 2005. Preservation of the endogenous antioxidant system of fish muscle by grape polyphenols during frozen storage. Eur. Food Res. Tech. 220, 514–519.

Pazos, M., Iglesias, J., Maestre, R., Medina, I., 2010. Structure–activity relationships of polyphenols to prevent lipid oxidation in pelagic fish muscle. J. Agric. Food Chem. 58, 11067–11074.

Pezeshk, S., Rezaei, M., Hosseini, H., 2011. Effects of turmeric, shallot extracts, and their combination on quality characteristics of vacuum-packaged rainbow trout stored at 4 ± 1 °C. J. Food Sci. 76, M387–M391.

Pivarnik, L.F., Faustman, C., Suman, S.P., Palmer, C., Richard, N.L., Ellis, P.C., DiLiberti, M., 2013. Quality assessment of commercially processed carbon monoxide-treated Tilapia fillets. J. Food Sci. 78, S902–S910.

Polat, A., Özogul, Y., Kuley, E., Özogul, F., Özyurt, G., Şimşek, A., 2013. Tocopherol content of commercial fish species as affected by microwave cooking. J. Food Biochem. 37, 381–387.

Porter, P.J., Kennish, J.M., Kramer, D.E., 1992. The effects of exsanguinations of sockeye salmon on the changes in lipid composition during frozen storage. In: Bligh, E.G. (Ed.), Seafood Science and Technology. Fishing News Books, Oxford, pp. 76–83.

Prabhasankar, P., Ganesan, P., Bhaskar, N., Hirose, A., Stephen, N., Gowda, L.R., Miyashita, K., 2009. Edible Japanese seaweed, wakame (*Undaria pinnatifida*) as an ingredient in pasta: chemical, functional and structural evaluation. Food Chem. 115, 501–508.

Qiu, C., Xia, W., Jiang, Q., 2013. High hydrostatic pressure inactivation kinetics of the endogenous lipoxygenase in crude silver carp (*Hypophthalmichthys molitrix*) extract. Int. J. Food Sci. Technol. 48, 1142–1147.

Qiu, X., Chen, S., Dong, S., 2014. Effects of silver carp antioxidant peptide on the lipid oxidation of sierra fish fillets (*Scomberomorus niphonius*) during frozen storage. J. Food Biochem. 38, 167–174.

Quitral, V., Donoso, M.L., Ortiz, J., Herrera, M.V., Araya, H., Aubourg, S.P., 2009. Chemical changes during the chilled storage of Chilean jack mackerel (*Trachurus murphyi*): effect of a plant-extract icing system. LWT - Food Sci. Technol 42, 1450–1454.

Raghavan, S., Kristinsson, H.G., 2014. Influence of processing on lipids and lipid oxidation in aquatic foods. In: Kristinsson, H.G. (Ed.), Antioxidants and Functional Components in Aquatic Foods. John Wiley & Sons Ltd., West Sussex, UK, pp. 43–94.

Raghavan, S., Kristinsson, H.G., 2008. Antioxidative efficacy of alkali-treated tilapia protein hydrolysates: a comparative study of five enzymes. J. Agric. Food Chem. 56, 1434–1441.

Raghavan, S., Hultin, H.O., 2005. Effect of carrier solvent polarity on selective incorporation of exogenous δ-tocopherol into muscle membranes. J. Muscle Foods 16, 117–125.

Reeder, B.J., Svistunenko, D.A., Sharpe, M.A., Wilson, M.T., 2002. Characteristics and mechanism of formation of peroxide-induced heme to protein cross-linking in myoglobin. Biochemistry 41, 367–375.

Ribas, L., Flos, R., Reig, L., MacKenzie, S., Barton, B., Tort, L., 2007. Comparison of methods for anaesthetizing Senegal sole (*Solea senegalensis*) before slaughter: stress responses and final product quality. Aquaculture 269, 250–258.

Richards, M.P., Dettmann, M.A., Grunwald, E.W., 2005. Pro-oxidative characteristics of trout hemoglobin and myoglobin: a role for released heme in oxidation of lipids. J. Agric. Food Chem. 53, 10231–10238.

Richards, M.P., Hultin, H.O., 2000. Effect of pH on lipid oxidation using trout hemolysate as a catalyst: a possible role for deoxyhemoglobin. J. Agric. Food Chem. 48, 3141–3147.

Richards, M.P., Hultin, H.O., 2002. Contributions of blood and blood components to lipid oxidation in fish muscle. J. Agric. Food Chem. 50, 555–564.

Richards, M.P., Hultin, H.O., 2003. Effects of added hemolysate from mackerel, herring and rainbow trout on lipid oxidation of washed cod muscle. Fish Sci. 69, 1298–1300.

Richards, M.P., Kelleher, S.D., Hultin, H.O., 1998. Effect of washing with or without antioxidants on quality retention of mackerel fillets during refrigerated and frozen storage. J. Agric. Food Chem. 46, 4363–4371.

Riggs, A., 1970. Properties of hemoglobins. In: Hoar, W.S., Randall, D.J., (Eds.), Fish Physiology. The Nervous System, Circulation, and Respiration, vol. IV. Academic Press, New York, pp. 209–252.

Romero, F.J., Ordoñez, I., Arduini, A., Cadenas, E., 1992. The reactivity of thiols and disulfides with different redox states of myoglobin. Redox and addition reactions and formation of thiyl radical intermediates. J. Biol. Chem. 267 (3), 1680–1688.

Rostamzad, H., Shabanpour, B., Kashaninejad, M., Shabani, A., 2010. Inhibitory impacts of natural antioxidants (ascorbic and citric acid) and vacuum packaging on lipid oxidation in frozen Persian sturgeon fillets. Iran. J. Fish Sci. 9, 279–292.

Ruff, N., FitzGerald, R.D., Cross, T.F., Teurtrie, G., Kerry, J.P., 2002. Slaughtering method and dietary α-tocopheryl acetate supplementation affect rigor mortis and fillet shelf-life of turbot *Scophthalmus maximus* L. Aquacult. Res. 33 (9), 703–714.

Ryter, S.W., Tyrrell, R.M., 2000. The heme synthesis and degradation pathways: role in oxidant sensitivity: heme oxygenase has both pro- and antioxidant properties. Free Radic. Biol. Med. 28, 289–309.

Sae-leaw, T., Benjakul, S., Gokoglu, N., Nalinanon, S., 2013. Changes in lipids and fishy odour development in skin from Nile tilapia (*Oreochromis niloticus*) stored in ice. Food Chem. 141, 2466–2472.

Saeed, S., Howell, N.K., 2001. 12-Lipoxygenase activity in the muscle tissue of Atlantic mackerel (*Scomber scombrus*) and its prevention by antioxidants. J. Sci. Food Agric. 81, 745–750.

Sakai, T., Ohtsubo, S., Minami, T., Terayama, M., 2006. Effect of bleeding on hemoglobin contents and lipid oxidation in the skipjack muscle. Biosci. Biotechnol. Biochem. 70, 1006–1008.

Sakai, T., Nasu, A., Habiro, A., 1998. Changes in protein carbonyls and malonaldehyde contents in stored fish meat. Fish Sci. 64, 495–496.

Sampels, S., Åsli, M., Vogt, G., Mørkøre, T., 2010. Berry marinades enhance oxidative stability of herring fillets. J. Agric. Food Chem. 58, 12230–12237.

Samson, E., Stodolnik, L., Zamrażania, W., Solenia, O., Aktywność, N., Tkanki, L., Ikry, M.I., 2001. Effect of freezing and salting on the activity of lipoxygenase of the muscle tissue and roe of Baltic herring. Acta Ichthyologica et Piscatoria 31.

Sánchez-Alonso, I., Borderías, J., Larsson, K., Undeland, I., 2007. Inhibition of hemoglobin-mediated oxidation of regular and lipid-fortified washed cod mince by a white grape dietary fiber. J. Agric. Food Chem. 55, 5299–5305.

Sannaveerappa, T., Cai, H., Richards, M.P., Undeland, I., 2014. Factors affecting the binding of trout HbI and HbIV to washed cod mince model system and their influence on lipid oxidation. Food Chem. 143, 392–397.

Sannaveerappa, T., Carlsson, N.-G., Sandberg, A.-S., Undeland, I., 2007. Antioxidative properties of press juice from herring (*Clupea harengus*) against hemoglobin (Hb) mediated oxidation of washed cod mince. J. Agric. Food Chem. 55, 9581–9591.

Sante-Lhoutellier, V., Aubry, L., Gatellier, P., 2007. Effect of oxidation on in vitro digestibility of skeletal muscle myofibrillar proteins. J. Agric. Food Chem. 55, 5343–5348.

Secci, G., Parisi, G., Dasilva, G., Medina, I., 2016. Stress during slaughter increases lipid metabolites and decreases oxidative stability of farmed rainbow trout (*Oncorhynchus mykiss*) during frozen storage. Food Chem. 190, 5–11.

Shantha, N.C., Decker, E.A., 1994. Rapid, sensitive, iron-based spectrophotometric methods for determination of peroxide values of food lipids. J. AOAC Int. 77, 421–424.

Shahidi, F., 2001. Headspace volatile aldehydes as indicators of lipid oxidation in foods. In: Rouseff, R.L., Cadwallader, K.R. (Eds.), Advances in Experimental Medicine and Biology, vol. 488. Plenum Publisher, New York, pp. 113–114.

Shaviklo, G.R., Thorkelsson, G., Arason, S., Sveinsdottir, K., 2012. Characteristics of freeze-dried fish protein isolated from saithe (*Pollachius virens*). J. Food Sci. Technol. 49, 309–318 2012.

Shewfelt, R.L., Hultin, H.O., 1983. Inhibition of enzymic and non-enzymic lipid perox-idation of flounder muscle sarcoplasmic reticulum by pretreatement with phospholipase A_2. Biochim. Biophys. Acta Gen. Subj. 751, 432–438.

Slabyj, B.M., Hultin, H.O., 1982. Lipid peroxidation by microsomal fractions isolated from light and dark muscles of herring (*Clupea harengus*). J. Food Sci. 47, 1395–1398.

Slabyj, B.M., Hultin, H.O., 1984. Oxidation of a lipid emulsion by a peroxidizing microsomal fraction from herring muscle. J. Food Sci. 49, 1392–1393.

Sohn, J.H., Ushio, H., Ishida, N., Yamashita, M., Terayama, M., Ohshima, T., 2007. Effect of bleeding treatment and perfusion of yellowtail on lipid oxidation in post-mortem muscle. Food Chem. 104, 962–970.

Sohn, J.H., Ohshima, T., 2010. Control of lipid oxidation and meat color deterioration in skipjack tuna muscle during ice storage. Fish Sci. 76, 703–710.

Soldatov, A.A., 2006. Organ blood flow and vessels of microcirculatory bed in fish. J. Evol. Biochem. Physiol. 42, 243–252.

Stodolnik, L., Stawicka, A., Szczepanik, G., Aubourg, S.P., 2005. Rancidity inhibition study in frozen whole mackerel (*Scomber scombrus*) following flaxseed (*Linum usitatissimum*) extract treatment. Grasas y Aceites 56, 198–204.

Syväoja, E.-L., Salminen, K., Piironen, V., Varo, P., Kerojoki, O., Koivistoinen, P., 1985. Tocopherols and tocotrienols in Finnish foods: fish and fish products. J. Am. Oil Chem. Soc. 62, 1245–1248.

Sugimura, R., Suda, M., Sho, A., Takahashi, T., Sashima, T., Abe, M., Miyashita, K., Kazuo, M., 2012. Stability of fucoxanthin in dried Undaria pinnatifida (Wakame) and baked products (Scones) containing wakame powder. Food Sci. Tech. Res. 18, 687–693.

Swoboda, P.A., Peers, K.E., 1977. Metallic odour caused by vinyl ketones formed in the oxidation of butterfat. The identification of octa-1,cis-5-dien-3-one. J. Sci. Food Agric. 28, 1019–1024.

Tatiyaborworntham, N., Richards, M.P., January 2015. The effect of phospholipase A_2 from pig pancreas on lipid oxidation in minced muscles from fish, poultry and pork. Natl. Provis. 22–24.

Tejada, M., Huidobro, A., 2002. Quality of farmed gilthead seabream (*Sparus aurata*) during ice storage related to the slaughter method and gutting. Eur. Food Res. Tech. 215, 1–7.

Thanonkaew, A., Benjakul, S., Visessanguan, W., Decker, E.A., 2006. The effect of metal ions on lipid oxidation, colour and physicochemical properties of cuttlefish (*Sepia pharaonis*) subjected to multiple freeze–thaw cycles. Food Chem. 95, 591–599.

Theodore, A.E., Raghavan, S., Kristinsson, H.G., 2008. Antioxidative activity of protein hydrolysates prepared from alkaline-aided channel catfish protein isolates. J. Agric. Food Chem. 56, 7459–7466.

Thiansilakul, Y., Benjakul, S., Park, S.Y., Richards, M.P., 2012. Characteristics of myoglobin and haemoglobin-mediated lipid oxidation in washed mince from bighead carp (*Hypophthalmichthys nobilis*). Food Chem. 132, 892–900.

Thiansilakul, Y., Benjakul, S., Richards, M.P., 2011. Effect of myoglobin from Eastern little tuna muscle on lipid oxidation of washed Asian sea bass mince at different pH conditions. J. Food Sci. 76, 242–249.

Tichivangana, J.Z., Morrissey, P.A., 1982. Lipid oxidation in cooked fish muscle. Int. J. Food Sci. Technol. 6, 157–163.

Tichivangana, J.Z., Morrissey, P.A., 1984. Factors influencing lipid oxidation in heated fish muscle systems. Ir. J. Food Sci. Technol. 8, 47–57.

Timm-Heinrich, M., Eymard, S., Baron, C.P., Nielsen, H.H., Jacobsen, C., 2013. Oxidative changes during ice storage of rainbow trout (*Oncorhynchus mykiss*) fed different ratios of marine and vegetable feed ingredients. Food Chem. 136, 1220–1230.

Tokur, B., Ozkütük, S., Atici, E., Ozyurt, G., Ozyurt, C.E., 2006. Chemical and sensory quality changes of fish fingers, made from mirror carp (*Cyprinus carpio* L. 1758), during frozen storage (−18 °C). Food Chem. 99, 335–341.

Tretsven, W.I., Patten, B.G., 1981. Effect of arterial incisions on the amount of bleeding and flesh quality of rainbow trout. Mar. Fish Rev. 43, 16–18.

Tseng, Y.C., Xiong, Y.L., Webster, C.D., 2005. The preservation of the quality of the muscle in frozen Australian red claw crayfish (*Cherax quadricarinatus*) by pre-storage anti-oxidant dipping treatments. Int. J. Food Sci. Technol. 40, 841–848.

Tuckey, N.P., Forster, M.E., Gieseg, S.P., 2009. Lipid oxidation is inhibited by isoeugenol exposure in Chinook salmon (*Oncorhynchus tshawytscha*) fillets during storage at 15 °C. J. Food Sci. 74, C333–C338.

Undeland, I., Ekstrand, B., Lingnert, H., 1997. Lipid oxidation in herring (*Clupea harengus*) light muscle, dark muscle and skin, stored separately or as intact fillets. J. Am. Oil Chem. Soc. 75 (5), 581–590.

Undeland, I., Ekstrand, B., Lingnert, H., 1998a. Lipid oxidation in minced herring (*Clupea harengus*) during frozen storage. Effect of washing and precooking. J. Agric. Food Chem. 46, 2319–2328.

Undeland, I., Stading, M., Lingnert, H., 1998b. Influence of skinning on lipid oxidation in different horizontal layers of herring (*Clupea harengus*) during frozen storage. J. Sci. Food Agric. 78, 441–450.

Undeland, I., Hall, G., Lingnert, H., 1999. Lipid oxidation in fillets of herring (*Clupea harengus*) during ice storage. J. Agric. Food Chem. 47, 524–532.

Undeland, I., Hall, G., Wendin, K., Gangby, I., Rutgersson, A., 2005. Preventing lipid oxidation during recovery of functional proteins from herring (*Clupea harengus*) fillets by an acid solubilization process. J. Agric. Food Chem. 53, 5625–5634.

Undeland, I., Hultin, H.O., Richards, M.P., 2002. Added triacylglycerols do not hasten hemoglobin-mediated lipid oxidation in washed minced cod muscle. J. Agric. Food Chem. 50, 6847–6853.

Undeland, I., Hultin, H.O., Richards, M.P., 2003. Aqueous extracts from some muscles inhibit hemoglobin-mediated oxidation of cod muscle membrane lipids. J. Agric. Food Chem. 51, 3111–3119.

Undeland, I., Kristinsson, H.G., Hultin, H.O., 2004. Hemoglobin-mediated oxidation of washed minced cod muscle phospholipids: effect of pH and hemoglobin source. J. Agric. Food Chem. 52, 4444–4451.

Van Der Zee, J., Barr, D.P., Mason, R.P., 1996. ESR spin trapping investigation of radical formation from the reaction between hematin and tert-butyl hydroperoxide. Free Radic. Biol. Med. 20, 199–206.

Van Durme, J., Goiris, K., De Winne, A., De Cooman, L., Muylaert, K., 2013. Evaluation of the volatile composition and sensory properties of five species of microalgae. J. Agric. Food Chem. 61, 10881–10890.

Vázquez, M., Torres, J.A., Gallardo, J.M., Saraiva, J., Aubourg, S.P., 2013. Lipid hydrolysis and oxidation development in frozen mackerel (*Scomber scombrus*): effect of a high hydrostatic pressure pre-treatment. Innovative Food Sci. Emerg. Technol. 18, 24–30.

Waagbo, R., Hamre, K., Bjerkas, E., Berge, R., Wathne, E., Lio, O., Torstensen, B., 2003. Cataract formation in Atlantic salmon, *Salmo salar* L, smolt relative to dietary pro- and antioxidants and lipid level. J. Fish Dis. 26, 213–229.

Wang, X., Quinn, P.J., 1999. The effect of α-tocopherol on the thermotropic phase behaviour of dipalmitoylphosphatidylethanolamine. Eur. J. Biochem. 264 (1), 1–8.

Wang, J.H., Liu, D.M., Zheng, A.N., Liu, Y.L., Chen, Q., Wang, F.X., Li, X.H., Yu, J., 2012. Preliminary purification and distribution of lipoxygenase in grass carp. Adv. Mater. Res. 550, 1468–1472.

Wang, T., Jónsdóttir, R., Kristinsson, H.G., Thorkelsson, G., Jacobsen, C., Hamaguchi, P.Y., Ólafsdóttir, G., 2010. Inhibition of haemoglobin-mediated lipid oxidation in washed cod muscle and cod protein isolates by *Fucus vesiculosus* extract and fractions. Food Chem. 123, 321–330.

Whang, K., Peng, I.C., 1988. Electron paramagnetic resonance studies of the effectiveness of myoglobin and its derivatives as photosensitizers in singlet oxygen generation. J. Food Sci. 53 (6), 1863–1865.

Weber, J., Bochi, V.C., Ribeiro, C.P., Victório, A.d. M., Emanuelli, T., 2008. Effect of different cooking methods on the oxidation, proximate and fatty acid composition of silver catfish (*Rhamdia quelen*) fillets. Food Chem. 106, 140–146.

Weers, J.J., Thomason, C.E., 1990. Internal Publication. Petrolite Corp. Research and Development, St Louis, USA.

Wetterskog, D., Undeland, I., 2004. Loss of redness (a*) as a tool to follow hemoglobin-mediated lipid oxidation in washed cod mince. J. Agric. Food Chem. 52, 7214–7221.

Wilson, R.R., Knowles, F.C., 1987. Temperature adaptation of fish hemoglobins reflected in rates of autoxidation. Arch. Biochem. Biophys. 255, 210–213.

Winterbourn, C.C., 1990. Oxidative reactions of hemoglobin. Meth. Enzym. 186, 265–272.

Wongwichian, C., Klomklao, S., Panpipat, W., Benjakul, S., Chaijan, M., 2015. Interrelationship between myoglobin and lipid oxidations in oxeye scad (Selar boops) muscle during iced storage. Food Chem. 174, 279–285.

Xie, Z., Baba, S.P., Sweeney, B.R., Barski, O.A., 2013. Detoxification of aldehydes by histidine-containing dipeptides: from chemistry to clinical implications. Chem. Biol. Interact. 202, 288–297.

Xing, Y., Yoo, Y., Kelleher, S.D., Nawar, W.W., Hultin, H.O., 1993. Lack of changes in fatty acid composition of mackerel and cod during iced and frozen storage. J. Food Lipids 1, 1–14.

Yagiz, Y., Kristinsson, H., Balaban, M., Marshall, M., 2007. Effect of high pressure treatment on the quality of rainbow trout (*Oncorhynchus mykiss*) and mahi mahi (*Coryphaena hippurus*). J. Food Sci. 72, C509–C515.

Yamaguchi, T., Sugimura, R., Shimajiri, J., Suda, M., Abe, M., Hosokawa, M., Miyashita, K., 2012. Oxidative stability of glyceroglycolipids containing polyunsaturated fatty acids. J. Oleo Sci. 61, 505–513.

Yarnpakdee, S., Benjakul, S., Nalinanon, S., Kristinsson, H.G., 2012a. Lipid oxidation and fishy odour development in protein hydrolysate from Nile tilapia (*Oreochromis niloticus*) muscle as affected by freshness and antioxidants. Food Chem. 132, 1781–1788.

Yarnpakdee, S., Benjakul, S., Kristinsson, H.G., Maqsood, S., 2012b. Effect of pretreatment on lipid oxidation and fishy odour development in protein hydrolysates from the muscle of Indian mackerel. Food Chem. 135, 2474–2482.

Yarnpakdee, S., Benjakul, S., Kristinsson, H.G., 2012c. Effect of pretreatments on chemical compositions of mince from Nile tilapia (*Oreochromis niloticus*) and fishy odor development in protein hydrolysate. Int. Aquatic Res. 4, 1–16.

Yarnpakdee, S., Benjakul, S., Kristinsson, H.G., 2014. Lipid oxidation and fishy odour in protein hydrolysate derived from Nile tilapia (*Oreochromis niloticus*) protein isolate as influenced by haemoglobin. J. Sci. Food Agric. 94, 219–226.

Yerlikaya, P., Gokoglu, N., 2010. Inhibition effects of green tea and grape seed extracts on lipid oxidation in bonito fillets during frozen storage. Int. J. Food Sci. Technol. 45, 252–257.

Younathan, M.T., Watts, B.M., 1959. Relationship of meat pigments to lipid oxidation. Food Res. 24, 728–733.

Zamora, R., Hidalgo, F.J., 2005. Coordinate contribution of lipid oxidation and Maillard reaction to the nonenzymatic food browning. Crit. Rev. Food Sci. Nutr. 45, 49–59.

Zhang, J., Wu, D., Liu, D., Fang, Z., Chen, J., Hu, Y., Ye, X., 2013. Effect of cooking styles on the lipid oxidation and fatty acid composition of grass carp (*Ctenopharynyodon idellus*) fillet. J. Food Biochem. 37, 212–219.

Zweytick, D., Athenstaedt, K., Daum, G., 2000. Intracellular lipid particles of eukaryotic cells. Biochim. Biophys. Acta 1469, 101–120.

Chapter 12

Oxidative Stability and Shelf Life of Crackers, Cookies, and Biscuits

Etsehiwot Gebreselassie and Hall Clifford
Department of Plant Sciences, North Dakota State University, Fargo, ND, USA

12.1 APPLICATION OF INFORMATION FROM CHAPTER

Cookies, crackers, and biscuits are made primarily with solid fats; however, there are still sufficient unsaturated fatty acids in these products that can undergo oxidation, resulting in an unacceptable product. A variety of methods are available to assess the oxidative stability of food products. A combination of primary (e.g., peroxide value) and secondary (e.g., headspace volatiles) oxidation product indicators should be used to assess oxidation of products. These methods coupled with sensory evaluations can provide an excellent predictor of shelf life stability. The oxidation evaluation should be conducted using multiple approaches. The approach to assessing shelf life includes storage of the products under various temperatures (20, 38, and 50 °C) and relative humidities (25%, 45%, and 65%). Storage of the product in the traditional packaging (i.e., material, gas flush, etc.) provides information relevant to the oxidation that occurs during storage in the package while assessing the shelf life of the product stored in open containers can provide information about product stability once the package is open, similar to what consumers would do once the product is purchased, stored, and opened for several days at home. Challenging the antioxidant systems during product development is necessary if one expects that the antioxidant will function correctly in a consumer product. Traditional synthetic antioxidants BHA and BHT have excellent carry-through properties and provide excellent oxidation protection for cookies, crackers, and biscuits. These products are suitable in neutral flavored products since these antioxidants carry little flavor. However, the move away from synthetic antioxidants has led to the incorporation of a multitude of natural antioxidants. The use of plant extracts that have a strong flavor may not be problematic in products that carry a flavoring (e.g., herb flavored cracker) since the product itself has a flavor.

Oxidative Stability and Shelf Life of Foods Containing Oils and Fats
http://dx.doi.org/10.1016/B978-1-63067-056-6.00012-4

In products with neutral flavors the use of flavored plant extracts is problematic from a flavor standpoint. The use of deodorized plant extracts or neutral flavored products, such as tocopherols could be viable options. The formation of Maillard reaction products (MRP) during baking also provides additional benefit from an antioxidant perspective.

12.2 INTRODUCTION TO CRACKERS, COOKIES, AND BISCUITS

The U.S. market for bakery goods had reached a value of US$67.3 billion in the year 2010, and it is expected to reach US$77.2 billion by 2016 (Anonymous, 2012). Cereal-based products are widely consumed and accepted by different age groups. They are also good sources of energy due to their high contents of carbohydrates, fats, and low water content (Skrbic and Cvejanov, 2011). In addition, these baked goods are being used to deliver nutrients, such as vitamins, minerals, phenolic compounds, and dietary fibers to consumers (Skrbic and Cvejanov, 2011). Apart from the nutrition, consumers expect their foods to remain safe and be of high quality at time of purchase, storage, and consumption (Kilcast and Subramaniam, 2000). To maintain quality, safety, and guarantee the continued growth of these baked goods, one must understand the factors that affect the stability/shelf life of these products.

Before proceeding with the main content of this chapter, the terms used for baked goods will be defined to avoid confusion. Different terms are used in different countries. The term biscuit in the United Kingdom and Europe refers generically to both cookies and crackers of various forms (Manley, 2008). Biscuit can be sweet or savory. Confectionary biscuits are generally sweet while savory biscuits are unsweetened but may have added salt or other seasoning for flavor. However, in the United States, Canada, and elsewhere, the word "biscuit" refers to a quick or leavened bread-like product, such as scones and American biscuits that are not typically sweet. Thin, salty, and unsweetened products are referred to as "crackers." The word "cookie," on the other hand, refers to a small cake made from a sweet-dough, which can have a texture, after baking, ranging from chewy to a dry-crunchy (Manley, 2008; Gwirtz et al., 2007). In this chapter, the terms used for baked goods refer to the U.S. definition.

Biscuits, crackers, and cookies are principally made from cereal flours. Some common cereal flours include wheat, barley, corn, and oats; however, wheat flour remains the most common flour (Walker and Li, 2008). The food industry trend to use whole grains has improved the nutrient content of these products. However, the use of whole grains can also contribute to increased product oxidation.

Wheat flour is the principal raw material used in the production of biscuits, crackers, and cookies (Gwirtz et al., 2007). One reason for this is due to the presence of two proteins in wheat, gliadin, and glutenin, which form gluten when combined with water during dough development. Gluten is a viscoelastic protein that contributes to the desired texture of crackers and cookies (Gwirtz et al., 2007). However, excess gluten can negatively impact product texture. Most biscuits, crackers, and cookies are made from soft-wheat flours with protein contents of 8.0–9.5%, 9–10%, and 7.5–9.5%, respectively. Use of such flours with lower protein content enhances spreading of these baked goods (Gwirtz et al., 2007). In parts of the world where wheat is not a major crop, use of composite flours is common (Seibel, 2007). The composite flours can contain wheat flour and native plant flours, such as cassava, soy, corn, rice, and buckwheat.

Other than flour, other essential ingredients are added into the baking process. For instance, the ingredients used to make biscuits include flour, high levels of fat/lipid, and low amounts of sugar, water, salt, and chemical leavening (Gwirtz et al., 2007). After mixing these ingredients, the dough is sheeted, cut into a desired shape, and baked. Traditionally, plastic fats such as shortening are used to achieve a flaky texture in biscuits (Daglioglu et al., 2004; Gwirtz et al., 2007). In addition, using a high fat content gives the biscuit crust a shiny appearance. To make crackers, flour, water, shortening, and yeast are mixed to form a sponge, which can sometimes be fermented, and then sheeted, layered, laminated, and cut into a desired shape. Cookies are baked products of dough with low water content, and higher amounts of sugar and shortening (Gwirtz et al., 2007). The way in which the lipid is incorporated into the products can impact stability. Crackers and biscuits generally use a process whereby the shortening is laminated, i.e., layered between dough layers, while in cookies the fat is mixed into the dough prior to sheeting and forming. The physical protection of the fat by protein and starch network will be different among the crackers, biscuits, and cookies. In general, the oxidative stability of finished products can be attributed to a number of factors such as ingredients used in the formulation, processing method, addition of antioxidants, and packaging method.

12.3 ROLE OF INGREDIENTS IN PRODUCT OXIDATIVE STABILITY

12.3.1 Flour

In order to detect deterioration in foods, multiple techniques can be applied. Sensory is one common technique used to quantify eating quality (e.g., appearance, aroma, flavor, and texture) using trained panels (Kilcast and Subramaniam, 2000). Peroxide values, free fatty acids, anisidine value, and volatile markers such as hexanal and pentanal are used to assess oxidative and hydrolytic rancidity (Kilcast and Subramaniam, 2000). For raw and processed cereals, volatile markers, more specifically hexanal is used as a good indicator of oxidation (Fritsch and Gale, 1977; Ekstrand et al., 1993; Rutgersson et al., 2000). Furthermore, the type and quality of raw material and ingredients used in formulations can affect the stability of the end product (Galić et al., 2009).

For baked goods, including biscuits, crackers, and cookies, cereal grains are used as raw materials. Knowledge of the lipid composition of the raw materials and changes that occur in the raw materials prior to use in products is beneficial in understanding oxidative stability of these cereal-based products (Hall, 2010). For instance, linoleic acid (LA) is a major fatty acid in almost all cereal grains. In wheat, corn, and barley, LA accounts 50–60% of the total lipid. Palmitic and oleic acids make up 10–25% and 12–22% of the total lipid in cereal grains, respectively (Barnes, 1983; Mano et al., 1990; Becker, 2007). Since LA makes up the majority of the total fatty acids, hexanal and pentane are good indicators of oxidation in cereal-based products (Frankel et al., 1981; Frankel, 2005).

Enzymes such as lipase and lipoxygenase are among numerous enzymes endogenous to wheat. During grain processing, disruption of cells facilitates the interaction of enzymes and lipid substrates, resulting in rancidity of the flour (Galliard, 1983). Decreases in total lipid and triacylglycerides (TAG) and increased free fatty acid (FFA) were observed during storage of wheat flour. Galliard (1983) concluded that the lipase likely caused the degradation of the TAG into FFA and that oxidation of the FFA led to carbonyl formation and thereby reduction in total lipids. Shearer and Warwick (1983) observed an increased FFA in flour over 48 months. They reported lower FFA contents in strong, i.e., high protein, flour than in weak, i.e., low protein, flour and more FFA in high moisture flour than in low moisture flour. Galliard (1983) observed an increased carbonyl index once FFA reached a maximum, and that off-flavor scores observed by panelists dramatically increased with increasing carbonyl formation. Similar to flour, a sensory panel observed that untreated wheat germ became rancid in 4 weeks; however, no rancidity was detected for 7 weeks after heat treatment (Sjovall et al., 2000). Sjovall et al. (2000) reported that the volatile compounds hexanal, α-pinene, 1-hexanol, and 3-carene formed during the storage of wheat germ likely contributed to rancidity detected by the sensory panel. Thus, the type of treatment applied to flour and wheat germ or the presence of oxidation precursor, e.g., FFA, hydroperoxides, can indirectly affect the final product by bringing oxidation precursors into the product.

Other cereal grains, such as barley, have been used for β-glucan enrichment in cookies and crackers (Casiraghi et al., 2006). However, when the traditional formulation is changed, the shelf life of the final product can also be affected. For instance, more hexanal formation was observed in treated barley, e.g., steeped and heat processed, than untreated barley (Rutgersson et al., 2000). The hexanal formation is likely caused by the enzyme lipoxygenase. Lipoxygenase is a common enzyme, in cereal grains, responsible for the oxidation of 1,4-pentadiene moieties in unsaturated fatty acids (Lehtinen et al., 2003). Thermal processing (<50 °C) is known to activate lipoxygenase, causing oxidation, and continued oxidation has also been observed during storage, likely caused by autooxidation initiated from the products of lipoxygenase (Rutgersson et al., 2000). Corn is another commonly used cereal grain. However, in some cases corn is subjected to a nixtamalization process, which involves the cooking of corn in alkali, before being ground and added into snack foods (Hall, 2010). Hexanal was found to be a good indicator of oxidation in corn flour (Lopez-Duarte and Vidal-Quintanar, 2009). Thermal processing of corn germ can inactivate lipoxygenase, but facilitates oxidation during storage; thus, the type of corn used can affect the stability of the end product (Gardner and Inglett, 1971). Rice as a raw material has become more common with the development of the gluten free industry. Shin et al. (1986) detected hexanal in cooked brown rice that was stored for 12 months at different temperatures (i.e., 5 and 35 °C). In addition, the use of rice bran can lead to hydrolytic and oxidative deterioration (Ramezanzadeh et al., 1999; Lam and Proctor, 2003). During a 50-day storage, hexanal and 2-pentylfuran were found to cause odor during the initial storage time, but near the end of the storage period, 2-nonenal was responsible for the oxidized odor (Lam and Proctor, 2003). Oats are common cereal grains used in baked goods, including biscuits, crackers, and cookies (Manley, 2008). However, oats are known to have high lipase activity, and inactivation of this enzyme is necessary to produce a shelf-stable product (Nasirulla et al., 1989). If not treated, lipid hydrolysis and bitter flavor can develop in the end product (Lehtinen et al., 2003). Molteberg et al. (1996) reported an increased level of hexanal (10-fold) and 2-pentylfuran (36-fold) in raw oat flour during a 42-day storage compared to heat treated oat flour. Although heat treatment inactivates lipoxygenase, inactivating this enzyme is not always recommended as it plays a positive role during dough development (Cumbee et al., 1997; Shiiba et al., 1991). Even after heat treatment, oat-containing products are more susceptible to rancidity than any other cereal-based products (Lehtinen et al., 2003). Lehtinen et al. (2003) observed that polar lipids in oat kernel were more susceptible to heat treatment than storage lipids. The high polar lipid content of oats compared to other cereals might account for the greater susceptibility of oat to oxidation.

Some parts of the world produce baked goods from locally grown crops. Okpala et al. (2012) were able to produce acceptable cookies made from germinated pigeon pea, fermented sorghum, and cocoyam flour. Germination is an inexpensive process that can have positive effect on the nutritional profile of cereals, pseudo cereals, and legumes. During this process, compounds (e.g., polyphenols and flavonoids) with antioxidant activities become more available, impeding oxidative deterioration (Hubner and Arendt, 2013; Alvarez-Jubete et al., 2010). For instance, germination of amaranth increased the free and bound phenolics by 206% and 226%, respectively, and total flavonoid content by 213%. Thus, the use of germinated amaranth flour in food products can be a good source of natural antioxidants. The increased level of these compounds during germination impedes lipid oxidation and unfavorable flavor changes (Maillard et al., 1996; Molteberg et al., 1996; Goupy et al., 1999).

12.3.2 Other Ingredients

Other than flours, ingredients incorporated into baked goods can also influence shelf life. For instance, fat is an essential ingredient of bakery goods, including biscuits, crackers, and cookies (Daglioglu et al., 2004). The type of fat used also plays a major role in the end product quality, including shape, sensory, and shelf life of the baked goods (Daglioglu et al., 2004; Manley, 2008). Other ingredients that are high in pro-oxidants enhance lipid oxidation. Pro-oxidants include transition metals, enzymes such as lipoxygenase, and photosensitizers. In addition, any ingredient that influences water activity can in turn affect lipid oxidation.

Lipid or fat sources include plastic fats (e.g., shortening or butter), flour oils, and dairy products added during the baking process (Daglioglu et al., 2004). For instance, increased temperature can cause solid fats to melt, and fats in the liquid state act as solvents, facilitating chemical reactions that include oxidation (Kilcast and Subramaniam, 2000). The polyunsaturated fatty acid (PUFA) of the melted fats become more accessible compared to their presence in solid form and will undergo greater oxidation resulting in higher peroxide values and unfavorable sensory qualities (Daglioglu et al., 2004). Kaddour et al. (2006) found that a healthy cracker formula with 10% rapeseed oil oxidized significantly more than a cracker with 20% saturated palm oil. Biscuit (i.e., short bread) fortified with high oleic sunflower oil had initially high peroxide values (>20 mEq/kg) that appeared to level off (35 mEq/kg) after 90 days of storage (Marconi et al., 2014). Use of saturated fats appear ideal from the oxidative stability perspective, but such substitution affects dietary saturated fat intact.

Frampton (1994) noted the achievement of oxidative stability through hydrogenation of unsaturated fatty acids. Although an ideal solution to the food manufacturer, consumption of hydrogenated fats and *trans* fats resulting from hydrogenation are linked to cardiovascular diseases (Katan et al., 1994). Recently, palm oil and palm fractions have replaced hydrogenated vegetable oils and are being used as fats for dough (Talbot, 2010). Rossell (1994) reported that palm oil was more shelf-stable than traditional fats, including coconut oil and beef tallow. The stability of baked goods can also be compromised by using ingredients with high lipid content. For instance, the high lipid content in oats decreases the shelf life of many oat-based products (Cognat et al., 2014).

Use of confectionery coatings and fillings in baked goods can influence shelf life. For example, chocolate is often integrated into cookies, biscuits, and crackers. Chocolate is mainly composed of cocoa butter, which contains both saturated (65%) and polyunsaturated (3%) fats (Talbot, 2010). Due to high saturation and presence of natural antioxidants (e.g., tocopherols), cocoa butter is very stable to oxidation. On the other hand, milk fat has similar degree of unsaturation as cocoa butter, but their fatty acid composition differs. The presence of short chain fatty acids in milk fat makes it more susceptible to oxidation than cocoa butter. Talbot (2010) also observed lower antioxidant levels in milk fat than cocoa butter. For producing bakery goods using milk chocolate, the presence of pro-oxidants, such as metals should be minimized (Talbot, 2010). In addition, milk chocolate can be a source of stale flavor in bakery goods. Subramaniam (2009) observed increased staleness sensory scores in milk chocolate stored over a 13-week period. However, this staleness could also come from oxidation products catalyzed via lipases and lipoxygenases (Afoakwa et al., 2008). Other ingredients incorporated into bakery products include nuts and nut pastes. Lipoxygenase and pro-oxidants (e.g., copper and iron) are responsible for the oxidation of nuts (Talbot, 2010). Food processing techniques, such as roasting and size reduction, are known to facilitate oxidation in nuts. Therefore, specification for oxidation markers in nuts must be considered if nut ingredients are to be used in cookies.

The presence of pro-oxidants in the raw material (e.g., white flour) can degrade lipid hydroperoxides. Fortification of white flour with iron is a common source of pro-oxidants and decreases the oxidative stability of the raw material (Barden and Decker, 2013). Other sources of pro-oxidants can come from processing equipment and tap water (Barden and Decker, 2013; Taylor, 1987). For oxidation to occur, a very small amount of iron (<50 ppb) is required (Waraho et al., 2011). Barden (2014) evaluated oxidation of crackers enriched with different iron concentrations. The concentrations of iron tested had no effect on the stability of the cracker. Barden (2014) proposed that the low-water environment in crackers might have contributed to the low mobility of iron, making it difficult to initiate oxidation via hydroperoxide decomposition. In contrast, heme iron promoted the oxidation of palm oil (Aleman et al., 2010). These authors reported that 500 mg/kg ascorbyl palmitate combined with 300 mg/kg citric acid controlled the oxidation of palm oil. In contrast, a chocolate filling typically used in soft baked filled biscuits with palm oil and ascorbyl palmitate and citric acid did not produce the same results. Aleman et al. (2014) found that no synergistic activity resulted from the combination of ascorbyl palmitate and citric acids in a chocolate filling model. They also noted that the age of the citric acid and the differences fatty acid might account for differences between the two heme iron studies. They also noted that concentration of 200–400 mg ascorbyl palmitate/kg prevented oxidation. The various effects observed by researchers suggest that the form of iron and the matrices contribute to the prooxidant activity or lack thereof. Frampton (1994) suggested the use of additives (e.g., citric acid) to chelate metals in oils and fats, minimizing the oxidation of fats. The addition of natural extracts to dough may also be a source of the metal chelators, which may contribute to the antioxidant activity of plant material as supported by Watanabe et al. (2010).

12.4 OXIDATIVE STABILITY OF COOKIES, CRACKERS, AND BISCUITS

12.4.1 Processing and Storage Effects

Lipid oxidation has been known to influence the shelf life of low-moisture foods, such as cookies and crackers (Labuza et al., 1970). The high surface area of crackers facilitates exposure to air, which leads to lipid oxidation (Bassiouny et al., 1990). However, Barden and Decker (2013) reported that crackers stored at high temperature (55 °C) remained relatively stable. The low moisture environment in products such as crackers impacts oxidation differently compared to high-moisture environments. Therefore, the observed stability of the cracker may be driven by water activity of the cracker more than the presence of prooxidants.

Lipid is an essential baking ingredient that gives baked goods (e.g., biscuits and cookies) their desired textural and dimensional characteristics (Smith, 1972; Smith et al., 2004; Kozlowska et al., 2014). However, retention of the lipid in their original state is important for maintaining shelf life. In Turkish biscuits, the fat composition includes palmitic, stearic, oleic, and linoleic acids (Daglioglu et al., 2000). Due to the use of hydrogenated vegetable oils, monounsaturated *trans*-fatty acids were also found to be predominant in these biscuits (Daglioglu et al., 2000). Although small but significant changes in peroxides occurred over 12 months in various cookies and crackers, fatty acids had minimal impact in the shelf life of different cookies and crackers as no changes in fatty acid composition occurred during the 12 months (Daglioglu et al., 2004). In contrast, Caponio et al. (2007) reported a decrease in polyunsaturated fatty acids, and an increase in triacylglycerol oligopolymers and oxidized triacylglycerols as supporting evidence for lipid degradation during biscuit making. Bhanger et al. (2008) also observed a reduction in unsaturated fatty acids in cookies containing rice bran extract. A significant increase in secondary oxidation products occurred between the kneading (i.e., dough development stage) step and cookie after baking (Caponio et al., 2008). The lack of increased primary oxidation products between these two steps suggests the decomposition of hydroperoxides (Caponio et al., 2008). Furthermore, the incorporation of air into the dough during kneading may have facilitated the oxidation process. The storage temperature may also contribute to observed differences in oxidation among researchers. Calligaris et al. (2007) observed dramatically different rates of oxidation during the storage of cookies at different temperatures and that peroxides could be predicted using a modified Arrhenius equation. These authors also used sensory evaluation to derive a model that predicted shelf life. The model (Eqn (1)) developed applies to a temperature range of 20–45 °C and is based on peroxide value and panelists rejection of the samples.

$$SL = \frac{(0.24\,(F\%) + 3.28) - PV_i}{k_0 C \exp^{-(E_a/RT)}} \qquad (1)$$

12.4.1.1 Antioxidant Addition

The consumption of foods that are rich in phytochemicals may reduce the incidence of chronic degenerative diseases, including cancer, cardiovascular, and neurodegenerative diseases (Liu, 2013). For this reason, the consumption of antioxidant-rich foods (e.g., fruits and vegetables) has increased (Dillard and German, 2000). Apart from the health benefits, natural antioxidants from plants have been added to bakery products to prevent oxidation and extend shelf life (Pokorny, 1991). The effectiveness of plant extracts as antioxidants has been shown in multiple studies (Cao et al., 1996; Vinson et al., 1998; Al-Saikhan et al., 1995).

The inclusion of spices, such as cinnamon, has been shown to increase the shelf life of crackers, cookies, and biscuits (Bassiouny et al., 1990; Reddy et al., 2005). Enriching cookies with spices (e.g., *Majorana syriaca*) impeded lipid oxidation and improved organoleptic properties (Al-Bandak and Oreopoulou, 2011). The antioxidant activity of *Majorana syriaca* was attributed to a number of compounds including flavonoids and flavonoid glycosides and rosmarinic acid (Al-Bandak and Oreopoulou, 2007). Spices and herbs are also considered good sources of phenolic compounds, making potential sources of antioxidants. Thus, utilizing spices like sage, rosemary, and thyme can prevent lipid oxidation, unwanted sensory changes, and nutritional degradation (El-Alim et al., 1999; Kozlowska et al., 2014). Kozlowska et al. (2014) verified the effect of thyme and rosemary extracts on fat stability of cookies. Cookies containing the spice extracts were stored for 21 days at 60 °C. Thyme extract, at 0.02% and 0.2% levels, decreased the formation of hydroperoxides in the cookies during the storage period. Kozlowska et al. (2014) attributed this oxidation stability to the phenolic compounds found in the thyme extract. Some of these phenolic compounds, including thymol, carvacrol, rosmarinic acid, and apigenin have the ability to scavenge free radicals (Kozlowska et al., 2014). Bassiouny et al. (1990) also reported similar oxidative stability of crackers enriched with plant extracts, such as marjoram, spearmint, peppermint, and basil. Ethyl ether extracts (0.02%) of spearmint controlled peroxide formation in soda crackers better than the marjoram, peppermint, and basil extract. None of the extracts were as effective as BHA (Bassiouny et al., 1990). Lipid oxidation products of linoleic and oleic acids were found to be responsible for the off-flavors in the cookies (Cognat et al., 2014). In addition, the high peroxide value in the oat-based product confirmed the presence of lipid oxidation. Interestingly, the addition of rosemary extract in the biscuits did not have any significant effect to the oxidative stability of the cookies (Cognat et al., 2014). Given the various oxidation outcomes, the use of spices and herbs as antioxidant sources warrants further investigation.

Reddy et al. (2005) evaluated the antioxidant activity of three plant extracts (i.e., Indian gooseberries, raisins, and drumstick leaves) in biscuits. Before adding these extracts, their antioxidant activity was determined; the drumstick leaves (*Moringa oleifera*) had stronger antioxidant activity than the Indian gooseberries (*Emblica officinalis*), raisins, and the synthetic antioxidants butylated hydroxy anisole (BHA) and butylated hydroxy toluene (BHT). Oxidative stability of cookies made with the three plant extracts and BHA was also evaluated. The cookie containing the drumstick leaf extract had the greatest lipid stability by the end of the 6-week storage period (Reddy et al., 2005). Reddy et al. (2005) also evaluated the antioxidant activity of the powder form of the plants. They observed that the powder form of all three plants had lower antioxidant activity than their respective solvent extracts. Bassiouny et al. (1990) also observed that solvent extract of marjoram, spearmint, peppermint, and basil were more effective than their powdered counterparts. The difference in activity is likely due to the distribution of the antioxidants in food products and the ability of the active chemical compound to diffuse from the plant matrix into the food product, or more specifically the lipid. The uniform distribution of antioxidants in food products is essential to attain the maximum benefits. Furthermore, addition of the plant extracts, in a solution form, to the lipid medium improves the distribution of the extracts compared to adding the extract directly to the dough (Reddy et al., 2005). Addition of the extract to lipid also concentrates the antioxidants in the material they are designed to protect, i.e., lipid, and thereby imparting protection against oxidation.

One of the major issues with spices and herbs is the intense flavor characteristics imparted into food products. Medicinal plants can be an alternative in cases where the plant has a relatively neutral flavor. *Cratoxylum formosum*, a plant indigenous to Thailand, ethanolic extract, controlled the formation of both peroxide values and TBAR values similar to tocopherol and BHT in rice crackers (Maisuthisakul et al., 2007). The dried powdered extract was less effective as expected. The high chlorogenic acid content was believed to be the primary antioxidant in the extract (Maisuthisakul et al., 2007). Misan et al. (2011) evaluated the antioxidant activity of a mixture of medicinal plants (i.e., Frangulae cortex 35%, Mentha piperitae 20%, Carvi fructus 20%, Petroselini fructus 25%) in a cookie. They found the extract followed a dose-dependent response where the 6% fortification prevented the oxidation of the cookie to a greater extent than the 2% and 4% concentration. A number of phenolic compounds were identified and the authors suggested that the combination of the phenolics acted synergistically. Green tea extracts effectively controlled the oxidation in both cookies and crackers. A 1% addition of green tea extract controlled hydroperoxide formation in cookies better than BHA while lower (i.e., 0.02%, 0.1%) green tea extracts were less effective but still inhibited hydroperoxide formation (Mildner-Szkudlarz et al., 2009). A slight reduction of the green tea catechins during baking (Sharma and Zhou, 2011) likely contributed to the observed effects in the Mildner-Szkudlarz et al. (2009) study. However, the green tea extracts were not effective in controlling secondary oxidation formation. Radočaj et al. (2014) found significantly more radical scavenging activity in gluten free crackers containing 2–8% decaffeinated green tea leaves compared to the hempseed flour control. In this study, the DPPH radical scavenging assay of the ground products was used to assess the antioxidant activity and thus additional research such as volatile analysis and peroxide value determination is needed to fully characterize the antioxidant potential of decaffeinated green tea leaves in crackers.

The stability of natural antioxidants can be compromised during food processing (Jonsson, 1991). During the baking (160–180 °C) of the cookies containing the raisins, drumstick leaves, and Indian gooseberries, no change in antioxidant activity was observed (Reddy et al., 2005). The temperature did not affect the stability of these antioxidants, but the short duration (10 min) of the baking process might not have been sufficient to decompose the antioxidants in the extracts. Fresh mango-peels are good sources of bioactive compounds, including polyphenols and carotenoids (Ajila et al., 2007). The antioxidant activity of mango-peel powder (MPP) was higher than the synthetic antioxidant, BHA. Further study of the MPP addition to soft-dough biscuits supported the in vitro antioxidant results (Ajila et al., 2008). Although minimal loses in polyphenol content is expected during the baking process due to high temperatures (160–180 °C), the total polyphenol content of the biscuit containing MPP increased from 540 to 4500 μg GAE/g (Ajila et al., 2008). In addition, the DPPH scavenging property of MPP fortified biscuit increased when the amount of MPP incorporated into the biscuit increased. This increased antioxidant activity was attributed to the increased levels of polyphenols and carotenoids in the dough as the amount of MPP added to the dough increased (Ajila et al., 2008). Ranhotra et al. (1995) reported a 23% reduction in β-carotene during the baking of crackers. They did observe a 34% increase in the *cis* forms; however, during cracker storage, minimal β-carotene loss occurred.

Byproducts of fruit processing has been evaluated as a functional ingredient in cookies. One example includes grape marc (i.e., grape residue after pressing) (Pasqualone et al., 2013). The grape marc consists of the skin, seeds, and stem of the grape, making it a good source of anthocyanins, catechins, glycosides of flavonols, and phenolic acids (Mazza and Francis, 1995; Negro et al., 2003). Fortifying cookies with the grape marc extract resulted in increased phenolic compounds and improved oxidative stability (Pasqualone et al., 2014). Using high-quality ingredients, such as extra virgin oil and semolina, also enhanced the antioxidant property of cookies without the grape marc extract (Pasqualone et al., 2014).

Antioxidant activities of seeds or seed coat extracts have been documented in lipid systems. However, less extensive literature is available for cookies and crackers. Vanilla bean extract prevented the hydroperoxide formation in sugar cookies slightly better than synthetic vanillin (Anuradha et al., 2010). These authors observed increased vanillin loss over a 60-day period and that the natural vanilla extract lost significantly less vanillin. Therefore, vanillin likely contributed to the antioxidant activity but other compounds in the natural vanilla extract also contributed to the stability of the cookies. Budryn and Nebesny (2013) observed comparable antioxidant activity for roasted and green coffee bean extracts in cookies. The 0.5% concentration for both extracts proved to be the most effective. The author suggested that polyphenolics were the active components considering that green coffee bean extract was obtained from beans that were not roasted. However, the cookies with roasted coffee beans had slightly lower, but not significant, peroxide value compared to the cookies with green coffee extract. This suggests that compounds formed during roasting may provide some benefit. de Camargo et al. (2014) reported improved stability of cookies fortified with peanut skins. As expected, oxidative stability of the cookies was better with increasing peanut skin concentrations.

Radical scavenging activity of ground extracted buckwheat crackers was significantly higher than that of a corresponding whole-wheat cracker (Sedej et al., 2011). The author also observed higher tocopherol, phenolic acids, and flavonoids, which they indicated were responsible for the radical scavenging activity. Bhanger et al. (2008) reported similar oxidative stability of cookies containing 500 mg/kg methanol extract of rice bran and 200 mg/kg BHT. Increasing the rice bran extract to 1000 and 2000 mg/kg improved oxidative stability cookie compared to the BHT treatment. Watanabe et al. (2010) reported that defatted rice bran was effective at controlling iron catalyzed oxidation of cookies fortified with different oils. One suggested antioxidant in rice bran includes oryzanol. Kumar et al. (2014) reported that oryzanol content remained stable over 120-day storage of a cookie prepared with approximately 300 mg oryzanol/100 g. Storage under various relative humidities did not affect oyzanol content; however, oxidative stability was affected by relative humidities above 44%. At these humidities, hydroperoxides decomposed and likely resulted in the formation of secondary oxidation product, which were not measured by the authors. In contrast to rice bran, a shortbread biscuit containing coarse ground barley flour did not protect the biscuit from oxidation (Verardo et al., 2011). One reason might relate to the large flour particle size that was incorporated into the cookie formula and the inability of the potential barley antioxidants to diffuse into the fat during preparation.

12.4.1.2 Maillard Reaction Products

In cereals, glycine is readily available and is likely one of the amino acids involved in the production of MRPs of baked or high temperature–treated cereal products. Rada-Mendoza et al. (2004) reported the presence of MRPs in cookies, crackers, and breakfast cereals, indicating a potential antioxidant source. Baking of butter cookies at 150 °C for 20–30 min produced MRPs that were sufficient to provide an antioxidant effect. Wijewickreme and Kitts (1998) reported, however, that α-tocopherol was more effective than MRPs in preventing the oxidation of flour dough.

Cognat et al. (2012) reported that pyrazine was present in an oat cookie. They also noted that as the cookie became rancid, the level of 2,6-dimethylpyrazine decreased. We have observed similar results in unpublished work on roasted flaxseed where several pyrazine decreased with increasing oxidation. These data support the important antioxidant activity of the pyrazines, a result of the Maillard browning reaction. The body of research supports the antioxidant activity of MRPs in numerous food and model systems (Bressa et al., 1996). In addition to added antioxidants, the oxidative stability of baked products such as crackers and cookies likely depends on MRPs.

12.4.1.3 Packaging

The purpose of packaging food is to protect the product from various degradation processes, such as oxidation and staling, and physical destruction. Protection from light and oxygen are the two main functions of cracker, cookie, and biscuit packaging (Manley, 1983). Metalized films or metalized-oriented polypropylene using nitrogen flush provides the most protection against light and oxidation, but the cost is generally prohibitive for many cracker, cookie, and biscuit products. Polyvinyl chloride–polyethylene (PVC/PE) copolymer packaging has low gas and water permeability and thus could be an alternative to metalized films in products provided that paperboard is used to provide support against physical damage (Galić et al., 2009). Oriented polypropylene can be used in products not requiring gas barrier properties. However, only gas barriers films such as PVC/PE can be used for products requiring modified atmospheric packaging (MAP), e.g., nitrogen flush. Lu and Xu (2010) reported that polyethylene terephthalate/aluminum/oriented nylon/polyethylene (PET/AL/ONY/PE) was the most effective packaging material at preventing the oxidation of cookies. Other high oxygen barrier packaging also protected the cookies against oxidation better than nonoxygen barrier film, i.e., oriented polypropylene/polyethylene copolymer. Packages with light barrier properties were also effective in controlling oxidation of cookies compared to no light barrier package (Lu and Xu, 2009). Cookie shelf life increased from approximately 32 days for cookies exposed directly to light (1.9 $\mu W/cm^2$) to over 76 days in light barrier packaging. Therefore, photooxidation was clearly an issue in cookies and demonstrates the need for protection from light.

In the past few decades, active packaging systems beyond MAP have been developed. The purpose of these types of packages is to provide a benefit beyond simply being an inert barrier (Berenzon and Saguy, 1998; Vermeiren et al., 1999). Two examples of active packaging include the addition of oxygen absorbers and films with impregnated antioxidants. Significantly lower peroxide values and hexanal contents were observed in crackers packaged with an iron-based oxygen absorber compared to a control package without an oxygen absorber (Berenzon and Saguy, 1998). These observations held true at both 25 and 35 °C storage environments.

REFERENCES

Afoakwa, E., Paterson, A., Fowler, M., Ryan, A., 2008. Flavor formation and character in cocoa and chocolate: a critical review. Crit. Rev. Food Sci. Nutr. 48, 840–857.

Ajila, C.M., Leelavathi, K., Prasada Rao, U.J.S., 2008. Improvement of dietary fiber content and antioxidant properties in soft dough biscuits with the incorporation of mango peel powder. J. Cereal Sci. 48, 319–326.

Ajila, C.M., Bhat, S.G., Prasada Rao, U.J.S., 2007. Valuable components of raw and ripe peels from two indian mango varieties. Food Chem. 102, 1006–1011.

Al-Bandak, G., Oreopoulou, V., 2007. Antioxidant properties and composition of *Majorana syriaca* extracts. Eur. J. Lipid Sci. Technol. 109, 247–255.

Al-Bandak, G., Oreopoulou, V., 2011. Inhibition of lipid oxidation in fried chips and cookies by *Majorana syriaca*. Int. J. Food Sci. Technol. 46, 290–296.

Aleman, M., Nuchi, C., Bou, R., Tres, A., Polo, J., Guardiola, F., Codony, R., 2010. Effectiveness of antioxidants in preventing oxidation of palm oil enriched with heme iron: a model for iron fortification in baked products. Eur. J. Lipid Sci. Technol. 112, 761–769.

Aleman, M., Bou, R., Tres, A., Polo, J., Codony, R., Guardiola, F., 2014. The effect of citric acid and ascorbyl palmitate in palm oil enriched with heme iron: a model for iron fortification in bakery products. Eur. J. Lipid Sci. Technol. 116, 300–310.

Al-Saikhan, M.S., Howard, L.R., Miller, J.C., 1995. Antioxidant activity and total phenolics in different genotypes of potato. J. Food Sci. 60, 341–343.

Alvarez-Jubete, L., Wijngaard, H., Arendt, E.K., Gallagher, E., 2010. Polyphenol composition and in vitro antioxidant activity of amaranth, quinoa, buckwheat and wheat as affected by sprouting and baking. Food Chem. 119, 770–778.

Anonymous, 2012. Agriculture and Agri-food Canada. Bakery Products in the United States Market Indicator Report , pp. 1–9. Accessed online http://www5.agr.gc.ca/resources/prod/Internet-Internet/MISB-DGSIM/ATS-SEA/PDF/6113-eng.pdf.

Anuradha, K., Naidu, M., Manohar, R., Indiramma, A., 2010. Effect of vanilla extract on radical scavenging activity in biscuits. Flavour Fragr. J. 25, 488–492.

Barden, L., 2014. Understanding Lipid Oxidation in Low-Moisture Food. University of Massachusetts-Amherst. Doctoral Dissertations, Paper 155.

Barden, L., Decker, E., 2013. Lipid oxidation in low-moisture food: a review. Crit. Rev. Food Sci. Nutr. http://dx.doi.org/10.1080/10408398.2013.848833.

Barnes, P.J., 1983. Non-saponifiable lipids in cereals. In: Lipids in Cereal Technology (Food Science and Technology: A Series of Monographs). Academic Press, London, England, pp. 33–55.

Bassiouny, S.S., Hassanien, F.R., El-Razik Ali, F., El-Kayati, S.M., 1990. Efficiency of antioxidants from natural sources in bakery products. Food Chem. 37, 297–305.

Becker, R., 2007. Fatty acids in food cereal grains and grain products. In: Chow, C.K. (Ed.), Fatty Acids in Foods and Their Health Implications, third ed. CRC/Taylor and Francis, Boca Raton, FL, pp. 303–316.

Berenzon, S., Saguy, I.S., 1998. Oxygen absorbers for extension of crackers shelf life. LWT Food Sci. Technol. 31, 1–5.

Bhanger, M., Iqbal, S., Anwar, F., Imran, M., Akhtar, M., Zia-Ul-Haq, M., 2008. Antioxidant potential of rice bran extracts and its effects on stabilisation of cookies under ambient storage. Int. J. Food Sci. Technol. 43, 779–786.

Bressa, F., Tesson, N., Dalla Rosa, M., Sensidoni, A., Tubaro, F., 1996. Antioxidant effect of Maillard reaction products: application to a butter cookie of a competition kinetics analysis. J. Agric. Food Chem. 44, 692–695.Budryn, G., Nebesny, E., 2013. Effect of green and roasted coffee antioxidants on quality and shelf life of cookies and chocolates. J. Food Proc. Preserv. 37, 835–845.

Calligaris, S., Manzocco, L., Kravina, G., Nicoli, M., 2007. Shelf life modeling of bakery products using oxidation indices. J. Agric. Food Chem. 55, 2004–2009.

Cao, G., Sofic, E., Prior, R.L., 1996. Antioxidant capacity of tea and common vegetables. J. Agri. Food Chem. 44, 3426–3431.

Caponio, F., Summo, C., Pasqualone, A., Bilancia, M., 2008. Effect of kneading and baking on the degradation of the lipid fraction of biscuits. J. Cereal Sci. 48, 407–412.

Caponio, F., Gomes, T., Pasqualone, A., Summo, C., 2007. Use of the high performance size exclusion chromatography analysis for the measurement of the degree of hydrolytic and oxidative degradation of the lipid fraction of biscuits. Food Chem. 102, 232–236.

Casiraghi, M.C., Garsetti, M., Testolin, G., Brighenti, F., 2006. Post-prandial responses to cereal products enriched with barley beta-glucan. J. Am. Coll. Nutr. 25, 313–320.
Cognat, C., Shepherd, T., Verall, S.R., Stewart, D., 2014. Relationship between volatile profile and sensory development of an oat-based biscuit. Food Chem. 160, 72–81.
Cognat, C., Shepherd, T., Verall, S.R., Stewart, D., 2012. Comparison of two headspace sampling techniques for the analysis of off-flavour volatiles from oat based products. Food Chem. 134, 1592–1600.
Cumbee, B., Hildebrand, D.F., Addo, K., 1997. Soybean flour lipoxygenase isozymes effects on wheat flour dough rheological and bread making properties. J. Food Sci. 62, 281–294.
Daglioglu, O., Tasan, M., Gecgel, U., Daglioglu, F., 2004. Changes in oxidative stability of selected bakery products during shelf life. Food Sci. Technol. Res. 10, 464–468.
Daglioglu, O., Tasan, M., Tuncel, B., 2000. Determination of fatty acid composition and total trans fatty acids of Turkish biscuits by capillary gas–liquid chromatography. Eur. Food Res. Technol. 211, 41–44.
de Camargo, A., Vidal, C., Canniatti-Brazaca, S., Shahidi, F., 2014. Fortification of cookies with peanut skins: effects on the composition, polyphenols, antioxidant properties, and sensory quality. J. Agric. Food Chem. 62, 11228–11235.
Dillard, C.J., German, J.B., 2000. Phytochemicals: nutraceuticals and human health. J. Sci. Food Agric. 80, 1744–1756.
Ekstrand, B., Gangby, I., Akesson, G., StoÈllman, U., Lingnert, H., Dahl, S., 1993. Lipase activity and development of rancidity in oats and oat products related to heat treatment during processing. J. Cereal Sci. 17, 247–254.
El-Alim, S.S.L.A., Lugasi, A., Hovari, J., Dworschak, E., 1999. Culinary herbs inhibit lipid oxidation in raw and cooked minced meat patties during storage. J. Sci. Food Agric. 79, 277–285.
Frampton, A., 1994. Prevention of Rancidity in Confectionery and Biscuits-A Hazard Analysis Critical Control Point (HACCP) Approach. In: Allen, J.C., Hamilton, R.J. (Eds.), Rancidity in Foods, third ed. Blackie Academic and Professional, Glasgow (Chapter 9).
Frankel, E., 2005. Lipid Oxidation, second ed. The Oily Press, PJ Barnes & Associates, Bridgewater, UK.
Frankel, E., Neff, W., Selke, E., 1981. Analysis of auto-oxidized fats by gas chromatography-mass spectrometry: VII volatile thermal-decomposition products of pure hydroperoxides from auto-oxidized and photosensitized oxidized methyl oleate, linoleate, and linolenate. Lipids 16, 279–285.
Fritsch, C., Gale, J., 1977. Hexanal as a measure of rancidity in low fat foods. J. Am. Oil Chem. Soc. 54, 225–228.
Galić, K., Ćurić, D., Gabrić, D., 2009. Shelf life of packaged bakery goods—a review. Crit. Rev. Food Sci. Nutr. 49, 405–426.
Galliard, T., 1983. Enzymic degradation of cereal lipids. In: Lipids in Cereal Technology (Food Science and Technology: A Series of Monographs). Academic Press, London, England, pp. 111–147.
Gardner, H., Inglett, G., 1971. Food products from corn germ-enzyme activity and oil stability. J. Food Sci. 36, 645–648.
Goupy, P., Hugues, M., Boivin, P., Amiot, M., 1999. Antioxidant composition and activity of barley (*Hordeum vulgare*), and malt extracts and of isolated phenolic compounds. J. Sci. Food Agric. 79, 1625–1634.
Gwirtz, J.A., Willyard, M.R., McFall, K.L., 2007. Wheat flour products in North America. In: Future of Flour—A Compendium of Flour Improvement. Agrimedia GmbH, Clenze, Deutschland, pp. 295–308.
Hall III, C., 2010. Oxidation of cereals and snack products. In: Decker, E., McClements, J., Elias, R. (Eds.), Oxidation in Foods and Beverages and Antioxidant Applications. Management in

Different Industry Sectors, vol. 2. Woodhead Publishing Series in Food Science, Technology and Nutrition, Ch. 11. pp. 369–390.

Hubner, F., Arendt, E.K., 2013. Germination of cereal grains as a way to improve the nutritional value: a review. Crit. Rev. Food Sci. Nutr. 53, 853–861.

Jonsson, I., 1991. Thermal degradation of carotenoids and influence of their physiological functions. In: Freidman, M. (Ed.), Nutrition & Toxicological Consequences of Food Processing. Plenum Press, New York, NY, USA, pp. 75–82.

Kaddour, A., Grand, E., Barouh, N., Barea, B., Villeneuve, P., Cuq, B., 2006. Near-infrared spectroscopy for the determination of lipid oxidation in cereal food products. Eur. J. Lipid Sci. Technol. 108, 1037–1046.

Katan, M.B., Zock, P.L., Mensink, R.P., 1994. Effects of fats and fatty acids on blood lipids in humans: an overview. Am. J. Clin. Nutr. 60, 1017S–1022S.

Kilcast, D., Subramaniam, P., 2000. The Stability and Shelf life of Food, first ed. Woodhead Publishing Limited, Cambridge, England. pp. 1.

Kozlowska, M., Zbikowska, A., Gruczynska, E., Zontala, K., Poltorak, A., 2014. Effects of spice extracts on lipid fraction oxidative stability of cookies investigated by DSC. J. Therm. Anal. Calor. 118, 1697–1705.

Kumar, P., Manohar, R.S., Indiramma, A.R., Krishna, A.G., 2014. Stability of oryzanol fortified biscuits on storage. J. Food Sci. Technol. 51, 2552–2559.

Labuza, T.P., Tannenbaum, S.R., Karel, M., 1970. Water content and stability of low-moisture and intermediate-moisture foods. Food Technol. 543–550.

Lam, H.S., Proctor, A., 2003. Hydrolysis of acylglycerols and phospholipids of milled rice surface lipid during storage. In: Norman, R.J., Meullenet, J.-F., Moldenhauer, K.A.K., Lehtinen, P., Kiiliainen, K., Lehtomaki, I., Laakso, S. (Eds.), Effect of heat treatment on lipid stability in processed oats. J. Cereal Sci. 37, 215–221.

Liu, R.H., 2013. Dietary bioactive compounds and their health implications. J. Food Sci. 78, A18–A25.

Lopez-Duarte, A., Vidal-Quintanar, R., 2009. Oxidation of linoleic acid as a marker for shelf life of corn flour. Food Chem. 114, 478–483.

Lu, L.X., Xu, F., 2009. Effect of light-barrier property of packaging film on the photo-oxidation and shelf life of cookies based on accelerated tests. Packag. Technol. Sci. 22, 107–113.

Lu, L.X., Xu, F., 2010. Effects of initial oxygen concentration and oxygen-barrier property of film on fat oxidation rate of packed cookies. Eur. Food Res. Technol. 231, 347–351.

Lehtinen, P., Kiiliainen, K., Lehtomaki, K., Laakso, S., 2003. Effect of heat treatment on lipid stability in processed oats. J. Cereal Sci. 37, 215–221.

Maillard, M.N., Soum, M.H., Boivin, P., Berset, C., 1996. Antioxidant activity of barley and malt: relationship with phenolic content. LWT Food Sci. Technol. 29, 238–244.

Maisuthisakul, P., Gordon, M.H., Pongsawatmanit, R., Suttajit, M., 2007. Enhancing the oxidative stability of rice crackers by addition of the ethanolic extract of phytochemicals from *Cratoxylum formosum* dyer. Asia Pac. J. Clin. Nutr. 16 (Suppl. 1), 37–42.

Manley, D.J., 1983. Technology of Biscuits, Crackers, and Cookies. Ellis Horwood Series in Food Science and Technology. Chichester, West Sussex.

Manley, D., 2008. Technology of Biscuits, Crackers, and Cookies, third ed. Woodhead Publishing Limited, Cambridge, England.

Mano, Y., Ohnishi, M., Sato, H., Nakanishi, H., Maemoto, M., Ito, S., 1990. Compositions of lipid classes, fatty-acids and sterols in domestic rye grains. J. Jpn. Soc. Food Sci. Technol. 37, 338–345.

Marconi, O., Martini, R., Mangione, A., Falconi, C., Pepe, C., Perretti, G., 2014. Palatability and stability of shortbread made with low saturated fat content. J. Food Sci. 79, C469–C475.

Mazza, G., Francis, F.J., 1995. Anthocyanins in grapes and grape products. Crit. Rev. Food Sci. Nutr. 35, 341–371.

Mildner-Szkudlarz, S., Zawirska-Wojtasiak, R., Obuchowski, W., Goslinski, M., 2009. Evaluation of antioxidant activity of Green Tea extract and its effect on the biscuits lipid fraction oxidative stability. J. Food Sci. 74, S362–S370.

Misan, A., Mimica-Dukic, N., Sakac, M., Mandic, A., Sedej, I., Simurina, O., Tumbas, V., 2011. Antioxidant activity of medicinal plant extracts in cookies. J. Food Sci. 76, C1239–C1244.

Molteberg, E., Magnus, E., Bjorge, J., Nilsson, A., 1996. Sensory and chemical studies of lipid oxidation in raw and heat-treated oat flours. Cereal Chem. 73, 579–587.

Nasirullah, T.M., Krishnamurthy, M.N., Nasgaraja, K.V., 1989. Effect of stabilization on the quality characteristics of rice-bran oil. J. Am. Oil Chem. Soc. 66, 661–663.

Negro, C., Tommasi, L., Miceli, A., 2003. Phenolic compounds and antioxidant activity from red grape marc extracts. Bioresour. Technol. 87, 41–44.

Okpala, L., Okoli, E., Udensi, E., 2012. Physico-chemical and sensory properties of cookies made from blends of germinated pigeon pea, fermented sorghum, and cocoyam flours. Food Sci. Nutr. 1, 8–14.

Pasqualone, A., Bianco, A.M., Paradiso, V.M., Summo, C., Gambacorta, G., Caponio, F., 2014. Physio-chemical, sensory, and volatile profiles of biscuits enriched with grape marc extract. Food Res. Int. 65, 385 393.

Pasqualone, A., Bianc, A.M., Paradiso, V.M., 2013. Production trials to improve the nutritional quality of biscuits and to enrich them with natural anthocyanins. J. Food 11, 301–308.

Pokorny, J., 1991. Natural antioxidants for food use. Trends Food Sci. Technol. 2, 223–226.

Rada-Mendoza, M., Garcia-Banos, J., Villamiel, M., Olano, A., 2004. Study on nonenzymatic browning in cookies, crackers and breakfast cereals by maltulose and furosine determination. J. Cereal Chem. 39, 167–173.

Radočaj, O., Dimić, E., Tsao, R., 2014. Effects of hemp (*Cannabis sativa* L.) seed oil press-cake and decaffeinated green tea leaves (*Camellia sinesis*) on functional characteristics of gluten-free crackers. J. Food Sci. 79, C318–C325.

Ramezanzadeh, F.M., Rao, R.M., Windhauser, M., Prinyawiwatkul, W., Tulley, R., Marshall, W.E., 1999. Prevention of hydrolytic rancidity in rice bran during storage. J. Agri. Food Chem. 47, 3050–3052.

Ranhotra, G., Gelroth, J., Langemeier, J., Rogers, D., 1995. Stability and contribution of beta-carotene added to whole wheat bread and crackers. Cereal Chem. 72, 139–141.

Reddy, V., Urooj, A., Kumar, A., 2005. Evaluation of antioxidant activity of some plant extracts and their application in biscuits. Food Chem. 90, 317–321.

Rossell, J.B., 1994. Measurement of rancidity. In: Allen, J.C., Hamilton, R.J. (Eds.), Rancidity in Foods, third ed. Blackie Academic and Professional, Glasgow (Chapter 2).

Rutgersson, A., Toukkuri, V., Reinikainen, P., Lingnert, H., 2000. Influence of hydrothermal treatment on lipid oxidation in barley. Cereal Chem. 77, 407–413.

Sedej, I., Sakac, M., Mandic, A., Misan, A., Pestoric, M., Simurina, O., Canadanovic-Brunet, J., 2011. Quality assessment of gluten-free crackers based on buckwheat flour. LWT Food Sci. Technol. 44, 694–699.

Seibel, W., 2007. Composite flours. In: Future of Flour—A Compendium of Flour Improvement. Agrimedia GmbH, Clenze, Deutschland, pp. 193–198.

Sharma, A., Zhou, W., 2011. Stability study of green tea catechins during the biscuit making process. Food Chem. 126, 568–573.

Shearer, G., Warwick, M.J., 1983. The effect of storage on the lipids and bread making properties of wheat flour. In: Barnes, P.J. (Ed.), Lipids in Cereal Technology. Academic Press Inc., New York, pp. 253–267.

Shiiba, K., Negishi, Y., Okada, K., Nagao, S., 1991. Purification and characterization of lipoxygenase isozymes from wheat germ. Cereal Chem. 68, 115–122.

Shin, M., Yoon, S., Rhee, J., Kwon, T., 1986. Correlation between oxidative deterioration of unsaturated lipid and normal-hexanal during storage of brown rice. J. Food Sci. 51, 460–463.

Sjovall, O., Virtalaine, T., Lapvetelainen, A., Kallio, H., 2000. Development of rancidity in wheat germ analyzed by headspace gas chromatography and sensory analysis. J. Agric. Food Chem. 48, 3522–3527.

Skrbic, B., Cvejanov, J., 2011. The enrichment of wheat cookies with high-oleic sunflower seed and hull-less barley flour: impact on nutritional composition, content of heavy elements and physical properties. Food Chem. 124, 1416–1422.

Smith, J.P., Daifas, D.P., El-Khoury, W., Koukoutsis, J., El-Khoury, A., 2004. Shelf life and safety concerns of bakery products—a review. Crit. Rev. Food Sci. Nutr. 44, 19–55.

Smith, W.H., 1972. Hard Semi-Sweet Biscuits. Biscuits, Crackers and Cookies: Technology, Production and Management, vol. 1. Applied Science Publishers Ltd, London, pp. 466–473.

Subramaniam, P., 2009. Shelf life prediction and testing. In: Talbot, G. (Ed.), Science and Technology of Coated and Filled Chocolate, Confectionery and Bakery Products. Woodhead Publishing, Cambridge (Chapter 12), pp. 233-254.

Taylor, A., 1987. Effect of water quality on lipid oxidation. Food Sci. Technol. Today, 1, 158–159.

Talbot, G., 2010. Oxidation of confectionery products and biscuits. In: Decker, E., McClements, J., Elias, R. (Eds.), Management in Different Industry SectorsWoodhead Publishing Series in Food Science, Technology and Nutrition, vol. 2. Ch. 10, pp. 344–368.

Verardo, V., Riciputi, Y., Messia, M., Vallicelli, M., Falasca, L., Marconi, E., Caboni, M., 2011. Dietary fiber and flavan-3-ols in shortbread biscuits enriched with barley flours co-products. Int. J. Food Sci. Nutr. 62, 262–269.

Vermeiren, L., Devlieghere, F., van Beest, M., de Kruijf, N., Debevere, J., 1999. Developments in the active packaging of foods. Trends Food Sci. Technol. 10, 77–86.

Vinson, J.A., Hao, Y., Su, X., Zubik, L., 1998. Phenol antioxidant quantity and quality in foods: vegetables. J. Agri. Food Chem. 46, 3630–3634.

Walker, C.E., Li, J., 2008. Crops and their processing. In: Pyler, E. (Ed.), Baking Science and Technology, fourth ed. Kansas City, MO, pp. 567–600.

Waraho, T., McClements, D.J., Decker, E.A., 2011. Mechanisms of lipid oxidation in food dispersions. Trends Food Sci. Technol. 22, 3–13.

Watanabe, Y., Hashimoto, K., Omori, A., Uda, Y., Nomura, M., 2010. Suppressive ability of defatted rice bran against lipid oxidation in cookies containing iron. Biosci. Biotechnol. Biochem. 74, 262–265.

Wijewickreme, A., Kitts, D., 1998. Oxidative reactions of model Maillard reaction products and alpha-tocopherol in a fluor-lipid mixture. J. Food Sci. 63, 466–471.

Chapter 13

Packaging Technologies to Control Lipid Oxidation

John L. Koontz
FDA–CFSAN, Division of Food Processing Science and Technology, Bedford Park, IL, USA

13.1 INTRODUCTION

Prevention of lipid oxidation, and its deleterious effects on food quality, continues to be a fundamental goal of the food industry. The initial products of oxidation in foods containing unsaturated fatty acids are hydroperoxides, which are colorless, tasteless, and odorless. However, these hydroperoxides decompose into aldehydes, ketones, and carbonyls that produce undesirable odors and flavors associated with rancidity. The primary food quality issues influenced by lipid oxidation include the development of off-flavors, altered texture and color, and reduced nutritional value. Edible oil oxidation results in degradation of essential fatty acids and production of toxic compounds and oxidized polymers. Several studies suggest that oxidation of lipids from the diet may play a direct role in the development of chronic diseases in the human body (Kanazawa et al., 2002; Kanner, 2007; Riemersma, 2002).

The oxidative stability of a packaged food product can be greatly impacted by its interaction with the external environment including atmospheric O_2, light irradiation, and water vapor. Oxidation of edible oils can occur by either the mechanism of auto-oxidation or photosensitized oxidation, depending upon the type of reactive oxygen species present (Choe and Min, 2006). Atmospheric triplet oxygen ($^3O_2$) is a radical that reacts with lipid radicals and causes auto-oxidation, which is a free radical chain reaction. Auto-oxidation, as a result of O_2 permeation through the packaging material, limits the quality and shelf life of lipid-containing foods. Singlet oxygen ($^1O_2$) is an electrophilic nonradical produced during photosensitized oxidation of lipid in the presence of light, sensitizers, and atmospheric O_2. Photosensitized oxidation, as a result of ultraviolet (UV) and visible light transmission through the packaging, has a deleterious impact on food quality and shelf life. Several intrinsic factors affect the oxidative stability of edible oils, including composition of fatty acids and concentrations of transition metals, pigments, phospholipids, free fatty acids,

Oxidative Stability and Shelf Life of Foods Containing Oils and Fats
http://dx.doi.org/10.1016/B978-1-63067-056-6.00013-6

mono- and diacylglycerols, and antioxidants. Lipid oxidation reaction rates are also strongly dependent on the water activity (a_w) of a food system (Nelson and Labuza, 1992). In foods with a_w values of less than 0.3, the continued removal of water results in an accelerated rate of lipid oxidation. As water is added to food products with a_w values in the range of 0.4–0.7, oxidation rates generally increase likely due to an increase in reactant mobility such as transition metals and O_2. The rate of lipid oxidation then decreases in foods with a_w greater than 0.7. Permeation of water vapor through the packaging material alters the a_w of the contained food, and consequently may either increase or decrease the lipid oxidation rate depending upon the initial a_w of the food. The extent to which these primary mechanisms of oxidation occur within packaged foods is dependent upon the barrier properties of the selected packaging material.

Food packaging has traditionally been defined as a passive barrier with its goal to protect its contents from the adverse effects of the external environment, such as microorganisms, water vapor, gases, odors, light, dust, and mechanical forces. Polymer packaging can act as a barrier between the outside environment and the interior atmosphere of the package in order to control lipid oxidation and maintain food quality. Figure 13.1 illustrates the interactions that occur in foods contained in polymer packaging which may impact oxidative stability. These interactions include those between the headspace and the food product, between the headspace and the external environment, and between the food product and the external environment. The permeation of a small molecule, such as molecular O_2 or water vapor, through a polymer film involves the following four stages: (1) absorption of the permeating species onto the surface of the polymer; (2) dissolution of the gas or vapor into the polymer matrix; (3) diffusion through the wall along a concentration gradient; and (4) desorption

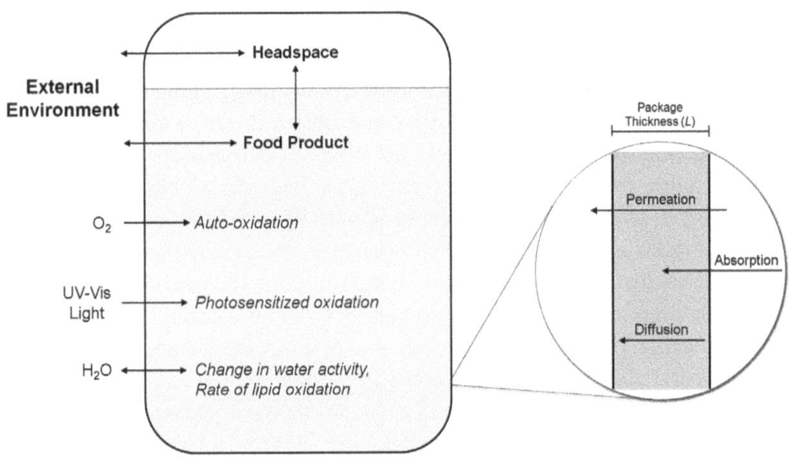

FIGURE 13.1 Mass and energy transfer through a hermetically sealed polymer package and product interactions with the external environment that affect the oxidative stability of foods.

from the other surface (Ashley, 1985). The passive barrier properties of polymers dictate their resistance to the sorption and diffusion of permeants, such as O_2 and water vapor.

Active packaging materials possess attributes beyond their basic physical barrier properties and play an active role in food preservation. The performance of package systems can be enhanced by active packaging in which subsidiary constituents are deliberately included in the packaging material to maintain or improve the sensory, safety, and quality aspects of the food.

Both commercialized packaging technologies and research publications provide the food industry with a variety of potential solutions. The main polymer packaging strategies currently available to inhibit oxidation of foods include use of high-barrier polymer films which limit O_2 permeation, use of modified atmospheres, protection against light exposure, and incorporation of O_2 and off-odor (aldehyde) scavengers as active components. Meanwhile, research has laid the foundation for development of innovative active packaging technologies that may aid in preventing lipid oxidation, maintaining nutritional quality, and extending the shelf life of packaged food. In such active systems, polymer packaging may serve as a reservoir or matrix from which active components, such as antioxidants, are delivered in a controlled manner into a food product. Antioxidants and metal-chelating agents have also been immobilized by covalent grafting onto polymer surfaces with the aim of controlling lipid oxidation in foods.

13.2 BARRIER PROPERTIES OF PACKAGING MATERIALS

The selection of suitable packaging materials for a particular food product requires optimization and typically centers on the barrier properties of the packaging material. Packaging acts as a passive barrier against the ingress of deteriorative vectors such as O_2 and water vapor, and simultaneously protects food from loss of desirable components such as moisture, carbon dioxide, and flavors and aromas. Foods can be classified according to their sensitivity to O_2 or water vapor and the appropriate packaging material can then be evaluated and selected for its ability to provide the degree of protection necessary to achieve a desired product shelf life. The degree of product protection from O_2 permeation required to achieve a 1-year shelf life at 25 °C has been estimated as a general guideline (Ashley, 1985). The most O_2-sensitive food products, such as canned meats, fish, milk, and vegetables, have a 1-year shelf life if O_2 gain is limited within the range of 1–5 ppm. Less sensitive products, such as snack foods, canned fruits, dried foods, and nuts, tolerate O_2 concentrations of up to 5–15 ppm for 1 year. Edible oils and fats, salad dressings, preserves, and pickles have a higher estimated acceptable O_2 gain in the range of 50–200 ppm during 1 year of storage. Regardless of a specific food's O_2 sensitivity, the highest food quality for maximum extension of shelf life will always be obtained by use of packaging materials with the highest barrier to O_2 and light.

Food packaging is manufactured using several different groups of materials, including glass, metals, paper, and thermoplastic polymers. The properties and performance characteristics of these various packaging materials vary across a wide range. Metal cans and glass containers can be regarded as practically impermeable with an absolute barrier to O_2, water vapor, and flavors and aromas, provided that the integrity of the metal can seams and glass container closures remain intact. Glass containers are inert to the food they hold, and have the benefit of optical transparency, thereby allowing consumers to visually evaluate the packaged food. On the other hand, the transparency also provides a lack of protection from light irradiation.

Aluminum foil has excellent barrier properties with a minimum thickness of at least 15 μm, but below this thickness, the frequency of pinhole formation is greater, thereby limiting its barrier properties (Lamberti and Escher, 2007). Aluminum foil is typically used in polymer laminates when insufficient barrier properties are the limiting factor for the shelf stability of food and also functions as a complete barrier against the influence of light. The aluminum layer in aluminum foil plastic laminate packages can be classified as rigid (>180 μm), semi-rigid (50–180 μm), or flexible (9–50 μm) (Lamberti and Escher, 2007). Thin coatings of metallic aluminum, aluminum oxide, or silicon oxide on plastic laminates can offer an alternative to aluminum foil laminates, but their barrier properties are somewhat less efficient due to their limited thicknesses, which are on the order of 50–100 nm. Paper and paperboard packaging materials are poor barriers to gases and water vapor and, as a result, are typically coated or laminated with thermoplastic polymers to improve the package barrier properties.

Unlike glass and metals, no single thermoplastic packaging polymer offers an absolute barrier to O_2 or water vapor. Despite this limitation, monolayer polymer structures have the ability to satisfy the barrier requirements for packaged foods in many cases. Other situations require combinations of several different types of polymers in the form of multilayer structures to provide an effective barrier for the intended shelf life of the packaged product while also having the ability to withstand processing conditions and achieve a good-quality seal. Multilayer polymer materials provide varying degrees of barrier protection, which are highly dependent on the molecular structure of the individual monolayers used in their manufacture. Several other factors affecting polymer permeability include environmental temperature and relative humidity, polymer free volume, polymer crystallinity, polymer chain orientation, and penetrant concentration (Dhoot et al., 2002). The selection of a barrier polymer for a specific application typically involves tradeoffs between permeation, mechanical, and aesthetic properties as well as economic and recycling considerations. The O_2- and water vapor–barrier properties of important polymer packaging materials for foods vary over several orders of magnitude as summarized in Figure 13.2. Transmission rate data from polymer film samples is extremely useful in initial material evaluations, research and development applications and prioritizing selections of candidates for a package. The O_2-transmission rate (OTR) and the water

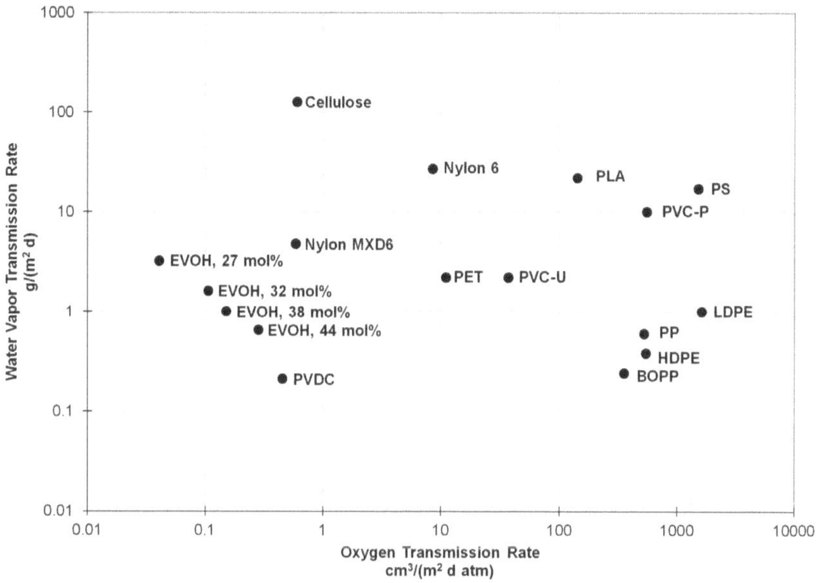

FIGURE 13.2 Transmittance for O_2 and water vapor of food packaging polymers at 23 °C, where permeability was normalized to 100-μm film thickness. Water vapor permeation was measured at 85% relative humidity (with the exception of nylon MXD6 at 90% RH). *Adapted from Langowski (2008).*

vapor transmission rate (WVTR) can vary by approximate factors of 40,000 and 600, respectively (Langowski, 2008). The final package configuration should also be analyzed, especially when developing shelf life predictions, since other variables such as the product volume-to-packaging surface area ratio can have a considerable impact on a product's oxidative stability.

It is difficult to provide an objective definition of a barrier polymer but a material that provides a sufficient barrier for a specific end-use requirement can be considered to be a barrier polymer for that purpose. The polyolefins are an important class of polymers which include low density polyethylene (LDPE), linear low-density polyethylene (LLDPE), high-density polyethylene (HDPE), and polypropylene (PP) and are generally characterized by their excellent barriers to water vapor, poor barrier to O_2, and good sealing properties. Polystyrene (PS) is highly permeable to O_2 and water vapor. As a result, one of its largest applications is in the packaging of fresh produce. Unplasticized poly(vinyl chloride) (PVC) has higher water vapor permeability but lower O_2 permeability than that of the polyolefins. The gas and moisture barrier properties of PVC are generally decreased as the material is plasticized dependent upon the type and content of plasticizer. Poly(ethylene terephthalate) (PET) is a polyester that has a much higher impermeability to gases but a lower impermeability to water vapor relative to the polyolefins. Nylons generally have low O_2 permeabilities when the films are dry but are highly permeable to water vapor due to its polar amide group. Nylon-6 is a polyamide commonly used in films that has relatively good O_2-barrier properties but poor water vapor barrier properties and is sensitive to relative humidity (RH).

A high-barrier food packaging material has a low O_2 permeability that has been generally defined as an OTR of less than 1 $cm^3/(m^2\,d\,atm)$ (López-Rubio, 2011). Ethylene–vinyl alcohol (EVOH) copolymer, poly(*meta*-xylene adipamide) (commercially known as nylon MXD6), or poly(vinylidene chloride) (PVDC) can be considered as very suitable selections for high O_2-barrier packaging film structures. EVOH copolymers offer superior barriers to O_2, water vapor, and aromas as well as demonstrate excellent processing properties. These properties have enabled polymer packaging containing EVOH barrier layers to replace many metal and glass packaging formats. Nylon MXD6 has better O_2-barrier properties than nylon-6 and PET at all humidities and is better than EVOH at 100% RH. Nylon MXD6 has significantly higher gas and moisture barrier properties than other nylons. PVDC copolymer has a unique combination of low permeability to water vapor and O_2 due to its crystallinity and low free volume in the amorphous phase (Robertson, 2013).

Figure 13.3 illustrates the effect of relative humidity on the O_2 permeability of important packaging polymers. The OTR of nylon-6 and nylon MXD6 increases considerably when environmental conditions are greater than 50% RH. As the ethylene content of EVOH copolymer increases, the O_2-barrier properties decrease and the moisture barrier properties improve. The O_2-barrier property

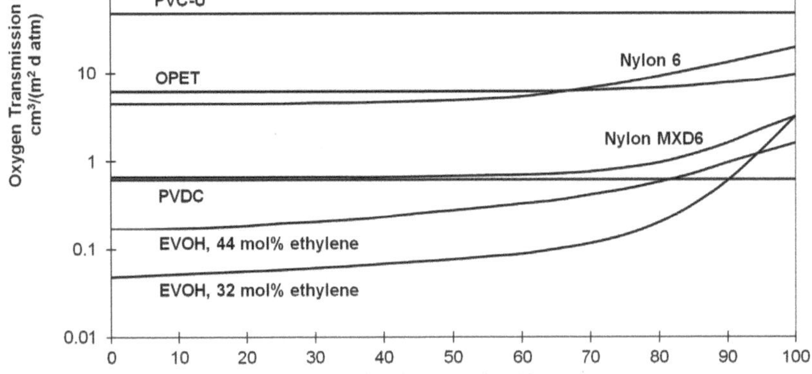

FIGURE 13.3 The effect of relative humidity on the O_2 transmittance of commercial polymers at 20 °C, where permeability was normalized to 100-μm film thickness. *Adapted from Kuraray America Inc. (2015).*

of EVOH is highly sensitive to moisture as shown by EVOH of 32 mol% ethylene having an increase in OTR of greater than one order of magnitude across the range from 0% to 100% RH. EVOH is often used as a barrier layer in laminate structures for thermal processing which require the EVOH layer to be embedded between two hydrophobic polymers for it to maintain its superior O_2-barrier properties. Current barrier technologies for polymer packaging, current trends influencing the use of polymer barrier packaging, and recent innovations in materials for high barrier flexible and rigid polymer packaging have been reviewed in extensive detail elsewhere (Lange and Wyser, 2003; Robertson, 2013). All active packaging technologies that function to inhibit lipid oxidation have the same general requirement of the presence of a passive high-O_2 barrier layer within the package structure.

13.3 FOOD–PACKAGING INTERACTIONS AND OXIDATIVE STABILITY

O_2 permeation resulting in auto-oxidation, UV–visible light transmission resulting in photosensitized oxidation, and water vapor permeation, which alters a product's water activity and subsequent lipid oxidation rate are potential packaging interactions that affect the oxidative stability of foods. The barrier properties of the selected packaging material determine the degree to which these oxidative mechanisms occur within packaged foods. The O_2 permeation of polymer materials is also dependent on O_2 concentration inside and outside the polymer package, which often provides a difference in partial pressures. With low O_2 content in the package headspace, a high OTR polymer generally leads to a net influx of O_2 molecules through the packaging.

The effect of a polymer's gas barrier properties on the oxidative stability of foods is well established with low O_2 permeability films resulting in extended shelf life. The oxidative stability of edible vegetable oils is very clearly affected by the O_2 permeability of monolayer polymer packaging materials. After 60 days of storage in the dark at 24 °C and 37 °C, vegetable oils packaged in bottles had oxidative stability rankings as follows: transparent glass > PVC = PET > PP > PS (Tawfik and Huyghebaert, 1999). PET and PVC gave a degree of oxidative stability which was closer to glass, while PS with the highest permeability rate caused a rapid deterioration in the quality of olive, sunflower, and palm oils. The oxidation of unsaturated linoleic acid in corn masa flour ($a_w = 0.45$) packaged in polyethylene pouches at 55 °C at 55% RH was very fast, with a loss of 71% when stored for 113 days (Márquez-Castillo and Vidal-Quintanar, 2011). In contrast, multilayer pouches of EVOH, nylon, and polyethylene offered significant oxidative protection under the same storage conditions with only a 12% loss of linoleic acid over 117 days. Corn masa flour packaged in coextruded EVOH pouches have a longer shelf life than masa flour packaged in polyethylene pouches due to EVOH having a much higher barrier for O_2 permeation than polyethylene.

UV–visible light transmission through polymer and glass packaging materials can result in photosensitized oxidation that deteriorates the quality of contained food products. Sunflower oil packaged without a headspace in 1-L bottles of PET and glass was exposed to fluorescent light producing significantly higher peroxide content than oil stored in the dark (Kucuk and Caner, 2005). After 9 months of storage, sunflower oil packaged in glass had mean peroxide values of 8.7 mEq O_2/kg in light and 0.8 mEq O_2/kg in the dark. Sunflower oil packaged in PET bottles had peroxide values of 13.9 mEq O_2/kg in light and 1.1 mEq O_2/kg in the dark after 9 months of storage. The effect of the moisture barrier properties of polymer packaging on the water activity of a food and its resultant impact on lipid oxidation rate has received little attention, but it is believed that it plays a minor role in affecting a product's oxidative stability compared to either O_2 permeation or light transmission.

13.4 MODIFIED ATMOSPHERE PACKAGING

Reduction of the O_2 concentration of a packaged food is a common method used to inhibit lipid oxidation. Modified atmosphere packaging (MAP) is the enclosure of food in a package whereby the atmosphere inside the package is modified or altered to provide an optimum atmosphere for increasing shelf life and maintaining food quality. Active modification of the atmosphere involves displacing the air with a controlled, desired mixture of gases (CO_2, O_2, N_2, CO, Ar), in a procedure generally referred to as gas flushing. Vacuum packaging can also be considered a form of MAP since removing most of the air has modified the atmosphere inside the package. O_2 is generally excluded or reduced to a concentration as low as possible since it promotes lipid oxidation, pigment oxidation, browning reactions, and growth of common spoilage microorganisms. There are a few exceptions where O_2 is required inside the package, such as for the respiration of fruits and vegetables or retention of color in red meat.

Essentially any concentration of O_2 is sufficient for promoting lipid oxidation because of the exponential nature of the diffusion-limited reaction involving the addition of O_2 to the alkyl radical (McClements and Decker, 2008). The auto-oxidation rate of edible oil at greater than 4% headspace O_2 content was independent of O_2 concentration and was directly dependent on the concentration of lipid (Labuza and Dugan, 1971). However, at headspace O_2 concentrations of less than 4%, the oxidation rate was solely dependent on O_2 concentration and was independent of lipid concentration (Karel, 1992). Below 0.5% headspace O_2, the oxidation rate constant of vegetable oil at 50 °C increased rapidly as the O_2 concentration was increased. However, at concentrations of 1–2% O_2, oxidation continued to increase but at a much slower rate (Andersson and Lingnert, 1999).

Complete removal and displacement of atmospheric O_2 in the package headspace and that dissolved or physically entrapped within the food matrix by an inert gas or vacuum remain difficult processes. The higher solubility of O_2 in the lipid phase of foods than in the aqueous phase also serves to complicate removal of dissolved O_2 (McClements and Decker, 2008). The effectiveness of MAP packaging is dependent on the inherent barrier properties of the polymer material and the efficiency of producing a hermetic seal after packing. The common high gas barrier structures have been used in MAP of nonrespiring foods, including laminates and coextruded films containing EVOH, PVDC, and nylons as a barrier layer. The inner seal layer is typically LDPE, which additionally provides a good moisture vapor barrier.

MAP has been suggested as a protective treatment to extend the shelf life of corn masa flour packaged in coextruded EVOH pouches at 55 °C under incandescent lighting (Márquez-Castillo and Vidal-Quintanar, 2011). The storage time to reach 10% oxidative consumption of unsaturated linoleic acid in corn flour was used as a threshold for evaluating the deterioration of product quality. A 10% loss of linoleic acid occurred within EVOH pouches under atmospheric air at 37 days and under vacuum treatment at 108 days, but was never attained in pouches under either CO_2 or N_2 atmospheres during 170 days of total storage time. MAP systems of EVOH–CO_2 or EVOH–N_2 showed the greatest inhibition of lipid oxidation of corn flour, achieving a 6-month shelf life without any direct addition of antioxidants, vacuum treatment, or storage in darkness. Sliced sausages packed in high-barrier polymer laminate films and stored for 60 days of light exposure under 100% N_2 atmosphere had higher lipid oxidative stability than those packed under vacuum as evaluated by thiobarbituric acid reactive substances (TBARS) and cholesterol oxidation products (Zanardi et al., 2002). Both MAP and vacuum packaging extend the shelf life of sausages by removal of molecular O_2. However, low residual levels of O_2 cannot be avoided in either case. Storage in atmospheres modified with N_2 to reduce O_2 levels below 10% significantly reduced auto-oxidation in whole and kernel hazelnuts (Martín et al., 2001). No major differences in oxidative rancidity of hazelnuts were observed at O_2 concentrations of 1%, 5%, or 10%. However, there was a significant difference in rancidity, as determined by peroxide values and concentrations of dienes, in hazelnuts stored under O_2 atmospheres of 10% or less and hazelnuts stored under 20% O_2 atmospheres.

The protective effect of MAP at reducing lipid oxidation rates is variable across different food applications. This is likely due to a number of factors, including the improper selection of gas barrier polymer materials, the lack of package characterization, variable residual O_2 concentrations after gas flushing, and numerous different methods to assess oxidation in foods which make comparable evaluations challenging. For example, use of a N_2 atmosphere reduced conjugated diene formation in peeled, raw almonds compared with those packaged in air after 150 days of storage, but no protective effect was observed in roasted almonds (Sanchez-Bel et al., 2011). In addition, N_2 atmosphere had no effect on hydroperoxide formation or sensory scores for organoleptic quality in either roasted or peeled almonds. However, it is important to highlight that the researchers used PP/PE bags, which have high gas permeability, for packaging of almonds, and the N_2 atmosphere was only modified to attain levels of 2–3% O_2 and 0.2–0.3% CO_2. Potato crisps stored under an air atmosphere for 3 months experienced a loss of polyunsaturated fatty acids (PUFA), while those stored under N_2 and vacuum atmospheres retained higher levels of PUFAs (Sanches Silva et al., 2004). All potato crisps were packaged with the same materials in this study, but the barrier properties of the packaging film were not identified. While there was no significant difference between N_2 atmosphere and vacuum in terms of oxidative stability, the vacuum treatment resulted in product deformation unlike MAP. Sliced cheese packaged in a PS container with a polyester barrier layer and lid under MAP conditions of 25% CO_2 and 75% N_2 (initial residual O_2 content of 0.4%) showed no significant difference in oxidative stability after a 21-day continuous light exposure as determined by peroxide value (Kristensen et al., 2000). However, sensory evaluation of odor and taste attributes presented significant differences between MAP-packaged cheeses exposed to and protected from fluorescent light.

13.5 ACTIVE FOOD PACKAGING

Active packaging systems offer a variety of additional complementary tools in the effort to control lipid oxidation by removing unwanted compounds from the food (dissolved and headspace O_2, transition metals, or off-flavors/odors), adding a desirable compound to the food (antioxidant), or preventing degradative UV–visible light from impacting the food (UV absorber, pigment). Robertson (2013) defined active packaging as "packaging in which subsidiary constituents have been deliberately included in or on either the packaging material or the package headspace to enhance the performance of the package system." The performance of such deliberately interactive packaging should maintain or improve the sensory, safety, and quality aspects of the food. The normal application of MAP is passive as opposed to active packaging, unless there is some way in which the package or its components actively affect the internal gas atmosphere by means other than normal permeation through polymer films. The design and production of a functional and efficient active flexible package can be difficult to realize because of the complexity of such systems.

The array of active packaging technologies available to prevent lipid oxidation all require a high gas-barrier packaging structure that is hermetically sealed in order to have the potential to function properly and provide optimal benefits to the contained food. Since oxidation is a free radical chain reaction, an extremely proactive approach must be taken to prevent the initiation step, where formation of a low concentration of free radicals occurs to begin the oxidative process. The most O_2-sensitive food products to be packaged in flexible polymer packaging require hurdle technology to achieve their greatest oxidative stability and extension of shelf life. A combination of complementary approaches using a passive high-O_2 barrier, modified atmosphere, and active packaging systems (O_2 scavengers, UV absorbers, antioxidant controlled release, transition metal chelation, and aldehyde scavengers) that have a high capacity to prevent radical initiation and hydroperoxide decomposition is needed to maintain the oxidative stability of such foods. Active polymer packaging systems have the goal of more effectively limiting oxidation, maintaining nutritional quality, inhibiting the formation of harmful oxidation products, and extending the shelf life of foods.

13.5.1 O_2-Scavenging Systems

The typical residual O_2 levels achievable by MAP or vacuum packaging are in the 0.3–3.0% range. Since essentially any concentration of O_2 is sufficient for promoting lipid oxidation, additional technologies are needed to reduce this residual O_2 content within a package to an absolute minimum. O_2 scavengers are materials incorporated into package structures that chemically combine with O_2 from the inner package environment, including the food product itself through diffusion, for effective removal of O_2 (Brody et al., 2001d). Despite the numerous publications on active packaging over the last 30 years, there are only a few systems currently on the market which can be considered commercially significant (Robertson, 2013). Ferrous iron-based compounds, ascorbic acid and its derivatives, catechol, organometallic ligands, oxidative enzymes, unsaturated hydrocarbons, and specific nylons are the most important classes of O_2 scavengers.

Active barrier packaging materials have a finite O_2-absorption capacity; therefore, they require protection from O_2 in the external environment so that their O_2-absorption capacity is not wasted. The properties and characteristics of commercially available O_2-scavenging systems along with their manufacturers have been summarized in detail (Pereira de Abreu et al., 2012). O_2-scavenging packaging systems can be classified into two categories: (1) sachets or labels containing O_2-absorbing compounds placed inside packages as a component and (2) O_2-absorbing materials with active compounds incorporated directly into or on the packaging materials.

13.5.1.1 Sachets and Labels

The most widely used O_2 absorbers are contained within small sachets holding a combination of various iron-based powders and catalysts that scavenge O_2 by preferentially oxidizing within the food package. The sachets are microperforated or highly permeable to gases and water vapor, resistant to damage, and compatible with food. Ferrous iron–based scavengers are by far the most effective O_2-scavenger material in commercial use. Powdered iron is used to provide a large surface area for reaction and water is required for the O_2 absorbents to function. In some cases water is added during sachet manufacture, while in others moisture must be adsorbed from the food before O_2 can be absorbed. One type can be stored dry without absorbing O_2 until exposed to a_w greater than 0.85, while another requires storage in a low or zero O_2 atmosphere since they begin to react immediately. O_2 is irreversibly converted to a stable oxide by iron powder with the overall reaction as follows:

$$Fe + \frac{3}{4}O_2 + 1\frac{1}{2}H_2O \rightarrow Fe(OH)_3$$

The use of iron powder enables the reduction of O_2 concentration in the package headspace to less than 0.01% (100 ppm).

Characterization of O_2 absorbers can be accomplished by the two main properties of absorption capacity and absorption rate constant (Robertson, 2013). The absorption capacity of 1 g of iron is equal to approximately 300 mL of O_2. Commercial O_2 sachets are produced in many different sizes with the ability to consume 20–2000 mL of O_2. The absorption rate is often the property of most importance in terms of food quality, but few studies evaluating this parameter have been performed. In PVDC pouches containing air at 0% RH, a commercial O_2-scavenger sachet did not experience significant changes in its scavenging capacity of about 110 mL of O_2 at temperatures of 5 and 35 °C (Charles et al., 2006). However, the absorption rate constant increased with increasing temperature with values of 0.027/h at 5 °C and 0.210/h at 35 °C. An individual O_2-scavenger sachet in a PVDC pouch containing air at 20 °C and 0% RH absorbed about 9% of its maximal O_2 volume after 1 h and reached about 78% of capacity after 10 h, which emphasized the need to rapidly place self-activated scavengers inside the package followed immediately by sealing.

Ferrous iron O_2 scavengers were commercially developed by Mitsubishi Gas Chemical Company under the trade name Ageless® but are also available from several other suppliers under various trade names (Brody et al., 2001b). According to the manufacturer, Ageless® sachets are able to maintain packages of snack foods at less than 0.1% O_2 over a 3-month shelf life at ambient temperature. The selection of absorbent type and size depends on many factors including the desired shelf life of the food, properties of the food (a_w, dissolved O_2, size, shape, weight), initial O_2 level in the package headspace, and the O_2 permeability of the packaging material. For example, a sachet containing 1 g of ferrous iron has the capacity to control O_2 permeation for 84 days for a 76-μm-thick polyester package with an OTR of 4 $cm^3/(m^2\,d)$ and a surface area of 1 m^2 (Brody et al., 2001b). This calculation assumes that the oxidation rate of ferrous iron to the ferric state is faster than the diffusion rate of O_2 into the package.

While sachets and pads can be a highly efficient form of active packaging, they have some disadvantages. They should not be used in vacuum packs or flexible film packages where the film may cling to the sachet and separate it from areas where it is needed to perform its function. Sachets can be adhered to the inner wall of polymer packaging having greater rigidity, or labels containing the O_2 absorber can be affixed to the inner package wall to overcome this problem. Sachets and labels are also not suitable for liquid foods or high humidity applications, and the iron-based varieties typically are incompatible when metal detectors are installed on packaging lines. The potential accidental ingestion of the contents of an iron-based sachet by the consumer has been suggested as a detriment to even further widespread commercialization. O_2-absorbing adhesive labels attached to the inside of the package have helped to solve the problem of possible sachet ingestion. Metallized self-adhesive labels hold a dispersion of iron powder in microporous polymer strips, but the O_2-absorbing capacity of these active labels is limited to around 100 mL due to less availability of iron for reaction with O_2 compared to sachets. It is also important to note that the use of sachets or other O_2-scavenging systems to create an O_2-free atmosphere could present a food safety risk from pathogenic anaerobic bacteria, such as *Clostridium botulinum,* for low-acid foods under high a_w and unrefrigerated conditions (Skinner and Reddy, 2006).

13.5.1.2 O_2-Absorbing Materials

The incorporation of O_2-absorbing materials directly onto or into the polymer packaging structure can present an attractive alternative to sachet and label O_2 scavengers. O_2-scavenging polymers offer several potential advantages over sachets including capability for use under retort conditions, elimination of food product distortion that may occur when a sachet is in contact with food, and increased production efficiency and convenience from simplified packaging and filling processes. The versatility of O_2-absorbing polymers is demonstrated by the ability to vary their layer thickness or blend composition to match the O_2 concentration to be removed. O_2-absorbing systems can be incorporated into package structures and various films with the required reactive components dispersed within the polymer matrix or sandwiched between film layers. The speed and capacity of O_2-absorbing films are considerably lower than O_2 scavenging sachets, which is a primary limitation to their use on a broader scale. The activation or triggering mechanism of O_2-absorbing films should ideally be stable in atmospheric air and not begin consuming O_2 until the food is packaged, which presents an additional challenge. Exposure to light irradiation or elevated moisture have served as the main approaches used to activate O_2-absorbing films, however, storage of O_2-absorbing materials in high-barrier bags with a positive pressure of nitrogen is also used to prevent O_2 absorption until the package forming process.

Polymer-based O_2 scavengers can be extruded into a multilayer packaging film structure as an active layer. O_2-scavenging polymer structures in which the reactive component is not in direct contact with the package headspace are limited by the time required for O_2 to diffuse to the reaction centers. Not only is the O_2 permeability of the polymer reduced at lower temperatures, but also the scavenging reaction rate is reduced with decreasing temperatures. An idealized multilayer packaging structure containing an integrated active layer of an O_2 scavenger is illustrated in Figure 13.4. In this example, the moisture-sensitive EVOH barrier layer is protected from water by LDPE as the hydrophobic structural layer. The passive high gas barrier (EVOH) layer serves to protect

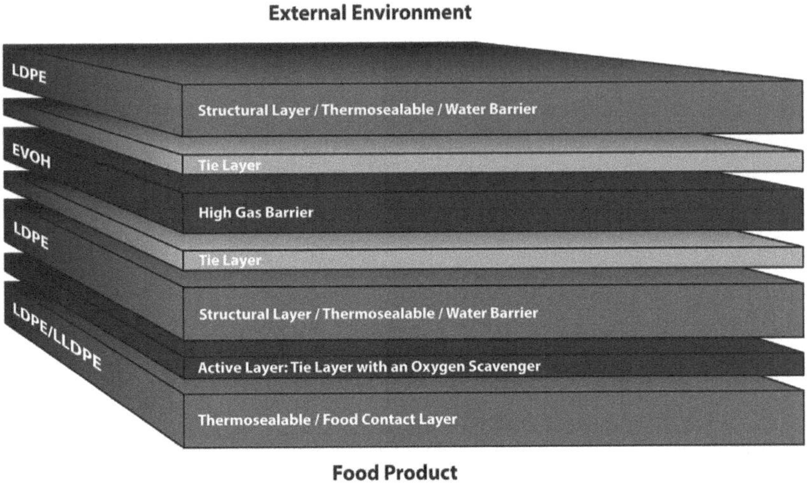

FIGURE 13.4 Schematic ideal O_2-scavenging multilayer packaging film structure with some common types of structural and barrier polymers.

the O_2-scavenging layer from exhaustion by O_2 in the external environment, thereby conserving the O_2-absorption capacity of the scavenging layer for the food product and package headspace. The high O_2-permeability properties of the food-contact LDPE layer are used to minimize any reduction in the O_2-scavenging reaction rate of the active layer, while LDPE still provides excellent heat sealing properties.

Many of the polymeric O_2-scavenging systems are proprietary technologies with the majority of their characteristics and structures described only in patent literature. For example, one of the first polymeric O_2-scavenging systems developed to provide a total barrier to O_2 was a PET/nylon MXD6/cobalt structure (Brody et al., 2001b). The cobalt salt of an organic acid catalyzes the reaction of nylon MXD6 with the O_2 as it permeates through the wall of the package. In comparison, a reference 1-L PET bottle has an OTR of $3.5\,cm^3/(m^2\,d)$ at 22 °C, while a 1-L PET/nylon MXD6 (4%)/cobalt bottle has an OTR of less than $0.04\,cm^3/(m^2\,d)$ at 22 °C. The very low OTR continues for a period of nearly 2 years under these conditions. The O_2-scavenging capacity is dependent on the specific composition of the nylon MXD6 and cobalt blend and the wall thickness of the package structure. The application of this approach has been limited by the development of more cost-effective methods of providing an O_2 barrier in PET bottles.

Polymeric O_2-scavenging systems have been developed that absorb O_2 in modified atmosphere packages and serve as total O_2 barriers upon activation. A transparent, moisture-independent, polymer-based O_2-scavenging film was developed composed of an oxidizable polyolefinic polymer, a photoinitiator, and a catalyst (Brody et al., 2001a). The oxidizable polymer functioned to bind molecular O_2, the photoinitiator absorbed UV radiation and provided the energy to begin the reaction, and the catalyst aided in increasing the rate of the scavenging reaction. The O_2-scavenging layer was inactive until activation by exposing the film to UV radiation on the packaging line. Upon activation, the O_2-scavenging reaction proceeded until either the headspace O_2 was consumed or the film capacity was exhausted.

Ascorbic acid and its derivatives readily react with and remove atmospheric O_2 upon their conversion to dehydroascorbic acid. This oxidation or hydrogen transfer reaction is catalyzed by the presence of metal ions such as iron. Ascorbic acid O_2 scavengers have been used commercially by their incorporation into the closure liners of polymeric bottles (Brody et al., 2001b). The use of enzymes, most commonly glucose oxidase plus catalase, as O_2 scavengers incorporated into polymer material has been investigated. Since many foods may have minimal contact with the package surface, it has been recommended that the scavenging reaction occur separately from the food by placing the enzymes in liquid form, including glucose as the reactant, between two polymer layers (López-Rubio et al., 2004). In this system, the outer polymer layer serves as a high gas barrier while the inner food contact layer is highly gas permeable.

Many problems in using polymer-based O_2 scavengers result from the difficulty of guaranteeing the functional performance of the package scavenging activity over time while also maintaining the mechanical and optical properties of the packaging. The shelf life of extra virgin olive oil and sunflower oil packaged in monolayer PET bottles loaded with 1% and 5% of an O_2 scavenger (OXISTOP™) and stored at room temperature under diffuse lighting conditions has been studied (Sacchi et al., 2008). The effect of the O_2 scavenger on dissolved O_2 content of vegetable oil was observed during the first 3 months of storage. PET bottles with 5% O_2 scavenger showed a dissolved O_2 content in oils, which was lower than that of oils packaged in PET with 1% O_2 scavenger and neat PET over the entire 6-month storage period. However, the O_2 scavenger material was found to have no significant influence on the quality and organoleptic characteristics of olive oil.

Active PET monolayer films were formed by cast extrusion of PET with the O_2 scavenger Amosorb DFC 4020 (Galdi et al., 2008). The O_2-scavenging component consisted of a transition metal as cobalt salt, which catalyzes unsaturated hydrocarbon dienes linked to a polyester backbone. The barrier properties of active PET films demonstrated the effectiveness of the O_2 scavenger at concentrations of 1, 3, and 5% in reducing the O_2 permeability of PET. PET with 5% O_2 scavenger had an O_2-scavenging capacity in the range of 0.81–0.86 cm^3/g, but its activity was limited to approximately 26 h. However, O_2-scavenging capacity and lifetime are strictly dependent on the scavenger concentration in the polymer matrix and on the film thickness. This study used active PET with a reduced film thickness of approximately 30 μm, while the O_2 scavenger was blended with PET for producing monolayer active bottles with wall thicknesses of about 250 μm. Active PET films displayed a significant ability to reduce the oxidative effect of color variation of fresh apple slices. The presence of an O_2 scavenger (Amosorb®) in 300-mL PET bottles was able to better maintain the quality and authenticity attributes of extra virgin olive oil over 13 months of storage (Cecchi et al., 2010). The protective role of the O_2 scavenger against O_2 permeation was confirmed by a reduced level of primary and secondary oxidation products in olive oil compared with that found in neat PET bottles stored under the same conditions. The decline in antioxidant activity during storage of olive oil was also reduced by the active barrier.

The effectiveness of an iron-based O_2 absorber incorporated into a blend of LDPE and LLDPE ($ABSO_2RB$ sealant) within an opaque laminate packaging structure of PET/Al foil/$ABSO_2RB$ sealant/PE was investigated for its ability to maintain and extend shelf life of a hot-filled cheese-spread as an MRE (meal-ready-to-eat) component (Gomes et al., 2009). The O_2-absorbing laminate was efficient in reducing headspace O_2 concentration from 20.4% to 6.8% within 24 h, and to less than 0.5% in 11 days of storage at 27 °C. Throughout the 1-year storage of packaged cheese spread, headspace O_2 levels remained below 0.5% and the development of rancidity in the product was reduced. The benefits of using O_2-scavenging polymers on retaining nutrition and prolonging shelf life of high-fat, liquid-based products have clearly been demonstrated. Another iron-based O_2 scavenger (SHELFPLUS™ O_2 2400) was found to compensate for the free diffusion of O_2 through a 10-μm pinhole defect in the sealing layer, but not through 17-μm defects (Sängerlaub et al., 2013).

Polymer-based light-sensitive films have been developed that use energy transfer to convert triplet O_2 to highly reactive singlet oxygen ($^1O_2$), which is absorbed irreversibly by the film (Maloba et al., 1996). Photosensitive dyes, eosin and curcumin, were added to the singlet-O_2 scavenger, polyfuryloxirane (PFO), and blended with ethyl cellulose to form cast films of 20% PFO in ethyl cellulose. The O_2-scavenging PFO/ethyl cellulose film delayed increases of primary and secondary oxidation products in sunflower oil as shown in Figure 13.5 by peroxide value and hexanal concentration, respectively. Many O_2-scavenging

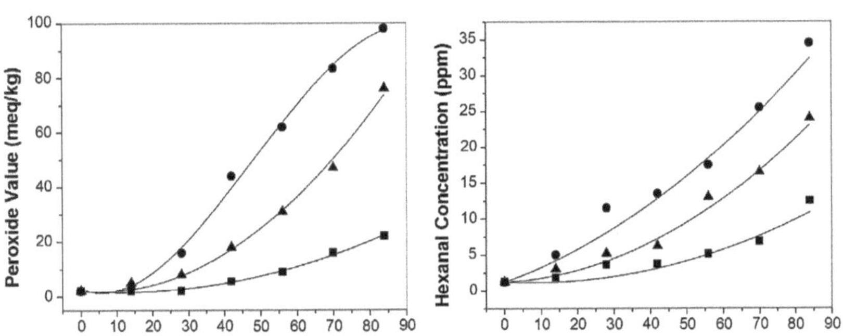

FIGURE 13.5 Peroxide values and development of headspace hexanal for sunflower oil stored at 23 °C. ●, sunflower oil control; ▲, sunflower oil with 0.02% BHT; ■, sunflower oil stored with 20% polyfuryloxirane (PFO) in ethyl cellulose O_2-scavenging film. *Adapted from Maloba et al. (1996).*

systems, such as iron-based scavengers, reduce residual O_2 levels in a package to less than 0.1% relatively slowly, taking from 1 to 4 days. Headspace O_2 levels can be reduced to less than 1% with an irradiation period of 1–2 h using PFO/ethyl cellulose film. The high scavenging rate of PFO/ethyl cellulose films offers the potential for some food packages to be scavenged of most of their O_2 while on the filling line. In this study, the film was an insert in the food package, but its greatest potential use would be integrated as a layer in a laminate structure. Figure 13.5 illustrates the overall goal of each active packaging system reviewed in this chapter to delay the initiation step of auto-oxidation or photo-oxidation and its deteriorating effects on food quality in order to extend food product shelf life.

13.5.2 Light Screening Pigments and UV Absorbers

The optical properties of a packaging material relate to the degree of light penetration and the subsequent effect of that transmission. Shorter wavelength radiation, such as UV light (200–400 nm), is more energetic than visible light (400–750 nm) and, therefore, often more detrimental to a food's oxidative stability and nutrient quality. Visible light of shorter wavelengths has also caused more detrimental effects on photosensitized oxidation of lipids than longer wavelength visible light (Lennersten and Lingnert, 2000; Sattar et al., 1976). Photo-oxidative protection of specific foods can be provided by packaging materials depending on several factors, including (a) inherent absorption characteristics of the material, (b) material thickness, (c) material processing conditions, and (d) addition of UV–visible light absorbing compounds to the material (Mortensen et al., 2004). Packaging materials with a high barrier to UV–visible light, such as metal, paperboard, and aluminum foil–polymer laminates, will always be preferred to maintain optimal food quality from photo-oxidation. However, consumers continue to place great value on the ability to view the product through the package. This same optical clarity allows visible and often UV light to impact the product, which has the detrimental effect of initiating photo-oxidation. The UV–visible light transmission of glass and thermoplastics are important properties since $^1O_2$ oxidation occurs in the presence of light.

Clear flint glass transmits nearly all visible light above 400 nm with a decreasing degree of UV light transmission to almost a complete block as wavelengths approach 300 nm (Maury et al., 2010). Transmission may be controlled by incorporation of color additives such as metallic oxides, sulfide, or selenides (Robertson, 2013). Amber glass provides light protection by allowing less than 10% incident radiation across the wavelength range of 290–450 nm. Antique Green (dark brown/green) and Classic Green (darker green) glass both reduce the amount of visible light reaching the product within the bottle, although they also allow a small amount of UV light transmission (Maury et al., 2010). The UV cut-off wavelength for the darker bottles is greater than for the lighter colored bottles, however, all bottle types allow UV transmission of different intensities around 350 nm. Olive oil packaged in clear glass bottles lead to photo-oxidation and reduced shelf life, while the use of green glass decreased the rate of photo-oxidation (Kiritsakis et al., 2002).

Polymer packaging materials transmit similar intensities of visible light; however, they offer varying degrees of protection against damaging UV wavelengths. LDPE, HDPE, and PP provide practically no light barrier with almost complete transmittance in the UV–visible region, while PET absorbs UV radiation at wavelengths below around 300 nm (Coltro et al., 2003; Lennersten and Lingnert, 2000). Polymer materials can be incorporated with UV absorbers to block UV wavelengths without affecting the transmission of visible light or optical clarity of the package. UV absorbers are compounds that minimize photo-oxidation by absorption of incident UV energy and accompanying quenching of photoexcited chromophores. The ideal UV absorber should absorb all terrestrial UV radiation, but still allow transmittance of all radiation with wavelengths greater than 400 nm. Commercial UV absorbers typically have an absorption spectrum that does not tail significantly into the visible region which would result in a yellowing of the polymer. Different classes of commercial UV absorbers are used including phenolic derivatives of benzophenone, benzotriazole, and 1,3,5-triazine and nonphenolic UV absorbers represented by oxanilide, malonate, and bifunctional malonate/hindered amine stabilizer (Pospíšil and Nešpůrek, 2008). Pigments, such as titanium dioxide, may also be added into polymer materials to increase light scattering and, therefore, reduce transmittance of light especially at wavelengths shorter than 400 nm (Mortensen et al., 2004).

Light-induced lipid oxidation of mayonnaise without packaging material was accelerated most by irradiance at 365 nm, followed by 405 and 435 nm, with no effect at wavelengths longer than 470 nm (Lennersten and Lingnert, 2000). Mayonnaise stored in PET exposed to cool white fluorescent lighting showed higher hexanal concentration than that stored in polyethylene naphthalate (PEN) after 41 days. The UV cut-off of PEN near 380 nm acts as a UV protective filter similar to that of addition of a UV absorber to PET by extending the UV cut-off of PET from around 320 nm. The irradiance peak at 365 nm was determined to cause greater lipid oxidation in mayonnaise stored in PET, which allowed approximately 40% transmittance at 365 nm, than that stored in PEN, which only allowed around 1% transmittance at that wavelength. However, even though PEN effectively blocked the UV contribution from the fluorescent light exposure, the mayonnaise still became oxidized indicating that visible light also had a considerable effect in this food system.

PET bottles used for packaging edible oil that contained 0.08% of the UV absorber Tinuvin 326 (benzotriazole class) promoted a 90% reduction of light transmission at 360 nm relative to a neat PET bottle (Coltro et al., 2003). In Figure 13.6, the light protective UV cut-off of PET at about 320 nm is extended into the long UV wavelength of near 380 nm with use of different concentrations of a UV absorber. Further improvements in light barrier were expected to

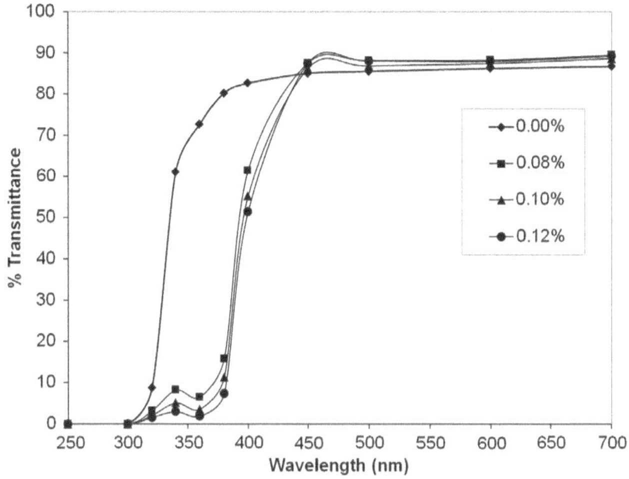

FIGURE 13.6 UV–visible light transmission spectra of poly(ethylene terephthalate) (PET) bottles (200 μm average thickness) with different concentrations of UV absorber Tinuvin 326. *Adapted from Coltro et al. (2003).*

be small at concentrations greater than 0.11% Tinuvin 326. To protect soybean oil in transparent packages from photo-oxidation, Tinuvin 326 was dispersed within the regrind layer of coextruded multilayered PP packages containing an EVOH barrier layer (Pascall et al., 1995). Soybean oil in packages with 0.3% Tinuvin 326 experienced less photo-oxidation than oil in packages without any UV absorber when exposed to fluorescent light at 21 °C and 35 °C for 35 days. The UV absorber Tinuvin 234 (benzotriazole class) added to PET bottles was found to be effective in reducing the sensory changes of soybean oil and enhancing its oxidative stability during 6 months of storage at 25 °C under fluorescent lighting (De Azeredo et al., 2003). Tinuvin 234 was added at concentrations ranging from 0.05 to 0.20% in PET, but 0.1% was the minimum concentration required to achieve adequate oil stability during 6 months of storage.

Transparent polymer materials with UV absorbers can be used to display light-sensitive products on retail shelves by offering nearly complete protection from UV radiation. However, it is important to remember that many foods contain natural pigment compounds, such as riboflavin and chlorophylls, which act as efficient photosensitizers by absorbing visible light and initiating photo-oxidation without the food ever having received any UV light exposure. For example, many researchers have shown that exposure to visible light causes a significant increase in light oxidation in milk due to riboflavin and other photosensitizers (van Aardt et al., 2001; Webster et al., 2009). While transparent PET bottles containing a UV absorber provided some protection against light oxidation of milk, it was not extensive enough to extend the shelf life longer than milk packaged in neat PET bottles (Mestdagh et al., 2005).

The UV absorber did not completely exclude wavelengths above 400 nm which are detrimental to riboflavin. Photo-oxidation occurs when vegetable oils are exposed to light through the action of natural photosensitizers such as chlorophyll. Extra virgin olive oil packaged in clear PET bottles and exposed intermittently to light at 22 °C was found to retain its quality characteristics for a period of less than 3 months (Pristouri et al., 2010). Addition of a UV absorber in the PET bottles provided only partial light protection to the olive oil with a shelf life of 3 months while complete protection from light using aluminum foil extended the shelf life to at least 9 months. The retention of olive oil quality was found to be most affected by light, followed by the container headspace and OTR of the packaging material. If visible light with wavelengths that are damaging to a photosensitizer is not completely blocked by the packaging material, it can have the ability to photo-oxidize and degrade the organoleptic quality of the contained food.

13.5.3 Antioxidant Controlled-Release

Antioxidants have traditionally been used as direct food additives in a single initial dose to protect food against generation of free radicals and lipid oxidation. Antioxidants significantly extend the shelf life of foods containing lipids susceptible to oxidation such as vegetable oils, animal fats, flavorings, spices, cereals, nuts, processed meats, and snack products. Butylated hydroxyanisole (BHA) and butylated hydroxytoluene (BHT) while being classic lipid antioxidants in food have also been incorporated into polyolefin polymers to inhibit oxidation of the polymer material itself. The direct addition of antioxidants to food in one large initial dose is limited by the potential for rapid depletion of the antioxidants, in addition to very high initial concentrations which may unintentionally degrade product quality. Under certain conditions, compounds with antioxidant activity may exhibit pro-oxidant action (Lankin et al., 1999; Takahashi, 1995). Natural antioxidants, including phenolic antioxidants and tocopherols, have been shown to undergo loss of activity and become pro-oxidants at high concentrations (Jadhav et al., 1996). The motivation for controlled-release antioxidant packaging is to deliver one or more active components from the package material to an O_2-sensitive food at a specific optimized rate throughout the product shelf life. The continuous replenishment of antioxidants to the food from the packaging material which serves as a reservoir, offers the potential to eliminate the direct addition of antioxidants into the food. Controlled-release antioxidant packaging must be used on an unoxidized food to minimize free radical initiation since it is a futile process to add antioxidants to an already oxidized substrate.

The ultimate challenge of controlled-release packaging is matching the rate of antioxidant diffusion from the food contact layer with the requirements of the specific food to maintain oxidative stability. Active compounds can be incorporated into food packaging for the release of nonvolatile antioxidants into the food or volatile antioxidants into the headspace or atmosphere surrounding the food by a controlled mechanism of diffusion. The selected antioxidant and packaging material should be compatible in order to achieve a homogeneous distribution, and the partition coefficients of the antioxidant should favor its release from the package to the food or headspace. After mass transfer into the food, the effectiveness of the antioxidant will be determined by its solubility characteristics. Therefore, it is critical that the type of antioxidant be selected as a function of the type of food to be packaged. Nonpolar antioxidants are generally more suitable for foods with high lipid content and polar antioxidants are generally more appropriate for more polar foods. However, the "antioxidant paradox" must also be considered where hydrophilic free radical scavengers (FRS) may be more effective antioxidants than hydrophobic FRS in bulk oils, while hydrophobic FRS may be more effective in emulsified oils (Porter, 1993). The often complex customization required for the development of antioxidant-release packaging systems will present new challenges for large-scale package manufacturing.

The ability of antioxidants incorporated into flexible packaging materials to delay lipid oxidation in food products has been known for many years. Rho et al. (1986) showed that coating the inner surface of a polyethylene package with *t*-butylhydroquinone (TBHQ) extended the shelf life of deep-fried instant noodles to twice that of an equal amount of TBHQ added directly into the frying oil. Hoojjat et al. (1987) demonstrated the effectiveness of HDPE film loaded with BHT in delaying lipid oxidation of an oat flake cereal due to mass transfer of the antioxidant from the package to product by an evaporation/sorption mechanism. The cereal contained in HDPE pouches with a higher BHT level (0.32%) showed greater oxidative stability than cereal stored in HDPE pouches with low BHT content (0.02%) (Miltz et al., 1988). BHT could not be detected in HDPE pouches after 6 weeks of storage at 39 °C, but 19% of the BHT originally present in the packaging material was found to be adsorbed on the cereal. BHT and BHA-treated PE films in contact with vegetable oils have been shown to have a stabilizing effect on stored oils against oxidative degradation (Sharma et al., 1990). An additional application of BHT-incorporated PE film was the inhibition of lipid oxidation in both fish muscle and oil (Huang and Weng, 1998). The two main mechanisms of incorporating antioxidants into polymer packaging materials are coating by solution casting (Figure 13.7(A)) and melt compounding followed by extrusion (Figure 13.7(B)).

13.5.3.1 Antioxidant Compounding

Antioxidant additives can be directly incorporated into the polymer matrix in the melt by mixing during the compounding process. Cast or blown films containing compounded antioxidant are formed as a result of an extrusion process. This technique is preferred by packaging converters since most conventional flexible packaging is formed either completely or partially by extrusion. The main potential

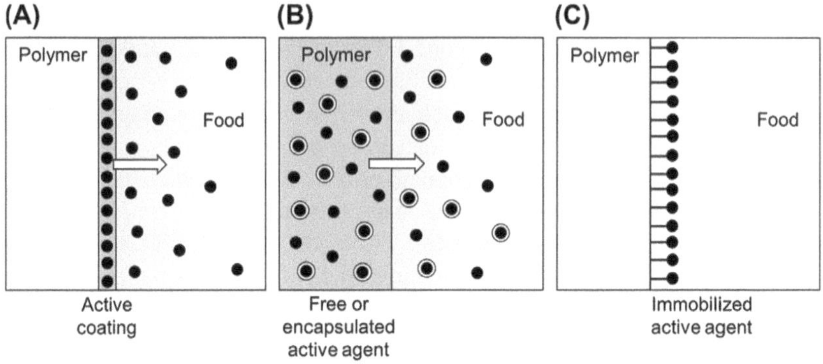

FIGURE 13.7 Antioxidant active packaging systems: (A) coating an active layer to the polymer surface; (B) incorporating free or encapsulated active agent into the polymer matrix; and (C) covalent immobilization of the active agent on the polymer surface. *Adapted from Tian et al. (2013b).*

limitation of this manufacturing process is that the antioxidant additive is exposed to high temperature, shearing force, and high pressure which could impact its active loading concentration and performance.

Original active film development with synthetic antioxidant release has shifted to the use of natural antioxidants, such as α-tocopherol, in packaging materials due to significant consumer and industry demand. The presence of a cereal product contained within an active package was found to affect the release of α-tocopherol from LDPE pouches (Wessling et al., 2000a). LDPE pouches containing oatmeal lost approximately 400 mg/g of α-tocopherol during 4 weeks of storage at 40 °C, but LDPE pouches stored without food had retained 900 mg/g of α-tocopherol during the same period. The different characteristics of food, such as contact phase properties, fat, alcohol, and acid content, influence the retention of α-tocopherol within the LDPE packaging material (Wessling et al., 1999). In PP film, α-tocopherol retention was found to be nearly complete; however, a significant loss of α-tocopherol from LDPE film was observed when the film was in contact with mayonnaise, cream, and low-fat milk. α-Tocopherol levels of 0.1% and 1.0% have been used in LDPE films with the dual function of protecting the polymer from oxidative degradation during processing and delaying the onset of oxidation in the packaged food (Wessling et al., 2000b). High levels of α-tocopherol in LDPE films show a positive effect on the oxidation stability of linoleic acid emulsions at low temperatures, but may have detrimental effects on LDPE polymer characteristics (Wessling et al., 2000b). The shelf life of corn oil packaged in LDPE bags containing α-tocopherol (2% and 4%) was maintained for approximately 16 weeks compared to 12 weeks for oil in contact with neat LDPE film at 30 °C (Graciano-Verdugo et al., 2010). A detailed inventory of packaging films containing α-tocopherol and various natural antioxidants has recently been catalogued (Sanches-Silva et al., 2014).

Hydrophilic EVOH films containing quercetin (1% and 5%) and catechin (0.5% and 2%) were obtained by cast extrusion (López-de-Dicastillo et al., 2010a). Catechin-containing films showed greater release than quercetin in aqueous and alcoholic food simulants, but the extent and kinetics of release were dependent on the compatibility between the antioxidant and the food simulant. Fried peanuts packaged in EVOH films with 1% quercetin and 0.5% and 2% catechin under air atmosphere at 37 °C provided a great deal of oxidative protection (López-de-Dicastillo et al., 2012b). The first significant increment in hexanal concentration from peanuts stored in these active films was observed at 25 days, while those stored in neat EVOH had an equivalent level of hexanal at close to 5 days of storage. Films with 5% quercetin offered less oxidative protection than at the 1% level, indicating some degree of pro-oxidant activity at high antioxidant loading concentrations. In contrast, these EVOH films with 5% quercetin performed as the most efficient antioxidant packaging films in contact with sunflower oil as determined by lowest hydroperoxide content in the oil over 150 days of storage at 37 °C. The release profiles of quercetin into lipid simulant (95% ethanol) have been shown to be strongly influenced by polymer hydrophobicity (Chen et al., 2012). The initial release rate was much faster, the concentration released was much higher, and the equilibrium time was much shorter for EVOH compared to LDPE. These studies highlight the complex design required for antioxidant controlled-release packaging that involves the proper selection of an antioxidant that is compatible with both the polymer matrix and the type of food product to be packaged.

Active films containing 4% of a rosemary extract or 4% of an oregano extract enhanced the oxidative stability of lamb steaks packaged in modified atmosphere and displayed under illumination at 1 °C (Camo et al., 2008). Oregano-containing films were significantly more efficient than those with rosemary, having an effect similar to that of direct addition of rosemary extract. Carvacrol, a component of oregano essential oil, was incorporated into HDPE sheets at the 1% and 2% levels by extrusion and compression molding (Peltzer et al., 2009). The release of carvacrol into lipid and aqueous simulants was demonstrated indicating that this antioxidant may have activity across a wide range of food products. Sesamol, a component from sesame seed and oil, was evaluated for its antioxidant activity with 0.1% and 0.2% initial loading in co-extruded multilayer structures of LLDPE, HDPE, and ethylene–vinyl acetate (EVA) (Zhu et al., 2013). Sesamol films extended the induction period of linoleic acid oxidation by eight times at 23 °C compared with the control film. In oat cereal packaged with sesamol films after 12 months of storage at 23 °C, hexanal formation was reduced by 59% compared with cereal packaged with control films. Volatile antioxidants, such as sesamol and BHT, have the ability to provide distinct advantages for food products that have little direct contact with the inner surface of the packaging. An important consideration of antioxidant controlled-release systems is whether the natural antioxidants impart color to both the packaging and the food, such as quercetin and catechin, or alter the organoleptic properties of the food, such as essential oils.

Another proposed strategy to provide oxidative protection in a food is the use of binary antioxidants in packaging materials with complementary release rates, where one antioxidant would release at an appropriate fast rate to provide short-term/intermediate inhibition and the other antioxidant would release at an appropriate slow rate to provide intermediate/long-term inhibition (Chen et al., 2012). Polymer blend films consisting of both hydrophilic and hydrophobic polymers can be designed to match the properties of the antioxidant to achieve a moderate release rate that is maintained for a prolonged period of time. The range of diffusivities of the common packaging polymers also enables the selection of optimal antioxidant release rates for specific foods which will provide the longest shelf life. The addition of an HDPE layer as a coextruded film with LDPE as the base layer containing carotenoids modified antioxidant release by slowing the release rate and increasing the time to reach equilibrium (Colín-Chávez et al., 2013, 2014). The release rate of natural antioxidants from active PP films has been modified using vegetable oil plasticizers and different block copolymers as polymer chain extenders and plasticizers to develop different release profiles for a specific antioxidant into food simulants (Castro López et al., 2012; Castro López et al., 2013; López-de-Dicastillo et al., 2013a).

Cyclodextrin (CD) may be incorporated into polymer packaging materials to change the release rate and enhance the stability of encapsulated antioxidants. CDs are cyclic oligosaccharides consisting of six (α-CD), seven (β-CD), eight (γ-CD), or more glucose units that have the ability to form inclusion complexes with guest molecules, resulting in an encapsulation process on the molecular scale. The release of additive from polymer matrices with added CDs can be reduced by (a) increasing its molecular weight during complexation, which results in lowered diffusivity; (b) lowering the concentration of diffusible species by forming poorly soluble complexes; (c) reducing the concentration of diffusible species by forming complexes in which the host is covalently bound to the polymer; or (d) functioning as cross-linking agents and reducing polymer mesh size (Bibby et al., 2000). CD inclusion complexes with the natural antioxidants, α-tocopherol and quercetin, have been incorporated into LLDPE films and the release rate of both free and complexed antioxidant from the active film into coconut oil was determined (Koontz et al., 2010a, 2010b). α-Tocopherol–β-CD and quercetin–γ-CD complexes were compounded at the 1.5% level into LLDPE followed by compression molding into films. The limited solubility of quercetin in LLDPE did not allow release from the film into coconut oil due to phase segregation of the quercetin in the bulk polymer. After 28 days, the total release of α-tocopherol from LLDPE into coconut oil at 30 °C was 70% in its free form, while only 8% in its CD-complexed form. The mechanism by which α-tocopherol was released was modified due to its inclusion inside the β-CD cavity within the LLDPE matrix as indicated by its diffusion coefficient decreasing by two orders of magnitude. A one order of magnitude decrease in diffusion coefficient was calculated for an α-tocopherol–β-CD complex compared with free α-tocopherol during release from LDPE film into 95% ethanol as a fatty food stimulant (Siró et al., 2006). CD inclusion complexes of antioxidants within polymers offer another strategy to effectively target delivery of an optimal antioxidant concentration to a food to achieve prolonged activity, resulting in the extended shelf life of foods.

13.5.3.2 Antioxidant Coating

Antioxidant compounds can also be coated or physically adsorbed onto the food contact surface of packaging materials for their sustained release into food products to inhibit oxidative reactions. The process of solution casting involves dissolving both the antioxidant and polymer into an appropriate solvent followed by application of the solution to the packaging surface as substrate and evaporating the solvent system. Compatibility must exist between the substrate and coating polymers for good adhesion properties. The surface of the film substrate may be treated to promote adhesion by application of primers, UV irradiation, corona discharge, and atmospheric pressure plasma (Contini et al., 2014). One of the main benefits of coating by solution casting is that heat-sensitive antioxidants can be incorporated into film with minimum loss of activity. Although active films produced by coating onto a polymer substrate have been used in film development, this method does not lend itself to preferred commercial production processes such as extrusion.

Various studies have been performed on coated antioxidant release rates and their ability to provide enhanced oxidative stability to food systems. Mangiferin, a major bioactive compound in mango extract exhibiting strong antioxidant capacity, has been coated onto EVA film by a solvent casting process (Boonnattakorn et al., 2015). The release rate of mangiferin increased with increasing vinyl acetate contents of 12%, 18%, 25%, and 40%, which was likely due to decreasing polymer crystallinity. Natural antioxidants extracted from barley husks that were solution-coated onto LDPE films were demonstrated to have efficacy in reducing the rate of lipid hydrolysis and increasing oxidative stability of salmon during frozen storage (Pereira de Abreu et al., 2010). Commercial citrus extract, a mixture of bioflavonoid antioxidants, was applied in solution by a nebulizer coating procedure onto PET trays (Contini et al., 2012). Cooked turkey meat slices stored on PET trays coated with citrus extract at 4 °C for 4 days had significantly lower TBARS and hexanal values than those of product stored on uncoated trays.

Active antioxidant films can also be produced by solution casting a coating onto glass, ceramic, or polytetrafluoroethylene plates and subsequent removal from the substrate to achieve free-standing monolayer films. Active films were produced by incorporation of ascorbic acid, ferulic acid, quercetin, and green tea extract into an EVOH matrix by solution casting for the improvement of lipid stability in brined sardines (López-de-Dicastillo et al., 2012a). EVOH films containing 5% green tea extract were most effective at slowing hydroperoxide formation and reducing malondialdehyde concentration in sardines. In contrast, 5% ascorbic acid films failed to prevent lipid oxidation in packaged sardines and even exhibited some pro-oxidant effects. Green tea extract has also been incorporated into chitosan by solvent casting to produce active films with enhanced radical scavenging activity in addition to improved mechanical and barrier properties (Siripatrawan and Harte, 2010).

13.5.4 Covalent Immobilization of Antioxidants

Active packaging with covalent immobilization of functional compounds (Figure 13.7(C)) offers a unique approach to impart antioxidant activity into food systems without the use of direct additives in the product formulation (Tian et al., 2013b). Antioxidant controlled-release packaging is dependent on the mass transfer of active compounds to food products for its effectiveness. Controlled-release systems may be limited by their eventual loss of activity upon depletion of antioxidants from the polymer matrix. Covalent immobilization technologies offer the potential to enable active packaging to possess a longer duration of activity. The covalent immobilization of functional compounds onto the food contact surface of packaging is not likely to affect the mechanical and optical properties of the bulk polymer as is sometimes the case for active compounds within the polymer matrix. However, immobilized antioxidants require direct contact with the food product to impart their functionality, which limits the potential application of this technology to liquid and semi-liquid foods.

Commonly used commercial food packaging polymers have an inert surface which must be pretreated to introduce reactive sites on the polymer surface for the subsequent attachment of functional compounds. The main surface modification techniques developed to perform surface functionalization include wet chemical oxidation, ionized gas treatment, and UV irradiation (Tian et al., 2013b). UV irradiation has been reported as the preferred method for surface functionalization on the polymer surface. Long-wave UV light in the presence of a photoinitiator can be used to convert light energy into useful chemical energy by abstracting hydrogen from the polymer surface to generate free radicals. The free radicals formed on the polymer surface then have the ability to initiate the graft polymerization of functional monomers from the polymer surface.

Different types of antioxidants have been covalently immobilized for the development of antioxidant packaging materials. Caffeic acid was immobilized onto a PP film surface though the covalent binding of caffeoyl chloride on a modified polymeric surface of PP film photografted with hydroxyethyl methacrylate as monomer (PP-*g*-HEMA) (Arrua et al., 2010). PP-*g*-HEMA-caffeic acid films showed good protective activity against ascorbic acid oxidation in real samples of orange juice. Gallic acid and catechin were covalently inserted into chitosan polymer chains via a free-radical grafting procedure

(Curcio et al., 2009). The antioxidant activity of chitosan was increased after its functionalization with both antioxidant compounds. The antioxidant effectiveness of gallic acid–grafted chitosan films has been evaluated on ground peanut powder during 15 weeks of storage at 50 °C (Schreiber et al., 2013). Ground peanuts packaged in grafted gallic acid–chitosan had significantly better quality than in LDPE as indicated by lower levels of conjugated trienes, peroxide value, and TBARS. Grafted gallic acid–chitosan pouches had lower levels of accumulated end-products of lipid oxidation compared to both neat chitosan and LDPE pouches, which showed that gallic acid maintained its activity while grafted to chitosan during the entire 15-week storage. Green tea extract, as a source of flavonoid antioxidants, was immobilized onto PP through the incorporation of maleic anhydride grafted PP to develop radical scavenging materials and to reduce antioxidant release to the food during microwave heating and sterilization conditions (López et al., 2013b).

Inhibition of lipid oxidation in oil-in-water emulsions containing free fatty acids by the presence of ethylenediamine tetraacetic acid (EDTA) indicates that the most prevalent mechanism for the pro-oxidant activity of free fatty acids is the attraction of transition metals to the surface of the emulsion droplet, where they can interact with lipid and accelerate oxidation (Waraho et al., 2009). Transition metal–promoted oxidative degradation reactions occur at metal concentrations naturally occurring in foods, which are often in the low ppb range (Goddard et al., 2012). In Figure 13.8, poly(acrylic acid) (PAA) grafted onto PP films (PP-*g*-PAA) was developed to chelate transition metals with the goal of inhibiting such metal-promoted lipid oxidative reactions (Tian et al., 2012a). In the first reaction step, the photoinitiator benzophenone was covalently grafted to the PP surface. This benzophenone surface initiation step ensures that the grafted PAA is covalently bound to PP film and not merely a coating of cross-linked PAA. In the second step, the graft polymerization of the acrylic acid monomer from the PP surface was initiated to form PP-*g*-PAA. Carboxylic acid groups on each acrylic acid polymer chain graft are able to bind ferrous iron (Fe^{2+}) at a 2:1 theoretical ligand to metal binding ratio. PP-*g*-PAA films have been shown to

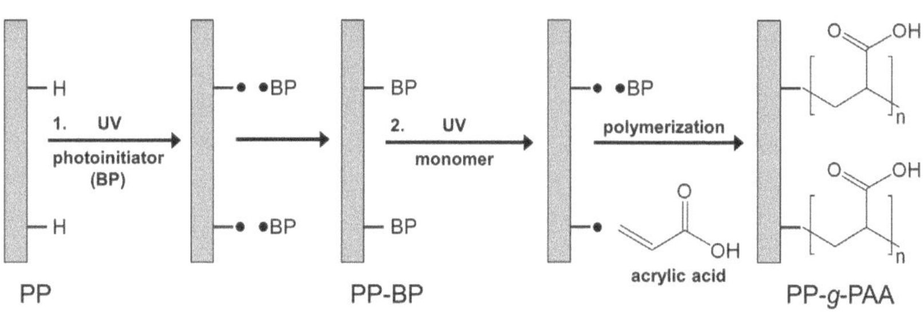

FIGURE 13.8 Two-step photoinitiated graft polymerization of acrylic acid monomer onto the polypropylene (PP) surface with benzophenone (BP) as photoinitiator to form covalently immobilized metal-chelating active packaging materials. *Adapted from Tian et al. (2012a).*

be effective at inhibiting the oxidation of oil-in-water emulsions by chelation of pro-oxidant ferrous ions. The oxidative stability of a soybean oil-in-water emulsion at 37 °C was extended from 2 days in neat PP films to 9 days in PP-*g*-PAA films as indicated by the development of hexanal. These PP-*g*-PAA films were reported to have an equivalent chelating activity of 72.2 μmol/L EDTA, which is well above the minimum EDTA concentration of 2.5 μmol/L required to dramatically decrease lipid oxidation. Iron-chelating active packaging offers the potential to allow for the removal of the synthetic additive EDTA from product formulations while still maintaining product quality.

The strongest activity of PP-*g*-PAA films to prevent oxidative reactions in an oil-in-water emulsion was at pH 7.0, and the activity decreased with decreasing pH (Tian et al., 2014). The ability to control iron-promoted oxidative reactions in acid and acidified foods was investigated by use of a biomimetic iron chelating active packaging material (Tian et al., 2013a). Poly(hydroxamic acid) (PHA) was selected as a biomimetic metal chelator to graft from a PP film surface by photoinitiated surface graft polymerization. PP-*g*-PHA films demonstrated the ability to control lipid oxidation in an acidified soybean oil-in-water emulsion system at pH 3.0. The performance of PP-*g*-PHA films was even more efficient than EDTA at inhibiting the formation of volatile oxidation products. The iron chelating activity of PP-*g*-PAA films can be tailored to adapt the functional characteristics of chelating polymer materials for specific applications of active packaging by adjusting the graft polymerization conditions (Roman et al., 2014). Graft chain length and density had a significant effect on the chelating activity of PP-*g*-PAA both in terms of overall iron-binding capacity as well as ligand to metal binding ratio. PAA has also been used as an iron chelator to impart antioxidant capacity to LDPE films by its covalent immobilization to the LDPE surface (Tian et al., 2012b). Due to their effective chelation of ferrous ions, these active chelating films would not be suitable for packaging of foods that serve as important nutritional sources of iron or other metals.

13.5.5 Aldehyde-Scavenging Packaging

The principle source of off-flavors developed by lipid oxidation is hydroperoxides, which are unstable and readily decompose to form aliphatic aldehydes, ketones, and alcohols. Many of these secondary oxidation products have undesirable odors with particularly low odor thresholds that tend to impart a rancid taste to foods. For example, hexanal as the main decomposition product of linoleic acid has reported odor thresholds of 120 ppb in vegetable oil (Buttery et al., 1973), 9 ppb in water (Ahmed et al., 1978), and 0.05 ppb in air (Blank, 2002). Such low odor thresholds mean that only a minute amount of the total lipid in a food needs to be oxidized in order to produce an objectionable off-odor. Therefore, lipid-containing foods can be organoleptically unacceptable long before any substantial loss of their nutritional and functional properties as a result of oxidative degradation. Active polymer packaging can selectively absorb volatile aldehydes from the oxidation of lipids in foods to remove such undesirable off-odors of rancidity from inside the package. Selective scalping can be accomplished by incorporation of absorbents based on reactive aldehyde chemistry or incorporation of CD into polymers.

DuPont has performed extensive research on incorporating aldehyde scavengers into package materials to remove the end-products of lipid oxidation, such as hexanal and heptanal, leading to a series of patents (Brodie and Visioli, 1994; Brody et al., 2001c). Aldehyde-scavenging packaging to remove hexanal from a product involves hexanal dissolution into a polyolefin polymer matrix, hexanal diffusion through the polymer matrix, and hexanal reaction with polyethylene imine (PEI) (Del Nobile et al., 2002). The product of the reaction of dissolved hexanal and PEI is the hexanal chemically grafted to the PEI backbone. The aldehyde scavenger can be incorporated into several different flexible package structures including as a layer in multilayer polymer structure, into the polymer used in a cap liner, and into barrier coextrusions for food products such as snacks, crackers, cookies, and cereals (Brody et al., 2001c). The effectiveness of the scavenger at reducing aldehyde concentration in the headspace has been demonstrated in packages of peanut butter, instant coffee, and snack products.

Nylon MXD6, D-sorbitol, and α-CD have been tested as aldehyde-scavenging agents in PET film blends at the 5% level (Suloff et al., 2003). These aldehyde-scavenging films sorbed 2–10 times higher concentrations of aldehydes than neat PET films in an acidified aqueous model solution. Aldehyde-scavenging films demonstrated selective sorption, preferring smaller molecular weight aldehydes to larger aldehydes for the homologous series from the 2-carbon acetaldehyde to the 6-carbon hexanal. Nylons and D-sorbitol react with aldehydes by nucleophillic addition to form imines and acetals, respectively. CDs form inclusion complexes with aldehydes through weak intermolecular forces, including hydrophobic interactions and van der Waals forces. α-CD scavenges acetaldehyde through a hydrogen bonding and size-enclosure mechanism, with each molecule of α-CD being able to accommodate one molecule of acetaldehyde (Mrozinski et al., 2013). β-CD has been incorporated at 10%, 20%, and 30% levels in an EVOH copolymer with a 44% molar percentage of ethylene by extrusion with glycerol as a dispersion aid to develop active packaging for the inclusion or retention of nonpolar organic compounds (López-de-Dicastillo et al., 2010b). EVOH films containing β-CD have increased sorption capacity that was used to reduce the presence of aldehydes from the headspace of packaged fried peanuts (López-de-Dicastillo et al., 2011). Films containing β-CD significantly reduced the hexanal concentration of peanuts, achieving a 50% decrease throughout a 1–5 week period compared to the control without EVOH film. At 10 weeks, however, no differences were observed between treatments indicating that the scavenging capacity of the β-CD films had been exhausted.

The previously highlighted active packaging technologies are more proactive approaches at controlling lipid oxidation by inhibiting the initiators of the oxidative reaction and are generally a more preferred approach compared to the reactionary scavenging of oxidation products. Alternatively, aldehyde-scavenging packaging can also provide beneficial effects to food quality by its ability to remove off-flavor compounds that migrate from the packaging material to the food product. For example, acetaldehyde is a thermal degradation product of PET formed during polymerization, processing, and forming of packaging materials. Aldehyde-scavenging packaging will not interrupt or inhibit oxidation reactions, however, it has the capability to remove undesirable secondary oxidation products. Lipid-containing foods have the potential to remain organoleptically acceptable for extended periods of time when stored in such active packaging before there is substantial loss of nutritional or functional properties. It is critical to emphasize that those active materials which scavenge undesirable oxidation products should not mislead or pose a health risk to the consumer by masking spoiled or deteriorated food.

13.6 REGULATORY PATHS TO SAFE COMMERCIALIZATION

The food and packaging industries will increasingly consider bringing to market the various active packaging technologies that may limit oxidation and extend the shelf life of lipid-containing foods in an effort to deliver the greatest potential of high product quality to consumers. The complex nature of many types of active packaging offers new technical considerations for companies seeking regulatory approval, therefore, it is important to review how current U.S. Food and Drug Administration (FDA) regulations apply to new food contact substances.

13.6.1 FDA Regulatory Programs

U.S. FDA's current premarket regulatory programs for food additives and ingredients include the food additive petition (FAP), which is the traditional petition process that is very time intensive, threshold of regulation (TOR), food contact substance notification (FCN), and generally recognized as safe (GRAS) notification programs. Direct food additives are substances deliberately added to a food to accomplish an intended technical effect and are authorized via the petition process, resulting in the publication of a regulation specifying the identity and use conditions of the food additive. The TOR program is limited to only substances used in food packaging or processing equipment that do not show any intended effect on the food itself and have a dietary concentration below 0.5 ppb. These are not the full criteria for the acceptance of a TOR submission; additional criteria are found in 21 CFR 170.39.

The FCN program was intended to replace the FAP and TOR programs as the primary means by which FDA authorizes the use of food additives that are food contact substances. A food contact substance (FCS) is "any substance intended for use as a component of materials used in manufacturing, packing, packaging, transporting, or holding food if such use is not intended to have a technical effect in such food." These materials were formerly referred to as indirect and secondary direct food additives. FCSs are authorized for use via the FCN program, unless circumstances warrant use of the petition process. FCSs are materials that unintentionally become components of food, such as components of packaging materials, and have a technical effect in a food contact material, but not in the food itself. FCS notifications are effective only for the specific manufacturer, the specific FCS, and the specified conditions of use. Other manufacturers cannot market products containing the same FCS without filing their own notification. FDA recommends that specific chemistry information should be submitted in an FCN, TOR, or FAP for an FCS, which includes the material identity and composition, method of manufacture, intended use, intended technical effect, migration data and analytical methods, and consumer exposure.

Substances whose uses are generally recognized as safe by qualified experts are not required by law to receive FDA premarket approval. In the absence of FDA premarket approval, a manufacturer must conclude that the use of a substance is GRAS and is therefore exempt from the FDA premarket approval requirements. Once a manufacturer makes a GRAS conclusion for the use of a substance, the manufacturer may submit a GRAS notice to FDA under the agency's GRAS Notification Program. FDA then evaluates the data and information in the GRAS notice to assess whether the data and information demonstrate safety and general recognition of safety of the manufacturer's use of the substance. Following this evaluation, FDA issues a letter stating whether the agency has questions regarding the manufacturer's GRAS conclusion. After making a GRAS conclusion, a manufacturer may also choose not to submit a GRAS notice to FDA (a so-called GRAS self-determination).

13.6.2 FDA Perspective on Active Packaging Materials

The regulatory authority given to FDA is sufficiently robust and flexible to accommodate the various active packaging formats and configurations, while maintaining its safety standard of a reasonable certainty of no harm. FDA's premarket safety review does not focus on the relative effectiveness of the food contact material (but data demonstrating the FCS achieves its intended technical effect are a component of an FCN), but rather focuses on the safety of chemical components that may be expected to migrate from the packaging into food. Therefore, it is irrelevant whether packaging materials are classified as active, passive, or intelligent (Song and Hepp, 2005). Active packaging for controlled release of antioxidants into food for the intended technical effect of delaying the onset of lipid oxidation may be considered either a direct additive or GRAS substance and should follow the FAP program or GRAS Notification program, respectively. Nonmigratory active packaging by covalent grafting of metal chelators for the intended technical effect of binding trace metals that promote lipid oxidation in the food without intentional migration may be considered an indirect additive and should follow the FCN program. Polymer-based O_2-scavenging films and UV light absorbers are considered indirect additives and should likewise follow the FCN program.

The FDA encourages innovation in active packaging applications that may limit oxidation, maintain nutritional quality, and extend the shelf life of foods. FDA reminds developers that such materials are subject to the premarket approval requirements of the Federal Food, Drug, and Cosmetic Act if components of the package are reasonably expected to become components of food or affect the characteristics of food as a result of their intended use. FDA provides several online guidance documents on ingredients, additives, and packaging that are intended for industry use (U.S. Food and Drug Administration, 2015). FDA recommends that developers meet with the agency before submitting an FAP, GRAS notice, or FCN to prevent an expenditure of resources on experiments that may not provide adequate data. Before initiating studies for a submission, consultation with FDA regarding the design of experimental protocols that are likely to address FDA's concerns will result in a faster time to market for active packaging technologies.

LIST OF ABBREVIATIONS

BHA Butylated hydroxyanisole
BHT Butylated hydroxytoluene
BOPP Biaxially oriented polypropylene
CD Cyclodextrin
EVA Ethylene–vinyl acetate
EDTA Ethylenediamine tetraacetic acid
EVOH Ethylene-vinyl alcohol copolymer
FAP Food additive petition
FCN Food contact substance notification
FCS Food contact substance
FRS Free radical scavenger
g Grafted
GRAS Generally recognized as safe
HDPE High density polyethylene
HEMA Hydroxyethyl methacrylate
LDPE Low density polyethylene
LLDPE Linear low density polyethylene
MAP Modified atmosphere packaging
MRE Meal-ready-to-eat
Nylon Polyamide
Nylon MXD6 Poly(*meta*-xylene adipamide)
$^1O_2$ Singlet oxygen
$^3O_2$ Atmospheric triplet oxygen
OTR Oxygen transmission rate
PAA Poly(acrylic acid)

PEN Poly(ethylene naphthalate)
PET Poly(ethylene terephthalate)
PFO Polyfuryloxirane
PHA Poly(hydroxamic acid)
PP Polypropylene
PLA Poly(lactic acid)
PS Polystyrene
PUFA Polyunsaturated fatty acids
PVC-P Poly(vinyl chloride), plasticized
PVC-U Poly(vinyl chloride), unplasticized
PVDC Poly(vinylidene chloride)
RH Relative humidity
TBARS Thiobarbituric acid reactive substances
TBHQ t-butylhydroquinone
TOR Threshold of regulation
WVTR Water vapor transmission rate

REFERENCES

Ahmed, E.M., Dennison, R.A., Dougherty, R.H., Shaw, P.E., 1978. Flavor and odor thresholds in water of selected orange juice components. J. Agric. Food Chem. 26, 187–191.

Andersson, K., Lingnert, H., 1999. Kinetic studies of oxygen dependence during initial lipid oxidation in rapeseed oil. J. Food Sci. 64, 262–266.

Arrua, D., Strumia, M.C., Nazareno, M.A., 2010. Immobilization of caffeic acid on a polypropylene film: synthesis and antioxidant properties. J. Agric. Food Chem. 58, 9228–9234.

Ashley, R.J., 1985. Permeability and plastics packaging. In: Comyn, J. (Ed.), Polymer Permeability. Chapman & Hall, London, pp. 269–308.

Bibby, D.C., Davies, N.M., Tucker, I.G., 2000. Mechanisms by which cyclodextrins modify drug release from polymeric drug delivery systems. Int. J. Pharm. 197, 1–11.

Blank, I., 2002. Gas chromatography-olfactometry in food aroma analysis. In: Marsili, R. (Ed.), Flavor, Fragrance, and Odor Analysis, second ed. CRC Press, Boca Raton, pp. 297–331.

Boonnattakorn, R., Chonhenchob, V., Siddiq, M., Singh, S.P., 2015. Controlled release of mangiferin using ethylene vinyl acetate matrix for antioxidant packaging. Packag. Technol. Sci. 28, 241–252.

Brodie, V., Visioli, D.L., February 8, 1994. Aldehyde scavenging compositions and methods relating thereto. US 5284892.

Brody, A.L., Strupinsky, E.R., Kline, L.R., 2001a. Oxygen scavenging in 2000 and beyond. In: Active Packaging for Food Applications. CRC Press, Boca Raton, pp. 65–86.

Brody, A.L., Strupinsky, E.R., Kline, L.R., 2001b. Oxygen scavenger systems. In: Active Packaging for Food Applications. CRC Press, Boca Raton, pp. 31–64.

Brody, A.L., Strupinsky, E.R., Kline, L.R., 2001c. Odor removers. In: Active Packaging for Food Applications. CRC Press, Boca Raton, pp. 107–118.

Brody, A.L., Strupinsky, E.R., Kline, L.R., 2001d. Oxygen scavengers. In: Active Packaging for Food Applications. CRC Press, Boca Raton, pp. 11–30.

Buttery, R.G., Guadagni, D.G., Ling, L.C., 1973. Flavor compounds: volatilities in vegetable oil and oil-water mixtures. Estimation of odor thresholds. J. Agric. Food Chem. 21, 198–201.

Camo, J., Beltrán, J.A., Roncalés, P., 2008. Extension of the display life of lamb with an antioxidant active packaging. Meat Sci. 80, 1086–1091.

Castro López, Mdel, M., Dopico García, S., Ares Pernas, A., López Vilariño, J.M., González Rodríguez, M.V., 2012. Effect of PPG-PEG-PPG on the tocopherol-controlled release from films intended for food-packaging applications. J. Agric. Food Chem. 60, 8163–8170.

Castro López, Mdel, M., López de Dicastillo, C., López Vilariño, J.M., González Rodríguez, M.V., 2013. Improving the capacity of polypropylene to be used in antioxidant active films: incorporation of plasticizer and natural antioxidants. J. Agric. Food Chem. 61, 8462–8470.

Cecchi, T., Passamonti, P., Cecchi, P., 2010. Study of the quality of extra virgin olive oil stored in PET bottles with or without an oxygen scavenger. Food Chem. 120, 730–735.

Charles, F., Sanchez, J., Gontard, N., 2006. Absorption kinetics of oxygen and carbon dioxide scavengers as part of active modified atmosphere packaging. J. Food Eng. 72, 1–7.

Chen, X., Lee, D.S., Zhu, X., Yam, K.L., 2012. Release kinetics of tocopherol and quercetin from binary antioxidant controlled-release packaging films. J. Agric. Food Chem. 60, 3492–3497.

Choe, E., Min, D.B., 2006. Mechanisms and factors for edible oil oxidation. Compr. Rev. Food Sci. Food Saf. 5, 169–186.

Colín-Chávez, C., Soto-Valdez, H., Peralta, E., Lizardi-Mendoza, J., Balandrán-Quintana, R., 2013. Diffusion of natural astaxanthin from polyethylene active packaging films into a fatty food simulant. Food Res. Int. 54, 873–880.

Colín-Chávez, C., Soto-Valdez, H., Peralta, E., 2014. Diffusion of carotenoids from mono and bilayer polyethylene active packaging into soybean oil. Food Packag. Shelf Life 1, 170–178.

Coltro, L., Padula, M., Saron, E.S., Borghetti, J., Buratin, A.E.P., 2003. Evaluation of a UV absorber added to PET bottles for edible oil packaging. Packag. Technol. Sci. 16, 15–20.

Contini, C., Katsikogianni, M.G., O'Neill, F.T., O'Sullivan, M., Dowling, D.P., Monahan, F.J., 2012. PET trays coated with citrus extract exhibit antioxidant activity with cooked turkey meat. LWT Food Sci. Technol. 47, 471–477.

Contini, C., Katsikogianni, M.G., O'Neill, F.T., O'Sullivan, M., Boland, F., Dowling, D.P., Monahan, F.J., 2014. Storage stability of an antioxidant active packaging coated with citrus extract following a plasma jet pretreatment. Food Bioprocess Technol. 7, 2228–2240.

Curcio, M., Puoci, F., Iemma, F., Parisi, O.I., Cirillo, G., Spizzirri, U.G., Picci, N., 2009. Covalent insertion of antioxidant molecules on chitosan by a free radical grafting procedure. J. Agric. Food Chem. 57, 5933–5938.

De Azeredo, H.M.C., Faria, J.A.F., Da Silva, M.A.A.P., 2003. The efficiency of TBHQ, β-carotene, citric acid, and Tinuvin 234® on the sensory stability of soybean oil packaged in PET bottles. J. Food Sci. 68, 302–306.

Del Nobile, M.A., Buonocore, G.G., Fava, P., Piergiovanni, L., 2002. Modeling of hexanal sorption kinetic in an aldehydes scavenger film intended for food packaging applications. J. Food Sci. 67, 2687–2691.

Dhoot, S.N., Freeman, B.D., Stewart, M.E., 2002. Barrier polymers. In: Encyclopedia of Polymer Science and Technology, vol. 5. John Wiley & Sons, pp. 198–263.

Galdi, M.R., Nicolais, V., Di Maio, L., Incarnato, L., 2008. Production of active PET films: evaluation of scavenging activity. Packag. Technol. Sci. 21, 257–268.

Goddard, J.M., McClements, D.J., Decker, E.A., 2012. Innovative technologies in the control of lipid oxidation. Lipid Technol. 24, 275–277.

Gomes, C., Castell-Perez, M.E., Chimbombi, E., Barros, F., Sun, D., Liu, J.D., Sue, H.-J., Sherman, P., Dunne, P., Wright, A.O., 2009. Effect of oxygen-absorbing packaging on the shelf life of a liquid-based component of military operational rations. J. Food Sci. 74, E167–E176.

Graciano-Verdugo, A.Z., Soto-Valdez, H., Peralta, E., Cruz-Zárate, P., Islas-Rubio, A.R., Sánchez-Valdes, S., Sánchez-Escalante, A., González-Méndez, N., González-Ríos, H., 2010. Migration of α-tocopherol from LDPE films to corn oil and its effect on the oxidative stability. Food Res. Int. 43, 1073–1078.
Hoojjat, P., Hernandez, R.J., Miltz, J., 1987. Mass transfer of BHT from high density polyethylene film and its influence on product stability. J. Packag. Technol. 1, 78–81.
Huang, C.-H., Weng, Y.-M., 1998. Inhibition of lipid oxidation in fish muscle by antioxidant incorporated polyethylene film. J. Food Process. Preserv. 22, 199–209.
Jadhav, S.J., Nimbalkar, S.S., Kulkarni, A.D., Madhavi, D.L., 1996. Lipid oxidation in biological and food systems. In: Madhavi, D.L., Deshpande, S.S., Salunkhe, D.K. (Eds.), Food Antioxidants: Technological, Toxicological, and Health Perspectives. Marcel Dekker, New York, NY, pp. 5–63.
Kanazawa, A., Sawa, T., Akaike, T., Maeda, H., 2002. Dietary lipid peroxidation products and DNA damage in colon carcinogenesis. Eur. J. Lipid Sci. Technol. 104, 439–447.
Kanner, J., 2007. Dietary advanced lipid oxidation endproducts are risk factors to human health. Mol. Nutr. Food Res. 51, 1094–1101.
Karel, M., 1992. Kinetics of lipid oxidation. In: Schwarzberg, H.G., Hartel, R.W. (Eds.), Physical Chemistry of Foods. Marcel Dekker, New York, pp. 651–668.
Kiritsakis, A., Kanavouras, A., Kiritsakis, K., 2002. Chemical analysis, quality control and packaging issues of olive oil. Eur. J. Lipid Sci. Technol. 104, 628–638.
Koontz, J.L., Marcy, J.E., O'Keefe, S.F., Duncan, S.E., Long, T.E., Moffitt, R.D., 2010a. Polymer processing and characterization of LLDPE films loaded with α-tocopherol, quercetin, and their cyclodextrin inclusion complexes. J. Appl. Polym. Sci. 117, 2299–2309.
Koontz, J.L., Moffitt, R.D., Marcy, J.E., O'Keefe, S.F., Duncan, S.E., Long, T.E., 2010b. Controlled release of α-tocopherol, quercetin, and their cyclodextrin inclusion complexes from linear low-density polyethylene (LLDPE) films into a coconut oil model food system. Food Addit. Contam. Part A 27, 1598–1607.
Kristensen, D., Orlien, V., Mortensen, G., Brockhoff, P., Skibsted, L.H., 2000. Light-induced oxidation in sliced Havarti cheese packaged in modified atmosphere. Int. Dairy J. 10, 95–103.
Kucuk, M., Caner, C., 2005. Effect of packaging materials and storage conditions on sunflower oil quality. J. Food Lipids 12, 222–231.
Kuraray America Inc. Barrier to Oxygen. http://www.eval.eu/us/eval-properties/barrier-to-oxygen.aspx (accessed October 2015).
Labuza, T.P., Dugan, L.R., 1971. Kinetics of lipid oxidation in foods. CRC Critic. Rev. Food Technol. 2, 355–405.
Lamberti, M., Escher, F., 2007. Aluminium foil as a food packaging material in comparison with other materials. Food Rev. Int. 23, 407–433.
Lange, J., Wyser, Y., 2003. Recent innovations in barrier technologies for plastic packaging—a review. Packag. Technol. Sci. 16, 149–158.
Langowski, H.-C., 2008. Permeation of gases and condensable substances through monolayer and multilayer structures. In: Piringer, O.G., Baner, A.L. (Eds.), Plastic Packaging, second ed. Wiley-VCH, Weinheim, pp. 297–347.
Lankin, V.Z., Tikhaze, A.K., Konovalova, G.G., Kozachenko, A.I., 1999. Concentration-dependent inversion of antioxidant and prooxidant effects of β-carotene in tissues *in vivo*. Bull. Exp. Biol. Med. 128, 930–932.
Lennersten, M., Lingnert, H., 2000. Influence of wavelength and packaging material on lipid oxidation and colour changes in low-fat mayonnaise. Lebensmittel-Wissenschaft und Technologie 253–260.

López-de-Dicastillo, C., Catalá, R., Gavara, R., Hernández-Muñoz, P., 2011. Food applications of active packaging EVOH films containing cyclodextrins for the preferential scavenging of undesirable compounds. J. Food Eng. 104, 380–386.
López-de-Dicastillo, C., Alonso, J.M., Catalá, R., Gavara, R., Hernández-Muñoz, P., 2010a. Improving the antioxidant protection of packaged food by incorporating natural flavonoids into ethylene-vinyl alcohol copolymer (EVOH) films. J. Agric. Food Chem. 58, 10958–10964.
López-de-Dicastillo, C., Gallur, M., Catalá, R., Gavara, R., Hernández-Muñoz, P., 2010b. Immobilization of β-cyclodextrin in ethylene-vinyl alcohol copolymer for active food packaging applications. J. Membr. Sci. 353, 184–191.
López-de-Dicastillo, C., Gómez-Estaca, J., Catalá, R., Gavara, R., Hernández-Muñoz, P., 2012a. Active antioxidant packaging films: development and effect on lipid stability of brined sardines. Food Chem. 131, 1376–1384.
López-de-Dicastillo, C., Pezo, D., Nerín, C., López-Carballo, G., Catalá, R., Gavara, R., Hernández-Muñoz, P., 2012b. Reducing oxidation of foods through antioxidant active packaging based on ethyl vinyl alcohol and natural flavonoids. Packag. Technol. Sci. 25, 457–466.
López-de-Dicastillo, C., Ares Pernas, A., Castro López, Mdel, M., López Vilariño, J.M., González Rodríguez, M.V., 2013a. Enhancing the release of the antioxidant tocopherol from polypropylene films by incorporating the natural plasticizers lecithin, olive oil, or sunflower oil. J. Agric. Food Chem. 61, 11848–11857.
López-de-Dicastillo, C., Castro-López, Mdel, M., López-Vilariño, J.M., González-Rodríguez, M.V., 2013b. Immobilization of green tea extract on polypropylene films to control the antioxidant activity in food packaging. Food Res. Int. 53, 522–528.
López-Rubio, A., Almenar, E., Hernandez-Muñoz, P., Lagarón, J.M., Catalá, R., Gavara, R., 2004. Overview of active polymer-based packaging technologies for food applications. Food Rev. Int. 20, 357–387.
López-Rubio, A., 2011. Ethylene-vinyl alcohol (EVOH) copolymers. In: Lagarón, J.-M. (Ed.), Multifunctional and Nanoreinforced Polymers for Food Packaging. Woodhead Publishing, Oxford, pp. 261–284.
Maloba, F.W., Rooney, M.L., Wormell, P., Nguyen, M., 1996. Improved oxidative stability of sunflower oil in the presence of an oxygen-scavenging film. J. Am. Oil Chem. Soc. 73, 181–185.
Márquez-Castillo, A., Vidal-Quintanar, R.L., 2011. Improvements in the shelf life of commercial corn dry masa flour (CMF) by reducing lipid oxidation. J. Food Sci. 76, C236–C241.
Martín, M.B.S., Fernández-García, T., Romero, A., López, A., 2001. Effect of modified atmosphere storage on hazelnut quality. J. Food Process. Preserv. 25, 309–321.
Maury, C., Clark, A.C., Scollary, G.R., 2010. Determination of the impact of bottle colour and phenolic concentration on pigment development in white wine stored under external conditions. Anal. Chim. Acta 660, 81–86.
McClements, D.J., Decker, E.A., 2008. Lipids. In: Damodaran, S., Parkin, K.L., Fennema, O.R. (Eds.), Fennema's Food Chemistry, fourth ed. CRC Press, Boca Raton, pp. 155–216.
Mestdagh, F., De Meulenaer, B., De Clippeleer, J., Devlieghere, F., Huyghebaert, A., 2005. Protective influence of several packaging materials on light oxidation of milk. J. Dairy Sci. 88, 499–510.
Miltz, J., Hoojjat, P., Han, J.K., Giacin, J.R., Harte, B.R., Gray, I.J., 1988. Loss of antioxidants from high-density polyethylene. In: Hotchkiss, J.H. (Ed.), ACS Symposium Series, vol. 365. American Chemical Society, Washington, DC, pp. 83–93.
Mortensen, G., Bertelsen, G., Mortensen, B.K., Stapelfeldt, H., 2004. Light-induced changes in packaged cheeses—a review. Int. Dairy J. 14, 85–102.

Mrozinski, B.A., Kim, Y.-W., Lofgren, E.A., Jabarin, S.A., 2013. Chemistry of the interactions of acetaldehyde scavengers for poly(ethylene terephthalate). J. Appl. Polym. Sci. 130, 4191–4200.
Nelson, K.A., Labuza, T.P., 1992. Relationship between water and lipid oxidation rates. In: St Angelo, A.J. (Ed.), Lipid Oxidation in Food. American Chemical Society, Washington, DC, pp. 93–103.
Pascall, M.A., Harte, B.R., Giacin, J.R., Gray, J.I., 1995. Decreasing lipid oxidation in soybean oil by a UV absorber in the packaging material. J. Food Sci. 60, 1116–1119.
Peltzer, M., Wagner, J., Jiménez, A., 2009. Migration study of carvacrol as a natural antioxidant in high-density polyethylene for active packaging. Food Addit. Contam. Part A 26, 938–946.
Pereira de Abreu, D.A., Paseiro Losada, P., Maroto, J., Cruz, J.M., 2010. Evaluation of the effectiveness of a new active packaging film containing natural antioxidants (from barley husks) that retard lipid damage in frozen Atlantic salmon (*Salmo salar* L.). Food Res. Int. 43, 1277–1282.
Pereira de Abreu, D.A., Cruz, J.M., Paseiro Losada, P., 2012. Active and intelligent packaging for the food industry. Food Rev. Int. 28, 146–187.
Porter, W.L., 1993. Paradoxical behavior of antioxidants in food and biological systems. In: William, G.M. (Ed.), Antioxidants: Chemical Physiological, Nutritional and Toxicological Aspects. Princeton Scientific, pp. 93–122.
Pospíšil, J., Nešpůrek, S., 2008. Polymer additives. In: Piringer, O.G., Baner, A.L. (Eds.), Plastic Packaging, second ed. Wiley-VCH, Weinheim, pp. 63–88.
Pristouri, G., Badeka, A., Kontominas, M.G., 2010. Effect of packaging material headspace, oxygen and light transmission, temperature and storage time on quality characteristics of extra virgin olive oil. Food Control 21, 412–418.
Rho, K.L., Seib, P.A., Chung, O.K., Chung, D.S., 1986. Retardation of rancidity in deep-fried instant noodles (ramyon). J. Am. Oil Chem. Soc. 63, 251–256.
Riemersma, R.A., 2002. Analysis and possible significance of oxidised lipids in food. Eur. J. Lipid Sci. Technol. 104, 419–420.
Robertson, G.L., 2013. Food Packaging: Principles and Practice, third ed. CRC Press, Boca Raton. pp. 703.
Roman, M.J., Tian, F., Decker, E.A., Goddard, J.M., 2014. Iron chelating polypropylene films: manipulating photoinitiated graft polymerization to tailor chelating activity. J. Appl. Polym. Sci. 131, 39948.
Sacchi, R., Savarese, M., Del Regno, A., Paduano, A., Terminiello, R., Ambrosino, M.L., 2008. Shelf life of vegetable oils bottled in different scavenging polyethyleneterephthalate (PET) containers. Packag. Technol. Sci. 21, 269–277.
Sanches-Silva, A., Costa, D., Albuquerque, T.G., Buonocore, G.G., Ramos, F., Castilho, M.C., Machado, A.V., Costa, H.S., 2014. Trends in the use of natural antioxidants in active food packaging: a review. Food Addit. Contam. Part A 31, 374–395.
Sanches Silva, A., López Hernández, J., Paseiro Losada, P., 2004. Modified atmosphere packaging and temperature effect on potato crisps oxidation during storage. Anal. Chim. Acta 524, 185–189.
Sanchez-Bel, P., Egea, I., Pretel, M.T., Flores, F.B., Romojaro, F., Martínez-Madrid, M.C., 2011. Roasting and packaging in nitrogen atmosphere protect almond var. Guara against lipid oxidation. Food Sci. Technol. Int. 17, 529–540.
Sängerlaub, S., Gibis, D., Kirchhoff, E., Tittjung, M., Schmid, M., Müller, K., 2013. Compensation of pinhole defects in food packages by application of iron-based oxygen scavenging multilayer films. Packag. Technol. Sci. 26, 17–30.
Sattar, A., deMan, J.M., Alexander, J.C., 1976. Effect of wavelength on light induced quality deterioration of edible oils and fats. Can. Inst. Food Sci. Technol. J. 9, 108–113.

Schreiber, S.B., Bozell, J.J., Hayes, D.G., Zivanovic, S., 2013. Introduction of primary antioxidant activity to chitosan for application as a multifunctional food packaging material. Food Hydrocoll. 33, 207–214.

Sharma, G.K., Madhura, C.V., Arya, S.S., 1990. Interaction of plastic films with foods. II. Effect of polyethylene and polypropylene films on the stability of vegetable oils. J. Food Sci. Technol. 27, 328–331.

Siripatrawan, U., Harte, B.R., 2010. Physical properties and antioxidant activity of an active film from chitosan incorporated with green tea extract. Food Hydrocoll. 24, 770–775.

Siró, I., Fenyvesi, É., Szente, L., De Meulenaer, B., Devlieghere, F., Orgoványi, J., Sényi, J., Barta, J., 2006. Release of alpha-tocopherol from antioxidative low-density polyethylene film into fatty food simulant: influence of complexation in beta-cyclodextrin. Food Addit. Contam. 23, 845–853.

Skinner, G.E., Reddy, N.R., 2006. Hazards associated with *Clostridium botulinum* in modified atmosphere packaged fresh fish and fishery products. In: Otwell, W.S., Kristinsson, H.G., Balaban, M.O. (Eds.), Modified Atmospheric Processing and Packaging of Fish: Filtered Smokes, Carbon Monoxide, and Reduced Oxygen Packaging. Blackwell Publishing, Ames, IA, pp. 163–192.

Song, Y.S., Hepp, M.A., 2005. US food and drug administration approach to regulating intelligent and active packaging components. In: Han, J.H. (Ed.), Innovations in Food Packaging. Elsevier, Amsterdam, pp. 475–481.

Suloff, E.C., Marcy, J.E., Blakistone, B.A., Duncan, S.E., Long, T.E., O'Keefe, S.F., 2003. Sorption behavior of selected aldehyde-scavenging agents in poly(ethylene terephthalate) blends. J. Food Sci. 68, 2028–2033.

Takahashi, O., 1995. Haemorrhagic toxicity of a large dose of α-, β-, γ- and δ-tocopherols, ubiquinone, β-carotene, retinol acetate and L-ascorbic acid in the rat. Food Chem. Toxicol. 33, 121–128.

Tawfik, M.S., Huyghebaert, A., 1999. Interaction of packaging materials and vegetable oils: oil stability. Food Chem. 64, 451–459.

Tian, F., Decker, E.A., Goddard, J.M., 2012a. Control of lipid oxidation by nonmigratory active packaging films prepared by photoinitiated graft polymerization. J. Agric. Food Chem. 60, 7710–7718.

Tian, F., Decker, E.A., Goddard, J.M., 2012b. Development of an iron chelating polyethylene film for active packaging applications. J. Agric. Food Chem. 60, 2046–2052.

Tian, F., Decker, E.A., Goddard, J.M., 2013a. Controlling lipid oxidation via a biomimetic iron chelating active packaging material. J. Agric. Food Chem. 61, 12397–12404.

Tian, F., Decker, E.A., Goddard, J.M., 2013b. Controlling lipid oxidation of food by active packaging technologies. Food Funct. 4, 669–680.

Tian, F., Decker, E.A., McClements, D.J., Goddard, J.M., 2014. Influence of non-migratory metal-chelating active packaging film on food quality: impact on physical and chemical stability of emulsions. Food Chem. 151, 257–265.

U.S. Food and Drug Administration. Ingredients, Additives, GRAS & Packaging Guidance Documents & Regulatory Information. http://www.fda.gov/Food/GuidanceRegulation/GuidanceDocuments-RegulatoryInformation/IngredientsAdditivesGRASPackaging/default.htm (accessed October 2015).

van Aardt, M., Duncan, S.E., Marcy, J.E., Long, T.E., Hackney, C.R., 2001. Effectiveness of poly(ethylene terephthalate) and high-density polyethylene in protection of milk flavor. J. Dairy Sci. 84, 1341–1347.

Waraho, T., Cardenia, V., Rodriguez-Estrada, M.T., McClements, D.J., Decker, E.A., 2009. Prooxidant mechanisms of free fatty acids in stripped soybean oil-in-water emulsions. J. Agric. Food Chem. 57, 7112–7117.

Webster, J.B., Duncan, S.E., Marcy, J.E., O'Keefe, S.F., 2009. Controlling light oxidation flavor in milk by blocking riboflavin excitation wavelengths by interference. J. Food Sci. 74, S390–S398.

Wessling, C., Nielsen, T., Leufvén, A., Jägerstad, M., 1999. Retention of α-tocopherol in low-density polyethylene (LDPE) and polypropylene (PP) in contact with foodstuffs and food-simulating liquids. J. Sci. Food Agric. 79, 1635–1641.

Wessling, C., Nielsen, T., Giacin, J.R., 2000a. Antioxidant ability of BHT- and α-tocopherol-impregnated LDPE film in packaging of oatmeal. J. Sci. Food Agric. 81, 194–201.

Wessling, C., Nielsen, T., Leufvén, A., 2000b. The influence of α-tocopherol concentration on the stability of linoleic acid and the properties of low-density polyethylene. Packag. Technol. Sci. 13, 19–28.

Zanardi, E., Dorigoni, V., Badiani, A., Chizzolini, R., 2002. Lipid and colour stability of Milano-type sausages: effect of packing conditions. Meat Sci. 61, 7–14.

Zhu, X., Schaich, K., Chen, X., Yam, K., 2013. Antioxidant effects of sesamol released from polymeric films on lipid oxidation in linoleic acid and oat cereal. Packag. Technol. Sci. 26, 31–38.

Index

Note: Page numbers followed by "f" indicate figures, "t" indicate tables.

B

E

F

G

H

I

P

Q

R

T

U

V

W

X

Z

FSC
www.fsc.org
MIX
Papier | Fördert
gute Waldnutzung
FSC® C083411

Kolibri 360 GmbH
Ferdinand-Jühlke-Straße 7
99095 Erfurt, Deutschland
produktsicherheit@kolibri360.de